Modern Carbonyl Chemistry

Edited by Junzo Otera

Related Titles from Wiley-VCH

H.-G. Schmalz (Ed.)
Organic Synthesis Highlights IV
2000. Softcover. ISBN 3-527-29916-5

F. Diederich / P.-J. Stang (Eds.)
Templated Organic Synthesis
2000. Hardcover. ISBN 3-527-29666-2

K. C. Nicolaou / E. J. Sorensen
Classics in Total Synthesis
1996. Hardcover. ISBN 3-572-29284-5
1996. Softcover. ISBN 3-572-29231-4

Modern Carbonyl Chemistry

Edited by Junzo Otera

WILEY-VCH

Weinheim · New York · Chichester · Brisbane · Singapore · Toronto

Edited by:
Prof. Dr. Junzo Otera
Department of Applied Chemistry
Okayama University of Science
Ridai-cho
Okayama 700-0005
Japan

Reprint 2001

Library of Congress Card No. applied for.

A cataloque record for this book is available from the British Library.

Die Deutsche Bibliothek – Cataloguing-in-Publication Data
A catalogue record for this book is available from Der Deutschen Bibliothek
ISBN 3-527-29871-1

Printed on acid-free and chlorine-free paper.

Composition: K+V Fotosatz GmbH, D-64743 Beerfelden
Printing: Strauss Offsetdruck, D-69509 Mörlenbach
Bookbinding: Wilh. Osswald & Co., D-67433 Neustadt/Weinstr.

Printed in the Federal Republic of Germany

Preface

The chemical transformations of carbonyl groups have played a central role in organic chemistry. The ease with which these groups undergo nucleophilic additions has led to a variety of synthetically useful reactions. Most significantly, the emergence of new reactions has not only enabled progress to be made in carbonyl chemistry itself but exerted a great influence on other fields. In this sense, the (Barbier-)Grignard reaction was the first epoch-making event. The discovery of this reaction dates back almost a century, but still holds the status of one of the most fundamental reactions in modern synthetic chemistry due to its facile availability and versatility. Furthermore, the contribution of this reaction to organometallic chemistry is, in particular, worthy of note. Before the discovery of the Grignard reaction, a considerable number of organometallic compounds and their reactions had already been known. However, the Grignard reaction clearly exemplified the synthetic potential of organometallic compounds for the first time.

The next milestone appeared in the 1950s in the context of the development of asymmetric reactions. Various stereochemical reactions induced by facial discrimination of the carbonyl group have always been pivotal in this field. Cram's rule inspired an explosion of studies on diastereoselective reactions followed by enantioselective versions. The recent outstanding progress in the non-linear effect of chirality or asymmetric autocatalysis heavily relies on the carbonyl addition reactions. Thanks to these achievements, natural products chemistry has enjoyed extensive advancement in the synthesis of complex molecules. It is no exaggeration to say that we are now in a position to be able to make any molecules in as highly selective a manner as we want.

The activation of the carbonyl group by Lewis acids was another leap made in the 1960s as typified by Mukaiyama-aldol reaction. In sharp contrast to the conventional carbonyl addition reactions that had been run under basic conditions, this new method allowed the addition of various nucleophiles under acidic conditions with high chemo- and stereocontrol and, consequently, the scope of the carbonyl addition reaction was extensively expanded. The Lewis acid-promoted allylation with allylmetals and ene reaction also received as much attention as the aldol-type reaction. It should be further pointed out that the catalytic versions of asymmetric reactions, which represent one of the most exciting topics in recent synthetic chemistry, owe their development strongly to the Lewis acid activation protocol. The design of a variety of chiral ligands for metals has produced luxuriant fruits in this field.

The inversion of reactivity (*Umpolung*) by the use of masked carbonyls was another achievement of modern synthetic chemistry, yet the direct generation of acyl anions has been the latest subject of focus. Further remarkable reactivity variation was introduced by radical chemistry, which had for long time been regarded as a technique in which control of stereochemistry is difficult. Nevertheless, the recent innovation in this field allowed novel stereoselective reactions to proceed under neutral conditions.

Needless to say, it is important for synthetic chemists to devise new reactions. However, it is important as well, especially in the light of yielding to the growing demands for green chemistry, to improve the reaction conditions. The change of the reaction media is one of such treatments, on which intensive attention is now concentrated. The use of water in place of organic solvents is of particular interest in terms of both economical and ecological aspects. Reaction without solvent is an ultimate goal, and apparently a solid state reaction is a means to this end. These technologies have been successfully applied to a variety of carbonyl reactions.

As such, "modern carbonyl chemistry" has made vast and profound progress in manifold aspects. Unfortunately, however, these achievements are rather dispersed, and hence it is not easy for us to survey them comprehensively. This book was undertaken to overcome such inconvenience by collecting relevant subjects together. To my great pleasure, this was successfully realized through contributions by leading chemists in this field. I wish to express my sincere thanks to these authors, who agreed to share this book despite their tight schedules.

Okayama, March 2000 Junzo Otera

Contents

List of Authors

Neil G. Almstead
245 Roger Adams Laboratory, Box 18
Department of Chemistry
University of Illinois
600 S. Mathews Avenue
Urbana
IL 61801
USA

Erick M. Carreira
ETH-Zürich
Organic Chemistry Laboratory
Universitätsstrasse 16
8092 Zürich
Switzerland

Sherry R. Chemler
Department of Chemistry
University of Michigan
Ann Arbor
MI 48109
USA

Cameron J. Cowden
Merck Sharp and Dohme Research Laboratories
Department of Process Research
Hertford Road
Hoddesdon EN11 9BU
United Kingdom

James M. Coxon
Department of Chemistry
University of Canterbury
Private Bag 4800
Christchurch
New Zealand

Scott E. Denmark
Department of Chemistry
University of Illinois
245 Roger Adams Laboratory, Box 18
600 S. Mathews Avenue
Urbana
IL 61801
USA

Gregory C. Fu
Department of Chemistry
Massachusetts Institute of Technology
Room 18-411
77 Massachusetts Avenue
Cambridge
MA 02139-4307
USA

Keiji Iwamoto
Department of Applied Chemistry
Faculty of Engineering
Osaka University
Suita
Osaka 565-0871
Japan

Shu Kobayashi
Graduate School of Pharmaceutical Sciences
University of Tokyo
Hongo
Bunkyo-ku
Tokyo 113-0033
Japan

Mitsuo Komatsu
Department of Applied Chemistry
Graduate School of Engineering
Osaka University
Suita
Osaka 565-0871
Japan

Richard T. Luibrand
Department of Chemistry
California State University
Hayward
CA 94542
USA

Kei Manabe
Graduate School of Pharmaceutical Sciences
University of Tokyo
Hongo
Bunkyo-ku
Tokyo 113-0033
Japan

Keiji Maruoka
Department of Chemistry
Graduate School of Science
Hokkaido University
Sapporo 060-0810
Japan

Koichi Mikami
Department of Chemical Technology
Tokyo Institute of Technology
Ookayama
Meguro-ku
Tokyo 152-8552
Japan

Shinji Murai
Department of Applied Chemistry
Faculty of Engineering
Osaka University
Suita
Osaka 565-0871
Japan

Satoshi Nagayama
Graduate School of Pharmaceutical Sciences
University of Tokyo
Hongo
Bunkyo-ku
Tokyo 113-0033
Japan

Takashi Ooi
Department of Chemistry
Graduate School of Science
Hokkaido University
Sapporo 060-0810
Japan

Ian Paterson
Chemical Laboratory
University of Cambridge
Lensfield Road
Cambridge CB2 1EW
United Kingdom

William R. Roush
Department of Chemistry
University of Michigan
Ann Arbor
MI 48109
USA

Ilhyong Ryu
Department of Applied Chemistry
Graduate School of Engineering
Osaka University
Suita
Osaka 565-0871
Japan

Susumu Saito
Graduate School of Engineering
Nagoya University
Chikusa
Nagoya 464-8603
Japan

Mukund P. Sibi
Department of Chemistry
North Dakota State University
Fargo
ND 58105-5516
USA

Tara R. Ternes
Department of Chemistry
North Dakota State University
Fargo
ND 58105-5516
USA

Fumio Toda
Department of Applied Chemistry
Faculty of Engineering
Ehime University
Matsuyama
Ehimc 790
Japan

Kiyoshi Tomioka
Graduate School of Pharmaceutical
Sciences
Kyoto University
Yoshida
Sakyo-ku
Kyoto 606-8501
Japan

Debra J. Wallace
Merck Sharp and Dohme Research
Laboratories
Department of Process Research
Hertford Road
Hoddesdon EN11 9BU
United Kingdom

Hisashi Yamamoto
Graduate School of Engineering
Nagoya University
CREST, (JST)
Furo-cho
Chikusa
Nagoya 464-8603
Japan

1 Carbonyl-Lewis Acid Complexes

Takashi Ooi and Keiji Maruoka

1.1 Introduction

Carbonyl is one of the commonest functional groups to be found in organic compounds and is often encountered in biomolecules such as peptides and lipids, playing a crucial role in organizing three-dimensional intermolecular associations [1]. For organic chemists, in turn, it is always an extremely useful functional group for manipulation, and either Brønsted acids or Lewis acids are routinely employed for its activation. Since Lewis acid complexation of carbonyl compounds can have a dramatic effect on the rates and selectivities of reactions at carbonyl centers [2] as we have witnessed in the real explosion of new methodologies over the last two decades [3], carbonyl-Lewis acid complexations play a fundamental role in organic and bioorganic chemistry [4]. Concrete discussion on this issue inevitably requires a deep understanding of the nature of both carbonyl functionality and metal-centered Lewis acids, and to identify key parameters of Lewis acid function would be a tremendous task, requiring a knowledge of each of the components that would contribute to the overall reactivity and selectivity of Lewis acid-mediated reactions. Such comprehension is thought to be a necessary prerequisite, especially when trying to determine the origin of their stereochemical outcome, because it is clear that the conformational preferences of the carbonyl-Lewis acid complex are ultimately responsible for determining the stereochemical course of the reactions [4].

When considering factors that may influence the reactivity and conformation of carbonyl-Lewis acid complexes, primary attention should be given to the modes of coordination of Lewis acids to carbonyl groups, analyzing the exact location of the Lewis acid with respect to its carbonyl ligand. There are several different possible modes of coordination. First, purely electrostatic interaction can be considered, in which the metal is situated at the negative end of the C=O dipole [C–O–M = 180° (**A**)]. The second possibility is the coordination of the metal to one of the lone pairs on the carbonyl oxygen, with the metal being in the nodal plane of the C=O π-bond (**B**). Although facial differentiation would not be feasible in this case due to the planar alignment, the Lewis acid will be *syn* or *anti* to particular substituents (R^1). The third, a bent non-planar mode of bonding results from movement of the metal out of the carbonyl π nodal plane (**C**), which is often adopted by the bulky Lewis acids. The last mode is a η^2 coordination of metal to the C=O π bond, where the carbonyl is formally the donor, but back-bonding into the C=O π^* orbital occurs (**D**). This mode has been reported for transition metal complexes [5]. The metal center is out of the carbonyl plane and blocks a particular face, possibly directing the attack of the nucleophile on the opposite face. However, the two substituents (R^1 and R^2) would not be differentiated. So far, 1:2 carbonyl-Lewis acid complexes (**E1**) in which two individual Lewis acids interact simultaneously with the carbonyl oxygen atom are unknown. However, the double coordination of carbonyl compounds by bidentate Lewis acids (**E2**) should have useful chemical consequences, including enhanced reactivity of carbonyls [6].

MLn O R1 R2 A; MLn O R1 R2 B; MLn O R1 R2 C; O MLn R1 R2 D; LnM MLn O R1 R2 E1; LnM MLn O R1 R2 E2

In the meantime, the effect of Lewis acid coordination on the inherent conformational preferences adjacent to the carbonyl group should be discussed. In particular, the answer to how is the *s-cis*/*s-trans* equilibrium of α,β-unsaturated carbonyls affected by Lewis acid complexation must enable the exposure of one or the other face of the unsaturated system to a given nucleophile, which will surely be of great importance when dealing with enantioselective reactions.

MLn O R3 *s-cis* ⇌ MLn O R3 *s-trans*

Additionally, Lewis acid complexes of carbonyl compounds bearing heteroatom-containing functionality (X) in appropriate proximity are an interesting subject to be addressed. Such chelate-type carbonyl-Lewis acid complex formation is generally a favorable process, and can bring an enhancement of reactivity and selectivity by the effective activation of the carbonyl moiety compared to the non-chelation case, implying considerable utility in organic synthesis [7].

X O R4 * R5 —MLn→ [Ln M X O * R4 R5] —:Nu→ X OH Nu * * R4 R5

The first chapter on these carbonyl-Lewis acid complexes uses information on (1) theoretical study, (2) NMR study, and (3) X-ray crystallographic study. For the rest, the subjects of chelate complexes and bidentate Lewis acid complexes, mainly featuring recent advances, are discussed.

1.2 Theoretical Study of Carbonyl-Lewis Acid Complexes

Reetz and co-workers determined the structure of the benzaldehyde-boron trifluoride adduct by X-ray crystallography in 1986 (Fig. 1-1) [8], which shows that the Lewis acid is placed 1.59 Å from the carbonyl oxygen, along the direction of the oxygen lone pair and *anti* to the phenyl substituent.

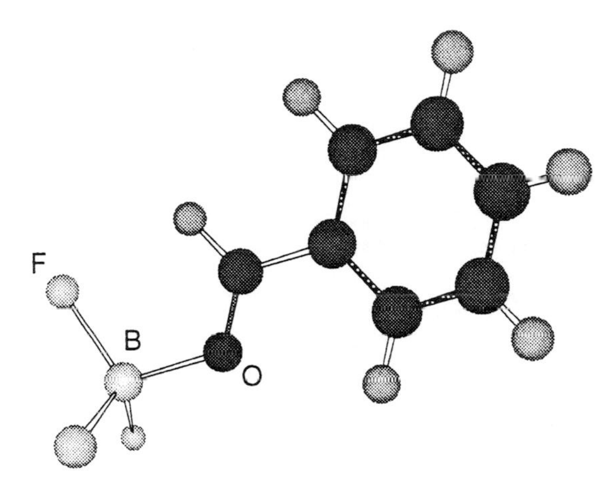

Figure 1-1. Crystal structure of BF_3-benzaldehyde complex.

They also performed MNDO calculations in order to understand the bonding of aldehyde-BF_3 adducts, employing acetaldehyde as a model substrate [8]. Minimum energy geometries revealed that the *anti* form (**1**), in which the C–C–O–B skeleton lies in a common plane, turned out to be more stable than the *syn* isomer (**3**) by 1.8 kcal/mol. Although the linear form (**2**) represents the least-energy transition state for internal *anti/syn* isomerization, it is not a minimum on the energy surface (Scheme 1-1).

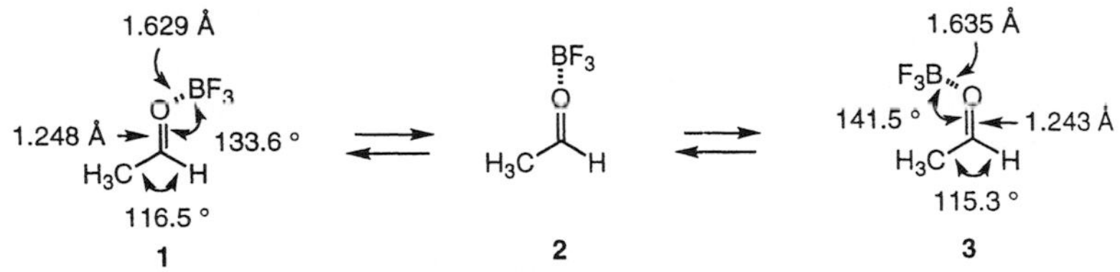

1.629 Å; 1.248 Å; 133.6 °; 116.5 ° — **1** — ΔH_f -297.07 kcal/mol

2 — -291.79

1.635 Å; 141.5 °; 1.243 Å; 115.3 ° — **3** — -295.27

Scheme 1-1. MNDO-optimized parameters.

They further investigated the fact that BF_3 complexation lowers the energy of the π^*_{co} orbital, making the carbonyl more susceptible to nucleophilic attack [9] and the fact that the extent of LUMO lowering is a little greater in the *anti* complex than in the *syn* isomer. Eventually, the coefficient of the orbital at the carbonyl C atom in π^*_{co} increases in magnitude upon complexation, a phenomenon which also enhances the ease of nucleophilic addition (Fig. 1-2).

Wiberg studied rotational barriers in formaldehyde, propanal and acetone coordinated to Lewis acids such as BF_3 or $AlCl_3$, where all complexes were found to prefer bent geometries. For formaldehyde complexes, linear structures are 6.10 kcal/mol higher in energy and out-of-plane structure (π-complexes) even higher [10].

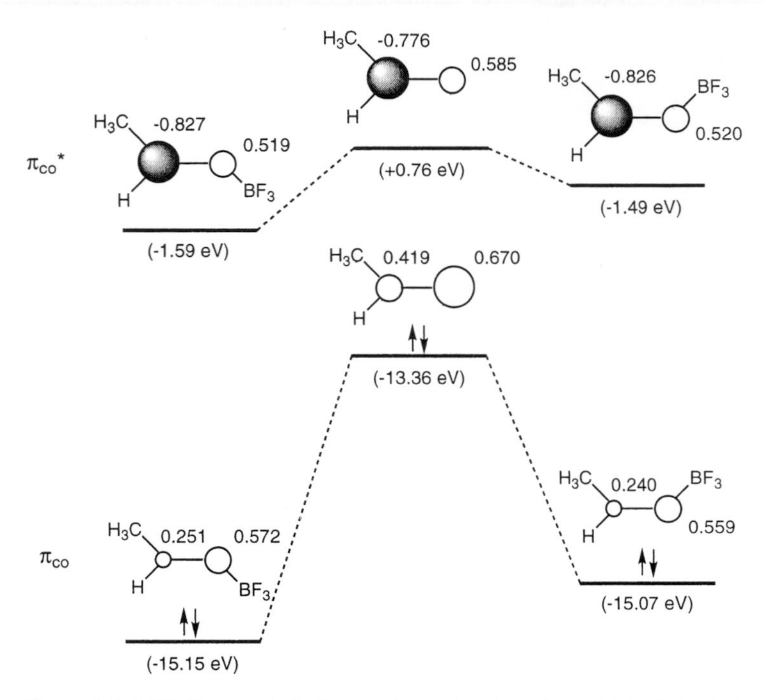

Figure 1-2. MNDO-computed effects of complexation of acetaldehyde by BF_3.

Quite recently, Ren, Cramer and Squires presented an interesting experimental approach for evaluating acidity and electrophilicity of simple aldehydes and ketones when complexed to Lewis acids by employing acetaldehyde-BF_3 complex as a representative case [11].

$$\underset{\mathbf{4}}{CH_3CH{=}\overset{\oplus}{O}{-}\overset{\ominus}{B}F_3} \xrightarrow{\Delta H_{diss}(\mathbf{4})} CH_3CH{=}O + BF_3$$

$$\Delta H_{acid}(\mathbf{4}) \downarrow \qquad \qquad \uparrow -\Delta H_{acid}(CH_3CHO)$$

$$H^{\oplus} + \underset{\mathbf{5}}{H_2C{=}CH{-}O{-}\overset{\ominus}{B}F_3} \xrightarrow{\Delta H_{diss}(\mathbf{5})} H^{\oplus} + H_2C{=}CH{-}O^{\ominus} + BF_3$$

Scheme 1-2. Thermochemical cycle to derive the gas-phase acidity of the complex **4**.

A thermochemical cycle that can be used to derive the gas-phase acidity (ΔH_{acid}) of the CH_3CHO-BF_3 complex is illustrated in Scheme 1-2. The increase in acidity of the α-proton in CH_3CHO-BF_3 relative to free CH_3CHO is equal to the difference in the BF_3 binding energies of the neutral aldehyde and the enolate ion. Thus, it is possible to determine the gas-phase acidity of the complex from BF_3 bond strengths and the absolute acidity of CH_3CHO. Energy-resolved collision-induced dissociation (CID) in a flowing aftergrow-triple quadrupole apparatus [12] was used to measure the BF_3 binding energy of $[CH_2CHO\text{-}BF_3]^-$, and 61±3 kcal/mol was obtained as an average value for the 298 K dissociation enthalpy of **5**. Since ab initio calculations carried out at the MP2, B3LYP, CBS and CCSD(T) level of theory all point to a value of 11±2 kcal/mol for the complexation energy of CH_3CHO-BF_3, the BF_3 bond strength in the enolate is 50 kcal/mol greater than that in the neutral complex. Combining the result with the gas-phase acidity of CH_3CHO, ΔH_{acid} (CH_3CHO) = 365.8±2.2 kcal/mol, affords a value for the acidity of complex **4** of 316±4 kcal/mol. Theoretical estimates for ΔH_{acid}(**4**) obtained from B3LYP/6-31+G(d) and CBS-4 calculations, which are 315.7 kcal/mol and 314.7 kcal/mol respectively, are in excellent agreement with the value. Therefore, BF_3 coordination has essentially transformed the carbonyl compounds into gas-phase superacids, molecules with Brønsted acidities comparable to or greater than that of sulfuric acid (ΔH_{acid} = 309 kcal/mol).

By employing hydride ion affinity (HIA, the enthalpy change for the reaction: $RH^- \rightarrow R + H^-$) [13] of a carbonyl group as a useful thermochemical correlate, the extent to which the complexation of CH_3CHO to BF_3 makes the carbonyl electrophilic in the gas phase was also clearly demonstrated.

Mikami et al. reported highly enantioselective ene-type reaction with trihaloacetaldehydes (**7**) catalyzed by the chiral titanium complex (**6**), where the significant influence of the halogen substituents not only on the product ratio but also on the enantiomeric excesses of homoallylic (**8**) and allylic (**9**) alcohol products was thought to be of mechanistic interest (Scheme 1-3) [14].

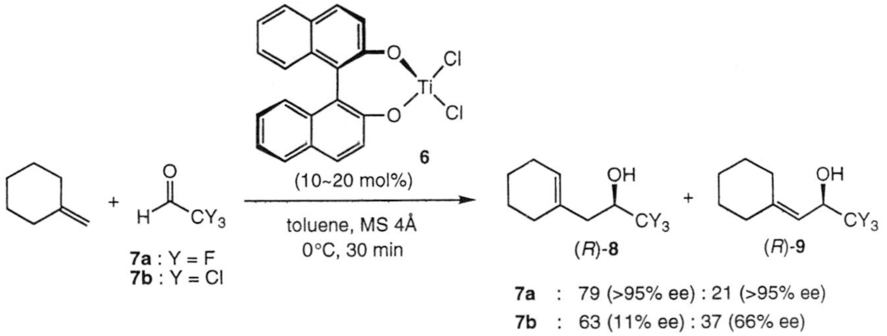

6 (10~20 mol%)
toluene, MS 4Å
0°C, 30 min

7a : Y = F
7b : Y = Cl

(*R*)-**8** (*R*)-**9**

7a : 79 (>95% ee) : 21 (>95% ee)
7b : 63 (11% ee) : 37 (66% ee)

Scheme 1-3. Catalytic asymmetric ene-type reaction with chiral titanium catalyst.

They estimated the ene reactivity of trihaloaldehydes on the basis of the atomic charge and LUMO energy level by running an MO calculation on the aldehyde-H^+ complexes as a model of aldehyde/chiral Lewis acid complexes [15]. The results from semi-empirical (MNDO and PM3) and ab initio (6-31G**) calculations are listed in Table 1-1.

Table 1-1. Computational analysis of trihaloaldehyde-H^+ complexes.

		Fluoral (**7a**)-H^+	Chloral (**7b**)-H^+	Acetaldehyde-H^+
Ab initio (RHF/6-31G**)	LUMO (eV)	–5.40	–4.88	–4.09
	C_1 charge	+0.61	+0.64	+0.70
PM3	LUMO (eV)	–8.64	–7.83	–7.42
	C_1 charge	+0.36	+0.39	+0.44
MNDO	LUMO (eV)	–8.55	–8.14	–7.37
	C_1 charge	+0.36	+0.42	+0.42

The frontier orbital interaction between HOMO of the ene component and the LUMO of the carbonyl enophile is the primary interaction. Thus, the fluoral (**7a**) complex with the lower LUMO energy level is considered to be more reactive enophile species, giving mainly the homoallylic alcohols (**8**) and the chloral (**7b**) complex with the higher LUMO energy level is the less reactive one. On the other hand, the chloral complex bears the greater positive charge at the carbonyl carbon (C_1) and hence is the more reactive in the cationic reaction, eventually leading to the allylic alcohols (**9**). Notably, the data revealed that the positive charge at the carbonyl carbon of the simple acetaldehyde is greater than in the case of the fluoral (**7a**). Accordingly, the ene reactivity of aldehydes is determined in terms of the balance of LUMO energy level vs. electron density on the carbonyl carbon (C_1).

Branchadall and co-workers studied the effect of Lewis acid ($AlCl_3$) catalysis on the Diels-Alder reactions of methyl (*Z*)-(*S*)-4,5-(2,2-propylidenedioxy)pent-2-enoate (**10**) with cyclopentadiene, which usually exhibits a high level of *syn-endo* selectivity under the influence of the Lewis acid, at the B-LYP/6-31G* level [16]. The most stable conformation of the complex **10**-$AlCl_3$ revealed the significant difference with the structure of uncomplexed molecule, i.e., *s-trans* arrangement of the carbonyl group with respect to the carbon-carbon double bond (Fig. 1-3).

The comparison of the potential energy barriers with those corresponding to the uncatalyzed reaction shows that the $AlCl_3$ enables a drastic lowering of the barriers. Generally, the catalyst stabilizes with preference of the *endo* transition state over the *exo* one and the *syn* over the *anti*. This fact leads to an increase of both *syn/anti* and *endo/exo* selectivity (Table 1-2).

By carrying out a partition of the potential energy barriers into the distortion energy and the interaction energy, contributions of the steric and electronic effects to the $AlCl_3$-catalyzed Diels-Alder reaction were precisely discussed [16].

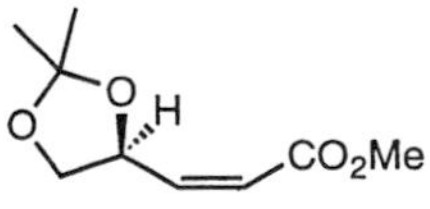

10

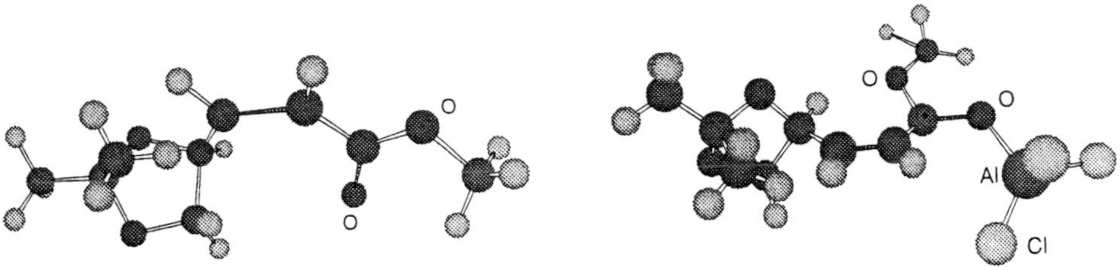

10 10-$AlCl_3$

Figure 1-3. Geometries of **10** and complex **10**-$AlCl_3$ obtained at the AM1 (B-LYP/6-31G*) level.

Table 1-2. Potential energy barriers computed at the B-LYP/6-31G* level for the AM1 (B-LYP) geometries of the transition state of the reaction of **10** with cyclopentadiene.

Uncatalyzed	TS	ΔE [a])	$AlCl_3$-catalyzed	TS	ΔE [a])
	syn-endo	26.3		*syn-endo*	16.9
	syn-exo	27.1		*syn-exo*	18.9
	anti-endo	28.0		anti-endo	19.3
	anti-exo	28.6		*anti-exo*	21.0

[a]) In kcal/mol BSSE correction included.

1.3 NMR Study of Carbonyl-Lewis Acid Complexes

NMR studies have been able to provide a lot of information on the structure of carbonyl-Lewis acid complexes in solution including stoichiometry and conformation, which are of direct importance with respect to the stereochemical outcome of the reactions involving these complexes.

In conjunction with their X-ray crystallographic study of benzaldehyde-BF_3 complex, Reetz and co-workers investigated the structure in solution using a heteronuclear Overhauser experiment, where the fluorine atoms were irradiated, leading to a 5% enhancement of the aldehydic proton resonance, but causing no affect on the aromatic protons [4b]. Although the result suggests that BF_3 is situated *anti* to the phenyl ring as observed in the solid state, it does not rule out a small amount of the *syn* isomer which may be in equilibrium with the *anti* isomer.

In the studies of $SnCl_4$ or $BF_3{\cdot}OEt_2$-promoted intramolecular addition of allylic organometallic reagents to aldehydes to clarify the origin of the stereoselectivity,

Denmark and co-workers disclosed that the crystal structure of the 4-*tert*-butylbenzaldehyde (**11**)-$SnCl_4$ complex shows a 2:1 stoichiometry with two nonequivalent aromatic aldehyde *cis* to one another around the octahedrally coordinated tin atom (Fig. 1-4) and further carried out variable-temperature NMR experiments with this complex [17].

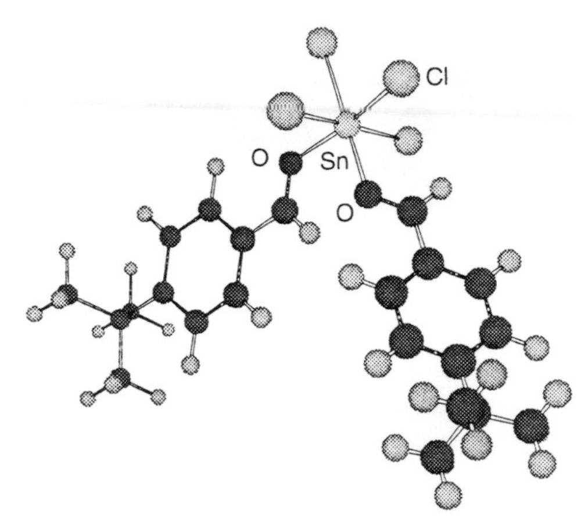

Figure 1-4. Crystal structure of 4-*tert*-butylbenzaldehyde (**11**)-$SnCl_4$ complex.

Table 1-3. Chemical shift differences on complexation.[a]

Temp (°C)	$(\mathbf{11})_2SnCl_4$			$(\mathbf{12})_2SnCl_4$		
	Complex	Neutral	$\Delta\delta$[b]	Complex	Neutral	$\Delta\delta$[b]
–100	199.72	192.61	7.11		195.38	
–80	199.76	192.46	7.30		195.16	
–60	199.77	192.29	7.48	202.51	194.89	7.62
–40	199.75	192.13	7.62	202.45	194.62	7.83
–20	199.67	191.97	7.70	202.36	194.35	8.01
0	199.44	191.79	7.65	202.25	194.10	8.15
20	198.86	191.63	7.23	202.07	193.81	8.26
40				201.75	193.58	8.17

[a]) All complexes were prepared with 2 equiv of the aldehyde and 1 equiv of $SnCl_4$.
[b]) $\Delta\delta=\delta$ (complex)–δ (neutral); positive numbers are downfield shifts.

A compilation of the resonances for the neutral and complexed 4-*tert*-butylbenzaldehyde (**11**) as well as the chemical shift differences ($\Delta\delta$) of the carbonyl carbon from –100 to 40 °C are shown in Table 1-3. Actually, the chemical shift differences were temperature independent, and the complex (in CD_2Cl_2) did not show any indication of dissociation as the temperature increased. Surprisingly, even at 120 °C (in toluene-d_8), the complex with $SnCl_4$ showed no signs of dissociation, indicating the strong basicity of **11** as a carbonyl ligand. A similar tendency was observed in the complexation of (*E*)-2-heptenal (**12**) with $SnCl_4$, as also shown in Table 1-3, though it appeared to dissociate slightly around 40 °C (in toluene-d_8). Denmark also established the conformation of α,β-enal unit by the use of difference NOE measurement, and the results are collected in Table 1-4, which also includes the data for BF_3 complex.

Table 1-4. The NOE data for (*E*)-2-Heptenal (**12**).[a]

Complex	Temp[b] (°C)	NOE (saturate/observe, %)			
		H^a/H^b	H^c/H^a	H^b/H^a	H^b/H^c
12	–95	14.0	16.4	0	0
(**12**)$_2$-$SnCl_4$[c]	–95	–29.2	–38	0	0
(**12**)-BF_3[d]	–95	–3.4	–3.3	0	0

[a]) In CD_2Cl_2.
[b]) Calibrated probe temperature.
[c]) Complex formed by addition of 0.5 equiv of $SnCl_4$ to the aldehyde.
[d]) Complex formed by addition of BF_3(g) (1.0 equiv) to the aldehyde.

Irradiation of H^a in the $SnCl_4$ complex resulted in a strong –29.2% NOE to H^c. When H^b was irradiated, no NOE to either H^a or H^c was detected. Irradiation of H^c again led to a strong NOE of –38.0% to H^a. These indicate that the complex of $SnCl_4$ and (*E*)-2-heptenal is primarily in the *s-trans* conformation in solution. Although the magnitude of the NOE observed with BF_3 (g) was appreciably lower than that observed with $SnCl_4$ due to the lower molecular weight, a similar trend of the NOE data was observed in the BF_3-aldehyde complex, confirming the *s-trans* conformation in solution (Fig. 1-5).

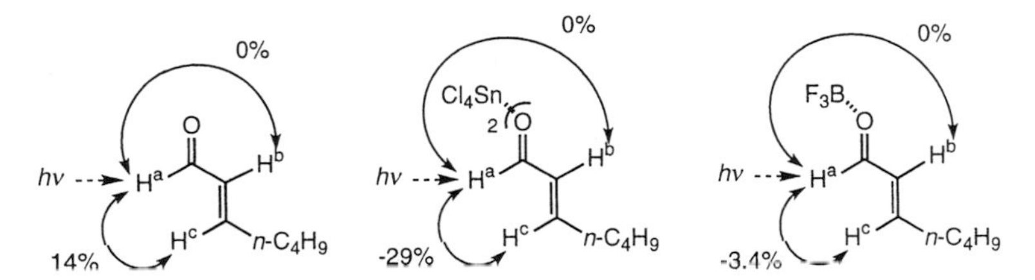

Figure 1-5. NOE observed for uncomplexed (*E*)-2-heptenal (**12**) and complexes with $SnCl_4$ and BF_3.

Corey and co-workers performed ^{1}H NMR molecular dynamics and NOE studies of the 2-methylacrolein-BF_3 complex in CD_2Cl_2 and demonstrated that the *s-trans* structure of the complex predominates in the solution at 185 K [18].

The spectroscopic evidence for the structure of carbonyl-Lewis acid complexes in solution is obviously quite relevant for elucidation of the transition states of Lewis acid-promoted stereoselective reactions. Yamamoto, Ishihara and Gao investigated the boron-substituent-dependent enantioselectivity of the chiral CAB-catalyzed asymmetric Diels-Alder reaction toward obtaining mechanistic information on the conformational preferences in α,β-enals in the transition state assembly of the reaction [19]. As summarized in Table 1-5, the Diels-Alder reaction of cyclopentadiene with methacrolein under catalysis by CAB gave the (2*R*)-enantiomer as the major product regardless of the steric feature of the boron-substituent (R), suggesting that this substrate appears to favor *s-trans* conformation in the transition state assembly of the catalytic Diels-Alder reaction. On the other hand, the stereochemistry of the reaction of cyclopentadiene with acrolein and crotonaldehyde was dramatically reversed by altering the structure of the boron-substituent (R). A sterically bulky aryl boron-substituent such as the *o*-phenoxyphenyl group may cause the active capacity between the boron-substituent and 2,6-diisopropoxyphenyl moiety to decrease. Therefore, the *s-trans*-coordinated α-nonsubstituted acrolein which would lead to the (2*R*)-product changed so that it favored the *s-cis* conformation inversely, thereby diminishing the steric size and giving the (2*S*)-product.

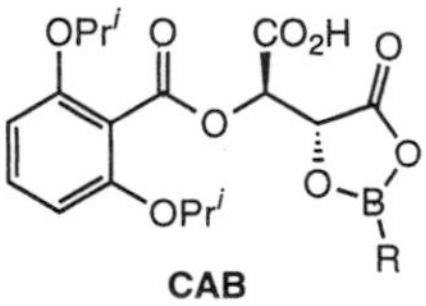

CAB

Table 1-5. Asymmetric Diels-Alder reaction of α,β-enal with cyclopentadiene catalyzed by CAB. [a])

CAB (R)	% ee (Confign)		
	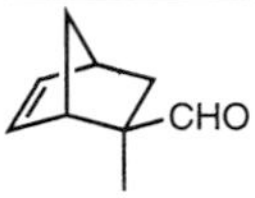 CHO	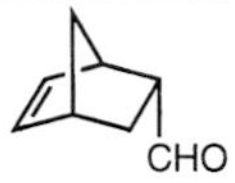 CHO	 CHO
$C_4H_9C{\equiv}C$	64 (*R*)	58 (*R*)	42 (*R*)
H	87 (*R*)	47 (*R*)	2 (*S*)
Ph	80 (*R*)	10 (*S*)	37 (*S*)
3,5-$(CF_3)_2C_6H_3$		3 (*S*)	59 (*S*)
o-$PhOC_6H_4$	93 (*R*)	57 (*S*)	67 (*S*)

[a]) The reaction was carried out in propionitrile for several hours using 10~20 mol% of CAB and cyclopentadiene (3 equiv) at −78 °C.

This hypothesis was supported by the difference NOE measurement of the CAB-complexed methacrolein and crotonaldehyde in CD_2Cl_2 at −95~−75 °C. Irradiation of H^a uniformly resulted in a strong NOE to H^c and no NOE to H^b and H^d and irradiation of H^c led to a strong NOE to H^a, indicating that the complex of methacrolein with CAB is primarily in the *s-trans* conformation independently of the boron substituent (Table 1-6). However, all the NOE data results for crotonaldehyde in Table 1-7 clearly reveal that (1) uncomplexed crotonaldehyde is primarily in the *s-trans* conformation, (2) the crotonaldehyde complexes with CAB

($R = C_4H_9C{\equiv}C, H$) is in the *s-trans* conformation, and (3) the crotonaldehyde complexed with CAB ($R = 3,5\text{-}(CF_3)_2C_6H_3$, *o*-$PhOC_6H_4$) is in the *s-cis* conformation.

Table 1-6. The NOE data for methacrolein complexed with CAB.

Complex	Temp [a] (°C)	NOE (saturate/observe, %)			
		H^a/H^b	H^a/H^c	H^a/H^d	H^c/H^a
Methacrolein only	–95	0	6.3	0	18
Methacrolein-CAB (R=H) [b]	–95	0	–10	0	6.3
Methacrolein-CAB (R=*o*-$PhOC_6H_4$) [b]	–75	0	–22	0	–33

[a] Calibrated probe temperature.
[b] Complex formed by addition of 0.72 equiv of the aldehyde to CAB.

Table 1-7. The NOE data for crotonaldehyde complexed with CAB.

Complex	Temp [a] (°C)	NOE (saturate/observe, %)			
		H^a/H^b	H^a/H^c	H^b/H^a	H^c/H^a
Crotonaldehyde only	–95	0	5.4	–	–
Crotonaldehyde-CAB ($R = C_4H_9C{\equiv}C$) [b]	–75	0	6	–	13
Crotonaldehyde-CAB (R=H) [b]	–95	0	18	–	–
Crotonaldehyde-CAB ($R = 3,5\text{-}(CF_3)_2C_6H_3$) [b]	–75	–32	0	–48	–
Methacrolein-CAB (R=*o*-$PhOC_6H_4$) [b]	–75	–14	0	–18	–

[a] Calibrated probe temperature.
[b] Complex formed by addition of 0.72 equiv of the aldehyde to CAB.

1.4 X-ray Crystallographic Study of Carbonyl-Lewis Acid Complexes

Attracted by the wealth and accuracy of the information provided by X-ray crystallographic analysis, organic chemists utilize this powerful tool to infer mechanistic and conformational hypotheses based on structural data [20], though it seems implausible that static, crystalline species could reveal any information regarding the dynamics of transition states. This attitude has been particularly fruitful in the study of weak intermolecular forces such as hydrogen bonding [21] and Lewis acid-base interactions [22, 23], and obviously carbonyl-Lewis acid complexation is not the exception.

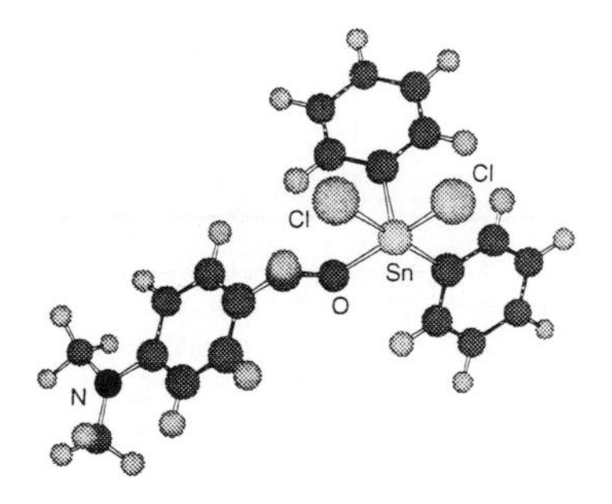

Figure 1-6. Crystal structure of Ph_2SnCl_2-*p*-NMe_2-C_6H_4CHO complex.

In addition to the studies by Reetz and Denmark already described in the previous sections, several aldehyde-Lewis acid complexes have been reported. In all cases, the Lewis acid is located *trans* to the aldehyde residue and coordinate in a σ-fashion.

The *p*-(dimethylamino)benzaldehyde-Ph_2SnCl_2 complex adopts the trigonal-bipyramidal structure, and such 1:1 complexation is common for complexes of alkyl and aryl tins (Fig. 1-6) [24].

Hersh and co-workers reported that catalysis for the Diels-Alder reactions of cyclopentadiene with α,β-unsaturated enones was induced by $(Me_3P)(CO)_3(NO)W$-$FSbF_5$. In order to probe the mechanism of the Diels-Alder catalysis, a single-crystal X-ray diffraction study of $[(Me_3P)(CO)_3(NO)(acrolein)W]^+SbF_6^-$ was carried out, and σ-type coordination was found to be present in the solid state [25]. This structure provides clear evidence for the preferred *s-trans* conformation of the α,β-unsaturated aldehyde unit, which would be expected by theoretical studies (Fig. 1-7).

Along with a great deal of effort devoted to the development of chiral Lewis acids during the last decade [26], defining the conformation about the dative bond between substrate and Lewis acid is one of the important issues encountered in the design of an effective catalyst [23].

Ln*M O=C R H F

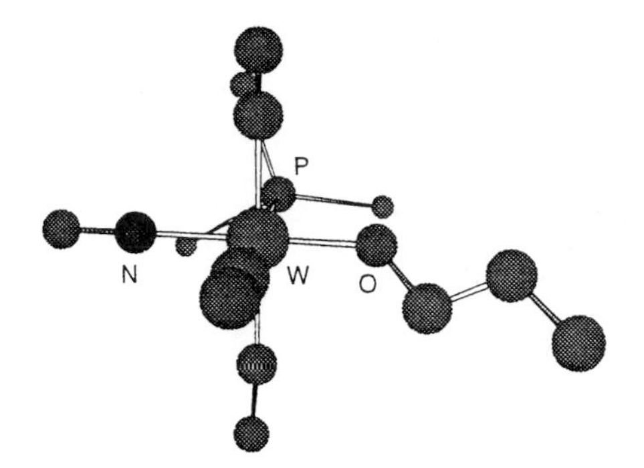

Figure 1-7. Crystal structure of $[(Me_3P)(CO)_3(NO)(acrolein)W]^+$.

Fu and co-workers recently reported a new approach to restricting the rotational degree of freedom (see complex **F**), which focuses on the development of Lewis acids that bear an empty σ-symmetry and an empty π-symmetry orbital as illustrated in Fig. 1-8 [27]. These vacant orbitals can simultaneously accept electron from an oxygen lone pair and from the π-system of a carbonyl group. The distinguishing feature of this approach is the π-symmetry interaction, which at once organizes the Lewis acid-base complex and activates the carbonyl toward nucleophilic addition.

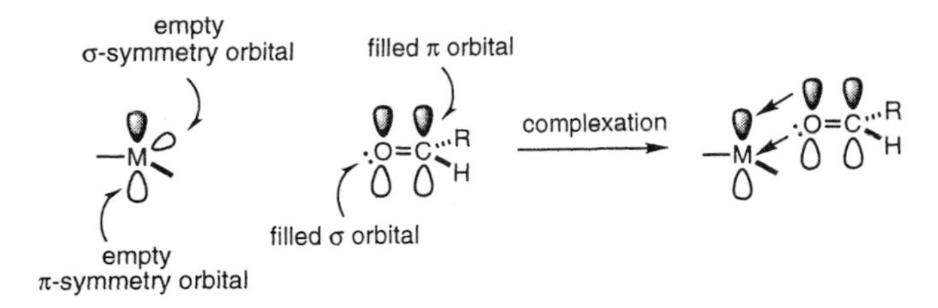

Figure 1-8. Simultaneous σ and π donation by a carbonyl group to a divalent Lewis acid.

Treatment of boracycle **13** [28] with $[(MeCN)_3Cr(CO)_3]$ in THF gave air- and moisture-sensitive $[(\eta^6\text{-borabenzene-THF})Cr(CO)_3]$ (**14**) that reacts with 3-(dimethylamino)acrolein to afford $[(\eta^6$-borabenzene-(3-(dimethylamino)acrolein) $Cr(CO)_3]$ (**15**). The crystal structure of **15** displays the following features, each of which is typical for aldehyde-Lewis acid complexes: (1) the Lewis acidic atom lies in the plane of the carbonyl group (B–O4–C7–C8 = –176°); (2) the Lewis acid binds *syn* to the hydrogen of the aldehyde rather than *anti*; (3) the Lewis acid-oxygen-carbon angle is roughly 120° (B–O4–C7 = 123°) (Scheme 1-4).

The coplanarity of the borabenzene ring and the α,β-unsaturated aldehyde is particularly noteworthy, a conformation that allows interaction between π-symmetry orbitals of the two fragments. The bond lengths in **15** are indicative of an unusually strong interaction between the Lewis acid and carbonyl moiety, since the

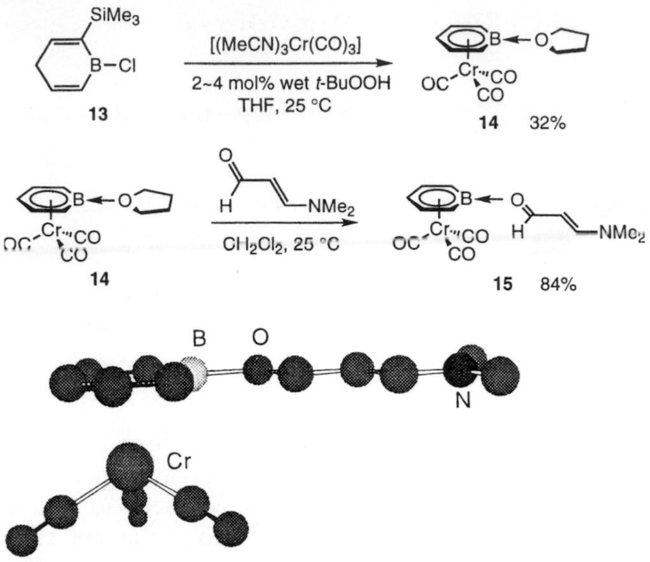

Scheme 1-4. Preparation of [(η^6-borabenzene-(3-(dimethylamino)acrolein)Cr(CO)$_3$] (**15**) and its crystal structure.

B–O length is extremely short (1.451 Å) for a dative bond, approaching the value found for the covalent B–O bond of Na[(η^6-borabenzene-OMe)Cr(CO)$_3$] (1.400 Å). The C=O and $C_{carbonyl}$–C_α bonds in **15** have comparable lengths, indicating that the π electrons of 3-(dimethylamino)acrolein are delocalized effectively upon complexation to [(η^6-borabenzene)Cr(CO)$_3$] (Table 1-8) [29].

Table 1-8. Bond lengths [Å] in bound and free 3-aminoacroleins.

Parameter	**15**	Free 3-aminoacrolein
B–O	1.451(5)	–
C=O	1.316(4)	1.23
$C_{carbonyl}$-C_α	1.345(5)	1.41

1.5 Carbonyl-Lewis Acid Chelation Complexes

1.5.1 With Transition Metal Elements

A high degree of stereoselectivity can be realized under chelation control, where an oxygen atom of an ether function (or more generally a Lewis base) in the α-, β- or possibly γ-position of carbonyl compounds can serve as an anchor for the metal center of a Lewis acid. Since Cram's pioneering work on chelation control in Grignard-type addition to chiral alkoxy carbonyl substrates [30], a number of studies on related subjects have appeared [31], and related transition state structures have been calculated [32]. Chelation control involves Cram's cyclic model and requires a Lewis acid bearing two coordination sites (usually transition metal-centered Lewis acids).

Reetz and co-workers reported that the $TiCl_4$-promoted addition of allyltrimethylsilane to 2-methoxycyclohexanone (**16**) produced a single diastereomer resulting from an equatorial attack. This remarkable selectivity can be nicely accounted for by the chelation of 2-methoxycyclohexanone (**16**) with $TiCl_4$, which was spectroscopically supported by ^{13}C NMR measurement, i.e., the uniform downfield shift of the carbonyl carbon atom, the α-carbon atom and the methoxy carbon atom of **16** (Scheme 1-5) [33].

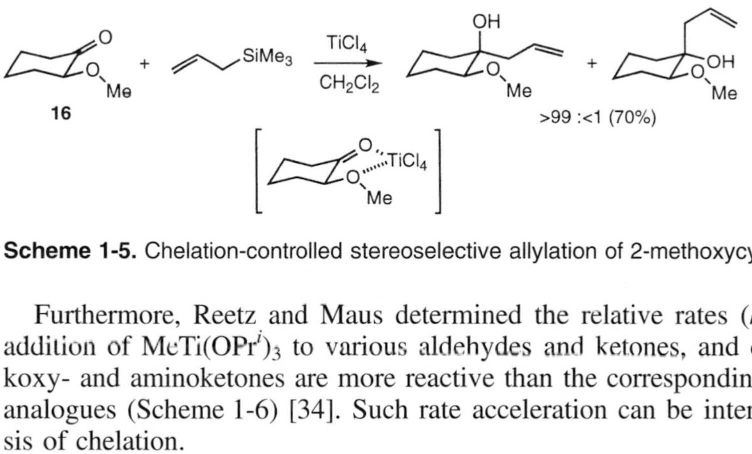

O Me **16** + SiMe$_3$ $\xrightarrow[CH_2Cl_2]{TiCl_4}$ OH O Me + OH O Me

>99 :<1 (70%)

[O TiCl$_4$ O Me]

Scheme 1-5. Chelation-controlled stereoselective allylation of 2-methoxycyclohexanone.

Furthermore, Reetz and Maus determined the relative rates (k_{rel}-values) of the addition of $MeTi(OPr^i)_3$ to various aldehydes and ketones, and claimed that α-alkoxy- and aminoketones are more reactive than the corresponding heteroatom-free analogues (Scheme 1-6) [34]. Such rate acceleration can be interpreted on the basis of chelation.

Panek and co-workers demonstrated that the reaction of (*S*)-2-benzyloxypropanal (**17**) with the allylic silane (*S*)-**18** under the influence of $BF_3{\cdot}OEt_2$ gave the *syn* homoallylic alcohol **19** with excellent level of Felkin induction [35]. Interestingly, the high level of *syn* selectivity was also observed in the condensation of **17** and (*R*)-**18**. On the other hand, the reaction of **17** with (*S*)-**18** in the presence of $TiCl_4$ produced *anti* homoallylic alcohol **20** almost exclusively, whereas the reaction with (*R*)-**18** promoted by $TiCl_4$ afforded *syn* homoallylic alcohol **21**. Presumably, the reactions proceeded through a Cram chelate transition state model

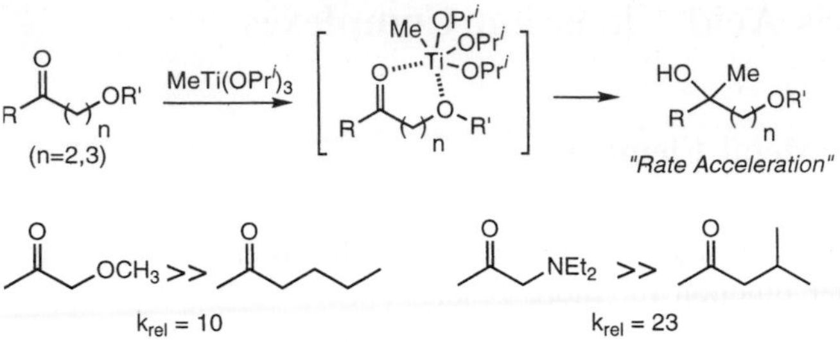

O
R OR'
n
(n=2,3)
$MeTi(OPr^i)_3$
Me OPr^i
OPr^i
O Ti OPr^i
O
R R'
n
HO Me
R OR'
n
"Rate Acceleration"

O O O O
OCH_3 >> NEt_2 >>
k_{rel} = 10 k_{rel} = 23

Scheme 1-6. Rate acceleration by chelate formation.

[30]. The set of experiments suggest that the stereochemistry of the emerging hydroxy group appears to be affected by chirality of aldehydes while the absolute stereochemical relationships are dictated by the configuration of the C-SiR_3 bond (Scheme 1-7) [36].

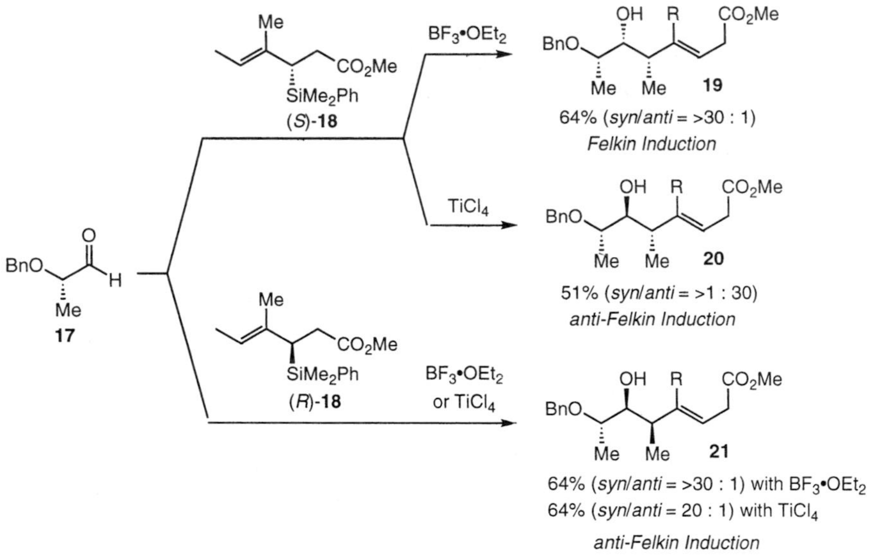

Me
CO_2Me
$SiMe_2Ph$
(*S*)-**18**
$BF_3•OEt_2$
OH R CO_2Me
BnO
Me Me **19**
64% (*syn*/*anti* = >30 : 1)
Felkin Induction
$TiCl_4$
OH R CO_2Me
BnO
Me Me **20**
51% (*syn*/*anti* = >1 : 30)
anti-Felkin Induction
O
BnO H
Me
17
Me
CO_2Me
$SiMe_2Ph$
(*R*)-**18**
$BF_3•OEt_2$
or $TiCl_4$
OH R CO_2Me
BnO
Me Me **21**
64% (*syn*/*anti* = >30 : 1) with $BF_3•OEt_2$
64% (*syn*/*anti* = 20 : 1) with $TiCl_4$
anti-Felkin Induction

Scheme 1-7. Double stereodifferentiation in the Lewis acid-promoted crotylation of (*S*)-2-benzyloxypropanal with chiral allylic silanes.

The $TiCl_4$-mediated Mukaiyama aldol reactions between π-allyltricarbonyliron lactone complexes and chiral aldehydes were well documented by Ley and coworkers [37]. (*R*)-Trimethylsilyl enol ether **23** (>96% ee) was prepared from the methyl ketone complex **22** by treatment with Me_3SiOTf/Et_3N in CH_2Cl_2 and this was then reacted with (*R*)- and (*S*)-2-benzyloxypropanal **24** under the influence of $TiCl_4$ in CH_2Cl_2 at –78 °C. Although the reactions proceeded very slowly and apparent hydrolysis of the silyl enol ether occurred, the aldol products **25** and **26** were isolated in excellent diastereoselectivity in both cases (Scheme 1-8). Interestingly, the diastereofacial preference of the present aldol reaction was governed almost entirely by the aldehyde, the inherent *re* face preference of the silyl enol ether having no significant effect.

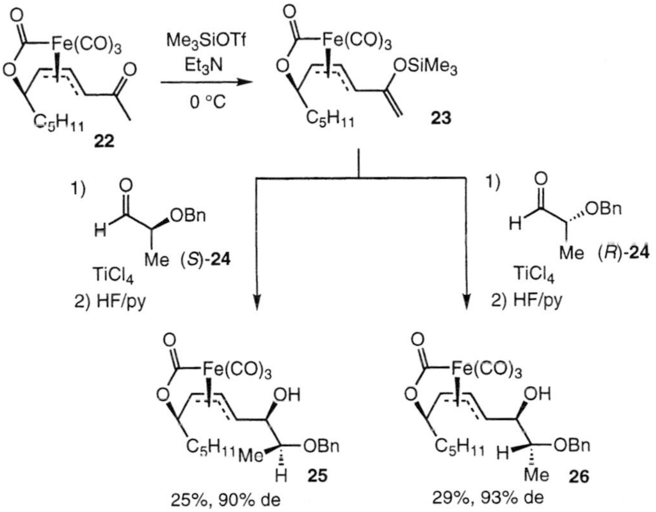

Scheme 1-8. Reaction of silyl enol ether complex **23** with α-benzyloxy aldehydes.

C_2-symmetric bis(oxazolinyl)pyridine (pybox)-Cu (II) complex **27** has been shown to catalyze highly enantioselective Mukaiyama aldol reactions between (benzyloxy)acetaldehyde and silyl ketene acetals by Evans and co-workers as exemplified in Scheme 1-9 [38]. Here, the requirement for a chelating substituent on the aldehyde partner is critical to catalyst selectivity, as α-(*tert*-butyldimethylsiloxy)acetaldehyde gave lower enantioselectivity (56% ee). In addition, β-(benzyloxy)propionaldehyde provided the racemic product, indicating a strict requirement for a five-membered catalyst-substrate chelate.

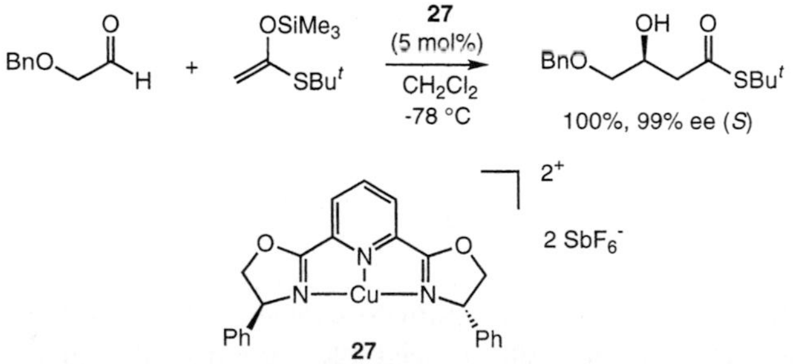

Scheme 1-9. C_2-Symmetric copper (II) complex catalyzed enantioselective aldol addition of silyl ketene acetal to benzyloxy acetaldehyde.

In the stereochemical model of the catalyst-aldehyde chelate complex, the square pyramidal complex **28** [39], the *re* aldehyde enantioface is shielded by the ligand phenyl group exposing the *si* enantioface to nucleophilic attack (Fig. 1-9). Since enantioselective formation of (*S*)-β-hydroxy esters is observed (*si* face attack), the absolute stereochemistry of the products is consistent with the proposed coordination model.

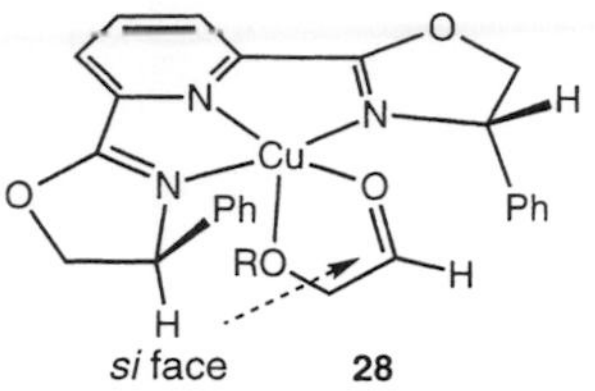

Figure 1-9. Stereochemical model of the catalyst-α-alkoxy aldehyde complex.

1.5.2 With Main-group Metal Elements

Although Lewis acids having main-group metals such as boron and aluminum (trivalent compounds; **G**) have been widely used in synthetic organic chemistry [40], they have been believed to act only as non-chelating Lewis acids through the formation of the corresponding tetracoordination complexes **H** with Lewis bases (L) [4]. Aside from several recent examples of neutral pentacoordinate, trigonal-bipyramidal complexes **I** [41, 42], little is known about another, synthetically more important pentacoordinate chelate-type complexes **J** [43], and its nature still remains to be studied.

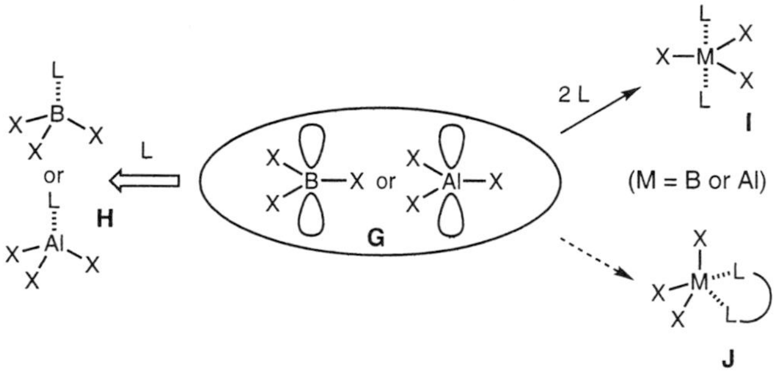

By taking the high affinity of boron and aluminum to oxygen into consideration [44], the authors investigated chelation-induced selective reduction of α-methoxy ketone **29** and its deoxy analog **30** with Bu_3SnH in the presence of several Lewis acids. Treatment of an equimolar mixture of **29** and **30** with a commonly used chelating Lewis acid, $TiCl_4$ and Bu_3SnH in toluene at –78 °C gave rise to a mixture of α-methoxy alcohol **31** accompanied by **32**. Under similar reaction conditions, reduction of **29** and **30** (1 : 1 ratio) with Me_3Al or $(C_6F_5)_3B$ [45] afforded α-methoxy alcohol **31** as a sole isolable product. The result implies the preferable formation of chelating pentacoordinate **K** (MX_3=Me_3Al or $(C_6F_5)_3B$) rather than a tetracoordinate **L** as illustrated in Scheme 1-10 [46].

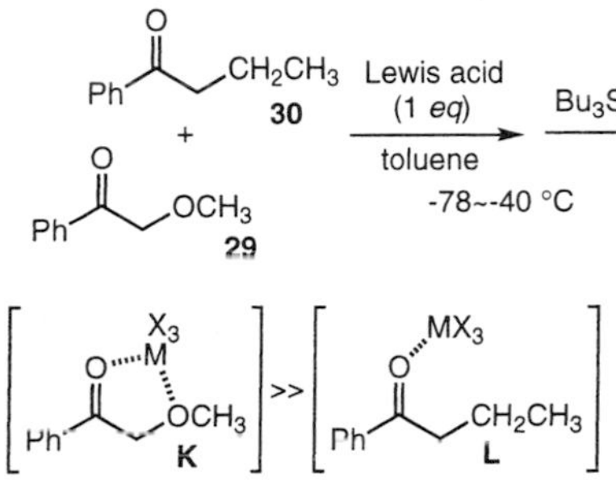

Lewis acid	**31** / **32**
$TiCl_4$	7:1 (85%)
Me_3Al	>20 : <1 (76%)
$(C_6F_5)_3B$	>20 : <1 (72%)

Scheme 1-10. Discriminative reduction of α-alkoxy ketone.

Moreover, $(C_6F_5)_3B$-promoted reduction of simple α-substituted ketone **33a** with Bu_3SnH gave a mixture of diastereomeric alcohols **34**, whereas chelation-controlled reduction of α-methoxy-α-methyl ketone **33b** with $(C_6F_5)_3B/Bu_3SnH$ afforded single diastereomer **35** exclusively (Scheme 1-11) [46].

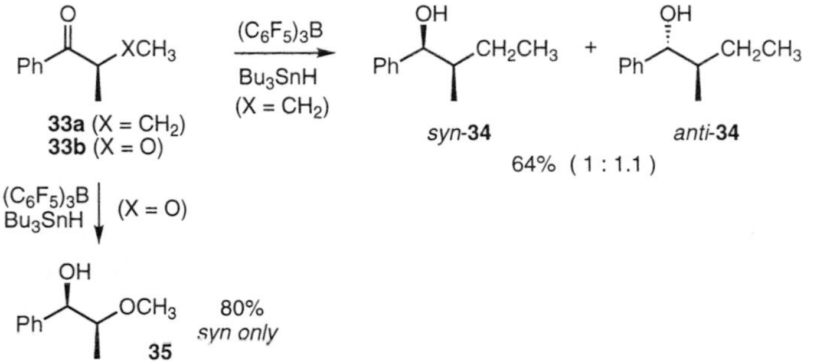

Scheme 1-11. $(C_6F_5)_3B$-promoted diastereoselective reduction via a pentacoordinate chelate-type intermediate.

Rate acceleration provided by the chelate formation was further observed in a discrimination experiment between two isomeric alkoxycarbonyl compounds **36** and **37** (X=OMe; R^1=Pr^i or H) as shown in Table 1-9 (entries 1-4). Thus, chelation-induced selective reduction of *o*-methoxyisobutyrophenone **36** (X=OMe; R^1=Pr^i) was observed to furnish *o*-methoxyphenyl carbinol **38** (X=OMe; R^1=Pr^i; R^2=H) preferentially with $(C_6F_5)_3B$ and Me_3Al. The $(C_6F_5)_3B$ and Me_3Al-promoted discriminative allylation of an equimolar mixture of *o*- and *p*-anisaldehyde,

36 and **37** (X=OMe; R^1=H) with allyltributyltin afforded *o*-methoxy homoallylic alcohol **38** (X=OMe; R^1=H; R^2=CH_2CH=CH_2) almost exclusively.

The chemoselective *o*-allylation of 2-methoxyphenyl-1,5-dicarboxaldehyde (**40**) appears feasible in the presence of organoboron Lewis acid (Scheme 1-12) [46].

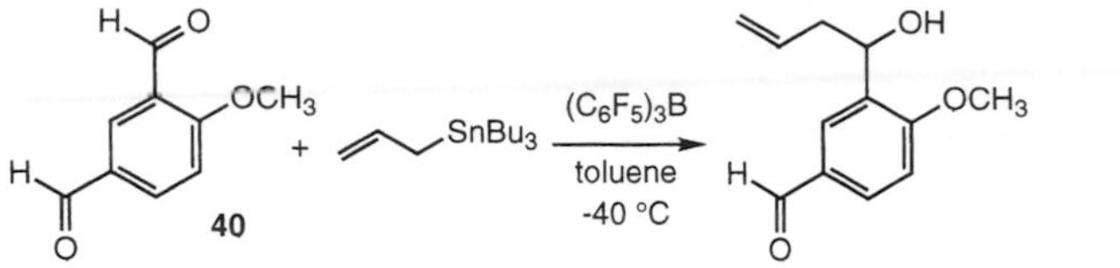

$(C_6F_5)_3B$, toluene, -40 °C; 40 + allyl-$SnBu_3$ → 70% (*o*-/*p*-allylation = 13 : 1)

Scheme 1-12. Chemoselective allylation of aldehyde carbonyls with $(C_6F_5)_3B$ and allyltributyltin.

In addition to its oxygenophilicity, aluminum has a high affinity toward fluorine [44], which enables selective alkylation of fluorocarbonyl compounds with organoaluminum reagents based on the pentacoordinate chelate formation as also shown in Table 1-9 [47]. Treatment of an equimolar mixture of 2-fluorobenzaldehyde **36** (X=F; R^1=H) and 4-fluorobenzaldehyde **37** (X=F; R^1=H) in toluene at –78°C with $Me_2AlC{\equiv}CPh$ resulted in formation of two different propargyl alcohols **38** and **39** (X=F; R^1=H, R^2=$C{\equiv}CPh$) in a ratio of 9.2:1 (entry 5, Table 1-9). The selectivity is lowered by switching the metals of PhC≡C–M from Al to Mg, Ti, and to Li (entries 6–8). A similar metal effect is observed with BuC≡C–M (M=$AlMe_2$ or Li) (entries 9 and 10). The high affinity of aluminum to fluorine compared to other halogens is evident from the discrimination experiment between chloro analogs with $Me_2AlC{\equiv}CPh$ (entry 11).

The advantage of aluminum reagents over other metal reagents was also seen in the Lewis acid-promoted reactions of fluoro carbonyl compounds with other alkylating agents (entries 12–22, Table 1-9) [48]. Indeed, Me_3Al-promoted selective allylation of an equimolar mixture of 2- and 4-fluorobenzaldehydes with allyltributyltin afforded the homoallylic alcohol **38** (X=F; R^1=H, R^2=CH_2CH=CH_2) almost exclusively (entry 12). A similar tendency is also observable in the selective reduction of *o*-fluorophenyl ketone **36** (X=F; R^1=Pr^i) over the *p*-fluoro analog with Me_3Al/Bu_3SnH (entry 17). Unsatisfactory results were obtained with Ti, Mg, Li, Si reagents in terms of chemical yield and selectivity (entries 13–16 and 20–22).

Based on these findings, the authors have found that high *anti*-selectivity [49] is achieved in the aldol reactions of fluoro aldehydes with ketene silyl acetals in the presence of Me_3Al (Scheme 1-13) [48]. For instance, Me_3Al-induced reaction of *o*-fluorobenzaldehyde (**40a**) with a substituted ketene silyl acetal gave rise to a mixture of fluoro *β*-hydroxy esters **41a** and **42a** with high diastereoselectivity (16:1), probably due to the effective fixation of the carbonyl moiety, while the selectivity was dramatically lowered when other common Lewis acids such as $BF_3{\cdot}OEt_2$, $TiCl_4$ and Me_3SiOTf [50] were used. In contrast, however, *o*-anisaldehyde (**40b**) and benzaldehyde (**40c**) exhibited moderate selectivity (5.3:1).

Table 1-9. Chemoselective functionalization of *o*- and *p*-substituted carbonyl derivatives.

36 + **37** → reagent or LA / Nu, toluene, -78~-40 °C → **38** + **39**

(LA = Lewis acid; Nu = nucleophile)

Run	Substrate (**36** and **37**)	Reagent Lewis acid/nucleophile	Product (**38** and **39**)	Ratio (**38**:**39**)
1	X=OMe; R^1=Pr^i	$(C_6F_5)_3/Bu_3SnH$	R^2=H	>20:1
2		Me_3Al/Bu_3SnH		11:1
3	X=OMe; R^1=H	$(C_6F_5)_3B/CH_2$=$CHCH_2SnBu_3$	R^2=CH_2CH=CH_2	>20:1
4		Me_3Al/CH_2=$CHCH_2SnBu_3$		>20:1
5	X=F; R^1=H	PhC≡$CAlMe_2$	R^2=C≡CPh	9.2:1
6		PhC≡CMgBr		3.7:1
7		PhC≡$CTiCl(OPr^i)_2$		2.7:1
8		PhC≡CLi		1.8:1
9		BuC≡$CAlMe_2$	R^2=C≡CBu	7.1:1
10		PhC≡CLi		1.1:1
11	X=Cl; R^1=H	PhC≡$CAlMe_2$	R^2=C≡CPh	2.4:1
12	X=F; R^1=H	Me_3Al/CH_2=$CHCH_2SnBu_3$	R^2=CH_2CH=CH_2	31:1
13		$TiCl_2(OPr^i)_2/CH_2$=$CHCH_2SnBu_3$		9.8:1
14		$MgBr_2/CH_2$=$CHCH_2SnBu_3$		7.1:1
15		$LiClO_4/CH_2$=$CHCH_2SnBu_3$		4.1:1
16		$SiCl_4/CH_2$=$CHCH_2SnBu_3$		3.8:1
17	X=F; R^1=Pr^i	Me_3Al/Bu_3SnH	R^2=H	34:1
18		Et_3Al/Bu_3SnH		26:1
19		Me_2AlCl/Bu_3SnH		7.3:1
20		$SiCl_4/Bu_3SnH$		2.3:1
21		$TiCl_2(OPr^i)_2/Bu_3SnH$		1.9:1
22		$MgBr_2/Bu_3SnH$		1.9:1

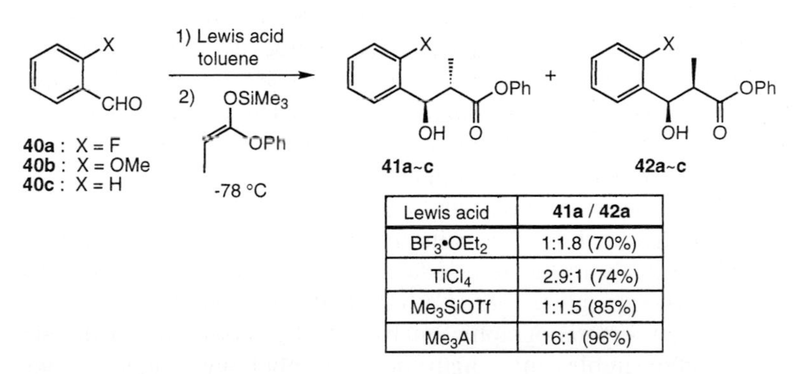

Lewis acid	**41a / 42a**
BF_3•OEt_2	1:1.8 (70%)
$TiCl_4$	2.9:1 (74%)
Me_3SiOTf	1:1.5 (85%)
Me_3Al	16:1 (96%)

Scheme 1-13. *anti*-Selective aldol reaction based on Al-F interaction.

1.6 Carbonyl-Bidentate Lewis Acid Complexes

1.6.1 Basic Study

The two principal modes of coordination of carbonyls to metal-centered Lewis acids are σ-bonding (**B**) and π-bonding (**D**) [51]. The former mode is generally preferred with main-group Lewis acids. In addition, simultaneous coordination to carbonyl groups with two metals of type (**E1**) would alter the reactivity and selectivity of the carbonyl substrates. Examples of such double coordination with two main-group metals are rare despite its theoretical, mechanistic, and synthetic importance, simply because of the high preference for the single coordination mode (**B**) even in the presence of excess Lewis acids, and hence the nature of such di-σ-bonding (**E1**) has remained an elusive phenomenon [52].

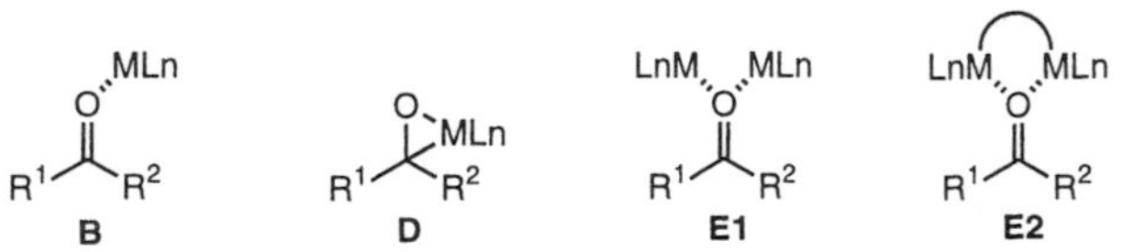

Wuest and co-workers introduced phenylenedimercury dichloride (**43**) as a bidentate Lewis acid forming a 1 : 1 complex of type **E2** with dimethylformamide in which the carbonyl oxygen atom is bonded to both atoms of mercury at once [53]. Furthermore, they discovered that crystallization of the more strongly Lewis-acidic bis(trifluoroacetate) **44** [54] from dimethylformamide or diethylformamide produces complexes in which **44** and amide are present in a 2 : 3 molar ratio. X-ray crystallographic analysis of the complex with diethylformamide revealed the remarkable feature of the structure, i.e., the binding of the amide whose oxygen atom interacts simultaneously with four Lewis-acidic atoms of mercury, creating the unprecedented partial structure as shown in Fig. 1-10 [55]. Each 1,2-phenylenedimercury unit forms one short and one long dative Hg–O bond. The two shortest bonds are those that lie closest to the carbonyl plane, while the two longest bonds lie in an approximately orthogonal plane.

The same group have shown that in an intramolecular case, where the carbonyl was tethered by two aluminums, both the main-group Lewis acids coordinated simultaneously and symmetrically to the central ketone as depicted in Scheme 1-14 [56]. By carrying out low-temperature (–85 °C) ^{13}C NMR [the upfield shift of the carbonyl carbon] and X-ray crystallographic analysis [the carbon-oxygen double bond (1.34(2) Å) is comparable in length to the phenoxy single bonds (1.32(2) Å)], they concluded that the corresponding resonance hybrid must contribute despite a partial loss of aromaticity.

The double-coordination ability of 1,8-naphthalenediylbis(dichloroborane) (**45**) has been elucidated by Oh and Reilly using 2,6-dimethylpyranone (**46**) as a carbo-

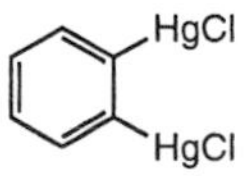

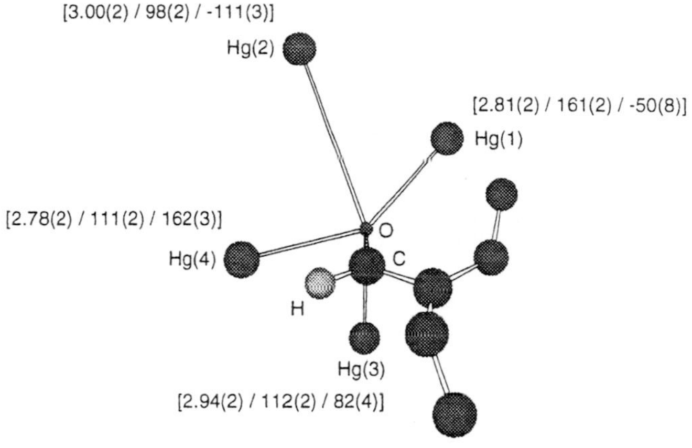

Figure 1-10. Partial structure of **44**-diethylformamide complex.

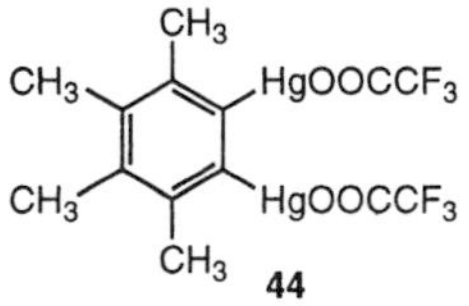

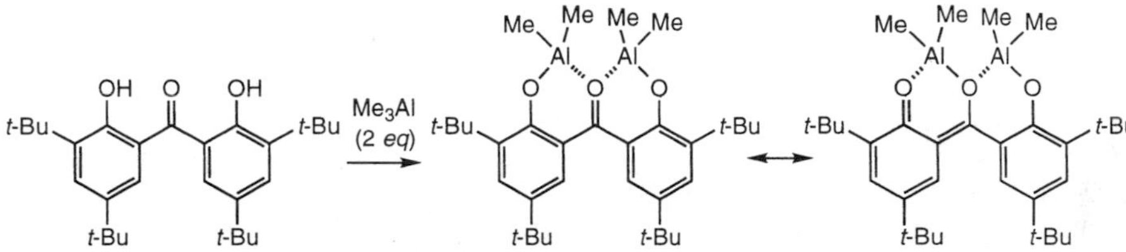

Scheme 1-14. Intramolecular simultaneous coordination of two aluminums to ketone carbonyl.

nyl substrate. A titration experiment was performed with **45** and **46**, and monitored by ^{1}H NMR, which revealed that there were no sharp titration points. The spectrum of a 1:2 mixture of **45** and **46** indicates that all ketone **46** is coordinated, suggesting the existence of a complex of type **48** that is in equilibrium with complex **47**. With one equivalent of **46**, there is a significant amount of complex **48**, while the signal due to complex **48** is totally suppressed with excess Lewis acid **45** and the sharp signals corresponding to complex **47** is clearly observed (Scheme 1-15) [57].

The authors have designed modified bis(organoaluminum) reagent **49** for the efficient simultaneous coordination toward carbonyls (type **E2**), and successfully elucidated its reactivity and selectivity in the typical synthetic transformations [58].

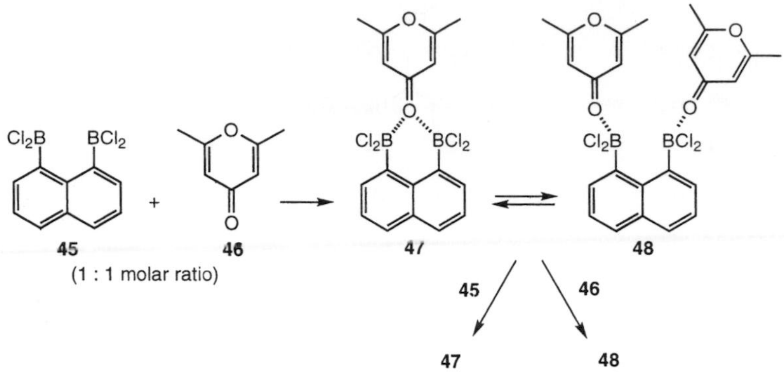

Scheme 1-15. 1,8-Naphthalenediylbis(dichloroborane) (**45**) as bidentate Lewis acid.

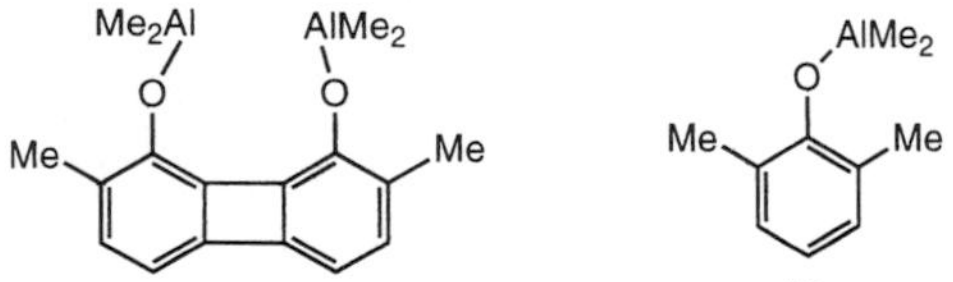

49 **50**

Initial complexation of 5-nonanone with the (2,7-dimethyl-1,8-biphenylenedioxy)bis(dimethylaluminum) (**49**) (1.1 equiv) in CH_2Cl_2 and subsequent reaction of Bu_3SnH at low temperature gave rise to the corresponding 5-nonanol in high yield. In marked contrast, however, reduction of 5-nonanone with Bu_3SnH in the presence of monodentate organoaluminum reagent **50** under similar reaction conditions afforded 5-nonanol in only 6% yield. These results clearly demonstrate that the bidentate Lewis acid **49** strongly enhances the reactivity of ketone carbonyl toward hydride transfer via the double electrophilic activation of carbonyl moiety. A similar tendency is observed in the acetophenone carbonyl reduction (Scheme 1-16).

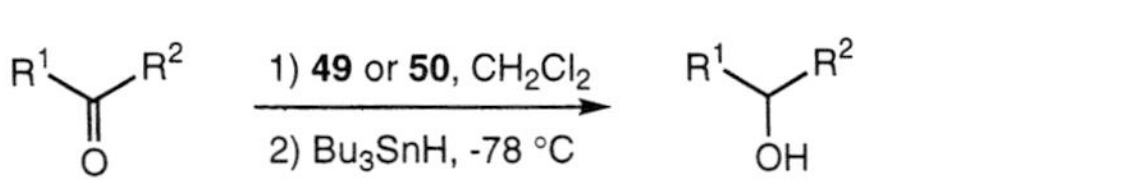

$R^1 = R^2 = Bu$: 86% with **49** (6% with **50**)
$R^1 = Ph, R^2 = Me$: 91% with **49** (9% with **50**)

Scheme 1-16. Double electrophilic activation by bidentate aluminum Lewis acid **49** in the reduction of ketones.

Further, the Mukaiyama aldol reaction of 1-(trimethylsiloxy)-1-cyclohexene and benzaldehyde was effected with the bidentate **49**, giving the aldol products (*erythro/threo* = 1 : 3) in 87% yield, though its monodentate counterpart **50** showed no evidence of reaction under similar conditions (Scheme 1-17).

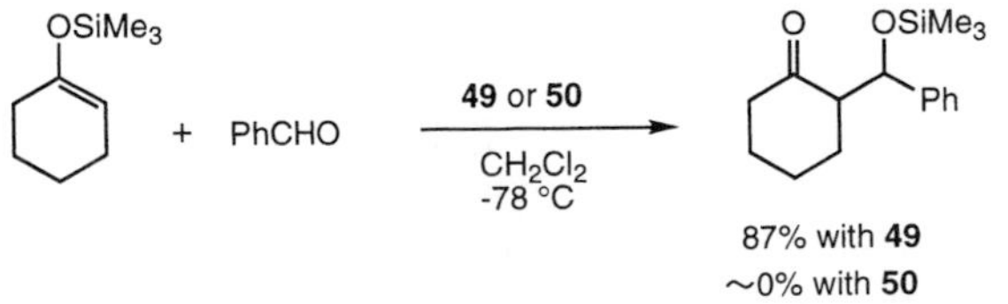

Scheme 1-17. Reactivity of **49** in the Mukaiyama aldol reaction.

Spectroscopic evidence for the double coordination and activation behavior of the bidentate **49** was obtained by low-temperature ^{13}C NMR spectroscopy using DMF as a carbonyl substrate. The 75 MHz ^{13}C NMR measurement of the 1:1 monodentate **50**-DMF complex (**M**) in $CDCl_3$ at –50 °C showed that the original signals of DMF carbonyl at δ 162.66 shifted downfield to δ 164.05. In contrast, the 1:1 bidentate **49**-DMF chelation complex under similar conditions undergoes a further downfield shift for the DMF carbonyl (δ 165.62), implying the strong electrophilic activation of the DMF carbonyl by the intervention of double coordination complex (**N**). Addition of one more equiv of DMF to the 1:1 bidentate **49**-DMF complex gives two signals at δ 163.71 and δ 165.63 in a ratio of about 1:1, suggesting the equilibrium between the coordination complex (**O**) and the double coordination complex (**N**) (Scheme 1-18).

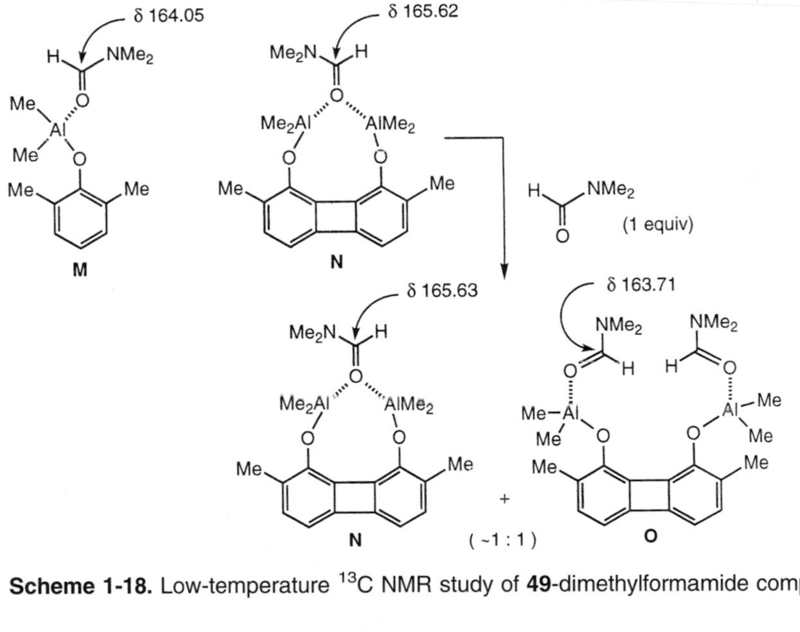

Scheme 1-18. Low-temperature ^{13}C NMR study of **49**-dimethylformamide complex.

Another interesting feature of the bidentate Lewis acid **49** in organic synthesis is the regio- and stereocontrolled Michael addition of silyl ketene acetals to α,β-unsaturated ketones as acceptors. Reaction of benzalacetone and silyl ketene acetal **51** with dimethylaluminum aryloxides of type **52** gave rise to a mixture of Michael adducts **53** and **54** almost exclusively, where the *Z* selectivity decreased with increase in the steric size of a phenoxy ligand in **52**. Indeed, switching the phenoxy group to 2,6-xylenoxy, 2,6-diisopropylphenoxy, and 2,6-di-*tert*-butylphenoxy groups, the *Z* selectivity decreased from 80:20 to 70:30, 67:33, and 33:67, respectively. Based on the experimental findings, the stereochemical outcome of the *Z*-isomeric Michael adduct **53** is interpreted for by the preferable complex (**P**) formation of benzalacetone with sterically less hindered **52a** or **50**. With more hindered **52b** or **52c**, the coordination complex (**R**) is then favored rather than the sterically congested complex (**Q**), thereby increasing the formation of *E*-isomeric Michael adduct **54**. In the ultimate case, bidentate **51** can be utilized to obtain *E*-isomeric **54** as a major product via the complex (**S**) formation with *s-trans* conformation (Scheme 1-19).

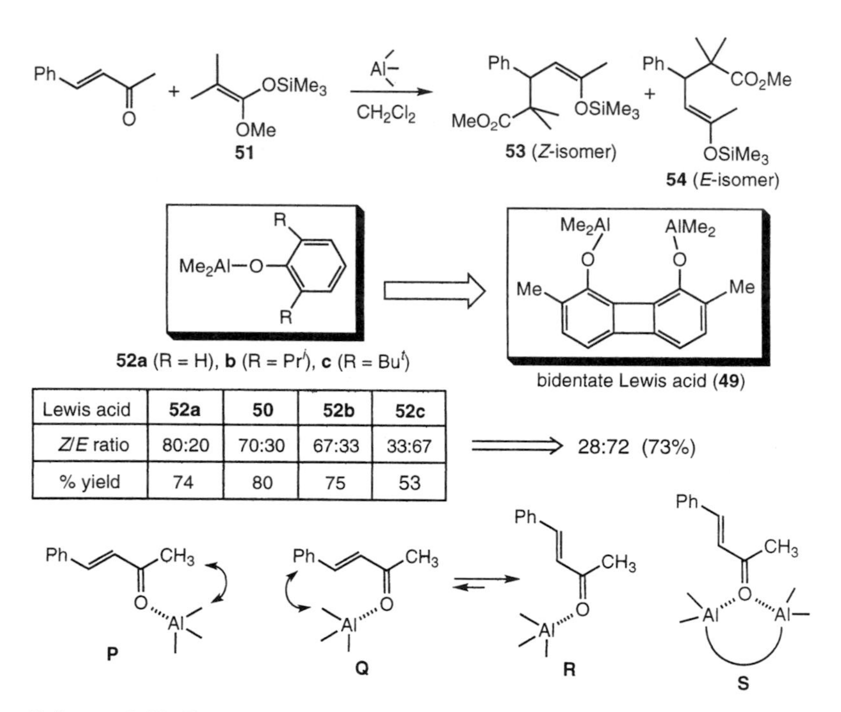

Lewis acid	52a	50	52b	52c
Z/*E* ratio	80:20	70:30	67:33	33:67
% yield	74	80	75	53

Scheme 1-19. Stereocontrolled Michael addition of silyl ketene acetal to benzalacetone.

The authors devised the new bidentate titanium catalyst (anthraquinone-1,8-dioxy)bis(triisopropoxytitanium) (**55**) and utilized it for the simultaneous coordination of carbonyl substrates [59]. Comparison of the reactivity and selectivity with the corresponding monodentate titanium catalyst **56** in several synthetic examples genuinely demonstrates the high double-activation ability of **55** toward carbonyls under catalytic conditions as illustrated in Scheme 1-20.

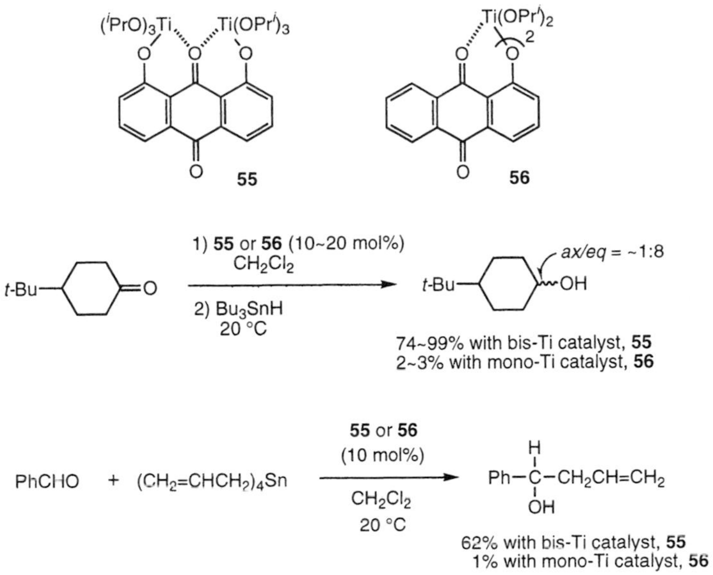

(^{i}PrO)$_3$Ti Ti(OPr^i)$_3$ **55** Ti(OPr^i)$_2$ 2 **56**

t-Bu–C$_6$H$_{10}$=O → 1) **55** or **56** (10~20 mol%) CH_2Cl_2 2) Bu_3SnH 20 °C → t-Bu–C$_6$H$_{10}$–OH (ax/eq = ~1:8)

74~99% with bis-Ti catalyst, **55**
2~3% with mono-Ti catalyst, **56**

PhCHO + $(CH_2{=}CHCH_2)_4Sn$ → **55** or **56** (10 mol%), CH_2Cl_2, 20 °C → Ph–CH(OH)–$CH_2CH{=}CH_2$

62% with bis-Ti catalyst, **55**
1% with mono-Ti catalyst, **56**

Scheme 1-20. Double activation ability of bidentate titanium catalyst **55** in the reduction and allylation of carbonyl substrates.

1.6.2 Synthetic Aspect

Although bidentate Lewis acids still remain poorly studied, it is increasingly difficult to dismiss them as esoteric reagents of mere academic interest because truly efficient and useful synthetic applications have recently appeared. The authors reported a new catalytic Meerwein-Ponndorf-Verley reduction [60, 61] system based on the bidentate Lewis acid chemistry [62]. Treatment of benzaldehyde with (2,7-dimethyl-1,8-biphenylenedioxy)bis(diisopropoxyaluminum) (**57**) at room temperature instantaneously produced the reduced benzyl alcohol almost quantitatively (entry 2, Table 1-10). Moreover, even with 5 mol% of the catalyst **57** the reduction proceeds quite smoothly at room temperature to furnish benzyl alcohol in 81% yield after 1 h (entry 3, Table 1-10). This remarkable efficiency can be ascribed to the double electrophilic activation of carbonyls by the bidentate aluminum catalyst (Scheme 1-21).

Other selected examples are summarized in Table 1-10. In addition to aldehydes, both cyclic and acyclic ketones can be reduced equally well. *sec*-Phenethyl alcohol (**59**, R=Ph) as hydride source works more effectively than *i*-PrOH. Based on this finding, the asymmetric MPV reduction of unsymmetrical ketones [63] with chiral alcohol in the presence of catalyst **58** was examined. Treatment of 2-chloroacetophenone (**60**) with optically pure (*R*)-(+)-*sec*-phenethyl alcohol (1 equiv) under the influence of catalytic **58** afforded (*S*)-(+)-2-chloro-1-phenylethanol (**61**) with moderate asymmetric induction (82%, 54% ee). Switching chiral alcohols from (*R*)-(+)-*sec*-phenethyl alcohol to (*R*)-(+)-*α*-methyl-2-naphthalenemethanol and (*R*)-(+)-*sec*-*o*-bromophenethyl alcohol further enhanced the optical yields of **61** in 70 and 82% ee, respectively [62].

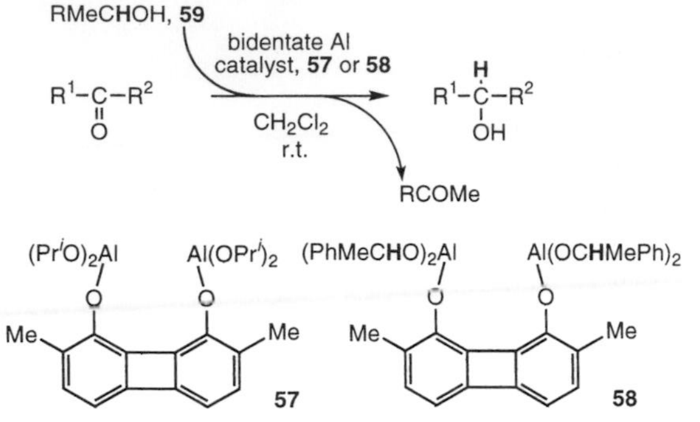

Scheme 1-21. Catalytic Meerwein-Ponndorf-Verley reduction of carbonyl compounds with bidentate aluminum alkoxides.

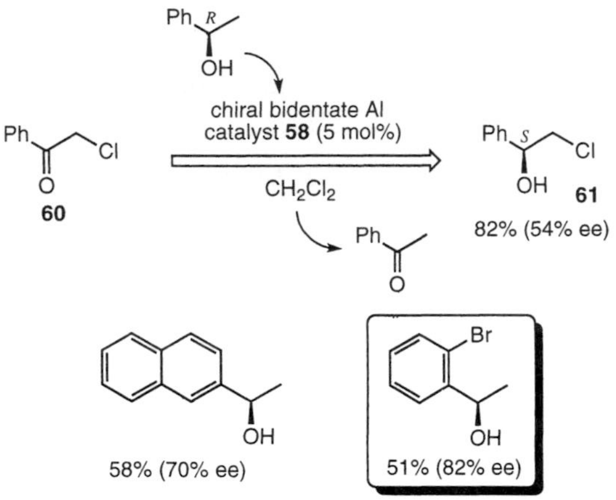

Scheme 1-22. Asymmetric MPV reduction of unsymmetrical ketones.

Furthermore, the authors have successfully developed a highly accelerated Oppenauer oxidation system [64, 65] using a bidentate aluminum catalyst. This modified catalytic system effectively oxidizes a variety of secondary alcohols to the corresponding ketones as shown in Scheme 1-23. For example, reaction of (2,7-dimethyl-1,8-biphenylenedioxy)bis(dimethylaluminum) (**49**, 5 mol%) with carveol (**62**) in the presence of 4 Å molecular sieves, and subsequent treatment with pivalaldehyde (3 equiv) yielded carvone (**63**) in 91% yield. Under the oxidation conditions, cholesterol (**64**) was converted to 4-cholesten-3-one (**65**) in 75% yield (91% yield with 5 equiv of *t*-BuCHO) [62].

A simultaneous reduction/oxidation sequence of hydroxy carbonyl substrates in the Meerwein-Ponndorf-Verley reduction can be accomplished by use of a catalytic amount of (2,7-dimethyl-1,8-biphenylenedioxy)bis(dimethylaluminum) (**49**). This represents an efficient hydride transfer from the *sec*-alcohol moiety to the remote carbonyl group and, due to its insensitivity to other functionalities, should find vast potential in the synthesis of complex polyfunctional molecules including both natural

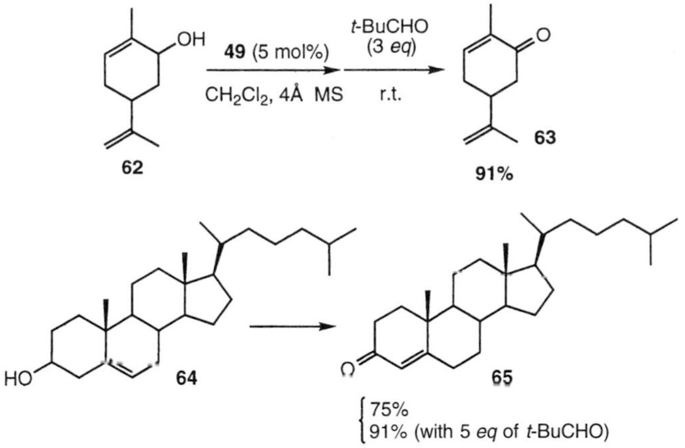

Scheme 1-23. Efficient, catalytic Oppenauer oxidation using bidentate aluminum catalyst.

and unnatural products. Treatment of hydroxy aldehyde **66** with **49** (5 mol%) in CH_2Cl_2 at 21 °C resulted in formation of hydroxy ketone **67** in 78% yield. As expected, the use of 25 mol% of **49** enhanced the rate and the chemical yield was increased to 92%. It should be noted that the present reduction/oxidation sequence is highly chemoselective, and can be utilized in the presence of other functionalities such as esters, amides, *tert*-alcohols, nitriles and nitro compounds as depicted in Scheme 1-24 [66].

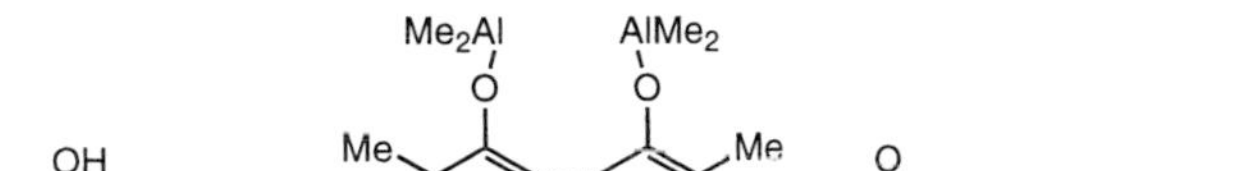

Scheme 1-24. Chemoselective simultaneous reduction/oxidation of hydroxy carbonyl substrates.

Table 1-10. Catalytic MPV reduction of carbonyl substrates with bidentate Al catalyst[a])

Entry	Substrate	Al reagent	Hydride source	Conditions	Yield, %
1	PhCHO	$Al(OPr^i)_3$ (1 eq)	*i*-PrOH (1 eq)	r.t., 2 h	10
2		**57** (1 *eq*)	*i*-PrOH (1 eq)	r.t., 1 min	>99
3		**57** (5 mol%)	*i*-PrOH (1 eq)	r.t., 1 h	81
4		**57** (5 mol%)	*i*-PrOH (3 eq)	r.t., 1 h	96
5	$PhCH(CH_2)_2C{=}O$	$Al(OPr^i)_3$ (1 eq)	*i*-PrOH (1 eq)	r.t., 2 h	trace
6		**57** (5 mol%)	*i*-PrOH (1 eq)	r.t., 1 h	91
7		**57** (5 mol%)	*i*-PrOH (1 eq)	r.t., 2 h	99
8	$PhC(=O)CH_2Cl$	$Al(OPr^i)_3$ (1 eq)	*i*-PrOH (1 eq)	r.t., 2 h	N.R.
9		**57** (5 mol%)	*i*-PrOH (1 eq)	r.t., 2 h	75
10		**57** (5 mol%)	*i*-PrOH (1 eq)	r.t., 10 h	89
11		**58** (5 mol%)	PhMeCHOH (1 eq)	r.t., 2 h	>99
12	$CH_3(CH_2)_8COCH_3$	$Al(OPr^i)_3$ (1 eq)	*i*-PrOH (1 eq)	r.t., 5 h	N.R.[b])
13		**57** (5 mol%)	*i*-PrOH (1 eq)	r.t., 5 h	52
14		**58** (5 mol%)	PhMeCHOH (1 eq)	r.t., 5 h	73
15		**58** (5 mol%)	PhMeCHOH (3 eq)	r.t., 5 h	89
16	$PhCH{=}CHCOCH_3$	**57** (5 mol%)	*i*-PrOH (1 e)	r.t., 5 h	31[c])
17		**58** (5 mol%)	PhMeCHOH (6 eq)	r.t., 5 h	70[c])

[a]) The MPV reduction of carbonyl substrates was effected with several Al catalysts under the given reaction conditions.
[b]) N.R. = No reaction.
[c]) Yields of 1,2-reduction products.

References

1. Dugas, H. Ed. *Bioorganic Chemistry, 3rd edn.*, Springer, New York, 1996.
2. (a) Heathcock, C.H. *Asymmetric Synthesis*, Morrison, J.P., Ed.; Academic, New York, 1984; Vol. 3, p 111. (b) Reetz, M.T. *Angew. Chem., Int. Ed. Engl.* **1984**, *23*, 556. (c) Oppolzer, W. *Angew. Chem., Int. Ed. Engl.* **1984**, *23*, 876. (d) ApSimon, J.W.; Collier, T.L. *Tetrahedron* **1986**, *42*, 5157. (e) Corey, E.J.; Bakshi, R.K.; Shibata, S. *J. Am. Chem. Soc.* **1987**, *109*, 5551.
3. The development of mild nucleophilic reagents for carbon-carbon bond formation is a typical example: (a) Calas, R.; Dunogues, J.; Deleris, G.; Pisciotti, F. *J. Organomet. Chem.* **1974**, *69*, C15. (b) Deleri, G.; Dunogues, J.; Calas, R. *J. Organomet. Chem.* **1975**, *93*, 43. (c) Hosomi, A.; Sakurai, H. *Tetrahedron Lett.* **1976**, 1295.
4. (a) Yamaguchi, M. In *Comprehensive Organic Synthesis, Vol. 1, Addition to C-X-Bonds, Part 1*; Schreiber, S.L., Ed.; Pergamon Press: Oxford, 1991; Chapter 1.11. (b) *Selectivities in Lewis Acid Promoted Reactions*; Schinzer, D., Ed.; Kluwer Academic Publishers: Dordrecht, 1989. (c) Santelli, M.; Pons, J.-M. *Lewis Acids and Selectivity in Organic Synthesis*; CRC Press, Boca Raton, 1995.
5. (a) Brunner, H.; Wachter, J.; Bernal, I.; Creswick, M. *Angew. Chem., Int. Ed. Engl.* **1979**, *18*, 861. (b) Sacerdoti, M.; Bertolasi, V.; Gilli, G. *Acta Crystallogr., Sect. B* **1980**, *36*, 1061. (c) Poll, T.; Metter, J.O.; Helmchen, G. *Angew. Chem., Int. Ed. Engl.* **1985**, *24*, 112. (d) Fernandez, J.M.; Emerson, K.; Larsen, R.H.; Gladysz, J.A. *J. Am. Chem. Soc.* **1986**, *108*, 8268. (e) Mendez, N.Q.; Arif, A.M.; Gladysz, J.A. *Angew. Chem., Int. Ed. Engl.* **1990**, *29*, 1473.
6. See section 1.6.
7. Reetz, M.T.; Maus, S. *Tetrahedron* **1987**, *43*, 101. See also ref 2b.
8. Reetz, M.T.; Hullmann, M.; Massa, W.; Berger, S.; Rademacher, P.; Heymanns, P. *J. Am. Chem. Soc.* **1986**, *108*, 2405.
9. Houk, K.N.; Strozier, R.W. *J. Am. Chem. Soc.* **1973**, *95*, 4094.

10. LePage, T.J.; Wiberg, K.B. *J. Am. Chem. Soc.* **1988**, *110*, 6642.
11. Ren, J.; Cramer, C.J.; Squires, R.R. *J. Am. Chem. Soc.* **1999**, *121*, 2633.
12. Marinelli, P.J.; Paulino, J.A.; Sunderlin, L.S.; Wenthold, P.G.; Poutsma, J.C.; Squires, R.R. *Int. J. Mass Spectrom. Ion Processes* **1994**, *130*, 89.
13. Bartmess, J.E. *Mass Spectrom. Rev.* **1989**, *8*, 297.
14. Mikami, K.; Yajima, T.; Terada, M.; Uchimaru, T. *Tetrahedron Lett.* **1993**, *34*, 7591.
15. A semi-empirical and ab initio MO study on fluoroketones: Linderman, R. J.; Jamois, E. A. *J. Fluor. Chem.* **1991**, *53*, 79.
16. Sbai, A.; Branchadell, V.; Ortuno, R.M.; Oliva, A. *J. Org. Chem.* **1997**, *62*, 3049.
17. (a) Denmark, S.E.; Henke, B.R.; Weber, E. *J. Am. Chem. Soc.* **1987**, *109*, 2512. (b) Denmark, S.E.; Almstead, N.G. *J. Am. Chem. Soc.* **1993**, *115*, 3133.
18. Corey, E.J.; Loh, T.-P.; Sarshar, S.; Azimioara, M. *Tetrahedron Lett.* **1992**, *33*, 6945.
19. Ishihara, K.; Gao, Q.; Yamamoto, H. *J. Am. Chem. Soc.* **1993**, *115*, 10413.
20. (a) Seebach, D. *Angew. Chem., Int. Ed. Engl.* **1988**, *27*, 1624. (b) Boche, G. *Angew. Chem., Int. Ed. Engl.* **1989**, *28*, 277.
21. (a) Ceccarelli, C.; Jeffrey, G.A.; Taylor, R. *J. Mol. Struct.* **1981**, *70*, 255. (b) Newton, R.; Jeffrey, G.A.; Takagi, S. *J. Am. Chem. Soc.* **1979**, *101*, 1997.
22. Chakrabarti, P.; Dunitz, J.D. *Helv. Chim. Acta* **1982**, *65*, 1482.
23. Shambayati, S.; Crowe, W.E.; Schreiber, S.L. *Angew. Chem., Int. Ed. Engl.* **1990**, *29*, 256.
24. Mahadevan, C.; Seshasayee, M.; Kothiwal, A.S. *Cryst. Struct. Commun.* **1982**, *11*, 1725.
25. Honeychuck, R.V.; Bonnesen, P.V.; Farahi, J.; Hersh, W.H. *J. Org. Chem.* **1987**, *52*, 5293.
26. Recent reviews: (a) Ishihara, K.; Yamamoto, H. *Advances in Catalytic Processes*; JAI, Greenwich, CT, 1995; p 29. (b) Deloux, L.; Srebnik, M. *Chem. Rev.* **1993**, *93*, 763. (c) Mikami, K.; Shimizu, M. *Chem. Rev.* **1992**, *92*, 1021. (d) Narasaka, K. *Synthesis* **1991**, 1.
27. (a) Amendola, M.C.; Stockman, K.E.; Hoic, D.A.; Davis, W. M.; Fu, G.C. *Angew. Chem., Int. Ed. Engl.* **1997**, *36*, 267. (b) Tweddell, J.; Hoic, D.A.; Fu, G.C. *J. Org. Chem.* **1997**, *62*, 8286.
28. Hoic, D.A.; Wolf, J.R.; Davis, W.M.; Fu, G.C. *Organometallics* **1996**, *15*, 1315.
29. X-ray diffraction studies of free 3-aminoacroleins: (a) Kulpe, S.; Schulz, B. *Krist. Tech.* **1979**, *14*, 159. (b) Bai, C.; Yu, Z.; Fu, H.; Tang, Y. *Jiegou Huaxue* **1984**, *3*, 65.
30. (a) Cram, D.J.; Kopecky, K.R. *J. Am. Chem. Soc.* **1959**, *81*, 2748. (b) Leitereg, T.J.; Cram, D.J. *J. Am. Chem. Soc.* **1968**, *90*, 4019.
31. Nogradi, M. In *Stereoselective Synthesis*; VCH, Weinheim, 1987; p 160.
32. (a) Frenking, G.; Kohler, K. F.; Reetz, M. T. *Tetrahedron* **1993**, *49*, 3971. (b) Frenking, G.; Kohler, K. F.; Reetz, M. T. *Tetrahedron* **1993**, *49*, 3983.
33. Reetz, M. T.; Kesseler, K.; Schmidtberger, S.; Wenderoth, B.; Steinbach, R. *Angew. Chem., Int. Ed. Engl.* **1983**, *22*, 989.
34. Reetz, M.T.; Maus, S. *Tetrahedron* **1987**, *43*, 101.
35. (a) Cherest, M.; Felkin, H.; Prudent, N. *Tetrahedron Lett.* **1968**, 2199. (b) Anh, N.T.; Eisenstein, O. *Nouv. J. Chem.* **1977**, *1*, 61.
36. (a) Jain, N.F.; Cirillo, P.F.; Pelletier, R.; Panek, J.S. *Tetrahedron Lett.* **1995**, *36*, 8727. (b) Jain, N.F.; Takenaka, N.; Panek, J.S. *J. Am. Chem. Soc.* **1996**, *118*, 12475.
37. Ley, S.V.; Cox, L.R.; Worrall, J.M. *J. Chem. Soc., Perkin Trans. 1* **1998**, 3349.
38. (a) Evans, D.A.; Murry, J.A.; Kozlowski, M.C. *J. Am. Chem. Soc.* **1996**, *118*, 5814. (b) Evans, D.A.; Kozlowski, M.C.; Murry, J.A.; Burgey, C.S.; Campos, K.R.; Connell, B.T.; Staples, R.J. *J. Am. Chem. Soc.* **1999**, *121*, 669.
39. Five-coordinated Cu(II) complexes exhibit a strong tendency toward either square pyramidal or trigonal bipyramidal geometries: Hathaway, B.J. In *Comprehensive Coordination Chemistry*; Wilkinson, G. Ed.; Pergamon Press, New York, 1987; Vol. 5, Chapter 53.
40. Negishi, E. *Organometallics in Organic Synthesis*; John Wiley & Sons, New York, 1980.
41. Lee, D.Y., Maetin, J.C. *J. Am. Chem. Soc.* **1984**, *106*, 5745.
42. (a) Heitsch, C. W.; Nordman, C.E.; Parry, P.W. *Inorg. Chem.* **1963**, *2*, 508. (b) Palenick, G. *Acta Crystallogr.* **1964**, *17*, 1573. (c) Beattie, I.R.; Ozin, G.A. *J. Chem. Soc. A*, **1968**, 2373. (d) Bennett, F.R.; Elms, F.M.; Gardiner, M.G.; Koutsantonis, G.A.; Raston, C.L.; Roberts, N.K. *Organometallics* **1992**, *11*, 1457. (e) Muller, G.; Lachmann, J.; Rufinska, A. *Organometallics* **1992**, *11*, 2970. (f) Fryzuk, M.D.; Giesbrecht, G.R.; Olovsson, G.; Rettig, S.J. *Organometallics* **1996**, *15*, 4832.

43. Maruoka, K.; Ooi, T. *Chem. Eur. J.* **1999**, *5*, 829.
44. Lide, D.R. *CRC Handbook of Chemistry and Physics; 78th edn., CRC Press, New York, 1997/1998.*
45. Utility of $(C_6F_5)_3B$ as Lewis acids: (a) Ishihara, K.; Hanaki, N.; Yamamoto, H. *Synlett* **1993**, 577. (b) Parks, D.J.; Piers, W.E. *J. Am. Chem. Soc.* **1996**, *118*, 9440.
46. Ooi, T.; Uraguchi, D.; Kagoshima, N.; Maruoka, K. *J. Am. Chem. Soc.* **1998**, *120*, 5327.
47. Ooi, T.; Kagoshima, N.; Maruoka, K. *J. Am. Chem. Soc.* **1997**, *119*, 5754.
48. Ooi, T.; Kagoshima, N.; Uraguchi, D.; Maruoka, K. *Tetrahedron Lett.* **1998**, *39*, 7105.
49. For a recent example: Ghosh, A.K.; Onishi, M. *J. Am. Chem. Soc.* **1996**, *118*, 2527.
50. Noyori, R.; Murata, S.; Suzuki, M. *Tetrahedron* **1981**, *37*, 3899.
51. (a) Harman, W.D.; Fairlie, D.P.; Taube, H. *J. Am. Chem. Soc.* **1986**, *108*, 8223. (b) Huang, Y.-H.; Gladysz, J. *J. Chem. Educ.* **1988**, *65*, 298. (c) Klein, D. P.; Dalton, D. M.; Mendez, N. Q.; Arif, A. M.; Gladysz, J. *J. Organomet. Chem.* **1991**, *412*, C7.
52. Bidentate Brønsted acids: (a) Hine, J.; Linden, S.-M.; Kanagasabapathy, V.M. *J. Org. Chem.* **1985**, *50*, 5096. (b) Hine, J.; Ahn, K. *J. Org. Chem.* **1987**, *52*, 2083. (c) Kelly, T.R.; Meghani, P.; Ekkundi, V.S. *Tetrahedron Lett.* **1990**, *31*, 3381.
53. Beauchamp, A.L.; Olivier, M.J.; Wuest, J.D.; Zacharie, B. *Organometallics* **1987**, *6*, 153.
54. Nadeau, F.; Simard, M.; Wuest, J.D. *Organometallics* **1990**, *9*, 1311.
55. Simard, M.; Vaugeois, J.; Wuest, J.D. *J. Am. Chem. Soc.* **1993**, *115*, 370.
56. Sharma, V.; Simard, M.; Wuest, J.D. *J. Am. Chem. Soc.* **1992**, *114*, 7931.
57. (a) Reilly, M.; Oh, T. *Tetrahedron Lett.* **1995**, *36*, 217. (b) Reilly, M.; Oh, T. *Tetrahedron Lett.* **1995**, *36*, 221. See also: Reilly, M.; Oh, T. *Tetrahedron Lett.* **1994**, *35*, 7209.
58. Ooi, T.; Takahashi, M.; Maruoka, K. *J. Am. Chem. Soc.* **1996**, *118*, 11307.
59. Asao, N.; Kii, S.; Hanawa, H.; Maruoka, K. *Tetrahedron Lett.* **1998**, *39*, 3729.
60. Review: Wilds, A.L. *Org. React.* **1944**, *2*, 178.
61. Recent improvements of MPV reduction: (a) Kagan, H.; Namy, J. *Tetrahedron* **1986**, *42*, 6573. (b) Huskens, J.; De Graauw, C.; Peters, J.; Van Bekkum, H. *Recl. Trav. Chim. Pays-Bas* **1994**, 1007. (c) Barbry, D.; Torchy, S. *Tetrahedron Lett.* **1997**, *38*, 2959. (d) Akamanchi, K.; Varalakshmy, N.R. *Tetrahedron Lett.* **1995**, *36*, 3571. (e) Akamanchi, K.; Varalakshmy, N.R.; Chaudhari, B.A. *Synlett* **1997**, 371.
62. Ooi, T.; Miura, T.; Maruoka, K. *Angew. Chem., Int. Ed. Engl.* **1998**, *37*, 2347.
63. (a) Morrison, J.D.; Mosher, H.S. *Asymmetric Organic Reactions*; American Chemical Society: Washington, D. C., 1976; p 160. (b) Evans, D.A.; Nelson, S.G.; Gagne, M.R.; Muci, A. R. *J. Am. Chem. Soc.* **1993**, *115*, 9800.
64. Oppenauer, R. V. *Rec. Trav. Chim.* **1937**, *56*, 137.
65. Recent modification: Akamanchi, K.G.; Chaudhari, B.A. *Tetrahedron Lett.* **1997**, *38*, 6925.
66. Ooi, T.; Itagaki, Y.; Miura, T.; Maruoka, K. *Tetrahedron Lett.* **1999**, *40*, 2137.

2 Carbonyl Recognition

Susumu Saito and Hisashi Yamamoto

2.1 General Scope

The carbonyl recognition systems involved in enzymes [1] and their subsequent mimics [2] utilize multiply arranged hydrogen bondings to complex carbonyl functionalities. Enzymes discriminate between molecules with respect to the capacity of their active sites to form specific complexes with their substrates to promote chemical reactivity and selectivity. Unlike these complicated but precise networks extended by weak hydrogen bondings, Lewis acids have characteristic coordination bonding and thus selectively bind their substrates using a metal, single binding site. Since the appearance of the first report of the formation of crystalline complexes of BF_3 and aromatic aldehydes [3], Lewis acids have found a prominent role in organic synthesis through their action on carbonyl-containing compounds. They have been used as effective catalysts for a number of C–C bond formations; however, the true origins of the "Lewis acid effect" are still poorly understood. As mentioned in the previous chapter, a role of Lewis acid is to make a complex with each carbonyl substrate in a distinctive coordination fashion [3]. Lewis acids initially differentiate between the steric and electronic characters of substituents attached to a carbonyl carbon, subsequently altering one of the lone pairs of the carbonyl oxygen (Scheme 2-1). Thereafter, a selective coordination might result even with relatively small Lewis acids such as BF_3 [4]. Accordingly, most Lewis acids prefer complexation within the π-nodal plane of the carbonyl group, although some exceptions were found.

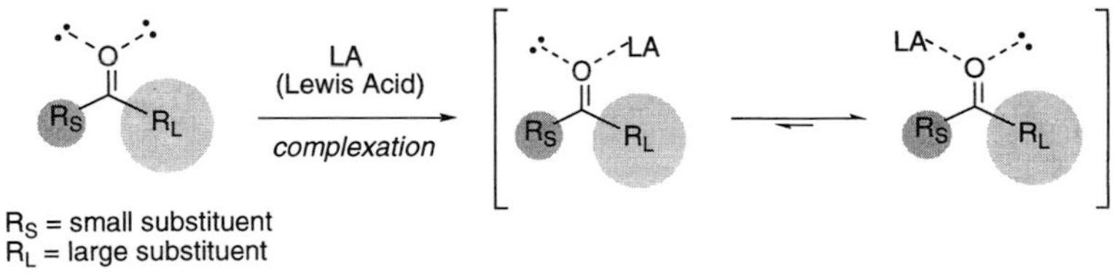

R_S = small substituent
R_L = large substituent

Scheme 2-1.

During the last two decades, a great deal of attention has been devoted to the discrimination between structurally and/or electronically similar substrates. Conventional Lewis acids such as $TiCl_4$, $SnCl_4$, $AlCl_3$, BF_3 etc. showed poor chemoselectivity upon complexation due to their high reactivities. In principle, two reasonable approaches by Lewis acids have been developed to address these challenges. One is by adjusting the electronic properties of a Lewis acidic metal (often by decreasing Lewis acidity) or just by changing one metal to another; another solution is by developing designer Lewis acids [5], namely designing the ligand required for the attachment of a specific (often bulky) shape to Lewis acids. This is the main subject of the present chapter. Both manipulations employ an ingenious idea to make a specific complex between a metal and a carbonyl base as they form preferential interactions, and are potentially useful for the chemoselective functionalization of either stable or labile molecules.

It is widely accepted that the carbonyl reactivity toward nucleophiles increases in the order aldehyde>ketone>ester>amide [6]. This reactivity order is simply based on the extent to which each carbonyl carbon is sterically and electronically activated. However, reactivities might change when these carbonyl substrates are subjected to a Lewis acid. It is generally assumed that the coordination capability of the carbonyl oxygen to Lewis acids is the means by which Lewis acids activate carbonyl substrates. Thus, in some respects, the reaction rate parallels the Lewis basicity of the carbonyls. Furthermore, the reactivity of a carbonyl substrate depends on the reaction type as well as the Lewis acid employed. Special care must be taken in assessing the relationship between the relative reaction rate, the relative Lewis basicity, and the inherent carbonyl reactivity of each substrate. It is instructive to take a look at the following example (Schemes 2-2 and 2-3; Fig. 2-1).

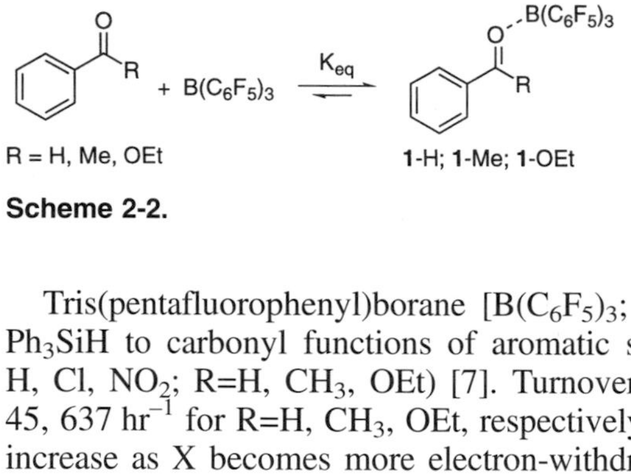

+ $B(C_6F_5)_3$ $\xrightleftharpoons{K_{eq}}$ $B(C_6F_5)_3$

R = H, Me, OEt   **1**-H; **1**-Me; **1**-OEt

Scheme 2-2.

Tris(pentafluorophenyl)borane [$B(C_6F_5)_3$; 1–4 mol%] catalyzes the addition of Ph_3SiH to carbonyl functions of aromatic substrates *p*-X-C_6H_4(C=O)R (X=CH_3, H, Cl, NO_2; R=H, CH_3, OEt) [7]. Turnover numbers for X=H substrates are 19, 45, 637 hr^{-1} for R=H, CH_3, OEt, respectively. The reaction rates of hydrosilylation increase as X becomes more electron-withdrawing. However, the strength of substrate binding to $B(C_6F_5)_3$ is in the opposite order (equilibrium constants (K_{eq}) for Scheme 2-2: 2.1×10^4, 1.1×10^3, and 1.9×10^2 for X=H, Me, OEt, respectively). It appears that the role of the Lewis acid is not to activate the carbonyl substrate and that liberation of "free" $B(C_6F_5)_3$ is required for productive reaction. The mechanism is highly likely to be consistent with the scenario shown in Scheme 2-3. Through these pathways, the paradoxical observation that esters are reduced more rapidly than aldehydes and ketones, despite being more weakly bound to the Lewis acid, is accommodated. Note that the basicity of the substrate is, however, relevant in competition reactions. When a 1 : 1 mixture of PhCHO and $PhCO_2Et$ was subjected to this hydrosilylation, more basic PhCHO was selectively reduced. The transition state **TS-1** was proposed to explain this unusual nucleophilic/electrophilic catalysis (Fig. 2-1). It should be emphasized that the relative reaction rate which is measured independently for each substrate is entirely distinct from that obtained from such a competition experiment.

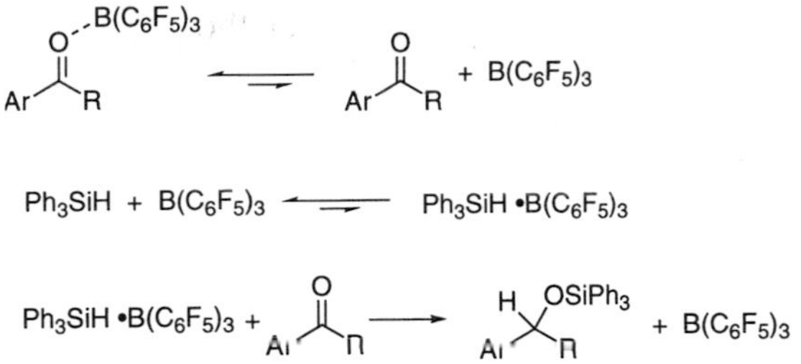

Scheme 2-3.

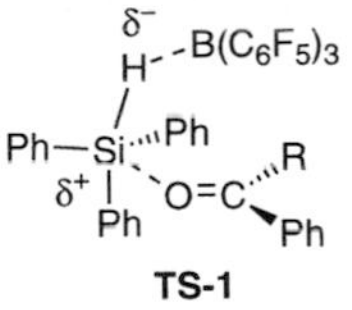

Figure 2-1

Methylaluminum bis(2,6-di-*tert*-butyl-4-methylphenoxide) (MAD) [8] and aluminum tris(2,6-diphenylphenoxide) (ATPH) [9] are bulky Lewis acids that are readily obtainable from Me_3Al and the corresponding phenol (Fig. 2-2). Compared with the above example, where $B(C_6F_5)_3$ forms the Lewis acid-base complexes reversibly, the bulky aluminum reagents coordinate with carbonyl substrates essentially irreversibly owing to high oxophilicity. For instance, the induced ^1H NMR chemical shifts of the ATPH-PhCHO complex ($1{\times}10^{-1}$ M) do not change at all even at high dilution ($5{\times}10^{-3}$ M).

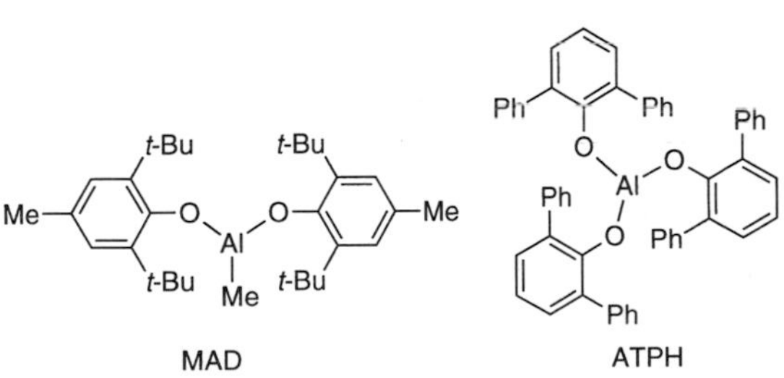

Figure 2-2

In the presence of different carbonyl substrates, ATPH is assumed to kinetically favor the formation of a stable complex with a sterically less-hindered substrate (or sometimes with an electronically more donating substrate). In other words, carbonyl groups on the outside of a molecule bind to the Lewis acid rather tightly, while those on the inside of a molecule will not form stable complexes [10]. In cases where the stable complexes are formed, the rate of the thermodynamic exchange with a free substrate seems very slow. By taking these characteristic advantages of the complexation with bulky aluminum reagents, an otherwise sterically more-hindered substrate is recognized as a relatively less-hindered counterpart. This leads to chemoselective functionalization of the originally more hindered one by activated nucleophiles (i.e. organolithiums, organomagnesiums, etc.). On the contrary, rather deactivated nucleophiles [i.e. ketene silyl acetals (KSA), silyl enol ethers, allyl silanes, allyl stannanes, etc.] react only when carbonyl substrates are activated with a Lewis acid, so that in this case the less-hindered species are functionalized chemoselectively (Scheme 2-4). The carbonyls complexed with the bulky Lewis acids are thus sterically deactivated, and are instead electronically activated (Fig. 2-3). It is easy to discuss and interpret a certain origin of chemoselectivity by taking into consideration the irreversibility of the aluminum reagents. In contrast, the reversible complexation as exemplified by the use of $B(C_6F_5)_3$ renders chemoselective outcomes, especially on treatment with activated nucleophiles, more difficult to predict. The general principle of these ideas is well demonstrated in this chapter.

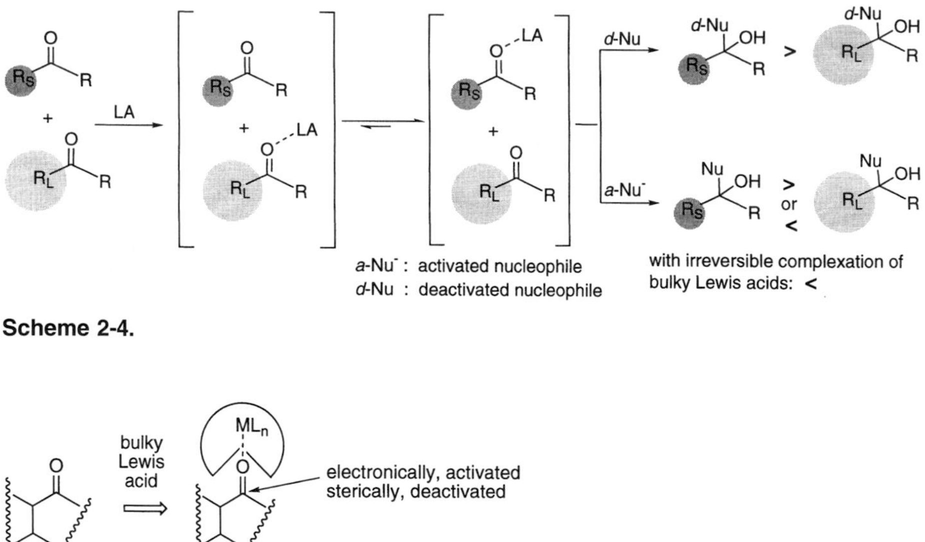

Scheme 2-4.

Figure 2-3

2.2 Recognition of Carbonyl Substrate with Bulky Lewis Acid

2.2.1 Design, Preparation and Availability of Bulky Aluminum Reagent

Most aluminum reagents in solution exist in dimeric, trimeric, or higher oligomeric structures [11]. In contrast, MAD and ATPH are monomeric in organic solvents. Lewis acidity of these reagents decreases with the coordination of more electron-donating aryloxides compared to alkyl aluminum compounds, but this can be compensated by loosening of the aggregation. MAD, ATPH, and methylaluminum bis(2,6-diphenylphenoxide) (MAPH) [12] are readily prepared by treatment of Me_3Al with a corresponding amount of the phenol in toluene (or in CH_2Cl_2) at room temperature for 0.5–1 hour with rigorous exclusion of air and moisture. Methane gas evolves spontaneously. These reagents can be used without further purification for all of the reactions described here. The reactivity of a phenol toward R_3Al largely depends on the steric limitations of phenol and R (Scheme 2-5) [13]. For example, the reaction of Me_3Al with 1 equiv of 2,6-di-*tert*-butyl-4-methylphenol (BHT-H) does not yield the monoaryloxide complex but instead the bis(phenoxide) MAD. This result has been explained in terms of ligand exchange reactions [14]. In contrast, the reaction of BHT-H with $Al(i\text{-}Bu)_3$ and $Al(t\text{-}Bu)_3$ in a 1 : 1 molar ratio yields the mono-aryloxides $Al(i\text{-}Bu)_2(BHT)$ and $Al(t\text{-}Bu)_2(BHT)$, respectively. The stable monomeric monophenoxides can thus be isolated when sterically more demanding trialkylaluminums are employed. On the other hand, treatment of 3 equiv of BHT-H with Me_3Al in CH_2Cl_2 at room temperature under argon results in the generation of bisphenoxide MAD together with 1 equiv of the unreacted phenol. In contrast, 3 equiv of 2,6-diphenylphenol completely reacts with 1 equiv of Me_3Al to produce the tris(phenoxide)

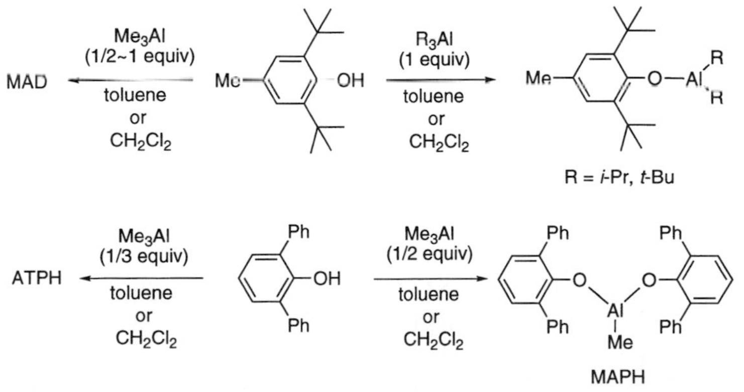

Scheme 2-5.

ATPH (Scheme 2-5). Aluminum tris(2,6-di-*tert*-butyl-4-methylphenoxide) (ATD) [15, 16] was first reported by Barron and co-workers. They exploited a two-step synthesis of ATD starting from $LiAlH_4$ as an aluminum source. Apparently the structural modification of aluminum aryloxides that leads to numerous variants offers real advantages over conventional Lewis acids regarding preparation and handling.

2.2.2 Selective Coordination with Carbonyl Oxygen

As described in Chapter 1, there are several different possible modes for the coordination of Lewis acids to carbonyl groups (Fig. 2-4). One is a purely electrostatic interaction, in which the metal is situated at the negative end of the C=O dipole, where C–O–Al=180° (**A**). The second possibility is the coordination of the metal to one of the lone pairs on the carbonyl oxygen, with the metal being in the nodal plane of the C=O π-bond (**B**). This mode of bonding can be seen in a large number of the X-ray crystal structures of Lewis acid-base complexes [3]. The third, a bent non-planar mode of bonding results from movement of the metal out of the carbonyl π nodal plane (**C**). The bulky Lewis acid complexation with ester or amide sometimes prefers this mode [17]. The last mode is a η^2 coordination of a metal to the C=O π bond in which the carbonyl π orbital is the donor, accompanied by back-bonding into C=O π^* orbital (**D**). Although this mode has been reported for transition metals [18], it does not seem likely for main-group elements. Unfortunately, doubly coordinated termolecular complexes in mode **E1** are presently unknown. In contrast, Wuest and co-workers recently reported that closely related doubly coordinated bimolecular complexes **E2** can be formed when two sites of Lewis acidity are joined to create single bidentate reagents [19]. The X-ray crystal structure of the bidentate dimercury complex of *N*,*N*-diethylacetamide also support viability of this mode of complexation [20]. Furthermore, as well documented in the previous chapter, the double coordination of carbonyl compounds by bidentate Lewis acids increases the carbonyl reactivity.

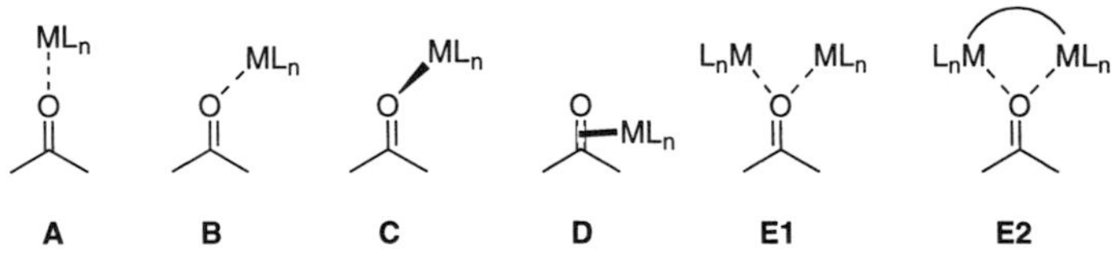

A B C D E1 E2

Figure 2-4

Since the aluminum reagents ATPH and MAD are considerably bulky, it is envisaged that the binding of incoming carbonyl bases would be hampered by significant steric repulsions due to the bulk of phenol ligands. However, the X-ray crystal structure of the *N*,*N*-dimethylformamide (DMF)-ATPH complex [21] disclosed that the three arene rings of ATPH form a propeller-like arrangement around the aluminum center, and hence ATPH has a cavity with C_3 symmetry. Furthermore, the X-ray crystal structure of the benzaldehyde-ATPH complex [4] shows that the cavity surrounds the carbonyl substrate upon complexation with slight distortion from C_3 symmetry. These two carbonyl substrates are effectively and tightly encapsulated into the cavity (Fig. 2-5).

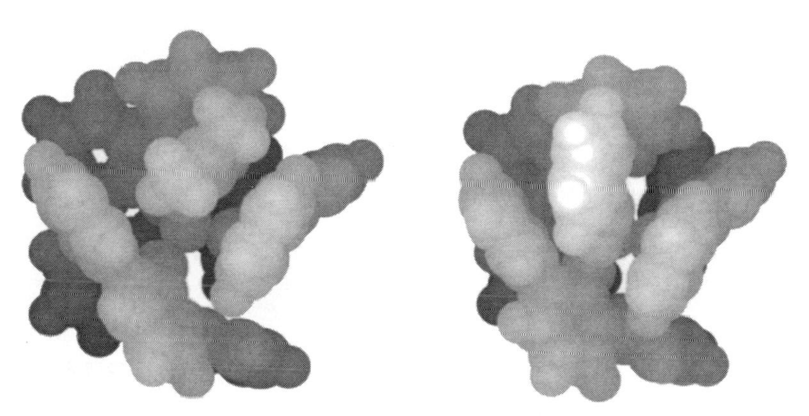

Figure 2-5

A particularly notable structural feature of these aluminum-carbonyl complexes is the Al–O–C angles and Al–O distances (Table 2-1 and Fig. 2-5), which confirm that the size and the shape of the cavity changes flexibly depending on the substrate. The angle values θ and ϕ shed light on the coordination mode of carbonyls, and several results obtained from the X-ray crystal structures of bulky Lewis acid-base complexes are worthy of comment (Table 2-1). With a few exceptions in which the coordination of metal is slightly deviated from the π nodal plane (i.e. the mode **C** with $\phi > 10°$: entries 4, 5, 8 and 11), the ϕ values show that the carbonyl is coordinated with the metal in mode **B** (entries 1–3, 6, 7, 9 and 10). The θ values vary diversely with differing carbonyl substrates, i.e., the coordination seems inherent to each substrate. However, as shown in Chapter 1, there are some general rules expected from a series of these distinctive θ values. The bulky Lewis acid-base complexes also stay within these rules. For example, crotonaldehyde and methyl crotonate prefer *anti, s-trans* and *syn, s-trans* conformations, respectively, by complexation with ATPH (entries 6 and 8). These results are in accord with the coordination bias of conjugated carbonyl compounds with relatively small Lewis acids (Fig. 2-6) [3]. However, in cases where esters were employed, it is of interest that a diverse set of θ and ϕ values is observed (entries 1, 8 and 11): e.g., $\theta = 168.4°$, being approximate to mode **A** (entry 1). It should be pointed out that the methyl group of methyl crotonate adopts the *syn* conformation to the carbonyl axis (entry 8; Fig. 2-6), consistent with the location of the methyl and ethyl groups of the benzoates (entries 1 and 11). Apparently, the coordination mode of organic carbonyls to bulky aluminum and borane reagents is quite flexible, and steric effects predominate.

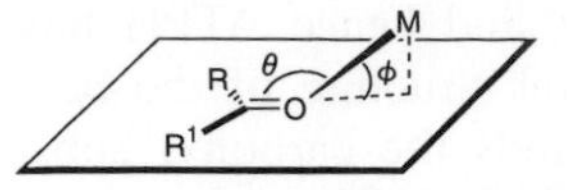

Table 2-1. The X-ray crystal structure properties of bulky Lewis acid-base complexes*.

Entry [ref.]	Lewis acid-base complex	Selected properties
1 [22]		R = Ph; R^1 = OMe θ = 168.4(3)° ϕ = 3.2° Al-O(sp^2): 1.851 (7) Å
2 [23]		R = R^1 = Ph θ = 144(6)° ϕ = 6.9° Al-O(sp^2): 1.903 (6) Å
3 [22]		R = H; R^1 = *t*-Bu θ = 136.0(3)° ϕ = 8.1° Al-O(sp^2): 1.920 (5) Å
4 [24]		R = CH-R'; R^1 = -(C=O)-R" θ = 139.7(3)° ϕ = 16.8° Al-O(sp^2): 1.923 Å
5 [21]		R = H; R^1 = NMe_2 θ = 138.1(3)° ϕ = 11.0(3)° Al-O(sp^2): 1.795 Å
6 [25]		R = H; R^1 = CH=CHMe θ = 145.9° ϕ = 6.1(8)° Al-O(sp^2): 1.829(2) Å *anti, s-trans*
7 [4]		R = H; R^1 = Ph θ = 135.7(9)° ϕ = 2.34(7)° Al-O(sp^2): 1.856(5) Å
8 [26]		R = CH=CHMe; R^1 = OMe θ = 143.8(3)° ϕ = 19.9(7)° Al-O(sp^2): 1.808(6) Å *syn, s-trans*
9 [7]		R = H; R^1 = Ph θ = 126.7(5)° ϕ = 4.6° B-O(sp^2): 1.610 (8) Å
10 [7]		R = Me; R^1 = Ph θ = 133.6(3)° ϕ = 4.2° B-O(sp^2): 1.576 (5) Å
11 [7]		R = Ph; R^1 = OEt θ = 138.2(4)° ϕ = 15.6° B-O(sp^2): 1.594 (5) Å

*) Some θ and ϕ values are cited from the CD-ROM data base supplied by the Cambridge Crystal Data Centre (CCDC).

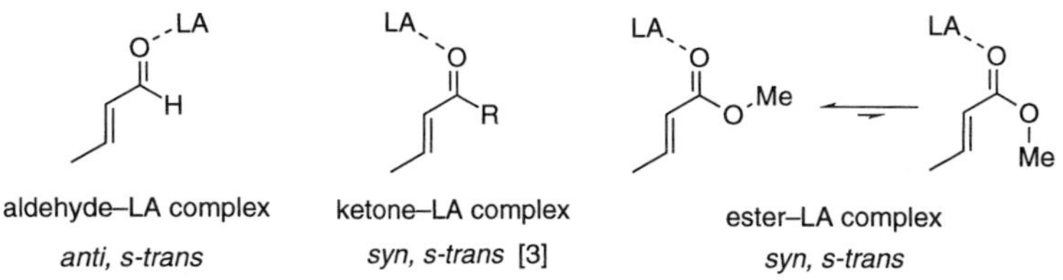

aldehyde–LA complex *anti, s-trans* ketone–LA complex *syn, s-trans* [3] ester–LA complex *syn, s-trans*

Figure 2-6

According to this information, the cavity of ATPH should be able to differentiate between carbonyl substrates which, once accepted into the cavity, should exhibit unprecedented reactivity and selectivity under the steric and electronic environment of the arene rings. ^{1}H NMR measurement of crotonaldehyde-ATPH complex (300 MHz, CD_2Cl_2) revealed that the original chemical shifts of the aldehydic proton (H_a) at δ 9.50, and the α- and β-carbon protons (H_b and H_c) at δ 6.13 and δ 6.89, were significantly shifted upfield to δ 6.21, δ 4.92 and δ 6.40, respectively. The largest $\Delta\delta$ value of H_a of 3.29 ppm suggests that the carbonyl is effectively shielded by the arene rings of the cavity. This observation is in contrast to the resonance frequencies of the crotonaldehyde-Et_2AlCl complex at −60 °C (H_a, δ 9.32; H_b, δ 6.65; H_c, δ 7.84) and those of crotonaldehyde complexes with other ordinary Lewis acids [27]. Similar shift changes into the upfield region are also seen in other ATPH-carbonyl complexes. In contrast, MAD, while adequately sterically encumbered, is more structurally flexible than ATPH, and can adopt various dynamic conformations depending on the incoming carbonyl substrate as shown by the X-ray crystal structures. MAD thus can contribute for the purpose of either protection or activation of a carbonyl group depending on the reaction type employed, whereas ATPH acts preferably as a carbonyl protector in cases where the reactive nucleophiles are employed. In fact, the Michael addition of BuMgCl to cinnamaldehyde occurred more effectively with ATPH than with MAD, which gave rise to a considerable amount of 1,2-adduct (Scheme 2-6) [21].

ATPH, toluene; BuMgCl, ether, -78 °C; [1,4] [1,2]; *1,4-adduct* + *1,2-adduct* 99% (90:10); MAD, toluene; 95% (7:93)

Scheme 2-6.

Astonishing behavior of ATPH in a carbonyl recognition event was highlighted in the kinetic generation of more substituted enolates of unsymmetrical dialkyl ketones [28]. Most likely, ATPH prefers coordination with one of the lone pairs of carbonyl oxygen *anti* to the more sterically hindered α carbon of unsymmetrical ketones. As a consequence, the less hindered *α*-site is surrounded and efficiently shielded by the steric bulk of ATPH, thus hampering the trajectory of the nucleophilic attack of LDA (Scheme 2-7). Subsequent treatment with methyl triflate (MeOTf) gave rise to regioselective formation of the more-substituted *α*-carbon (94% yield, regioselectivity=99:1). MAD demonstrated less efficiency with respect to the regioselectivity (ca. 1:1).

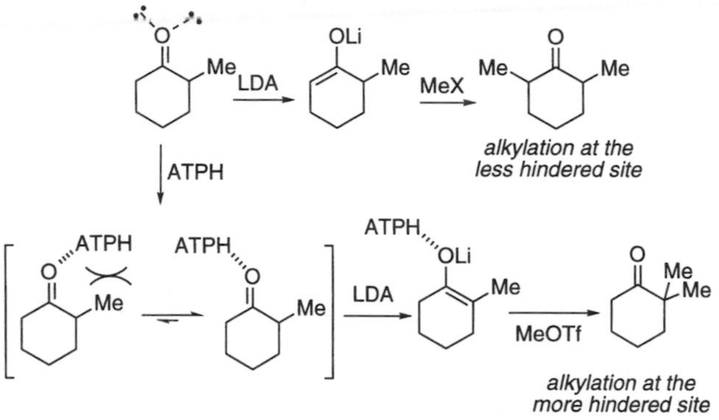

Scheme 2-7.

2.3 Chemoselective Functionalization of Different Carbonyl Group

There exist three possible modes for the functionalization of carbonyl compounds (Fig. 2-7). These are categorized into the reactions of: (1) activated nucleophiles that promote inter- or intramolecular substituent transfer (**F1**), (2) deactivated nucleophiles with Lewis acid assisting (**F2**), and (3) activated nucleophiles with Lewis acid assisting (**F3**), what we call the "amphiphilic activation system" [29]. These three courses of reactions are also interpretable by the mode of activation of either nucleophile or carbonyl substrate, or both. In general, chemoselective functionalization of carbonyls utilizing modes **F1**–**F3** has been realized by varying metals (M) to change nucleophilic potential of their ligands or by adjusting both electronic and steric properties of a metal. Furthermore, in some instances, two of the modes operate simultaneously to achieve chemoselectivity. It is easy to functionalize a more labile molecule chemoselectively, and a number of examples in which metal hydrides reduced carbonyls have been reported. The reader can refer to other specialized reviews on this subject [30]. We focus here mainly on representative carbon-carbon forming reactions.

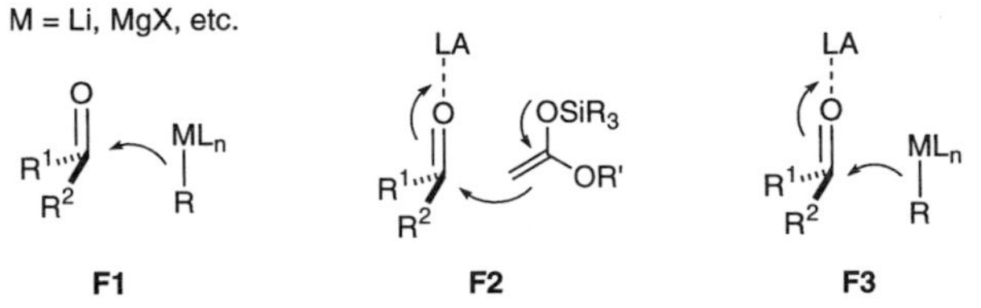

M = Li, MgX, etc.

F1 F2 F3

Figure 2-7

2.3.1 Aldehyde vs Ketone

There is no doubt that aldehydes are more reactive than ketones toward nucleophiles. However, both carbonyl substrates are functionalized by activated nucleophiles e.g. RLi or RMgX (X = halogen), with poor chemoselectivity. For example, benzaldehyde is not a dominant species to be alkylated in the coexistence of acetophenone. Reetz and co-workers addressed these difficulties by systematic studies on ligand effects in carbonyl addition reactions of RMgL (L = relatively bulky ligand) [31]. Upon reacting a 1:1 mixture of PhCHO and PhCOMe with 1 equiv of RMgL in a competition experiment, the aldehyde reacted essentially exclusively to form adduct **2-H** (Table 2-2, entries 2–4).

One of the important and powerful methods for controlling chemoselectivity of carbanions consists in titanation with reagents of the type $RTiL_3$ (L=Cl, OR′, NR'_2, etc.) (Scheme 2-8; Table 2-1) [32]. These chemoselective behaviors of RMgL and $RTiL_3$ have been proposed to originate from the bulky ligands L at the metals. Other typical results related to this subject are listed in Table 2-1. The studies to induce high chemoselectivity are particularly focused on various organometallics which are used for the transfer of allylic groups (entries 5–12, Table 2-2). The Bu_4Pb-$TiCl_4$ [33] and organoiron×metal salts [34] reagents also proved to be effective in the competition experiments (Scheme 2-9).

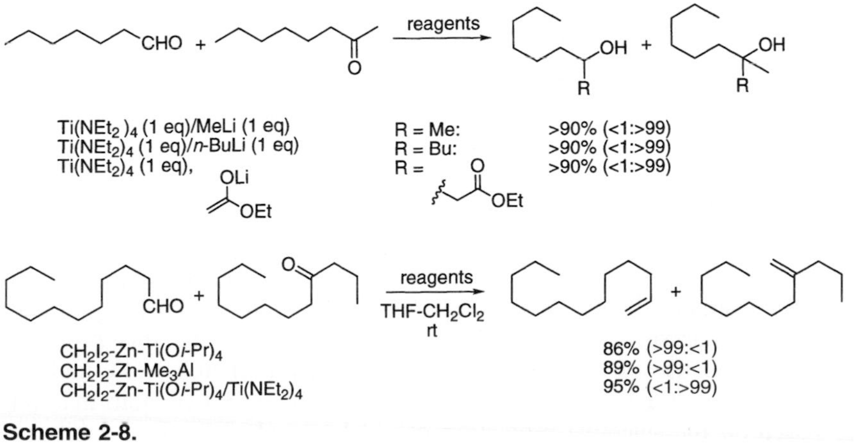

reagents	R	yield (ratio)
$Ti(NEt_2)_4$ (1 eq)/MeLi (1 eq)	R = Me:	>90% (<1:>99)
$Ti(NEt_2)_4$ (1 eq)/*n*-BuLi (1 eq)	R = Bu:	>90% (<1:>99)
$Ti(NEt_2)_4$ (1 eq), $CH_2=C(OLi)OEt$	R = CH_2CO_2Et	>90% (<1:>99)

reagents, THF-CH_2Cl_2, rt

reagents	yield (ratio)
CH_2I_2-Zn-Ti(O*i*-Pr)$_4$	86% (>99:<1)
CH_2I_2-Zn-Me_3Al	89% (>99:<1)
CH_2I_2-Zn-Ti(O*i*-Pr)$_4$/Ti(NEt$_2$)$_4$	95% (<1:>99)

Scheme 2-8.

PhCHO + PhCOMe ⟶ **2-H** + **2-Me**

Table 2-2. Chemoselective alkylation of a mixture of PhCHO and PhCOMe.

Entry	Reagent(s)	Solvent	Products	Yield (2-H:2-Me)	Ref.
1	BuMgBr	THF	R = Bu	88% (85:15)	[31]
2	BuMgOTf	THF		79% (91:9)	[31]
3	BuMgOTs	THF		76% (93:7)	[31]
4	$BuMgOSO_2C_6H_2(CH_3)_3$	THF		68% (95:5)	[31]
5	$CH_2{=}CHCH_2Ti(O\textit{i}\text{-}Pr)_3$	THF	R = allyl	(84:16)	[32]
6	$CH_2{=}CHCH_2Ti(O\textit{i}\text{-}Pr)_4MgCl$	THF		(98:2)	[32]
7	$BiCl_3/Zn$, $CH_2{=}CHCH_2Br$	THF		>99% (>99:<1)	[43]
8	$BiCl_3/Fe$ $CH_2{=}CHCH_2Br$	THF		80% (>99:<1)	[43]
9	$Sc(OTf)_3$, $(CH_2{=}CHCH_2)_4Ge$	$CH_3NO_2\text{-}H_2O$		96% (>99:<1)	[44]
10	HCl, $(CH_2{=}CHCH_2)_4Sn$	$THF\text{-}H_2O$		>85% (>99.98:<0.002)	[45]
11	$Pd(PPh_3)_4/Zn$, $CH_3CH{=}CHCH_2OAc$	dioxane	R = $CH(CH_3)CH{=}CH_2$	70% (>99:<1)	[46]
12	$R_2B\text{-}CH{=}C{=}CH_2$	Et_2O	R = $CH_2C{\equiv}CH$	>99% (97:3)	[47]
13	MeLi	CH_2Cl_2	R = Me	>80% (4:1)	[38]
14	MeMgI	CH_2Cl_2		>80% (22:1)	[38]
15	MAD/MeLi	CH_2Cl_2		64% (6:1)	[38]
16	MAD/MeMgI	CH_2Cl_2		79% (>99:<1)	[38]
17	$Me_2AlNMe/MeLi$	CH_2Cl_2		>99% (1:7)	[38]
18	$MoOCl_3(THF)_2/MeLi$ (2 eq) (4 eq)	THF		82% (95:5)	[48]
20	$MeTi(O\textit{i}\text{-}Pr)_3$	Et_2O		(>99:<1)	[32]

The reagent combination of $Bu_2Sn(OTf)_2$/2-stanna-1,3-dithiolane led to a similar bias of chemoselectivity (Scheme 2-10) [35].

Contrary to the greater ease of the transformations of aldehydes compared with ketones, the achievement of the opposite chemoselectivity has been a significant challenge. Two similar strategies have been applied to overcome this problem. One is the combined use of two different modes of **F1** (Scheme 2-11), and the other is relevant to mode **F1** together with mode **F3** (Scheme 2-12). Both manipulations are based on the selective protection of aldehyde which is more labile to-

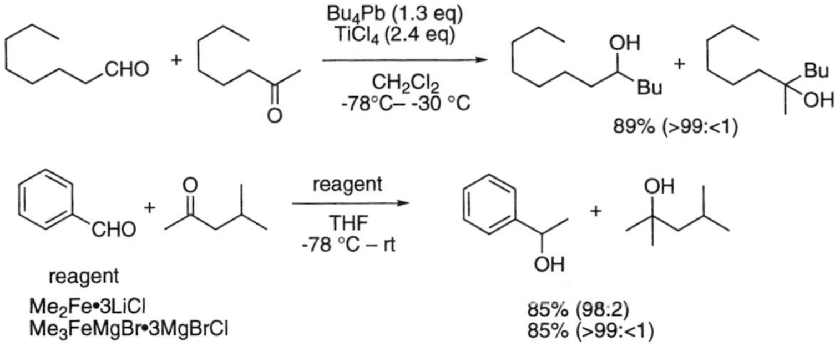

CHO + O → Bu_4Pb (1.3 eq), $TiCl_4$ (2.4 eq), CH_2Cl_2, -78°C– -30 °C → OH, Bu + Bu, OH

89% (>99:<1)

CHO + O → reagent, THF, -78 °C – rt → OH + OH

reagent

$Me_2Fe{\bullet}3LiCl$ 85% (98:2)

$Me_3FeMgBr{\bullet}3MgBrCl$ 85% (>99:<1)

Scheme 2-9.

Bu_2Sn(S, S) (1.2 eq)

CHO (**3**) + O → $Bu_2Sn(OTf)_2$, (30 mol%), Cl–Cl, 0 °C → S, S + S, S

74% (>99:<1)

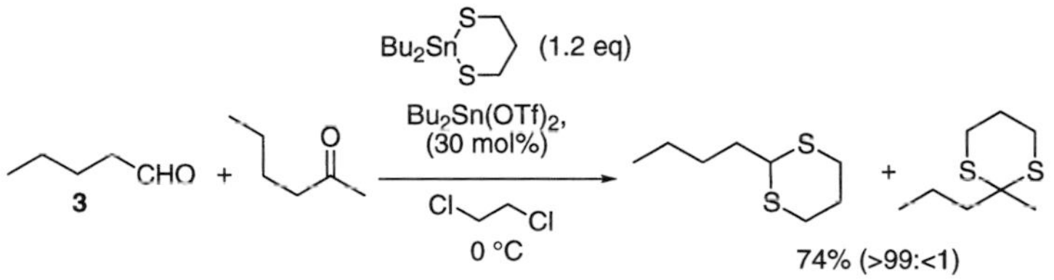

Scheme 2-10.

ward the first reagent employed. Consequently, the second reagent, an activated nucleophile, reacts selectively with a more stable carbonyl counterpart which remains inert to the first reagent. To this end, the amino ate complex of (allyl) $Ti(NMe_2)_4MgCl$ was originally devised [36]. The relative mobility of the ligands on this reagent increases in the order of NMe_2>allyl. Subsequently, a related and general procedure was developed in which $Ti(NR_3)_4$ was used as an *in situ* protecting reagent (Scheme 2-8) [37]. This scope was expanded to other metal amides such as $R^1_2AlNR_2$ [38] as well as the TMSOTf-Me_2S system (Scheme 2-11) [39]. The reactions are highly likely to proceed via the protected intermediates in the forms of **I-1**, **I-2**, and **I-3**, respectively (Scheme 2-11).

MAPH is another possibility available for the chemoselective functionalization of ketones (Scheme 2-12) [12]. The alkylation involved in mode **F3** is almost inoperative because the aldehydic carbon is sterically, i.e., kinetically, deactivated by the cooperation of the MAPH ligands, thereby presumably forming a sort of sandwich structure (**I-4**). The reaction rate of RLi (R=Bu, Ph, etc.) with ketones consequently becomes faster.

An entirely different approach employs organotin reagents. $(C_6F_5)_2SnBr_2$ prefers selective activation of ketones over aldehydes when treated with a ketene silyl acetal (KSA) (Schemes 2-13 and 2-14) [40]. TMSOTf and $Sc(OTf)_3$ also showed a moderate to high level of preference for ketones with exposure to KSA. The novel tin reagent systems $Bu_2Sn(OTf)_2/Me_3SiCN$ and $Bu_2Sn(OTf)_2/Et_3SiH$ with the coexistence of Me_3SiOMe afforded the selective methyl ether formation at ketonic carbons (Scheme 2-13) [41].

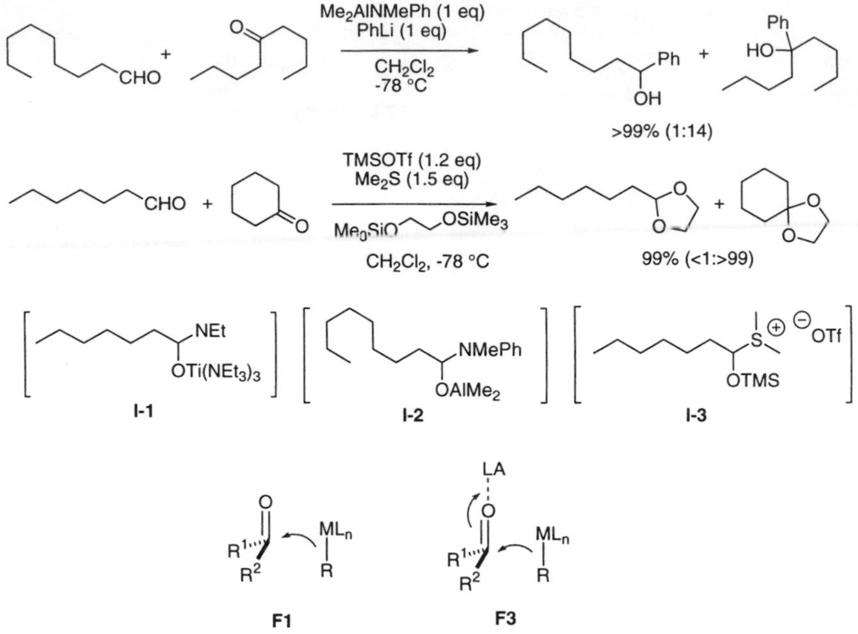

Scheme 2-11.

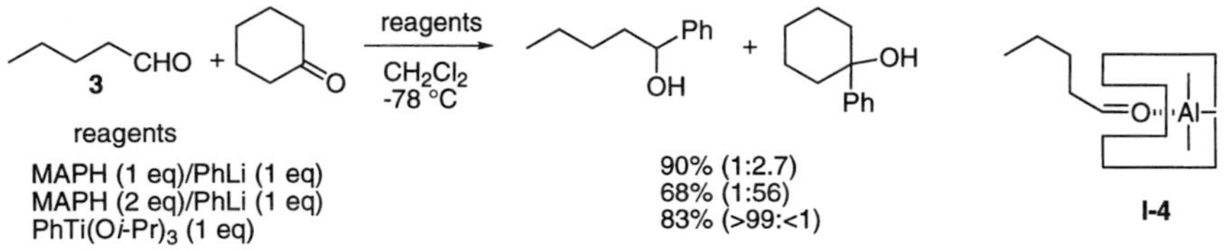

Scheme 2-12.

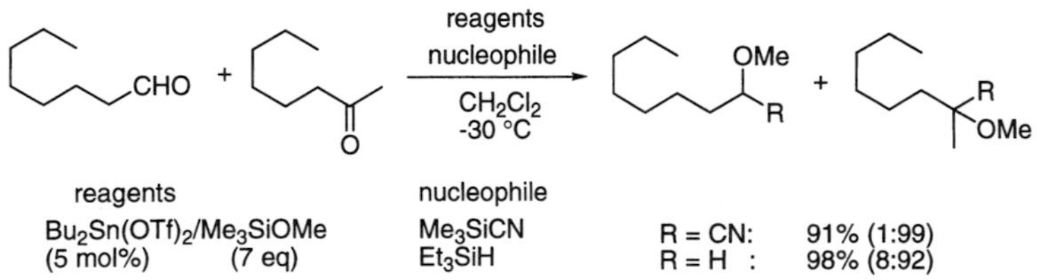

Scheme 2-13.

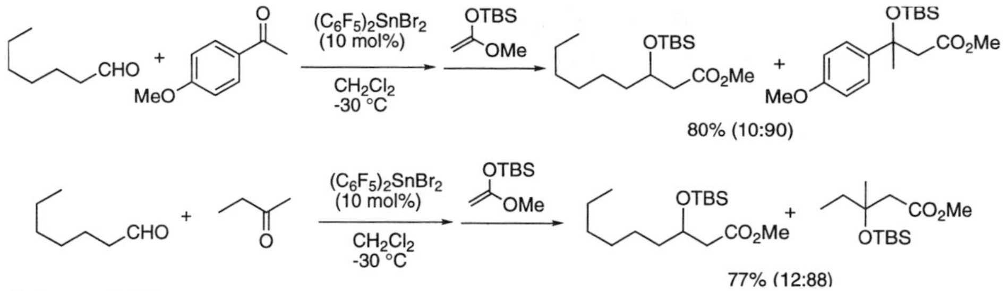

Scheme 2-14.

The neighboring group participation was invoked for the preferential capture of a ketonic function of keto aldehyde substrates by enol silyl ethers in TMSOTf-catalyzed reactions [42]. TMSOTf initially coordinates with the aldehyde, and the ketonic oxygen attacks intramolecularly the activated aldehydic carbonyl, thus resulting in the domino-type activation of ketonic functions (Scheme 2-15).

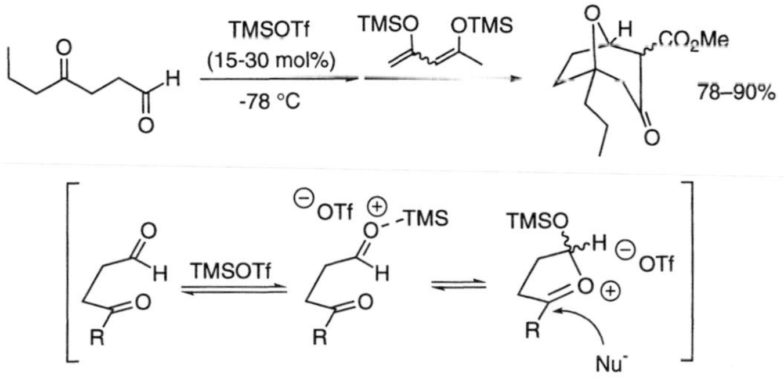

Scheme 2-15.

2.3.2 Ketone vs Ketone

2.3.2.1 Saturated vs Saturated

Discrimination between two different saturated ketones is by use of an idea similar to that described above for the molecular recognition of the aldehyde vs ketone. Additional advantages of Ti-ate complexes were emphasized (Scheme 2-16) [36]. Cyclic ketones are allylated chemoselectively with the ate complex in the form of (allyl)Ti(i-OPr)$_4$MgCl [36] in the presence of acyclic ketones. Cyclic ketones are, in principle, more reactive than acyclic ketones. In this case as well, amino ate complex (allyl)Ti(NMe$_2$)$_4$MgCl underwent the reaction with essentially complete reversal of chemoselectivity [36]. Even more difficult discrimination was encountered between six- and five-membered ketones, while tetraallylstannane was found to be a superb reagent (Scheme 2-17) [45]. These general trends were also ascertained by dithianation with the combined reagents of $Bu_2Sn(OTf)_2$ and 2-stanna-1,3-dithiolane (Scheme 2-18) [35].

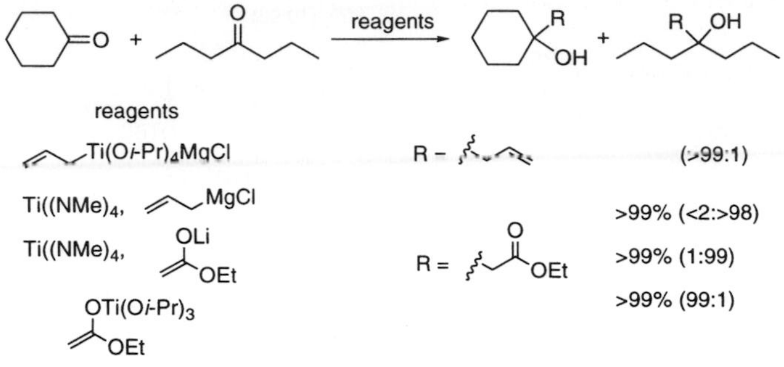

Scheme 2-16.

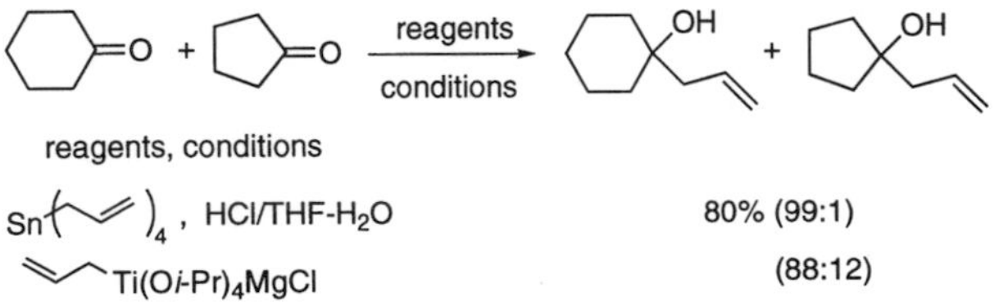

Scheme 2-17.

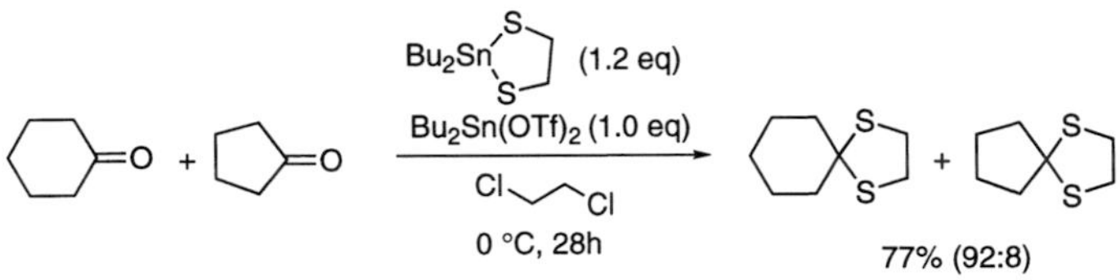

Scheme 2-18.

The chelation effect of $MeCrCl_2(THF)_3$ enhances the rate acceleration, thus leading to preferred conversion of α-hydroxy ketones into methylated products (Scheme 2-19) [49].

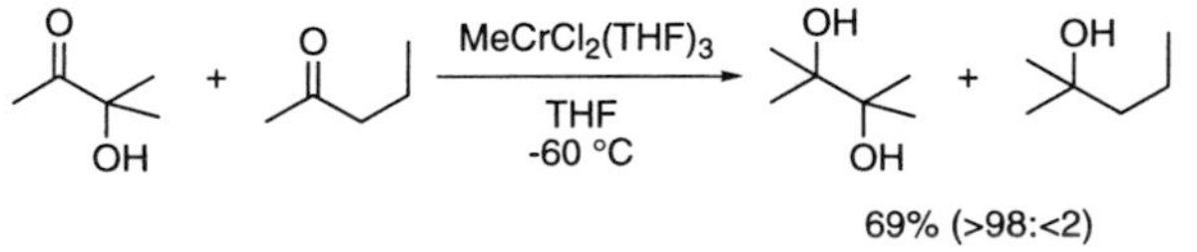

Scheme 2-19.

Among a series of bulky aluminum reagents, MAD and ATPH form coordination complexes with less hindered or more basic ketones preferentially, where MAD allows the activation of ketone carbonyls, whereas ATPH stabilizes, for subsequent nucleophilic alkylations (Scheme 2-20) [50]. These results are in contrast to the MAD-DIBAL reduction system in which MAD serves as an effective stabilizer for sterically less-hindered ketones (Scheme 2-21) [51].

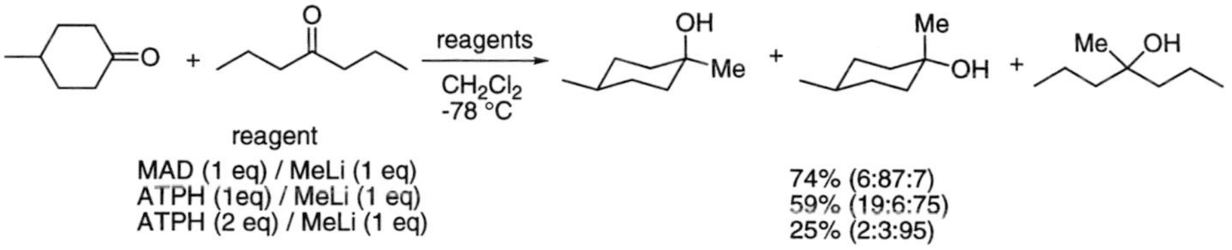

reagents; CH_2Cl_2, -78 °C

reagent	
MAD (1 eq) / MeLi (1 eq)	74% (6:87:7)
ATPH (1eq) / MeLi (1 eq)	59% (19:6:75)
ATPH (2 eq) / MeLi (1 eq)	25% (2:3:95)

Scheme 2-20.

MAD (1 eq); CH_2Cl_2, -78 °C; 99 : 1; DIBAL-H (1 eq), -78 °C; 66% (1:10)

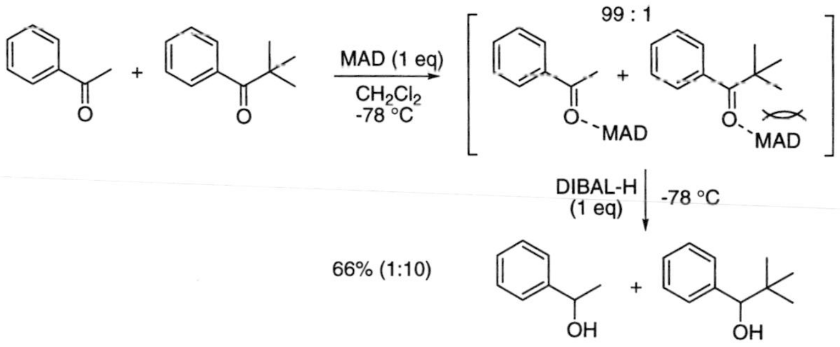

Scheme 2-21.

2.3.2.2 Saturated vs conjugated

Conjugated ketones (enones) form more stable complexes with Lewis acids compared to saturated ketones (alkanones). This leads to preferential functionalization of enones when $Eu(fod)_3$ [52] and $Bu_2Sn(OTf)_2$ [35] are subjected to chemical transformations (Schemes 2-22 and 2-23). The exposure of the thiolamine (Scheme 2-23) to a mixture of enone and alkanone units in the steroid also shows similar preference for the reaction with the enone [53].

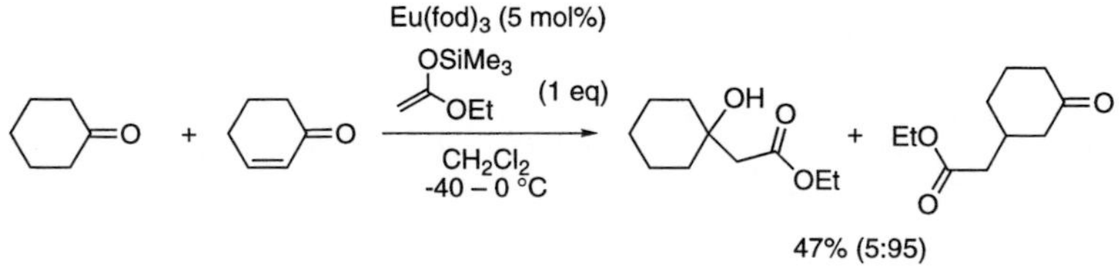

$Eu(fod)_3$ (5 mol%); $OSiMe_3$/OEt (1 eq); CH_2Cl_2, -40 – 0 °C; 47% (5:95)

Scheme 2-22.

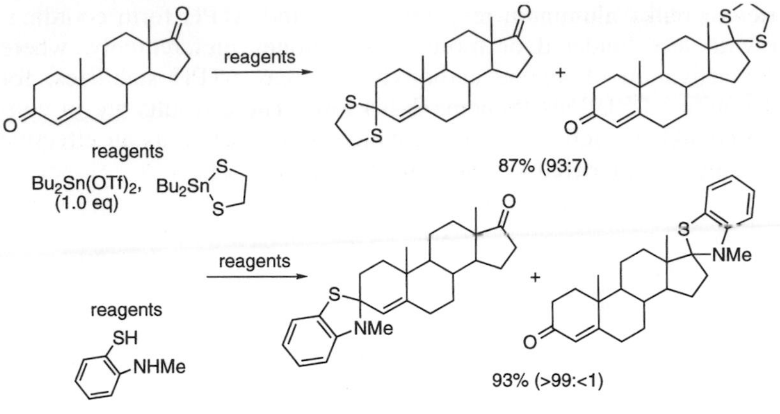

reagents

reagents: $Bu_2Sn(OTf)_2$ (1.0 eq), Bu_2Sn(S–S)

87% (93:7)

reagents: 2-(methylamino)thiophenol (SH, NHMe)

93% (>99:<1)

Scheme 2-23.

Marko and co-workers recently found a unique property of triorganothallium compounds (TOT): their ability to react preferentially with enones (Scheme 2-24) [54]. One of the most striking observations is that selectivity of the process increases as the substrate becomes more conjugated and the intrinsic reactivity of the carbonyl function decreases. This chemical event can be interpreted by the single electron transfer from the thallium species to enones, followed by recombination of the resulting radical species **I-5** and **I-6**.

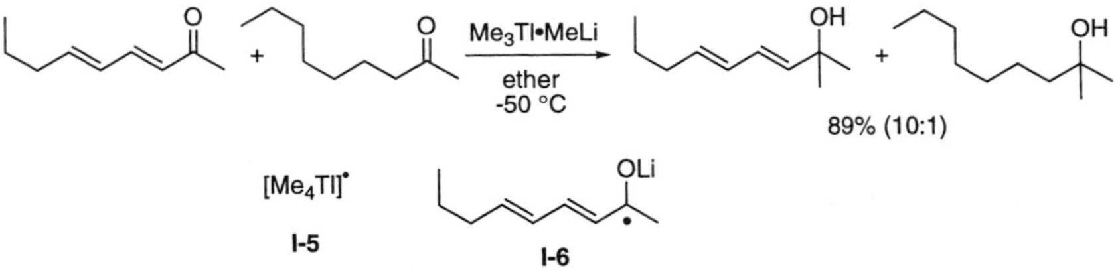

$Me_3Tl{\bullet}MeLi$, ether, -50 °C

89% (10:1)

$[Me_4Tl]^{\bullet}$ **I-5**

OLi **I-6**

Scheme 2-24.

Although most systems allow selective transformations of enones over alkanoes, the $2MoOCl_3(THF)_2$-4MeLi combination affords exceptional chemoselectivity in the homologation of alkanoic functions (Scheme 2-25) [48].

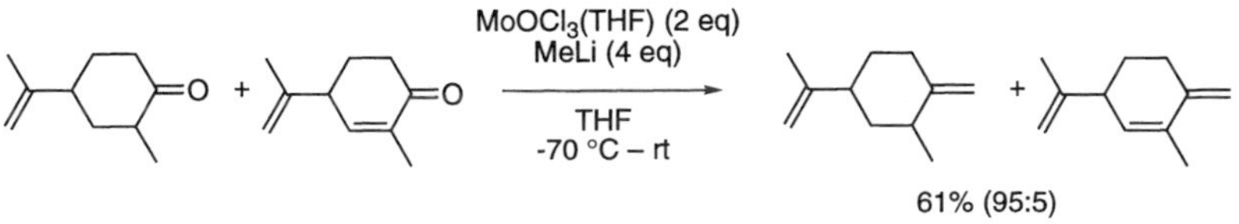

$MoOCl_3(THF)$ (2 eq), MeLi (4 eq), THF, -70 °C – rt

61% (95:5)

Scheme 2-25.

2.3.3 Aldehyde vs Aldehyde

2.3.3.1 Saturated vs saturated

The discrimination between different aldehyde carbonyls with a Lewis acid is rather difficult owing to their high reactivity and inherent coordination aptitude to these acids. The complexation with aldehydes is consistent with *anti* to the substituent of an aldehydic carbon [3], hence steric factors have less influence on the discrimination events (Fig. 2-8).

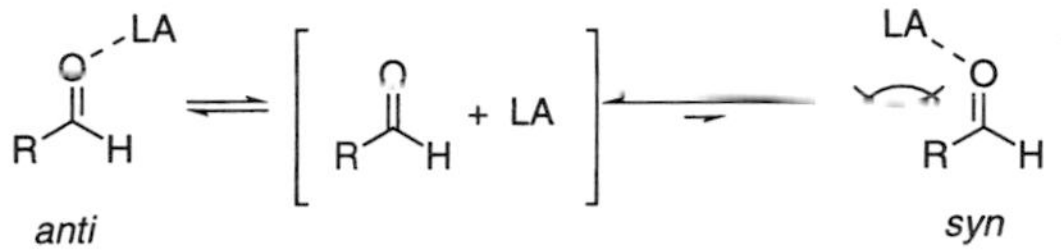

anti *syn*

Figure 2-8

Furthermore, saturated aldehydes are somewhat less basic than saturated ketones or esters, resulting in reversible complexation even with bulky aluminum reagents. However, whether the equilibrium [Lewis acid + base ⇌ Lewis acid-base complex] is reversible or irreversible, the selective functionalization of more labile or sterically less-encumbered aldehydes is facile using bulky or mild Lewis acids.

Table 2-3. Eu(dppm)$_3$-catalyzed chemoselective aldolization.

Entry	Aldehydes	KSA	Temp (°C)	Aldols yield (adols ratio)
1	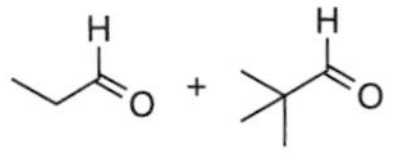	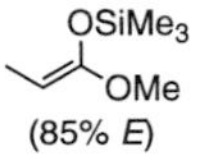 OSiMe$_3$ OMe (85% *E*)	0	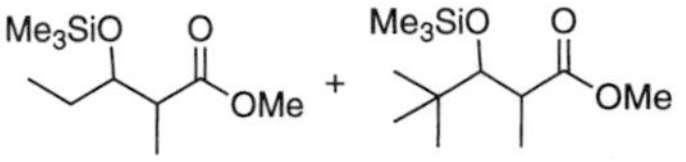 61% (>99:<1)
2	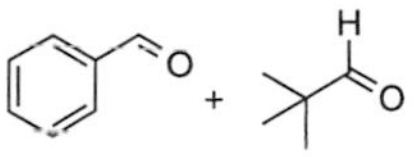	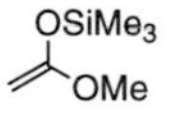 OSiMe$_3$ OMe	-40	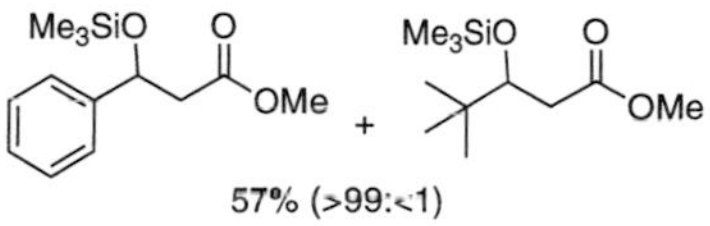 57% (>99:<1)
3	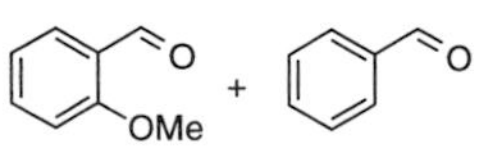 OMe	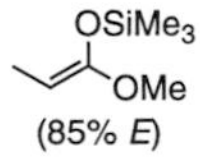 OSiMe$_3$ OMe (85% *E*)	-78	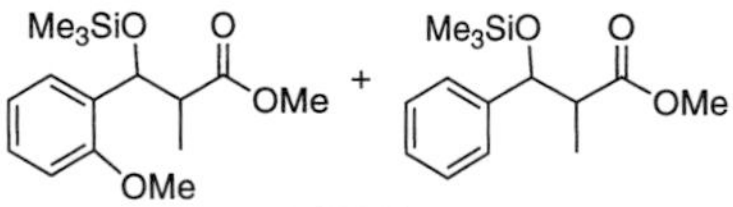 78% (95:5)
4	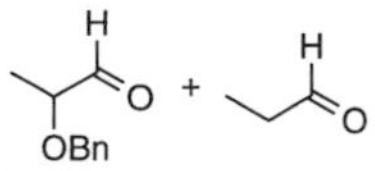 OBn	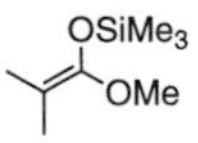 OSiMe$_3$ OMe	-40	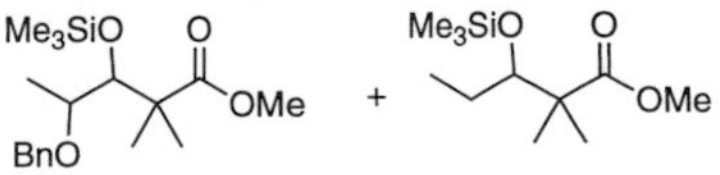 80% (>99:<1)

The Eu-catalyst Eu(dppm)$_3$ provides a remarkable level of chemoselectivity but is only effective for the Mukaiyama-aldol reaction of aldehydes with several ketene silyl acetals (KSA) (Table 2-3) [55]. When ketones and aldehydes are treated, respectively, with KSA and ketone-derived silyl enol ethers, no reaction results. The rate enhancement by chelation control (entry 4, Table 2-3) is intriguing. This is a feature common to other Lewis acids such as $TiCl_4$ [56] or $LiClO_4$ [57].

For comparison, ATPH can be used for this kind of differentiation more efficiently in the hetero-Diels-Alder reaction [58]. With ATPH, silyl enol ether **Si-1** exhibited adequate potential for the aldolization unlike the observed poor reactivity with KSA alone using the above Eu-catalyst (Scheme 2-26; Table 2-4). Even the β-substituents of the aldehydes can be differentiated (entries 4 and 5, Table 2-4). The hetero-atom-containing aldehyde was effectively discriminated, showing non-chelation ability of ATPH (entry 6, Table 2-4). When aldehydes are encapsulated in the ATPH cavity, the hitherto small steric effects turned out in these cases to be dominant. The importance of the effect of the cavity was illustrated further by a comparative experiment with bulky MAD (**5**:**6**=3.7:1).

Table 2-4. Chemoselective functionalization between two different aldehydes with ATPH[a].

Entry	Aldehydes	Hetero-Diels-Alder adducts yield (ratio)	Aldol adducts (ratio)	Allylated adducts (ratio)
1	CHO + CHO			82% (>99:<1)
2	CHO + CHO			77% (24:1)
3	CHO + CHO	61% (30:1)		
4	CHO + CHO	64% (63:1)	63% (91:1)	
5	CHO + CHO	81% (31:1)	61% (27:1)	
6	CHO + Ph O CHO	70% (>99:<1)		

[a]) Unless otherwise specified, an equimolar mixture of two different aldehydes in CH_2Cl_2 was treated with Danishefsky diene or 2-trimethylsiloxy)propene in the presence of ATPH (1 ~ 2 eq) at –78 °C. Allylation was conducted using allyltributylstannane 5 mol% of ATPH at –78 °C.

MAPH is able to stabilize formaldehyde [59], which is also applicable to the selective protection of sterically less-hindered aldehyde carbonyls (Scheme 2-27) [12]. This characteristic advantage led to the opposite sense of chemoselectivity: selective nucleophilic alkylation of sterically more-hindered aldehydes with RLi (R=Ph, Bu, etc.) reagents.

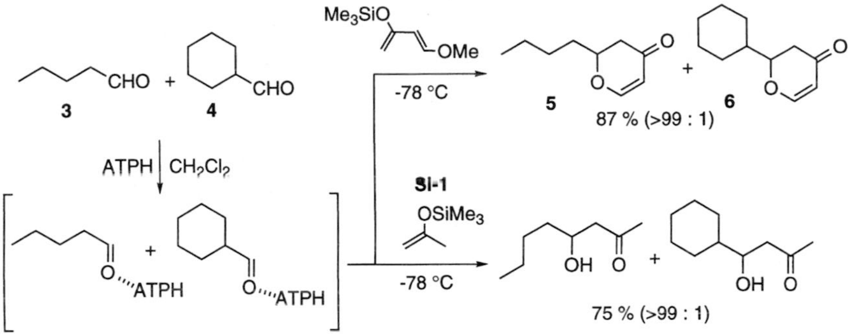

Scheme 2-26.

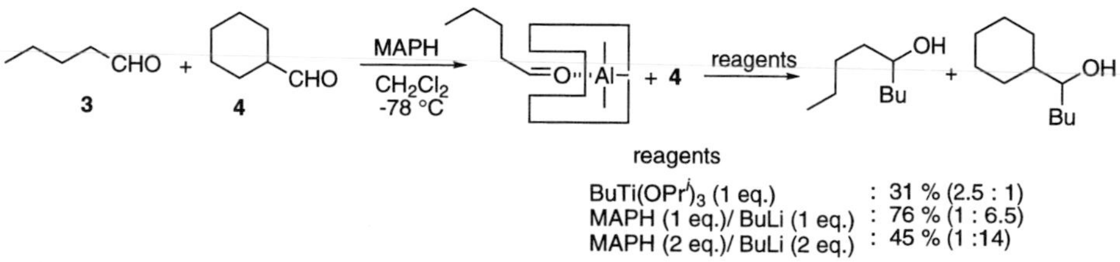

reagents

BuTi(OPr^i)$_3$ (1 eq.) : 31 % (2.5 : 1)
MAPH (1 eq.)/ BuLi (1 eq.) : 76 % (1 : 6.5)
MAPH (2 eq.)/ BuLi (2 eq.) : 45 % (1 :14)

Scheme 2-27.

2.3.3.2 Saturated vs conjugated

In general, conjugated aldehydes make more stable complexes with Lewis acids than saturated aldehydes. This coordination tendency allows selective activation of conjugated aldehyde which leads to chemoselective conversion of these species (entry 2, Table 2-5). There are, however, some exceptions. When a 1 : 1 mixture of benzaldehyde and cyclohexanecarboxaldehyde (**4**) was exposed to Me_3SiCN in the presence of Me_2AlCl, the silylcyanation of **4** predominated (entry 1, Table 2-5) [60].

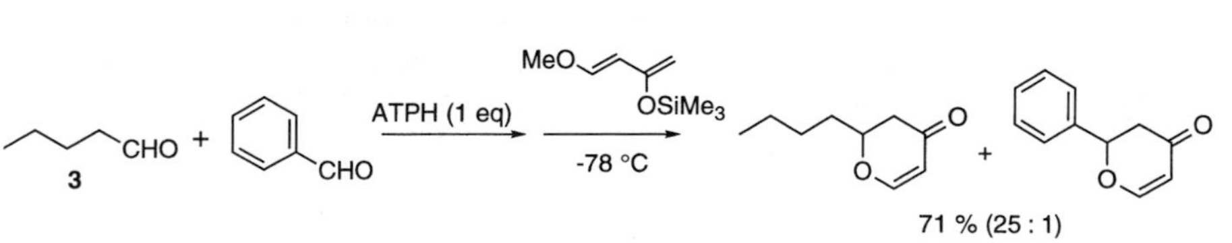

Scheme 2-28.

Interpretation of the discrimination events using ATPH is rather complicated. The carbonyl of PhCHO is highly likely to be stabilized by the phenyl groups of the cavity of ATPH. The induced ^{1}H NMR chemical shifts of the ATPH-*o,o′,p*-trimethylbenzaldehyde (**7**) complex (toluene-d_8, 300 MHz) showed considerable upfield shift of the H_a (aldehydic proton) resonance (δ 8.08 ppm; originally, 10.33 ppm). Thus, aldehydes having a linear alkyl chain (relatively small aldehyde) are selectively functionalized by deactivated nucleophiles in the presence of PhCHO (Scheme 2-28). On the contrary, saturated aldehydes having an α-branched alkyl chain (relatively large aldehyde) are more sterically deactivated on complexation with ATPH, consequently leading to the reversal of chemoselectivity, i.e., functionalization of PhCHO. These contradictory results can be explained by the synergetic steric effect that is extended by the interplays of the substituents of aldehydes and the bulk of ATPH (Fig. 2-9). It was demonstrated that 5 mol% of ATPH was productive enough for the chemoselective allylation with the allylstannane reagent (entry 3, Table 2-5) [61].

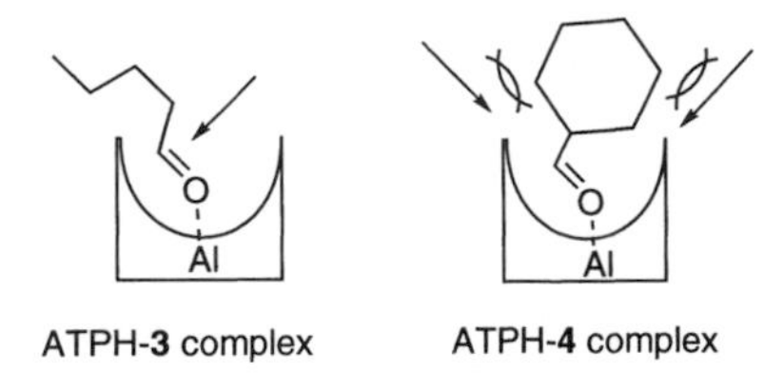

Figure 2-9

Table 2-5. Chemoselective functionalization of cyclohexanecarboxaldehyde vs PhCHO.

Entry	Reagents	Temp (°C)	Products yield (ratio)
	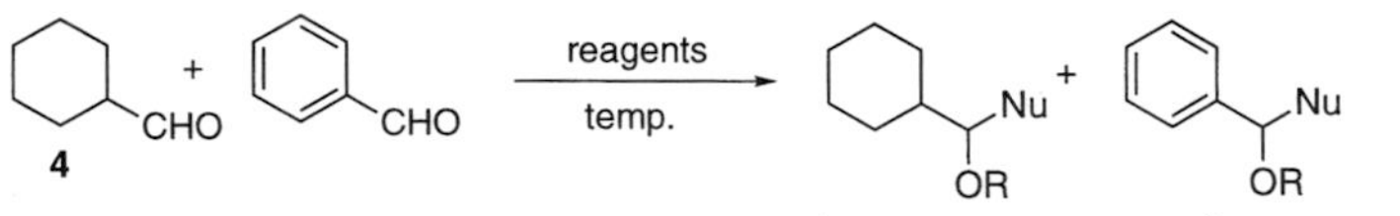 **4** (CHO) + PhCHO	reagents, temp.	cyclohexyl-CH(OR)Nu + Ph-CH(OR)Nu
1	Me_2AlCl, Me_3SiCN	-78 °C	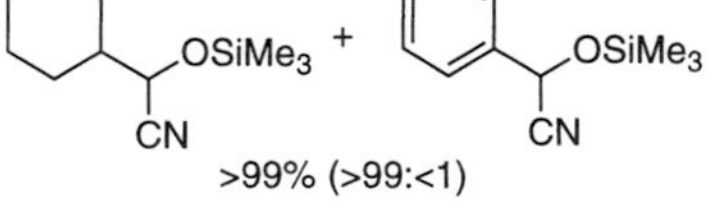 cyclohexyl-CH(CN)OSiMe₃ + Ph-CH(CN)OSiMe₃ >99% (>99:<1)
2	Me_2AlCl, 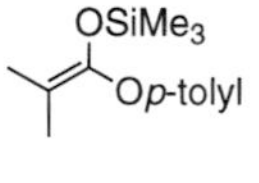 OSiMe₃ / O*p*-tolyl	-78 °C	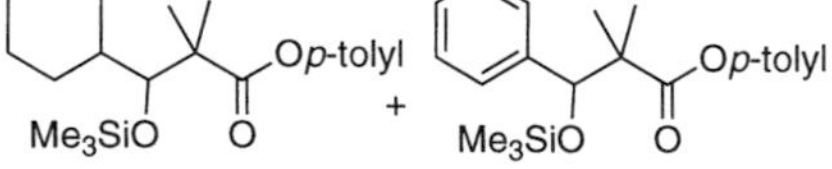 Me₃SiO / O*p*-tolyl / O + Me₃SiO / O*p*-tolyl / O >99% (<1:>99)
3	ATPH (5 mol%), 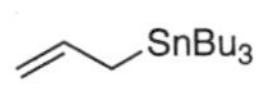 SnBu₃	-40 °C	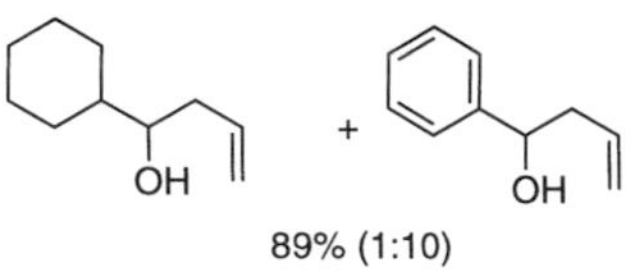 OH + OH 89% (1:10)

2.3.3.3 Conjugated vs conjugated

The coordination ability of conjugated aldehydes plays a crucial role in evaluating the reactivity of each substrate: the more basic the carbonyl of an aldehyde is, the more reactive the aldehyde becomes. $Eu(dppm)_3$ [55] and $Bu_2Sn(OTf)_2$ [62] selectively activate one aldehyde over another along this line (entry 3, Table 2-3; entries 1–3, Table 2-6). Note that cinnamaldehyde is more reactive than PhCHO using $Bu_2Sn(OTf)_2$. The activation with ATPH is more sensitively influenced by the steric environment of conjugated aldehydes (entry 4, Table 2-6) [61].

Table 2-6. Chemoselective functionalization of two different conjugated aldehydes.

Entry	Conjugated aldehydes	Reagents	Products yield (ratio)
1	PhCHO + MeO-C6H4-CHO	$Bu_2Sn(OTf)_2$, CH2=C(OTBS)OEt	Ph-CH(OSiMe3)CH2CO2Et + MeO-C6H4-CH(OSiMe3)CH2CO2Et; 83% (<1:>99)
2	PhCHO + NC-C6H4-CHO	$Bu_2Sn(OTf)_2$, CH2=C(OTBS)OEt	Ph-CH(OSiMe3)CH2CO2Et + MeO-C6H4-CH(OSiMe3)CH2CO2Et; 54% (>99:<1)
3	PhCH=CHCHO + PhCHO	$Bu_2Sn(OTf)_2$, CH2=C(OTBS)OEt	PhCH=CH-CH(OSiMe3)CH2CO2Et + Ph-CH(OSiMe3)CH2CO2Et; 79% (97:3)
4	PhCHO + **7** (mesityl-CHO)	ATPH (5 mol%), CH2=CHCH2SnBu3	Ph-CH(OH)CH2CH=CH2 + Mes-CH(OH)CH2CH=CH2; 83% (>99:<1)

2.3.4 Aldehyde vs Acetal

Alkanones and alkanals did not react with enol silyl ethers in the presence of TMSOTf or a TMSOTf/2,6-di-*t*-butylpyridine mixture (Scheme 2-29) [63]. In contrast, the totally opposite selectivity was reported for the competition reaction between a ketone (or aldehyde) and an acetal in the TMSOTf-catalyzed aldol reaction with the KSA of methyl acetate (silyl=TBS group) [64]. The chemoselective preference for ketone over acetal in this KSA-aldol system also arose with $(C_6F_5)_2SnBr_2$, $Sc(OTf)_3$, Bu_3SnClO_4, and $TrClO_4$ (Table 2-7) [64]. It has been pointed out that these preferences for ketone and aldehyde are unique to KSA. Bu_3SnClO_4, $Bu_2Sn(ClO_4)_2$, and $Bu_2Sn(OTf)_2$ are excellent activators of aldehydes over acetals or ketals, although the latter two tin reagents were later demonstrated to be inefficient promoters of ketonic carbonyls [65].

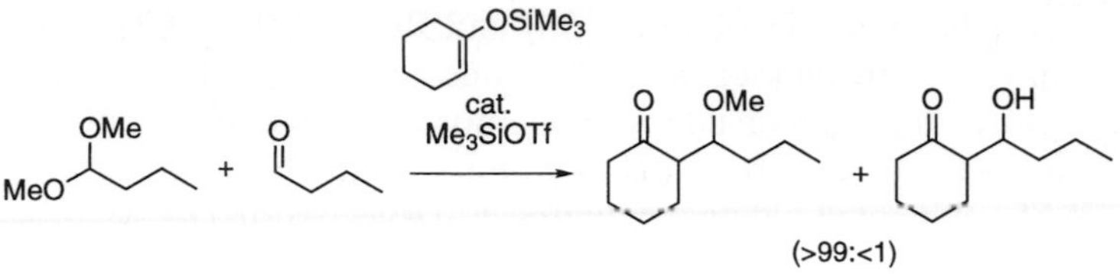

OMe, MeO + O → ($OSiMe_3$, cat. Me_3SiOTf) → O OMe + O OH

(>99:<1)

Scheme 2-29.

Table 2-7. Chemoselective functionalization of carbonyls over acetals or ketals.

Entry	Aldehyde + acetal (or ketal)	Reagents	Products yield (ratio)
1	PhCHO 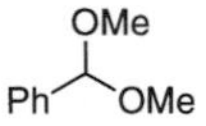 + Ph(OMe)OMe	Bu_3SnClO_4, 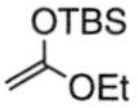 OTBS / OEt	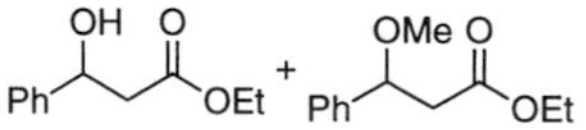 Ph, OH O, OEt + Ph, OMe O, OEt — 94% (100:0)
2		$Bu_2Sn(ClO_4)_2$, 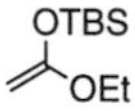 OTBS / OEt	84% (100:0)
3	PhCHO 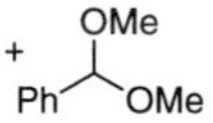 + Ph(OMe)OMe	$Bu_2Sn(OTf)_2$, 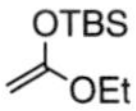 OTBS / OEt	62% (>99:<1)
4	PhCHO 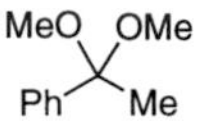 + MeO OMe, Ph Me	$Bu_2Sn(OTf)_2$, 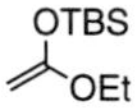 OTBS / OEt	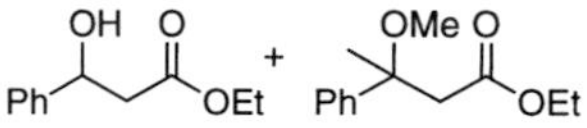 Ph, OH O, OEt + Ph, OMe O, OEt — 55% (>99:<1)
5	PhCOMe 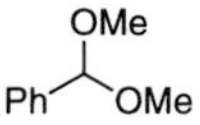 + Ph(OMe)OMe	$(C_6F_5)_2SnBr_2$ 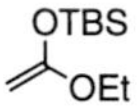 OTBS / OEt	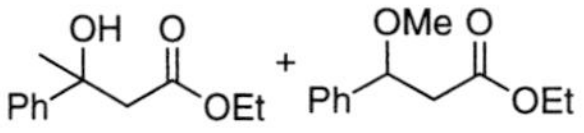 Ph, OH O, OEt + Ph, OMe O, OEt — 91% (100:0)
6	PhCOMe 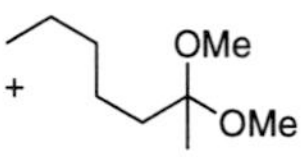 + OMe, OMe	$(C_6F_5)_2SnBr_2$ 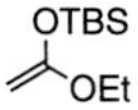 OTBS / OEt	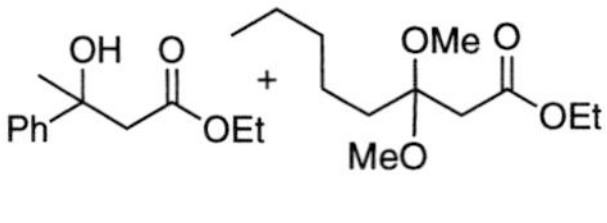 Ph, OH O, OEt + OMe O, OEt, MeO — 96% (100:0)
7	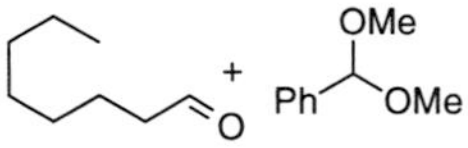 O + Ph(OMe)OMe	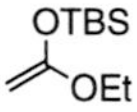 OTBS / OEt, Bu_3SnClO_4	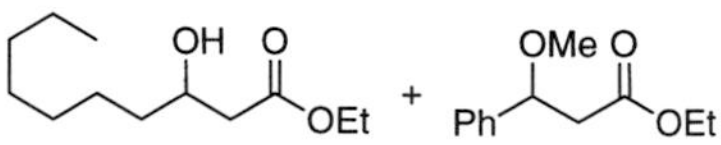 OH O, OEt + Ph, OMe O, OEt — 78% (100:0)
8	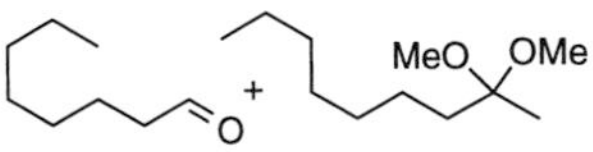 O + MeO OMe	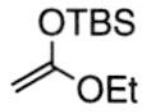 OTBS / OEt, Bu_3SnClO_4	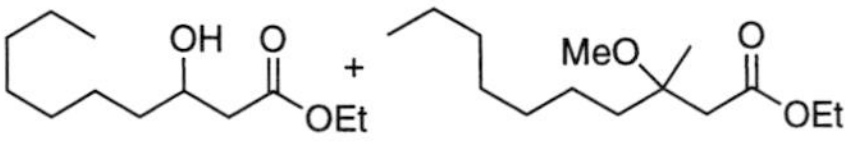 OH O, OEt + MeO, O, OEt — 64% (100:0)

Exceptions are aromatic aldehydes attached to an electron-withdrawing group (EWG). They have low reactivity comparable to that of acetals and ketals. The competition experiments between benzaldehyde derivatives having an electron-donating group (EDG) and EWG support the importance of Lewis basicity of substrates for the chemoselective complexation and activation of a carbonyl unit: the reactivity toward KSA increases in the order *p*-MeO-PhCHO > PhCHO > *p*-CN-PhCHO. It is likely that KSA inherently reacts with aldehydes more readily than with acetals, whereas enol silyl ethers are compatible with acetals and ketals. With this information, parallel differentiated recognition was achieved, i.e., the one-pot chemical transformation on separate reaction sites within a molecule [66]. Carbonyl groups react preferentially with KSA, while enol silyl ethers afford aldolization at acetal counterparts in the presence of ca. 36 mol% of $(C_6F_5)_2SnBr_2$ (Scheme 2-30).

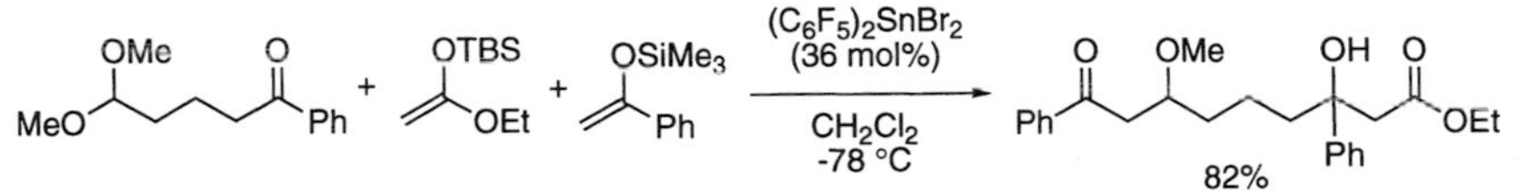

MeO, OMe, O, Ph + OTBS, OEt + OSiMe$_3$, Ph → $(C_6F_5)_2SnBr_2$ (36 mol%), CH_2Cl_2, -78 °C → Ph, O, OMe, OH, O, Ph, OEt 82%

Scheme 2-30.

A new concept was introduced in which the bidentate coordination of type **E2** facilitated the chemoselective conversion of carbonyl functions [67]. Distinct from the above results obtained with the typical tin-Lewis acids, this by **8** doubly activating system is suited even for the selective aldolization of aldehydes with enol silyl ethers. KSA and allylstannane reagents showed similar bias of preferential transformation of aldehydic carbonyls (Table 2-8).

Table 2-8. Chemoselective functionalization of carbonyls over acetals.

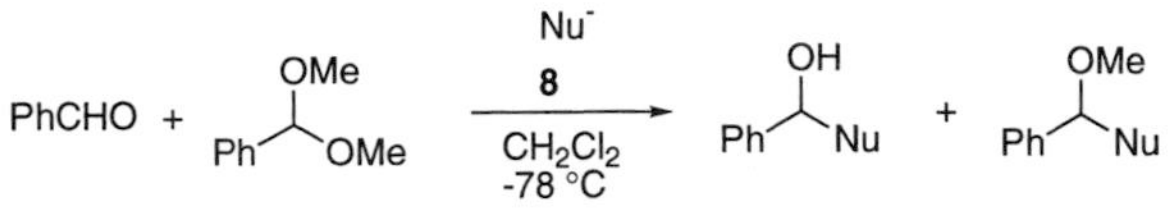

PhCHO + Ph(OMe)OMe → Nu⁻, **8**, CH_2Cl_2, -78 °C → Ph(OH)Nu + Ph(OMe)Nu

Entry	Nucleophile (Nu⁻)	Products	yield (ratio)
1	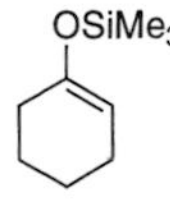 OSiMe$_3$	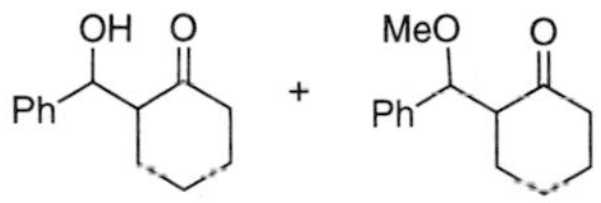 OH O, Ph + MeO O, Ph	84% (97:3)
2	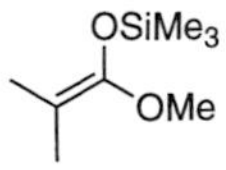 OSiMe$_3$, OMe	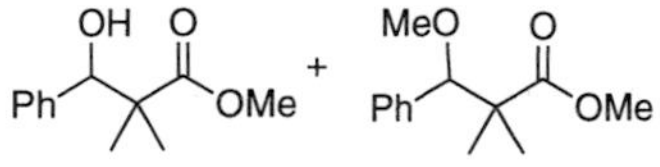 OH O, Ph, OMe + MeO O, Ph, OMe	86% (>99:<1)
3	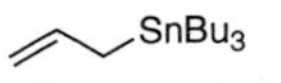 SnBu$_3$	 OH, Ph + MeO, Ph	97% (90:10)

Me$_2$AlO, OAlMe$_2$, Me, Me

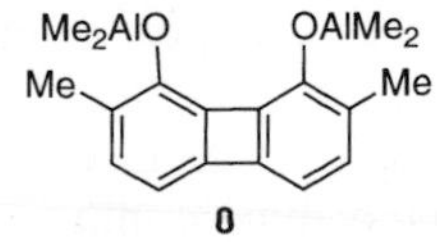

8

Under basic conditions such as those using MeLi, acetals are generally unreactive. In contrast, $MeTiCl_3$, which has a more Lewis-acidic metal center, can catalyze the selective methylation at an acetal moiety (Scheme 2-31) [68].

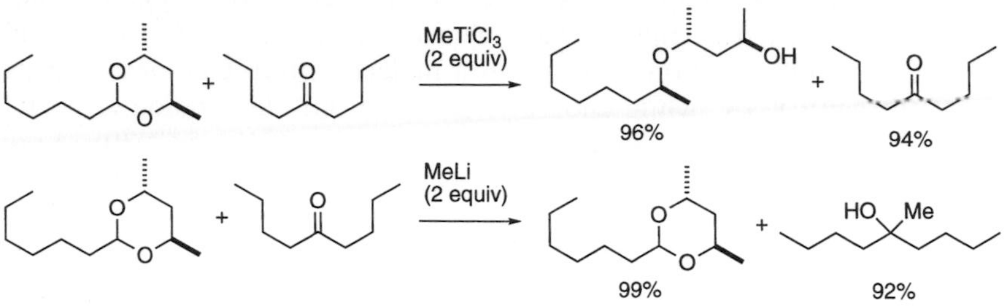

Scheme 2-31.

The silyl chloride-$InCl_3$ system proved to be effective for the chemoselective aldolization of aldehydic units (Scheme 2-32) [69].

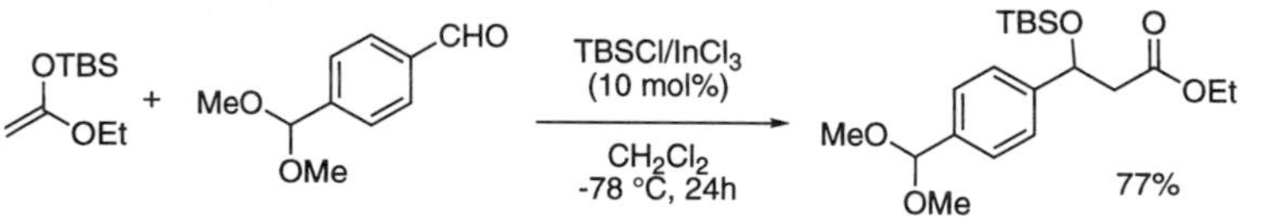

Scheme 2-32.

2.3.5 Aldehyde vs Aldimine

While selective reaction of aldehydes takes place with the typical Lewis acids $TiCl_4$, $SnCl_4$, TMSOTf, etc., lanthanide triflates [$Ln(OTf)_3$] are unique Lewis acids that change the reaction course dramatically: aldimine reacts selectively in the coexistence of aldehydes [70]. Among a series of $Ln(OTf)_3$ tested, $Yb(OTf)_3$ exhibited the most prominent chemoselectivity in addition to high chemical yields. The silyl enol ethers of ketones, allyltributylstannane and Me_3SiCN are all applicable as chemoselective nucleophiles (Table 2-9). Preferential formation of $Yb(OTf)_3$-aldimine complexes was postulated by ^{13}C NMR spectral analysis in the presence of PhCHO and *N*-benzylideneaniline.

This unique behavior of $Yb(OTf)_3$ was extended to a novel three-component coupling of aldehyde, amine, and several nucleophiles in organic solvents (Scheme 2-34) [71]. The aldimine formation is much faster than the nucleophilic addition to aldehydes. Synthetically, this *in situ* tandem transformation proved to be of significant value for the one-pot synthesis of β-lactam skeletones (Scheme 2-35) [72].

$Yb(OTf)_3$ and $Sc(OTf)_3$ are water-stable Lewis acids [73], and when combined with a surfactant such as SDS, the three-component coupling also proceeded ef-

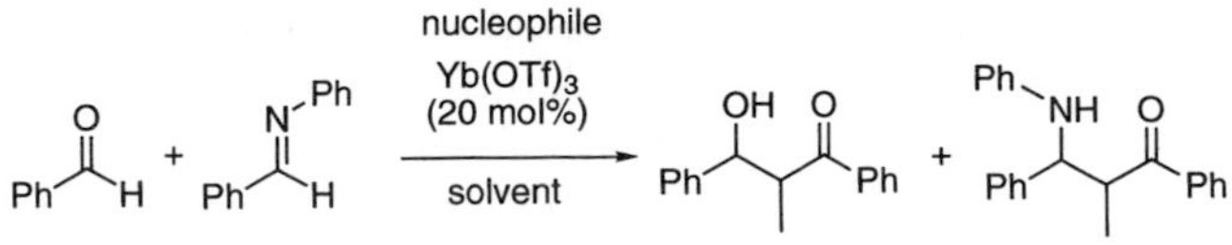

Scheme 2-33.

Table 2-9. Chemoselective functionalization of aldimine.

Entry	Nucleophile	Solvent	Temp. (°C)	Products yield (ratio)
1	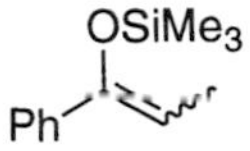	CH_2Cl_2	-23	94% (2:98)
2	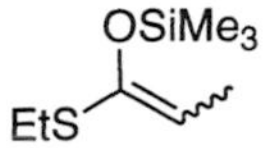	EtCN	-45	93% (<1:>99)
3	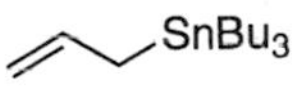 $SnBu_3$	EtCN	-45	81% (<1:>99)
4	Me_3SiCN	EtCN	-45	83% (<1:>99)

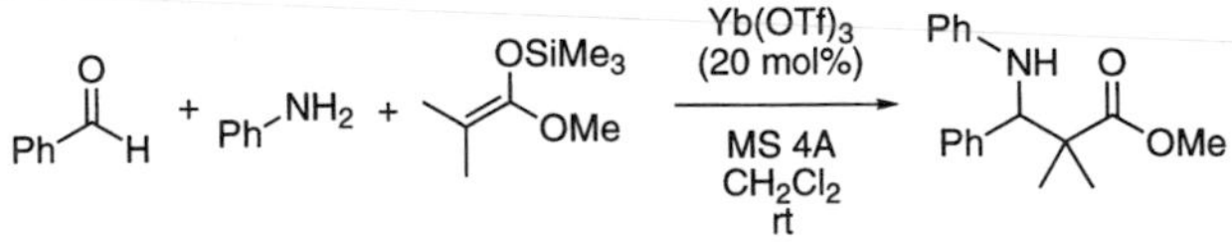

Scheme 2-34.

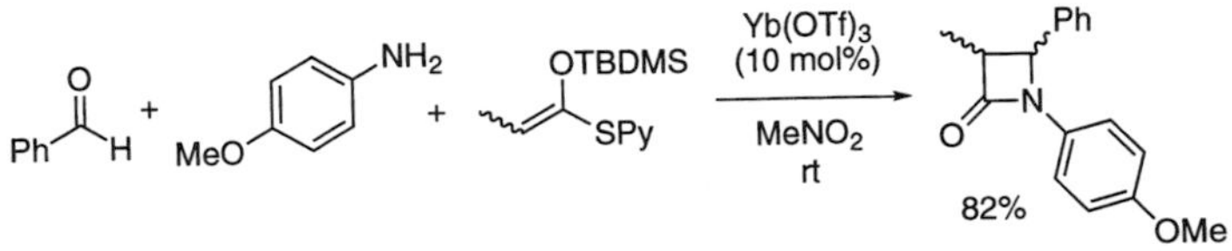

Scheme 2-35.

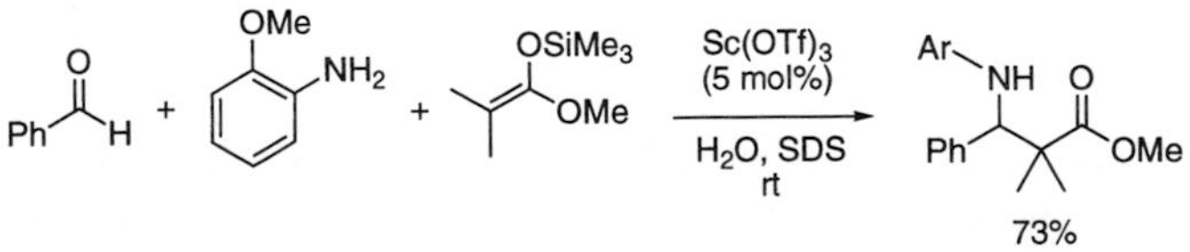

Scheme 2-36.

fectively in water (Schemes 21-36 and 2-37) [74]. Although catalytic amounts of $Sc(OTf)_3$ are good promoters for the Mukaiyama-type aldol reaction of aldehydes with silyl enol ethers in the SDS-water system, aldimines are selectively functionalized with the coexistence of aldehydes. There is an exceptional case where SDS is not required when vinyl ethers [75] and Bu_3SnCN [76] are employed (Schemes 2-38 and 2-39). It was later presented that allyltrimethylgermanium is also a compatible allylating reagent with the $Sc(OTf)_3$-catalyzed three-component coupling in $MeNO_2$ (Scheme 2-37) [77]. $Cu(OTf)_2$ is also a comparable reagent that is useful for the tandem transformation of aldehydes to β-aminocarbonyl compounds [78].

Allylating reagent	Conditions	Solvent	Product yield
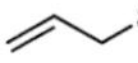 allyl-$SnBu_3$	SDS, $Sc(OTf)_3$ (20 mol%)	H_2O, rt	83%
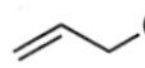 allyl-$GeEt_3$	$Sc(OTf)_3$ (10 mol%)	$MeNO_2$, rt	81%

Scheme 2-37.

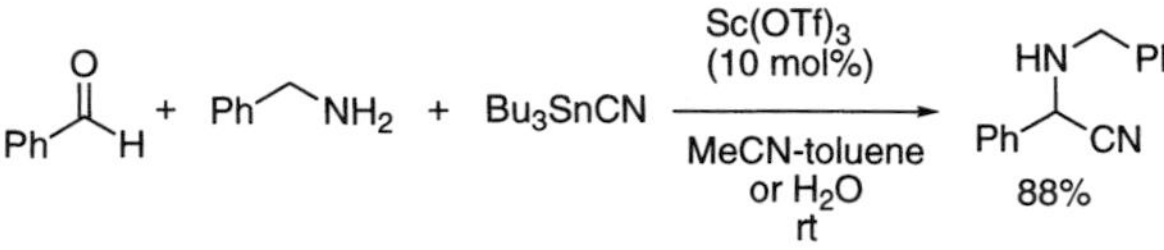

Ph-CHO + PhCH_2NH_2 + Bu_3SnCN → ($Sc(OTf)_3$ (10 mol%); MeCN-toluene or H_2O, rt) 88%

Scheme 2-38.

aq. HCHO + 4-Cl-$C_6H_4NH_2$ + 2-methoxypropene → ($Yb(OTf)_3$ (10 mol%); THF-H_2O, rt) 92%

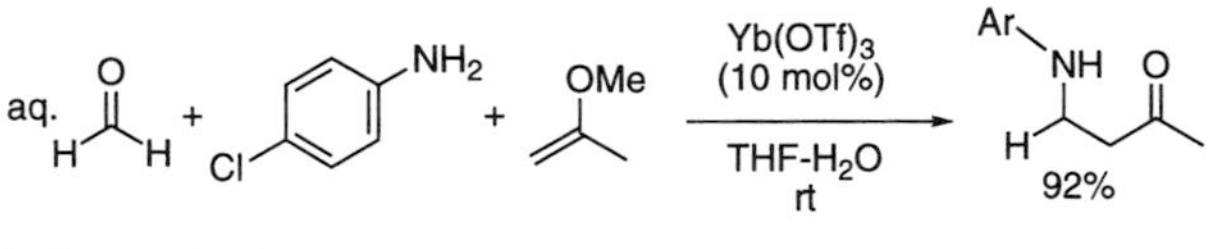

Scheme 2-39.

Recent investigation of bis-π-allylpalladium **I-7**, generated from allyltributylstannane and $PdCl_2(PPh_3)_2$ (Scheme 2-40), afforded the chemoselective functionalization of aldimines over aldehydes (Scheme 2-41) [79]. This unprecedented selectivity can be explained by the difference of the coordination ability between N and O atom to the transition metal. In general, the N-atom can coordinate to a transition metal more strongly.

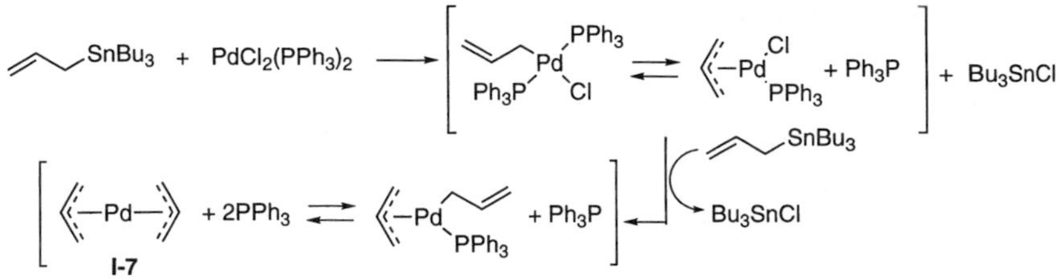

Scheme 2-40.

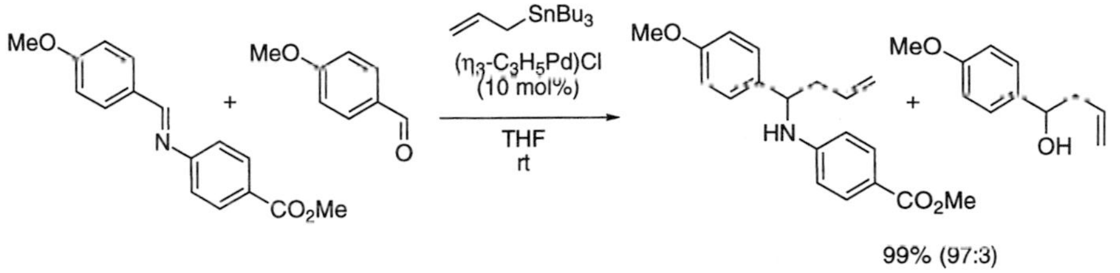

Scheme 2-41.

2.3.6 Ester vs Ester

The use of MAD to distinguish two different ester carbonyls resulted in success (Scheme 2-42) [80]. The reaction of *t*-butyl methyl fumarate **9** with 1.1 equiv of MAD in CH_2Cl_2 at –78 °C gave MAD-fumarate complex **I-8**. The exclusively formed structure of **I-8** was rigorously established by low-temperature ^{13}C NMR spectroscopy. The Diels-Alder reaction of complex **I-8** with cyclopentadiene at –78 °C in toluene gave, after 1 h, cycloadduct **10** predominantly with *endo* orientation of the methoxycarbonyl group. Thus, the methyl ester coordinated with MAD led to high *endo*-selectivity. Even methyl and isopropyl or methyl and ethyl can be fairly well discriminated with MAD.

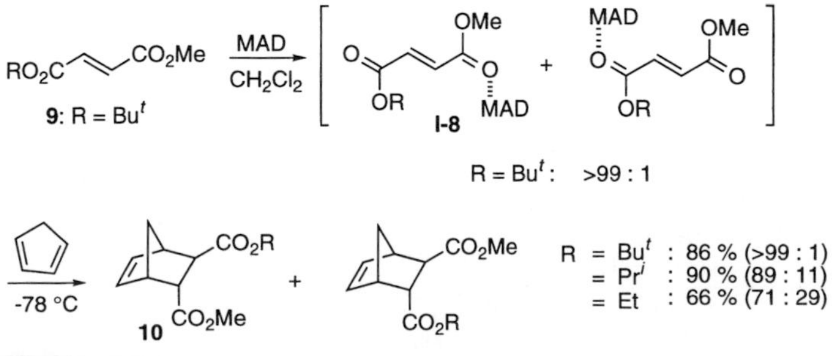

Scheme 2-42.

2.4 Carbonyl Recognition in Asymmetric Synthesis

The recognition of chiral carbonyl compounds, especially the substrates having a chiral stereogenic center at the α-carbon, have been a main subject of research for long time [81], since Curtin [82] and Cram [83] initially proposed a structural model for the diastereoselective addition to these molecules. There are in-depth studies related to this subject including chiral enolate addition, i.e., double asymmetric synthesis [84]. In this section, therefore, some other impressive topics of current interest are discussed briefly.

It is reasonable to anticipate that certain chiral ketones may discriminate between racemic organoaluminum reagents by diastereoselective complexation: preferential formation of one of the diastereomers (Schemes 2-43 and 2-44). Indeed, the Lewis-acidic enantiomer **11** that *in situ* remained intact promoted the asymmetric hetero-Diels-Alder reaction of several aldehydes with substituted Danishefsky diene **12** in high enantioselectivity. The so-called concept of "chiral poisoning" of one of two active enantiomers triggers the selective and relative activation of another enantiomer [85]. Similar approaches using this strategic "chiral poisoning" for asymmetric synthesis have also been reported [86].

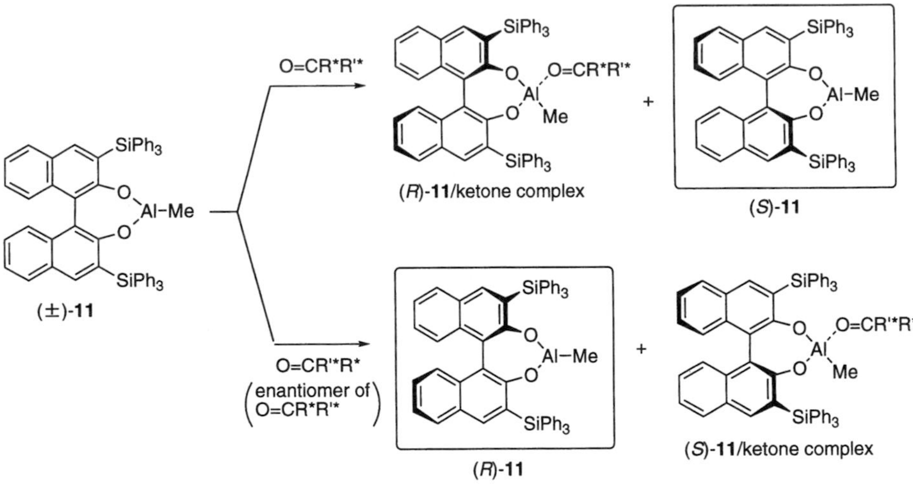

(±)-**11** → O=CR*R'* → (*R*)-**11**/ketone complex + (*S*)-**11**

(±)-**11** → O=CR'*R* (enantiomer of O=CR*R'*) → (*R*)-**11** + (*S*)-**11**/ketone complex

Scheme 2-43.

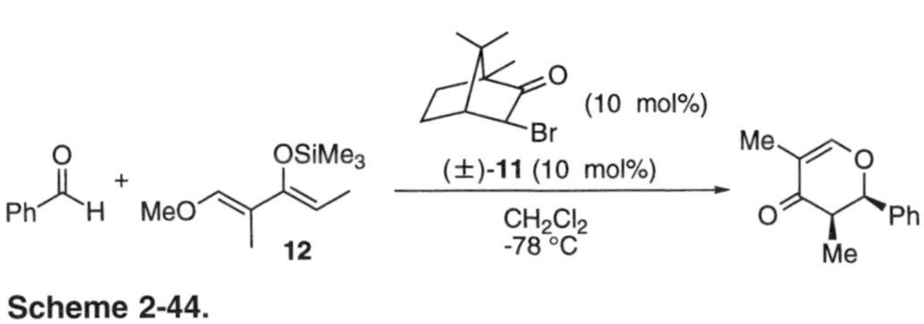

PhCHO + **12** (MeO–CH=C(Me)–C(OSiMe$_3$)=CH–Me) → (bromocamphor ketone, 10 mol%), (±)-**11** (10 mol%), CH_2Cl_2, -78 °C → dihydropyranone (Me, Ph, Me substituted)

Scheme 2-44.

The aluminum catalyst complexed with the ligand *S*-VAPOL mediates Diels-Alder reactions, exhibiting an autoinduction which is due to cooperative interaction of the product with the catalyst to generate a more selective catalytic species (Scheme 2-45). The ee% gradually increased as the reaction time lengthened. In a proposed intermediate of pentacoordinated aluminum complex **I-9**, the cycloadduct is recognized as a complementary ligand, leading to the high degree of asymmetric induction. The acrylate is activated effectively by this hybridized complex [87].

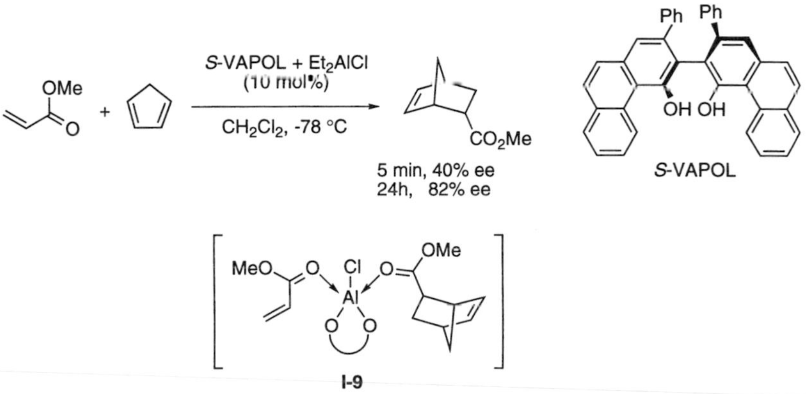

OMe + [cyclopentadiene] → S-VAPOL + Et_2AlCl (10 mol%), CH_2Cl_2, -78 °C → CO_2Me

5 min, 40% ee
24h, 82% ee

Ph Ph OH OH *S*-VAPOL

MeO O Cl Al O O OMe O

I-9

Scheme 2-45.

Kinetic resolution of racemic compounds is a representative chiral recognition event participating in fields of both carbonyl and non-carbonyl chemistry [88]. Under certain chiral circumstances, substrate enantiomers S_R and S_S react at different rates, k_R and k_S, yielding enantiomeric products P_R and P_S. This kind of usual kinetic resolution has an inherent disadvantage: the maximum yield of either enantiomer produced is 50%. In contrast, dynamic kinetic resolution (Scheme 2-46) is involved in the reaction where racemic substrates possess an epimerizable chiral stereogenic center, and thus can possibly give the maximum yield of 100%. Noyori, Kitamura, and co-workers reported on this subject in the asymmetric reduction of *β*-ketoesters by BINAP-Ru catalyst derivatives and H_2 (Scheme 2-46). (*S*)-**13** is recognized as a preferable substrate in the kinetic reduction, the rate of which is appreciably faster than that of (*R*)-**13** ($k_S/k_R = 15$) [89]. In addition, the rate constant of the equilibrium (k_{inv}) must be much larger than k_S. (*S,R*)-**14** was indeed generated exclusively.

Very recently, a spectacular improvement in carbonyl reduction chemistry was achieved by Noyori, Ohkuma, and co-workers (Scheme 2-47). They demonstrated that the three-components reduction system, BINAP (**15**)-Ru(II), 1,2-diamine and KOH is crucial for the chemoselective reduction of ketone carbonyls with H_2 in the presence of olefinic units. The extremely efficient reaction conditions [Ru(II) catalyst, 0.5 mol%; H_2, 1 ~ 8 atm; 28 °C; *i*-PrOH-toluene] render the reduction quite practical, bringing about almost quantitative yields as well as absolute che-

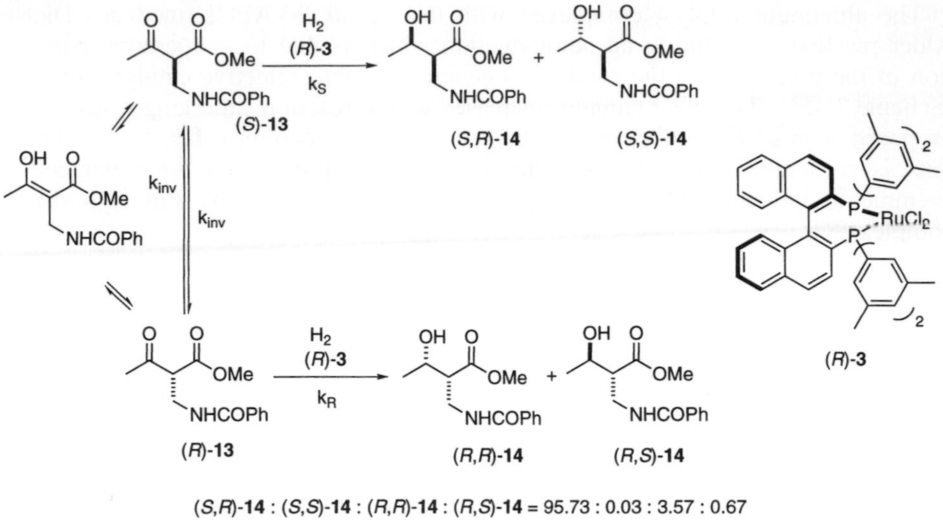

Scheme 2-46.

moselectivities [90]. Although *i*-PrOH is a critical solvent for the enhanced efficacy, the authors emphasized that the hydride source is H_2 and not from transfer hydrogenation. In connection with this pioneering discovery, the enantioselective version was put forth using similar synergetic effects of optically active **15** and 1,2-ethylenediamine derivative **16** [91, 92].

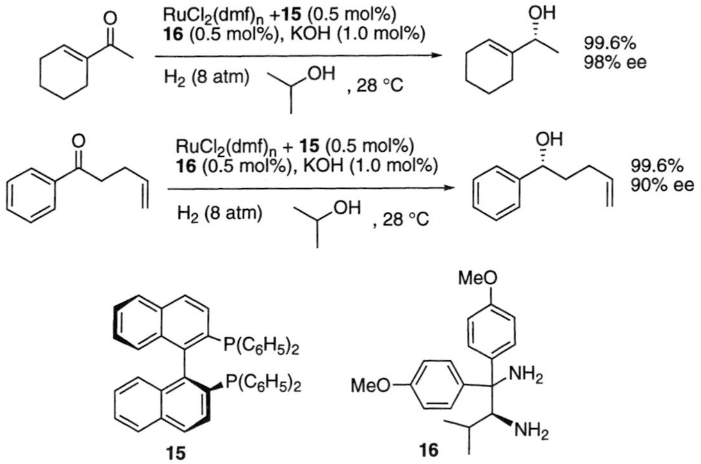

Scheme 2-47.

2.5 Closing Remarks

The development of chemoselective transformations that obviate a need for tedious protection-deprotection sequences is a central dogma for the efficient synthesis of natural and other complex organic molecules. The strategy utilizing Lewis acids is a powerful candidate to address this problem. Screening of metal variations to estimate the reactivity and selectivity of Lewis acids provides a short pathway to the requisite chemoselectivity. In addition, the potential effects of metal ligands on chemoselectivity are well demonstrated by many examples. Specifically, imparting a specific shape to Lewis acids is a powerful tool for the achievement of the selective complexation with a specific substrate as illustrated by the use of ATPH and MAD. Designer Lewis acids thus show attractive and potential usage in discriminating structurally and electronically similar carbonyl substrates. However, more efficient systems are still needed, and the mechanistic aspects of each chemoselective transformation remain to be established. Thus, the search for new and practical approaches for the design of Lewis acids remains a challenge in selective organic synthesis.

References

1. Fessner, W.-D.; Walter, C. In *Bioorganic Chemistry*; Schmidtchen, F. P., Ed.; Springer: Berlin, **1997**, pp 97.
2. (a) Rebek, J. Jr. *Angew. Chem.* **1990**, *102*, 261; *Angew. Chem. Int. Ed. Engl.* **1990**, *29*, 245. (b) Pieters, R.J.; Huc, I.; Rebek, J. Jr. *Chem. Eur. J.* **1995**, *1*, 183. (c) Dugas, H. Ed. *Bioorganic Chemistry, 3rd edn.*, Springer: New York, **1996**.
3. (a) Schreiber, S. L. In '*Comprehensive Organic Synthesis*' Trost, B. M.; Fleming, I. Ed.; Pergamon Press, Oxford, **1991**, Vol. 1, pp. 283. (b) Shambayashi, S.; Crowe, W.E.; Schreiber, S.L. *Angew. Chem.* **1990**, *102*, 273; *Angew. Chem. Int. Ed. Engl.* **1990**, *29*, 256.
4. (a) Reetz, M.T.; Hüllmann, M.; Massa, W.; Berger, S.; Rademacher, P.; Heymanns, P. *J. Am. Chem. Soc.* **1986**, *108*, 2405. (b) Hartman, J.S.; Stilbs, P.; Forsén, S. *Tetrahedron Lett.* **1975**, 3497. (c) Corey, E.J.; Rohde, J.J.; Fischer, A.; Azimioara, M.D. *Tetrahedron Lett.* **1997**, *38*, 33. (d) Corey, E.J.; Loh, T.-P.; Sarshar, S.; Azimioara, M. *Tetrahedron Lett.* **1992**, *33*, 6945. (e) Denmark, S.E.; Almstead, N.G. *J. Am. Chem. Soc.* **1993**, *115*, 3133.
5. (a) Saito, S.; Yamamoto, H. *Chem. Commun.* **1997**, 1585. (b) Yamamoto, H.; Yanagisawa, A.; Ishihara, K.; Saito, S. *Pure Appl. Chem.* **1998**, *70*, 1507. (c) Yamamoto, H.; Saito, S. *Pure Appl. Chem.* **1999**, in press.
6. Pine, S.H. Ed. *Organic Chemistry, 5th edn.*, McGraw-Hill, New York, **1987**, chap. 8 and 9.
7. Parks, D.J.; Piers, W. E. *J. Am. Chem. Soc.* **1996**, *118*, 9440.
8. (a) Maruoka, K.; Nagahara, S.; Yamamoto, H. *J. Am. Chem. Soc.* **1990**, *112*, 6225. (b) Maruoka, K.; Nagahara, S.; Yamamoto, H. *Tetrahedron Lett.* **1990**, *31*, 5475. (c) Ooi, T.; Maruoka, K. *Org. Synth. Chem. Jpn.* **1996**, *54*, 200.
9. (a) Maruoka, K.; Imoto, H.; Yamamoto, H. *J. Am. Chem. Soc.* **1994**, *116*, 12115. (b) Maruoka, K.; Saito, S.; Yamamoto, H. *J. Am. Chem. Soc.* **1995**, *117*, 1165. (c) Saito, S.; Yamamoto, H. *J. Org. Chem.* **1996**, *61*, 2928. (d) Saito, S.; Ito, M.; Maruoka, K.; Yamamoto, H. *Synlett.* **1997**, 357. (e) Ooi, T.; Hokke, K.; Maruoka, K. *Angew. Chem.* **1997**, *109*, 1230; *Angew. Chem. Int. Ed. Engl.* **1997**, *36*, 1181. (f) Saito, S.; Shiozawa, M.; Ito, M.; Yamamoto, H. *J. Am. Chem. Soc.* **1998**, *120*, 813. (g) Ooi, T.; Kondo, K.; Maruoka, K. *Angew. Chem.* **1998**, *110*, 3213; *Angew. Chem. Int. Ed. Engl.* **1998**, *37*, 3039. (h) Saito, S.; Shiozawa, M.; Yamamoto, H. *Angew. Chem.* in press.

10. See reference 50.
11. Elschenbroich, C.; Salzer, A. Ed. *Organometallics, 2nd edn.*, VCH, Weinheim, **1992**, pp. 75.
12. Maruoka, K.; Saito, S.; Concepcion, A.B.; Yamamoto, H. *J. Am. Chem. Soc.* **1993**, *115*, 1183.
13. Healy, M.D.; Power, M.B.; Barron, A.R. *Coord. Chem. Rev.* **1994**, *130*, 63.
14. Shreve, A.P.; Mulhaupt, R.; Fultz, W.; Calabrese, J.; Robbins, W.; Ittel, S.D. *Organometallics*, **1988**, *7*, 409.
15. Healy, M.D.; Barron, A.R. *Angew. Chem.* **1992**, *104*, 939; *Angew. Chem. Int. Ed. Engl.* **1992**, *31*, 921.
16. Maruoka, K.; Imoto, H.; Yamamoto, H. *Synlett* **1995**, 719.
17. See Table 2 1.
18. Gladysz, J.A.; Boone, B.J. *Angew. Chem.* **1997**, *106*, 566; *Angew. Chem. Int. Ed. Engl.* **1997**, *36*, 550.
19. (a) Wuest, J.D. *Acc. Chem. Res.* **1998**, in press. (b) Vaugeois, J.; Simard, M.; Wuest, J.D. *Coord. Chem. Rev.* **1995**, *145*, 55.
20. Vaugeois, J.; Wuest, J.D. *J. Am. Chem. Soc.* **1998**, *120*, 13016.
21. Maruoka, K.; Imoto, H.; Saito, S.; Yamamoto, H. *J. Am. Chem. Soc.* **1994**, *116*, 4131.
22. Power, M.B.; Bott, S.G.; Clark, D.L.; Atwood, J.L.; Barron, A.R. *Organometallics* **1990**, *9*, 3086.
23. Power, M.B.; Bott, S.G.; Atwood, J.L.; Barron, J.L. *J. Am. Chem. Soc.* **1990**, *112*, 3446.
24. Quinkert, G.; Becker, H.; Grosso, M.; Dambacher, G.; Bats, J.W.; Durner, G. *Tetrahedron Lett.* **1993**, *34*, 6885.
25. CCDC registry number: CCDC-112322.
26. CCDC registry number: CCDC-112404.
27. Childs, R.F.; Mulholland, D.L.; Nixon, A. *Can. J. Chem.* **1982**, *60*, 801.
28. Saito, S.; Ito, M.; Yamamoto, H. *J. Am. Chem. Soc.* **1997**, *119*, 611.
29. Maruoka, K.; Itoh, T.; Sakurai, M.; Nonoshita, K.; Yamamoto, H. *J. Am. Chem. Soc.* **1988**, *110*, 3588.
30. Keinan, E.; Greenspoon, N. In *Comprehensive Organic Synthesis*, Trost, B.M.; Fleming, I. Ed.; Pergamon Press, Oxford, **1991**, Vol. 8, pp. 523.
31. Reetz, M.T.; Harmat, N.; Mahrwald, R. *Angew. Chem.* **1992**, *104*, 333; *Angew. Chem. Int. Ed. Engl.* **1992**, *31*, 342.
32. Reetz, M.T. Ed. *Organotitanium Reagents in Organic Synthesis*, Springer, Berlin, **1986**, pp. 75.
33. Yamamoto, Y.; Yamada, J.; Asano, T. *Tetrahedron* **1992**, *48*, 5587.
34. (a) Kauffmann, T.; Laarmann, B.; Meges, D. *Tetrahedron Lett.* **1990**, *31*, 507. (b) Kauffmann, T.; Laarmann, B.; Menges, D.; Neiteler, G. *Chem. Ber.* **1992**, *125*, 163.
35. Sato, T.; Otera, J.; Nozaki, H. *J. Org. Chem.* **1993**, *58*, 4971.
36. (a) Reetz, M.T.; Westermann, J.; Steinbach, R.; Wenderoth, B.; Peter, R.; Ostarek, R.; Maus, S. *Chem. Ber.* **1985**, *118*, 1421. (b) Reetz, M. T.; Wenderoth, B. *Tetrahedron Lett.* **1982**, *23*, 5259.
37. (a) Reetz, M.T.; Wenderoth, B.; Peter, R. *J. Chem. Soc., Chem. Commun.* **1983**, 406. (b) Okazoe, T.; Hibino, J.; Takai, K.; Nozaki, H. *Tetrahedron Lett.* **1985**, *26*, 5581.
38. Maruoka, K.; Araki, Y.; Yamamoto, H. *Tetrahedron Lett.* **1988**, *29*, 3101.
39. Kim, S.; Kim, Y.G.; Kim, D. *Tetrahedron Lett.* **1992**, *33*, 2565.
40. Chen, J.; Sakamoto, K.; Orita, A.; Otera, J. *J. Org. Chem.* **1998**, *63*, 9739.
41. Sato, T.; Otera, J.; Nozaki, H. *J. Am. Chem. Soc.* **1990**, *112*, 901.
42. (a) Molander, G.A.; Shubert, D.C. *J. Am. Chem. Soc.* **1987**, *109*, 6877. (b) Molander, G.A.; Cameron, K.O. *J. Org. Chem.* **1991**, *56*, 2617. (c) Molander, G.A.; Cameron, K.O.; *J. Am. Chem. Soc.* **1993**, *115*, 830.
43. Wada, M.; Ohki, H.; Akiba, K. *Tetrahedron Lett.* **1986**, *27*, 4771.
44. Akiyama, T.; Iwai, J. *Tetrahedron Lett.* **1997**, *38*, 853.
45. Yanagisawa, A.; Inoue, H.; Morodome, M.; Yamamoto, H. *J. Am. Chem. Soc.* **1993**, *115*, 10356.
46. Masuyama, Y.; Kinugawa, N.; Kurusu, Y. *J. Org. Chem.* **1987**, *52*, 3704.
47. Brown, H.C.; Khire, U.R.; Narla, G.; Racherla, U.S. *J. Org. Chem.* **1995**, *60*, 544.
48. Kauffmann, T.; Baune, J.; Fiegenbaum, P.; Hansmersmann, U.; Neiteler, C.; Papenberg, M.; Wieschollek, R. *Chem. Ber.* **1993**, *126*, 89.
49. Kauffmann, T.; Beirich, C.; Hamsen, A.; Möller, T.; Philipp, C.; Wingbermühle, D. *Chem. Ber.* **1992**, *125*, 157.

50. Maruoka, K.; Imoto, H.; Yamamoto, H. *Synlett* **1994**, 441.
51. Maruoka, K.; Araki, Y.; Yamamoto, H. *J. Am. Chem. Soc.* **1988**, *110*, 2650.
52. Hanyuda, K.; Hirai, K.; Nakai, T. *Synlett* **1997**, 31.
53. Chikashita, H.; Komazawa, S.; Ishimoto, N. *Bull. Chem. Soc. Jpn.* **1989**, *62*, 1215.
54. Markó, I. E.; Leung, C. W. *J. Am. Chem. Soc.* **1994**, *116*, 371.
55. Mikami, K.; Terada, M.; Nakai, T. *J. Org. Chem.* **1991**, *56*, 5456.
56. Reetz, M.T.; Raguse, B.; Marth, C.F.; Hügel, H.M.; Bach, T.; Fox, D.N.A. *Tetrahedron* **1992**, *48*, 5731.
57. Henry, K.J. Jr.; Grieco, P.A.; Jagoe, C.T. *Tetrahedron Lett.* **1992**, *33*, 1817.
58. Maruoka, K.; Saito, S.; Yamamoto, H. *Synlett* **1994**, 439.
59. Maruoka, K.; Concepcion, A.B.; Murase, N.; Oishi, M.; Hirayama, N.; Yamamoto, H. *J. Am. Chem. Soc.* **1993**, *115*, 3943.
60. Mori, A.; Ohno, H.; Inoue, S. *Chem. Lett.* **1992**, 631.
61. Marx, A.; Yamamoto, H. *Synlett* in press.
62. Chen, J.; Otera, J. *Synlett* **1997**, 29.
63. Murata, S.; Suzuki, M.; Noyori, R. *Tetrahedron* **1988**, *44*, 4259.
64. Otera, J.; Chen, J. *Synlett* **1996**, 321.
65. Chen, J.; Otera, J. *Tetrahedron* **1997**, *53*, 14275.
66. Chen, J.; Otera, J. *Angew. Chem.* **1998**, *110*, 96; *Angew. Chem. Int. Ed. Engl.* **1998**, *37*, 91.
67. Ooi, T.; Tayama, E.; Takahashi, M.; Maruoka, K. *Tetrahedron Lett.* **1997**, *38*, 7403.
68. Mori, A.; Maruoka, K.; Yamamoto, H. *Tetrahedron Lett.* **1984**, *25*, 4421.
69. Mukaiyama, T.; Ohno, T.; Han, J.S.; Kobayashi, S. *Chem. Lett.* **1991**, 949.
70. Kobayashi, S.; Nagayama, S. *J. Am. Chem. Soc.* **1997**, *119*, 10049.
71. Kobayashi, S.; Araki, M.; Yasuda, M. *Tetrahedron Lett.* **1995**, *36*, 5773.
72. Annunziata, R.; Cinquini, M.; Cozzi, F.; Molteni, V.; Schupp, O. *J. Org. Chem.* **1996**, *61*, 8293.
73. (a) Kobayashi, S.; Nagayama, S.; Busujima, T. *J. Am. Chem. Soc.* **1998**, *120*, 8287. (b) Kobayashi, S.; Wakabayashi, T.; Oyamada, H. *Chem. Lett.* **1997**, 831.
74. Kobayashi, S.; Busujima, T.; Nagayama, S. *Chem. Commun.* **1998**, 19.
75. Kobayashi, S.; Ishitani, H. *Chem. Commun.* **1995**, 1379.
76. Kobayashi, S.; Busujima, T.; Nagayama, S. *Chem. Commun.* **1998**, 981.
77. Akiyama, T.; Iwai, J. *Synlett* **1998**, 273.
78. Kobayashi, S.; Busujima, T.; Nagayama, S. *Synlett* **1999**, in press.
79. Nakamura, H.; Iwama, H.; Yamamoto, Y. *J. Am. Chem. Soc.* **1996**, *118*, 6641.
80. Maruoka, K.; Saito, S.; Yamamoto, H. *J. Am. Chem. Soc.* **1992**, *114*, 1089.
81. (a) Heathcock, C.H. In *Asymmetric Synthesis*, Morrison, J.D. Ed.; Academic Press, San Diego, **1984**, Vol. 3, pp. 111. (b) Eliel, E.L. In *Asymmetric Synthesis*, Morrison, J.D. Ed.; Academic Press, San Diego, **1984**, Vol. 2, pp. 125.
82. Curtin, D.Y.; Harris, E.E.; Meislich, E.K. *J. Am. Chem. Soc.* **1952**, *74*, 2901.
83. Cram, D.J.; Abd Elhafez, F.A. *J. Am. Chem. Soc.* **1952**, *74*, 5828.
84. Masamune, S.; Choy, W.; Petersen, J.S.; Sita, L.R. *Angew. Chem. Int. Ed. Engl.* **1985**, *24*, 1.
85. Maruoka, K.; Yamamoto, H. *J. Am. Chem. Soc.* **1989**, *111*, 789.
86. (a) Faller, J.W.; Parr, J. *J. Am. Chem. Soc.* **1993**, *115*, 804. (b) Faller, J.W.; Sams, D.W.I.; Liu, X. *J. Am. Chem. Soc.* **1996**, *118*, 1217.
87. Heller, D.O.; Goldberg, D.R.; Wulff, W.D. *J. Am. Chem. Soc.* **1997**, *119*, 10551.
88. (a) Kagan, H.B.; Fiaud, J.C. In *Topics in Stereochemistry*, Eliel, E. L.; Wilen, S.H. Ed.; Wiley & Sons, New York, **1988**, pp. 249. (b) Finn, M.G.; Sharpless, K.B. In *Asymmetric Synthesis*, Morrison, J.D. Ed.; Academic Press, San Diego, **1985**, Vol. 5, pp. 247
89. Kitamura, M.; Tokunaga, M.; Noyori, R. *J. Am. Chem. Soc.* **1993**, *115*, 144.
90. Ohkuma, T.; Ooka, H.; Hashiguti, S.; Ikariya, T.; Noyori, R. *J. Am. Chem. Soc.* **1995**, *117*, 2675.
91. Ohkuma, T.; Ooka, H.; Ikariya, T.; Noyori, R. *J. Am. Chem. Soc.* **1995**, *117*, 10417.
92. Ohkuma, T.; Koizumi, M.; Doucet, H.; Pham, T.; Kozawa, M.; Murata, K.; Katayama, E.; Yokozawa, T.; Ikariya, T.; Noyori, R. *J. Am. Chem. Soc.* **1998**, *120*, 13529.

3 Pinacol Coupling

Gregory C. Fu

3.1 Introduction

The pinacol coupling reaction is a powerful method for forming from two carbonyl groups a doubly functionalized carbon-carbon bond (Eq. 3.1). The discovery of this venerable process dates from a report by Fittig in the middle of the nineteenth century [1]. Because the pinacol coupling reaction has already been the subject of a number of reviews [2], this chapter will focus on studies published during the past decade, specifically:

- applications of the pinacol coupling reaction in the total synthesis of natural products,
- new families of reagents for the pinacol coupling reaction, and
- new catalytic protocols for the pinacol coupling reaction.

reducing agent

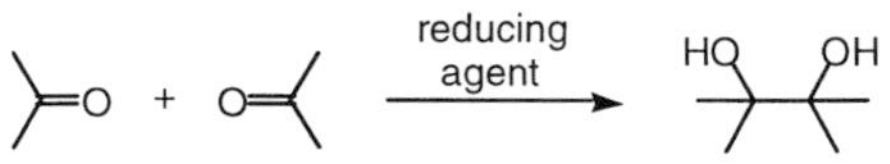

(3.1)

3.2 Background

Most pinacol coupling reactions are believed to proceed through the radical-radical coupling of ketyl anions, which are formed upon treatment of the carbonyl compound with a reducing agent (Fig. 3-1, pathway a). Depending on the reagent, other pathways, such as the addition of a ketyl anion to a carbonyl group (pathway b) or of a metallaoxirane to a carbonyl group (pathway c), may be followed.

The reducing agents that have been employed for the pinacol coupling of carbonyl compounds span the periodic table – for example, lithium [3], sodium [4], magnesium [5], cerium [6], samarium [7, 8], ytterbium [9], titanium [10, 11], vanadium [12], chromium [13], iron [14], zinc [15], and aluminum [16] reagents have been shown to be effective. The stereoselection of the pinacol coupling reaction has been investigated thoroughly for a number of reducing agents, and in certain cases excellent diastereoselectivity has been obtained [2].

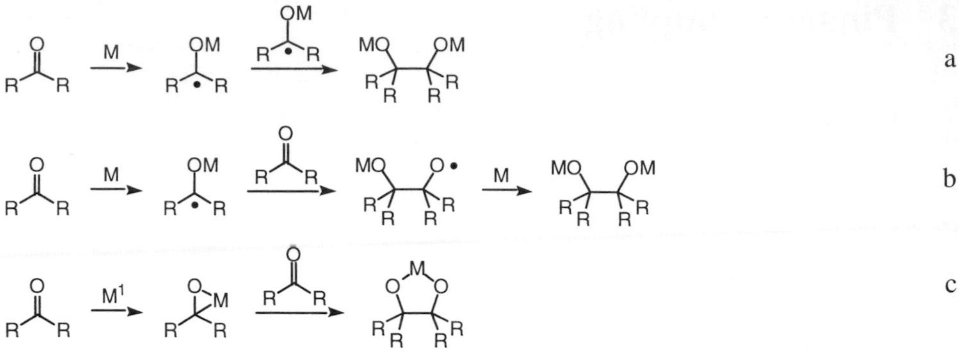

Figure 3-1. Some common pathways for pinacol coupling.

3.3 Applications of the Pinacol Coupling Reaction in the Total Synthesis of Natural Products

The frequency with which the pinacol coupling process is employed in the synthesis of complex molecules is a testament to its tremendous utility. In this section, recent applications to natural products total synthesis are surveyed. These "case studies" simultaneously attest to the triumphs made possible by the pinacol coupling reaction, provide an indication of the functional-group compatibility of the process, and point to challenges that remain to be solved.

Trehazolamine. Chiara has described a synthesis of trehazolamine, the aglycon of trehazolin, a powerful inhibitor of trehalase, wherein a pinacol cyclization provided the necessary cyclopentane framework (Eq. 3.2) [17]. Thus, treatment of the tetrabenzyl-protected keto aldehyde, derived from D-glucose, with SmI_2 furnished the two *cis* diols exclusively as a 1 : 1 mixture of isomers in excellent yield (>90%) [18]. Previous studies of five-membered ring formation mediated by SmI_2 have also reported good *cis/trans* selectivity, presumably due to the intervention of a samarium chelate.

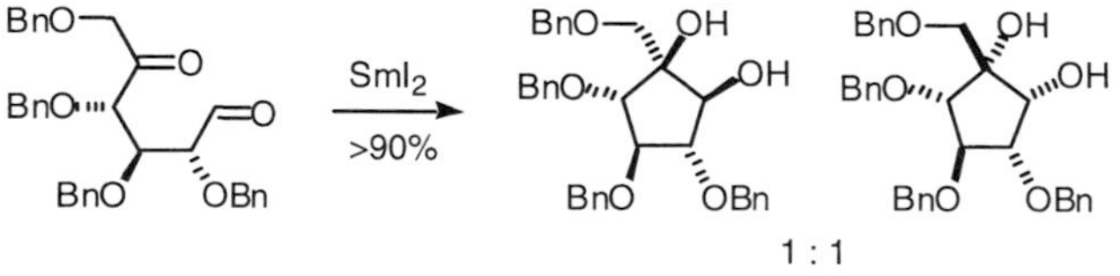

(3.2)

Caryose. Iadonisi has utilized a stereoselective intramolecular pinacol coupling during the course of a synthesis of caryose, a component of a cell wall lipopolysaccharide (Eq. 3.3) [19]. Treatment of the 1,5-ketoaldehyde with SmI_2 provided the isomeric *cis* diols in 64% yield and 93:7 diastereoselectivity. It is worth noting the similarity in structure but difference in stereoselection observed for this substrate compared with that employed by Chiara in the trehazolamine synthesis discussed above (Eq. 3.2).

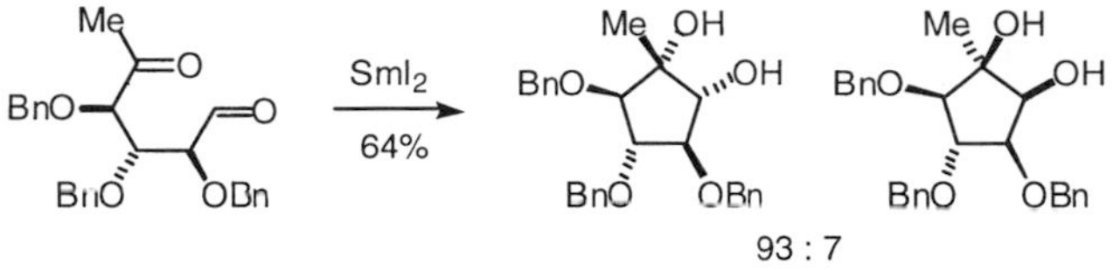

SmI_2, 64%; 93 : 7 (3.3)

Rocaglamide. Taylor has employed an intramolecular keto-aldehyde pinacol coupling reaction in a synthesis of rocaglamide, an anti-leukemia natural product (Eq. 3.4) [20]. Other reducing agents were either completely ineffective (e.g., Zn/TMSCl and Zn/$TiCl_4$) or significantly less efficient (e.g., Mg/Hg/$TiCl_4$ and $LiAlH_4$/$CpTiCl_3$) than SmI_2.

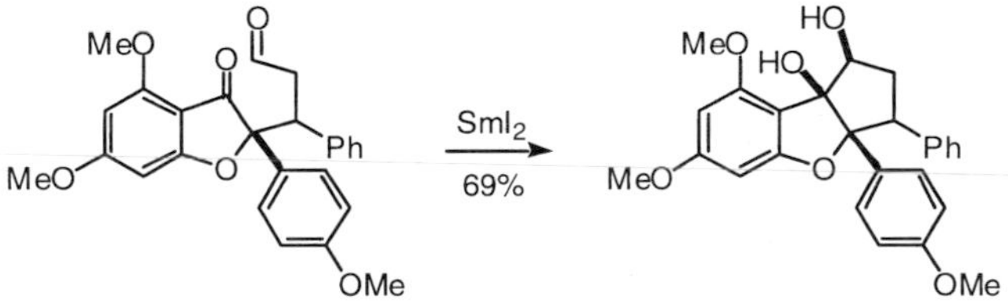

SmI_2, 69% (3.4)

L-*Chiro*-inositol. Chiara has reported a synthesis of L-*chiro*-inositol from D-sorbitol that employed a highly stereoselective intramolecular pinacol coupling reaction as the key step (Eq. 3.5) [21]. Thus, treatment of the illustrated dialdehyde with SmI_2 afforded the desired diastereomer with 94:6 stereoselection (neither of the other two diastereomers was detected) in >78% yield. The major reaction product was the one expected based upon previous studies of SmI_2-mediated pinacol cyclizations [22].

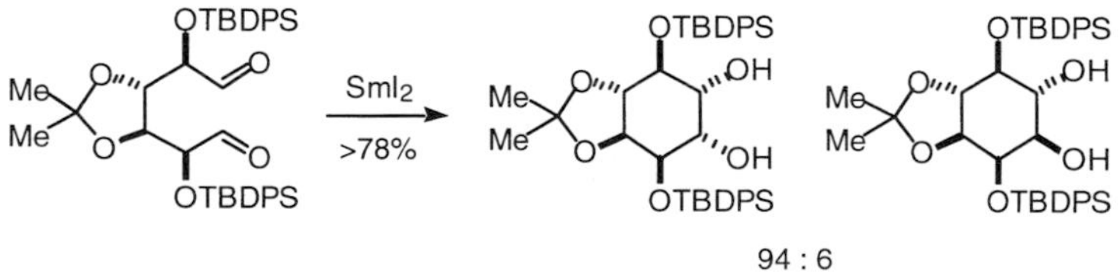

SmI_2, >78%; 94 : 6 (3.5)

Forskolin. Prange has described a formal total synthesis of forskolin, a diterpenoid that serves as an activator of adenylate cyclase, in which an intramolecular SmI_2-mediated pinacol coupling reaction was a pivotal transformation (Eq. 3.6) [23]. Only the desired diastereomer was produced in this process.

SmI_2, >95% (3.6)

(–)-Periplanone C. McMurry has reported an enantioselective synthesis of (-)-periplanone C, a sesquiterpene that serves as a sex pheromone for cockroaches, through a route involving a pinacol cyclization of a 1,10-keto aldehyde (Eq. 3.7) [24]. MM2 calculations based on a model for predicting the stereoselection of titanium-mediated pinacol coupling reactions were in qualitative, but not quantitative, agreement with the experimental results.

$TiCl_3(DME)_{1.5}$, Zn-Cu, >60%; + 3 diastereomers; 62 : 38 (3.7)

Sarcophytol B. McMurry has utilized a diastereoselective pinacol cyclization of a 1,14-dialdehyde in the final step of a five-step synthesis of sarcophytol B, a cembrane sesquiterpene (Eq. 3.8) [25]. This work established the relative stereochemistry of the natural product [26].

$TiCl_3(DME)_{1.5}$, Zn-Cu, 46% (3.8)

sarcophytol B

Isolobophytolide. McMurry has described a synthesis of isolobophytolide, a terpenoid component from a Pacific soft coral (Eq. 3.9) [27]. A key step in the synthesis was a pinacol cyclization of a 1,14-ketoaldehyde with $TiCl_3(DME)_{1.5}$/Zn-Cu. Unfortunately, all four possible stereoisomeric diol products were generated, with the desired isomer being formed in slight excess over the other three (21% : 19% : 11% : 7%).

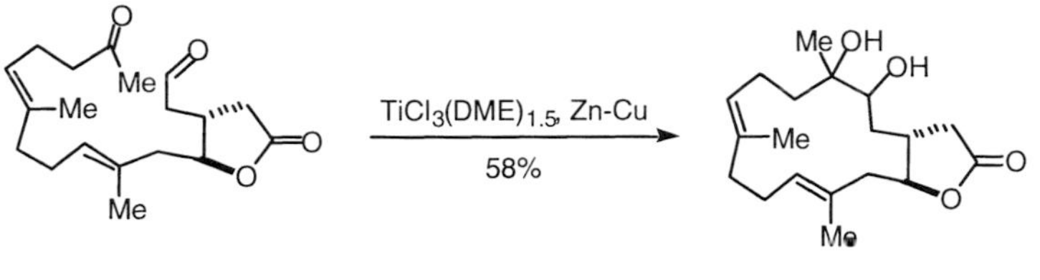

$TiCl_3(DME)_{1.5}$, Zn-Cu, 58% (3.9)

Palominol. Corey has reported the application of an intramolecular titanium-mediated pinacol coupling reaction to the synthesis of a 15-membered ring, en route to palominol, a marine diterpenoid that displays cytotoxicity toward the human colon cell line (Eq. 3.10) [28]. Slow addition of the keto aldehyde (32 h) to the titanium reagent furnished the cyclized product in 53% yield as a mixture of diastereomers (2.1 : 1).

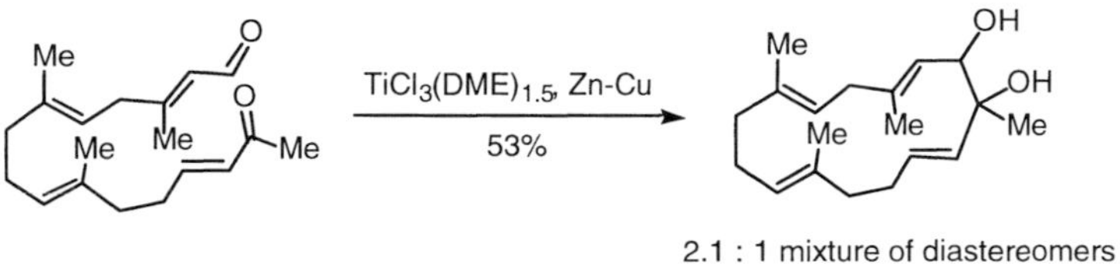

$TiCl_3(DME)_{1.5}$, Zn-Cu, 53% (3.10)

2.1 : 1 mixture of diastereomers

Shikonin. Torii has accomplished an efficient *intermolecular* pinacol coupling reaction, en route to the total synthesis of shikonin, a compound with antiinflammatory, antibacterial, and antitumor activity (Eq. 3.11) [29]. Because the aromatic aldehyde possessed a substituent capable of chelating to vanadium, Torii anticipated, based on precedent, that selective cross-coupling would be possible. The pinacol reaction proceeded in 73% yield and with good diastereoselectivity (5.5 : 1).

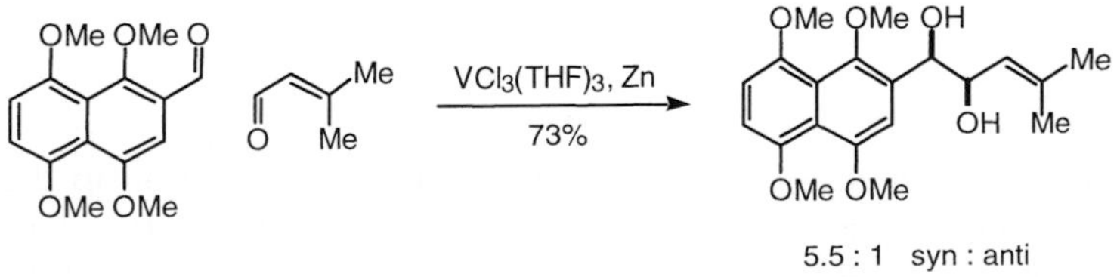

$VCl_3(THF)_3$, Zn, 73% (3.11)

5.5 : 1 syn : anti

Taxol. Nicolaou has reported the use of a pinacol cyclization to close the B ring of taxol, a diterpene of great interest due to its anticancer activity (Eq. 3.12) [30]. The reaction proceeded in modest yield (23%), with two of the major side products being a lactol and an olefin.

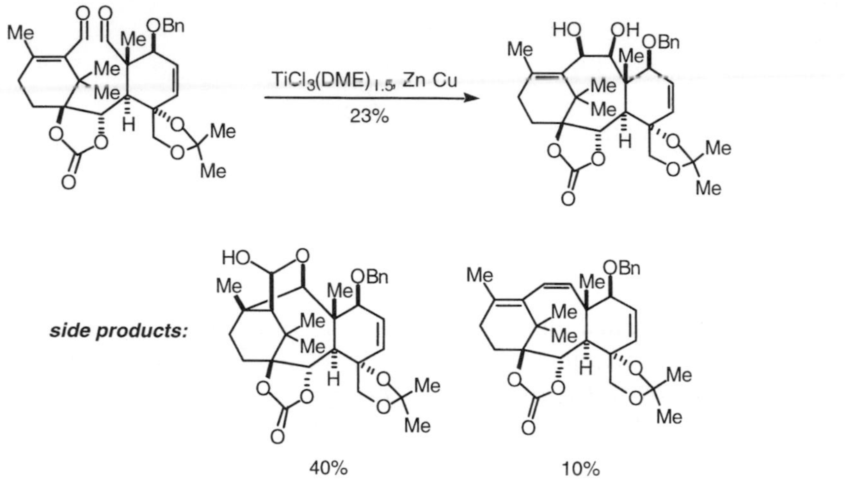

$TiCl_3(DME)_{1.5}$, Zn Cu

23%

side products:

40%   10%

(3.12)

Mukaiyama has described a total synthesis of taxol wherein pinacol cyclization of a diketone was employed to construct the A ring in good yield (Eq. 3.13) [31]. In this work, partially reduced alcohols and rearranged compounds were the major side products of the titanium-mediated coupling process.

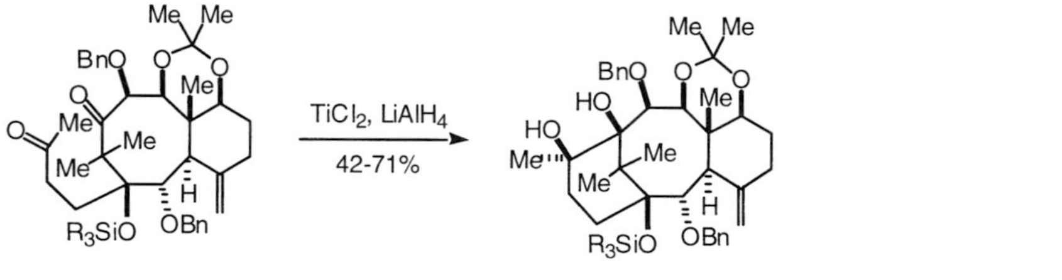

$TiCl_2$, $LiAlH_4$

42-71%

(3.13)

Summary

The utility of the pinacol coupling reaction is evident from its application to the synthesis of a diverse array of natural product targets. A wide range of ring sizes can be generated through intramolecular couplings, and the reaction conditions are compatible with a broad spectrum of functional groups.

In certain instances, excellent yields and stereoselectivities are observed in pinacol coupling reactions. However, it is clear from the reports described above that, in many instances, only modest yields and stereoselectivities are obtained. Indeed, it is a testament to the power of this transformation that the pinacol coupling is used in the synthesis of complex natural products (e.g., taxol) despite these limitations. The current deficiencies point to a number of opportunities for future work, including the development of both more efficient reagents and reagents that effectively control product stereochemistry in a predictable manner.

3.4 New Families of Reagents for the Pinacol Coupling Reaction

As indicated in Section 3.2 "Background", a diverse set of metals and metal complexes can effect the pinacol coupling of carbonyl groups. In the present section, newly discovered families of reagents for pinacol couplings will be described (modifications of existing families of reagents will not be discussed).

Gadolinium. Saussine has reported that gadolinium effects the pinacol coupling of benzophenone [32].

Indium. Kim has developed an indium-mediated pinacol coupling of aromatic aldehydes that proceeds in water or in water/*t*-BuOH (Eq. 3.14) [33]. Sonication greatly enhances the rate of the reaction. The diastereoselectivity of the pinacol coupling is variable, with d,l:meso ratios ranging from 6:1 to 1:2. Aliphatic aldehydes, as well as ketones, are inert to these conditions.

ArCHO —[In, sonication; H_2O/*t*-BuOH; 50-83%]→ ArCH(OH)CH(OH)Ar (3.14)

Manganese. Li and Chan [34] and Rieke [35] have independently reported that manganese reagents can accomplish the pinacol coupling of aromatic carbonyl compounds. In the Li/Chan study, reaction of an array of aromatic aldehydes proceeded in good to excellent yield in the presence of Mn/AcOH/H_2O, albeit with poor diastereoselectivity (Eq. 3.15). Under these conditions, aliphatic aldehydes are reduced to the corresponding alcohol, and ketones (aromatic or aliphatic) do not react.

Ar–CHO → (Mn/AcOH, H_2O, 62-92%) → Ar–CH(OH)–CH(OH)–Ar, ~1 : 1 mixture of diastereomers (3.15)

Manganese prepared according to the Rieke method has also been shown to serve as an effective reducing agent for the pinacol coupling of aromatic aldehydes, providing the 1,2-diol in yields comparable to Mn/AcOH (Eq. 3.16). Interestingly, in contrast to Mn/AcOH, the Rieke system affords good stereoselectivity for certain substrates, favoring formation of the d,l isomer. Aliphatic aldehydes do not undergo coupling to a significant extent when treated with Rieke manganese.

Ar–CHO → (Rieke Mn, 60-94%) → Ar–CH(OH)–CH(OH)–Ar, 1-13 : 1 (3.16)

Unlike Mn/AcOH, Rieke manganese efficiently couples aromatic ketones, furnishing good to excellent d,l selectivity (Eq. 3.17). As with Mn/AcOH, dialkyl ketones are not suitable substrates for Rieke Mn-mediated pinacol coupling.

Ar–CO–R → (Rieke Mn, 71-93%) → Ar–C(OH)(R)–C(OH)(R)–Ar, 3-99 : 1 (3.17)

These manganese-mediated pinacol processes are believed to proceed via single-electron transfer.

Neodymium. Saussine has reported that neodymium effects the pinacol coupling of benzophenone and of fluorenone [36].

Niobium. Szymoniak has developed a niobium-based method for the pinacol coupling of aliphatic aldehydes, aromatic aldehydes, and aromatic ketones (Eq. 3.18) [37]. In the presence of $NbCl_3$, intermolecular couplings proceed with consistently high diastereoselectivity. In many cases, the diol forms an acetal with the remaining aldehyde, and this is isolated at the end of the reaction. The stereoselection of this process is rationalized by a pathway involving complexation of an aldehyde to Nb(III) to produce a Nb(V) metallaoxirane, followed by insertion of a second aldehyde to furnish a niobium pinacolate. Presumably, steric considerations favor the *trans* orientation of the alkyl groups.

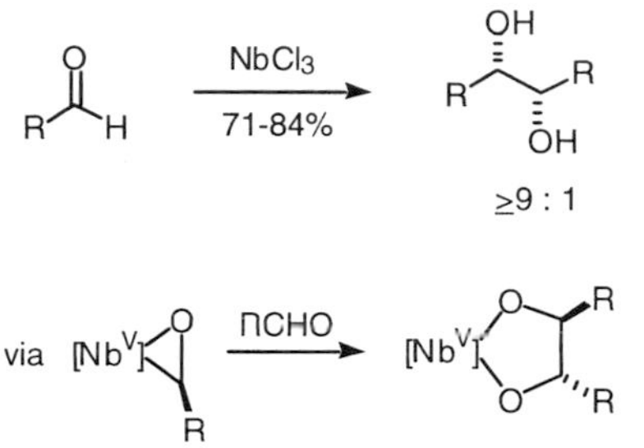
NbCl3
71-84%
OH
R
OH
≥9 : 1
via [Nb^V]
RCHO
R

(3.18)

In contrast, *intra*molecular pinacol couplings mediated by $NbCl_3$ proceed in lower yield and with essentially no diastereoselectivity (Eq. 3.19).

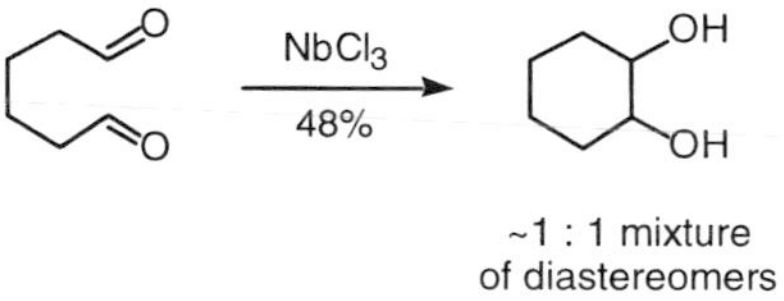
NbCl3
48%
OH
OH
~1 : 1 mixture
of diastereomers

(3.19)

Kammermeier and Jendralla have employed $NbCl_3$ in a gram-scale synthesis of a C_2-symmetric HIV-protease inhibitor (Eq. 3.20) [38]. Homocoupling of the illustrated dipeptide proceeds in modest yield and with good (>9 : 1 syn_1:syn_2) diastereoselectivity.

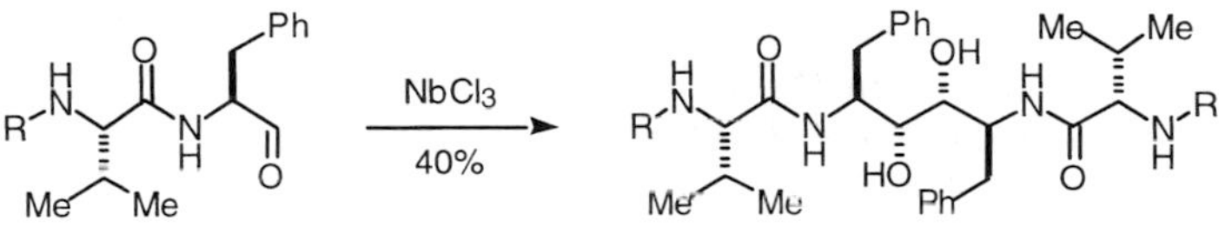
Ph
NbCl3
40%
Me
Me
OH
HO
Ph
R

(3.20)

Takai, Oshima, and Utimoto have noted that low-valent niobium ($NbCl_5$/Zn) reductively couples aldehydes and ketones [39].

Tantalum. Takai, Oshima, and Utimoto have also reported that $TaCl_5$/Zn effects the pinacol coupling of aliphatic aldehydes and aliphatic ketones in good yield (Eq. 3.21) [39].

Summary

A number of novel methods for effecting the pinacol coupling of carbonyl compounds have been described during the past decade, significantly expanding the already broad spectrum of reducing agents that can accomplish this powerful transformation. Since several of the new reagents (as well as certain reagents discovered earlier) are amenable to the development of chiral variants, it is anticipated that effective enantioselective reagents for pinacol coupling reactions will soon emerge.

3.5 New Catalytic Protocols for the Pinacol Coupling Reaction

One of the most exciting recent developments in the area of pinacol coupling is the discovery of a number of processes wherein the key coupling reagent is used as a catalyst and a second reducing agent (that does not itself effect pinacol coupling) is employed stoichiometrically [45]. In addition to their practical significance, these advances lay the necessary groundwork for efforts directed at the development of catalytic asymmetric pinacol coupling reactions.

Chromium. In 1998, Boland reported a chromium(II)-catalyzed pinacol coupling reaction of aromatic aldehydes and ketones (Eq. 3.28) [46–48]; aliphatic aldehydes are inert to these conditions. The turnover step relies upon a combination of manganese and Me_3SiCl to regenerate CrX_2 from an intermediate chromium(III) species (Fig. 3-3).

PhC(O)R —(cat. Cr(II), Mn/R_3SiX)→ Ph(R)C(OH)–C(OH)(R)Ph + Ph(R)C(OH)–C(OH)(R)Ph (3.28)

R = H (88%) 52 : 48
Me (57%) 48 : 52

As illustrated in Fig. 3-3, the CrX_2 catalyst may react initially with the carbonyl compound to generate a chromium ketyl. This chromium ketyl is then silylated by Me_3SiCl, generating a silicon ketyl and CrX_3. Dimerization of the silicon ketyl affords the pinacol adduct, and reduction of CrX_3 by Mn metal regenerates the CrX_2 catalyst.

With more bulky silylating agents, higher stereoselection is obtained, an observation consistent with a mechanism involving dimerization of silicon, rather than

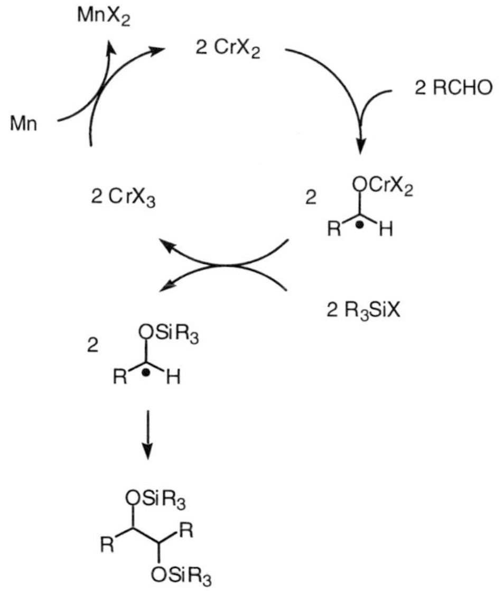

Figure 3-3. Possible pathway for chromium-catalyzed, manganese-mediated pinacol couplings.

chromium, ketyls (Fig. 3-3); unfortunately, with more bulky silicon chlorides, lower yields of the pinacol adduct are isolated. Also consistent with the postulate that silicon ketyls are coupling is the fact that the stereoselection is independent of the structure of the Cr(II) catalyst (e.g., $CrCl_2$ and $CrCp_2$ furnish the same diastereoselectivity).

With respect to the stoichiometric reducing agent, manganese is more effective than is zinc in these chromium-catalyzed couplings. In the absence of a chromium(II) catalyst, pinacol coupling by manganese/Me_3SiCl proceeds rather slowly.

Ruthenium. In 1995, Hidai reported that a diruthenium complex can catalyze the silylative dimerization of a wide variety of aromatic aldehydes (Eq. 3.29) [49, 50]. The analogous reaction of acetophenone proceeds more slowly and furnishes only 33% of the desired coupling product. It is postulated that this ruthenium-catalyzed

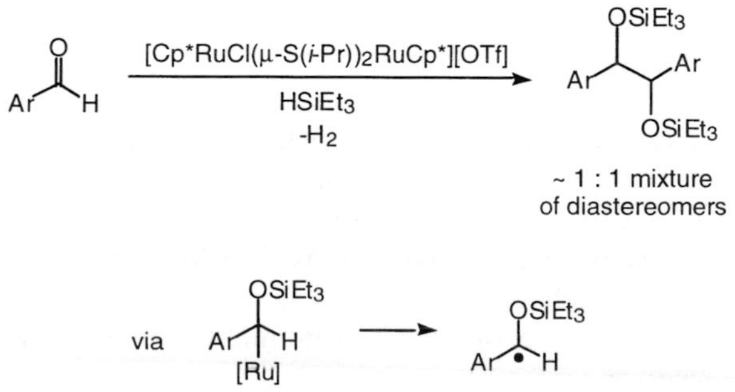

(3.29)

dimerization proceeds through the coupling of silicon ketyls, generated through homolysis of Ru-C bonds.

Samarium. In 1996, Endo determined that, in the presence of Mg/Me_3SiCl, carbonyl compounds can be reductively dimerized by a catalytic quantity of SmI_2 (Eq. 3.30) [51, 52]. Aliphatic and aromatic aldehydes and ketones can be coupled with this system. This catalytic process has recently been applied to the synthesis of hydroxyl-functionalized polymers [53].

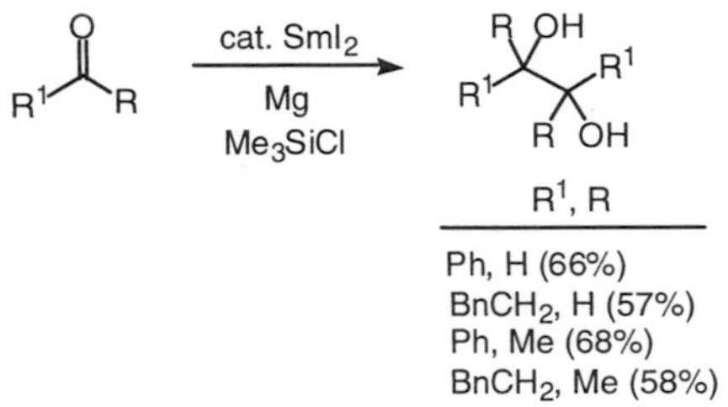
cat. SmI2
Mg
Me3SiCl
R1, R
Ph, H (66%)
BnCH2, H (57%)
Ph, Me (68%)
BnCH2, Me (58%)

(3.30)

The proposed mechanism for this samarium-catalyzed transformation is illustrated in Fig. 3-4. The pathway parallels that depicted in Fig. 3-3 for the chromium-catalyzed pinacol coupling reaction, with the exception of the timing of the silylation step. Based on the fact that pinacol reactions with catalytic and stoichiometric SmI_2 furnish the same diastereoselection, Endo believes that carbon-carbon bond formation precedes silylation of the alkoxide.

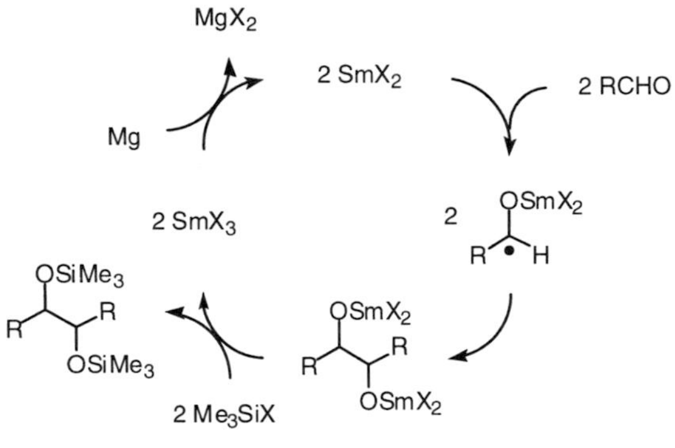
MgX2
Mg
2 SmX2
2 RCHO
2 SmX3
2 Me3SiX

Figure 3-4. Possible pathway for samarium-catalyzed, magnesium-mediated pinacol couplings.

In the absence of SmI_2 under otherwise identical conditions, Mg/Me_3SiCl does not effect the pinacol coupling of benzaldehyde. If Me_3SiCl is omitted from the reaction, then a complex mixture of products is observed, presumably due to inefficient regeneration of the Sm(II) catalyst.

Namy has recently described an alternative method for effecting SmI_2-catalyzed pinacol couplings (Eq. 3.31) [54]. Using mischmetall, an inexpensive alloy of the light lanthanides ($12/kg from Fluka), acetophenone can be reductively dimerized in 70% yield; in contrast to the Endo system, no Me_3SiCl is necessary. Carbon-carbon bond formation is presumed to involve coupling of samarium ketyls, based on identical diastereoselectivity in the presence of catalytic and stoichiometric SmI_2. In the absence of SmI_2, there is no reaction.

$$PhC(O)Me \xrightarrow[\text{mischmetall, 70\%}]{\text{cat. } SmI_2} Ph(Me)C(OH)\text{–}C(OH)(Me)Ph \quad 4:1 \tag{3.31}$$

Titanium. In 1988, Zhang reported that treatment of diaryl ketones with catalytic Cp_2TiCl_2 and stoichiometric *i*-BuMgBr results in reductive coupling (Eq. 3.32) [55]. Zhang speculated that a titanium ketyl is formed under these conditions.

$$ArC(O)Ar^1 \xrightarrow[i\text{-BuMgBr, 53-81\%}]{\text{cat. } Cp_2TiCl_2} Ar(Ar^1)C(OH)\text{–}C(OH)(Ar^1)Ar \tag{3.32}$$

via $ArC(O)Ar^1 \xrightarrow{Cp_2TiH} Ar\dot{C}(OTiCp_2H)Ar^1$

In 1997, Gansäuer reported that in the presence of a catalytic amount of Cp_2TiCl_2 and stoichiometric $Zn/Me_3SiCl/MgBr_2$, pinacol coupling of aromatic aldehydes proceeds in good yield and high diastereoselectivity (>10:1; Eq. 3.33) [56]. Because lower stereoselection is observed in the absence of $MgBr_2$, Gansäuer postulated that the illustrated trinuclear complex is a key intermediate in this process. Preferential formation of the *syn* diol is readily rationalized by this scheme.

$$ArCHO \xrightarrow[\text{Zn, } Me_3SiCl, MgBr_2, \text{ 78-91\%}]{\text{cat. } Cp_2TiCl_2} ArCH(OH)\text{–}CH(OH)Ar \quad >10:1 \text{ diastereoselectivity} \tag{3.33}$$

via $Cp_2Ti(\mu\text{-Cl})(\mu\text{-O}\dot{C}HR)Mg(\mu\text{-O}\dot{C}HR)(\mu\text{-Cl})TiCp_2$

Under the reaction conditions, Cp_2TiCl_2 should be reduced by zinc to Cp_2TiCl (Fig. 3-5). Addition of Cp_2TiCl to an aldehyde then generates a titanium ketyl, which dimerizes to furnish a titanium pinacolate. In what Gansäuer believes is the turnover-limiting step, the pinacolate reacts with Me_3SiCl to release the silylated pinacol adduct and to regenerate Cp_2TiCl_2.

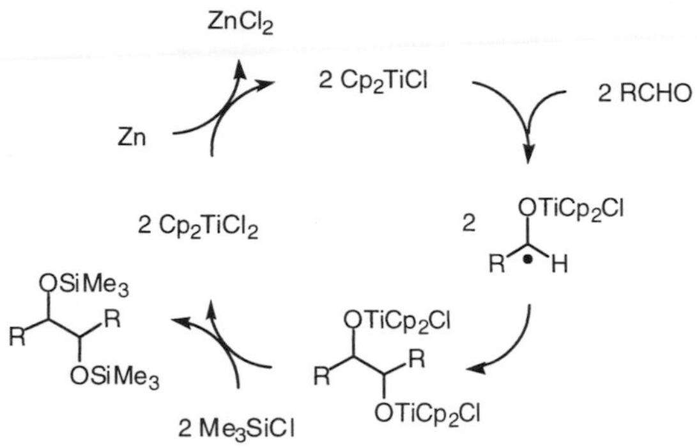

ZnCl2
Zn
2 Cp2TiCl
2 RCHO
2 Cp2TiCl2
2 OTiCp2Cl R • H
OSiMe3 R R OSiMe3
OTiCp2Cl R R OTiCp2Cl
2 Me3SiCl

Figure 3-5. Possible pathway for titanium-catalyzed, zinc-mediated pinacol couplings.

Gansäuer has established that Cp_2TiCl_2-catalyzed pinacol reactions that are run on a 0.5- or a 50-mmol scale provide identical results in terms of yield (%) and diastereoselectivity. Under otherwise identical conditions in the absence of Cp_2TiCl_2, a slower and non-stereoselective reductive dimerization is observed. More recently, Gansäuer has reported that *rac*-ethylenebis(η^5-tetrahydroindenyl)titanium dichloride provides higher diastereoselection (>20 : 1) than does Cp_2TiCl_2, and Hirao has determined that a catalytic Cp_2TiCl_2/Zn/Me_3SiCl system can effect the pinacol coupling of aliphatic aldehydes with modest to excellent de [57].

In 1997, Nelson described a related Ti(III) system (cat. $TiCl_3(THF)_3$; Zn/TMSCl) wherein additives can significantly affect both reactivity and stereoselectivity [58]. Thus, addition of substoichiometric quantities of protic compounds (e.g., catechol or 2,2′-biphenol) or Lewis-basic species (e.g., DMPU or DMF) results in ~5–10-fold acceleration relative to the parent system. *t*-BuOH proved to be the optimum additive among those that were surveyed. With this system, it is possible to efficiently accomplish the pinacol coupling of a broad spectrum of carbonyl compounds, including aliphatic and aromatic aldehydes and ketones (Eq. 3.34); no coupling product is observed in the absence of $TiCl_3(THF)_3$. In the case of aromatic, but not aliphatic, aldehydes, the diastereoselection can be enhanced by adding a catalytic quantity of 1,3-diethyl-1,3-diphenylurea.

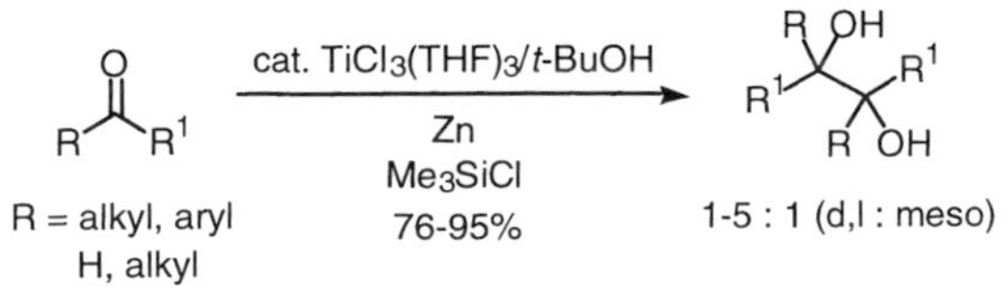

O
R R1
R = alkyl, aryl
H, alkyl
cat. TiCl3(THF)3/t-BuOH
Zn
Me3SiCl
76-95%
R OH
R1 R1
R OH
1-5 : 1 (d,l : meso)

(3.34)

Nelson has established that titanium-catalyzed intramolecular pinacol couplings are also possible (Eq. 3.35). For these processes, substantially improved yields are obtained if magnesium, rather than zinc, is employed as the stoichiometric reductant.

cat. $TiCl_3(THF)_3$/*t*-BuOH, Mg, Me_3SiCl, 74%; 8 : 1 (cis: trans) (3.35)

In 1999, Itoh reported that Cp_2TiPh is an effective catalyst for the inter- and intramolecular pinacol coupling of aldehydes [59]. In the case of 1,5- and 1,6-dialdehydes, very high *trans* selectivity is observed (Eqs. 3.36 and 3.37). This result is particularly interesting because many other reducing reagents have been shown to preferentially afford the *cis* isomer.

cat. Cp_2TiCl, Zn, Me_3SiCl, 40%; 99 : 1 (3.36)

t-Bu; cat. Cp_2TiCl, Zn, Me_3SiCl, 52%; *t*-Bu; one isomer (3.37)

In 1999, Cozzi and Umani-Ronchi described a diastereoselective intermolecular pinacol coupling of aromatic and aliphatic aldehydes in the presence of a catalytic quantity of $TiCl_4(THF)_2$/Schiff base (Eq. 3.38) [60]. Manganese is employed as the stoichiometric reductant; with the Cozzi/Umani-Ronchi system, zinc generally affords a lower yield of the diol. The reaction is believed to proceed via a pathway analogous to that illustrated in Fig. 3-5. The observations of Cozzi and Umani-Ronchi that the Schiff base affects reaction diastereoselectivity and increases the reaction rate bode well for studies of asymmetric variants. In an initial investigation, these workers obtained 10% ee in a reductive dimerization of benzaldehyde (Eq. 3.39).

At the same time as the Cozzi/Umani-Ronchi report, Nicholas also published a method for titanium-catalyzed pinacol couplings of aromatic and aliphatic aldehydes that employs manganese as the stoichiometric reducing agent; in this instance, the catalyst is Cp_2TiCl_2 [61]. Application of the enantiopure Brintzinger complex, (*R*,*R*)-ethylenebis(η^5-tetrahydroindenyl)titanium dichloride, to the reductive dimerization of benzaldehyde under these conditions furnishes the 1,2-diol with good diastereoselectivity (7 : 1 d,l: meso) and very promising enantioselectivity (60% ee; Eq. 3.40).

PhCHO → cat. $TiCl_4(THF)_2$/Schiff base, Mn, Me_3SiCl, 75% → PhCH(OH)CH(OH)Ph, 99 : 1 (d,l : meso)

Schiff base = (*t*-Bu, *t*-Bu, Ph, N, OH)

(3.38)

PhCHO → cat. $TiCl_4(THF)_2$/Schiff base, Mn, Me_3SiCl, 40% → PhCH(OH)CH(OH)Ph, 90 : 10 (d,l : meso), ***10% ee***

Schiff base = (*t*-Bu, *t*-Bu, N, OH, N, OH, *t*-Bu, *t*-Bu)

(3.39)

PhCHO → (R,R)-ethylenebis(η^5-tetrahydroindenyl)$TiCl_2$, Mn, Me_3SiCl → PhCH(OH)CH(OH)Ph, 7 : 1 (d,l : meso), 60% ee

(3.40)

Two drawbacks of the silyl-chloride-based strategy for catalyst regeneration (Figs. 3-3 to 3-5) are: (1) the need to hydrolyze the bis(silyl ether) adduct in order to furnish the coupling product as a diol, and (2) the turnover-limiting nature, in many instances, of the silylation step. In 1998, Gansäuer demonstrated that a proton-based approach to catalyst regeneration can address both of these issues (Eq. 3.41; Fig. 3-6) [62]. Through the addition of a weakly acidic proton source (specifically, an amine hydrochloride), the product is released from the catalyst as a diol through a rapid proton-transfer step. Cp_2TiCl_2 is also produced, ready to propagate the catalytic cycle (compare Figs. 3-5 and 3-6).

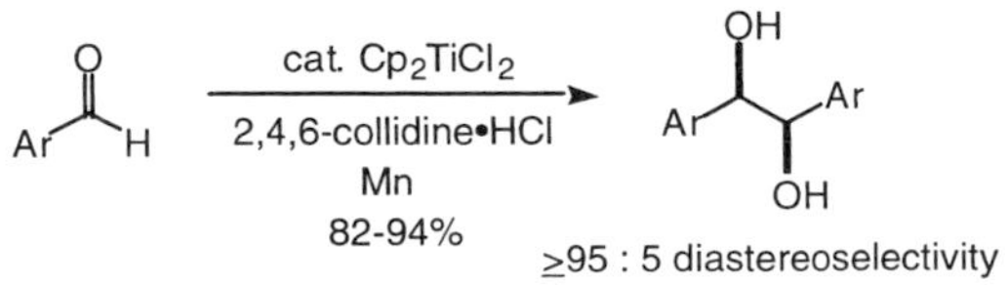

$$\text{ArCHO} \xrightarrow[\text{2,4,6-collidine•HCl, Mn, 82-94\%}]{\text{cat. } Cp_2TiCl_2} \text{ArCH(OH)CH(OH)Ar} \quad \geq 95:5 \text{ diastereoselectivity} \tag{3.41}$$

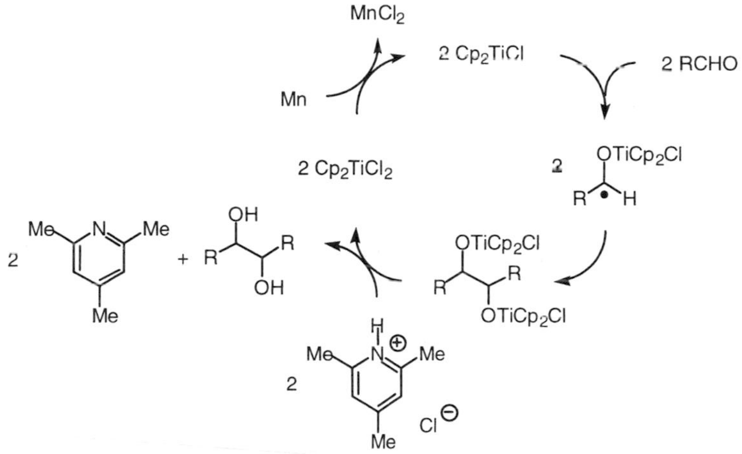

MnCl$_2$

Mn

2 Cp$_2$TiCl

2 RCHO

2 Cp$_2$TiCl$_2$

2 R(H)C•–OTiCp$_2$Cl

RCH(OTiCp$_2$Cl)CH(OTiCp$_2$Cl)R

2 2,4,6-collidine + RCH(OH)CH(OH)R

2 2,4,6-collidinium Cl$^{\ominus}$

Figure 3-6. Possible pathway for titanium-catalyzed pinacol couplings under protic conditions.

Gansäuer was able to apply this new catalytic process to the pinacol coupling of a variety of aromatic aldehydes in excellent yield and diastereoselectivity (Eq. 3.41). Other metals (e.g., zinc, magnesium, and aluminum) were inferior to manganese as the stoichiometric reducing agent.

Ephritikhine has also explored alternatives to silyl chlorides for liberating the coupling product from the initial metal pinacolate. In 1997 he reported that $AlCl_3$ may be employed in a $TiCl_4$-catalyzed, Li/Hg-mediated pinacol coupling of aliphatic aldehydes and ketones (Eq. 3.42) [63]; the diastereoselectivity of this process is modest ($\sim$2:1 for valeraldehyde). The proposed catalytic cycle is illustrated in Fig. 3-7.

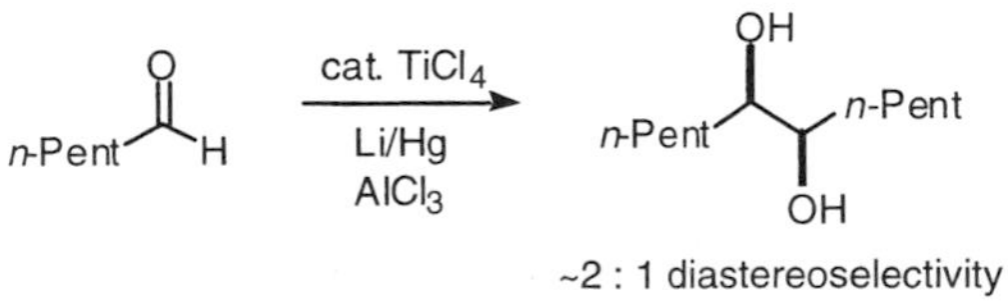

$$n\text{-PentCHO} \xrightarrow[\text{Li/Hg, } AlCl_3]{\text{cat. } TiCl_4} n\text{-PentCH(OH)CH(OH)}n\text{-Pent} \quad \sim 2:1 \text{ diastereoselectivity} \tag{3.42}$$

Uranium. Ephritikhine has reported a UCl_4-catalyzed variant of the $TiCl_4$-catalyzed process illustrated in Fig. 3-7 [63]. Both aliphatic aldehydes and aliphatic ketones are coupled by UCl_4 (cat.)/Li/Hg/$AlCl_3$ (reaction of acetone: 97% yield).

18. See also: (a) Adinolfi, M.; Barone, G.; Iadonisi, A.; Mangoni, L. *Tetrahedron Lett.* **1998**, *39*, 2021–2024. (b) Boiron, A.; Zillig, P.; Faber, D.; Giese, B. *J. Org. Chem.* **1998**, *63*, 5877–5882.
19. Adinolfi, M.; Barone, G.; Iadonisi, A.; Mangoni, L.; Manna, R. *Tetrahedron* **1997**, *53*, 11767–11780.
20. (a) Davey, A.E.; Schaeffer, M.J.; Taylor, R.J.K. *J. Chem. Soc., Chem. Commun.* **1991**, 1137–1139. (b) Dewey, A.E.; Schaeffer, M.J.; Taylor, R.J.K. *J. Chem. Soc., Perkin Trans. 1* **1992**, 2657–2666.
21. Chiara, J.L.; Valle, N. *Tetrahedron: Asymmetry* **1995**, *6*, 1895–1898.
22. For a stereoselective synthesis of *myo*-inositol using a SmI_2-mediated pinacol cyclization, see: Chiara, J.L.; Martin-Lomas, M. *Tetrahedron Lett.* **1994**, *35*, 2969–2972.
23. Anies, C.; Pancrazi, A.; Lallemand, J.-Y.; Prange, T. *Bull. Soc. Chem. Fr.* **1997**, *134*, 203–222.
24. McMurry, J.E.; Siemers, N.O. *Tetrahedron Lett.* **1994**, *35*, 4505–4508.
25. McMurry, J.E.; Rico, J.G.; Shih, Y.-n. *Tetrahedron Lett.* **1989**, *30*, 1173–1176.
26. For a synthesis of (–)-cembrene A using a titanium-mediated pinacol cyclization, see: Yue, X.; Li, Y. *Synthesis* **1996**, 736–740.
27. McMurry, J.E.; Dushin, R.G. *J. Am. Chem. Soc.* **1990**, *112*, 6942–6949.
28. Corey, E.J.; Kania, R.S. *Tetrahedron Lett.* **1998**, *39,* 741–744.
29. Torii, S.; Akiyama, K.; Yamashita, H.; Inokuchi, T. *Bull. Chem. Soc. Jpn.* **1995**, *68*, 2917–2922.
30. (a) Nicolaou, K.C.; Yang, Z.; Liu, J.J.; Ueno, H.; Nantermet, P. G.; Guy, R.K.; Claiborne, C.F.; Renaud, J.; Couladouros, E.A.; Paulvannan, K.; Sorensen, E.J. *Nature* **1994**, *367*, 630–634. (b) Nicolaou, K.C.; Yang, Z.; Liu, J.-J.; Nantermet, P.G.; Claiborne, C.F.; Renaud, J.; Guy, R.K.; Shibayama, K. *J. Am. Chem. Soc.* **1995**, *117*, 645–652.
31. Mukaiyama, T.; Shiina, I.; Iwadare, H.; Saitoh, M.; Nishimura, T.; Ohkawa, N.; Sakoh, H.; Nishimura, K.; Tani, Y.-i.; Hasegawa, M.; Yamada, K.; Saitoh, K. *Chem. Eur. J.* **1999**, *5*, 121–161.
32. Olivier, H.; Chauvin, Y.; Saussine, L. *Tetrahedron* **1989**, *45*, 165–169.
33. Lim, H.J.; Keum, G.; Kang, S.B.; Chung, B.Y.; Kim, Y. *Tetrahedron Lett.* **1998**, *39*, 4367–4368.
34. Li, C.-J.; Meng, Y.; Yi, X.-H.; Ma, J.; Chan, T.-H. *J. Org. Chem.* **1997**, *62*, 8632–8633.
35. Rieke, R.D.; Kim, S.-H. *J. Org. Chem.* **1998**, *63*, 5235–5239.
36. Olivier, H.; Chauvin, Y.; Saussine, L. *Tetrahedron* **1989**, *45*, 165–169.
37. (a) Szymoniak, J.; Besancon, J.; Moise, C. *Tetrahedron* **1992**, *48*, 3867–3876. (b) Szymoniak, J.; Besancon, J.; Moise, C. *Tetrahedron* **1994**, *50*, 2841–2848. See also: Kataoka, Y.; Takai, K.; Oshima, K.; Utimoto, K. *J. Org. Chem.* **1992** *57*, 1615–1618.
38. (a) Kammermeier, B.; Beck, G.; Jacobi, D.; Jendralla, H. *Angew. Chem., Int. Ed. Engl.* **1994**, *33*, 685–687. (b) Kammermeier, B.; Beck, G.; Holla, W.; Jacobi, D.; Napierski, B.; Jendralla, H. *Chem. Eur. J.* **1996**, *2*, 307–314.
39. See footnote 15 in: Kataoka, Y.; Takai, K.; Oshima, K.; Utimoto, K. *J. Org. Chem.* **1992**, *57*, 1615–1618.
40. Khan, R.H.; Mathur, R.K.; Ghosh, A.C. *Synth. Commun.* **1997**, *27*, 2193–2196.
41. (a) Hays, D.S.; Fu, G.C. *J. Am. Chem. Soc.* **1995**, *117*, 7283–7284. (b) Hays, D.S.; Fu, G.C. *J. Org.* **1998**, *63*, 6375–6381.
42. Villiers, C.; Adam, R.; Lance, M.; Nierlich, M.; Vigner, J.; Ephritikhine, M. *J. Chem. Soc., Chem. Commun.* **1991**, 1144–1145.
43. Ephritikhine, M.; Maury, O.; Villiers, C.; Lance, M.; Nierlich, M. *J. Chem. Soc., Dalton Trans.* **1998**, 3021–3027. See also: Maury, O.; Villiers, C.; Ephritikhine, M. *Angew. Chem., Int. Ed. Engl.* **1996**, *35*, 1129–1130.
44. Barden, M.C.; Schwartz, J. *J. Org. Chem.* **1997**, *62*, 7520–7521.
45. For early reports in this area, see: (a) Frainnet, E.; Bourhis, R.; Simonin, F.; Moulines, F. *J. Organomet. Chem.* **1976**, *105*, 17–31. (b) Inoue, H.; Fujimoto, N.; Imoto, E. *J. Chem. Soc., Chem. Commun.* **1977**, 412–413.
46. Svatos, A.; Boland, W. *Synlett* **1998**, 549–551.
47. For related work, see: Fürstner, A.; Shi, N. *J. Am. Chem. Soc.* **1996**, *118*, 12349–12357.
48. For earlier work on the use of TMSCl to facilitate reduction of metal-oxygen bonds, see: Fürstner, A.; Hupperts, A. *J. Am. Chem. Soc.* **1995**, *117*, 4468–4475.
49. Shimada, H.; Qu, J.-P.; Matsuzaka, H.; Ishii, Y.; Hidai, M. *Chem. Lett.* **1995**, 671–672.
50. For the analogous nickel-catalyzed process, see: Frainnet, E.; Bourhis, R.; Simonin, F.; Moulines, F. *J. Organomet. Chem.* **1976**, *105*, 17–31.

51. Nomura, R.; Matsuno, T.; Endo, T. *J. Am. Chem. Soc.* **1996**, *118*, 11666–11667.
52. For an electrochemical samarium-catalyzed pinacol coupling process, see: Leonard, E.; Dunach, E.; Perichon, J. *J. Chem. Soc., Chem. Commun.* **1989**, 276–277.
53. Brandukova-Szmikowski, N. E.; Greiner, A. *Acta Polym.* **1999**, *50*, 141–144.
54. Helion, F.; Namy, J.-L. *J. Org. Chem.* **1999**, *64*, 2944–2946.
55. Zhang, Y.; Liu, T. *Synth. Commun.* **1988**, *18*, 2173–2178.
56. (a) Gansäuer, A. *J. Chem. Soc., Chem. Commun.* **1997**, 457–458. (b) Gansäuer, A. *Synlett* **1997**, 363–364. (c) Gansäuer, A.; Moschioni, M.; Bauer, D. *Eur. J. Org. Chem.* **1998**, 1923–1927.
57. Hirao, T.; Hatano, B.; Asahara, M.; Muguruma, Y.; Ogawa, A. *Tetrahedron Lett.* **1998**, *39*, 5247–5248.
58. Lipski, T. A.; Hilfiker, M. A.; Nelson, S. G. *J. Org. Chem.* **1997**, *62*, 4566–4567.
59. Yamamoto, Y.; Hattori, R.; Itoh, K. *J. Chem. Soc., Chem. Commun.* **1999**, 825–826.
60. Bandini, M.; Cozzi, P. G.; Morganti, S.; Umani-Ronchi, A. *Tetrahedron Lett.* **1999**, *40*, 1997–2000.
61. Dunlap, M.; Nicholas, K. M. *Synth. Commun.* **1999**, *29*, 1097–1106.
62. (a) Gansäuer, A.; Bauer, D. *J. Org. Chem.* **1998**, *63*, 2070–2071. (b) Gansäuer, A.; Bauer, D. *Eur. J. Org. Chem.* **1998**, 2673–2676.
63. Maury, O.; Villiers, C.; Ephritikhine, M. *New J. Chem.* **1997**, *21*, 137–139.
64. (a) Hirao, T.; Asahara, M.; Muguruma, Y.; Ogawa, A. *J. Org. Chem.* **1998**, *63*, 2812–2813. (b) Hirao, T. *Synlett* **1999**, 175–181. See also: Hirao, T.; Hasegawa, T.; Muguruma, Y.; Ikeda, I. *J. Org. Chem.* **1996**, *61*, 366–367.

4 Modern Free Radical Methods for the Synthesis of Carbonyl Compounds

Ilhyong Ryu and Mitsuo Komatsu

4.1 Introduction

The synthesis of carbonyl compounds is of intrinsic importance in organic synthesis. Nevertheless, two decades ago, few people considered free radical reactions to be available as a method for the synthesis of such compounds. When confronted with the need to prepare carbonyl compounds, most researchers looked only for reliable ionic reactions or elegant metal-mediated or metal-catalyzed reactions, rather than risk exposing valuable substrates to radical reactions. Looking back on this time, such concerns were understandable, since only a few versatile and attractive methods existed for the synthesis of complex molecules. Perhaps acyl radical cyclization routes to cyclopentanones and cyclohexanones were the sole exception for which ionic and metal-assisted reactions cannot compete [1]. At present, however, the situation has changed. In response to a renaissance of new radical chemistry in organic synthesis over the past two decades [2], a large number of synthetically useful transformations leading to a wide range of carbonyl compounds have been developed, based on the use of modern free radical approaches. In the mid-1980s it was shown that cyclic ketones can be accessed by indirect methods via radical cyclization onto C–C triple bonds and C–O double bonds followed by oxidation. However, recent progress shows that cyclization to acylgermanes can be an efficient and direct method which obviates the tedious oxidation procedure.

Up until the end of the 1980s, radical carbonylation chemistry was rarely considered to be a viable synthetic method for the preparation of carbonyl compounds. In recent years, however, a dramatic change has occurred in this picture [3]. Nowadays, carbon monoxide has gained widespread acceptance in free radical chemistry as a valuable C1 synthon [4]. Indeed, many radical methods can allow for the incorporation of carbon monoxide directly into the carbonyl portion of aldehydes, ketones, esters, amides, etc. Radical carboxylation chemistry which relies on iodine atom transfer carbonylation is an even more recent development. In terms of indirect methods, the recent emergence of a series of sulfonyl oxime ethers has provided a new and powerful radical acylation methodology and clearly demonstrates the ongoing vitality of modern free radical methods for the synthesis of carbonyl compounds.

This chapter is designed to guide readers through the dramatic changes which occurred in just the past decade by showing selected but vivid examples of modern radical methods for the synthesis of carbonyl compounds. Examples are chosen mainly to present the new concepts which can be applied to further applications. Nevertheless, the reactions shown here will be sufficient to convince readers that the era has arrived in which organic chemists naturally consider free radical methods as a comparable choice among traditional ionic, metal-catalyzed processes as well as other synthetic methods.

4.2 Synthesis of Aldehydes

4.2.1 Radical Reduction of RCOX

One of the standard methods for the preparation of aldehydes involves the reduction of acid halides. A variety of stoichiometric reducing systems are available for this transformation, which include $NaAlH(OBu\text{-}t)_3$, $LiAlH(OBu\text{-}t)_3$, $NaBH(OMe)_3$. Catalytic hydrogenation with H_2 and Pd on carbon is also a popular method. In contrast, methods based on the radical reduction of acyl halides are synthetically less important. Radical reduction methods involve generation and subsequent hydrogen abstraction as key steps, which is complicated by decarbonylation of the intermediate acyl radicals. The first example in Scheme 4-1 shows that this competitive reaction is temperature dependent, where an acyl radical is generated from an acyl phenyl selenide via the abstraction of a phenylseleno group by tributyltin radical [5].

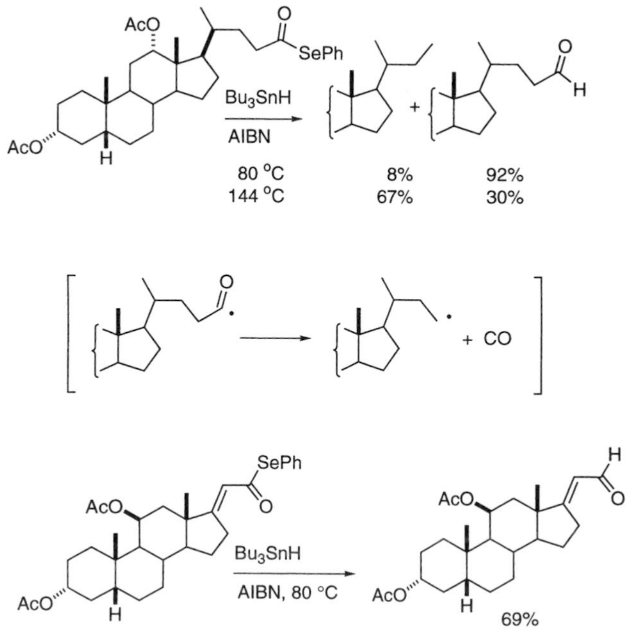

AcO, O, SePh, AcO, H, Bu_3SnH, AIBN, O, H, +

80 °C	8%	92%
144 °C	67%	30%

O, + CO

SePh, AcO, O, AcO, H, Bu_3SnH, AIBN, 80 °C, H, AcO, O, AcO, H, 69%

Scheme 4-1.

The radical reduction of acyl halides and related compounds, such as acyl chalcogenides, to aldehydes may find significance for primary, vinyl, and aromatic acyl radicals whose decarbonylation rates are significantly slower than those of the corresponding secondary and tertiary radicals [6]. In practice, however, this method is restricted to substrates which have serious incompatibilities with more traditional methods.

4.2.2 Radical Formylation and Hydroxymethylation with CO

The carbonylation of free radicals and the subsequent quenching of the resulting acyl radicals with hydrogen donor reagents has proven to be a useful route for the synthesis of aldehydes [7]. The premature quenching of alkyl radicals by tin hydride reagents and the course of decarbonylation of the acyl radicals can be suppressed by controlling the amount of carbon monoxide present via pressure adjustment [7]. The other reaction variable, tin hydride concentration, is also important, since decreasing the concentrations of tin hydride results in an increase of the ratio of carbon monoxide to tin hydride, so that the premature quenching of the radicals prior to carbonylation can be kept to a minimum.

Some results in Scheme 4-2 demonstrate that the formylation/reduction ratios are dependent on the hydrogen-donor ability of radical mediators. In the first equation, conventional tributyltin hydride is compared with Curran's ethylene-spaced fluorous tin hydride [8], and in the second equation with Chatgilialoglu's $(TMS)_3SiH$ [9]. Under identical reaction conditions, tributyltin hydride gave rise to a higher formylation/reduction ratio than ethylene-spaced fluorous tin hydride [10]. Therefore, a higher CO pressure and/or a higher dilution would be required for the fluorous analog to obtain results identical to tributyltin hydride. This result is consistent with a related kinetic study, which showed that the rate constant for primary alkyl radical trapping by fluorous tin hydride is about twice that for tributyltin hydride at 20 °C [11]. As shown in the second equation, $(TMS)_3SiH$, which has a relatively low H-donating ability [9], CO pressures can be significantly lowered, thus giving identical formylation/reduction ratios to those obtained with the use of tributyltin hydride [12].

Scheme 4-3 illustrates examples of radical formylation of several organic halides which contain an internal C–C double bond. The corresponding aldehydes were obtained from *trans* and *cis* 3-hexenyl bromides without loss of stereochemistry [7]. The 2,2-dimethyl-4-pentenyl radical undergoes double carbonylation to give a 4-oxo aldehyde in fair yield [13]. The use of tributylgermane as the mediator in this reaction results in the formation of a bicyclic lactone as the major product, for which atom transfer carbonylation is suggested as a possible reaction course (see below). On the other hand, the fourth reaction shows that 5-exo cyclization precedes carbonylation [10]. Free radical formylation can be extended to aromatic halides [14], but the carbonylation of organic halides, which gives stable radicals such as benzyl, allyl, α-keto radicals etc., appears difficult to achieve. For

n-C_9H_{19}Br + CO —[AIBN (10 mol%), R_3SnH (1.2-1.3 equiv), 80 °C, 2 h]→ n-C_9H_{19}CHO + n-C_9H_{20}

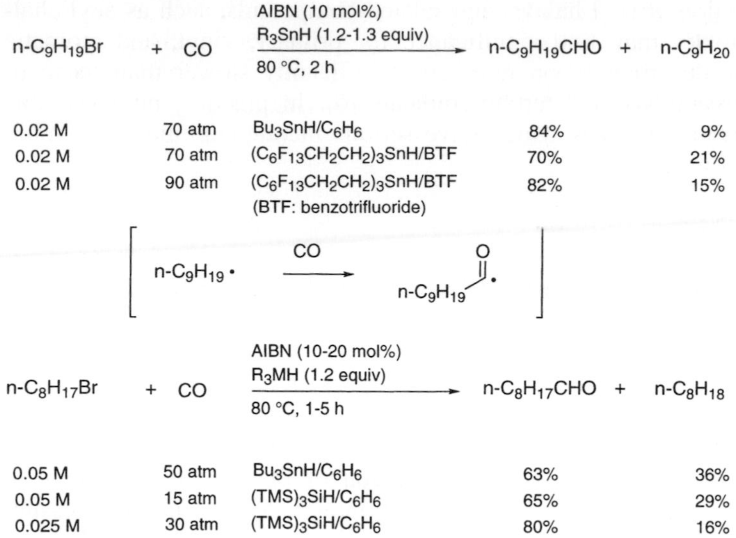

n-C_9H_{19}Br	CO		n-C_9H_{19}CHO	n-C_9H_{20}
0.02 M	70 atm	Bu_3SnH/C_6H_6	84%	9%
0.02 M	70 atm	$(C_6F_{13}CH_2CH_2)_3SnH$/BTF	70%	21%
0.02 M	90 atm	$(C_6F_{13}CH_2CH_2)_3SnH$/BTF	82%	15%

(BTF: benzotrifluoride)

[n-C_9H_{19}• —CO→ n-C_9H_{19}C(=O)•]

n-C_8H_{17}Br + CO —[AIBN (10-20 mol%), R_3MH (1.2 equiv), 80 °C, 1-5 h]→ n-C_8H_{17}CHO + n-C_8H_{18}

n-C_8H_{17}Br	CO		n-C_8H_{17}CHO	n-C_8H_{18}
0.05 M	50 atm	Bu_3SnH/C_6H_6	63%	36%
0.05 M	15 atm	$(TMS)_3SiH/C_6H_6$	65%	29%
0.025 M	30 atm	$(TMS)_3SiH/C_6H_6$	80%	16%

Scheme 4-2.

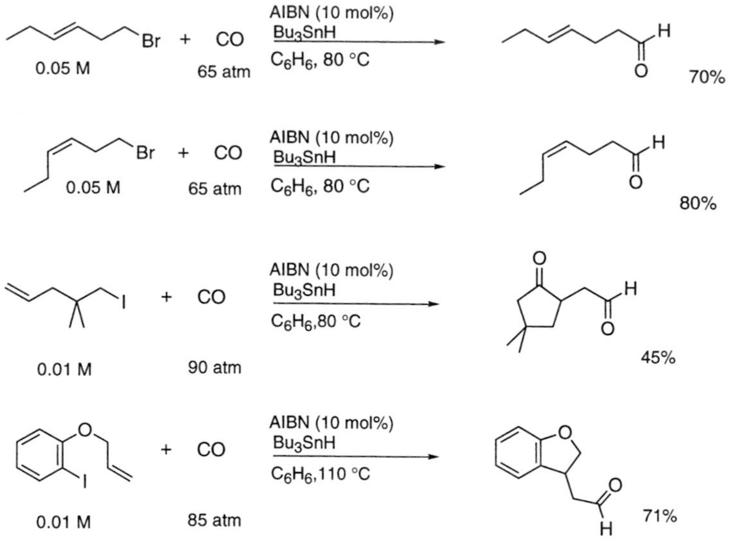

Br (0.05 M) + CO (65 atm) —[AIBN (10 mol%), Bu_3SnH, C_6H_6, 80 °C]→ aldehyde (H, O) 70%

Br (0.05 M) + CO (65 atm) —[AIBN (10 mol%), Bu_3SnH, C_6H_6, 80 °C]→ aldehyde (H, O) 80%

I (0.01 M) + CO (90 atm) —[AIBN (10 mol%), Bu_3SnH, C_6H_6,80 °C]→ cyclopentanone aldehyde (O, H, O) 45%

aryl iodide (O, I) (0.01 M) + CO (85 atm) —[AIBN (10 mol%), Bu_3SnH, C_6H_6,110 °C]→ dihydrobenzofuran aldehyde (O, O, H) 71%

Scheme 4-3.

the cases of allyl and benzyl halides, however, Pd-catalyzed carbonylation with tributyltin hydride can be a good alternative [15].

In a useful application, Kahne and Gupta reported the radical hydroxymethylation (formylation and in situ reduction) of organic halides, including sugar derivatives, using a catalytic amount of triphenylgermyl hydride in the presence of sodium cyanoborohydride as a reducing agent (the first equation in Scheme 4-4) [16]. The role of cyanoborohydride is to reproduce germyl hydride from the corresponding germyl iodide and for the in situ reduction of the aldehyde to the observed alcohol. Yamago, Yoshida and co-workers recently examined the photo and thermal reaction of tellurolglycosides with carbon monoxide [17]. Unfortunately, a CO insertion product was not formed (second equation). *a*-Alkoxy-substituted radicals, such as glycos-1-yl radical, are carbonylated under CO pressures with great difficulty, whereas ordinary alkyl radicals are readily carbonylated, and this is due, presumably, to the ready backward decarbonylation. On the other hand, as shown in the third equation, the insertion of 2,6-dimethylphenylisonitrile between the C–Te bond was successful.

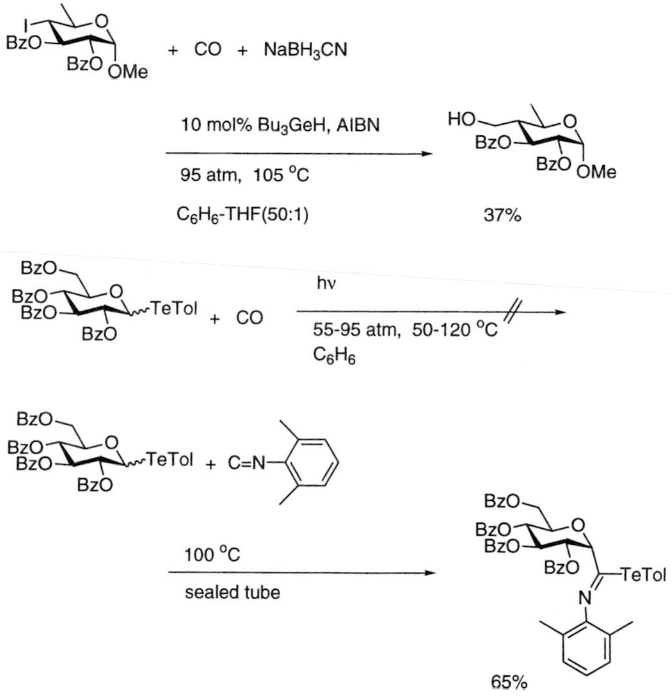

I, BzO, BzO, OMe + CO + $NaBH_3CN$

10 mol% Bu_3GeH, AIBN; 95 atm, 105 °C; C_6H_6-THF(50:1)

HO, BzO, BzO, OMe — 37%

BzO, BzO, BzO, BzO, TeTol + CO

hν; 55-95 atm, 50-120 °C; C_6H_6

BzO, BzO, BzO, BzO, TeTol + C=N–

100 °C; sealed tube

BzO, BzO, BzO, BzO, TeTol, N — 65%

Scheme 4-4.

Ryu, Curran and their co-workers have achieved the *fluorous* hydroxymethylation of organic halides using a catalytic quantity of a fluorous tin hydride [8] in the presence of sodium cyanoborohydride [10]. Interestingly, this fluorous reagent, as is usually the case for the related fluorous reactions [18], permits simple purification by a three-phase (aqueous/organic/fluorous) extractive workup. An example is given in Scheme 4-5. It should be noted that, unlike the cyclization-formylation sequence shown in the fourth equation in Scheme 4-3, the cyclization-hydroxymethylation sequence of the same substrate using a catalytic system was not feasible, since the acyl radical generated by the cyclization/carbonylation sequence adds to the C–C double bond of the starting substrate, giving an undesirable product. This was the result of the low concentration of the fluorous tin reagent in such a catalytic system.

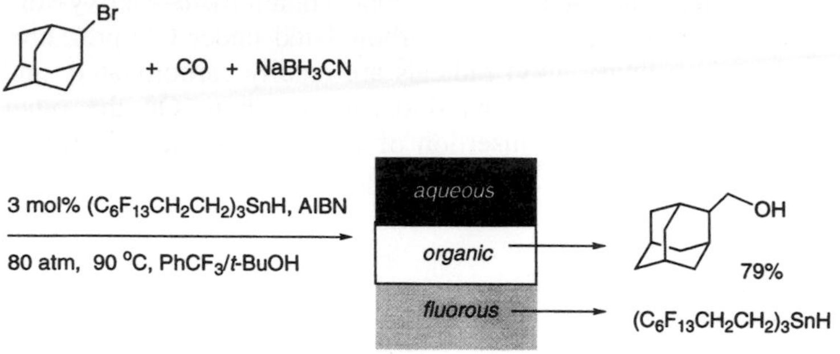

Br + CO + $NaBH_3CN$

3 mol% $(C_6F_{13}CH_2CH_2)_3SnH$, AIBN
80 atm, 90 °C, $PhCF_3$/t-BuOH

aqueous
organic
fluorous

OH
79%
$(C_6F_{13}CH_2CH_2)_3SnH$

Scheme 4-5.

The reducing ability of one-electron reducing reagents, such as zinc, can be used in the radical formylation of alkyl iodides, where one-electron reduction and proton quenching constitute the final step in the production of aldehydes [19]. Unfortunately, this zinc method involves competing reactions and a narrow scope in terms of substrates.

The stannylformylation of ordinary alkenes seems difficult to achieve, because of the reversibility of the stannyl radical addition step. That is to say, intermolecular CO trapping of the *β*-stannyl radical cannot compete well with the rapid reverse reaction which regenerates the tin radical and an alkene (Scheme 4-6).

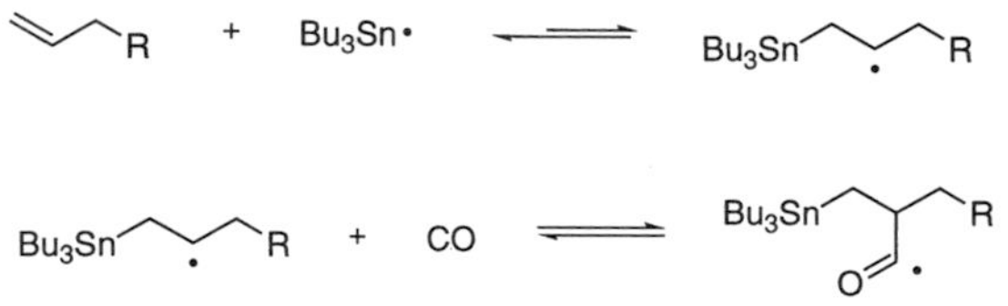

R + Bu_3Sn• ⇌ Bu_3Sn R

Bu_3Sn R + CO ⇌ Bu_3Sn R O

Scheme 4-6.

However, more rapid events, such as 5-exo cyclization and cyclopropylcarbinyl radical ring opening, can lead to radical formylation (Scheme 4-7). Thus, 1,6-dienes, when exposed to tin hydride/CO conditions, give fair yields of stannylformylation products via cyclization (Scheme 4-7) [20]. On the other hand, treatment of vinylcyclopropanes with CO under similar reaction conditions leads to stannylformylation products via a cyclopropylcarbinyl radical opening to a homoallylic radical (Scheme 4-7) [21].

Unlike tin radicals, thiyl radicals cannot abstract halogens from organic halides because of a mismatch in the polar effect but can add to C–C double and triple

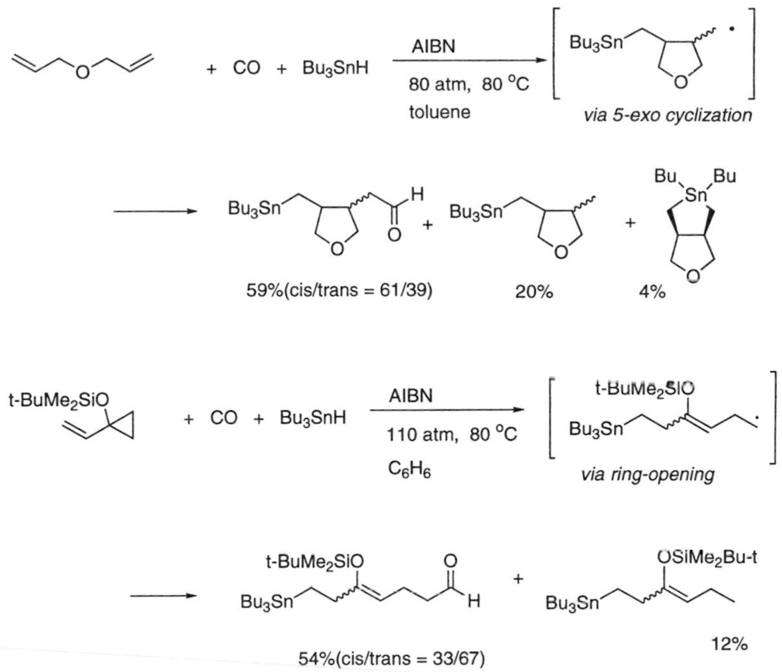

Scheme 4-7.

bonds. However, the addition of the thiyl radical to an alkene is reversible and, as a result, the carbonylation of the thus formed β-thio-substituted radical is highly inefficient. In the case of an alkyne, however, the formation of a stronger sp^2C–S bond overrides a similar problem. Vinyl radicals, which arise from the addition of thiyl radical to an alkyne, serve as viable intermediates which can participate in subsequent carbonylation. Indeed, Yoshida and co-workers reported that the thioformylation of terminal alkynes is an efficient reaction [22]. As illustrated in Scheme 4-8, thioformylation products are useful precursors of nitrogen- and sulfur-containing heterocycles. It is obvious that thioformylation in tandem reactions will have a great potential.

This thioformylation reaction of alkynes gave only *E*-isomers and this poses an intriguing mechanistic issue. To account for the stereoselectivity, two isomerization mechanisms need to be taken into consideration: one is *E/Z*-isomerization between the products induced by thiyl radicals and the other is the isomerization of α,β-unsaturated acyl radicals prior to H-abstraction, which is achieved via α-ketenyl radicals (Scheme 4-9). Interestingly, some recent examples clearly show that α,β-unsaturated acyl radicals can rapidly interconvert.

For example, the following example of tandem radical carbonylation, which was reported by Curran, Ryu, and their co-workers, suggests that the *E/Z* configurations of α,β-unsaturated acyl radicals are not directly associated with the subsequent radical cyclization (Scheme 4-10) [23]. The fact that the reaction was clean

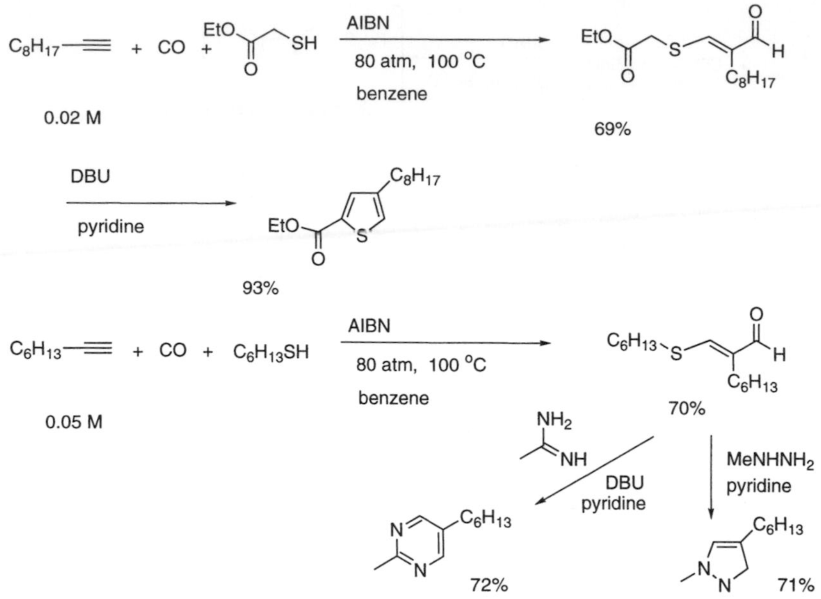

Scheme 4-8.

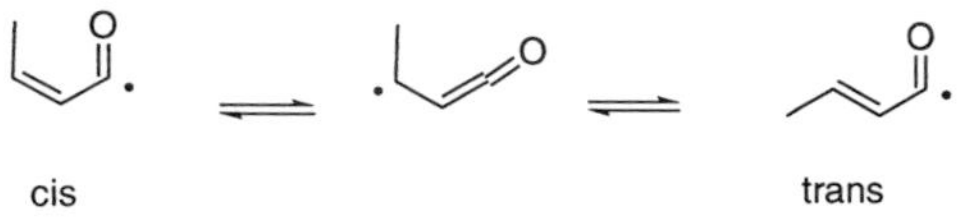

cis trans

Scheme 4-9.

and that the aldehyde anticipated via *trans* acyl radical was not formed suggests that *cis/trans* isomerization is very rapid. In contrast, as shown in the second equation, a pair of stereoisomeric sp^3-C–C(=O)$^\bullet$ radicals undergo different reaction modes, namely, formylation and cyclization, respectively.

Pattenden and co-workers recently succeeded in trapping an α-ketenyl radical by sequential 5-*exo*/5-*exo* cyclizations (Scheme 4-11) [24]. They further developed a series of interesting reactions based on this key isomerization reaction of α,β-unsaturated acyl radicals [25].

The combination of alkanes and CO is the simplest of all carbonylation methods which lead to aldehydes. Three research groups have reported the conversion of cyclohexane and carbon monoxide to cyclohexanecarboxaldehyde based on radical carbonylation by photolysis [26]. This alkane-formylation by radical carbonylations is conceptually important, but needs further improvement in order to be a useful synthetic method.

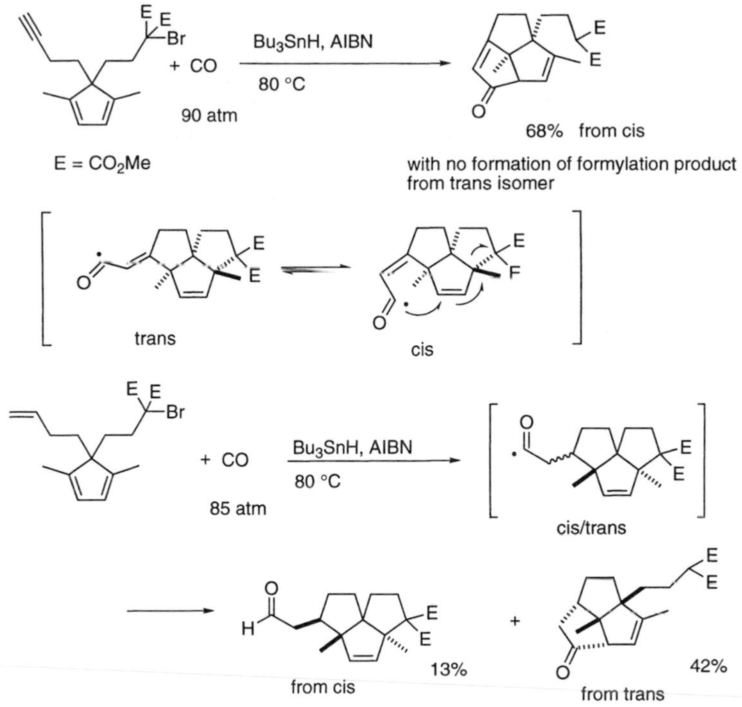

Scheme 4-10.

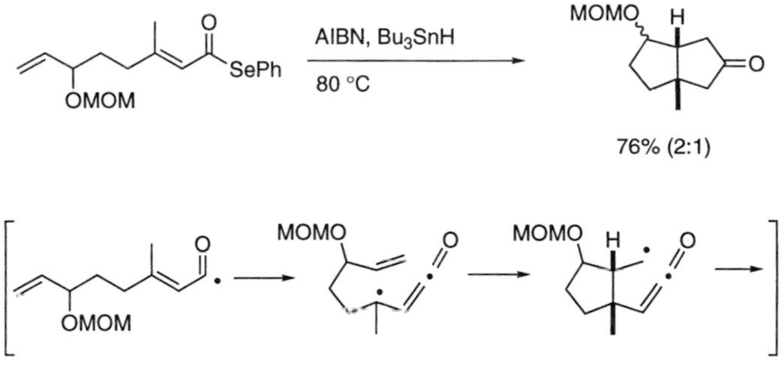

Scheme 4-11.

4.2.3 Radical Formylation of RX with a Sulfonyl Oxime Ether

Very recently, Kim and co-workers have reported a new useful radical acylation approach which uses a series of sulfonyl oxime ethers, some of which function as a viable radical C1 synthon. Unlike carbon monoxide and isonitriles, which operate as a radical acceptor/radical precursor type C1 synthon, sulfonyl oxime ethers complete the reaction by serving as a unimolecular chain transfer (UMCT) reagent which liberates phenyl- or alkylsulfonyl radical [27]. Whereas most of the oxime ethers are used for ketone synthesis (see below), the simplest formyl type of phenylsulfonyl oxime ether is a useful reagent for aldehyde synthesis [28]. Irradiation conditions are generally employed along with stoichiometric amounts of a reagent such as hexabutylditin or hexamethylditin, in order to propagate the radical chain. Aldoximes, produced as the primary products, were then hydrolyzed to aldehydes in high yields using a 30% HCHO solution in THF in the presence of a catalytic amount of HCl. Scheme 4-12 illustrates such examples. As seen from the second example, an application to a three-component coupling reaction is easy to achieve by this method.

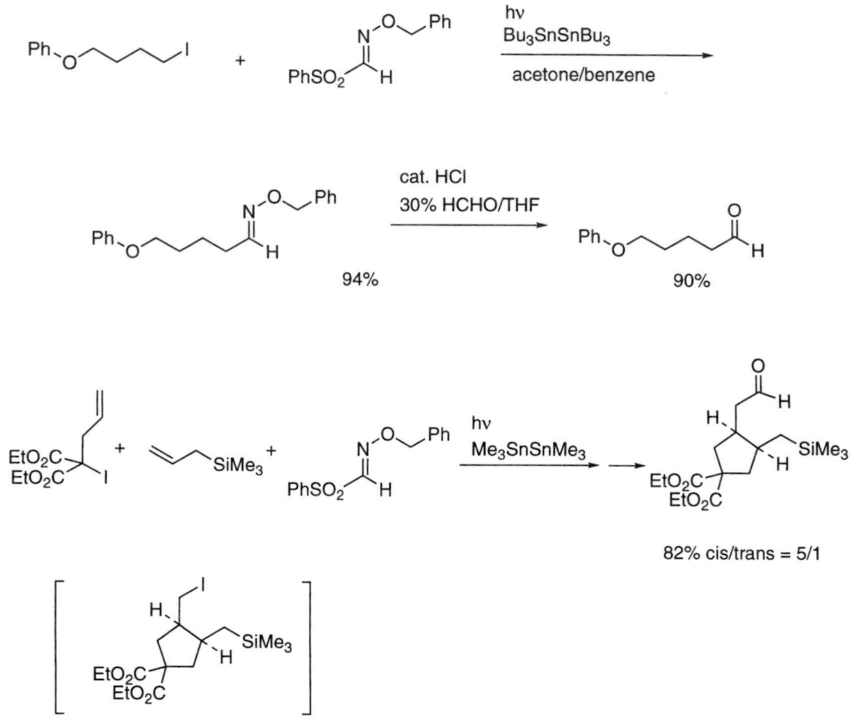

hv
$Bu_3SnSnBu_3$
acetone/benzene

cat. HCl
30% HCHO/THF

94%
90%

hv
$Me_3SnSnMe_3$

82% cis/trans = 5/1

Scheme 4-12.

It is noteworthy that this C1 reagent can be applied to a wide range of organic radicals involving rather stable radicals such as α-keto radical, α-alkoxyalkyl radical, and the benzyl radical for which the aforementioned radical formylation system with CO cannot be applied. The high reactivity of phenyl sulfonyl oxime ether is supported by kinetic studies [29]. The approximate rate constants for the addition of a primary alkyl radical to this phenyloxime ether was determined to be $k=9.6\times10^5$ M^{-1} s^{-1} at 25 °C, which is 1.8 times faster than the addition to acrylonitrile [30a, b] and ca. one order of magnitude faster than the addition to carbon monoxide [30c].

4.3 Synthesis of Ketones

A variety of indirect and direct methods are now available for ketone synthesis by radical reactions. Due to space limitations, this section will focus on selected topics, and only a few examples are shown for cases of frequently investigated approaches by acyl radical cyclizations. A recent review article on acyl radical chemistry provides a comprehensive survey of acyl radical cyclizations [1a].

4.3.1 Indirect Approaches by Cyclization onto CC, CN, and CHO

Kinetic work by Beckwith and Schiesser has shown that the 5-exo-dig cyclization of 5-hexynyl radical is faster than the corresponding 6-endo-dig cyclization (Scheme 4-13) [31]. Since the C–C double bond can be regarded as a latent carbonyl group, this type of cyclization serves as an indirect method for the synthesis of cyclopentanones. Using this technique, all products derived from well-known iodine atom-transfer cyclizations, which have been extensively investigated by the Curran group in the 1980s, can serve as the cyclic ketone precursors [32]. For example, the tricyclic product via tandem cyclization can be converted to the corresponding tricyclic ketone (Scheme 4-14) [32a]. The second example in Scheme 4-14, which was reported by Boger and Mathvink, shows an application of the concept to the synthesis of 9-methyldecalin-1,5-dione by tandem radical cyclization and ozonolysis of the resulting cyclized product [32d].

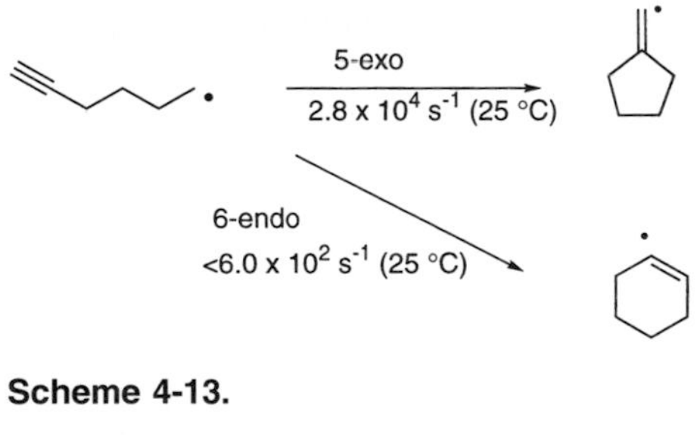

5-exo
$2.8 \times 10^4\ s^{-1}$ (25 °C)
oxidn.
6-endo
$<6.0 \times 10^2\ s^{-1}$ (25 °C)

Scheme 4-13.

Clive and co-workers employed the 5-exo-dig radical cyclization of a selenide as a key step in their total synthesis of (±)-fredericamycin A (Scheme 4-15) [33].

The same group also reported that cyclization onto a CN triple bond can be used to prepare cyclic ketones [34]. Secondary alcohols having a suitably located

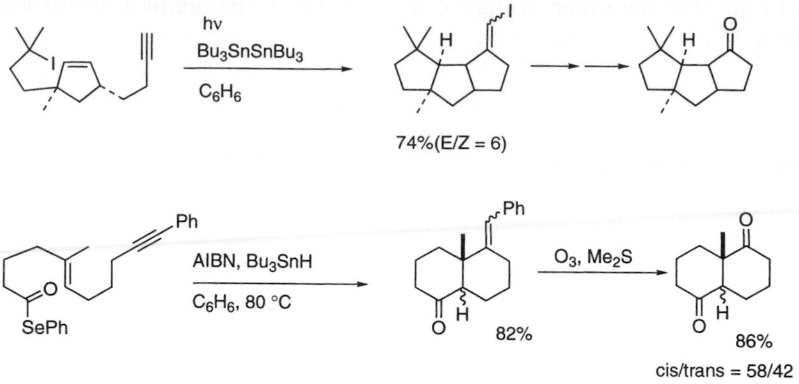

hv
$Bu_3SnSnBu_3$
C_6H_6
74%(E/Z = 6)

AIBN, Bu_3SnH
C_6H_6, 80 °C
82%
O_3, Me_2S
86%
cis/trans = 58/42

Scheme 4-14.

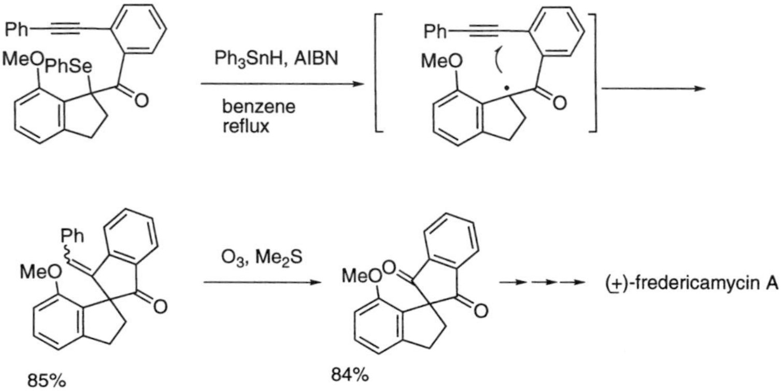

Ph_3SnH, AIBN
benzene
reflux

85%
O_3, Me_2S
84%
(+)-fredericamycin A

Scheme 4-15.

cyano group can be converted into the corresponding thiocarbonylimidazolides, and subsequent treatment with triphenyltin hydride and AIBN, followed by hydrolysis with aqueous acetic acid, gave good yields of cyclopentanones along with a small amount of uncyclized product.

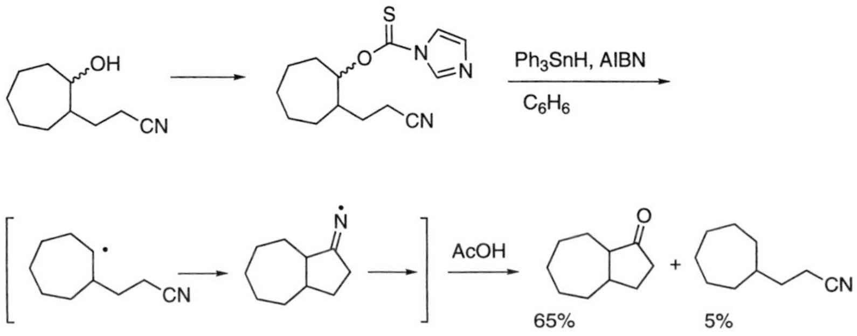

Ph_3SnH, AIBN
C_6H_6

AcOH
65%
5%

Scheme 4-16.

Tin hydride-mediated radical cyclizations onto a C–O double bond, coupled with subsequent oxidation, can be applied to ketone synthesis. Fraiser-Reid and co-workers demonstrated that the radical cyclization onto an aldehyde carbonyl group is particularly useful for the construction of cyclohexanol derivatives [35]. Examples shown in Scheme 4-17 well feature the cyclization. As Beckwith's kinetic work predicts (Scheme 4-18) [36], when this method is applied to a five-membered ring system, the rapid ring opening hinders the formation of cyclopentanols. Accordingly, only cyclohexanols can be reliably prepared using this approach.

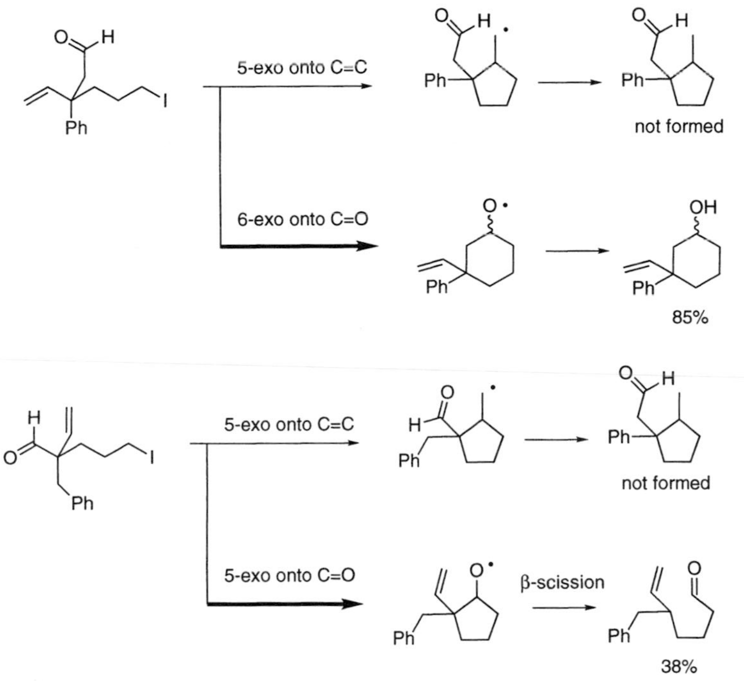

5-exo onto C=C

not formed

6-exo onto C=O

85%

5-exo onto C=C

not formed

5-exo onto C=O

β-scission

38%

Scheme 4-17.

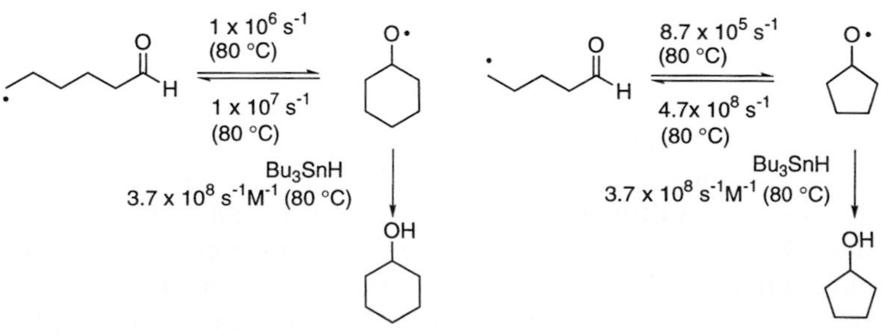

$1 \times 10^6\ s^{-1}$ (80 °C)

$1 \times 10^7\ s^{-1}$ (80 °C)

Bu_3SnH $3.7 \times 10^8\ s^{-1}M^{-1}$ (80 °C)

$8.7 \times 10^5\ s^{-1}$ (80 °C)

$4.7 \times 10^8\ s^{-1}$ (80 °C)

Bu_3SnH $3.7 \times 10^8\ s^{-1}M^{-1}$ (80 °C)

Scheme 4-18.

4.3.2 Carbonyl Addition/Elimination Approach

More straightforward methods are now available. A unique cyclic ketone synthesis using an acylgermane as the radical acceptor has been reported by Curran and co-workers [37]. The correct mechanistic inspection of Kiyooka's former work on photochemical isomerization of an unsaturated acylgermane to the corresponding β-germyl cyclopentanone [38] led the Curran group to the development of this new reaction scheme. Radicals add to acylgermanes and the rapid fragmentation of the resulting germylalkoxy radicals provides ketones and germyl radicals (Scheme 4-19). The germyl radicals, in turn, propagate the chain by the abstraction of a halogen atom. This acylgermane system can be reliably applied to cases where 5-*exo* and 6-*exo* cyclizations are possible. However, bimolecular reactions of alkyl radicals with acylgermanes are too slow to sufficiently propagate the radical chain [39].

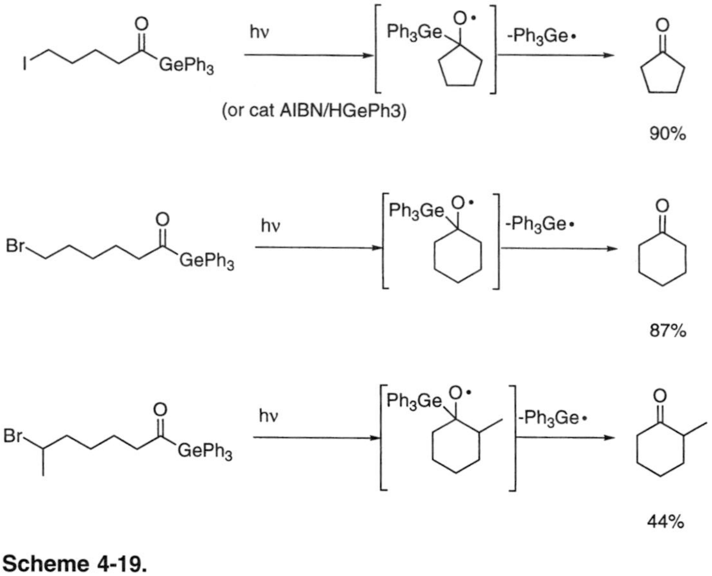

Scheme 4-19.

Acylsilanes also function as intramolecular radical acceptors, but, interestingly, a radical-Brook rearrangement takes place to give the silylated cycloalkanols (Scheme 4-20) [40].

A similar method for cyclic ketone synthesis via an intramolecular carbonyl addition/elimination strategy has been reported by Kim and Jon, who used acylsulfides and acylselenides as radical acceptors (Scheme 4-21) [41]. As has previously been observed in the cases of sulfonyl oxime ethers which liberate sulfonyl radicals, thiyl and selenenyl radicals react with ditin to propagate the chain. Generally acylselenides are more efficient substrates than acylsulfides due to the better leaving ability of the phenylseleno group. As shown in the third example in Scheme 4-21, a tandem cyclization leading to a tricyclic ketone has also been effected.

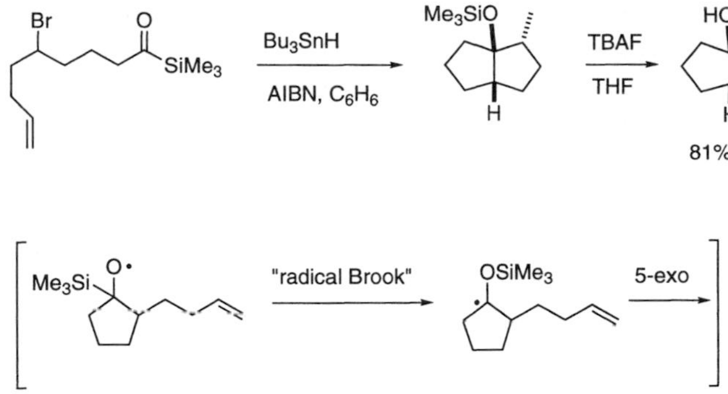

Scheme 4-20.

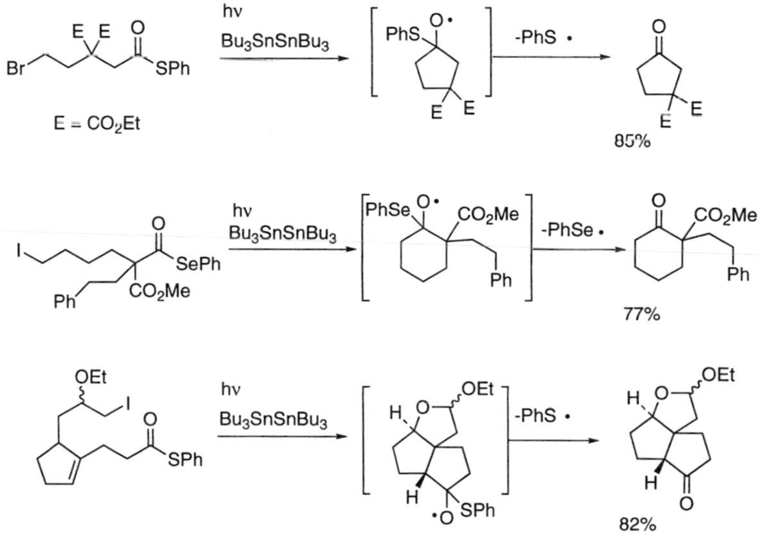

Scheme 4-21.

4.3.3 Radical Addition of RCHO, ACOX and Related Compounds to Alkenes

It has long been known that unsymmetrical ketones can be prepared by the reaction of aldehydes with alkenes under free-radical reaction conditions. Recently the revision of this chemistry has been reported by the Roberts group [42]. They introduced thiols as a *polarity reversal catalyst* for the addition of aldehydes to alkenes. Thiyl radicals are electrophilic, and therefore a polar S_H2 type transition state for the hydrogen transfer step from an aldehyde would be ideal in this situation. Indeed, the addition of aldehydes to a variety of alkenes can be effected by the addition of two portions of 5 to 10 mol% of thiols under mild reaction conditions (60 °C) using di-*tert*-butyl hyponitrite as an initiator (Scheme 4-22). Even in the case of electron-rich or ordinary alkenes, the propagation is effective, since the electrophilic thiyl radical abstracts hydrogen more favorably than nucleophilic alkyl radicals.

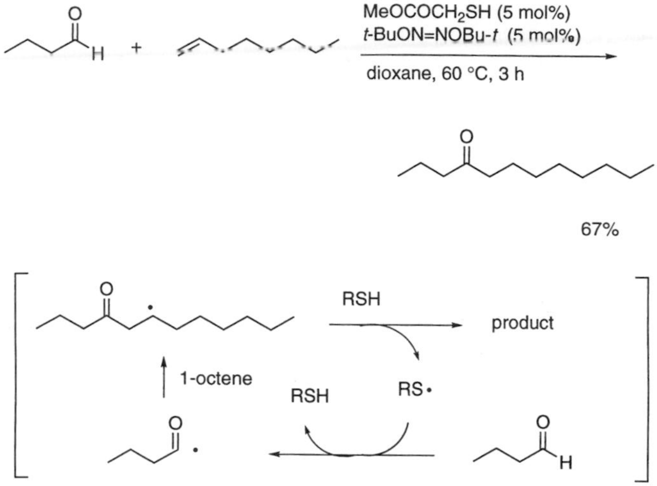

MeOCOCH$_2$SH (5 mol%)
t-BuON=NOBu-*t* (5 mol%)
dioxane, 60 °C, 3 h

67%

RSH
product
1-octene
RSH
RS•

Scheme 4-22.

Fuchs and Gong reported an acyl radical transfer reaction from aldehydes to acetylenic trifluoromethylsulfones to give acetylenic ketones [43]. In this case, trifluoromethyl radical, arising from the α-scission of trifluoromethylsulfonyl radical, abstracts the hydrogen of an aldehyde to form an acyl radical, which then propagates the chain.

Acyl radical sources, other than aldehydes, are also available. The group transfer addition of an acyl radical has been reported by Zard and co-workers, where *S*-acyl xanthates serve as acyl radical sources [44]. Crich and co-workers reported that an acyl radical, generated from an aromatic acyl telluride by photolysis, adds to an allylic sulfide which contains an ethoxycarbonyl group to form the corresponding β,γ-unsaturated ketones [45]. The addition pathway involves S_H2' type reaction with extrusion of a *tert*-butylthiyl radical.

Boger and Mathvink reported on the extensive synthesis of ketones based on the acyl transfer reaction of acyl selenides to alkenes using tin hydride as the radical mediator (Scheme 4-24) [46]. A radical arising from the addition of an acyl radical to alkenes abstracts hydrogen from the tin hydride with the liberation of a tin radical, thus creating a chain. The addition process is in competition with decarbonylation. In this regard, aroyl, vinylacyl, and primary alkylacyl radicals are most suitable for this reaction and secondary and tertiary acyl radicals are inferior.

Curran and Schwarz determined the optimal conditions for a similar acyl radical transfer reaction from acyl methyl selenide to methyl acrylate to achieve the

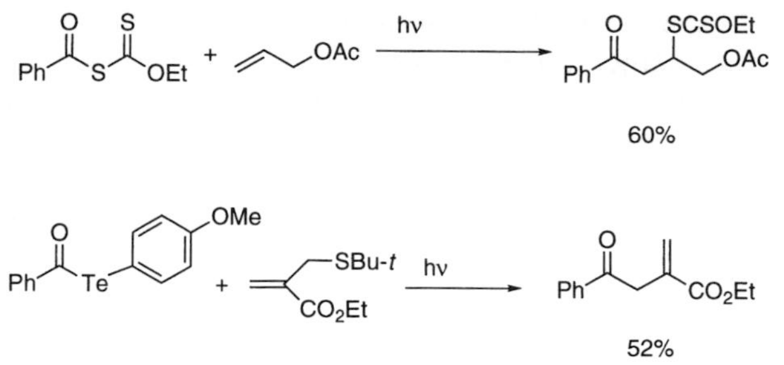

Scheme 4-23.

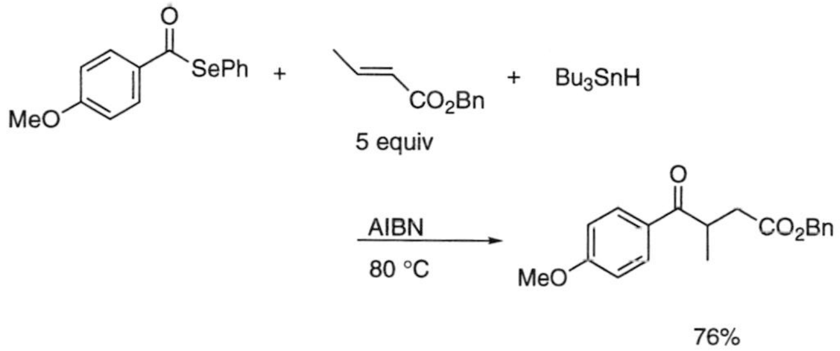

Scheme 4-24.

quantitative formation of the 4-keto ester [47]. Miyasaka and co-workers reported that acyl radicals generated from acyl selenides add to nucleoside-based enol ethers [48]. As seen in the example shown in Scheme 4-25, in order to compensate for the poor reactivity of the alkene, the acyl selenide and tin hydride are used in large excess in this case.

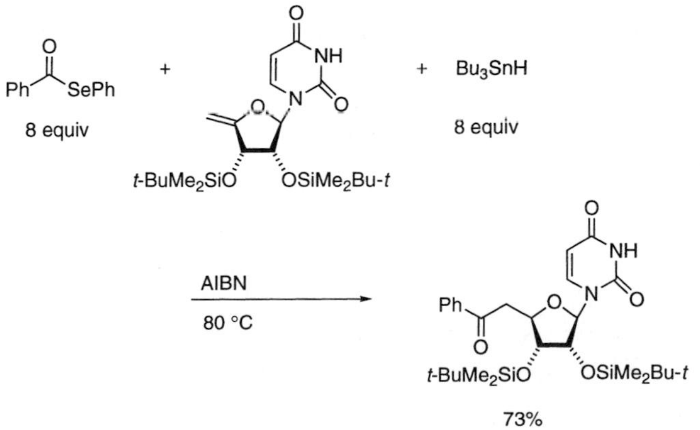

Scheme 4-25.

Sibi and Ji reported that acyl radicals, generated from acyl bromides, can participate in Lewis acid-mediated diastereoselective radical addition reactions (Scheme 4-26) [49]. Using triethylborane/O_2 as a radical initiator, the reaction was conducted at –78 °C.

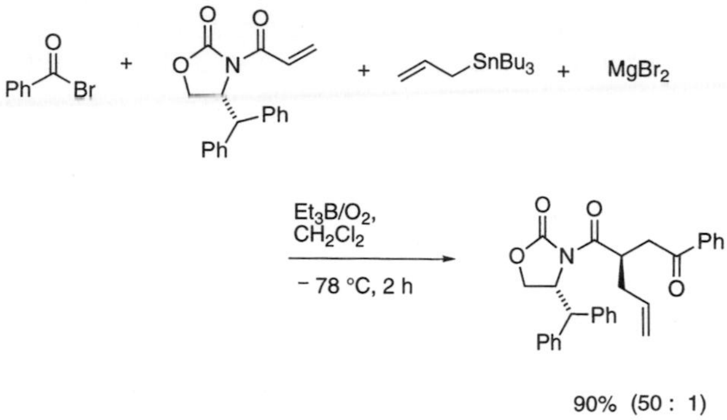

Scheme 4-26.

Narasaka and Sakurai found that chromium carbene complexes, when exposed to a copper(II) reagent, generate acyl radicals by a one-electron oxidation, and these then undergo addition to electron-deficient alkenes (Scheme 4-27) [50]. The resulting copper(I) species reduces the resulting radical to an anion, and subsequent protonation leads to the addition product. This redox type acyl radical transfer reaction works particularly well for aromatic acyl radical systems, for which decarbonylation is not a problem. Related work has also recently appeared [51].

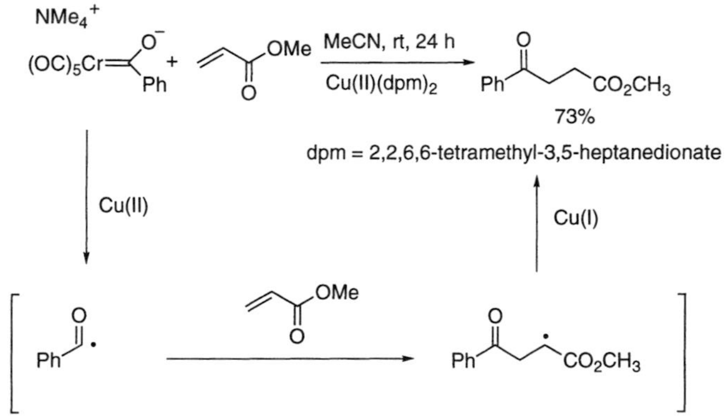

Scheme 4-27.

4.3.4 Radical Carbonylation with CO Incorporating Multi-Components

Using radical cascade reactions, carbon monoxide can be introduced directly into the carbonyl group of ketones. The tin hydride or $(TMS)_3SiH$-mediated radical coupling reaction of alkyl halides, CO, and alkenes permits the synthesis of unsymmetrical ketones, where the scope of alkenes is similar to that of the acyl selenide/alkene/tin hydride system but the scope of alkyl halides covers a wider range of aliphatic, aromatic, and vinylic halides [52]. An example is given in Scheme 4-28. To compete with the addition of the initial alkyl radical to the alkene and minimize premature quenching by radical mediators, a set of higher CO pressure and dilution conditions were used. In the case of the tin hydride-mediated reaction, an excess of alkene was used to suppress quenching of the acyl radical by tin hydride. However, the use of $(TMS)_3SiH$, a radical mediator, which is slower than tin hydride, enables the reaction to be carried out with nearly stoichiometric amounts of alkenes [12]. Because of this slower reduction, $(TMS)_3SiH$ can also mediate the carbonylation of free radicals at much lower CO pressures than are required for the tin hydride system.

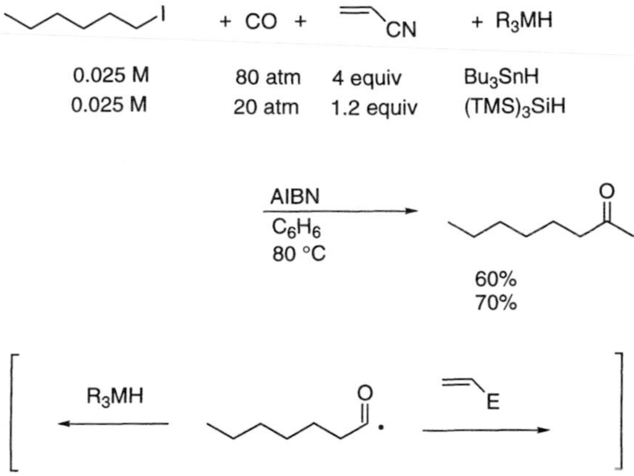

I	+ CO +	CN	+ R_3MH
0.025 M	80 atm	4 equiv	Bu_3SnH
0.025 M	20 atm	1.2 equiv	$(TMS)_3SiH$

AIBN, C_6H_6, 80 °C → product (O, CN): 60%, 70%

[← R_3MH — acyl radical — E →]

Scheme 4-28.

Unsaturated ketones can be readily synthesized by a three-component coupling reaction, comprised of alkyl halides, CO, and allyltin reagents [53]. Because of the slow direct addition of alkyl radicals to allyltin compounds [54], radical carbonylation with allyltin can be conducted at relatively low CO pressures to give good yields of β,γ-unsaturated ketones (Scheme 4-29).

Because of the nucleophilic nature of acyl radicals, in a mixed alkene system comprised of electron-deficient alkene and allyltin, they favor the electron-deficient alkene first and then the resulting product radicals, which have an electrophilic nature, then smoothly add to the allyltin (Scheme 4-30). This four-compo-

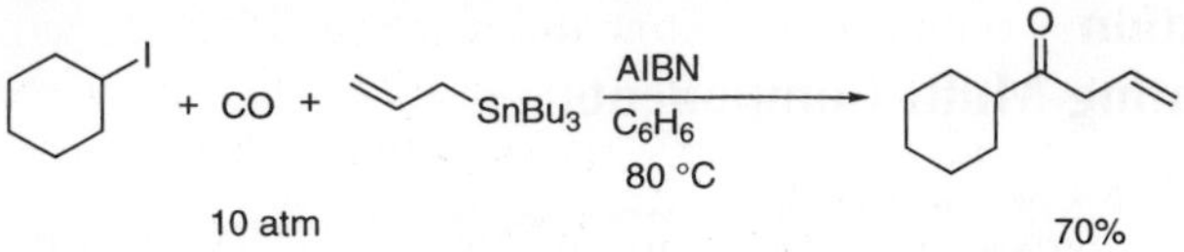

+ CO + SnBu3 AIBN C_6H_6 80 °C
10 atm 70%

Scheme 4-29.

nent coupling reaction provides a powerful radical cascade approach leading to β-functionalized δ,ε-unsaturated ketones, which are not readily accessible by other methods (Scheme 4-31) [55].

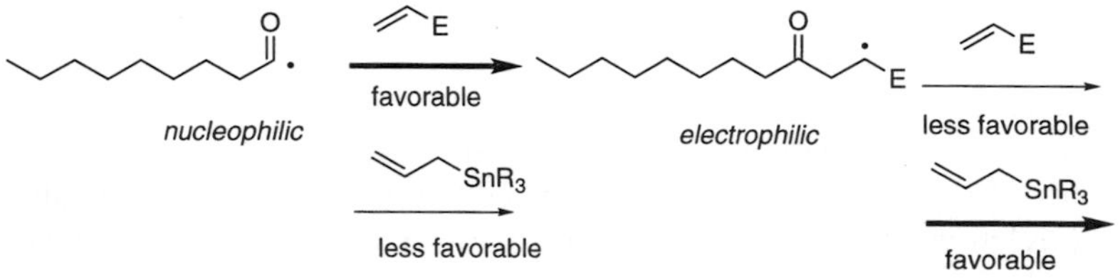

nucleophilic favorable / less favorable SnR_3 *electrophilic* less favorable / favorable

Scheme 4-30.

+ CO + + $SnBu_3$ 20 atm AIBN C_6H_6 74%

EtO + CO + OMe + $SnBu_3$ 20 atm AIBN C_6H_6 EtO OMe 67%

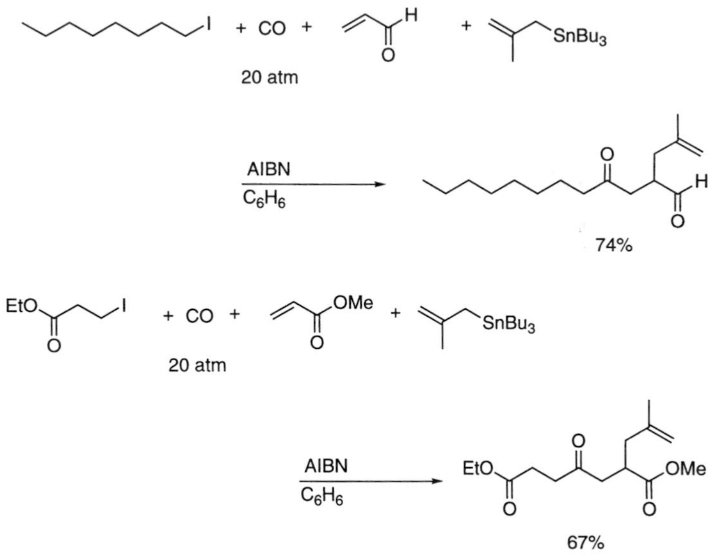

Scheme 4-31.

Propylene-spaced fluorous allyltin and methallyltin [56] proved particularly useful reagents for four-component coupling reactions, where alkyl halides, CO, alkenes and allyltin are combined in the given sequence [57]. Thus, slightly lower chain propagation abilities than those of the parent conventional tin reagents require fine tuning of the reaction conditions: higher concentrations are preferable to compensate for the modest chain propagation. After the reaction, BTF (benzotrifluoride) was removed by vacuum evaporation, and the resulting oil was partitioned into acetonitrile and FC-72 (perfluoroalkanes). Evaporation of the acetonitrile layer, followed by short column chromatography on silica gel, gave the pure product. The FC-72-layer contained fluorous tin compounds. As shown in the second example in Scheme 4-32, the alternative use of fluorous reverse phase silica gel (FRPS) [58] is ideal for the separation of products from tin compounds. The fluorous allyltin reagents were reproduced quantitatively by treatment of the tin residue with an ether solution of allyl and methallyl magnesium bromides and were then reused.

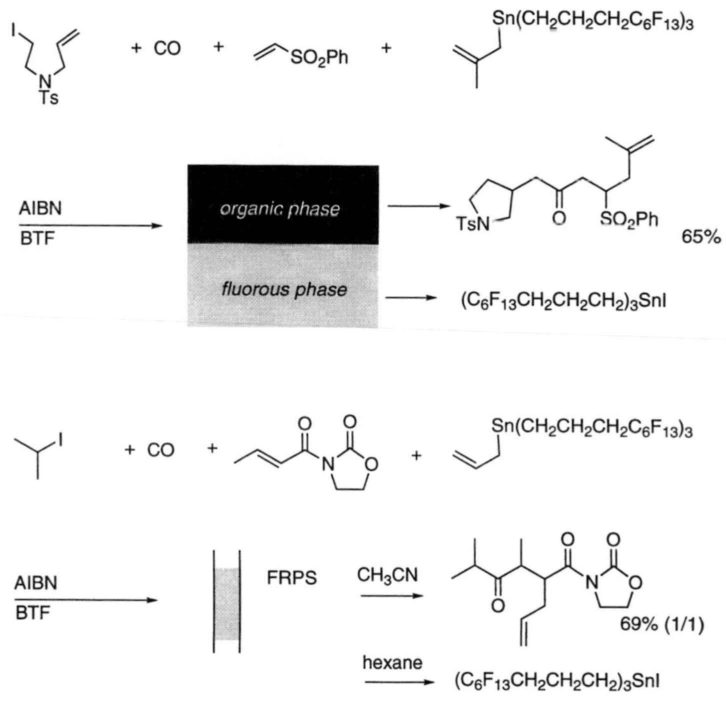

Scheme 4-32.

On the other hand, carbonylation via a one-electron reduction system by reducing metals can also lead to useful transformations [Scheme 4-33]. The first example in Scheme 4-34 shows that a zinc reduction system of pent-4-enyl iodide undergoes dual annulations, a [4+1] radical annulation with CO and a [3+2] anion annulation with alkenes, to give a bicyclo[3.3.0]octanol, in which four C–C bonds are produced [59]. An extension of the present strategy to the construction of bicyclo[3.2.1]octanol skeletons has also been successful with a system in which 6-*endo* cyclization is favored over 5-*exo* cyclization. In these two reactions three C–C bonds were created by radical reactions and one by anionic reaction.

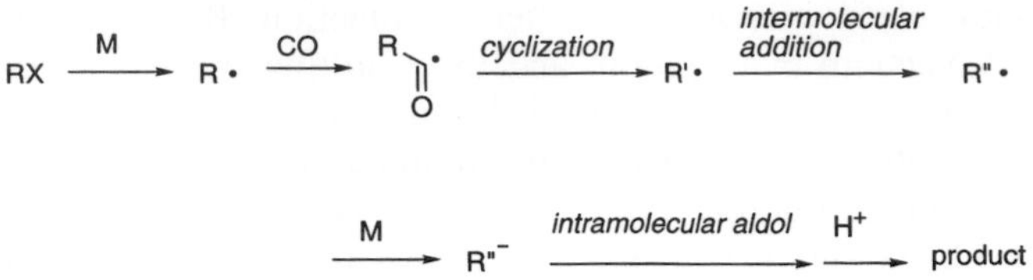

RX —M→ R• —CO→ RC(•)=O —cyclization→ R'• —intermolecular addition→ R''•

—M→ R''⁻ —intramolecular aldol→ —H⁺→ product

Scheme 4-33.

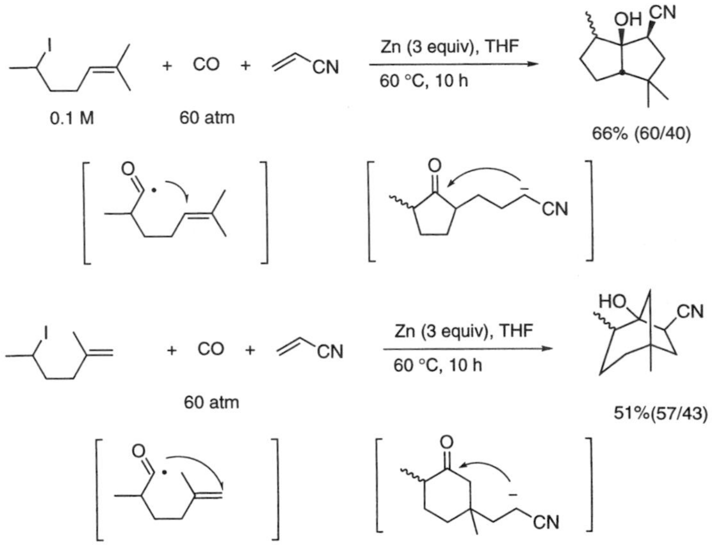

0.1 M + CO (60 atm) + CN, Zn (3 equiv), THF, 60 °C, 10 h; 66% (60/40)

60 atm; Zn (3 equiv), THF, 60 °C, 10 h; 51%(57/43)

Scheme 4-34.

Samarium diiodide when coupled with irradiation is a very reactive reducing reagent with respect to alkyl chlorides, whose oxidation potential is higher than those of the corresponding bromides and iodides. When such a reduction of an alkyl chloride was attempted under CO pressure [60], an unsymmetrical ketone was obtained, comprised of two molecules of alkyl chloride and two molecules of carbon monoxide. An α-hydroxy ketone, obtained via the dimerization of acylsamarium, is a likely precursor of the final product.

Among the wide range of cyclic ketone syntheses based on acyl radical cyclizations, the recently reported synthesis of 2-hydrazinocyclopentanones by Fallis and Brinza is noteworthy, since, unlike the corresponding cyclization onto a C–C double bond, the cyclization is selective in the 5-*exo* mode and no product arising from 6-*endo* cyclization was detected (Scheme 4-35) [61].

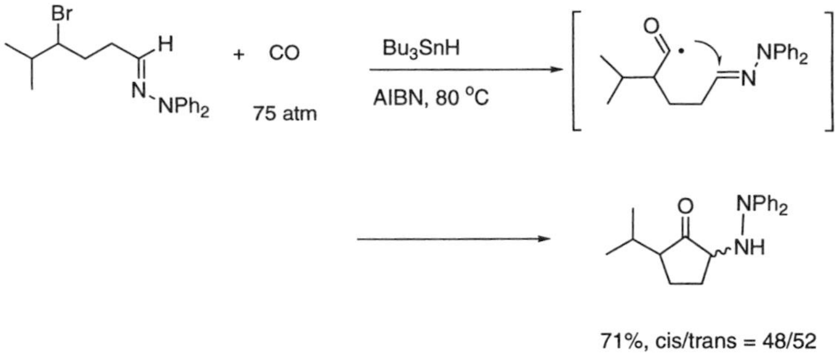

Scheme 4-35.

4.3.5 Radical Acylation with Sulfonyl Oxime Ethers

Sulfonyl oxime ethers, developed by Kim and co-workers, are useful reagents not only for the synthesis of aldehydes (Scheme 4-12) but for ketones as well. For example, methyl-substituted sulfonyl oxime ether can be used as a radical acetylation reagent [28]. More recently, Kim and Yoon developed a bissulfonyl oxime ether which is a very useful reagent for radical acylation [62]. Interestingly, this reagent serves as a new radical C1 synthon. However, unlike carbon monoxide and isonitriles, which serve as radical acceptor/radical precursor [63], this reagent serves as double geminal acceptor synthon (Scheme 4-36). By using sequential reaction procedures, this reagent allows for a double geminal alkylation reaction which leads to unsymmetrical ketones when coupled with the hydrolytic procedure with HCHO/THF/HCl (Scheme 4-37). As shown in the second example, the reaction is easily extendable to the synthesis of cyclic ketones, starting from ω-diiodo compounds. In a useful application of Kim's radical acylation approaches, α-keto esters can be synthesized efficiently, using a phenylsulfonyl methoxycarbonyl oxime ether [64].

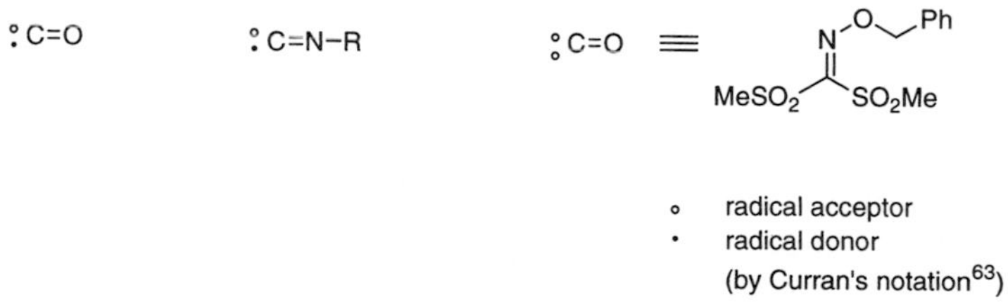

Scheme 4-36.

Radical carbonylation of an alkyl iodide in the presence of Kim's sulfonyl oxime ethers provides a new type of multi-component coupling reaction, and a typical example is given in Scheme 4-38 [65]. In this method, plural radical C1 synthons are consecutively combined.

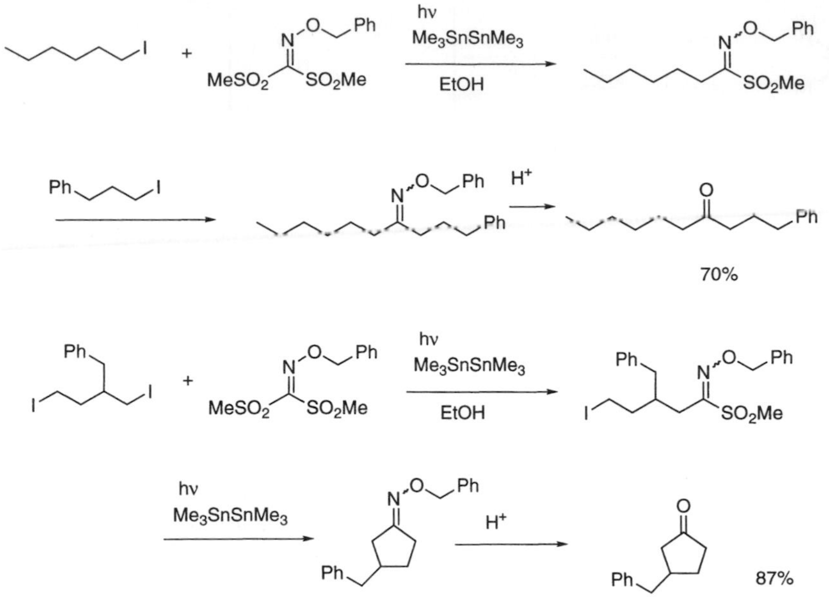

Scheme 4-37.

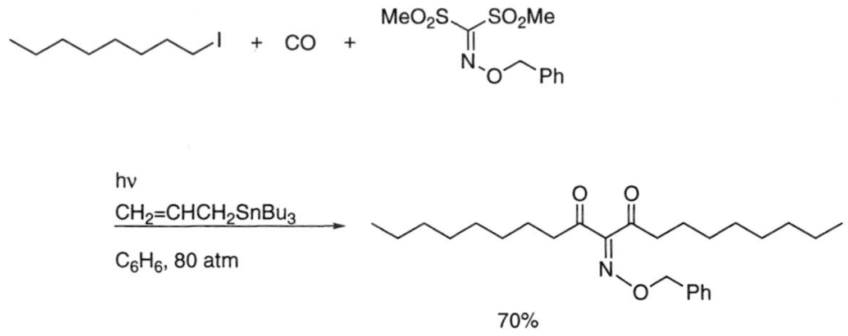

Scheme 4-38.

4.3.6 Acyl Radical Cyclization Approaches

The synthesis of cyclic ketones via acyl radical cyclizations represent by far the most frequently investigated approaches in the past two decades and such cyclizations are well discussed in a recent review [1a]. Two recent examples of the application to natural product synthesis are given here. A seven-membered ring, a key compound in the overall synthesis of (+)-confertin, was prepared by Shishido and co-workers, utilizing 7-*endo*-trig acyl radical cyclization which occurred in a highly efficient manner (Scheme 4-39) [66].

Scheme 4-40 details an application of the triple 6-*endo* cyclization, triggered by the formation of an acyl radical by Pattenden to the synthesis of the marine sponge metabolite spongian-16-one [67].

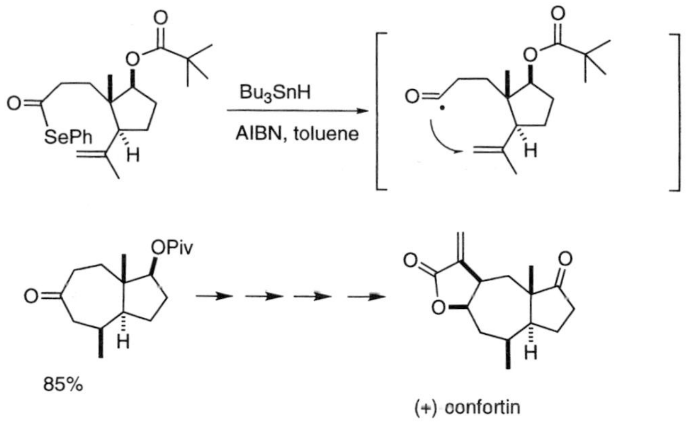

Scheme 4-39.

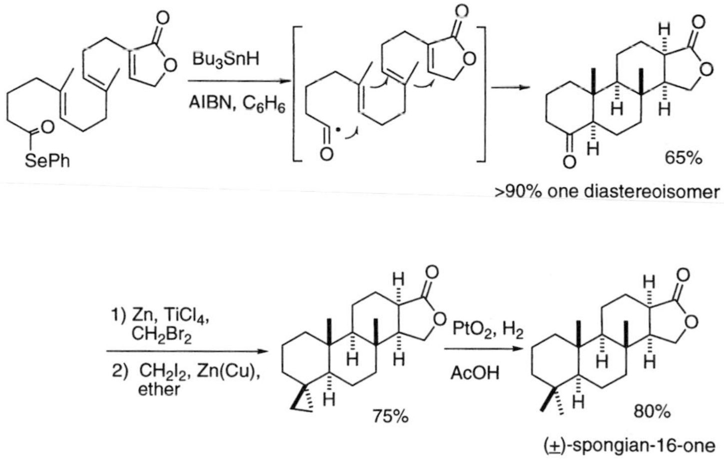

Scheme 4-40.

4.4 Synthesis of Carboxylic Acids and their Derivatives by Radical Reactions

It has long been known that autoxidation of aldehydes leads to carboxylic acids via a radical mechanism which involves the formation and oxidation of acyl radicals, leading to acyl cations via one-electron oxidation processes [68]. However, the recent topic in this field relates to the fact that many new synthetic methods for the synthesis of carboxylic acids derivatives have been developed which rely on the power of one-carbon homologative radical reactions.

4.4.1 Radical Carboxylation with CO_2

To our knowledge, only one report exists in which the formation of carboxylic acid by radical carboxylation with carbon dioxide has been documented. Curran and co-workers observed the formation of 9-anthracenecarboxylic acid in 10% yield, together with 71% yield of anthracene, when the radical reduction of 9-iodoanthracene with the ethylene-spaced fluorous tin hydride was run using supercritical CO_2 (90 °C, 280 atm) as the reaction media (Scheme 4-41) [69]. As demonstrated in this example, the CO_2 trapping reaction by radicals is not an efficient process and therefore is of limited synthetic utility. The rate constant for the addition is yet to be determined, but kinetic studies to date indicate that generally the decarboxylation of acyloxyl radicals is a rapid process [70].

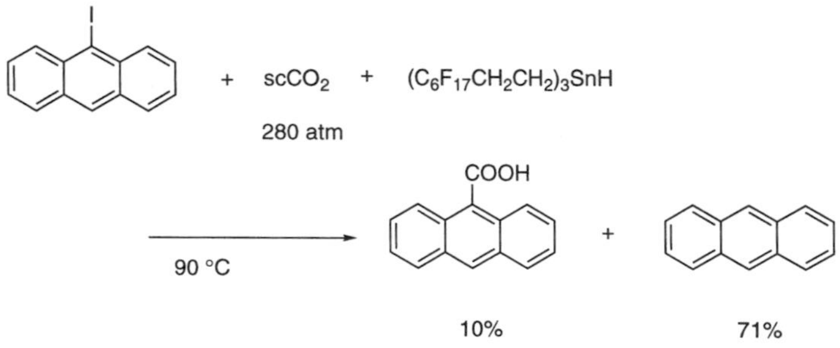

I

+ $scCO_2$ + $(C_6F_{17}CH_2CH_2)_3SnH$

280 atm

90 °C

COOH

+

10%

71%

Scheme 4-41.

4.4.2 Atom Transfer Carbonylation with CO

Although radical carboxylation with carbon dioxide is difficult to achieve, an equivalent reaction has been developed by using carbon monoxide instead of carbon dioxide. Recently developed atom transfer carbonylation reactions with carbon monoxide represents a promising approach. The general idea of the atom transfer carbonylation originated from the mechanistic consideration of unusual lactone formation in double carbonylation of 4-pentenyl iodide using a slow radical mediator such as tributylgermane or $(TMS)_3SiH$ [13].

When alkyl iodides and ROH were irradiated under CO pressure in the presence of a base such as potassium carbonate, good yields of carboxylic acid esters were obtained (Scheme 4-42) [71]. In the absence of a base, no carbonylation took place. The role of photo-irradiation is to initiate this hybrid radical/ionic reaction by effecting the homolysis of an R–I bond. The thermal initiation process involving allyltin and AIBN has also been found to be useful, as demonstrated by two examples of amide synthesis which are shown in Scheme 4-42 [72]. The likely mechanism involves (i) radical initiation via either irradiation or thermal initiation, (ii) radical chain propagation, composed of two reversible type radical reactions (carbonylation and iodine atom transfer) and (iii) ionic quenching to shift the equilibria of the reversible reactions. The same transformation was previously accomplished only by the procedure using a transition metal catalyst.

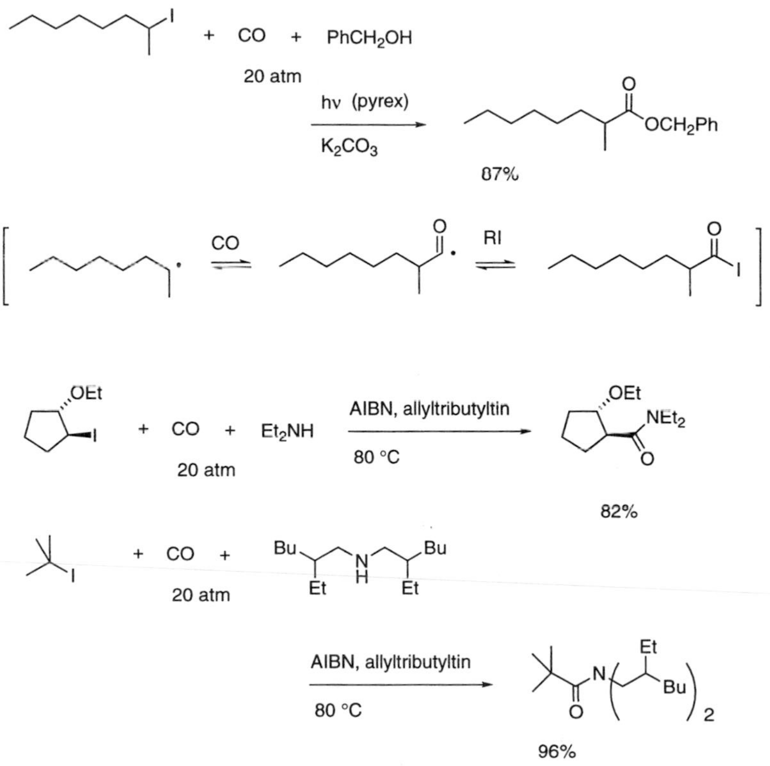

+ CO + $PhCH_2OH$
20 atm
hν (pyrex)
K_2CO_3
OCH_2Ph
87%

CO
RI

+ CO + Et_2NH
20 atm
AIBN, allyltributyltin
80 °C
OEt
NEt_2
82%

+ CO +
20 atm
AIBN, allyltributyltin
80 °C
96%

Scheme 4-42.

4.4.3 Group Transfer Carbonylation with CO

The photolysis of α-phenylselenoacetate and related compounds in the presence of an alkene and CO leads to acyl selenides via group transfer carbonylation. The mechanism of this three-component coupling reaction involves the addition of an α-(alkoxycarbonyl)methyl radical to an alkene, the trapping of the produced alkyl free radical by CO, and termination of the reaction by a phenylselenenyl group transfer from the starting material (Scheme 4-43) [73].

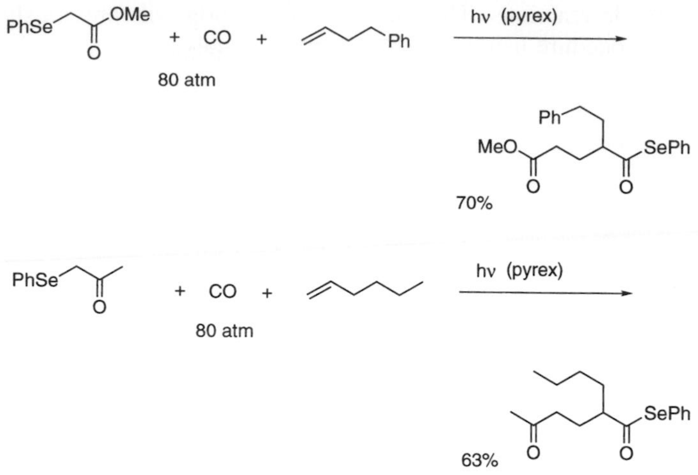

PhSe OMe + CO + Ph hν (pyrex)

80 atm

Ph MeO SePh

70%

PhSe + CO + hν (pyrex)

80 atm

SePh

63%

Scheme 4-43.

4.4.4 Oxidative Carboxylation with CO

Ryu and Alper found that manganese(III)-induced oxidative addition of carbonyl compounds to alkenes [74] can be combined successfully with radical carbonylations, which leads to fair to good yields of carboxylic acids [75]. Thus, a one-electron oxidation of an enolizable carbonyl compound, such as diethyl malonate, yields a malonyl radical, which then adds to an alkene and CO consecutively to form an acyl radical. The subsequent one-electron oxidation of an acyl radical leads to a carboxylic acid via an acyl cation (Scheme 4-44). Alper and Okuro reported that this system can be extended to include an alkyne as the substrate [76].

The oxidative carboxylation method reported recently from the Ishii group appears to be practical, since the key reagent NHPI(*N*-hydroxyphthalimide) is of catalytic use [77]. As an example, the carboxylation of adamantanes is shown in Scheme 4-45. Although the precise catalytic role of NHPI is presently unclear, the reaction may involve the generation of phthalimide-*N*-oxyl (PINO) from NHPI and O_2, followed by an abstraction of hydrogen from adamantane. The so-formed adamantyl radical then undergoes consecutive addition to CO and O_2 to form the adamantanecarbonylperoxy radical, which abstracts hydrogen from adamantane to form the peracid, a likely precursor of the carboxylic acid, along with the adamantyl radical.

4.4.5 Radical Carboxylation with Methyl Oxalyl Chloride

Kim and Jon recently reported a unique method for the synthesis of esters via a radical reaction of alkyl iodides with methyl oxalyl chloride under irradiation con-

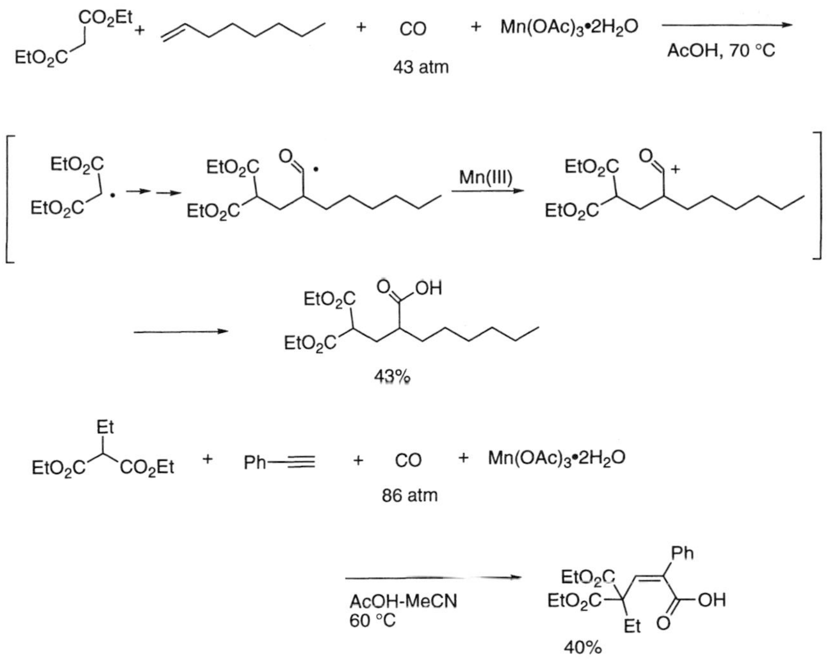

CO_2Et + + CO + $Mn(OAc)_3 \cdot 2H_2O$ → AcOH, 70 °C

EtO_2C 43 atm

EtO_2C EtO_2C Mn(III) EtO_2C

EtO_2C EtO_2C EtO_2C

EtO_2C OH

EtO_2C

43%

Et

EtO_2C CO_2Et + Ph— + CO + $Mn(OAc)_3 \cdot 2H_2O$

86 atm

Ph

AcOH-MeCN EtO_2C

60 °C EtO_2C OH

Et O

40%

Scheme 4-44.

O

N-OH

O (NHPI, 10 mol%)

+ CO + O_2

15 atm 1 atm $AcOH/(CHCl_2)_2$ 75% conversion

95 °C

COOH COOH

COOH + + COOH

56% 5% 8%

CO O_2

O O O

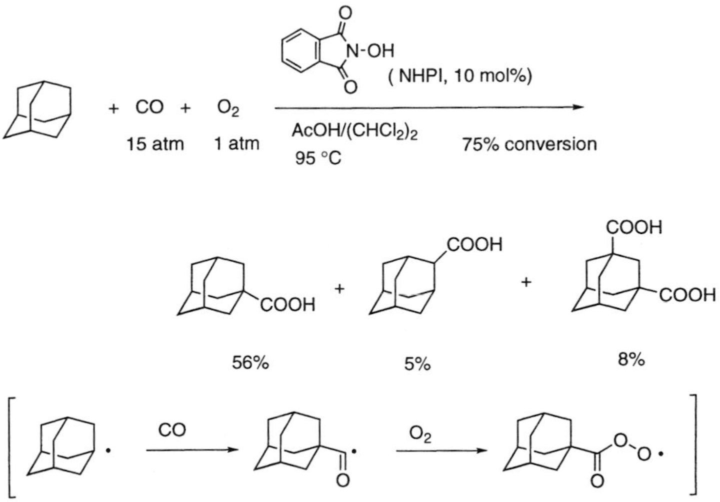

Scheme 4-45.

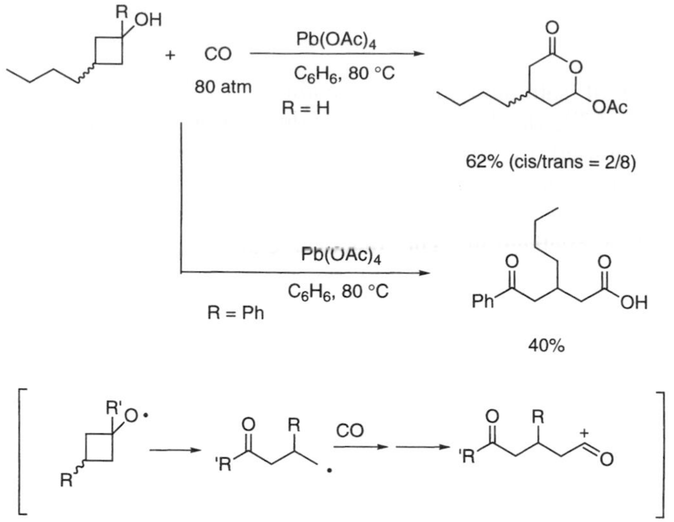

Scheme 4-48.

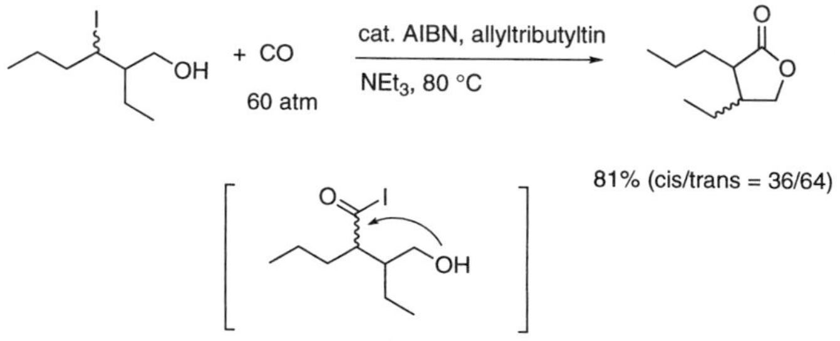

Scheme 4-49.

4+1 and 5+1 annulation protocols may be induced from alkyl bromides or selenides and tolerate the use of both aldimines and ketimines [88].

In view of the nucleophilic character usually exhibited by acyl radicals in their additions to C–C multiple bonds, these cyclizations are unusual, since the addition takes place at the more electron-rich nitrogen atom. However, reactions at higher concentrations of tin hydride clearly show that the 5-*exo* mode of cyclization is kinetically as well as thermodynamically favored. A further mechanistic possibility involves nucleophilic attack by the lone pair of the imine nitrogen electrons on the acyl radical, a process which is not a radical cyclization, but rather an ionic cyclization onto the carbonyl of an acyl radical.

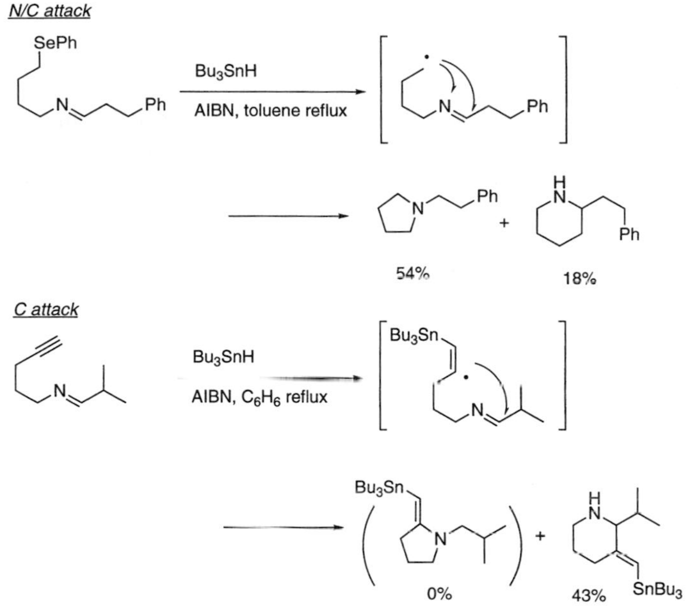

N/C attack

SePh

Bu$_3$SnH

AIBN, toluene reflux

Ph

54%

18%

C attack

Bu$_3$SnH

AIBN, C$_6$H$_6$ reflux

Bu$_3$Sn

SnBu$_3$

0%

43%

Scheme 4-50.

N attack

Br

+ CO

80 atm

Bu$_3$SnH

AIBN, 80 °C

81%

MeO

MeO

Br

+ CO

70 atm

Bu$_3$SnH

1,1'-azobis(cyclohexane-1-carbonitrile), 110 °C

49%

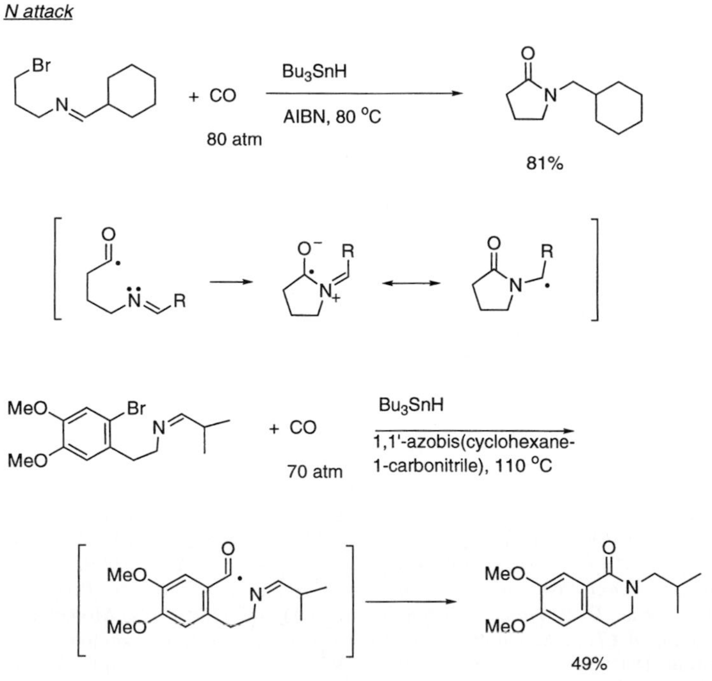

Scheme 4-51.

4.5.3 Thiolactones

The radical carbonylation of alkyl and aryl radicals and the cyclization of the resulting acyl radicals onto *tert*-butyl sulfides leads to the formation of γ-thiolactones with expulsion of the *tert*-butyl radical (Scheme 4-52) [89]. This process is applicable to a range of substituted 4-*tert*-butylthiobutyl bromides and iodides giving moderate to excellent yields of the corresponding thiolactones. Using acyl selenide/tin hydride chemistry and competition kinetic methods, the rate constant for the cyclization was determined to be 7.5×10^3 s^{-1} at 25 °C [89].

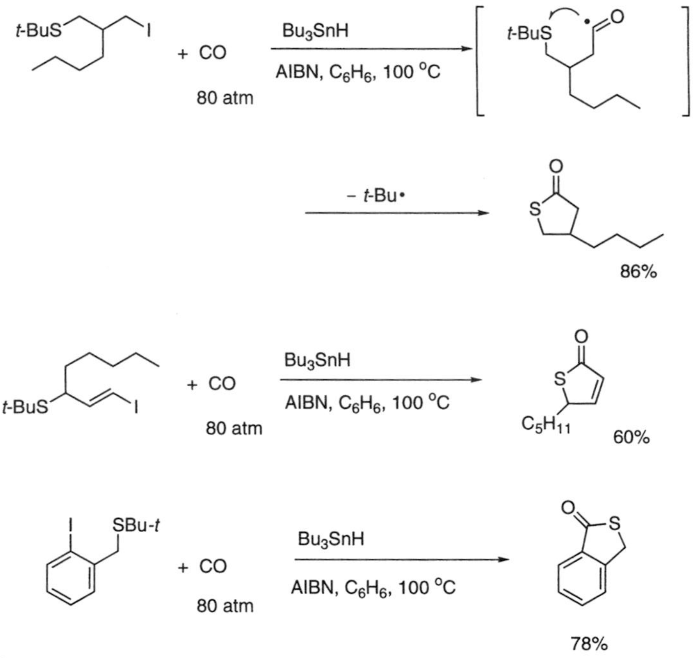

Scheme 4-52.

References

1. For recent reviews on acyl radicals, see: (a) Chatgilialoglu, C.; Crich, D.; Komatsu, M.; Ryu, I. *Chem. Rev.* **1999**, *99*, 1991. (b) Boger, D.L. *Israel J. Chem.* **1997**, *37*, 119. (c) Crich, D.; Yuan, H. In *Advances in Free Radical Chemistry*; Rawal, V.H., Ed.; Jai Press, New York, 1999; Vol. 2.
2. (a) Giese, B. *Radicals in Organic Synthesis: Formation of Carbon–Carbon Bonds*; Pergamon Press, Oxford, 1986. (b) Curran, D. P. *Synthesis*, **1988**, 417 (part 1); 489 (part 2). (c) Motherwell, W.B.; Crich, D. *Free Radical Chain Reactions in Organic Synthesis*, Academic, London, 1992. (d) Jasperse, C.P.; Curran, D.P.; Fevig, T.L. *Chem. Rev.* **1991**, *91*, 1237. (e) Beckwith, A.L.J.; Crich, D.; Duggan, P.J.; Yao, Q. *Chem. Rev.* **1997**, *97*, 3273. (f) Fallis, A.G.; Branza, I.M. *Tetrahedron* **1997**, *53*, 17543. (g) Curran, D.P.; Porter, N.A.; Giese, B. *Stereochemistry of Free*

Radical Reactions, VCH, Weinheim, 1996. (h) Sibi, M.P.; Porter, N.A. *Acc. Chem. Res.* **1999**, *32*, 163. (i) Baguley, P.A.; Walton, J.C. *Angew. Chem. Int. Ed.* **1998**, *37*, 3072.

3. Ryu, I.; Sonoda, N. *Angew. Chem. Int. Ed. Engl.* **1996**, *35*, 1050.
4. Ryu, I.; Sonoda, N.; Curran, D.P. *Chem. Rev.* **1996**, *96*, 177.
5. (a) Pfenninger, J.; Henberger, C.; Graf, W. *Helv. Chim. Acta* **1980**, *63*, 2328. (b) Pfenninger, J.; Graf, W. *Helv. Chim. Acta* **1980**, *63*, 1562.
6. Chatgilialoglu, C.; Ferreri, C.; Lucarini, M.; Pedrielli, P.; Pedulli, G. F. *Organometallics* **1995**, *14*, 2672.
7. Ryu, I.; Kusano, K.; Ogawa, A.; Kambe, N.; Sonoda, N. *J. Am. Chem. Soc.* **1990**, *112*, 1295.
8. Curran, D.P.; Hadida, S. *J. Am. Chem. Soc.* **1996**, *118*, 2531.
9. Chatgilialoglu, C. *Acc. Chem. Res.* **1992**, *25*, 188.
10. Ryu, I.; Niguma, T.; Minakata, S.; Komatsu, M.; Hadida, S.; Curran, D.P. *Tetrahedron Lett.* **1997**, *38*, 7883.
11. Horner, J.H.; Martinez, F.N.; Newcomb, M.; Hadida, S.; Curran, D. P. *Tetrahedron Lett.* **1997**, *38*, 2783.
12. Ryu, I.; Hasegawa, M.; Kurihara, A.; Ogawa, A.; Tsunoi, S.; Sonoda, N. *Synlett* **1993**, 143.
13. Tsunoi, S.; Ryu, I.; Yamasaki, S.; Fukushima, H.; Tanaka, M.; Komatsu, M.; Sonoda, N. *J. Am. Chem. Soc.* **1996**, *118*, 10670.
14. Ryu, I.; Kusano, K.; Masumi, N.; Yamazaki, H.; Ogawa, A.; Sonoda, N. *Tetrahedron Lett.* **1990**, *31*, 6887.
15. Baillargeon, V.P.; Stille, J.K. *J. Am. Chem. Soc.* **1983**, *105*, 7175.
16. Gupta, V.; Kahne, D. *Tetrahedron Lett.* **1993**, *34*, 591.
17. Yamago, S.; Miyazoe, H.; Goto, R.; Yoshida, J. *Tetrahedron Lett.* **1999**, *40*, 2347.
18. Curran, D.P. *Angew. Chem. Int. Ed.* **1998**, *37*, 1174.
19. Tsunoi, S.; Ryu, I.; Fukushima, H.; Tanaka, M.; Komatsu, M.; Sonoda, N. *Synlett* **1995**, 1249.
20. Ryu, I.; Kurihara, A.; Muraoka, H.; Tsunoi, S.; Kambe, N.; Sonoda, N. *J. Org. Chem.* **1994**, *59*, 7570.
21. Tsunoi, S.; Ryu, I.; Muraoka, H.; Tanaka, M.; Komatsu, M.; Sonoda, N. *Tetrahedron Lett.* **1996**, *37*, 6729.
22. Nakatani, S.; Yoshida, J.; Isoe, S. *J. Chem. Soc., Chem. Commun.* **1992**, 880.
23. Curran, D.P.; Sisko, J.; Balog, A.; Sonoda, N.; Nagahara, K.; Ryu, I. *J. Chem. Soc., Perkin. Trans. 1* **1998**, 1591.
24. Hayes, C.J.; Pattenden, G. *Tetrahedron Lett.* **1996**, *37*, 271.
25. (a) Pattenden, G.; Roberts, L. *Tetrahedron Lett.* **1996**, *37*, 4191. (b) Pattenden, G.; Roberts, L.; Blake, A.J. *J. Chem. Soc., Perkin Trans. 1* **1998**, 863.
26. (a) Ferguson, R.R.; Crabtree, R.H. *J. Org. Chem.* **1991**, *56*, 5503. (b) Boese, W.T.; Goldman, A.S. *Tetrahedron Lett.* **1992**, *33*, 2119. (c) Jaynes, B.S.; Hill, C.L. *J. Am. Chem. Soc.* **1995**, *117*, 4704.
27. Curran, D.P.; Xu, J.; Lazzarini, E. *J. Chem. Soc., Perkin Trans. 1* **1995**, 3049.
28. Kim, S.; Lee, I.Y.; Yoon, J.-Y.; Oh, D.H. *J. Am. Chem. Soc.* **1996**, *118*, 5138.
29. Kim, S.; Lee, I.Y. *Tetrahedron Lett.* **1998**, *39*, 1587.
30. (a) Citterio, A.; Arnoldi, A.; Minisci, F. *J. Org. Chem.* **1979**, *44*, 2674. (b) Giese, B.; Kretzschmar, G.; Meixner, J. *Angew. Chem. Int. Ed.* **1983**, *22*, 753. (c) Nagahara, K.; Ryu, I.; Kambe, N.; Komatsu, M.; Sonoda, N. *J. Org. Chem.* **1995**, *60*, 7384.
31. Beckwith, A.L.J.; Schiesser, C.H. *Tetrahedron* **1985**, *41*, 3925.
32. (a) Curran, D.P.; Chen, M.-H.; Kim, D. *J. Am. Chem. Soc.* **1986**, *108*, 2489. (b) Curran, D.P.; Chen, M.-H. *Tetrahedron Lett.* **1985**, *26*, 4991. (c) Curran, D.P. *Synthesis* **1988**, 489. (d) Boger, D.L.; Mathvink, R. J. *J. Am. Chem. Soc.* **1990**, *112*, 4003.
33. Clive, D.L.J.; Tao, Y.; Khodabocus, A.; Wu, Y.-J.; Angoh, A. G.; Bennet, S.M.; Boddy, C.N.; Bordeleau, L.; Kellner, D.; Middleton, D.S.; Nichols, C.J.; Richardson, S.R.; Vernon, P.G. *J. Am. Chem. Soc.* **1994**, *116*, 11275.
34. Clive, D.L.J.; Beaulieu, P.L.; Set, L. *J. Org. Chem.* **1984**, *49*, 1314.
35. Walton, R.; Fraser-Reid, B. *J. Am. Chem. Soc.* **1991**, *113*, 5791.
36. (a) Beckwith, A.L.J.; Hay, B.P. *J. Am. Chem. Soc.* **1989**, *111*, 2674. (b) Beckwith, A.L.J.; Hay, B.P. *J. Am. Chem. Soc.* **1989**, *111*, 230.
37. Curran, D.P.; Diederichsen, U.; Palovich, M. *J. Am. Chem. Soc.* **1997**, *119*, 4797.

Li + CO —(THF, -78 °C)→ O O 96.4 % (5.17)

$R^1Li + CHCl_2R^2 + CO \xrightarrow[2)\ 0\ °C,\ H_3O^+]{1)\ -100\ °C,\ 1\ h} R^1CH(OH)CCl_2R^2$ (5.18)

R^1 = Bu, ^{s}Bu, ^{t}Bu, C_6H_{13}
R^2 = H or aryl

$RLi + CH_3CN \xrightarrow[-110\ °C\ \text{or}\ -78\ °C]{CO} \xrightarrow{H_3O^+} RCH(OH)CH_2CN$ (5.19)

$^tBuC(O)TeBu \xrightarrow[-Bu_2Te]{BuLi,\ -105\ °C,\ THF\ /\ Et_2O} [^tBuC(O)Li] \xrightarrow{Me_3SiCl} PhC(O)SiMe_3$ 65 % (5.20)

$PhC(O)SiMe_3 + PhCH_2Br \xrightarrow[THF,\ 160\ °C,\ 3\ h]{KF,\ 18\text{-crown-6}} PhC(O)CH_2Ph$ 90 % (5.21)

$PhC(O)SiMe_3 + PhCHO \xrightarrow[THF]{TBAF} PhC(O)CH(OH)Ph$ 50 % (5.22)

Kambe reported a useful entry to acyllithium using an efficient lithium-tellurium exchange reaction (Eq. (5.20)) [12].

Walton and Ricci found that an acylsilane underwent desilylation by a naked fluoride ion to give an acyl anion, which was then trapped in situ by an alkyl halide (Eq. (5.21)) [27].

Later, Heathcock reported that a similar intermediate having an ammonium ion as the counter cation can be trapped by aldehydes or water (Eq. (5.22)) [28]. The issue of whether these anions are the free anions or of the ate complex type has not yet been resolved.

5.4 Intramolecular Conversion of Acyllithium

5.4.1 General

The complex outcomes which are commonly encountered in the reaction of alkyllithium with CO could be attributed to the exceedingly reactive nature of acyllithium. Therefore, novel methods which allow for the immediate conversion of the highly reactive acyllithium to a more stable but still reactive synthetic intermediate seem highly desirable. The requirements for such an intramolecular conversion in achieving high selectivity are as follows: (1) The intramolecular reaction should be faster than any other intermolecular reactions in which acyllithium intermediates participate. (2) The intramolecular conversion should give a new intermediate which contains no carbonyl function. Otherwise, strong nucleophiles such as acyllithium and alkyllithium present in the reaction mixture might rapidly attack such a functional group. Intramolecular conversions to give enolates, which are masked carbonyl functions, may be desirable. (3) Such a conversion should give a species which is not a dead end, but still potentially useful as a synthetic intermediate. (4) The new intermediate should be a discrete intermediate so that electrophiles could be added after its formation. It should be recalled that the major drawback of intramolecular reactions via acyllithium is the necessity of in situ reactions with electrophiles.

The major advances in this field have been brought about by the authors' group at Osaka University. As a strategy for attaining highly selective reactions via acyllithium, we employed intramolecular conversion. A number of tactics as shown in Scheme 5-2 and following sections were devised. It should be mentioned that these reactions are no longer 'acylation via acyllithium' but, rather, various type of enolate reactions derived from them.

5.4.2 Rearrangement

We examined the possibility of intramolecular conversion based on an anionic rearrangement [Scheme 5-2, (a)]. For group G, we selected silyl groups, since the 1,2-anionic rearrangement of organosilyl group is well known [29]. The results are given in Eqs. (5.23) and (5.24) [30, 31]. When silylmethyllithium was exposed to carbon monoxide (1 atm) at –78 °C in ether, the gradual absorption of carbon monoxide over a period of 2 h was observed. Quenching with Me_3SiCl gave an enediol disilyl ether as a major product (Eq. (5.23)) (33%, E/Z=50/50), and no desired product that was envisioned in Scheme 5-2 (a) was detected. A dramatic change occurred when this reaction was conducted at 15 °C in that the (1-siloxyvinyl)silane was produced in 86% yield. This indicates that the lithium enolate (R=H, G=Me_3Si) had been formed as the result of a silicon shift [(a) in Scheme 5-2]. Further examples of reactions of α-silylalkyllithium derivatives with carbon monoxide at 15 °C are given in Table 5-1.

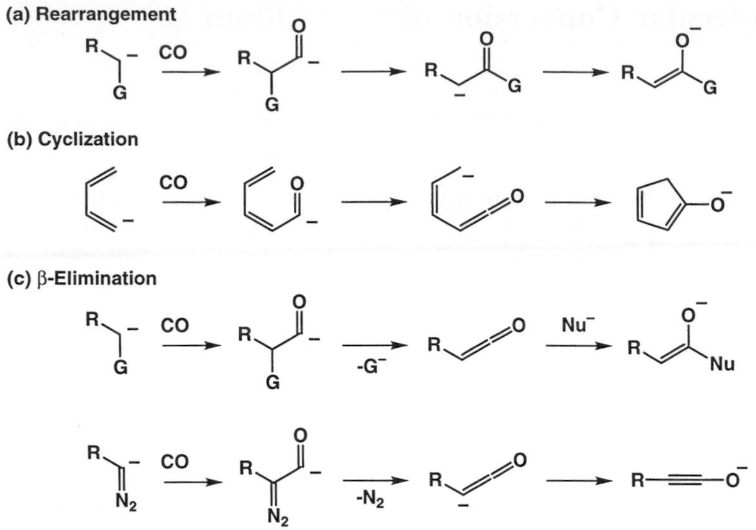

(a) Rearrangement

(b) Cyclization

(c) β-Elimination

Scheme 5-2. Tactics for intramolecular conversion of acyllithium.

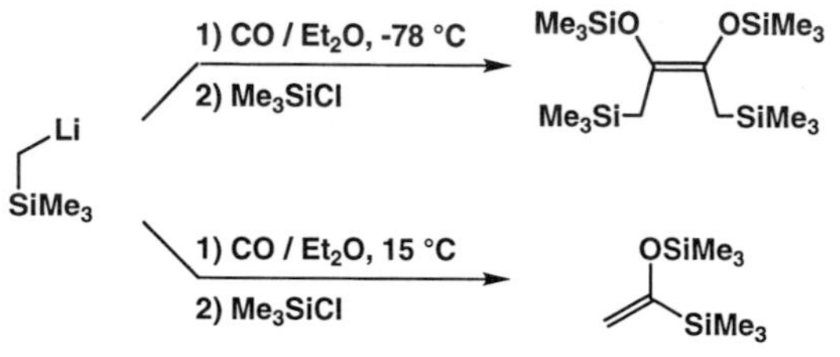

1) CO / Et_2O, -78 °C
2) Me_3SiCl (5.23)

1) CO / Et_2O, 15 °C
2) Me_3SiCl (5.24)

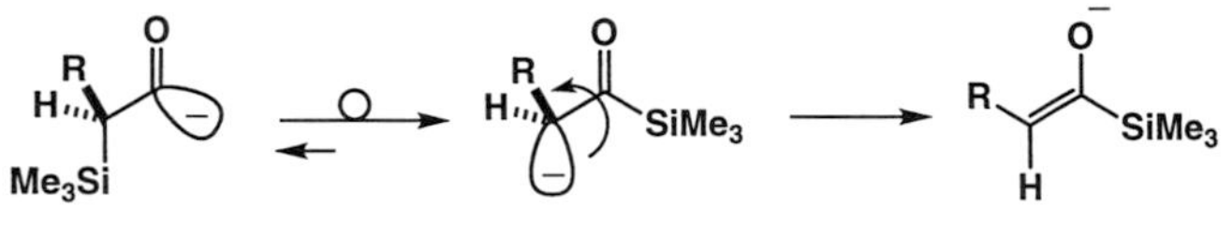

(5.25)

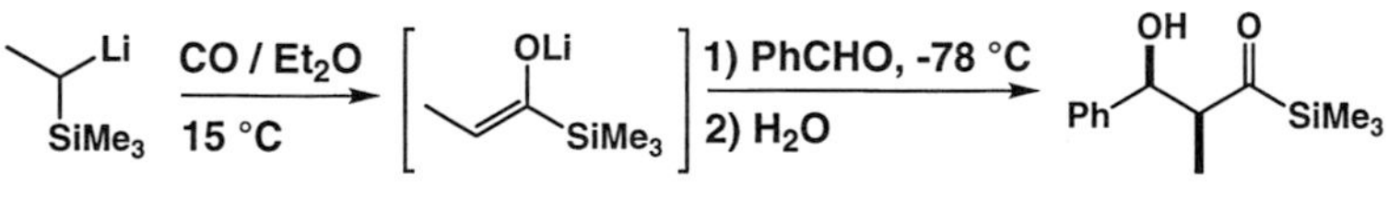

CO / Et_2O, 15 °C; 1) PhCHO, -78 °C 2) H_2O (5.26)

The preferential formation of *E* enolates deserves comment. The silicon shift would result in the negative charge in the plane perpendicular to that of the π-orbital of the carbonyl group, as depicted in Eq. (5.25). A subsequent 90° rotation around the C–C axis, in order to avoid steric congestion between the organosilicon group and R, would bring the negative charge into conjugation with carbonyl π-orbital to form the *E* enolates (Eq. (5.25)). The propionylsilane enolate, gener-

Table 5-1. Preparation of acylsilanes and their enol sily ethers.

Substrate	Solvent	Electrophile	Product	Yield; GLC, (isolated)
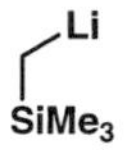 Li, $SiMe_3$	Et_2O	Me_3SiCl	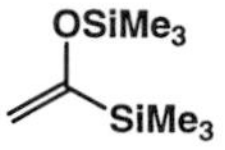 $OSiMe_3$, $SiMe_3$	86 (75)
		Me_3SiCl	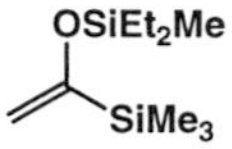 $OSiEt_2Me$, $SiMe_3$	80
		H_2O	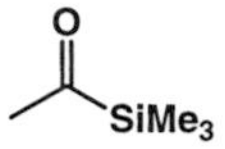 O, $SiMe_3$	72
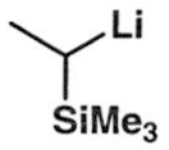 Li, $SiMe_3$	Et_2O	Me_3SiCl	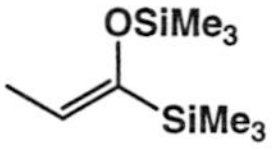 $OSiMe_3$, $SiMe_3$	88 ($E = 100$)
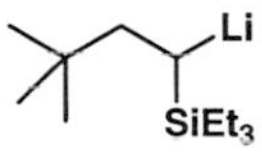 Li, $SiEt_3$	Et_2O	H_2O	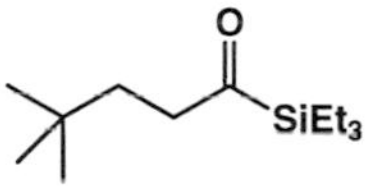 O, $SiEt_3$	73 (61)
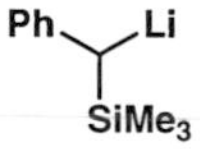 Ph, Li, $SiMe_3$	TMEDA	Me_3SiCl	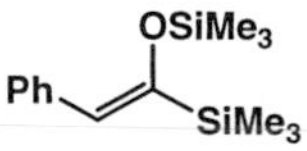 Ph, $OSiMe_3$, $SiMe_3$	92 ($E / Z = 89 / 11$)
	H_2O		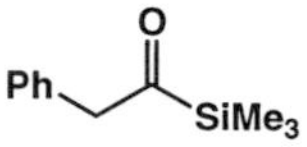 Ph, O, $SiMe_3$	94

ated in the stereochemically pure *E* form, then reacted with benzaldehyde to afford the *erythro* adduct as the major product (Eq. (5.26)) (52% yield, *erythro/threo*=93/7).

The allylsilane anion reacted with CO and underwent a similar rearrangement to provide a unique route to the dienolates (Eq. (5.27)) [31].

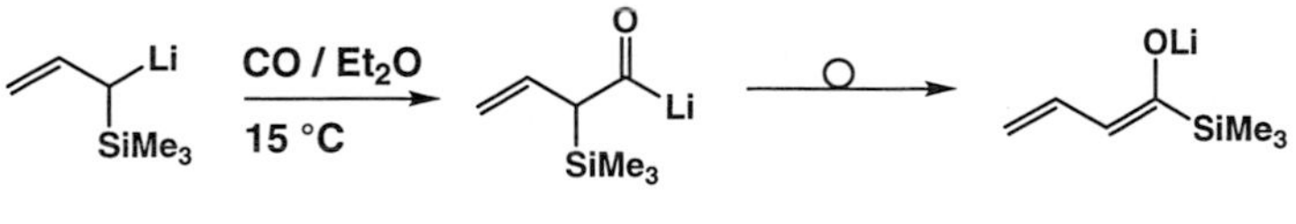

Li, $SiMe_3$ $\xrightarrow[15\ °C]{CO\ /\ Et_2O}$ O, Li, $SiMe_3$ ⟶ OLi, $SiMe_3$ (5.27)

5.4.3 Cyclization

Our second method involves the utilization of a cyclization reaction [Scheme 5-2(b)]. We examined reactions that generate the negative charge of an acyllithium on the termini of π-conjugation chains [32].

+ CO

(5.28)

$SiMe_3$ CO THF 15 °C, 2 h $SiMe_3$ Li O

Me_3Si OLi + $SiMe_3$ OLi

(5.29)

${}^tBuMe_2SiCl$ HMPA -78 °C Me_3Si $OSiMe_2{}^tBu$ + $SiMe_3$ $OSiMe_2{}^tBu$

59 % 15 %

The simplest example of such a system is depicted in Eq. (5.28). The lithium enolate of 2-(trimethylsilyl)cyclopropanone is formed by the reaction of [1-(trimethylsilyl)vinyl]lithium with carbon monoxide (Eq. (5.29)). Treatment of the vinyllithium with CO, at atmospheric pressure, in THF at 15 °C for 2 h followed by quenching with trimethylchlorosilane at –78 °C afforded a somewhat labile product, which decomposed during the usual hydrolytic work-up. Quenching with *tert*-butyldimethylchlorosilane/HMPA instead allowed isolation of the products. The major product was a silylated cyclopropane enolate. It is noteworthy that the overall sequence follows a formal [2+1]cycloaddition (Eq. (5.28)). The silylated allenolate was also formed as a by-product as the result of a 1,2-anionic silicon rearrangement.

Extension of one conjugation unit [Scheme 5-2(b)] also gave similar results [32]. The reaction of [2-phenyl-1-(silyl)vinyl]lithium with CO at 15 °C also underwent clean intramolecular transformation of the acyllithium, whereas cyclization to a five-membered ring was observed. Treatment of a THF solution of the vinyllithium with CO (1 atm) at 15 °C for 1.2 h followed by proton quenching yielded 2-(silyl)-1-indenol in 61% yield together with 3-(silyl)indanone (10%) (Eq. (5.30)). With the prolonged reaction time (24 h), 3-(silyl)indanone was obtained as a sole product in 60% isolated yield (70% GLC yield) after proton

Li $SiMe_3$ CO THF 15 °C H_2O 0 °C OH $SiMe_3$ + O $SiMe_3$

(5.30)

CO (15 °C, 1.2 h) 61 % 10 %
CO (15 °C, 24 h) 0 % 70 %

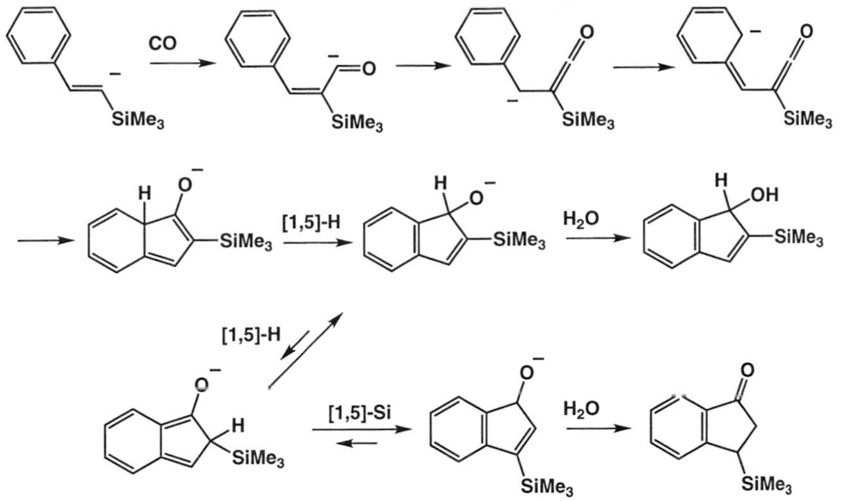

Scheme 5-3. Cyclization via acyllithium followed by [1,5]-H and [1,5]-Si shift.

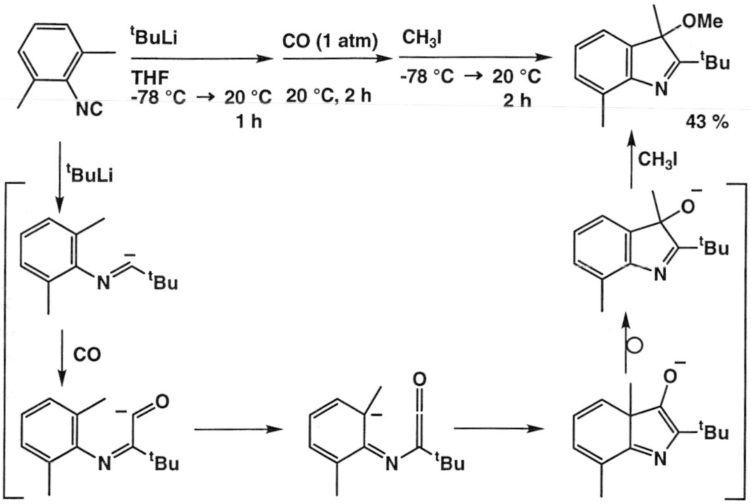

(5.31)

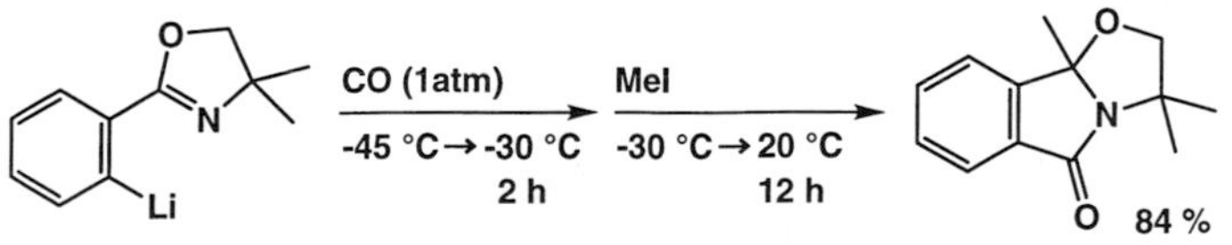

(5.32)

quenching. The proposed cyclization process leading to the alcohol and the ketone is shown in Scheme 5-3. The reaction of Eq. (5.30) involves cyclization from a dienyl anion. Similarly, 2-aza-dienyl anion (Eq. (5.31)) [33] and 4-aza-dienyl anion (Eq. (5.32)) [34] undergo cyclization via the corresponding acyllithium.

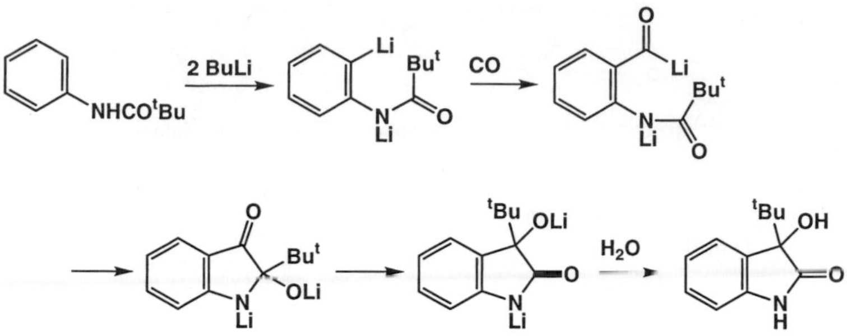

(5.33)

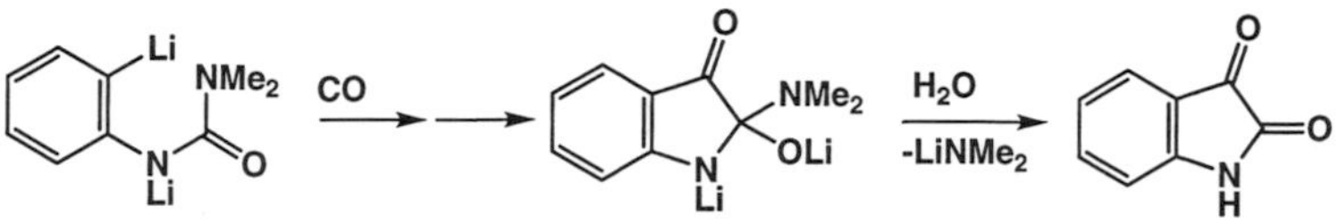

(5.34)

Smith reported cyclization reactions via acyllithium derived from a dianion and CO, which involved an alkyl group migration (Eq. (5.33)) [35]. They also reported a reaction similar to Eq. (5.33), but not involving alkyl migration (Eq. (5.34)) [36]. The cyclization of Eqs. (5.33) and (5.34) may proceed via ketenes or, alternatively, through direct intramolecular nucleophilic attack of the acyllithium on the amide carbonyl. Details of the mechanism are not known.

5.4.4 *β*-Elimination

As an intramolecular process for converting the very reactive acyllithium, we examined the possibility of *β*-elimination from an acyllithium having a suitable leaving group to afford a ketene as an intermediate [Scheme 5-2(c)]. To test this possibility, phenylthiomethyllithium, which is easily available via the deprotonation

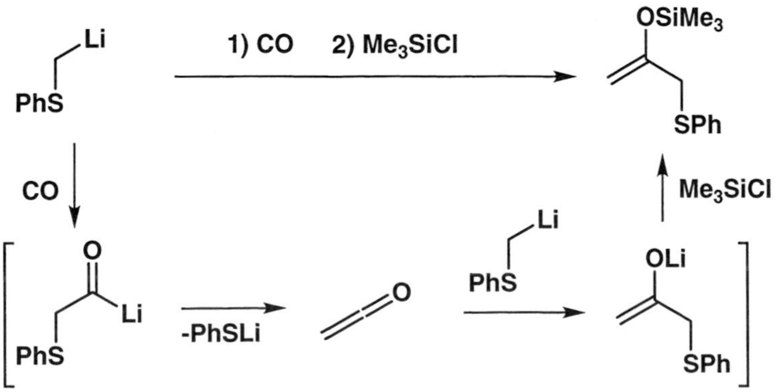

(5.35)

of thioanisole, was reacted with carbon monoxide at –20 °C for 1.5 h (Eq. (5.35)). Quenching with trimethylchlorosilane afforded an enol silyl ether in 48% yield, a result which indicated the formation of ketenes, as would be expected [37]. The reaction provides a clear demonstration of the validity of this method for the conversion of acyllithium.

Recently we reported another useful transformation which involves the intramolecular conversion of an acyllithium [38]. This reaction may be classified as a variation of the β-elimination or cyclic extrusion reaction. The reaction of a lithiated silyldiazomethane with CO underwent an extrusion of nitrogen from an acyllithium. The results not only provide a unique path to a lithium ynolate having a silyl group but also lead to a unique synthetic operation which enables 'ketenylation' (Eq. (5.36)). To a THF-hexane solution of trimethylsilyldiazomethane was added a hexane solution of BuLi (1.2 equiv) at –78 °C, and the mixture was stirred at the same temperature for 1 h. The mixture was then exposed to carbon monoxide at –78 °C and atmospheric pressure for 2 h. The addition of 1.1 equiv of triethylsilyl trifluoromethanesulfonate (–78 °C then 20 °C for 3 h) and workup with aqueous saturated NH_4Cl gave triethylsilyl(trimethylsilyl)ketene as the sole product in 85% yield. In the presence of Me_3Al, the ynolate thus generated can undergo ketenylation of epoxides (Eq. (5.37)). A similar ring-opening ketenylation of propylene imine does not require the use of Me_3Al and leads to a lactam (Eq. (5.38)). The examples shown above illustrate the potential of the use of intra-

Me3Si(Li)C=N2 —[CO (1 atm), -78 °C, 2 h]→ —[Et3SiOTf, -78 °C → 20 °C]→ (Me3Si)(Et3Si)C=C=O 85%

[CO; Me3Si(Li–C(=O))C=N+=N− → Me3Si … N, N, Li, O —[-N2]→ { Me3Si—≡—OLi ; Me3Si(Li)C=C=O } ; Et3SiOTf] (5.36)

Me3Si—≡—OLi —[1) Me3Al, -78 °C → 0 °C, 1 h; 2) cyclohexene oxide, -78 °C → 20 °C, 12 h]→ —[H+]→ SiMe3-substituted bicyclic lactone 93 % (5.37)

[SiMe3 ketene, O− → SiMe3, O−]

$$Me_3Si{-}C{\equiv}C{-}OLi \xrightarrow[-78\,°C\ \to\ 64\,°C,\ 6\,h]{\text{2-methyl-N-Ts-aziridine}} \xrightarrow{H^+} \text{3-}Me_3Si\text{-5-methyl-1-Ts-pyrrolidin-2-one}\quad 65\,\%\ (68{:}32) \tag{5.38}$$

molecular conversions to make use of exceedingly reactive acyllithium derivatives.

5.5 Reactions of Acyl Anions with Various Metals

5.5.1 Reactions with Mg

It has been reported, without giving details, in the early years of this century by Ferrario and Vinay [39] that some organomagnesium compounds reacted with CO. Later, Egorova reported [40] that the reaction of *t*-BuMgCl with CO gave an acyloin (Eq. (5.39)), while *i*-PrMgCl gave an olefin (Eq. (5.40)). Fischer [41] attempted to react Grignard reagents with CO under high pressure at high temperature and found that acyloins were formed from aromatic compounds (Eq. (5.41)) while olefins were obtained from aliphatic compounds (Eq. (5.42)). Similar results were also reported by Zelinski and Eidus [42, 43]. These reactions were, reportedly, thought to be possible only at higher temperatures and pressures [44]. Although in low yields, the formation of similar products at ambient temperature and pressure were observed by Ryang and Tsutsumi [45]. Louw found that the ad-

$$^tBuMgCl \xrightarrow{CO} \xrightarrow{H^+} {}^tBuCH(OH)C(=O){}^tBu \tag{5.39}$$

$$^iPrMgCl \xrightarrow{CO} \xrightarrow{H^+} {}^iPrCH{=}C(Me)_2 \tag{5.40}$$

$$PhMgBr \xrightarrow[110\,°C]{CO\ (100\ atm)} \xrightarrow{H^+} PhCH(OH)C(=O)Ph\quad 68\,\% \tag{5.41}$$

$$BuMgCl \xrightarrow[125\,°C]{CO\ (100\ atm)} \xrightarrow{H^+} BuCH{=}CHCH_2CH_2CH_3\quad 65\,\% \tag{5.42}$$

dition of HMPA facilitated the reaction of a Grignard reagent with CO under milder reaction conditions (room temperature and 30 atm of CO) [46].

5.5.2 Acylsodium

Only a few reports are available on reactions via acylsodium derivatives. Wanklyn [47] carried out a reaction of ethyl sodium, formed from Et_2Zn and metallic sodium, with CO and observed the formation of diethyl ketone (Eq. (5.43)). Later, Schluback prepared ethyl sodium from ethyl mercury and metallic sodium and then reacted it with CO. Triethylcarbinol was obtained in addition to diethyl ketone [48]. Ryang and Tsutsumi [49] reported the formation of different compounds for reactions using different solvents (Eqs. (5.44) and (5.45)).

$$\mathrm{EtNa} \xrightarrow{\mathrm{CO}} \mathrm{Et{-}C({=}O){-}Et} \tag{5.43}$$

$$\mathrm{PhNa} \xrightarrow[\mathit{THF}]{\mathrm{CO}} \mathrm{Ph{-}C({=}O){-}Ph} + \mathrm{Ph{-}C({=}O){-}CH(OH){-}Ph} \tag{5.44}$$

$$\mathrm{PhNa} \xrightarrow[\mathit{C_6H_6}]{\mathrm{CO}} \mathrm{Ph{-}C({=}O){-}C({=}O){-}Ph} \tag{5.45}$$

5.5.3 Acylcuprate

As has been described above, one of the drawbacks of acyllithium compounds as synthetic intermediates is that one cannot use them as discrete intermediates. In most (if not all) cases, the trapping reagent or reactant needs to be used in situ. This problem has been solved by derivatizing the organolithium compounds into cuprates prior to reacting them with CO. That a cuprate undergoes reaction with CO was first reported in 1972 by Schwartz, who reported the formation of a symmetrical ketone from Bu_2CuLi (Eq. (5.46)) [50]. In 1985, Seyferth at MIT found that a higher order cuprate gave a discrete acyl metal intermediate when exposed to CO and that this intermediate could be trapped by an enone which was added to the reaction mixture after the reaction with CO (Eq. (5.47)) [51]. The reaction

$$\mathrm{Bu_2CuLi} \xrightarrow[-22\ ^{\circ}\mathrm{C},\ 1\ \mathrm{h}]{\mathrm{CO(1atm)}} \xrightarrow{\mathrm{H_3O^+}} \underset{77\ \%}{\mathrm{BuC({=}O)Bu}} \tag{5.46}$$

$^{t}Bu_2Cu(CN)Li_2$ $\xrightarrow[-110\ °C]{CO(1atm)}$ (cyclohexenone) $\xrightarrow{NH_4OH / NH_4Cl}$ 78 % (5.47)

$Cu(CN)Li_2$ $\xrightarrow{CO\ (1atm)}$ [$Cu(CN)Li_2$] $\longrightarrow$ $\xrightarrow{H_3O^+}$ 73 % (5.48)

$\xrightarrow[CO(1atm)]{^{t}Bu_2Cu(CN)Li_2}$ $\xrightarrow{PhCOCl}$ 89 % (5.49)

$R_2Cu(CN)Li_2$ $\xrightarrow[-78\ °C \rightarrow r.t.]{CO\ (1atm),\ Bu_3P}$ $\xrightarrow{NH_4OH / NH_4Cl}$ $RC(O)CH(OH)R$ 65 - 85 % (5.50)

gave the alkylated enone which is not shown in Eq. (5.47) as a by-product. The formation of this by-product could be avoided by the use of *t*-BuCu(CN)Li instead of the dialkyl cuprate [52]. Such high order cuprate technology can also be applied to allyl-carbonyl group transfer (Eq. (5.48)) [53] and to 1,4-diketone synthesis (Eq. (5.49)) [54]. The higher order cuprate reacts with CO in the presence of Bu_3P to give a coupling product (Eq. (5.50)) [55].

5.5.4 Related Reactions Involving Other Metals

Jutzi reported the formation of enediol compounds from the reaction of trimethylsilyl lithium with CO (Eq. (5.51)) [56]. As shown above, they proposed a mechanism involving the dimerization of a lithioxy carbene, which is an acyllithium tautomer (Eq. (5.51)). A similar reaction suggests the possibility of further insertion

$$2\ Me_3SiLi + 2\ CO \longrightarrow 2\left[Me_3SiC(=O)Li \rightleftharpoons Me_3SiC(:)OLi\right] \longrightarrow \left[(Me_3Si)(LiO)C{=}C(OLi)(SiMe_3)\right] \xrightarrow{2\ Me_3SiCl} (Me_3Si)(Me_3SiO)C{=}C(OSiMe_3)(SiMe_3) \quad (5.51)$$

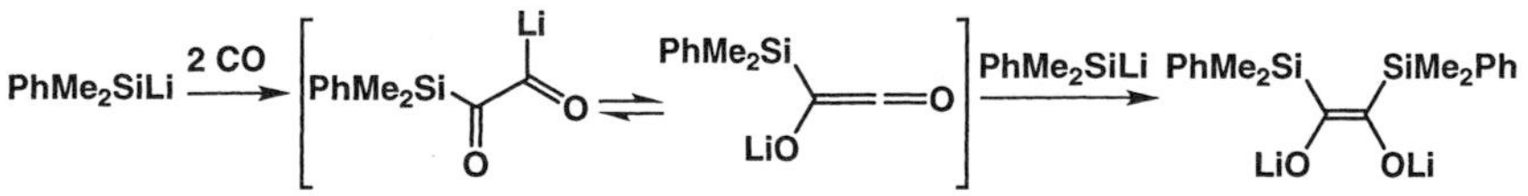

(5.52)

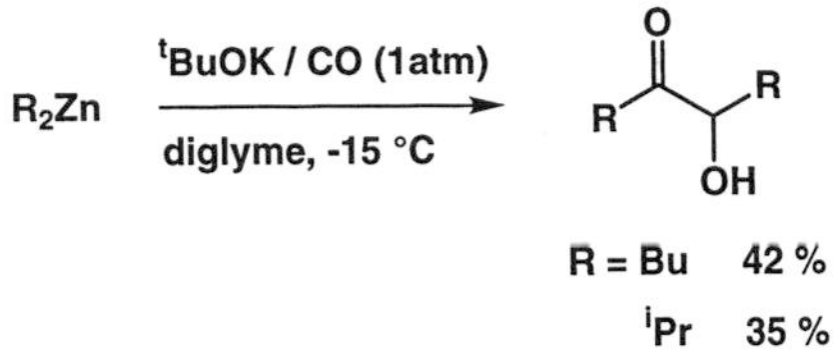

(5.53)

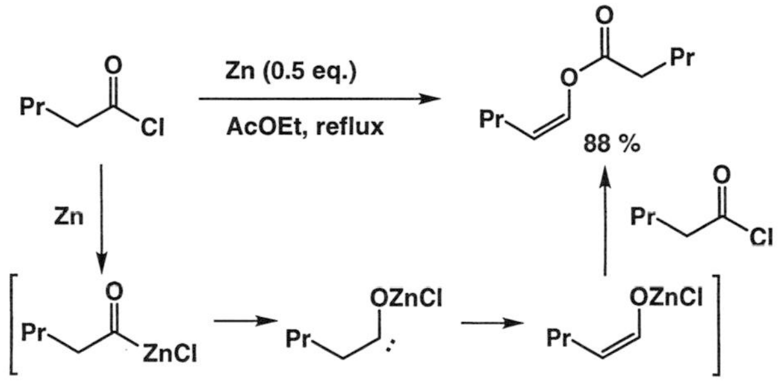

(5.54)

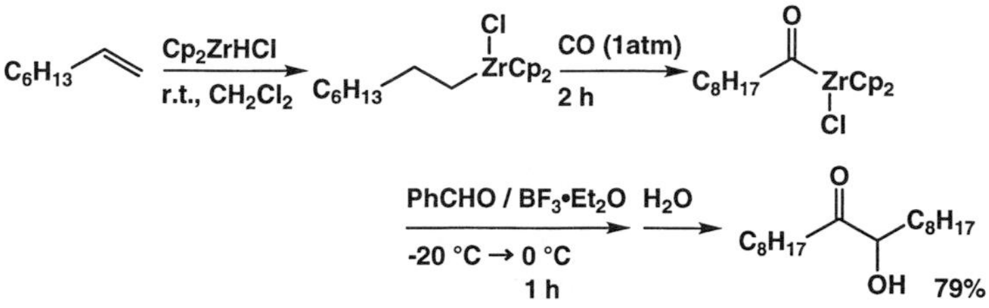

(5.55)

of CO into the acyllithium to give a dicarbonyl species followed by a ketene (Eq. (5.52)) [57]. It was reported that PhZnBr did not react with CO at ambient temperature and pressure [41]. Dialkylzinc, however, readily reacted with CO in the presence of *t*-BuOK to give an acyloin (Eq. (5.53)) [58].

The reaction of acid chlorides with zinc metal in an ethereal solvent usually resulted in esters arising from the acyl group and a portion of the solvent molecules [59, 60]. Later, Normant reported the formation of an enol ester and proposed the formation of an acylzinc followed by a 1,2-hydrogen rearrangement in zinc-oxy carbenoids (Eq. (5.54)) [61].

It is a well-established fact that acylzirconium compounds obtained by the hydrozirconation of alkenes or alkynes, followed by the carbonylation of the resulting intermediate, can be used for the synthesis of aldehydes or carboxylic acid derivatives [62]. Recently, it has been found that this intermediate is capable of re-

$$\mathrm{ArLi + Ni(CO)_4 \xrightarrow[\text{-78 °C, 3 h}]{Et_2O} \xrightarrow{H^+} ArC(=O)CH(OH)Ar} \qquad \text{Ar = phenyl 71 \%} \tag{5.56}$$

$$\mathrm{BuLi + Ni(CO)_4 \xrightarrow[\text{Et}_2\text{O, -78 °C, 20 h}]{PhCH{=}CHC(=O)CH_3} \xrightarrow{H^+} BuC(=O)CH(Ph)CH_2C(=O)CH_3} \qquad 92\ \% \tag{5.57}$$

$$\mathrm{Na_2Fe(CO)_4 \xrightarrow{RX\,/\,PPh_3} [RC(=O)Fe(CO)_3(PPh_3)]^-} \tag{5.58}$$

$$\mathrm{Na_2Fe(CO)_4 \xrightarrow{RX\,/\,CO \text{ or } RC(=O)Cl} [RC(=O)Fe(CO)_4]^-} \tag{5.59}$$

$$\mathrm{RC(=O)Cl \xrightarrow{2\,SmI_2} [RC(=O)SmI_2] \xrightarrow{RC(=O)Cl} RC(=O)C(=O)R} \tag{5.60}$$

$$\mathrm{[RC(=O)SmI_2] \xrightarrow{R'C(=O)R''} RC(=O){-}C(OH)R'R''} \tag{5.61}$$

acting with aldehydes in the presence of a Lewis acid such as $BF_3{\cdot}Et_2O$ (Eq. (5.55)) [63]. This reaction may be regarded as a direct nucleophilic acylation.

Ryang and Tsutsumi at Osaka University obtained an acyloin from alkyllithium and $Ni(CO)_4$ (Eq. (5.56)) [64]. Later, Corey and Hegedus applied this observation to an acyl group transfer reaction using nickel carbonyl (Eq. (5.57)) [65] and a cobalt carbonyl derivative [66].

Based on the observation made by Ryang and Tsutsumi [67] and also by Watanabe at Kyoto [68], Collman and others have developed a series of synthetic reactions via anionic acyliron complexes (Eqs. (5.58) and (5.59)) [69–71].

Acylsamarium species are believed to be formed as intermediates in the reaction of acyl chlorides with SmI_2, leading to diketones (Eq. (5.60)) or acyloins (Eq. (5.61)) [72]. Similarly, acylytterbium appears to be formed as an intermediate in the acylation reaction shown in Eq. (5.62)) [73]. It is also interesting that some lanthanoid and actinide complexes undergo nearly identical reactions to those observed in acyllithium chemistry. A trimethylsilylmethylthorium complex reacts with CO to give a thorium enolate of acylsilane (Eq. (5.63)) [74]. The reaction pattern is identical with that of the organolithium compound shown in Eq. (5.24).

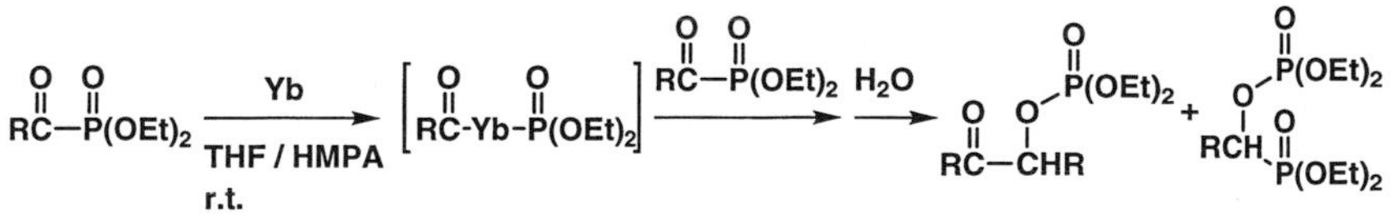

(5.62)

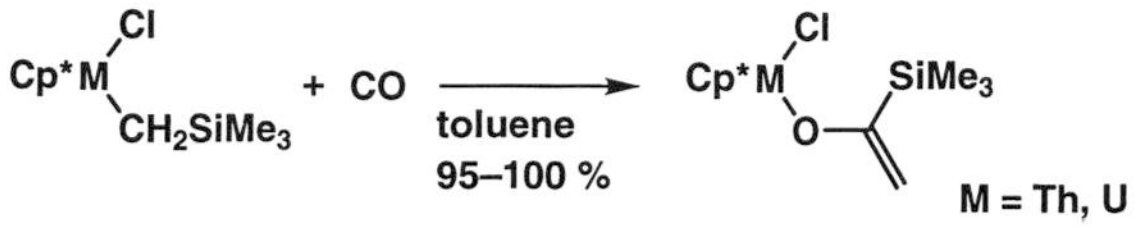

(5.63)

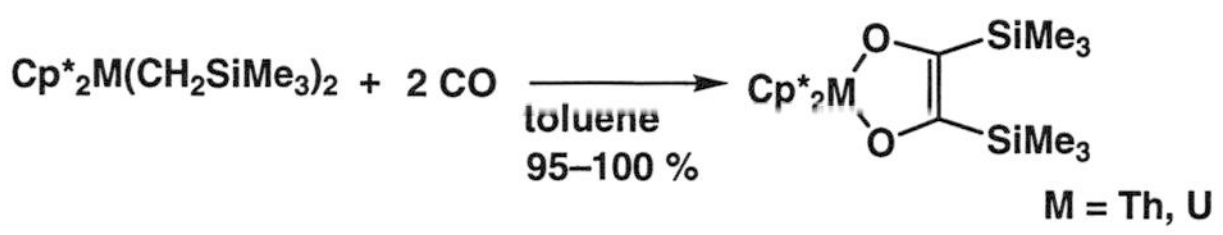

(5.64)

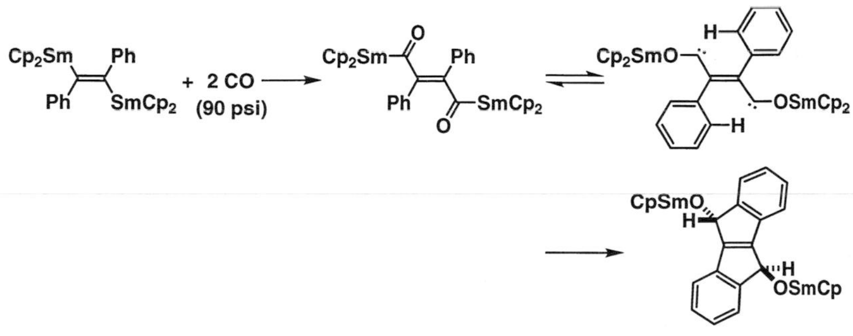

(5.65)

Bis(trimethylsilylmethyl)thorium gave an enediol compound (Eq. (5.64)) [74], which is again similar to the one observed for the lithium counterpart as shown in Eq. (5.23). Another example of the similarity of the acyllithium to the lanthanide can be seen in the reaction of a vinylsamarium complex with CO (Eq. (5.65)) [75]. Formally, the reaction (Eq. (5.65)) is almost identical to the reaction shown in Eq. (5.30) and those in Scheme 5-3 for a vinyllithium. The mechanisms suggested for each reaction are different, but none of these have yet been studied further in detail.

5.6 Structure of Acyllithium

The structure of acyllithium has not yet been established experimentally. A priori, three types of structures can be envisioned. These are acyllithium (or η^1-acyllithium), η^2-acyllithium and lithioxy carbene. For the case of lithioxy carbene, the linear (singlet) structure and the bent (triplet) are both possible. A theoretical study of the structure [76] indicates that the η^2-structure is more stable. Relationships between the reactivity of the acyllithium and its structure, especially that in solution, are interesting and deserve future study.

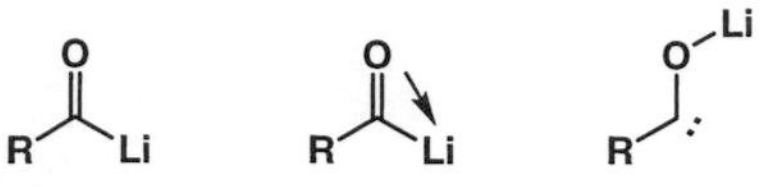

5.7 Carbamoyllithium

Carbamoyllithium is a nitrogen analog of acyllithium and a broadly defined carbonyl anion which has been the subject of extensive study.

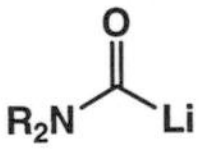

One of the characteristics of these species is that they are more stable than acyllithium. Examples, which are shown in Eqs. (5.66) to (5.70) [77–82] illustrate interesting features of carbamoyllithium. A detailed discussion, however, is beyond the scope of this review.

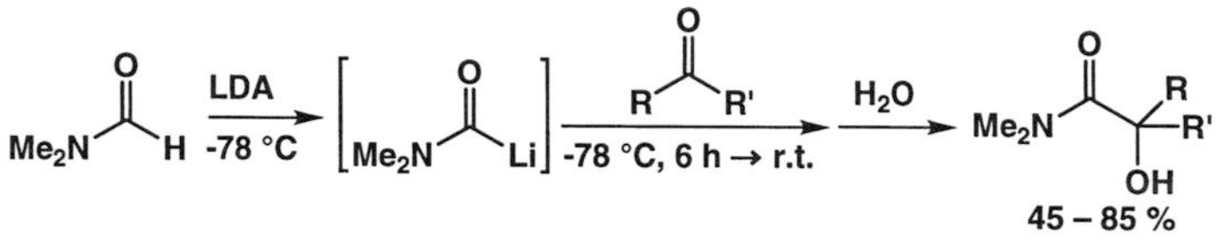

(5.66)

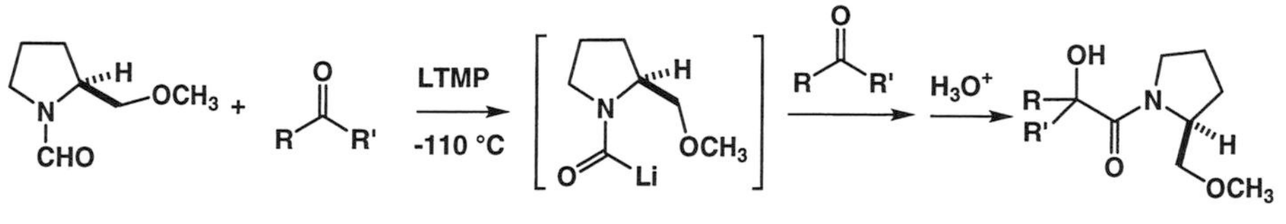

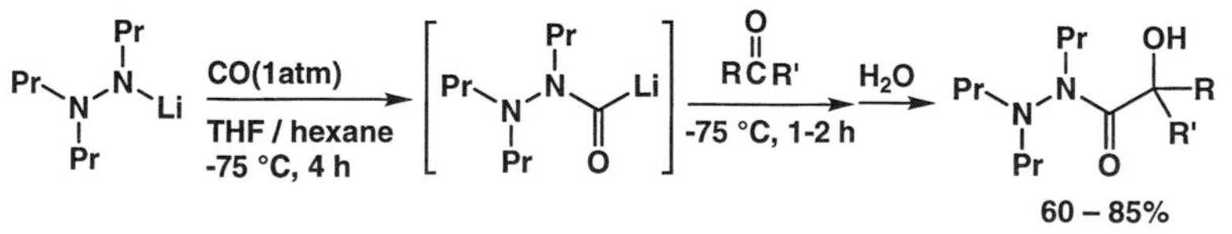

(5.67)

$$Et_2N\text{-}CO\text{-}TeBu \xrightarrow[-Bu_2Te]{BuLi,\ -110\ °C} [Et_2N\text{-}CO\text{-}Li] \xrightarrow[-105\ °C \rightarrow 25\ °C,\ 2\ h]{Ph_2CO} \xrightarrow{H^+} Et_2N\text{-}CO\text{-}C(OH)Ph_2\ \ 91\ \% \tag{5.68}$$

$$Ar\text{-}N(Li)SiR_3 \xrightarrow[20\ °C,\ 8\ h]{CO\ (1\ atm)} [Ar\text{-}N(SiR_3)\text{-}CO\text{-}Li \longrightarrow Ar\text{-}N(Li)\text{-}CO\text{-}SiR_3] \xrightarrow{MeI} Ar\text{-}N(Me)\text{-}CO\text{-}SiR_3\ \ 17-40\ \% \tag{5.69}$$

5.8 Conclusion

Acyllithium derivatives represent exotic molecules. In contrast to the acyl cation or acylinium ion, which is a well-documented species, acyllithium (or acylmagnesium etc.) derivatives are not discussed in depth in books on organic chemistry. As seen above, the rich chemistry of acyllithium is now available, and further development of the chemistry of this quite reactive and unique species is expected in the future.

We wish to acknowledge our able co-workers whose names are shown in the references. Without their collaboration this chapter would not have been possible.

References

1. For reviews see: Narayana, C.; Periasamy, M. *Synthesis* **1985**, 253. Nájera, C.; Yus, M. *Org. Prep. Proced. Int.* **1995**, *27*, 383. Warner, P. In *Comprehensive Organic Functional Group Transformations*; Katritzky, A.R.; Meth-Cohn, O.; Rees, C.W. Eds.; Pergamon Press, Oxford, **1995**; Vol. 5, pp. 435.
2. Wittig, G. *Angew. Chem.* **1940**, *53*, 241, footnote 58.
3. Ryang, M.; Tsutsumi, S. *J. Chem. Soc. Japan, Pure Chem. Soc. (Nippon Kagaku Zassi)* **1961**, *82*, 880; *Chem. Abstr.* **1963**, *58*, 12587a.
4. Ryang, M.; Tsutsumi, S. *Bull. Chem. Soc. Jpn.* **1962**, *35*, 1121; *Chem. Abstr.* **1962**, *57*, 11080g.
5. Jutzi, P.; Schröder, F.-W. *J. Organomet. Chem.* **1970**, *24*, 1.
6. Trzupek, L.S.; Newirth, T.L.; Kelly, E.G.; Nudelman, N.S.; Whitesides, G.M. *J. Am. Chem. Soc.* **1973**, *95*, 8118.
7. Nudelman, N.S.; Vitale, A.A. *Org. Prep. Proced.* **1981**, *13*, 144.
8. Nudelman, N.S.; Vitale, A.A. *J. Org. Chem.* **1981**, *46*, 4625.
9. Vitale, A.A.; Doctorovich, F.; Nudelman, N.S. *J. Organomet. Chem.* **1987**, *332*, 9.
10. Seyferth, D.; Weinstein, R.M. *J. Am. Chem. Soc.* **1982**, *104*, 5534.
11. Shiner, C.S.; Berks, A.H.; Fisher, A.M. *J. Am. Chem. Soc.* **1988**, *110*, 957.
12. Hiiro, T.; Morita, Y.; Inoue, T.; Kambe, N.; Ogawa, A.; Ryu, I.; Sonoda, N. *J. Am. Chem. Soc.* **1990**, *112*, 455.
13. Hiiro, T.; Mogami, T.; Kambe, N.; Fujiwara, S.-I.; Sonoda, N. *Synth. Commun.* **1990**, *20*, 703.
14. Kambe, N.; Inoue, T.; Sonoda, N. *Org. Synth.* **1993**, *42*, 154.
15. Seyferth, D.; Weinstein, R.M.; Wang, W.-L. *J. Org. Chem.* **1983**, *48*, 1146.

6 π-Facial Selectivity in Reaction of Carbonyls: A Computational Approach

James M. Coxon [1] and Richard T. Luibrand

6.1 Introduction

It is the aim of organic and computational chemists to be able to predict reaction rates for the various pathways and hence regio- and stereochemical outcomes. Organic chemists prefer simple 'arrow pushing' explanations, but have not yet found a universal rule, hence their growing respect for computational investigation. Computational chemists seek to quantify reaction rates through detailed knowledge of the shape of a potential energy surface and ultimately by understanding molecular dynamics. Organic chemists want to visualise and encapsulate this information in simple rules [1]. This chapter tells of this developing story for facial selection and is illustrated by discussion of a few selected examples of addition reactions to carbonyls.

The calculation of transition states and activation energies is now possible for even complex systems. High-level computational methods [2], facilitated by fast computers, and improvements in algorithms now give activation barriers which approximate to experimental values for reactions where the mechanism is known. However, when multiple conformations are possible the problem becomes expensive and time consuming. Inclusion of solvent in calculations is costly and complex and has only been possible in recent years [3].

An understanding of facial selection in aldehyde and ketone chemistry, the topic of this chapter, requires a knowledge of the reacting species and the mechanism for reaction. Addition to carbonyls can occur by a formally non-allowed $[_{\pi}2+_{\sigma}2]$ pathway or by an electrophile- or nucleophile-driven mechanism (Fig. 6-1) [4].

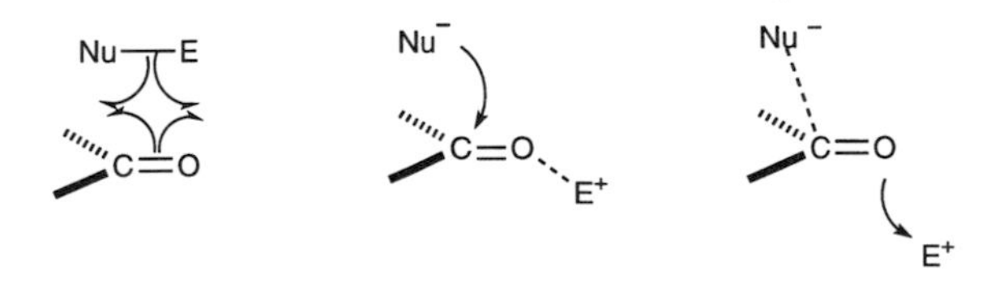

Figure 6-1. Addition to a carbonyl by a non allowed $[_{\pi}2+_{\sigma}2]$ concerted mechanism or by an electrophile- or nucleophile-driven mechanism.

[1] Dedicated to the memory of the late Professor David N. Kirk whose fascination with steroid reaction mechanisms and stereochemistry caught my imagination. I wish to thank D. Quentin McDonald, Rich Luibrand, Kendall Houk, and Alan P. Marchand. I acknowledge the Marsden Fund for research grants.

The energy difference of the HOMO of the nucleophile and the π^* LUMO of the carbonyl compared to the LUMO of the electrophile and the HOMO n-orbital of the carbonyl will be a factor in establishing whether the reaction is electrophile or nucleophile driven. In the case of a reaction catalysed by acid the reaction is considered to be electrophile driven and attack of the nucleophile occurs to the protonated carbonyl. A carbonyl, coordinated with a Lewis acid or cation (e.g. H^+, Li^+, Na^+, AlH_3 [5–7]) or uncoordinated, can be attacked by a neutral or anionic nucleophile. In the former case the nucleophile must bear an acidic hydrogen to allow for the formation of a neutral product [4]. Since reduction of aldehydes and ketones is exothermic the Hammond postulate dictates that the transition state is closer in energy and structure to the reactants than to the products.

Any explanation of facial selectivity must account for the diastereoselection observed in reactions of acyclic aldehydes and ketones and high stereochemical preference for axial attack in the reduction of sterically unhindered cyclohexanones along with observed substituent effects. A consideration of each will follow. Many theories have been proposed [8, 9] to account for experimental observations, but only a few have survived detailed scrutiny. In recent years the application of computational methods has increased our understanding of selectivity and can often allow reasonable predictions to be made even in complex systems. Experimental studies of anionic nucleophilic addition to carbonyl groups in the gas phase [10], however, show that this proceeds without an activation barrier. In fact Dewar [11] suggested that all reactions of anions with neutral species will proceed without activation in the gas phase. The "transition states" for reactions such as hydride addition to carbonyl compounds cannot therefore be modelled by gas phase procedures. In solution, desolvation of the anion is considered to account for the experimentally observed barrier to reaction.

6.2 Reduction in Acyclic Systems

One of the first and most satisfactory methods of predicting the most favoured diastereoisomeric transition state for nucleophilic attack at acyclic aldehydes and ketones was proposed by Cram [12] and is based on the size of the substituents to the carbonyl. He considered that the preferred conformation from which reaction occurs has the largest adjacent group antiperiplanar to the carbonyl (Fig. 6-2), and the nucleophile attacks from the side with the smaller (S) group.

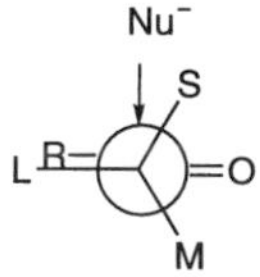

Figure 6-2. The Cram model for nucleophilic attack at acyclic carbonyl compounds.

When one of the substituents is a highly polar group (e.g. halogen) Cornforth [13] suggested that the polar group and oxygen atom stay as far apart as possible in order to minimise dipolar interactions: for example reduction of 2-chloro-1-deutero-2,3,3-trimethylbutanal with lithium di-butoxy aluminium hydride is consistent with this model (Fig. 6-3) [14].

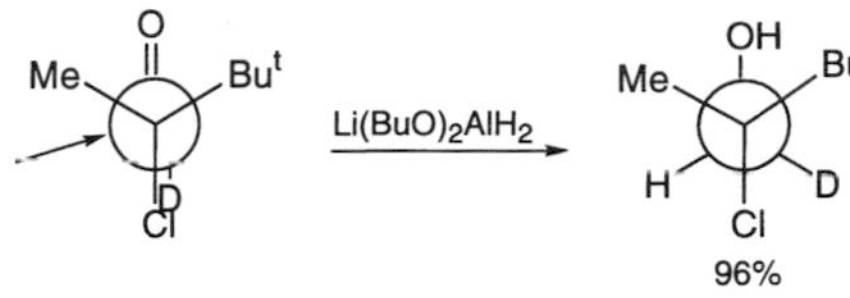

Figure 6-3. The Cornforth model for nucleophilic attack at acyclic carbonyl compounds containing a polar group.

Karabatsos [15] suggested that the incoming nucleophile approaches the carbonyl from the face with smallest group, but in contrast to Cram argued that the preferred conformation from which reaction occurs has the medium-sized substituent (M) eclipsed with the carbonyl (Fig. 6-4).

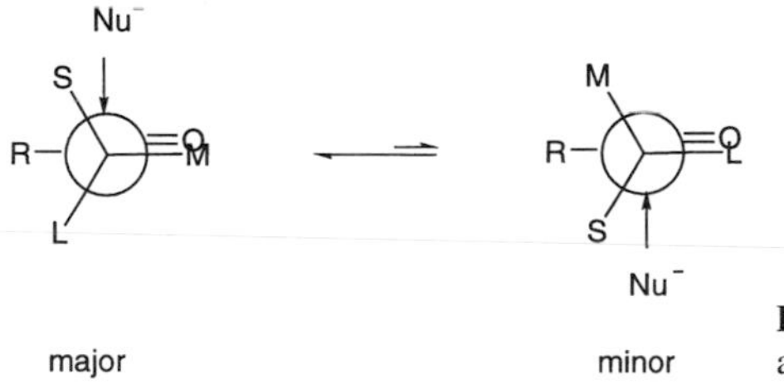

Figure 6-4. The Karabatsos model for nucleophilic attack at acyclic carbonyls.

Neither the Cram nor the Karabatsos model accounts for the effect of varying size of the carbonyl R group on the selectivity of reduction of acyclic ketones. To account for this Felkin [16] proposed that nucleophilic attack would proceed so as to minimise torsional strain in the transition state. He argued that the largest group, L, would be perpendicular to the carbonyl, and addition would occur *trans* to this group. By assuming that the interactions of the Small and Medium groups are greater with R than O (Fig. 6-5), the most favoured transition conformation was considered to have the Medium group positioned near the carbonyl oxygen. Early *ab initio* calculations (STO-3G) by Anh and Eisenstein [17] supported Felkin's proposal that transition states are favoured when nucleophilic attack occurs from an orientation antiperiplanar to an adjacent σ-bonded group. In this way the transition state is staggered and the Large group is *anti* to the incoming nucleophile.

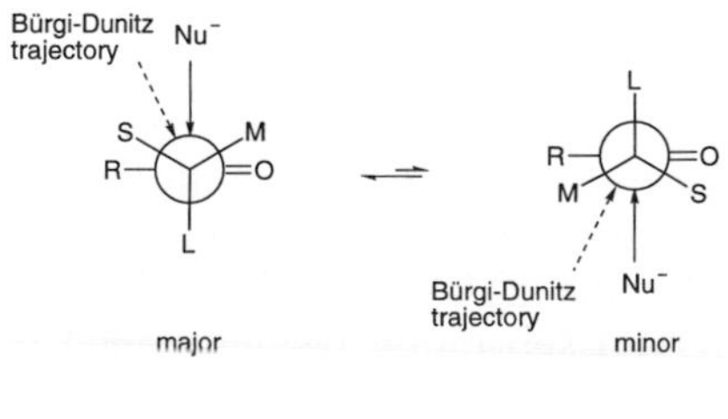

Figure 6-5. The Felkin model for nucleophilic attack at acyclic carbonyls.

Anh made a second contribution defining the importance of the Bürgi-Dunitz trajectory (an Nu-C-O angle of 105±5°) of nucleophilic attack at a carbonyl [18]. For the two possible conformations having the Large group perpendicular to the carbonyl the nucleophile will approach past the Small group rather than the Medium group (Fig. 6-5). This explanation does not require greater interactions on the Small and Medium groups with R than O even when R=H, as implicit in Felkin's proposal. Anh's theoretical study combined with Felkin's model has become known as the Felkin-Anh model. In the absence of overriding steric effects, if L is an electronegative substituent (e.g. Cl) with a low lying σ^*-orbital, the preferred face from which nucleophilic attack occurs is *anti* to the Cl, which will prefer to be orthogonal to the carbonyl at the transition state; the transition state is staggered and there is a favourable secondary orbital interaction of the nucleophile and the adjacent antiperiplanar σ-bond. By contrast, attack *syn* to the σ^*-orbital will result in eclipsing torsional interactions and unfavourable secondary orbital interactions between the nucleophile and σ^*-orbital. An antibonding secondary orbital and eclipsing interaction disfavours *syn*-periplanar attack. The most favourable transition state has the substituent with the lowest energy σ^*-orbital aligned antiperiplanar to the nucleophile and in the plane of the π^*-orbital. Application of the rule requires a knowledge of the energy of the σ^*-orbitals, often inversely related to the size of a group: Large<Medium<Small. The σ^*-orbital is stabilised by proximate electron-withdrawing groups. For the reaction of a series of acyclic aldehydes with lithium enolates [19], Heathcock concluded that the preference for substituents to be *anti* to the incoming nucleophile is in the order: MeO>*t*-Bu>Ph>*i*-Pr>Et>Me>H, a sequence determined by a balance of the σ^*-orbital energies and steric effects.

This Felkin-Anh model can be summarised as follows: (i) torsional strain is minimised in the transition state, (ii) the reaction occurs from a conformation with the Large group orthogonal to the carbonyl, and (iii) the trajectory of the nucleophile is *anti* to the Large group and past the Small group (Figs. 6-5 and 6-6). The second feature of this proposal can be rationalised by frontier orbital theory [20], which focuses on the stabilising HOMO/LUMO interactions between reactants which contribute most to the offset of torsional, steric and bending components of transition state energy.

Because of the principle of microscopic reversibility it is appropriate to consider frontier orbital analysis of the reaction in either direction. The Hammond postulate dictates that the more exothermic the reaction the more the transition state will reflect the starting geometry, and frontier orbital analysis of reactant orbitals is expected to be a better predictor of relative transition state orbital interactions than for an endothermic or a less exothermic process. Conversely, frontier orbital analysis of product orbitals in exothermic reactions would be a poorer predictor of transition state energy.

Interaction of the carbonyl π^*-orbital with an adjacent periplanar σ^*-orbital lowers the energy of the LUMO. This reduces the energy difference with the nucleophile HOMO (Fig. 6-6). The closer in energy that the π^*_{CO} is with an adjacent C–L σ^*-orbital the greater the lowering ($\Delta E'$) of the LUMO and the more favoured the reaction will be, i.e. the LUMO-HOMO separation becomes smaller.

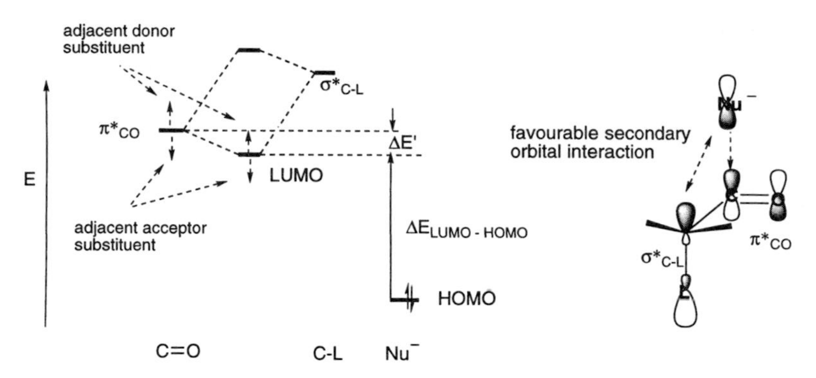

Figure 6-6. Interaction of the antibonding carbonyl orbital with an adjacent unoccupied σ^*-orbital.

Calculations have in fact shown that the energy of the LUMO is sensitive to the conformation about the C_α–C bond in an acyclic ketone [8]. In the absence of overriding steric considerations the lowest energy transition state will have the bond with the lowest lying σ^*-orbital orientated perpendicular to the plane of the carbonyl. The diastereoisomer favoured will result from attack from the opposite face of the carbonyl to the σ-bond with the lowest lying σ^*-orbital and involves a staggered transition state. Acceptor substituents increase reactivity consistent with the transition state acquiring negative charge in a reaction where the rate-determining step is addition of nucleophile [21]. Donor substituents on the carbonyl raise the energy of the π^*_{CO} orbital and reduce reactivity. Torsional interactions are minimised by reaction *anti* to the C–L bond, while *syn* attack results in eclipsing torsional interactions and is less favoured.

6.3 Reduction of Cyclohexanones

Because of the number of conformations that need to be considered for acyclic systems, cyclohexanones are somewhat simpler for analysis. However, even for these systems the situation is not easily amenable to isolating specific components of selectivity. Several explanations have been proposed over the years to account for the preference of axial attack of cyclohexanones by sterically unhindered nucleophiles ($LiAlH_4$, $NaBH_4$, AlH_3) [9]. Equatorial attack is favoured for sterically hindered cyclohexanones or reducing agents (Fig. 6-7).

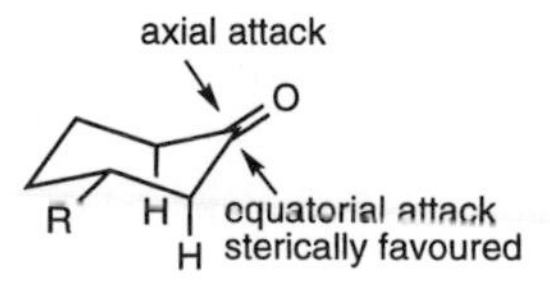

Figure 6-7. Axial and equatorial nucleophilic attack at cyclohexanone.

The first explanation for the preference for axial attack by hydride in conformationally rigid sterically unhindered cyclohexanones became known as "product development control" and was suggested to reflect a "late" transition state [22]. For hindered ketones steric interference with the nucleophile was considered to favour equatorial attack and this became known as "steric approach control" caused by an "early" transition state. Hudec [23] proposed that the preferred direction of approach to a carbonyl group is controlled by deviations in the angle by which the axis of the π^*-orbital of the carbonyl carbon atom is twisted thereby making the faces of the carbonyl diastereotopic.

The Felkin-Anh model has also been used to explain the preference for axial attack by nucleophiles on cyclohexanones and the effect of proximate substituents on facial selection. The *anti* periplanar geometry that Anh regarded as important in nucleophilic attack of carbonyl compounds is compromised by torsional strain in the reactions of cyclohexanones from the equatorial face. Felkin stated: "Whereas both torsional strain and steric strain can be simultaneously minimised in a reactant-like transition state when the substrate is acyclic... this is not possible in the cyclohexanone case. ...These reactions all proceed *via* reactant-like transition states. In the absence of polar effects, their steric outcome is determined by the relative magnitude of torsional strain and steric strain [in the axial and equatorial transition states]" [16].

An early approach to understanding the influence of electronic effects on π-facial diastereoselection was by Fukui [24] and focused on the analysis of ground-state properties of the reactants. The faces of the *p*-system were considered to be differentiated by mixing of *p*- and σ-orbitals resulting in a perturbation of the ground state HOMO and became known as non-equivalent orbital extension [25]. Klein [26] had earlier suggested that the LUMO of the carbonyl would be facially dissymmetric and influence facial selectivity, because cyclohexanone hyperconjugation of the π^*-orbital of the carbonyl with the adjacent axial C–H σ-bonds (shown in the Fig. 6-8 for only one C–H) results in non-equivalent orbital extension of the LUMO [27] shown simplistically in Fig. 6-8. This extension is regarded by Klein to favour axial addition of nucleophiles.

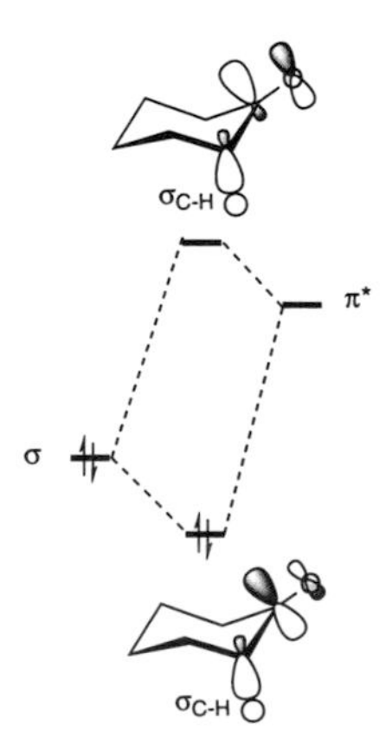

Figure 6-8. Interactions with an adjacent π-orbital causes uneven orbital extension of the LUMO.

Ab initio calculations [28] do show that orbital extension toward the axial side occurs not only for the LUMO but also for the HOMO [27]. This leads to extension of the orbital in the direction *trans* to the allylic bonds. In cyclohexanone, the ring distortion causes the C–H bond to be more eclipsed with the π^*-orbitals (O–C–C–H dihedral angle 115°) than the C–C bond (O–C–C–C dihedral angle 127°), resulting in orbital distortion in the axial direction. This model is consistent with the preference for axial nucleophilic attack with 1,3-dioxan-5-one derivatives (a preference even higher than for cyclohexanones) and equatorial attack for 1,3-dithian-5-one. In 1,3-dioxan-5-one the axial C–H bonds are in even better alignment with the carbonyl π^*-orbital than in cyclohexanone (O–C–C–H dihedral angle 80°). In 1,3-dithian-5-one ring distortions cause the C–S bonds to be more periplanar with the π orbital, causing equatorial orbital extension, and equatorial addition of hydride is favoured. This orbital extension could therefore be a factor to rationalise the known preference for axial attack observed in electrophilic attack of methylenecyclohexane.

Dannenberg [29] has attempted to provide a frontier orbital explanation for facial selectivity by a polarized π frontier orbital method. The new (polarized) orbitals are schematically represented as the combination of a p-orbital with two s-functions and have different coefficients associated with each face of each π center.

Cieplak [30], in a widely quoted qualitative approach [4–7], proposed that addition reactions to facially dissymmetric carbonyl systems are biased from steric and classical torsional control [31] by electron delocalisation of proximate-group electrons into the σ^*-orbital of the incipient (developing) carbon-nucleophile bond, i.e. the $\sigma^{*\ddagger}$-orbital. His hypothesis differs from the normal application of frontier orbital theory, which concerns the reactant or product orbitals, in an attempt to prejudge relative transition-state energies and does not consider interactions of substituents with specific transition-state orbitals. The Cieplak claim is that the controlling interaction in reactions "in cyclohexane-based systems" is "based on the concept of transition-state stabilization by electron donation into the vacant $\sigma^{*\ddagger}$ orbital associated with the incipient bond", nucleophilic attack being favoured *anti* to the better electron donating σ-bond. Cieplak [30] states: "During the axial attack of a reagent, the vacant orbital $\sigma^{*\ddagger}$ that develops along with the formation of the incipient bond interacts with the filled orbitals of the C(2)–H and C(6)–H bonds. During the equatorial attack, the $\sigma^{*\ddagger}$ orbital interacts with the filled orbitals of the ring bonds C(2)–C(3) and C(5)–C(6). The effect of steric hindrance is to favour the equatorial attack. The effect of the hyperconjugative σ assistance, however, favours the axial attack because the C–H bonds are better donors than the C–C bonds, and consequently the σ_{C-H}, $\sigma^{*\ddagger}$ stabilization energy is greater than the σ_{C-C}, $\sigma^{*\ddagger}$ stabilization". This result is illustrated diagrammatically in Fig. 6-9.

The generally accepted order of increasing σ-donor ability is $\sigma_{CO}<\sigma_{CN}<\sigma_{CCl}<\sigma_{CC}<\sigma_{CH}<\sigma_{CS}<\sigma_{CSi}$. However, sulfur- and silicon-containing substituents are better able to stabilise an *anti* transition state than an antiperiplanar C–C or C–H bond. The Cieplak argument is supported by the stereochemistry of reduction of C3-substituted cyclohexanones. An electron-withdrawing group at C3 en-

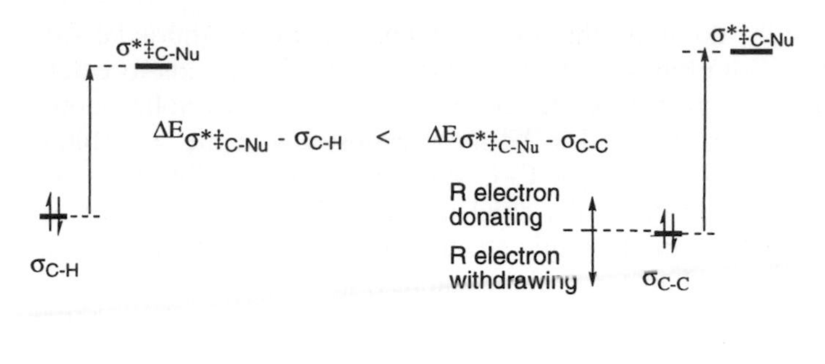

Figure 6-9. Cieplak model for axial and equatorial attack; a clear preference for axial addition is indicated [32].

hances axial attack; equatorial attack being disfavoured by a reduction in the electron-donating ability of the C2–C3 bond to the forming σ^*-orbital. On the other hand, when R is electron donating the antiperiplanar C2–C3 bond is better able to donate electrons to the σ^*-orbital of the forming equatorial carbon-nucleophile bond, and equatorial attack is promoted. The effect of donating and withdrawing substituents on the σ-C–C bond will influence the $\Delta E_{\text{LUMO-HOMO}}$, and this is also shown in Figure 6–9.

This theory is simple to apply and accounts for many experimental observations, with the proviso that there are likely to be situations where steric effects predominate. The Cieplak hypothesis is also supported by le Noble's work on 5-substituted adamantan-2-one; attack of nucleophiles occurs at the *syn* face if the 5-substituent is an electron-withdrawing group [33] and at the *anti* face if it is a donor (see later section) [34]. A serious problem with the Cieplak analysis, which is relevant to all systems where this hypothesis has been applied, concerns the large energy difference between the adjacent σ-orbital(s) and the $\sigma^{*\ddagger}$ orbital of the transition state, which will result in a minimum of orbital mixing and stabilisation/destabilisation, and this is discussed in more detail later [35–38].

We have recently argued [39] that facial selectivity in the Diels-Alder reactions of 5-substituted cyclopenta-1,3-dienes is a reflection of hyperconjugative effects; a frontier orbital analysis is shown in Fig. 6-10 [40]. The molecule adopts a conformation where the better electron-donating group (C–Me rather than C–X) hinders approach of the dienophile from that face. Overlap of the C–C σ-bond (a better donor than C–X) increases the energy of the diene ψ^2 HOMO. Furthermore overlap of the C–C σ^*-orbital with the diene LUMO ψ^3 reduces the energy of the

LUMO. These two effects reinforce and favour dienophile attack from the face *anti* to the better electron donor substituent, in this case the C–Me. In summary, the most electron-donating σ-bond (i.e., that which possesses the higher σ-orbital energy and/or the σ^*-orbital nearest in energy to the diene ψ^3 orbital) will block attack from that face, because as the reaction occurs the cyclopentadiene develops an envelope conformation and the electron-donating group (i.e. C–Me in the case discussed) is *pseudo-axial*. The HOMO/LUMO energy difference is less than for attack from the other face *anti* to C–X.

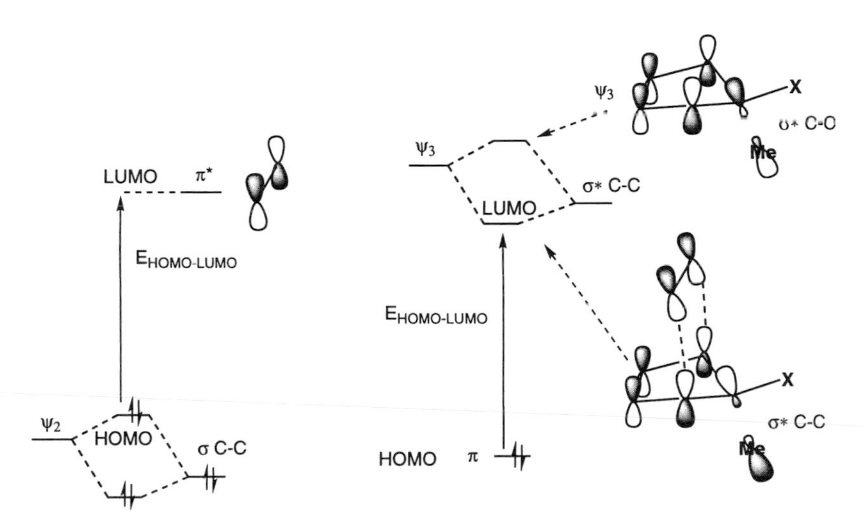

Figure 6-10. Hyperconjugative stabilization of the Diels-Alder reaction of substituted cyclopentadienes with ethylene *anti* to the better electron donor.

For axial and equatorial nucleophilic addition to cyclohexanone, the principle of microscopic reversibility dictates that frontier orbital analysis can be considered for addition of the nucleophile to the carbonyl or loss of nucleophile from the product. Since the reaction is considered to be exothermic the frontier orbital interaction that should best represent the transition energy is the orbital interaction of the nucleophile HOMO with the ketone π^* (LUMO) (Fig. 6-11).

It should be noted that Cieplak considers the interaction of antiperiplanar orbitals with the $\sigma^{*\ddagger}$-orbital, the LUMO orbital of the bond undergoing cleavage. A frontier orbital explanation for addition to cyclohexanone requires the energy of the carbonyl π^*-orbital to be facially differentiated by interaction with the adjacent C–C (shown in the Figure 6–11 as lower in energy than the C–H) and C–H bond(s) and to do so preferably without invoking orbital extension. The interaction depends not only on the energy of the C–H and C–C but on the relative orientation with the π^*-orbital. The interactions are therefore difficult to estimate without consideration of stereochemistry. If C–H is a better donor than C–C, and this is not without debate, and also assuming their overlap integrals are comparable, then the C–H raises the extended LUMO on the carbonyl more than for

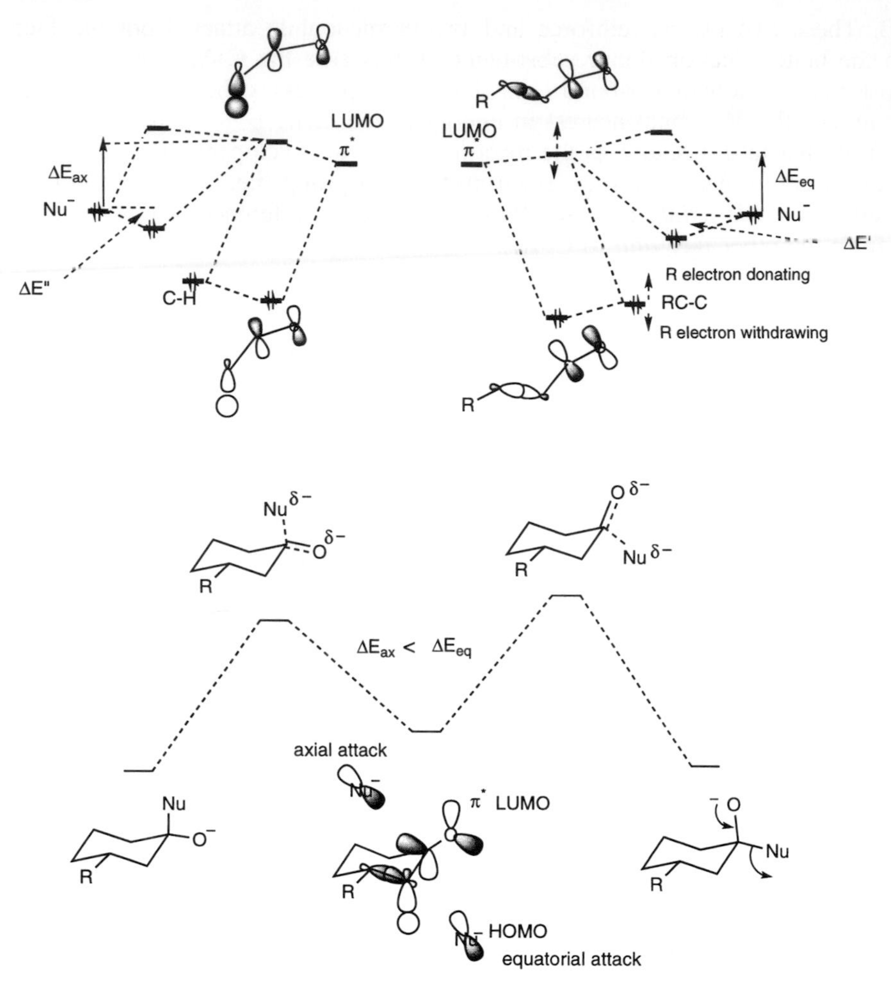

Figure 6-11. (top) Frontier molecular orbitals, (bottom) potential energy surface for axial and equatorial attack at 3-substituted cyclohexanones.

equatorial attack where the LUMO interacts with the lower energy adjacent C–Cs. The HOMO/LUMO energy difference for axial attack is thereby greater than for equatorial attack ($\Delta e_{ax} > \Delta E_{eq}$) (see Fig. 6-11 top), contrary to experiment. The effect of donating and withdrawing groups on an adjacent C–C bond is also shown in the Figure. All in all a frontier orbital analysis is difficult and not convincing, but may require a measure of the interaction energy (i.e. $\Delta E' > \Delta E''$) of the filled orbital formed from mixing the Nu^- HOMO and LUMO.

For the reverse reaction, namely the removal of Nu^-, the important frontier orbital interaction is between the C–Nu σ-orbital (the HOMO) and the C–Nu σ^*-orbital (the LUMO) (Fig. 6-12). A frontier orbital explanation requires a knowledge of how proximate bonds affect the energy of these orbitals. The orientation of these orbitals is relatively well known. It is known that electron-withdrawing R groups favour axial attack and electron-donating R groups equatorial attack.

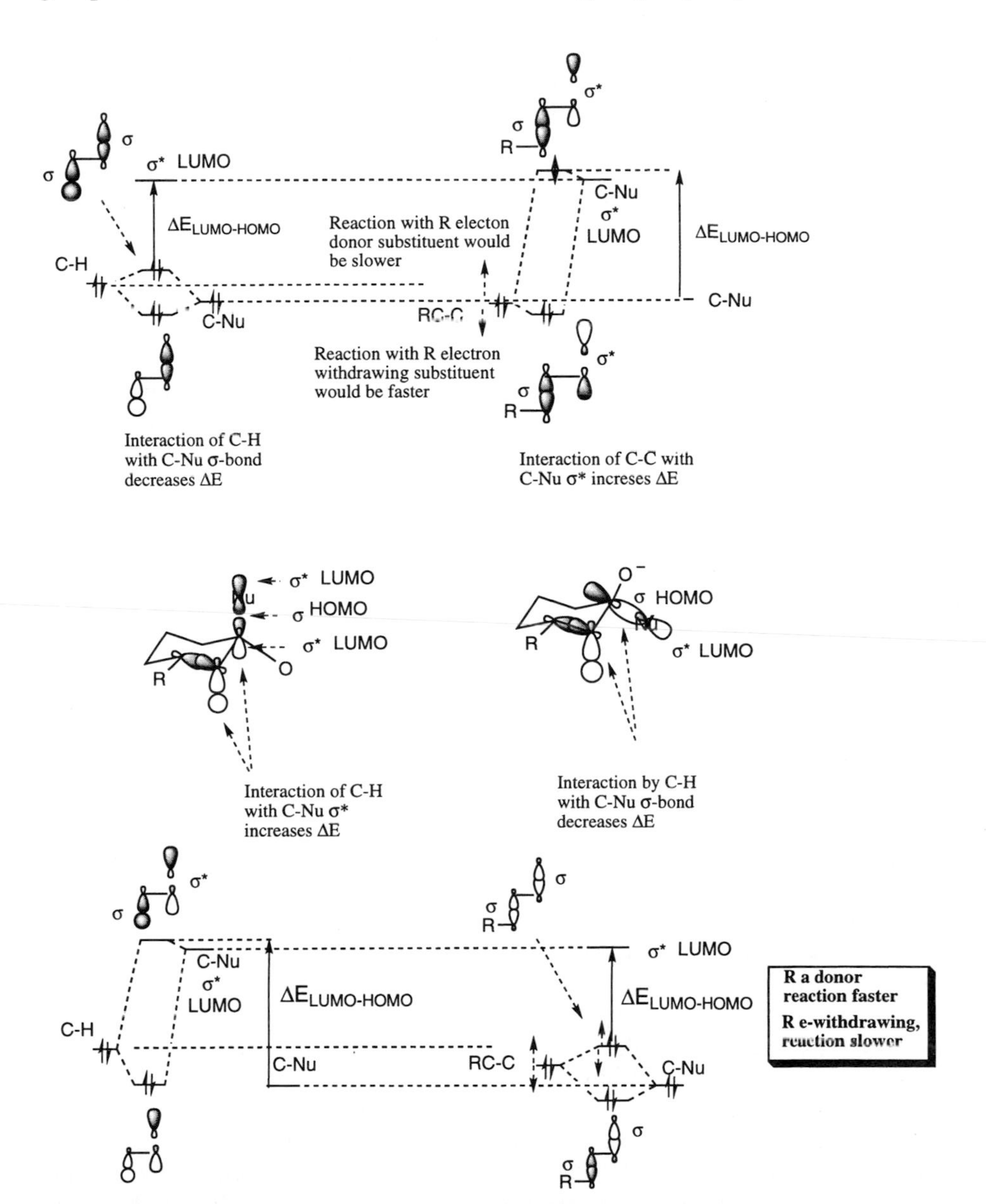

Figure 6-12. The effect of frontier orbital interactions of proximate C–H and RC–C bonds on $\Delta E_{\text{LUMO–HOMO}}$ for nucleophilic axial and equatorial attack at cyclohexanones.

The inherent preference for axial attack (steric, torsional and electronic) is modified by a C3R substituent. A C3 equatorial substituent has little effect on the frontier orbitals on the left hand side of the Figure, exerting an influence by mixing with the equatorial C–Nu σ and σ^* orbitals (right hand side of the Figure). The interaction of adjacent *anti* periplanar C–C bonds (C–C bonds being intermediate between C–Nu and C–H in donor ability) with the C–Nu σ- and σ^* orbitals is shown along with the influence of R substituents on the $\Delta E_{\text{LUMO-HOMO}}$. Because of the large energy difference between the RC–C σ-bond and the C–Nu σ^* orbital the interaction will be small (top right): an R-donor will increase the $\Delta E_{\text{LUMO-HOMO}}$ gap and slow the equatorial reaction contrary to experimental results. However the RC–C interaction with C–Nu σ-bond, and these orbitals are close in energy, decreases $\Delta E_{\text{LUMO-HOMO}}$ and thereby increases the favourableness of equatorial attack. The latter effect (highlighted in bold in the bottom right of Fig. 6-12) is the larger effect and is consistent with experimental values (electron-donating groups favour equatorial attack). The converse argument is true when R is electron-withdrawing: electron-withdrawing groups favour axial attack.

6.4 Mechanism of Carbonyl Reduction with AlH_3

The complex hydrides $LiAlH_4$ and $NaBH_4$ are reagents of choice for reduction of aldehydes and ketones; however, the mechanism of the reactions is not completely understood [9, 41]. $NaBH_4$ reacts rapidly in solution, although gas-phase theoretical calculations indicate significant activation energies. Calculations on the reduction of formaldehyde with borohydride show that the reaction takes place *via* two distinct steps; the first, an endothermic transfer of hydride to give borane and alkoxide ion [(E_a=41 kcal/mole (MNDO))], and the second, B–O bond formation [42]. An *ab initio* study located a product-like four-centre transition state (E_a=36 kcal/mole) for a one-step non-synchronous exothermic addition [43] and to avoid contradiction of the Hammond postulate was interpreted as a superposition of two reactions, initial hydride transfer from BH_4^- to formaldehyde, followed by BH_3 shift to the carbonyl oxygen atom. A concerted four-centre transition state for the $NaBH_4$ reduction of aldehydes and ketones in protic solvents is forbidden by orbital symmetry considerations and is not compatible with experimental evidence [44].

Several theoretical studies have considered LiH addition as a model for the more computationally difficult $LiAlH_4$ and $NaBH_4$ [45]. However, reaction of aldehydes and ketones with LiH seldom if ever leads to reduction [46]. AlH_3, while less commonly used than the complex boron and aluminium hydrides, is useful for reducing carbonyls [47] and therefore is a suitable model for computational study. Calculations [2, 5] show that gas-phase reduction of formaldehyde by AlH_3 occurs by formation of complex **1**, which rearranges *via* a four-centre transition state to form an aluminium methoxide product. Two conformational isomers of the complex have been located as energy minima. The more stable form **1b** [48] (by 1.1 kcal/mole, HF/3-21G*) has the hydrogen on aluminium eclipsing the carbonyl. The complex **1b** is separated from aluminium methoxide product **3** [49] by transition state **2** (activation enthalpies 3-21G* 16.7 kcal/mole; 6-31G* 15.0 kcal/mole; MP2/6-31G*//6-31G* 13.5 kcal/mole) (Fig. 6-13).

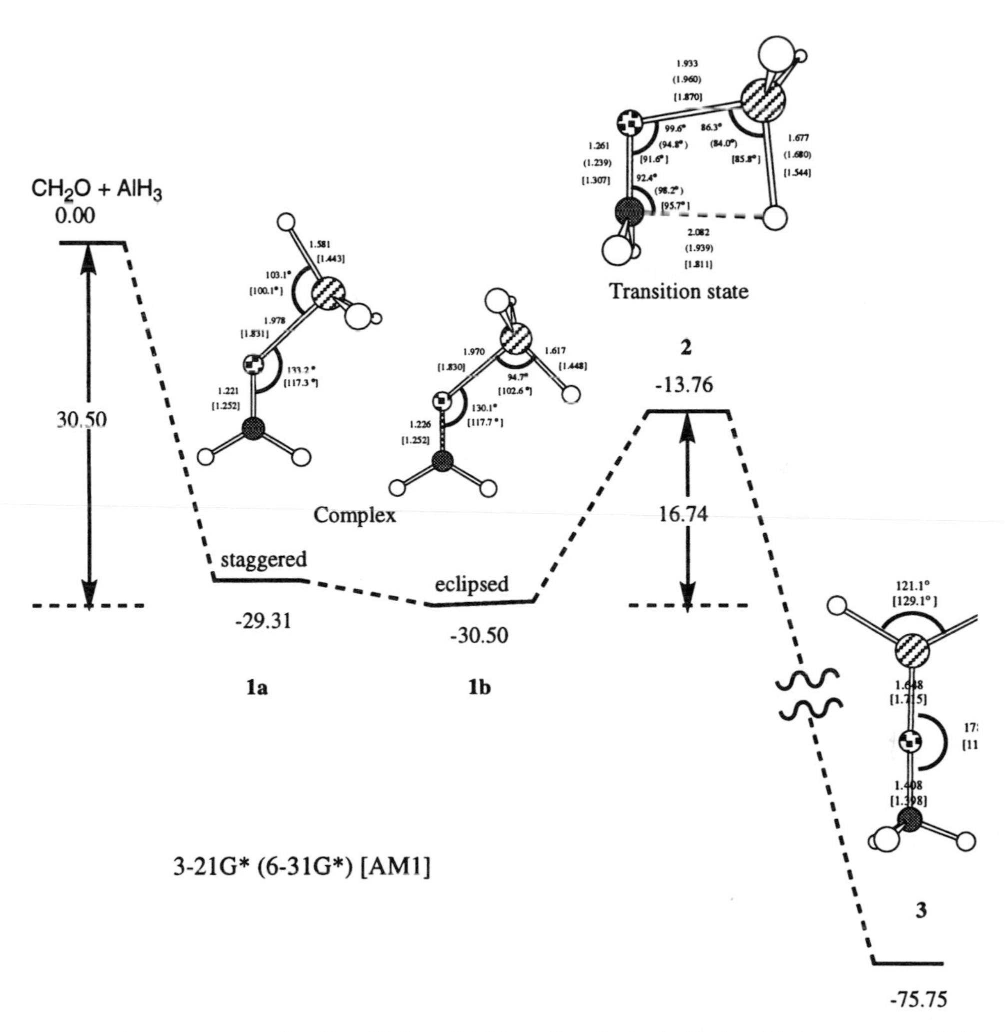

Figure 6-13. Reaction profiles for addition of AlH_3 to formaldehyde (kcal/mole) (structures at HF/3-21G* level).

The rate-determining step for reduction is the formation of the four-centre cyclic transition state **2**. The relatively short C–O and long H–C bond lengths indicate an early transition state consistent with the Hammond postulate. The obtuse H C O angle of attack (92.4°) is greater than the value observed for addition of

LiH to formaldehyde (88.7°, 3-21G), but less than the ideal Bürgi-Dunitz trajectory [18] (105±5°) due to the cyclic nature of the transition state. Geometry optimisation at the 6-31G* level shows an increase in this angle to 98.2°; the corresponding angle in LiH addition [45] is 88.8°. The H_{Nu}–C distance at the transition state is 0.8 Å shorter in the AlH_3 than LiH reaction, suggesting a slightly later transition state with AlH_3. The four-centre transition state has experimental support. Kinetic isotope effects in the AlH_3 reduction of benzophenone favour such a pathway. A Hammett plot for reaction of substituted benzophenones with AlH_3 is consistent with the rate-determining step involving nucleophilic attack at the carbonyl [50].

By removing the AlH_2^+ fragment from the four-centre saddle transition structure for the reaction of AlH_3 with formaldehyde and performing a single point calculation of the organic fragment (and separately AlH_2^+), the molecular orbitals that reflect the molecular orbitals for the reaction of H^- with formaldehyde at such a point in a gas phase reaction are obtained [5, 7]. It is not possible by normal methods to obtain such a structure as a stationary point since reaction of a neutral molecule with a charged species, as discussed earlier, results in a spontaneous reaction without a transition barrier. The orbitals corresponding to the transition orbitals are shown in Fig. 6-14. The node in the HOMO is between the carbon and oxygen. The facility for interaction with the σ*C–Nu and the σ*‡C–Nu orbitals will be dependent on the relative orientation of the donor and will be at a maximum when these orbitals are aligned *syn* or *anti* periplanar.

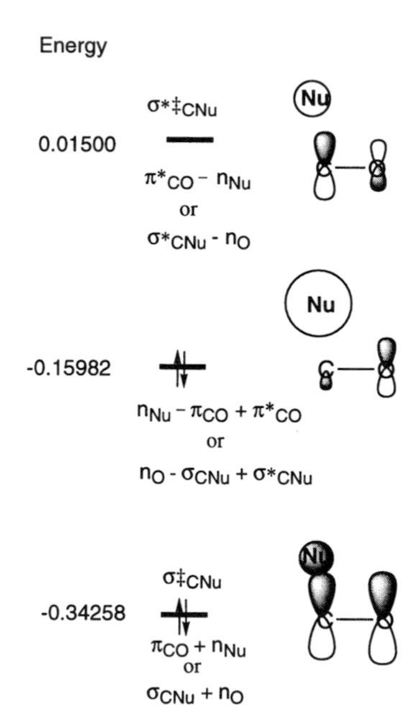

Figure 6-14. Molecular orbitals at the four-centre transition state geometry for AlH_3 addition to formaldehyde but after removal of AlH_2^+ (3-21G*).

6.5 Reduction of Cyclohexanone with AlH_3

The axial and equatorial transition structures for reduction of cyclohexanone [6] with AlH_3 (3-21G*) (Fig. 6-15) show incipient H_{Nu}–C bond distances of 1.891 and 1.889 Å. These are somewhat shorter than for LiH reduction (2.057 and 2.028 Å) [28, 51], indicative of a later transition state for AlH_3.

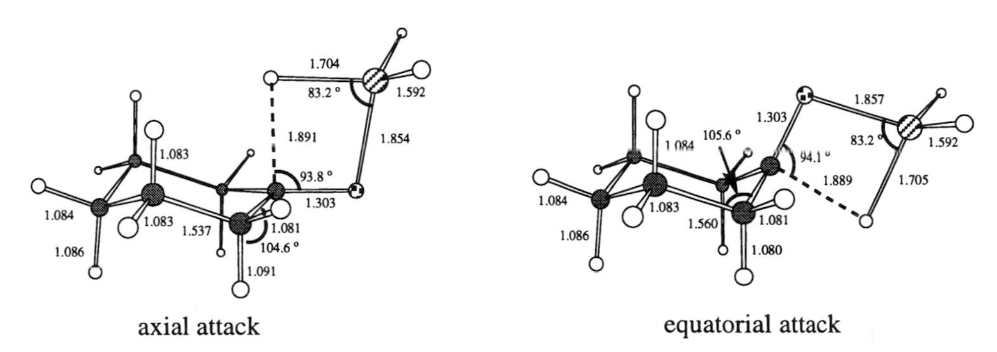

Figure 6-15. Transition states (3-21G*) for axial and equatorial AlH_3 addition to cyclohexanone.

Projections down the C_2–C_1 bond (Fig. 6-16a, b) show an O–C–C–H_{eq} dihedral angle in the axial and equatorial transition states of 34.4° and 30.2°, respectively, consistent with greater torsional strain in the latter, as first suggested by Felkin [52]. The ring is distorted as measured by the angle of the $C_1C_2C_6$–$C_2C_3C_5$ planes (Fig. 6-16c, d) and flattened to 142° at the transition state for axial and puckered to 119° for equatorial attack, compared with a value of 130° in cyclohexanone. The ring $C_6C_1C_2C_3$ dihedral angles (Fig. 6-16a, b) distort from 54° in cyclohexanone to 42° for axial and 65° for equatorial attack. The dihedral angles of the adjacent C–H and C–C bonds with the forming C–H_{Nu} bond are significant since they provide a measure of the ability of these bonds to participate in hyperconjugation. For axial attack an almost perfect antiperiplanar relationship [53] is calculated (H_{Nu}–C–C–H_{ax} dihedral 179°); the H_{Nu}–C–C–C dihedral for equatorial attack is 170° (Fig. 6-16c, d). Ring distortion provides a balance between minimising torsional strain and maximising orbital interactions with the adjacent σ-bonds (C–H for axial and C–C for equatorial attack). The calculated transition states of cyclohexanone reduction by AlH_3 show bond length changes and molecular orbitals consistent with hyperconjugative stabilisation.

Both semiempirical and *ab initio* calculations parallel experimental values for reduction of 4-*t*-butylcyclohexanone (Table 6-1). The transition state for axial attack is favoured over equatorial by 1.1 kcal/mole (3-21G* and AM1). This enthalpy difference can be separated into three components: the cyclohexanone fragment, the AlH_3 fragment, and an interaction enthalpy between these fragments. Their relative contributions are estimated by calculating the enthalpy of each fragment fixed at the transition state geometry. The axial cyclohexanone fragment is

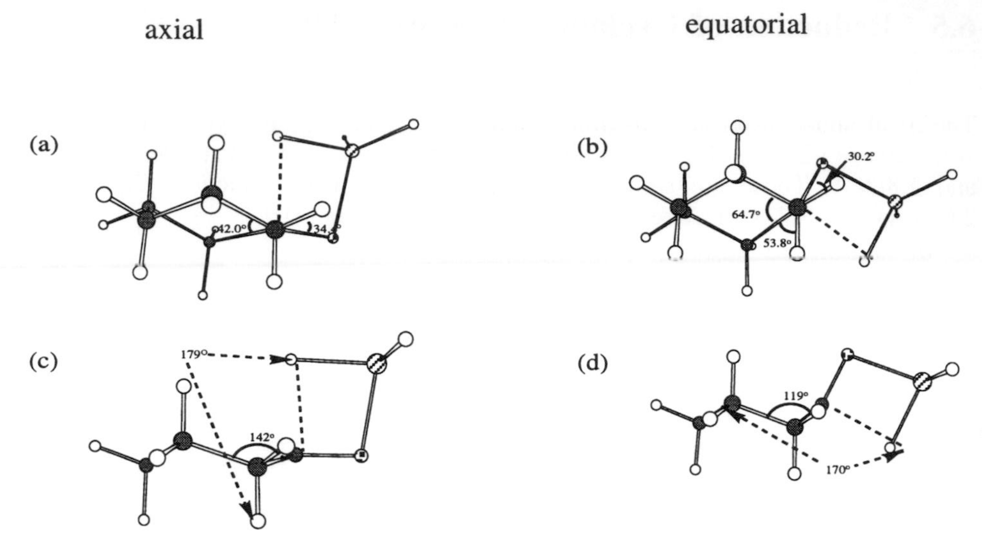

Figure 6-16. Dihedral angles in the transition states (3-21G*) for axial and equatorial AlH_3 addition to cyclohexanone. Relative energies are in kcal/mole.

more stable by 0.5 kcal/mole [54]. Enthalpies calculated for the cyclohexanone fragments are consistent with greater torsional strain in the equatorial attack transition state. The AlH_3 fragment shows a small enthalpy difference favouring equatorial attack (0.2 kcal/mole). The difference in energy between the transition states and the energies of the components at the transition state geometries, but at infinite separation, is a measure of the electronic interaction at the transition state, 0.8 kcal/mole favouring axial attack, and shows that electronic [55] as well as torsional effects are significant in determining facial selection in cyclohexanone reduction.

Hyperconjugative stabilisation, reported earlier for cyclohexanone as uneven orbital distribution [8, 56], exists in the transition states for AlH_3 reduction. Inspection of the molecular orbitals of the H_{Nu}–C bond in the axial transition state show a significant interaction with the antiperiplanar C–H bonds. The corresponding molecular orbitals for the equatorial transition state show a contribution from the antiperiplanar C–C bonds. This is further evidence that hyperconjugative stabilisation is a significant component of the electronic interaction enthalpy. The role of the attacking hydride in stabilisation of the transition state is shown by its removal; the enthalpy of the resulting frozen fragments results in favouring equatorial approach by 1.2 kcal/mole.

In each transition state, the bonds which are antiperiplanar to the forming H_{Nu}–C bond are elongated, axial C–Hs for axial attack, and C–C bonds for equatorial (Fig. 6-1). This effect has also been noted in calculations of the Diels-Alder transition states [39, 40]. Using the calculated length of the adjacent axial C–H bond of cyclohexanone (1.087 Å) as a reference, a further lengthening [58] to 1.091 Å (0.37%) in the axial transition state occurs, with no extension of the C_α–C_β bond.

Table 6-1. Reduction of 4-*t*-butylcyclohexanone (equatorial/axial alcohols).

Experimental	Theoretical [2]	(25°)[a]
$LiAlH_4$ 90/10 [57]	LiH (3-21G//3-21G) [27]	84/16
	LiH (6-31G*//3-21G) [27]	95/5
	LiH (MP2/6-31G*//3-21G) [28]	91/9
AlH_3 85/15 [57]	AlH_3 (AM1)	86/14
$NaBH_4$ 86/14 [9]	AlH_3(3-21G*//3-21G*)	87/13

[a] Boltzmann weighted distribution from the enthalpy difference.

The equatorial transition state shows a C_α–C_β bond extension to 1.560 Å (0.84%) compared with the calculated bond length of cyclohexanone (1.547 Å), with no extension of the C_α–H_{ax} bond. Similarly, the C_α–$C_{C=O}$ bond length shortened in both axial (1.509 Å, 0.4%) and equatorial structures (1.499 Å, 1.1%) compared to the corresponding calculated bond length in cyclohexanone (1.515 Å).

For AlH_3 addition to cyclohexanone the transition structures show lengthening of the σ-bonds antiperiplanar to the nucleophile [59]. Visualisation of the $p_{C=O}$ molecular orbital surfaces of the transition orbitals for axial addition reveals participation of the C_α–H_{ax} bond, but not C_α–C_β; the transition structure for equatorial addition shows C_α–C_β bond mixing, but not C_α–H_{ax} in the HOMO. The crystal structures of cyclohexanone complexes show bond lengthening and shortening consistent with this hyperconjugation [60]. The better antiperiplanar orientation of the nucleophile with the adjacent C–H bond in axial attack compared with the C–C bond for equatorial attack supports the suggestion that orbital alignment is important.

6.6 Reduction of Adamantanone with AlH_3

It is of considerable interest to establish the magnitude of electronic effects on facial selectivity. The symmetry of 2-adamantanone (**1** X=H) makes this structure ideal [61] to investigate electronic effects on transition state energy, since the faces of the carbonyl are little affected by steric and torsional effects with substitution at C5. The donor and acceptor ability of the four adjacent carbon-carbon bonds to the carbonyl can be varied without significantly altering the molecular structure.

The results of experiments on the reduction on 5-substituted adamantanones **1** and 5-azaadamantan-2-one N-oxide (**2**)[1] with $NaBH_4$ show that electron-withdrawing substituents favour attack by the complex hydride *syn* to the substituent or nitrogen [30, 34, 51] (Fig. 6-17). In the case of **2**, the effect is striking, with a *syn*/*anti* attack ratio of 96/4 for the formation of the *anti*/*syn* alcohols respectively (Table 6-1). Electron-donating substituents show a marginal preference for *anti* attack. Similarly, *syn* facial selectivity is found in free-radical reactions [63], thermal [64] and photo-cycloadditions [65], sigmatropic rearrangements [66], solvolytic reactions [51,67] and electrophilic additions [68] to methylene analogues of **1** which contain a C5 electron-withdrawing substituent.

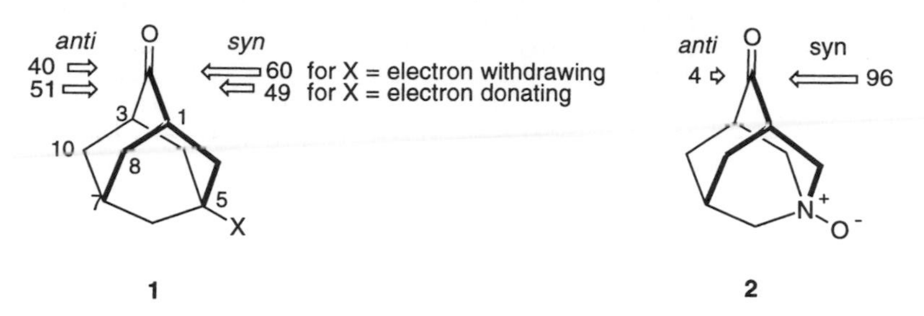

Figure 6-17. Facial selection observed in the reduction of 2-adamantanones with $NaBH_4$.

Figure 6-18 shows on the left the donor C–D bond followed by the frontier molecular orbitals of a nucleophile (Nu) and a carbonyl group. The orbitals of the transition state for nucleophilic addition can be constructed by mixing the reactant orbitals. The three orbitals corresponding to the transition structure can be correlated with reactant orbitals as follows: (1) the lowest orbital is the π orbital into which some of the Nu lone pair orbital has been mixed in a bonding fashion. The highest energy orbital is the anti-bonding admixture of the π^* orbital and the lower energy lone pair orbital. The central orbital is the Nu lone pair which has mixed with the π in an anti-bonding fashion and simultaneously with the π^* orbital in a bonding fashion. (2) Alternatively, the transition state orbitals can be correlated with the product orbitals shown on the right of the diagram. The σ and σ^* of the bond formed between nucleophile and the carbon of the carbonyl group and the lone pair on oxygen are the appropriate product orbitals which can be mixed to form the transition state frontier molecular orbitals.

Studies of adamantanone reduction have been interpreted by le Noble [8a, 8b, 9] as consistent with the Cieplak hypothesis, since the reaction occurs preferentially from the face opposite the more electron-rich σ-bond. The Cieplak hypothesis considers interactions of adjacent σ-bonds adjacent to the transition state with the $\sigma^{*\ddagger}$ orbital of the forming C–Nu bond (see Fig. 6-18). Figure 6-19 shows that the mixing of the adjacent antiperiplanar C–D (D=donor) with the $\sigma^{*\ddagger}_{C\text{–}Nu}$ of the transition state lowers the energy of the transition state.

This principle is applied in Fig. 6-20 to the reduction of substituted adamantanones, which shows that electron-donating groups decrease the HOMO/LUMO gap and favour reaction *anti* to the more donating adjacent C–C bonds. The reverse applies for electron-withdrawing substituents.

There are problems with the Cieplak hypothesis. A major problem is that donation by the donor σ-bonds to the $\sigma^{*\{}$ orbitals of the C–Nu forming bond would be expected to be small as these orbitals are markedly different in energy (Figs. 6-18, 6-19 and 6-20). Furthermore, while a donor substituent attached to the carbon α to the carbonyl will interact with the $\sigma^{*\ddagger}_{C\text{–}Nu}$ orbital to give stabilization

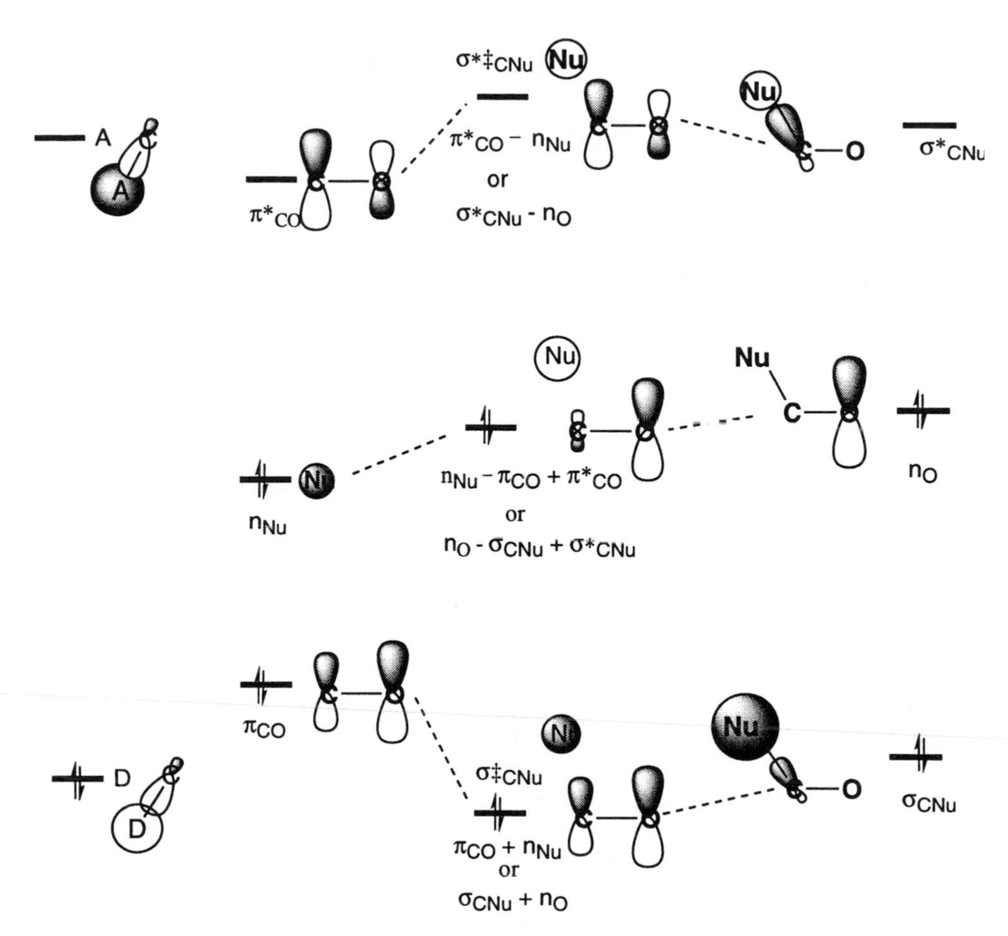

Figure 6-18. Molecular orbitals associated with nucleophilic addition to a carbonyl.

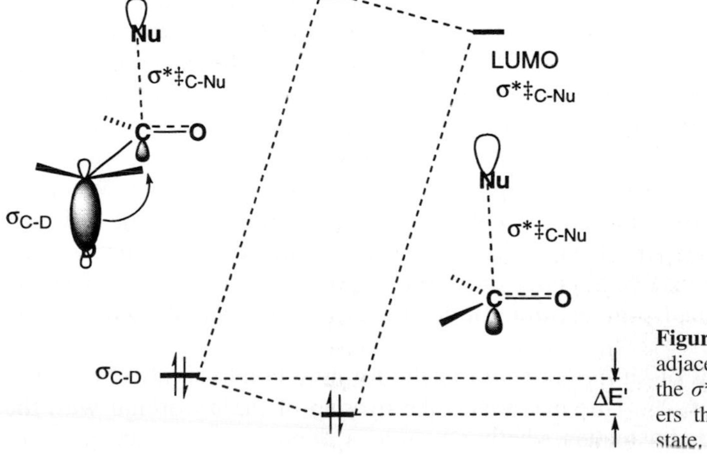

Figure 6-19. Donation by an adjacent antiperiplanar C–D into the $\sigma^{*\ddagger}_{C\text{–}Nu}$ of the transition lowers the energy of the transition state.

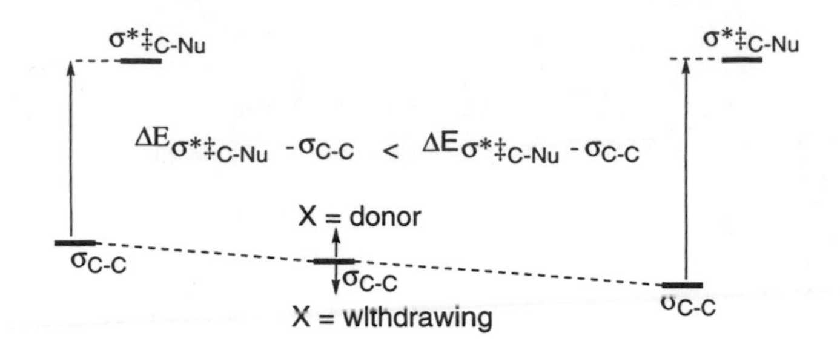

Figure 6.20. Cieplak preference for nucleophilic (reversion) addition *syn* to electron withdrawing and *anti* to electron-donating substituents.

(Fig. 6-19 and also Fig. 6-12), interaction with the $\sigma^{\ddagger}_{C-Nu}$ orbital will cause destabilisation. Furthermore, a donor will interact more strongly with the π^*_{CO} of the reactant carbonyl than the $\sigma^{*\ddagger}{}_{C-Nu}$ orbital (see Fig. 6-18), so that the transition state will be stabilised less than the reactants by a donor. This is in fact the origin of the deactivation of carbonyl groups by electron-donating substituents. The deactivation will be minimised when the σ_{CD} orbital is *anti*-periplanar to the $\sigma^{*\ddagger}_{C-Nu}$ orbital. This results in a staggered transition state, and one which has the nucleophile *anti* to the best donor. Looking from the adduct side, a donor substituent will interact strongly with the $\sigma^*{}_{C-Nu}$ orbital of the adduct, so, in the reverse reaction, the ability of the donor to accelerate the reaction will depend upon the relative energies of the $\sigma^*{}_{C-Nu}$ and the $\sigma^{*\ddagger}_{C-Nu}$ orbitals.

Why does a nucleophile like to attack *anti* to the D? The preferred *anti* arrangement of the geminal doubly occupied σ orbital of the donor and vacant σ^* or $\sigma^{*\ddagger}$ orbitals arises because the overlap is larger than that for the *syn* arrangement. At the same time, but not pointed out by Cieplak, there is less closed-shell repulsion between the donor orbital and the σ orbital in the *anti* arrangement than in the *syn*. In fact, these are two of the interactions which cause the transition state, or indeed any vicinal bonds, to prefer to be *anti* rather than *syn*. When all vicinal bonds on a substituted ethane are *syn*, an energy maximum, the eclipsed conformation, occurs. When all vicinal bonds on a substituted ethane are *anti*, an energy minimum obtains. By aligning the α-donor group *anti* to the $\sigma^{*\ddagger}$, Cieplak also aligns the two filled orbitals *anti*. Both of these operate simultaneously and in the same direction.

The usual alignment of an α-donor *anti* to the forming bond can also be understood in Figs. 6-18 to 6-20. This maximises the overlap of the σ^* orbital with the $\sigma^{\ddagger}$ orbital. An α bond, no matter whether it is to a donor or to an acceptor, always prefers to be *anti* to the forming bond to maximise stabilising filled-vacant orbital interactions and to minimise filled–filled orbital interactions. Le Noble has emphasised that an antiperiplanar alignment with the nucleophile may be more important than inherent electron-donating ability in determining participation of C–H or C–C in cyclohexanone.

These considerations show that the orbital interactions cited by Cieplak cannot be the controlling orbital interactions in nucleophilic additions. Why, then, is the Cieplak hypothesis nevertheless successful in so many cases? It is important first of all to acknowledge that there are a number of effects operating in nucleophilic additions, and no single one of them controls stereoselectivity in every case [69]. The question is to determine the relative magnitudes of these and to have some useful generalisations about which effects operate in different types of molecules.

Either the molecular orbital or the valence bond method may be used to analyse the interaction of substituents on the reactant, products, and transition states. Substituents which stabilise the transition state more than reactants will accelerate the reaction, while those which stabilise the reactant more than transition state will slow down the reaction. In valence bond theory, the reactants are represented by a nucleophile lone pair as an anion (Fig. 6-21, path a) and a neutral nucleophile (path b), and the carbonyl by the covalent and ionic resonance structures [70].

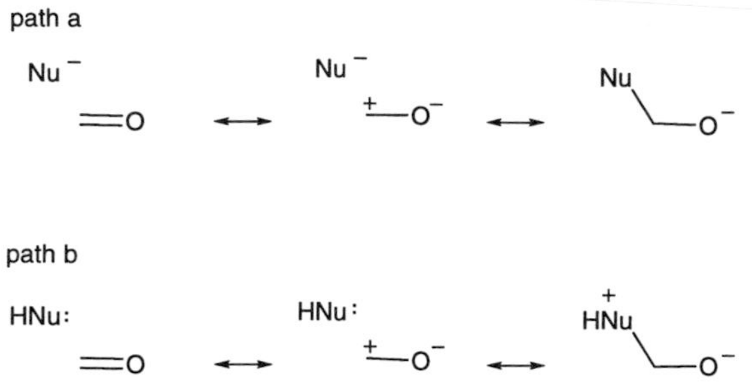

Figure 6-21. Valence bond representation of nucleophilic addition (Nu^- *path a* or HNu: *path b*) to a carbonyl.

The reactant may be considered as a polarised carbonyl bond, reflected in contributions of covalent and ionic resonance structures. As the reaction proceeds, the contribution of both of these is replaced by the structure at the right of the diagram. Because of the greater concentration of positive charge on carbon in the reactant, donor substituents stabilise reactants more than transition state. In summary, donor substituents deactivate carbonyls. Cieplak acknowledged that the reactivity effect of a donor might be different from its stereochemical effect.

6.7 Calculations on the Reduction of 5-Substituted Adamantanones with AlH_3

Nucleophilic attack at adamantanone is necessarily axial to one ring containing the carbonyl and equatorial to the other. The geometry of the transition structure at the reaction site is therefore between that associated with axial and that associated with equatorial addition to cyclohexanone. Torsional effects in the carbonyl-containing rings at the transition structure are compromised by the symmetry of the system, allowing electronic effects to dominate in determining facial selectivity. *Ab initio* and semiempirical calculations of the transition structures for reduction of 5-substituted adamantanones with AlH_3 show bond lengthening and molecular orbitals consistent with hyperconjugative stabilization. However it is not possible to correlate the extension with specific interactions of reactant orbitals. Both electronic and torsional effects contribute to facial selection.

Calculations [7] of the transition structures for the reaction of a series of 5-substituted adamantanones **1** and 5-azaadamantanone N-oxide **2** with AlH_3 [2, 71] closely parallel experimental results with $NaBH_4$ (Table 6-1). Electron-withdrawing groups at C-5 favour attack *syn* to the substituent and result in excess *anti* alcohol (Fig. 6-17). With the electron-donating trimethylsilyl group a marginal preference for *anti* attack is observed for $NaBH_4$. The reduction of the free amine **3** is an exception to the pattern that electron-donating groups favour *anti* attack (*syn*/*anti* attack 62/38). Semiempirical AM1 calculations predict a 33/67 ratio; however, if complexation occurs to give **4** the predicted ratio (60/40) is virtually identical to experiment ($NaBH_4$:62/38).

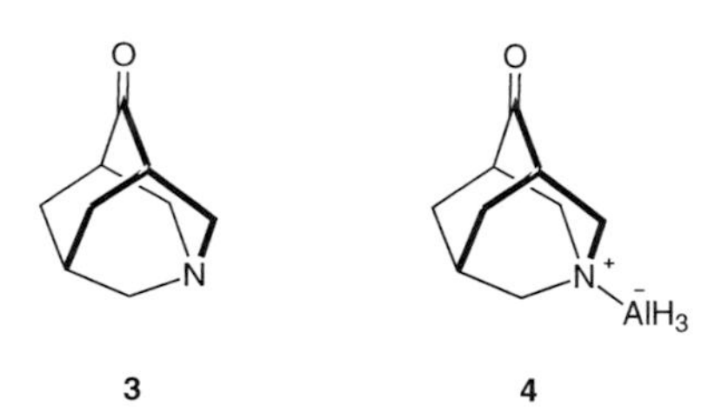

Semiempirical calculations for reduction of **2** do not reproduce experimental results, but *ab initio* calculations predict the high selectivity observed ($NaBH_4$ reduction is 96/4). The transition structure for AlH_3 addition to adamantanone shows flattening of the ring undergoing axial attack, i.e., the $C_{C=O}$ bends away from the direction of the approaching nucleophile from its initial position between the two bridgehead hydrogens in adamantanone as visualised by the angle defined by the $C_\alpha C_{C=O}C_{\alpha'}$–$C_\beta C_\alpha C_{\alpha'}$ planes, which is 135° in the ring undergoing axial attack (Fig. 6-22). The less constrained cyclohexanone AlH_3 transition structure flattens to a corresponding angle of 142° in axial attack results [6].

The internal C_α–$C_{C=O}$–$C_{\alpha'}$–$C_{\beta'}$ dihedral is also affected by the flattening and is 50° in the axial ring compared to 42° in the axial cyclohexanone transition struc-

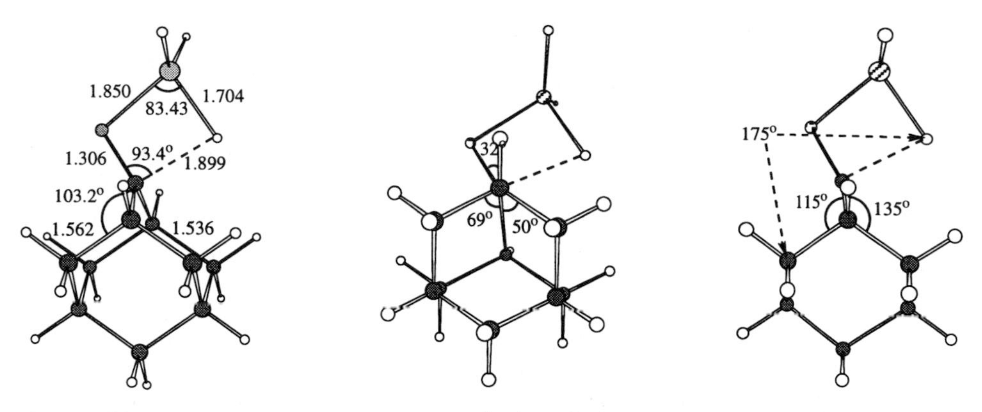

Figure 6-22. Transitions structure geometry for reduction of adamantanone (**1** X=H) with AlH_3.

ture. In adamantanone, the bending of the $C_{C=O}$ away from the nucleophile causes the ring undergoing equatorial attack to pucker, which also occurs in the equatorial attack transition structure of cyclohexanone. Ring puckering is greater for the adamantanone transition structure as measured by the 115° angle defined by the planes $C_\alpha C_{C=O}C_{\alpha'}$–$CC_\alpha C_{\alpha'}$ with an internal C_α–$C_{C=O}$-$C_{\alpha'}$–$C_{\beta'}$ dihedral angle of 69° compared to the corresponding values of 119° and 65° in the cyclohexanone equatorial transition structure. The O–$C_{C=O}$-C_α–H dihedral angle is 32°, compared to 34 and 30° in axial and equatorial cyclohexanone transition structures [6], indicative of a degree of torsional strain which is between the latter two structures. A measure of the ability of the adjacent σ-bonds to participate in a hyperconjugative interaction with the transition structure orbitals is given by the H_{Nu}–$C_{C=O}$–C_α–C_β dihedral angle of 175°, which is midway between the corresponding value for the axial (179°) and equatorial (170°) transition structures for cyclohexanone reduction. The N-oxide group in **2** flattens the nitrogen-containing ring, and the carbonyl group bends away from the nitrogen (Fig. 6-23). Since the positively charged N withdraws electrons from the proximate *syn* C_α–C_β bonds, the C_α–C_β bonds in the ring *anti* to the N have more electron density and interact better with the carbonyl.

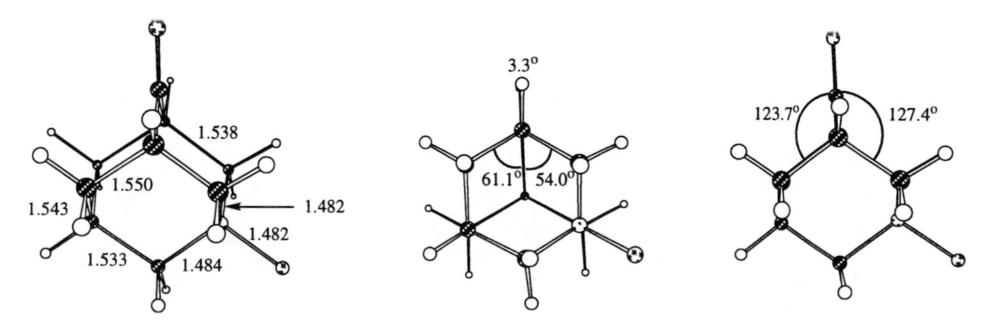

Figure 6-23. Geometry of **2** (3.21G*).

The ring flattening increases the ability for preferential hyperconjugation of the carbonyl with the electron-rich *anti* C_α–C_β bond by improving the *anti* periplanar relationship. The increased donor ability and better orbital alignment with the carbonyl orbitals results in uneven orbital extension and pyramidalization of the carbonyl [51]. The π-orbital extension occurs by interaction with the better aligned C_α–C_β bond which is *anti* to the nitrogen and also corresponds to the more electron-rich bond. The interaction is *anti*bonding, and extension toward the *syn* face results [72]. Relief of torsional strain may also play a role in the bending, since the bridgehead H–C_α bond no longer eclipses the carbonyl.

The calculated transition structures for both *syn* and *anti* attack at **2** (Fig. 6-24) are earlier along the reaction coordinate than the adamantanone transition structure, as measured by the longer H_{Nu}–$C_{C=O}$ distances (*syn* 1.933, *anti* 1.934, unsubstituted, 1.899 Å) and shorter Al–H_{Nu} lengths (*syn* 1.700, *anti* 1.699, unsubstituted 1.704 Å).

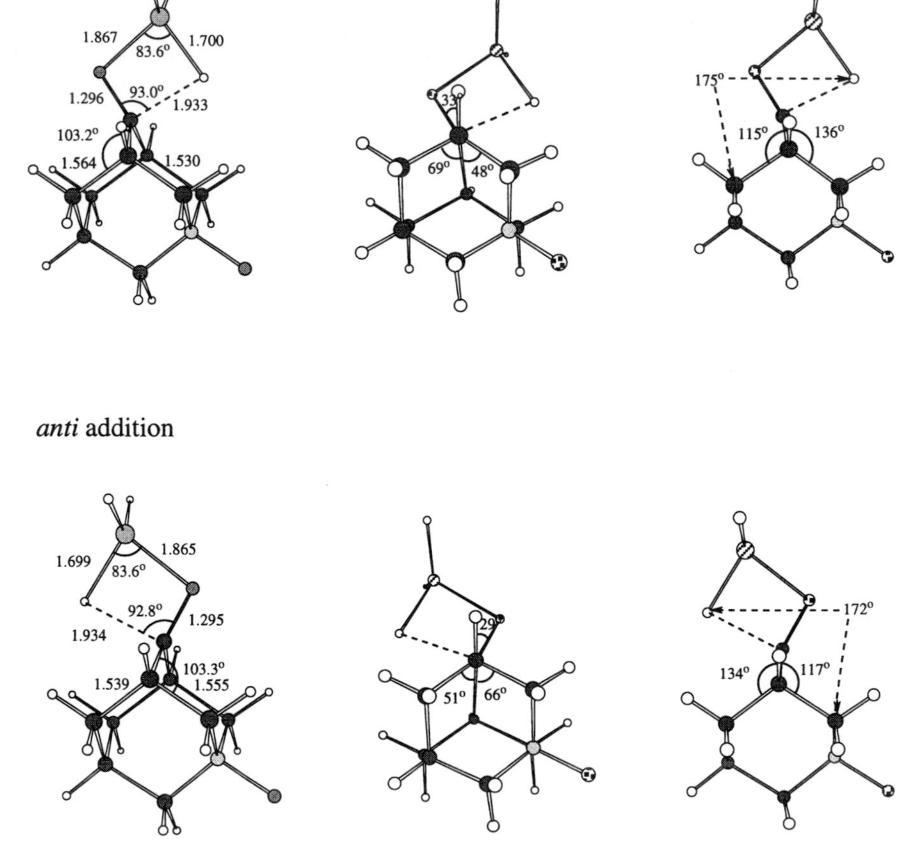

Figure 6-24. Transitions structure geometry for *syn* and *anti* reduction of **2** with AlH_3 (3-21G*).

For *syn* addition the transition structure exhibits minor additional flattening (but the same puckering) and a small increase of the O–$C_{C=O}$–C–H dihedral angle from 32 to 33°, indicative of slightly less torsional strain. By contrast, the transition structure for *anti* attack undergoes less ring flattening and less puckering than its unsubstituted counterpart and has a O–$C_{C=O}$–C–H dihedral angle of 29° consistent with an increase in torsional strain. The transition structure for *syn* attack of **2** is undiminished in its ability to achieve the antiperiplanar relationship (H_{Nu}–C–C–C dihedral angle 175°), but the *anti* isomer has a dihedral angle of 172°.

Syn addition is favoured by 1.2 kcal/mole, and this enthalpy difference can be factored into three components: distortion of the 5-azaadamantanone N-oxide skeleton to the transition structure geometry, AlH_3 distorted to the transition structure geometry and an electronic interaction enthalpy between these distorted fragments. The *syn* ketone fragment is more stable than the *anti* by 0.9 kcal/mole. Distortions of the AlH_3 fragment show a small enthalpy difference (0.2 kcal/mole favouring *anti* attack) and the residual enthalpy contribution is due to electronic interaction of the fragments (0.6 kcal/mole favouring *syn* attack).

Each of the calculated transition structures show elongation of the C–C bonds which are antiperiplanar to the incoming nucleophile. Using the calculated C_α–C_β bond length of adamantanone (1.546 Å) as a reference, the transition structures of adamantanone and the amine oxide **2** (*syn*) with AlH_3 show an extension of the antiperiplanar bonds to 1.562 (1.0%) and 1.564 (1.16%) respectively, with no lengthening (in fact a shortening) of the C_α–C_β bonds on the opposite face of the carbonyl (Figs. 6-22 and 6-24). The *anti* transition structure of **2** shows a corresponding C_α–C_β bond lengthening to 1.555 Å (0.58%). The $C_{C=O}$–C_α bond shortens from 1.514 to 1.500 Å (0.93%) and 1.502 Å (0.80%) for the parent and *syn* transition structures, respectively. The *anti* amine oxide transition structure shortens to 1.504 Å (0.66%). The $C_{C=O}$–C_α–C_β angles on the side away from the nucleophile decreased from 108.8° in adamantanone to 103.2° (0.54%) and 103.3° (0.54%) in the *syn* and *anti* transition structures.

The calculated transition structures for *syn* and *anti* addition of AlH_3 to **2** show a greater degree of carbonyl bending than is found for adamantanone. The degree of hyperconjugation is controlled by orbital overlap and electron availability. The H_{Nu}–$C_{C=O}$–C_α–C_β dihedral angles for the *syn* and *anti* transition structures are indicative of the ability of the C_α–C_β bonds to participate. In the *syn* and *anti* transition structures of the N-oxide the angles are 175° and 172° respectively, indicating a preferred geometry for the former. A measure of hyperconjugation at the transition structure is the degree of extension of the C_α–C_β bond length which is antiperiplanar to the approaching nucleophile. The *syn* transition structure exhibits the greater bond length extension (to 1.564 Å, 1.16% cf. *anti* to 1.555 Å, 0.58%). The greater hyperconjugation in the *syn* transition structure reflects the greater electron density in its antiperiplanar C_α–C_β bonds. The *anti* transition structure has both a poorer orbital alignment with the adjacent antiperiplanar C_α–C_β bonds which are poorer electron donors and exhibits more torsional strain than the *syn* transition structure as determined by the O–C–C–H dihedral angles (29° and 33° respectively). This is supported by the significant difference in enthalpy found in the azaadamantanone N-oxide (**2**) and AlH_3 components of the transition structure and the interaction enthalpy. The enthalpy difference of the ketone fragments corresponds with both the better hyperconjugative interactions in the *syn* transition structure and its smaller torsional strain. The better interaction enthalpy is consistent with the more favourable antiperiplanar relationship of the *syn* (175°) compared to the *anti* (172°).

An *ab initio* study of the 2-adamantyl cation [73] shows the "classical" structure (C_{2v} symmetry) as a transition structure, with the ground structure geometry having a significant bending of the C_α–C^+–C_α bridge toward one face of the cation (17.3°), the stabilisation resulting from the more parallel alignment of the carbonyl π orbitals with C_α–C_β bonds. Pyramidalization of the C^+–H bond occurs in the same direction (11.1°). The C_α–C_β bonds were unequal [1.542 and 1.603 Å (6-31G*)], with the longer bonds on the side closer to the C^+ bridge. This is also consistent with NMR data. The preferential hyperconjugation provides a lower energy2 structure than is obtained by the more extended double hyperconjugation of the transition structure cation. A molecular orbital surface shows that hyperconjugation of the C^+ with the adjacent C–C bond is almost entirely on one side of the molecule. Since hyperconjugation is more important with increasing positive charge, less bond lengthening will occur in the transition structure of AlH_3 reduction than in the 2-adamantyl cation, but more than would be found in the ketone if it were constrained to similar geometry. The geometry of ground state ketone **2** is distorted by bending the carbonyl to the same degree as is found in the transition structures for reduction (O–C–C–H constrained to 32.6°, as in Fig. 6-24), so that the resulting C_α–C_β bond lengths are 1.555 and 1.533 Å on the rings *anti* and *syn* to the nitrogen, respectively. The 1.555 Å bond length corresponds to 0.58% bond extension, compared to 1.564 Å (1.16% lengthening) found in the *syn* transition structure. Deforming ketone **2** by bending the carbonyl in the other direction to resemble the *anti* AlH_3 transition structure (O–C–C–H constrained to 29.1°, as in Fig. 6-24) showed no lengthening of the C_α–C_β bonds (1.543 Å *anti* and 1.545 Å *syn*) from the values of **2**. The degree of hyperconjugation depends on the extent of positive charge on the carbonyl carbon, the electron density in the participating bonds, and the degree of overlap. The surface of the incipient H_{Nu}–$C_{C=O}$ bonding molecular orbital shows hyperconjugation in the transition structure for reactions of **1** and **2**. The favoured *syn* amine oxide transition structure shows hyperconjugative participation of the more electron-rich C_α–C_β bond with the forming bond; the *anti* counterpart shows participation of the electron-deficient C_α–C_β bond and the amine oxide functional group, all of which are antiperiplanar. This alignment is consistent with NMR chemical shift and coupling evidence in adamantanones [74]. A wide range of chemical shifts and five-bond long-range coupling is observed [75] for derivatives of **5** but not **6**, indicative of optimal alignment for through-bond transmission in the former. A linear correlation between the C2 chemical shift and the magnitude of the deuterium isotope effect (change in chemical shift through four bonds) for several 5-deuterated derivatives of adamantanone **7** (X= =S; =O; =C(CN)$_2$; =CH$_2$; –O–CH$_2$–CH$_2$–O–; –Cl, –Cl; –Br, –Br; –S–CH$_2$–CH$_2$–S–; –H, –H; –CH$_2$–CH$_2$–) is interpreted as evidence that hyperconjugative interactions exist in the ground state in this system.

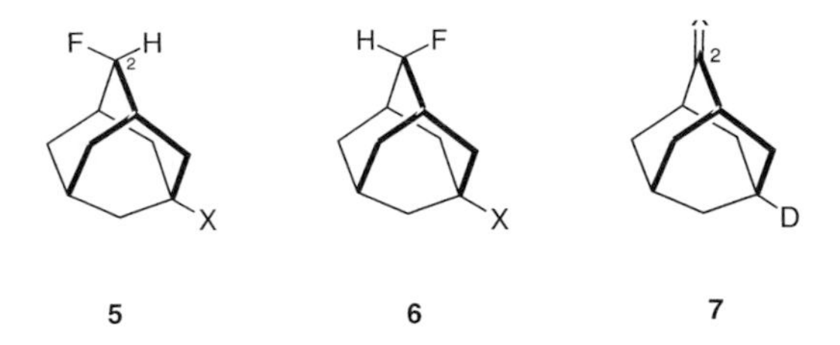

Electrostatic effects have been implicated as a stereoinductive factor in reductions which contain remote polar substituents [28, 51, 76], but they do not appear to represent a major contribution to the stabilization of the *syn* over *anti* transition structure from AlH_3 addition to **2**. The favoured *syn* transition structure for AlH_3 addition to **2** has a higher dipole moment (5.80 D) than the *anti* (4.61 D), indicating that there is less charge separation in the latter. If electrostatic effects played a significant role, a decrease in the stereoselection should be observed upon changing to a more polar solvent, and no decrease in stereoselection in $NaBH_4$ reduction upon changing from methanol to water or even saturated sodium chloride has been observed [51]. For reduction of 2-adamantanones **1** with AlH_3, bond length changes in a four-centre transition structure are consistent with hyperconjugative delocalization in the transition structure. The C_α–C_β bond which is antiperiplanar to the incoming nucleophile is lengthened, and ring distortions consistent with torsional strain minimisation and improvement of orbital alignment are present. These effects are also found in the *syn* and *anti* transition structures of 5-azaadamantanone N-oxide (**2**) with AlH_3. The greater bond lengthening occurs for *syn* addition consistent with the greater electron density in the adjacent periplanar bond and the more linear geometry of the bond and the nucleophile. The favoured (*syn*) structure has an alignment of the nucleophile with the more electron-rich C_α–C_β bonds which is closer to the ideal antiperiplanar orientation and with less torsional strain than is found in the transition structure for *anti* attack. The transition structure molecular orbital surfaces which involve the incipient bond show an interaction with the antiperiplanar C_α–C_β bonds. Electronic and torsional effects contribute to facial selection. The calculated structural changes of the transition states and their molecular orbitals are consistent with hyperconjugative delocalization of the C_α–H_{ax} bond for axial and the C_α–C_β bond for equatorial attack.

The growing ability of organic chemists to undertake computation reduces the need for simple generalisation of selectivity and offers the potential for calculation of activation barriers as a complementary tool in designing syntheses which require the prediction of facial or diastereoselection.

References

1. The Principle of *Conservation of Orbital Symmetry* for pericyclic reactions as enunciated by R.B. Woodward and R. Hoffmann comes closest to this description.
2. Gaussian 94: Revision A.1, Gaussian, Inc., Pittsburgh PA, Frisch, M.J.; Trucks, G.W.; Schlegel, H.B.; Gill, P.M.W.; Johnson, B.G.; Robb, M.A.; Cheeseman, J.R.; Keith, T.; Petersson, G.A.; Montgomery, J.A.; Raghavachari, H.; Al-Laham, M.A.; Zakrzewski, V.G.; Ortiz, J.V.; Foresman, J.B.; Cioslowski, J.; Stefanov, B.B.; Nanayakkara, A.; Challacombe, M.; Peng, C.Y.; Ayala, P.Y.; Chen, W.; Wong, M.W.; Andres, J.L.; Replogle, E.S.; Gomperts, R.; Martin, R.L.; Fox, D.J.; Binkley, J.S.; Defrees, D.J.; Baker, J.; Stewart, J.P.; Head-Gordon, M.; Gonzalez, C.; Pople, J.A. **1995**. SPARTAN (Version 2.1, Wavefunction, Inc., 18401 Von Karman, Irvine CA 92715). MOPAC: Dewar, M.J.S.; Zoebisch, E.G.; Healy, E. F.; Stewart, J.J.P. *J. Am. Chem. Soc.* **1985**, *107,* 3902.
3. Madura, J.D.; Jorgensen, W.L. *J. Am. Chem. Soc.*, **1986**, *108*, 2517. Jorgensen, W.L. *Acc. Chem. Res.*, **1989**, *22*, 184.
4. Coxon, J.M.; McDonald, D.Q. *Tetrahedron* **1992**, *48*, 3353.
5. Coxon, J.M.; Luibrand, R.T. *Tetrahedron Lett.* **1993**, *34*, 7093.
6. Coxon, J.M.; Luibrand, R.T. *Tetrahedron Lett.* **1993**, *34*, 7097.
7. Coxon, J.M.; Houk, K.N.; Luibrand, R.T. *J. Org. Chem.,* **1995**, *60*, 418.
8. (a) Li, H.; le Noble, W.J. *Recl. Trav. Chim. Pays-Bas*. **1992**, *111*, 199. (b) Frenking, G.; Köhler K.F.; Reetz, M.T. *Tetrahedron* **1991**, *47*, 8991.
9. Wigfield, D.C. *Tetrahedron* **1979**, *35*, 449.
10. Kleingeld, J.C.; Nibbering, N.M.M.; Grabowski, J.J.; DePuy, C. H.; Fukada, E.K.; McIver, R.T. *Tetrahedron Lett.*, **1982**, *23*, 4755. Johlman, C.L.; White, R.L.; Sawyer, D.T.; Wilkins, C.L. *J. Am. Chem. Soc.*, **1983**, *105*, 2091.
11. Dewar, M.J.S.; Storch, D.M. *J. Chem. Soc., Chem. Commun.*, **1985**, 94.
12. Cram, D.J.; Elhafez, F.A. *J. Am. Chem. Soc.* **1952**, *74*, 5828.
13. Cornforth, J.W.; Cornforth, R.H.; Mathew, K.K. *J. Chem. Soc.* **1959**, 112.
14. Blackett, B.N.; Coxon, J.M.; Hartshorn, M.P.; Richards, K.E. *Aust. J. Chem.* **1970**, *23*, 2077.
15. Karabatsos, G.J. *J. Am. Chem. Soc.* **1976**, *89*, 1367.
16. Cherest, M.; Felkin, H.; Prudent, N. *Tetrahedron Lett.* **1968**, *18*, 16.
17. Anh, N.T.; Eisenstein, O. *Nouv. J. Chim.* **1977**, *1*, 61. (H^- attack on MeCHClCHO and EtCHMe-CHO (STO-3G)).
18. Bürgi, H.B.; Dunitz, J.D.; Shefter, E.J. *J. Am. Chem. Soc.* **1973**, *95*, 5065. Bürgi, H.B.; Dunitz, J.D.; Lehn, J.M. *Tetrahedron*. **1974**, *30*, 1563. Bürgi, H.B.; Lehn, J.M.; Wipff, G. *J. Am. Chem. Soc.* **1974**, *96,* 1956.
19. Lodge, E.P.; Heathcock, C.H. *J. Am. Chem. Soc.* **1987**, *109*, 3353.
20. For reaction of an electrophile and a nucleophile it is the interaction of the HOMO of the nucleophile and the LUMO of the electrophile that results in the more important frontier orbital interaction, consistent with donation of electron density of the nucleophile to the electrophile.
21. Norman, R.O.C.; Coxon, J.M. *Principles of Organic Synthesis*, 3rd Edn., Blackie, Chapman and Hall, London, **1993**, p 88.
22. Dauben, W.G.; Fonken, G.S.; Noyce, D.S. *J. Am. Chem. Soc.* **1956**, *78*, 2579.
23. Giddings, M.R.; Hudec, J. *Can. J. Chem.*, **1981**, *59*, 459.
24. Inagaki, S.; Fujimoto, H.; Fukui, K. *J. Am. Chem. Soc.*, **1976**, *98*, 4054.
25. For a recent application of the "orbital mixing rule" see Ishida, M.; Beniya, Y.; Inagaki, S.; Kato, S. *J. Am. Chem. Soc.*, **1990**, *112*, 8980.
26. Klein, J. *Tetrahedron Lett.* **1973**, *44*, 4307.
27. Frenking, G.; Kohler, K.F.; Reetz, M.T. *Angew. Chem. Int. Ed. Engl.* **1994**, *30,* 1146.
28. Wu, Y.-D.; Houk, K.N.; Paddon-Row, M.N. *Angew. Chem. Int. Ed. Engl.* **1992**, *31*, 1019.
29. Huang, X.L.; Dannenberg, J.J. *J. Amer. Chem. Soc.* **1993**, *115*, 6017. Huang, X.L.; Dannenberg, J.J.; Duran, M.; and Bertrán. *J. Amer. Chem. Soc.* **1993**, *115*, 4024.
30. (a) Cieplak, A.S. *J. Amer. Chem. Soc.* **1981**, *103*, 4540. (b) Cieplak, A.S; Tait, B.D.; Johnson, C.R. *J. Amer. Chem. Soc.* **1989**, *111*, 8447. (c) Cieplak, A.S.; Tait, B.D.; Johnson, C.R. *J. Am. Chem. Soc.*, **1987**, *109*, 5875.

31. Schleyer, P. v. R. *J. Am. Chem. Soc.*, **1967**, *89*, 701.
32. The energy level of the $\sigma^{*\ddagger}_{C-Nu}$ orbitals for axial and equatorial addition are unknown. The Cieplak hypothesis requires that ΔE_{axial} ($\sigma^{*\ddagger}_{C-Nu}-\sigma_{C-X}$) < $\Delta E_{equatorial}$ ($\sigma^{*\ddagger}_{C-Nu}-\sigma_{C-C}$).
33. Li, H.; le Noble, W. J. *Tetrahedron Lett.*, **1990**, *31*, 4391.
34. Xie, M.; le Noble, W. J. *J. Org. Chem.*, **1989**, *54*, 3836.
35. Macaulay, J. B.; Fallis, A. G. *J. Am. Chem. Soc.*, **1990**, *112*, 1136.
Equatorial attack to the carbonyl would be stabilised by the C2–C3 and C5–C6 bonds which are antiperiplanar to the developing σ-bond with the nucleophile, but since these C–C bonds are considered poorer electron donors than C–H bonds, equatorial addition is less favoured.
37. A more classical torsional effect between adjacent σ-bonds was first delineated by von Schleyer and also favours axial addition vs equatorial attack since for the former an unfavourable torsional interaction is avoided [31].
38. The more bulky a nucleophile the more important attack from the more open equatorial face becomes.
39. Coxon, J. M.; Froese, R. J.; Ganguly, B.; Marchand, A. P.; Morokuma, K. *Syn Lett*, **1999**, 1681.
40. Reaction occurs faster at the face of the diene which has the antiperiplanar orientation of the best donor ligand and the forming ring bonds in the transition state, Macaulay, J. B.; Fallis, A. G. *J. Am. Chem. Soc.* **1990**, *112*, 1136. Coxon, J. M.; McDonald, D. Q. *Tetrahedron Lett.* **1992**, *48*, 651. Coxon, J. M.; McDonald, D. Q.; Steel, P. J. In *Advances in Detailed Reaction Mechanisms*, Coxon, J. M., Ed. JAI Press: Greenwich, CT; Vol. 3, **1994**, pp. 131–166.
41. Yamataka, H.; Hanafusa, T. *J. Am. Chem. Soc.* **1986**, *108*, 6643. Kayser, M. M.; Eliev, S.; Eisenstein, O. *Tetrahedron Lett.* **1983**, *24*, 1015. ApSimon, J. W.; Collier, T. L. *Tetrahedron* **1986**, *42*, 5157.
42. Dewar, M. J. S.; McKee, M. L. *J. Am. Chem. Soc.* **1978**, *100*, 7499.
43. Eisenstein, O.; Schlegel, H. B.; Kayser, M. M. *J. Org. Chem.* **1982**, *47*, 2886. Eisenstein, O.; Kayser, M.; Roy, M.; McMahon, T. B. *Can. J. Chem.* **1985**, *63*, 281.
44. Wigfield, D. C.; Gowland, F. W. *J. Org. Chem.* **1980**, *45*, 653. Wigfield, D. C.; Gowland, F. W. *Tetrahedron Lett.* **1976**, *17*, 3373. Ayres, D. C.; Kirk, D. N.; Sawdaye, R. *J. Chem. Soc.* (B), **1970**, 1133.
45. Kaufmann, E.; Schleyer, P. von R.; Houk, K. N.; Wu, Y. D. *J. Am. Chem. Soc.* **1985**, *107*, 5560. Calculation of the reaction pathway for LiH addition has been a pragmatic approach to understanding reduction with $LiAlH_4$.
46. A specially activated preparation of LiH has been reported, but it leads to a different reaction. Klusener, P. A. A.; Brandsma, L.; Verkruijsse, H. D.; Schleyer, P. von R.; Friedl, T.; Pi, R. *Angew. Chem. Int. Ed. Engl.* **1986**, *25*, 465 and references therein.
47. Norman, R. O. C.; Coxon, J. M. *Principles of Organic Synthesis*, 3rd Edn., Blackie, Chapman and Hall, London, **1993**, p. 175, 651. Ashby, E. C.; Boone, J. R. *J. Org. Chem.* **1976**, *41*, 2890. Calculations on the reduction of ketones by BH_3 have been reported. Masamune, S.; Kennedy, R. M.; Peterson, J. S.; Houk, K. N.; Wu, Y.-D. *J. Am. Chem. Soc.* **1986**, *108*, 7474.
48. The binding energy for **1a** (3-21G*, –29.3 kcal/mole) decreases at higher levels of theory, MP2/6-31G*//6-31G*, –21.0; MP3/6-31G*//6-31G*, –20.3 kcal/mole. The C–O–Al bond angle is sensitive to the basis set. Without consideration of the d orbitals on Al (3-21G) the value is 121.6° which increases to 130.1° (3-21G*) and 125.6° (6-31G*)° when the polarization functions are included. LePage, T. J.; Wiberg, K. B. *J. Am. Chem. Soc.* **1988**, *110*, 6642.
49. *Ab initio* calculations show a C–O–Al bond angle of **3** of 178.2° (3-21G*). A similar angle occurs for CH_3OAlH_2 (3-21G*, 176.2°), Barron, A. R.; Dobbs, K. D.; Francl, M. M. *J. Am. Chem. Soc.* **1991**, *113*, 39. At higher level (HF/6-31+G) where the ionic character in CH_3OAlH_2 is accounted for the angles reduces to 140°. Experimental values for sterically hindered C–O–Al bond angles in $AlMe(BHT)_2$ are 140.5° and 146.8°, Shreve, A. P.; Mulhaupt, R.; Fultz, W.; Calabrese, J.; Robbins, W.; Ittel, S. D. *Organometallics* **1988**, *7*, 409. The Al–O bond lengths observed from X-ray crystallographic studies (1.687 and 1.689 Å) are shorter than typical Al–O bonds (1.8–2.0 Å). The unusual obtuse bond angle and short Al–O distance are the result of interactions between the aluminium vacant *p*-orbital and a lone pair of electrons on oxygen.
50. Yamataka, H.; Hanafusa, T. *J. Org. Chem.* **1988**, *53*, 772. A kinetic study of the addition of lithium tri-*t*-butoxyaluminium hydride to alkylcyclohexanones and *p,p′*-disubstituted benzophenones has found support for a four-centred transition state which, unlike borohydride addition, is allowed by utilization of the orbitals on aluminium [44].

51. (a) Wu, Y.-D.; Tucker, J.A.; Houk, K.N. *J. Am. Chem. Soc.* **1991**, *113*, 5018. (b) Wu, Y. D.; Houk, K.-N. *J. Am. Chem. Soc.* **1987**, *109*, 908. (c) Hahn, J.M.; le Noble, W.J. *J. Am. Chem. Soc.* **1992**, *114*, 1916. (d) Cheung, C.K.; Tseng, L.T.; Lin, M.-H.; Srivastava, S.; le Noble, W.J. *J. Am. Chem. Soc.* **1987**, *109*, 1598; *ibid.* **1987**, *109*, 7239. (e) Mikami, K.; Shimizu, M. In *Advances in Detailed Reaction Mechanisms,* Coxon, J.M., Ed. JAI Press: Greenwich, CT; Vol. 3, **1994**, pp. 45.
52. Cherest, M.; Felkin, H. *Tetrahedron Lett.* **1968**, *18*, 2205.
53. The importance of an antiperiplanar approach, and ring flattening required to achieve it for axial attack in cyclohexanones was first noted by Anh [17]. Anh, N. T. *Top. Curr. Chem.* **1980**, *88*, 145.
54. The difference in strain of the cyclohexanone fragments is greater for AlH_3 addition than for LiH addition (3-21G*) [28, 51].
55. This is consistent with a significant electronic interaction contribution to the transition state enthalpy difference.
56. Houk has explained this in terms of secondary orbital interactions of the most eclipsed allylic sigma bonds with the $\pi^*_{C=O}$LUMO [28, 51].
57. Guyon, R.; Villa, P. *Bull. Soc. Chim. France* **1977**, 145.
58. In cyclohexanone C_α–H_{ax} bonds are already somewhat extended due to their interaction with the carbonyl.
59. In axial attack the C_α–H_{ax} bond is aligned with the $\pi_{C=O}$ orbital (H_{Nu}–$C_{C=O}$–C_α–C_β dihedral angle 179°). The C_α–C_β bond aligns better in the equatorial attack transition structure (H_{Nu}–$C_{C=O}$–C_α–C_β dihedral 170°).
60. Laube, T.; Hollenstein, S. *J. Am. Chem. Soc.* **1992**, *114*, 8812.
61. 2-Adamantanone is more rigid than cyclohexanone and less able to distort to achieve an optimal transition state geometry. The system avoids the problem present in studies of cyclohexanone of comparison of C–C and C–H.
62. Gung, B.W.; Wolf, M.A. *J. Org. Chem.* **1996**, *61*, 232.
63. Bodepuri, V.R.; le Noble, W.J. *J. Org. Chem.* **1991**, *56*, 5874.
64. Tsai, T.-L.; Chem, W.-C.; Yu, C.-H.; le Noble, W.J.; Chemg W.-S. *J. Org. Chem.* **1998**, *64*, 1099.
65. Chung, W.-S.; Turro, N.J.; Srivastava, S.; Li, H.; le Noble, W.J. *J. Am. Chem. Soc.* **1988**, *110*, 7882.
66. Mukherjee, A.; Schulman, E.M.; le Noble, W.J. *J. Org. Chem.* **1992**, *57*, 3120, 3126. Lin, M.H.; Watson, W.H.; Kashyap, R.P.; le Noble, W.J. *J. Org. Chem.* **1990**, *55*, 3597. Lin, M.H.; le Noble, W.J. *J. Org. Chem.* **1989**, *54*, 998.
67. Xie, M.; le Noble, W.J. *J. Org. Chem.* **1989**, *54*, 3839–3841. le Noble, W.J.; Chiou, D.-M.; Okaya, Y. *Tetrahedron Lett.* **1978**, *22*, 1961.
68. Srivastav, S.; le Noble, W.J. *J. Am. Chem. Soc.* **1987**, *109*, 5874.
69. Li, H; le Noble, W.J. *Recl. Trav. Chim. Pays-Bas* **1992**, *111*, 8447. Fundamental to the Cieplak hypothesis is the question of which *a* σ-bond will better participate, C–H or C–C. Houk has concluded that C–C bonds are more electron donating than C–H.
70. Frenking, G.; Köhler, K.F.; Reetz, M.T. *Tetrahedron* **1991**, *47*, 9005.
71. Le Noble has pointed out a similarity between Cieplak transition state stabilization and Winstein's proposal of σ assistance in the formation of carbocations [8a, 51]. In the Cieplak model, neighbouring σ-electrons delocalise into the $\sigma^{*\ddagger}$-orbital which forms with the nucleophile; in Winstein's carbocation model, the σ-electrons delocalise into a vacant *p* orbital.
72. The uneven orbital extension is observed by plotting the value of the LUMO coefficient at a given distance from the atom onto an electron density surface, as supported by SPARTAN.
73. Dutler, R.; Rauk, A.; Sorensen, T.S.; Whitworth, S.M. *J. Am. Chem. Soc.* **1989**, *111*, 9024.
74. Vinkovic, V.; Mlinaric-Majerski, K.; Marinic, Z. *Tetrahedron Lett.* **1992**, *33*, 7441.
75. Adcock, W.; Trout, N.A. *J. Org. Chem.* **1991**, *56*, 3299. Adcock, W.; Coope, J.; Shiner, V.J. Jr.; Trout, N.A. *J. Org. Chem.* **1990**, *55*, 1411. Adcock, W.; Krstic, A.R.; Duggan, P.J.; Shiner, V.J. Jr.; Coope, J.; Ensinger, M.W. *J. Am. Chem. Soc.*, **1990**, *112*, 3140.
76. Electrostatic interactions are considered to be the dominant controlling factor in the determination of stereoselection in the reduction of 3-fluorocyclohexanones and substituted 7-norbornanones. Paddon-Row, M.N.; Wu, Y.-D.; Houk, K.N. *J. Am. Chem. Soc.* **1992**, *114*, 10638.

7 Engineered Asymmetric Catalysis

Koichi Mikami

7.1 Introduction

Asymmetric catalysis of organic reactions is one of the most important topics in modern science and technology [1]. This technique affords a high proportion of the enantio-enriched product and a small proportion of waste material by taking advantage of a chiral catalyst [2]. Since the addition reaction to carbonyl compounds plays a central role in organic synthesis, asymmetric catalysis will in this case provide a powerful methodology for the catalytic asymmetric synthesis of target molecules. Highly promising candidates for such asymmetric catalysts are metal complexes bearing chiral ligands.

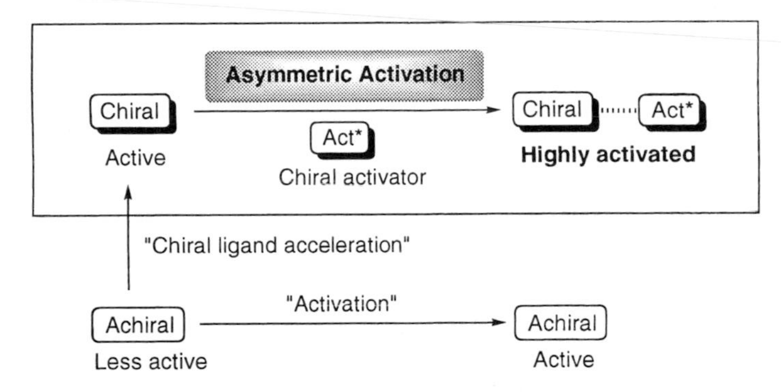

Figure 7-1. Asymmetric activation.

In homogeneous asymmetric catalysis, Sharpless et al. have emphasized the significance of “chiral ligand acceleration” [3] of an asymmetric catalyst from an achiral pre-catalyst via ligand exchange with a chiral ligand (Fig. 7-1). In heterogeneous asymmetric catalysis, the term “chiral modification” [4] is coined for the process in modifying an achiral heterogeneous catalyst, particularly on the surface, with a “chiral modifier”, namely “chiral ligand” (Fig. 7-1). However, a modifier is often reported to interact preferentially with a substrate [5] rather than the achiral heterogeneous catalyst surface [6].

The asymmetric catalysts thus prepared can be further evolved into a highly activated catalyst by engineering with chiral activators (Fig. 7-1). The term ‘asymmetric activation’ can be proposed for this process in an analogy to the activation process of an achiral reagent or catalyst to provide an activated but achiral one (e.g. an activated zinc reagent) [7]. This asymmetric activation process is particularly useful in racemic catalysis through selective activation of one enantiomer of the racemic catalysts.

While non-racemic catalysts thus developed can generate non-racemic products with or without the "non-linear relationship" in enantiomeric excesses between catalysts and products [8], racemic catalysts inherently give only a racemic mixture of chiral products. Recently, a strategy whereby a racemic catalyst is selectively deactivated by a chiral molecule has been reported to yield non-racemic products (Fig. 7-3). However, we have reported a conceptually opposite strategy to asymmetric catalysts in which a chiral activator selectively activates one enantiomer of a racemic chiral catalyst (Fig. 7-2). The advantage of this activation strategy over the deactivation counterpart is that the activated catalyst can produce a greater enantiomeric excess (x_{act}% ee) in the products, even with the catalytic amount of the chiral activator to the chiral catalyst, than can the enantiomerically pure catalyst on its own (x% ee).

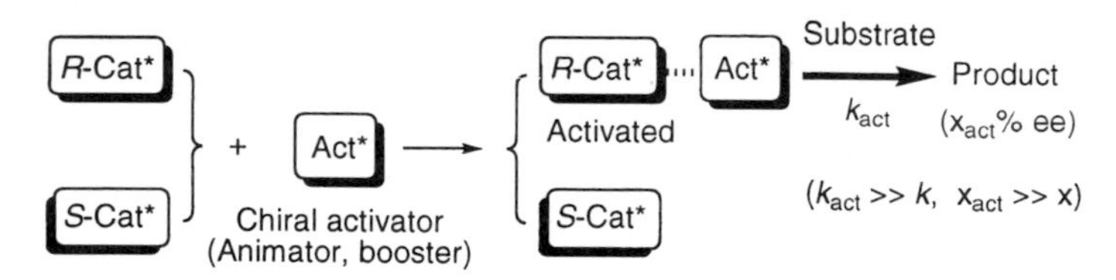

Figure 7-2. Asymmetric activation of racemic catalyst.

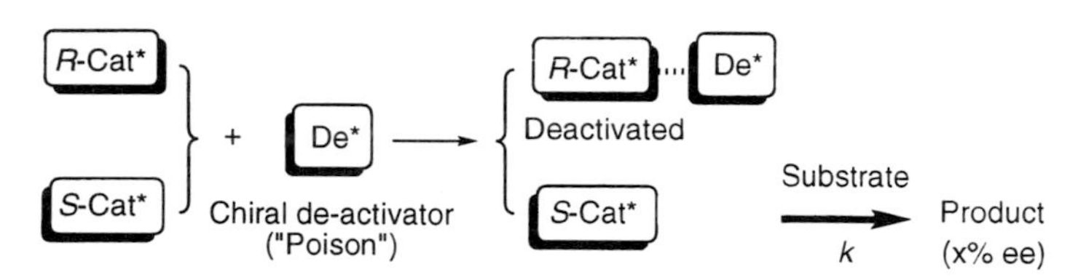

Figure 7-3. Asymmetric deactivation of racemic catalyst.

7.2 'Positive Non-Linear Effect' of Non-Racemic Catalysts

A chiral catalyst is not necessarily prepared from an enantio-pure ligand, because deviation from the linear relationship, namely "non-linear effect (NLE)", is sometimes observed between the enantiomeric purity of asymmetric catalysts and the optical yields of the products (Fig. 7-4) [8–11, 13, 14, 16, 17, 19–26]. Amongst other examples, the convex deviation which Kagan [9] and Mikami [10] independently refer to as positive non-linear effect (abbreviated as (+)-NLE) has attracted current attention, achieving a higher level of asymmetric induction than the enantio-purity of the non-racemic (partially resolved) catalysts [11].

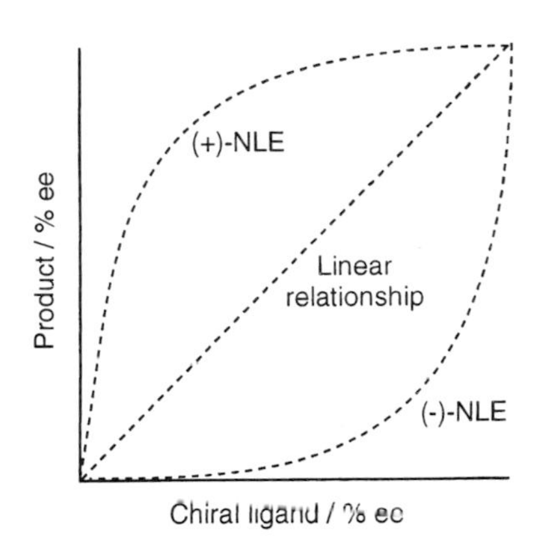

Figure 7-4. Relationship between the enantiomeric purity of chiral ligands and the optical yield of products.

Oguni has reported "asymmetric amplification" [12] ((+)-NLE) in an asymmetric carbonyl addition reaction of dialkylzinc reagents catalyzed by chiral aminoalcohols such as 1-piperidino-3,3-dimethyl-2-butanol (PDB) (Eq. (7.1)) [13]. Noyori et al. have reported a highly efficient aminoalcohol catalyst, (2*S*)-3-*exo*-(dimethylamino)isoborneol (DAIB) [14] and a beautiful investigation of asymmetric amplification in view of the stability and lower catalytic activity of the hetero-chiral dimer of the zinc aminoalcohol catalyst than the homo-chiral dimer (Fig. 7-5). We have reported a positive non-linear effect in a carbonyl-ene reaction [15] with glyoxylate catalyzed by binaphthol (binol)-derived chiral titanium complex (Eq. (7.2)) [10]. Bolm has also reported (+)-NLE in the 1,4-addition reaction of dialkylzinc by the catalysis of nickel complex with pyridyl alcohols [16].

PhCHO + Et_2Zn → [(−)-PDB (*t*Bu–CH(OH)*–CH$_2$–N-piperidine), 10.7% ee (2 mol%); hexane, −10 °C (96%)] → PhCH(OH)Et, 82% ee (7.1)

Ph(Me)C=CH_2 + H–CO–CO_2Me → [$Br_2Ti(O\textit{i}Pr)_2$ (1.0 mol%)/ binaphthol (OH, OH) 33.0% ee (1.0 mol%); MS 4A; CH_2Cl_2, −30 °C (92%)] → Ph–C(=CH_2)–CH_2–CH(OH)–CO_2Me, 91.4% ee (7.2)

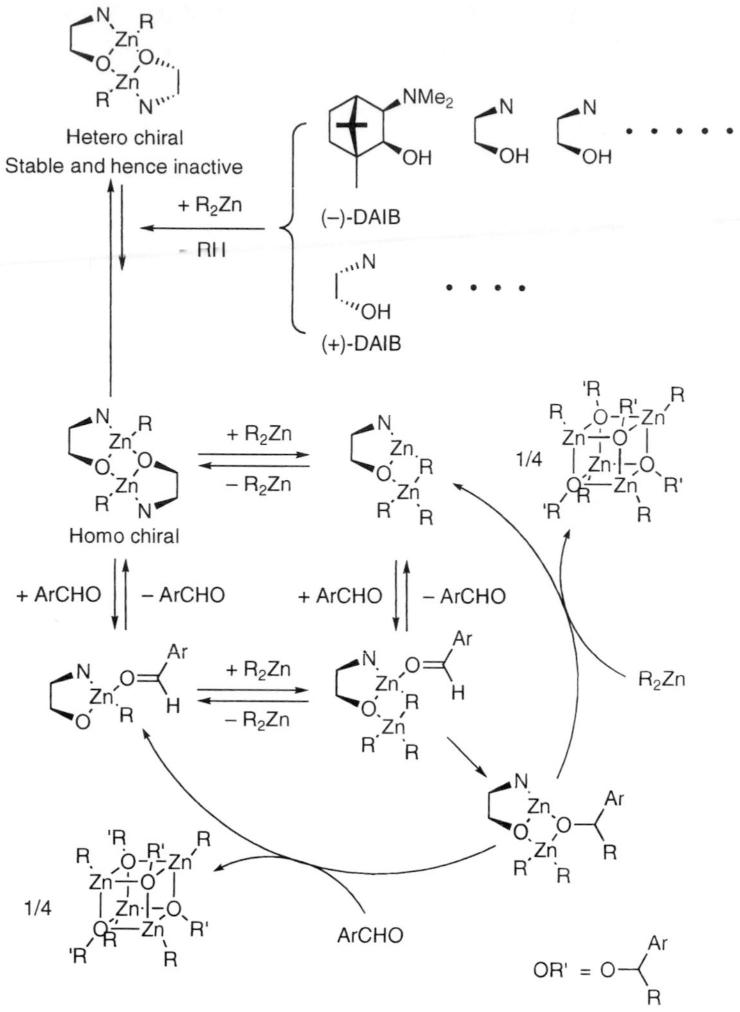

Figure 7-5. Mechanism of asymmetric amplification.

Significant levels of (+)-NLE are also observed in the asymmetric catalysis by cationic complexes bearing mer (meridional) tridentate ligands. An aqua complex exhibits a remarkable (+)-NLE (Eq. (7.3)). This chiral mer-Ni complex is prepared from $Ni(ClO_4)_2{\cdot}6H_2O$ and 4,6-dibenzofurandiyl-2,2′-bis(4-phenyloxazoline) (DBFOX/Ph) as a tridentate ligand [17]. Two mechanisms are involved in this (+)-NLE: the irreversible formation of heterochiral [(*R,R*)/(*S,S*)] 2:1 ligand:metal complexes and the water-bridged heterochiral oligomerization of 1:1 ligand: metal complexes.

Evans has studied the asymmetric catalysis of carbon-carbon bond forming reactions with C_2 symmetric bisoxazoline-Cu(II) complexes [18, 19]. The (+)-NLE observed in asymmetric aldol reactions catalyzed by the bis(oxazolinyl)pyridine

(PYBOX)-Cu complex (Eq. (7.4)) [19] is explained as a result of the relative stabilities of the heterochiral [(*S*,*S*)/(*R*,*R*)] and homochiral [(*S*,*S*)/(*S*,*S*)] ligand complexes with metal in a ratio of 2 : 1.

$Ni(ClO_4)_2{\bullet}6H_2O$ (20 mol%)
(*R*,*R*)-DBFOX/Ph
20% ee (20 mol%)
CH_2Cl_2, –40 °C
(95%)

96% ee (97% *endo*)

(*R*,*R*)-DBFOX/Ph

(7.3)

PYBOX-Cu
25% ee (10 mol%)
CH_2Cl_2, -78 °C

BnO, H, OTMS, S*t* Bu, OH, O

74% ee

2+

2 SbF_6^-

(*S*,*S*)-PYBOX-Cu

(7.4)

The negative non-linear effect [abbreviated as (–)-NLE] [9, 10] stands, in turn, for the opposite phenomenon of concave deviation (Fig. 7-4) [9b, 11l, m, n, z, ad, ag, ah, 20]. The partially resolved catalyst provides the product in % ee lower than calculated by linearity. In conjugate addition reaction of organocopper reagent, an interesting shape of NLE is reported. The (+)-NLE is observed at higher % ee of the chiral ligand, but (–)-NLE at lower % ee (Eq. (7.5)) [21]. Tanaka et al. have proposed a dimeric structure of the catalyst. However, Kagan has suggested a tetrameric complex as the reactive species by mathematical simulation of a four-chiral-ligands system (Fig. 7-6) [9a].

Significantly, the mode of preparation of a catalyst sometimes determines not only the presence or the absence of a non-linear effect (NLE) but also the direction (positive or negative) thereof [11l, af, ag, 20, 22, 23]. When the MS-free bi-

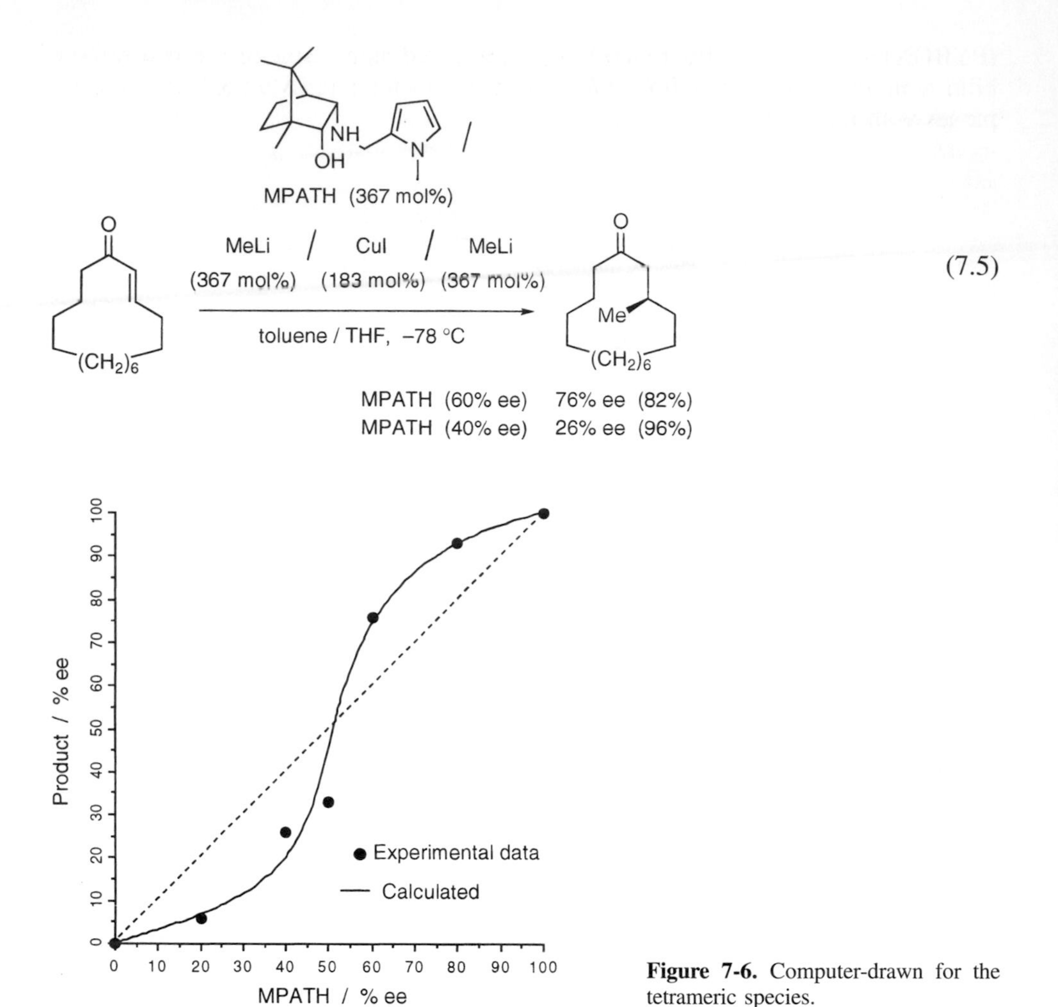

Figure 7-6. Computer-drawn for the tetrameric species.

nol-Ti catalyst (**1**′) is prepared from partially resolved binol and $Cl_2Ti(OiPr)_2$ in the presence of MS 4A, which is filtered off prior to the reaction (Fig. 7-7, Table 7-1) [22], a (+)-NLE is observed in the asymmetric Diels-Alder reaction (Run 1). The combined use of enantio-pure (*R*)-**1**′ and racemic (±)-**1**′ catalysts in a ratio of 1:1 results in a similar (+)-NLE (Run 2). In contrast, mixing enantio-pure MS-free binol-Ti catalysts, (*R*)- and (*S*)-**1**′ in a ratio of 3:1 leads to a linearity (Run 3). However, an MS-free catalyst obtained by mixing (*R*)- and (*S*)-**1**′ catalysts in the same ratio of 3:1, in the presence of MS, which is filtered off prior to the reaction, shows a (+)-NLE (Run 4). Moreover, in dichloromethane, the combined use of (*R*)- and (*S*)-**1**′ catalysts (3:1), even without prior treatment with MS, exhibits a (+)-NLE (Run 7). These experimental facts can be explained in the case that the complex consists of oligomers which do not interconvert in the absence of MS in toluene but do interconvert in dichloromethane (Runs 3 vs. 4 and 7).

When the reaction is carried out in the presence of MS, however, a (–)-NLE is observed (Run 5), because MS acts as an achiral catalyst for the Diels-Alder reaction (Run 6).

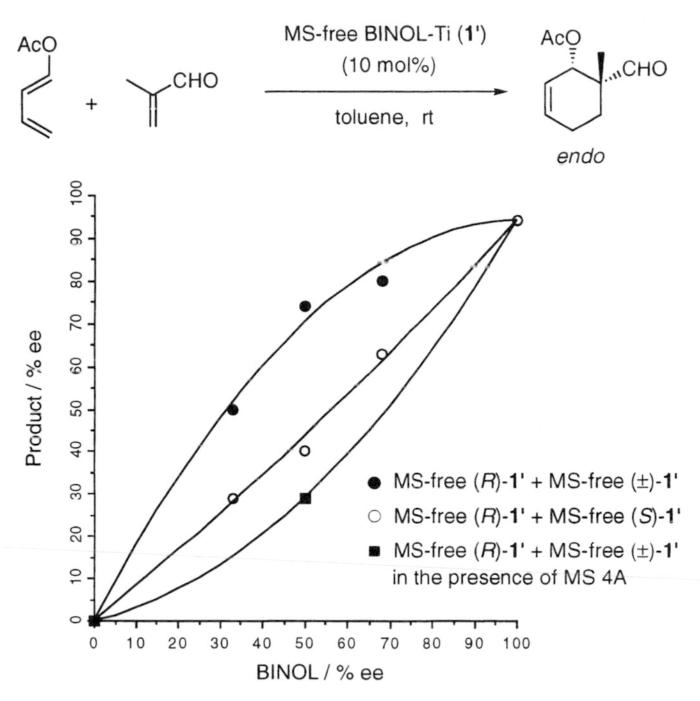

Figure 7-7. (+)-, (–)-NLE, and linear relationships depending on the catalyst preparation.

Table 7-1. NLE in asymmetric Diels-Alder reaction of 1-acetoxy-1,3-butadiene and methacrolein catalyzed by MS-free binol-Ti (**1'**).

Run	Preparation of MS-free binol-Ti (**1'**)	Ee [%] of **1'**	Yield [%]	*Endo* [%]	Ee [%]
1	Partially resolved binol (52% ee) and $Cl_2Ti(OiPr)_2$	52	41	98	76
2	MS-free (*R*)-**1'** and MS-free (±)-**1'** (1:1)	50	50	99	74
3	MS-free (*R*)-**1'** and MS-free (*S*)-**1'** (3:1)	50	62	99	40
4	MS-free (*R*)-**1'** and MS-free (*S*)-**1'** (3:1) in the presence of MS 4A which was filtered prior to the reaction	50	67	99	60
5	MS-free (*R*)-**1'** and MS-free (±)-**1'** (1:1) in the presence of MS 4A	50	62	95	29
6	Without MS-free catalyst (**1'**) in the presence of MS 4A	–	20	–	–
7	MS-free (*R*)-**1'** and MS-free (*S*)-**1'** (3:1) in CH_2Cl_2	50	52	99	53

Keck has reported that binol-derived titanium catalyst prepared in the presence or absence of MS 4A showed a (+)-NLE or a linearity, respectively [23].

Kagan has reported NLE as an indicator for distinction of closely related chiral catalysts (Eq. (7.6)) [20]. In asymmetric oxidation of sulfides with hydroperoxides promoted by chiral DET-Ti complexes, a wide diversity of titanium species is observed just by minor modifications of the catalyst preparation step. Stoichiometric use of a 1:4 mixture of $Ti(O\textit{i}Pr)_4$ and DET exhibits (–)-NLE. An addition of *i*PrOH to this mixture, e.g. a 1:4:4 mixture of $Ti(O\textit{i}Pr)_4$, DET, and *i*PrOH, provides (+)-NLE, while catalytic use of this ternary system leads to the disappearance of NLE.

DET-Ti

	DET % ee		
$Ti(O\textit{i}Pr)_4$: (*R*,*R*)-DET	50% ee	26% ee	(–)-NLE
= 1:4 (100 mol%)	100% ee	82% ee	
$Ti(O\textit{i}Pr)_4$: (*R*,*R*)-DET : *i*PrOH	50% ee	74% ee	(+)-NLE
= 1:4:4 (100 mol%)	100% ee	90% ee	
$Ti(O\textit{i}Pr)_4$: (*R*,*R*)-DET : *i*PrOH	50% ee	39% ee	Linear
= 1:4:4 (10 mol%)	100% ee	89% ee	

(7.6)

The study of NLE in asymmetric catalysis is useful in getting mechanistic insight and information about the active species involved in the catalytic cycle and their behavior in solution [24]. Jacobsen has used NLE as a mechanistic probe for the asymmetric ring opening of epoxides with trimethylsilyl azide catalyzed by the chiral Cr(salen) complex (Eq. (7.7)) [25]. The observation of significant (+)-NLE coupled with a second-order kinetic dependence on the Cr(salen) catalyst leads to a mechanistic proposal for simultaneous activation of both the epoxide and the azide by two different Cr(salen) complexes. On the basis of this cooperative mechanism they designed dimeric analogues of the Cr(salen) complex. Covalent linkage of the Cr(salen) complex unit with suitable tether length and position resulted in catalysts more reactive by 1–2 orders of magnitude than the monomeric analogues without any loss of enantioselectivity.

On the basis of NLE studies coupled with kinetic analyses, Denmark has disclosed that the mechanism of the rate acceleration by chiral phosphoramides in asymmetric aldol reactions of trichlorosilyl enolates with aldehydes stemmed from the ionization of the enolate by the basic phosphoramides (Eq. (7.8)) [26]. Sterically demanding phosphoramides (R=Ph) exhibit a linear relationship, through binding to the enolate in a 1:1 fashion and the resulting pentacoordinated cationic siliconate. In contrast, sterically less demanding phosphoramides (R=Me) with (+)-NLE can bind in a 2:1 fashion to result in the hexacoordinated cationic siliconate.

(salen)Cr

+ $TMSN_3$ → HO, N_3

Monomeric complex (1 mol%, 24 h, 100% yield) 93% ee
Dimeric analogue (n = 5) (0.05 mol%, 24 h, 100% yield) 93% ee

Dimeric analogue

(7.7)

Recently, Blackmond has demonstrated [27] that in these nonlinear catalytic systems a detailed analysis of the experimental reaction rate can give an independent confirmation of the mathematical models developed by Kagan [9a].

$OSiCl_3$ + PhCHO

(10 mol%)

CH_2Cl_2, –78 °C

(7.8)

Thus, consideration of the kinetic behavior of nonlinear catalytic reactions can provide valuable mechanistic insight into the NLE by comparison of the prediction of the models.

7.3 Auto-Catalysis

Another aspect of NLE is "asymmetric autocatalysis" as an event following symmetry breaking in nature. On the origin of chirality in nature, two major mechanisms have been proposed [28]. (1) Chance mechanism: To generate an optically active molecule just by chance followed by self-replication. (2) Determining mechanism to favor the one enantiomers: Some physicochemical elements are the determining factors to provide a non-equivalence of enantiomers. Non-conservation of parity may be due to weak interactions leading to small difference in energy between two enantiomeric forms [29]. A solid chiral adsorbent such as quartz can be another factor [30, 31]. Circularly polarized light [32] and geophysical fields such as rotation of the earth and magnetic fields [33] have long been proposed as a determining factor.

The non-equivalence of enantiomers through the spontaneous breaking of mirror-symmetry in nature is amplified by asymmetric autocatalytic reaction [34], e.g. Frank's "spontaneous asymmetric synthesis" [35, 36] (Fig. 7-8). Alberts and Wynberg have reported in "enantioselective autoinduction" that chiral lithium alkoxide products may be involved in the reaction to increase the enantioselectivity (Eq. (7.9)) [37]. The product % ee however does not exceed the level of catalyst % ee. In asymmetric hydrocyanation catalyzed by cyclic dipeptides, the (*S*)-cyanohydrin product complexes with the cyclic peptide to increase the enantioselectivity in the (*S*)-cyanohydrin product, the reaction going up to 95.8% ee (Eq. (7.10)) [38]. In the presence of achiral amine, (*R*)-1-phenylpropan-1-ol catalyzed carbonyl-addition reaction of diethylzinc has been reported to show lower % ee than that of the catalyst employed [39].

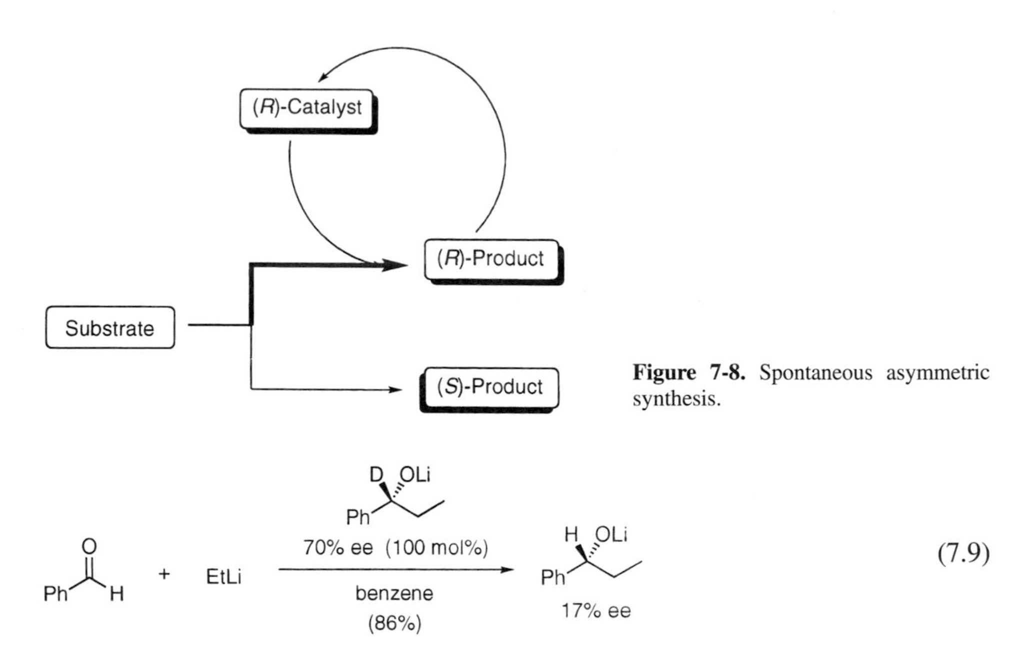

Figure 7-8. Spontaneous asymmetric synthesis.

Soai has reported the remarkable example of asymmetric autocatalysis in carbonyl-addition reactions of diisopropylzinc [40–43, 45]. Usually, zinc alkoxide forms an inactive tetramer. However, the use of pyridyl aldehyde as a substrate to give pyridyl alcohol product can loop the catalytic cycle without formation of the inac-

PhO-C₆H₄-CHO + HCN → (cyclo-dipeptide catalyst, 100% ee (2.2 mol%); toluene, 5 °C) → PhO-C₆H₄-CH(OH)CN (7.10)

	Time (h)	
In the absence of cyanohydrin	0.5	34.4% ee (*S*) (21%)
	1	66.2% ee (*S*) (39%)
	2	91.6% ee (*S*) (92%)
	4	92.0% ee (*S*) (94%)
In the presence of cyanohydrin 92.0% ee (*S*) (8.8 mol%)	0.5	95.8% ee (*S*) (55%)

tive tetramer [40]. In this autocatalytic system the product % ee does not exceed the level of catalyst % ee [41], while a chiral quinolyl alcohol as a catalyst instead of the pyridyl counterpart gives the product without any loss of enantio-purity [42 a]. A (+)-NLE is also found in the quinolyl alcohol as a catalyst [42 b]. A significant improvement of (+)-NLE is achieved by Soai in a similar carbonyl-addition reaction, however, to pyrimidyl aldehyde [43]. Starting from the (*S*)-alcohol in 2% ee (20 mol%), the 1st reaction provides the (*S*)-alcohol in 10% ee. The 4th reaction provides 88% ee via 57 and then 81% ees (Eq. (7.11)) [43 a]. Soai has also investigated an enantioselective autoinduction in the reduction of α-amino ketones with lithium aluminium hydride modified with a chiral 1,2-amino alcohol and an achiral amine [44]. He has also demonstrated amplification of a fairly small non-equivalence of enantiomers on the basis of asymmetric autocatalysis [45].

pyrimidine-5-carbaldehyde + (i-Pr)$_2$Zn → (zinc alkoxide catalyst (20 mol%); toluene, 0 °C) → pyrimidyl alcohol (7.11)

	Catalyst	Mixture of catalyst and product	Product
1st	2% ee	10% ee (46%)	16% ee (26%)
2nd	10% ee	57% ee (75%)	74% ee (55%)
3rd	57% ee	81% ee (80%)	89% ee (60%)
4th	81% ee	88% ee (75%)	90% ee (55%)
5th	88% ee	88% ee (70%)	88% ee (59%)

Thus, a little non-equivalence of enantiomers caused by symmetry breaking can be amplified through asymmetric autocatalysis to a large enantiomeric non-equivalence in molecules as found in nature.

7.4 'Asymmetric Deactivation' of Racemic Catalysts

Whilst non-racemic catalysts can generate non-racemic products with or without the NLE, racemic catalysts (0% ee) inherently produce only racemic (0% ee) products. A strategy whereby a racemic catalyst is enantiomer-selectively deactivated by a chiral molecule as a "catalyst poison" has recently been reported to yield non-racemic products (Fig. 7-3) [46–48]. A unique resolution of racemic CHIRAPHOS has been attained with a chiral iridium complex to give a deactiva-

(±)-CHIRAPHOS (2 mol%)

$R^* = CO_2^-$ (menthyl)

BF_4^- (1.2 mol%)

$[(nbd)_2Rh]^+ BF_4^-$ (0.8 mol%)

(7.12)

Inactive complex + Active catalyst

H_2, CH_2Cl_2

87% ee

cf. (*R*,*R*)-$[(chiraphos)Rh(nbd)]^+$ 90% ee

0% ee (10 mol%) / (10 mol%) H_3O^+

CH_2Cl_2, −78 °C (97%)

(7.13)

	80 (82% ee)	:	20
cf. (*S*)-Cat.	90 (95% ee)	:	10

(±)-[{(chiraphos)Rh}$_2$]$^{2+}$ / (6.7 mol%)

Ph_2PO ... NMe_2 ... SMe (4.7 mol%)

MeO_2C ... CO_2Me → (H_2, THF) MeO_2C ... CO_2Me

(7.14)

49% ee

cf. (*R*,*R*)-[{(chiraphos)Rh}$_2$]$^{2+}$ >98% ee

$RuCl_2$[(±)-binap](dmf)$_n$ / (0.3 mol%)

(1*R*,2*S*)-ephedrine (3 mol%)

H_2 (10 atm)

CH_2Cl_2 / MeOH

(7.15)

	>95% ee (23%)	(77%)
cf. $RuCl_2$[(*S*)-binap](dmf)$_n$	>95% ee (40%)	(60%)

tion form, leading to a chiral rhodium complex in association with the remaining enantiomer of CHIRAPHOS [46]. This process eventually results in a non-racemic hydrogenation product (Eq. (7.12)). A racemic aluminium reagent has been discriminated using chiral unreactive ketones to yield hetero Diels-Alder products with the remaining enantiomer of the aluminium reagent (Eq. (7.13)) [47]. More recently, "chiral poisoning" [48, 49] has been named for such a *deactivating* strategy in the context of a similar hydrogenation reaction by the asymmetric catalysis of the same CHIRAPHOS-Rh complex (Eq. (7.14)) [48 a, b]. A chiral amino alcohol, (1*R*,2*S*)-ephedrine, is also employable as a poison in the kinetic resolution of cyclic allylic alcohols using racemic binap (Eq. (7.15)) [48b, c]. However, the level of asymmetric induction does not exceed the level attained by the enantio-pure catalyst (Fig. 7-3).

Enantiomerically pure diisopropoxytitanium tartrate can also be used as a poison for racemic binaphthol-derived titanium diisopropoxide (**2**) (Eqs. (7.16) and (7.17)) [48d, e]. The % ee of the product increases with an increase in the amount of DIPT employed.

$Ti(O\textit{i}Pr)_4$ (30 mol%) / (±)-BINOL (20 mol%) / (–)-DIPT (X mol%); MS 4A, CH_2Cl_2

PhCHO + allyl-$SnBu_3$ → PhCH(OH)CH$_2$CH=CH$_2$ (7.16)

(–)-DIPT		
15 mol%	39% ee	(40%)
20 mol%	81% ee	(47%)
30 mol%	91% ee	(63%)

$Cl_2Ti(O\textit{i}Pr)_2$ (30 mol%) / (±)-BINOL (20 mol%) / (–)-DIPT (30 mol%); MS 4A, CH_2Cl_2 (87%)

methylenecyclopentane + HC(O)CCl_3 → (7.17)

OH/CCl_3 (cyclopentenyl)		OH/CCl_3 (cyclopentylidene)
64% ee		24% ee
91	:	9

7.5 'Asymmetric Activation' of Racemic Catalysts

An alternative but conceptually opposite strategy has been reported for asymmetric catalysis by racemic catalysts. A *chiral activator* selectively activates one enantiomer of a racemic chiral catalyst. A higher level of catalytic efficiency by more than two orders of magnitude ($k_{act} > k \times 10^2$), in addition to a higher enantioselectivity, might be attained than that achieved by an enantio-pure catalyst (x_{act}% ee > x% ee) (Fig. 7-2).

The ene reaction is one of the simplest ways for C–C bond formation, which converts readily available olefins with "C–H bond activation" at an allylic site by allylic transposition of the C=C bond into more functionalized products. The ene reaction encompasses a vast number of variants in terms of the enophile used [15b, 50]. Amongst others, the ene reactions of carbonyl enophiles, aldehydes in particular, which we refer to as 'carbonyl-ene reactions' [15], should in principle constitute a more efficient alternative to the carbonyl addition reaction of allylmetals for stereocontrol [51].

Catalysis of carbonyl-ene reaction with racemic binolato-Ti(O*i*Pr)$_2$ (**2**) achieves extremely high enantio-selectivity by adding another diol for the enantiomer-selective activation (Eq. (7.18) and Table 7-2) [52]. Significantly, a remarkably high enantioselectivity (89.8% ee, *R*) can be achieved using just a 0.5 equimolar amount (5 mol%) of (*R*)-binol activator added to a *racemic* (±)-binolato-Ti(O*i*Pr)$_2$ complex (**2**) (10 mol%). There is no spiro (binolato)$_2$Ti complex formation and ligand exchange reaction between (*S*)-binolato ligand with an additional (*R*)-binol observed within a reasonable reaction time.

(±)-BINOLato-Ti(O*i* Pr)$_2$ (**2**)
(10 mol%)

Chiral activator
(5 mol%)

toluene
0 °C, 1 h

(*R*)

(7.18)

The activation of the enantio-pure (*R*)-binolato-Ti(O*i*Pr)$_2$ catalyst (**2**) is also synthetically useful by further addition of (*R*)-binol (Eq. (7.19) and Table 7-3). The reaction proceeded quite smoothly to provide the carbonyl-ene product in higher chemical yield (82.1%) and enantioselectivity (96.8% ee) than those without additional binol (94.5% ee, 19.8%) (Run 2 vs. 1). Comparing the results of enantiomer-selective activation of the racemic catalyst (89.8% ee, *R*) (Table 7-2, Run 4) with those of the enantio-pure catalyst [with (96.8% ee, *R*) or without activator (94.5% ee, *R*)], the reaction catalyzed by the (*R*)-binolato-Ti(O*i*Pr)$_2$/(*R*)-binol complex (**2'**) is calculated to be 26.3 times as fast as that catalyzed by the (*S*)-binolato-Ti(O*i*Pr)$_2$ (**2**) in the racemic case (Fig. 7-9a). Indeed, kinetic studies show that the reaction catalyzed by the (*R*)-binolato-Ti(O*i*Pr)$_2$/(*R*)-binol complex (**2'**) is 25.6 ($=k_{act}/k$) times as fast as that catalyzed by the (*R*)-binolato-Ti(O*i*Pr)$_2$ (**2**). These results imply that the racemic (±)-binolato-Ti(O*i*Pr)$_2$ (**2**) and half-molar amount of (*R*)-binol assemble preferentially into the (*R*)-binolato-Ti(O*i*Pr)$_2$/(*R*)-bi-

Table 7-2. Enantiomer-selective activation of racemic binolato-Ti(O*i*Pr)$_2$ (**2**).

Run	Chiral activator	Yield [%]	Ee [%]
1	None	5.9	0
2	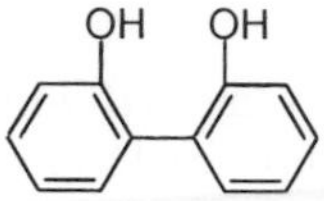 2,2'-biphenol (OH OH)	20	0
3	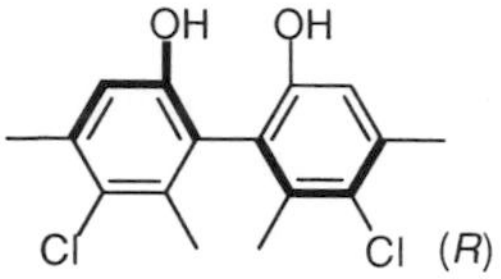 (R)-chloro-methyl-substituted biphenol (OH OH, Cl, Cl) (R)	38	80.8
4	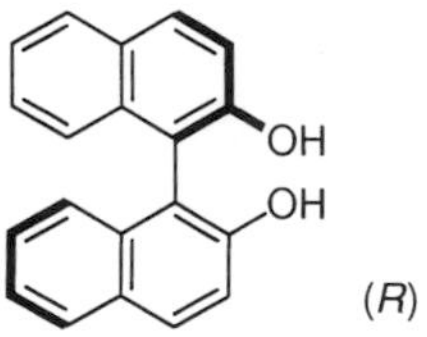 (R)-binol (OH, OH) (R)	52	89.8
5 [a]		35	80.0

[a]) 2.5 mol% of (*R*)-binol was used as a chiral activator.

nol complex (**2′**) and unchanged (*S*)-binolato-Ti(O*i*Pr)$_2$ (**2**). In contrast, the enantiomeric form of the additional chiral ligand [(*S*)-binol] activates the (*R*)-binolato-Ti(O*i*Pr)$_2$ (**2**) to a smaller degree (Run 3), thus providing the carbonyl-ene product in lower optical (86.0% ee, *R*) and chemical (48.0%) yields than (*R*)-binol does.

Another possibility is explored using racemic binol as an activator (Run 4). Racemic binol is added to the (*R*)-binolato-Ti(O*i*Pr)$_2$ (**2**), giving higher yield and enantioselectivity (95.7% ee, 69.2%) than that obtained by the original catalyst (*R*)-binolato-Ti(O*i*Pr)$_2$ (**2**) without additional binol (94.5% ee, 19.8%) (Run 4 vs 1). Comparing the results (95.7% ee, *R*) by the racemic activator with those of enantio-pure catalyst, (*R*)-binolato-Ti(O*i*Pr)$_2$/(*R*)-binol (**2′**) (96.8% ee, *R*) or (*R*)-binolato-Ti(O*i*Pr)$_2$/(*S*)-binol (86.0% ee, *R*) (Run 4 vs. 2 and 3), the reaction catalyzed by the (*R*)-binolato-Ti(O*i*Pr)$_2$/(*R*)-binol complex (**2′**) is calculated to be 8.8 times as fast as that catalyzed by the (*R*)-binolato-Ti(O*i*Pr)$_2$/(*S*)-binol (Fig. 7-9b). Kinetic studies show that the reaction catalyzed by the (*R*)-binolato-Ti(O*i*Pr)$_2$/(*R*)-binol complex (**2′**) is 9.2 ($=k_{act}/k'_{act}$) times as fast as that catalyzed by the (*R*)-binollato-Ti(O*i*Pr)$_2$/(*S*)-binol.

The great advantage of asymmetric activation of the racemic binolato-Ti(O*i*Pr)$_2$ complex (**2**) is highlighted in a catalytic version (Table 7-2, Run 5). High enantioselectivity (80.0% ee) is obtained by adding less than the stoichiometric amount (0.25 molar amount) of additional (*R*)-binol. A similar phenomenon on enantiomer-selective activation has been observed in aldol (Eq. (7.20)) [53] and

(*R*)-BINOLato-Ti(O*i*Pr)$_2$ (**2**)
(10 mol%)

BINOL (10 mol%)

Ph–C(=CH$_2$)CH$_3$ + H–C(=O)–CO$_2$*n*Bu → (toluene, 0 °C, 1 h) → (*R*)-product (OH, CO$_2$*n*Bu) (7.19)

Table 7-3. Asymmetric activation of enantio-pure (*R*)-binolato-Ti(O*i*Pr)$_2$ (**2**).

Run	binol	Yield [%]	Ee [%]
1	None	19.8	94.5
2	(*R*)-binol	82.1	96.8
3	(*S*)-binol	48.0	86.0
4	(±)-binol	69.2	95.7

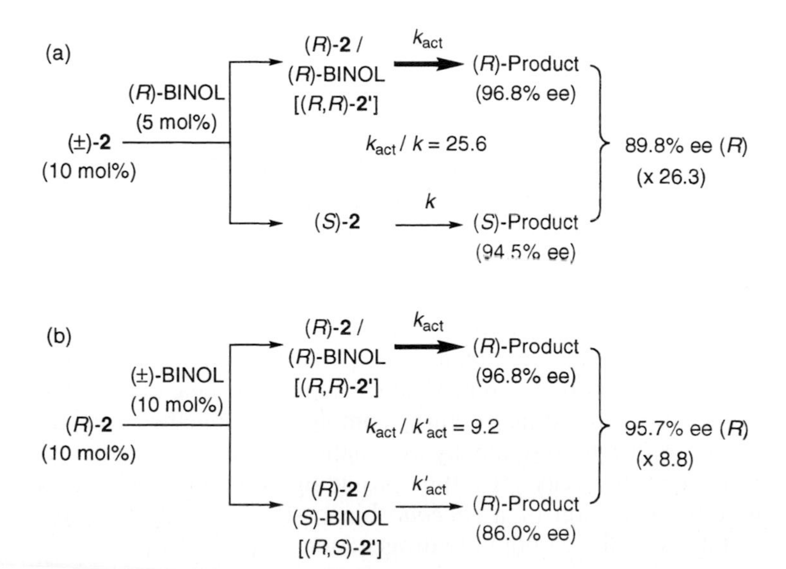

Figure 7-9. Kinetic feature of asymmetric activation of binolato-Ti(O*i*Pr)$_2$.

hetero Diels-Alder reactions (Eq. (7.21)) [54] catalyzed not only by a racemic but also by an enantiomerically pure binolato-Ti(O*i*Pr)$_2$ catalyst (**2**). Asymmetric activation of the (*R*)-binolato-Ti(O*i*Pr)$_2$ (**2**) by (*R*)-binols is essential to provide higher levels of enantioselectivity than those attained by the enantio-pure binolato-Ti(O*i*Pr)$_2$ catalyst (**2**) (5% ee) in the hetero Diels-Alder reaction of glyoxylates with the Danishefsky diene (Eq. (7.21)).

(*R*)-BINOLato-Ti(O*i* Pr)$_2$ (**2**) (10 mol%); (*R*)-BINOL (10 mol%); *t*BuS(OSiMe$_3$)C=CH$_2$ + H(C=O)C$_8$H$_{17}$ → toluene, 0 °C, 4 h → HCl / MeOH → *t*BuS(C=O)CH$_2$CH(OH)C$_8$H$_{17}$ (*S*)

97% ee (66%)
Without (*R*)-BINOL 91% ee (53%)

(7.20)

(*R*)-BINOLato-Ti(O*i* Pr)$_2$ (**2**) (10 mol%); (*R*)-BINOL (10 mol%); OMe, *t*BuMe$_2$SiO diene + H(C=O)CO$_2$*n*Bu → toluene, 0 °C, 4 h ~ 8 h → TFA → dihydropyranone CO$_2$*n*Bu (*R*)

84% ee (50%)
Without (*R*)-BINOL 5% ee (40%)

(7.21)

Activation of the (*R*)-binolato-Ti(O*i*Pr)$_2$ (**2**) by highly acidic and sterically demanding alcohols as achiral rather than chiral activators is also effective to provide higher levels of enantioselectivity than those attained by the parent enantio-pure binolato-Ti(O*i*Pr) catalyst (**2**) in the Mukaiyama aldol reaction of silyl enol ethers (Eq. (7.22)) [55].

Catalytic asymmetric hydrogenation has been shown to be one of the most efficient processes for the asymmetric functional group transformation of organic molecules. Noyori et al. have reported a remarkable example of enantioselective catalysis by the *enantio-pure* RuCl$_2$(binap)(dmf)$_n$ complex (**3**) together with an enantio-pure diamine and KOH to provide hydrogenation products of carbonyl compounds with high enantioselectivity [56], thus providing an opportunity for us to examine an asymmetric activation of a *racemic* binaps-RuCl$_2$ catalyst (**3**) for the enantioselective catalysis of the carbonyl hydrogenation (Eq. (7.23)) [57].

(*R*)-BINOLato-Ti(O*i* Pr)$_2$ (**2**)
(10 mol%)

ArOH (10 ~ 20 mol%)
MS 4A

$OSiMe_3$, *t* BuS + H, C_8H_{17} → toluene, 0 °C, 4 h → HCl / MeOH → *t* BuS, O, OH, C_8H_{17} (*S*)

(7.22)

ArOH			
2-naphthol	(20 mol%)	93% ee	(55%)
pentafluorophenol	(20 mol%)	97% ee	(40%)
pentafluorophenol	(10 mol%)	97% ee	(62%)
phenol	(10 mol%)	96% ee	(61%)
Without ArOH		91% ee	(53%)

Racemic BINAPs-$RuCl_2$ (**3**) / (*S,S*)-DPEN / KOH, *i* PrOH

Ar(C=O)CH$_3$ + H_2 → Ar-CH(OH)-CH$_3$ *

(±)-BINAPs (PAr_2, PAr_2); Ph, Ph, H_2N, NH_2 (*S,S*)-DPEN

(7.23)

a: Ar = 4-methylphenyl (TolBINAP)
b: Ar = 3,5-dimethylphenyl (DM-BINAP)
c: Ar = phenyl (BINAP)

The hydrogenation was performed in a mixture of racemic $RuCl_2$(tolbinap)(dmf)$_n$ (**3a**) [58] or $RuCl_2$(dmbinap)(dmf)$_n$ (**3b**) [59], an enantio-pure diamine such as (*S,S*)-1,2-diphenylethylenediamine [(*S,S*)-dpen] [60] or the (*R,R*)-enantiomer, and KOH in a ratio of 1:1:2, in modification of the reported procedure with the enantio-pure $RuCl_2$(binap)(dmf)$_n$ (**3c**) (Table 7-4).

A chiral diamine leads to a non-racemic hydrogenation product, supporting the importance of chirality in the diamine activator for selective activation of one enantiomer of the (±) $RuCl_2$(tolbinap) catalyst (**3a**) (Runs 2 vs 3). Thus, the asym-

AA AN

Table 7-4. Asymmetric activation of racemic binaps-$RuCl_2$ catalyst (**3**) by enantio-pure dpen [a]).

Run	3	Ketone	*T* [°C]	*t* [h]	Yield [%]	Ee [%]
1 [b])	(*R*)-**3a**	**AA**	28	18	2	29 (*S*)
2 [b])	(±)-**3a**	**AA**	28	18	<1	0
3	(±)-**3a**	**AA**	28	18	28	80 (*R*)
4	(±)-**3a**	**AA**	80	10	99	80 (*R*)
5	(*R*)-**3a**	**AA**	80	10	99	81 (*R*)
6	(*S*)-**3a**	**AA**	80	10	91	40 (*R*)
7	(±)-**3b**	**AN**	28	4	99	80 (*R*)
8	(±)-**3b**	**AN**	–35	7	95	90 (*R*)
9 [c])	(±)-**3b**	**AN**	–35	7	90	90 (*R*)
10	(*S*)-**3b**	**AN**	28	4	99	>99 (*R*)
11	(*R*)-**3b**	**AN**	28	4	99	56 (*S*)

[a]) Under H_2 (8 atm) atmosphere. Ketone : **3** : (*S,S*)-dpen : KOH=250 : 1 : 1 : 2. [b]) In the absence of (*S,S*)-dpen. [c]) 0.5 Molar amount of (*S,S*)-dpen per (±)-**3b** was used. **AN** : **3B** : dpen : KOH=250 : 1 : 0.5 : 2.

metric activation of the chiral $RuCl_2$(tolbinap) catalysts (**3a**) by the chiral diamine affords higher levels of asymmetric induction and catalytic activity than those attained by the enantio-pure catalyst (**3a**) alone (Runs 1 vs 3) even when starting from the *racemic* mixture of **3a**. The enantioselectivity thus obtained by the (±)-$RuCl_2$(tolbinap) complex (**3a**) and (*S,S*)-dpen is very close to that obtained by the matched pair [61] of (*R*)-$RuCl_2$(tolbinap) (**3a**)/(*S,S*)-diamine complex as exemplified (*R*)-binaps-$RuCl_2$(**3**)/(*S,S*)-dpen [(*R*)/(*S,S*)-**A**] (Runs 4 vs 5 and 6). However, the matched pair is dramatically changed over on going from 9-acetylanthracene (**AA**) to 1′-acetonaphthone (**AN**) (Runs 7 ~ 11); in the latter case, (*S*)-binaps-$RuCl_2$ (**3**)/(*S,S*)-dpen complex (**B**) is the more enantioselective combination than (*R*)-binaps-$RuCl_2$ (**3**)/(*S,S*)-dpen (**A**) to provide (*R*)-(+)-product in higher % ee (Runs 10 vs 11) (Fig. 7-10).

The dichotomous sense in enantioselectivity is determined by the ratio and catalytic activity (turnover frequency) of mono- or dihydrido binaps-RuHX/dpen complexes (X=H or Cl) [62], **A**′ and **B**′, which are derived from the diastereomeric complexes **A** and **B**, respectively, under the hydrogenation conditions (Fig. 7-10). It should be noted here that the catalytic activity critically depends on the nature of the carbonyl substrates. Interestingly, the use of a catalytic amount of diamine affords an equally high level of enantioselectivity to that obtained by an equimolar amount of diamine (Runs 9 vs 8). Indeed, the ^{31}P NMR spectrum

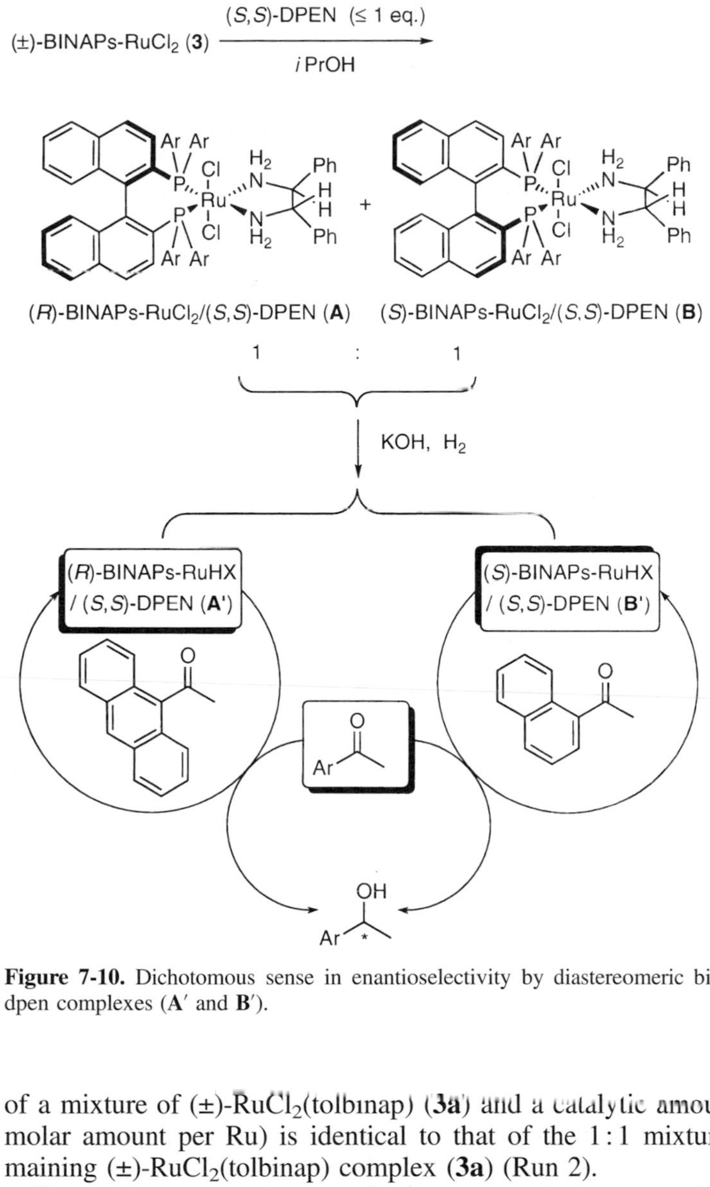

Figure 7-10. Dichotomous sense in enantioselectivity by diastereomeric binaps-RuHX (X=H or Cl)/dpen complexes (**A**′ and **B**′).

of a mixture of (±)-$RuCl_2$(tolbinap) (**3a**) and a catalytic amount of (*S*,*S*)-dpen (0.5 molar amount per Ru) is identical to that of the 1:1 mixture, except for the remaining (±)-$RuCl_2$(tolbinap) complex (**3a**) (Run 2).

The asymmetric activation phenomena can be interpreted through a continuum from the preferential complexation with the one enantiomer catalyst selectively giving the single activated diastereomer to the 1:1 complexation giving the activated diastereomeric mixture (1:1) in which the catalyst efficiency (turnover frequency) depends critically on the substrates employed.

For simplicity, the formation of the activated complexes can be discussed starting from the complexation of the chiral activator with racemic parent catalyst in monomeric form, following the thermodynamic and/or kinetic features (Fig. 7-11).

(1) Under equilibrium conditions between the activated catalyst and the parent catalyst (Fig. 7-11 a), the ratio of the activated diastereomeric catalysts depends on their thermodynamic stability. (2) Under non-equilibrium conditions, the ratio reflects the relative rate of the reaction of the enantiomeric catalyst with the chiral activator (Fig. 7-11 b). Of course, the use of 1.0 equivalent of the activator per the parent catalyst falls into a 1:1 mixture of the diastereomeric complexes. The kinetic or thermodynamic features described above are more apparent under the treatment with less than 1.0 equivalent of the activator.

Even with 0.5 equivalent of the activator, once a 1:1 mixture is formed, relative activity of these activated diastereomeric catalysts to the substrate is the factor to determine the outcome in terms of enantioselectivity of the asymmetric reaction. In other words, the turnover efficiency of these activated diastereomers should be dependent on the complex with the substrate used.

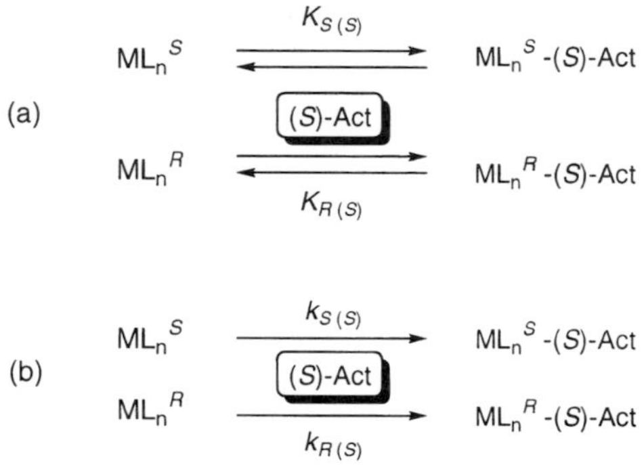

(a)
$ML_n^S \underset{}{\overset{K_{S(S)}}{\rightleftharpoons}} ML_n^S$-(*S*)-Act
(*S*)-Act
$ML_n^R \underset{K_{R(S)}}{\rightleftharpoons} ML_n^R$-(*S*)-Act

(b)
$ML_n^S \xrightarrow{k_{S(S)}} ML_n^S$-(*S*)-Act
(*S*)-Act
$ML_n^R \xrightarrow[k_{R(S)}]{} ML_n^R$-(*S*)-Act

Figure 7-11. Formation of activated diastereomeric catalysts under thermodynamic (a) or kinetic (b) conditions.

Therefore, the most crucial step is the catalytic asymmetric reaction with the substrate. The logarithm of the relative rate is varied from, for example, 0.01 to 100 (Fig. 7-12). Let us examine the case that the one activated diastereomeric complex provides the product in 100% ee (*R*) and that the other diastereomer provides the opposite enantiomeric product in 50% ee (*S*). Even when two activated diastereomer complexes are formed in 1:1 ratio, more than 98% ee of the product can be established in the case where the relative rate of the two activated diastereomers is 100 (log $K_{rel\text{-}act}=2$).

In the case that the relative rate with the two activated diastereomers is 100 (Fig. 7-13), more than 90% ee of the product can be attained even in the –75% de (12.5%) presence of the favorable diastereomer (dotted line); thermodynamically unstable and hence catalytically more active complexes may be found [63]. Similar phenomena can be drawn in a different way for the 1:1 formation of diastereomers. The relative rate of 14 (log $K_{rel\text{-}act}=1.15$) is sufficiently high to provide more than 90% ee of the desired product (Fig. 7-12).

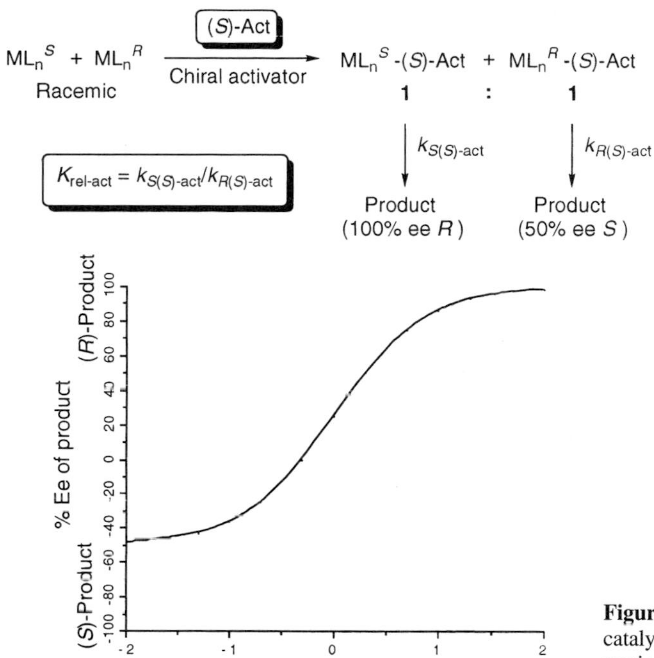

Figure 7-12. Asymmetric reaction catalyzed by activated 1:1 diastereomeric mixtures.

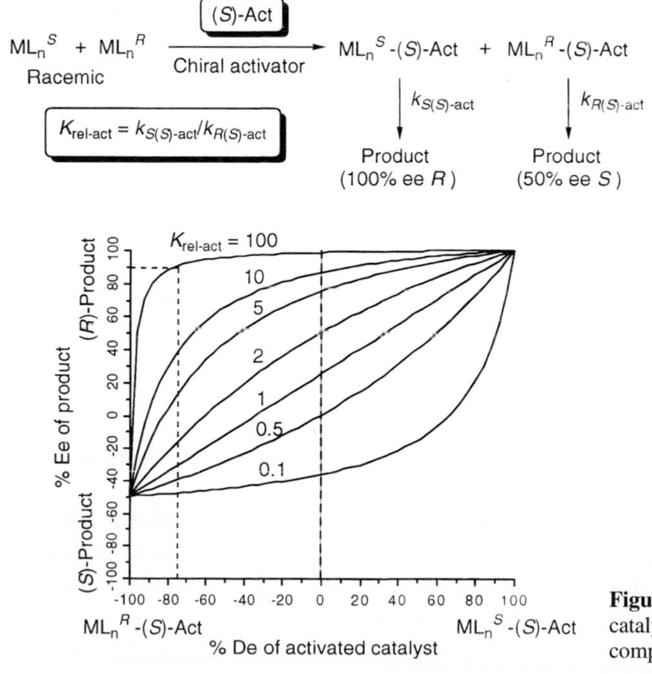

Figure 7-13. Asymmetric reaction catalyzed by activated diastereomeric complexes.

7.6 Asymmetric Activation of 'Pro-Atropisomeric' Catalysts

An advanced strategy for "asymmetric activation" can be seen in using chirally flexible ligands that achieves higher enantioselectivity than that attained by chirally rigid and hence racemic ligands. As described above, combination of a racemic binaps-$RuCl_2$ (**3**) species even with a 0.5 equimolar amount of an enantiomerically pure diamine gives a 1:1 mixture of two diastereomeric binaps-$RuCl_2$ (**3**)/dpen complexes. When the chirally rigid binaps is replaced by a flexible [64] and 'pro-atropisomeric' bipheps [65], diastereomeric complexes are formed, in principle, in unequal amounts (Fig. 7-14) [66]. When the major diastereomer shows higher chiral efficiency than that of the minor isomer, this strategy becomes more effective than the use of similar but chirally rigid analogues.

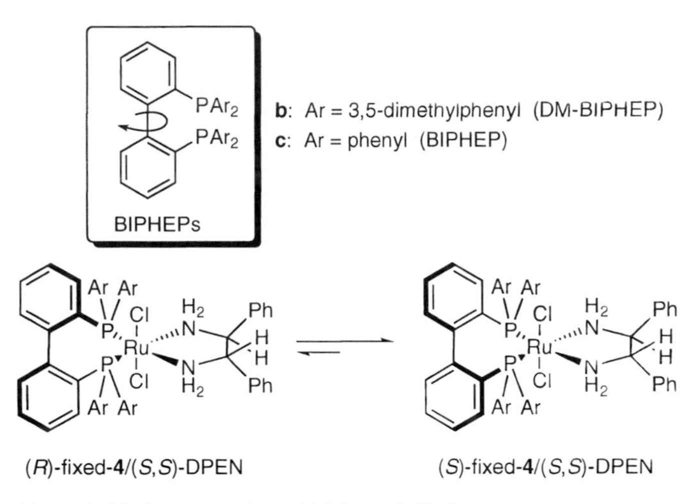

Figure 7-14. Stereomutation of bipheps-$RuCl_2$/dpen complexes.

The initially formed mixture of (*S*)- and (*R*)-$RuCl_2$(dmbiphep) (**4b**)/(*S,S*)-dpen in [D_8]2-propanol ($CDCl_3$:$(CD_3)_2CDOD$ = 1:2), when allowed to stand at room temperature (or at 80 °C), was found to give a 1:3 mixture of the (*R*)-**4b**/(*S,S*)-dpen-major diastereomers (Fig. 7-14). The equilibration occurred readily because of the conformational flexibility of bipheps-$RuCl_2$ (**4**)/diamine complexes. The dichloro complexes may be further converted to active mono- or dihydrido Ru species under hydrogenation conditions [62].

The significant effect of the conformationally flexible bipheps-$RuCl_2$ (**4**)/diamine complexes can be seen in hydrogenation (Table 7-5) of 1′-acetonaphthone (**AN**) (Run 1) in comparison with the enantioselectivity obtained using the (±)-$RuCl_2$(dmbiphep) (**3b**)/(*S,S*)-dpen complex (Run 2).

Table 7-5. Pro-atropisomeric biphep ligand for enantioselective hydrogenation [a]).

Run	Ketone	**4** or **3**	H_2 [atm]	*T* [°C]	*t* [h]	Yield [%]	Ee [%]
1	**AN**	**4b**	8	28	4	>99	84
2 [b])	**AN**	(±)-**3b**	8	28	4	>99	80
3	**AN**	**4b**	40	–35	12	>99	92
4 [b])	**AN**	(±)-**3b**	40	–35	7	>99	89
5	**AA**	**4c**	8	80	10	>99	70
6 [b])	**AA**	(±)-**3c**	8	80	10	>99	78

[a]) bipheps-$RuCl_2$ (**4**) / (*S,S*)-dpen in 2-propanol was pre-heated at 80 °C for 30 min. Ketone : **4** or **3** : (*S,S*)-dpen : KOH = 250 : 1 : 1 : 2. [b]) Without pre-heating operation.

Further increase in enantioselectivity was attained at a lower reaction temperature (Run 3). The enantioselectivity by the $RuCl_2$(dmbiphep) (**4b**)/(*S,S*)-dpen was higher than that by the (±)-$RuCl_2$(dmbinap) (**3b**)/(*S,S*)-dpen complex at the same low temperature and high pressure (Run 4). Thus, (*R*)-1-(1-naphthyl)ethanol was obtained with 92% ee in quantitative yield. $RuCl_2$(dmbiphep) (**4b**)/(*S,S*)-dpen was also useful in the reduction of *o*-methylacetophenone.

Brunner has reported the use of (*S,S*)-dpen to control the chirality of octahedral Co(II)$(acac)_2$/diamine complex. This chiral Co(II) complex catalyzes the Michael addition reaction to give the product in up to 66% ee at –50 °C (Eq. (7.24)) [67]. The (*S,S*)-configuration of the dpen ligand gives rise to the Δ-conformation of [Co$(acac)_2$(*S,S*)-dpen] complex to be thermodynamically more stable than the Λ-conformation [68]. The formation of (*R*)-product is rationalized on the basis of the assumptions that the ketone and ester carbonyl groups occupy the axial and equatorial positions, respectively, and that methyl vinyl ketone is directed to the *si*-face of the complexed ketoester via a hydrogen bonding of the vinyl ketone oxygen to one of the NH_2 groups.

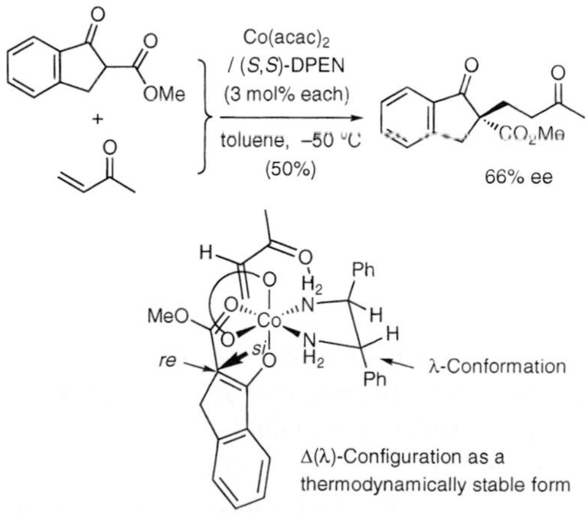

(7.24)

Self-organization of ligands in multi-component titanium catalysts [69] with conformationally flexible biphenols [52] is also found in the enantioselective glyoxylate-ene reaction [60] to give the significantly high enantioselectivity (Eq. (7.25)) [70]. Some molecular modeling was reported that the hexacoordination of the titanium atom would make the central titanium atom a center of chirality and that the Λ isomer is more favorable than the Δ isomer (Fig. 7-15).

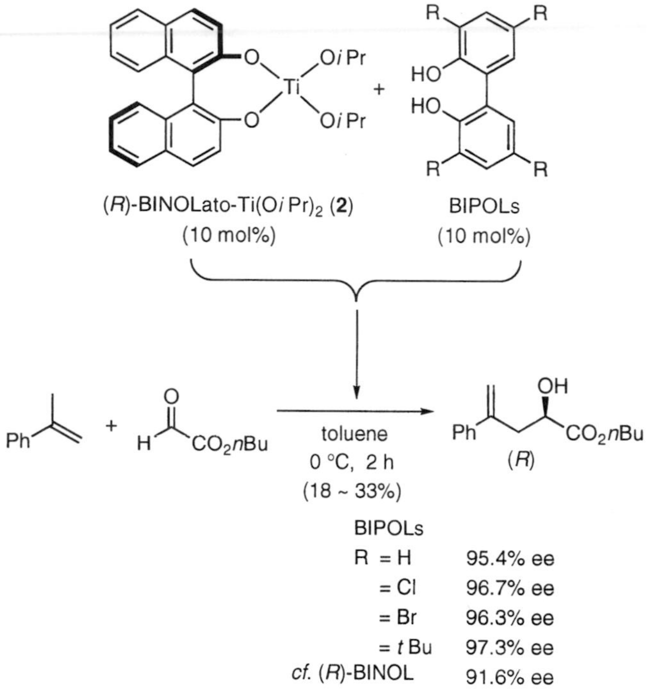

(R)-BINOLato-Ti(O*i* Pr)$_2$ (**2**) (10 mol%) + BIPOLs (10 mol%)

Ph–C(=CH$_2$)CH$_3$ + H–CO–CO_2nBu → (toluene, 0 °C, 2 h, (18 ~ 33%)) → (R)-product

(7.25)

BIPOLs	
R = H	95.4% ee
= Cl	96.7% ee
= Br	96.3% ee
= *t* Bu	97.3% ee
cf. (R)-BINOL	91.6% ee

(S)-BIPOL (R = *t* Bu)
(R)-BINOLato
Λ
-1.81 kcal

(R)-BINOLato
(S)-BIPOL (R = *t* Bu)
Δ
0.00 kcal

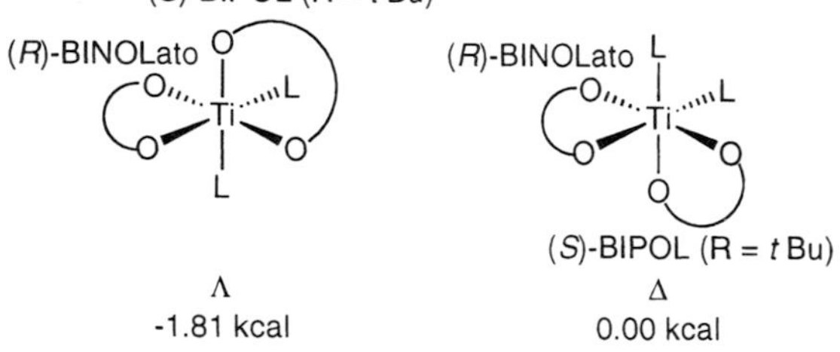

Figure 7-15. Energy differences of Λ and Δ isomers of (*R*)-binolato-Ti(O*i*Pr)$_2$/binol or bipol (R=*t*Bu) (L=*i*PrO).

Katsuki has studied asymmetric epoxidation of non-functionalized olefins catalyzed by chiral Mn(salen) complex. Recently they proposed that the ligands of Mn(salen) complexes take non-planar stepped conformation and the direction of the folding ligands is strongly related to the sense of chirality in the asymmetric epoxidation (Eq. (7.26)) [71]. On the basis of this proposal, conformational control of achiral Mn(salen) complex in two enantiomeric forms can be achieved by the use of chiral axial ligands (**AL***) [72]. The achiral Mn(salen) complex gave the epoxides in high enantio-purity in the presence of chiral bipyridine *N,N'*-dioxide as an axial ligand (**AL***).

(salen)Mn (4 mol%) / **AL*** (5 mol%), PhIO, MS 3A, $ClCH_2CH_2Cl$, −20 °C (90%); 83% ee (7.26)

(salen)Mn (+)-**AL***

Chiral *ansa*-metallocene complexes have become useful catalysts in asymmetric polymerization reactions [73]. While enantio-resolution of *ansa*-metallocene racemates cannot yield more than 50% of a particular enantiomer, the readily accessible racemate of a biphenyl-bridged metallocene complex (we abbreviate to 'biphecp'-M: M = Ti, Zr) has been quite recently reported to give enantio-pure *ansa*-titanocene and -zirconocene complexes through binol-induced asymmetric trans-

(*R*)-BIPHECp-Ti-(*R*)-BINOLato → 1) MeLi 2) HCl gas → (*R*)-BIHPECp-$TiCl_2$ (0.1 mol%); *n*BuLi (0.2 mol%); 148 atm H_2, toluene, 80 °C (96%); 98% ee (7.27)

formation (Fig. 7-16) [74]. The biphenyl-bridged complex (*R*)-biphecp-$TiCl_2$ in the presence of *n*BuLi resulted in an efficient asymmetric catalyst of imine hydrogenation (Eq. (7.27)).

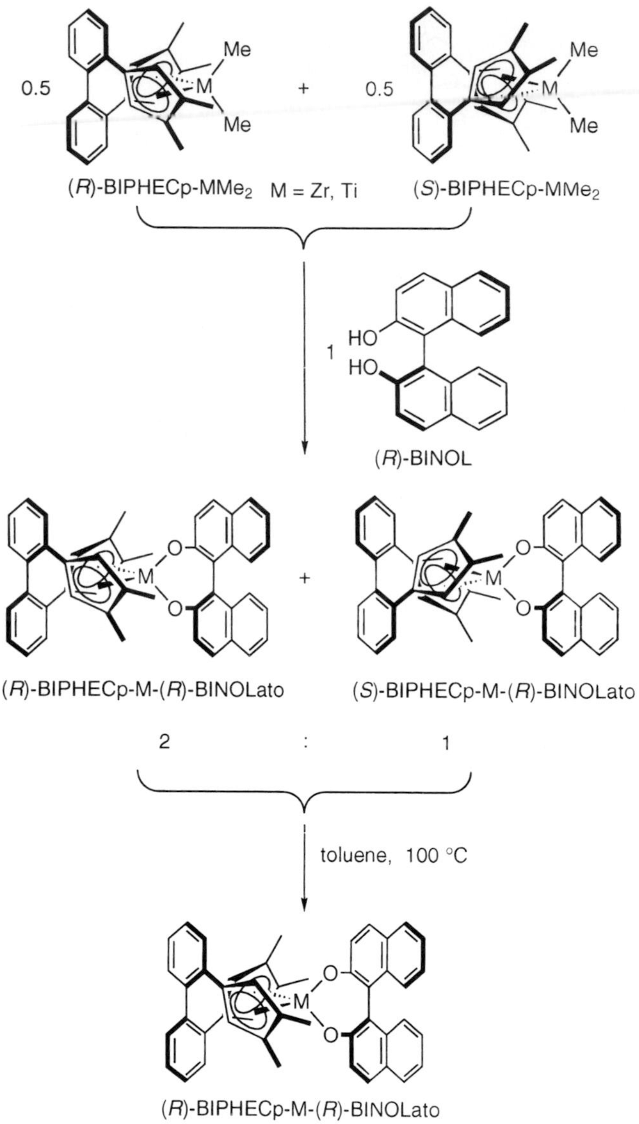

0.5 (*R*)-BIPHECp-MMe_2 + 0.5 (*S*)-BIPHECp-MMe_2 M = Zr, Ti

1 (*R*)-BINOL

(*R*)-BIPHECp-M-(*R*)-BINOLato + (*S*)-BIPHECp-M-(*R*)-BINOLato

2 : 1

toluene, 100 °C

(*R*)-BIPHECp-M-(*R*)-BINOLato

Figure 7-16. binolL-induced asymmetric transformation of biphenyl-bridged metallocene complexes.

Thus, the chirally rigid ligands can be replaced by flexible and hence 'pro-atropisomeric' ligands to give preferentially the favorable diastereomer with higher chiral efficiency than does the minor isomer. This strategy with flexible and 'pro-atropisomeric' ligands becomes more effective than the use of structurally similar but chirally rigid ligands. Therefore, the 'asymmetric activation' could provide a general and powerful strategy for the use of not only atropisomeric and hence racemic ligands but also chirally flexible and 'pro-atropisomeric' ligands without enantio-resolution.

7.7 High-Throughput Screening of Chiral Ligands and Activators

Combinatorial chemistry has been well recognized as a useful strategy for the discovery and optimization of drugs, metal complexes and solid-state materials [75]. Between the split-and-mix and parallel-matrix methodologies for combinatorial chemistry, the latter is more employable for lead optimization, wherein the high throughput screening (HTS) is an essential technique for tuning a variety of modulations [76]. However, a limited number of investigations have so far been reported even on chiral ligand optimization for coordination complexes [77]. Using *chiral* HPLC or GC analyses, it takes a long time to separate enantiomeric products and then to determine the enantioselectivity of the reactions. The application of a circular dichroism (CD)-based detection system in HPLC to *non*-chiral (*achiral*) stationary phases allows the simultaneous monitoring of the CD signal ($\Delta\varepsilon$), the absorption (ε) and their ratio ($g = \Delta\varepsilon/\varepsilon$) which is termed the dissymmetry or anisotropy factor. The g factor is independent of concentration and is linearly related to the enantiomeric excess [78]. With this technique, the % ee of the product could be determined within minutes without separation of the enantiomeric products using *chiral* stationary phases. Therefore, application of HPLC-CD provides a 'super high throughput screening (SHTS)' system for finding the most effective catalyst through asymmetric activation [79]. Chiral catalysts obtained via ligand exchange with chiral ligands (**L^{1}***, **L^{2}***, – –) may further evolve in parallel combination with chiral activators (**A^{1}***, **A^{2}***, – –) into the most catalytically active and enantioselective activated catalyst (Fig. 7-17).

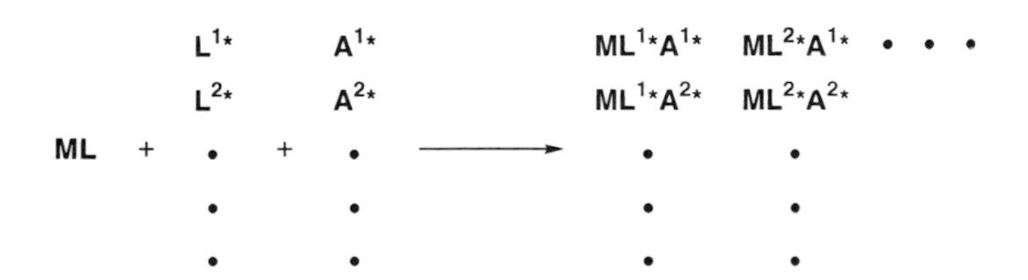

Figure 7-17. General principle for the creation of a catalyst system by asymmetric activation.

The super high throughput screening (SHTS) of the parallel solution library of activated catalysts is demonstrated in terms of the chiral ligands (**L**^{1}*, **L**^{2}*, – –) and activators (**A**^{1}*, **A**^{2}*, – –) for diol-Zn catalysts in the addition of diethylzinc to aldehydes by using HPLC-CD. Amongst asymmetric catalysis of C–C bond-forming reactions, enantioselective addition of diorganozinc reagents to aldehydes constitutes one of the most important and fundamental asymmetric reactions [1 c, 8 e, 80]. Since its initial report by Oguni [81], various chiral ligands including *β*-amino alcohols have been used for this type of reaction. However, less attention has been paid to C_2 symmetric binaphthols (binols) [82] despite their wide application as chiral ligands for B- [83], Al- [84], Ti- [85], Zr- [86], and Ln-catalysts [87] in enantioselective aldol, ene reactions and so forth, because of their lower catalytic activity and enantioselectivity for the organozinc addition reaction [88]. Only very recently, some modified binols [89], have been reported to be effective, but the simple binol itself is less effective in the reaction [88].

It is reasonable to assume that the active catalyst species is a monomeric zinc alkoxide in the addition of diethylzinc to aldehydes; the cleavage of the higher aggregates could result in an activation of the overall catalyst system (Fig. 7-18) [11 p, 90]. The addition of a chiral nitrogen activator for the activation of binols-zinc catalyst systems should be one of the most efficient ways because of its strong coordinating ability to the zinc cation to facilitate the alkyl transfer. As a result, a monomeric zinc complex is expected to be formed in a similar manner to that of chiral salen-zinc [91]. Furthermore, a bi-molecular combination of chiral activators with the diol-zinc complexes should be more convenient than the uni-molecular combination. Thus, the primary combinatory library of chiral ligands (**L**^{1}*–**L**^{5}*) and chiral activators (**A**^{1}*–**A**^{5}*) is initially examined, from which the lead can be further optimized for the next generation of the chiral ligands and activators (Fig. 7-19).

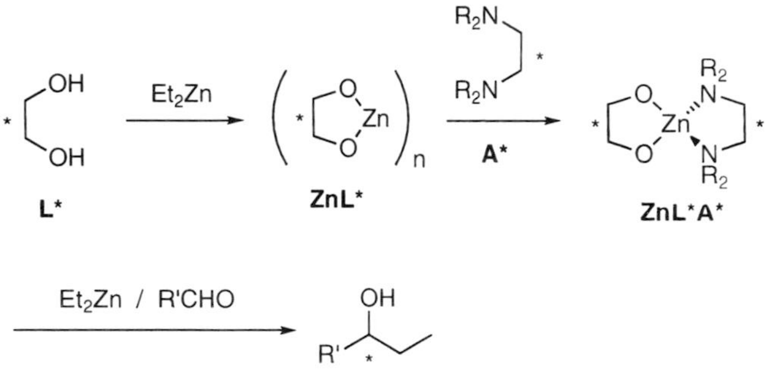

Figure 7-18. Asymmetric activation of chiral diol-zinc catalysts by chiral nitrogen ligands.

The activation effect was really observed in random screening of chiral ligand/activator combinations. Enantioselectivity of the reaction is also increased by matched combination of diol ligands and nitrogen activators. The replacement of 3- and 3′-positions with bulky phenyl groups, 3,3′-diphenyl-1,1′-bi-2-naphthol (**L**^{5}*), may further prevent the aggregation of binolato-Zn and increase the enantioselectivity, because of their steric demand to provide (*S*)-1-phenylpropanol in up to 65% ee and quantitative yields.

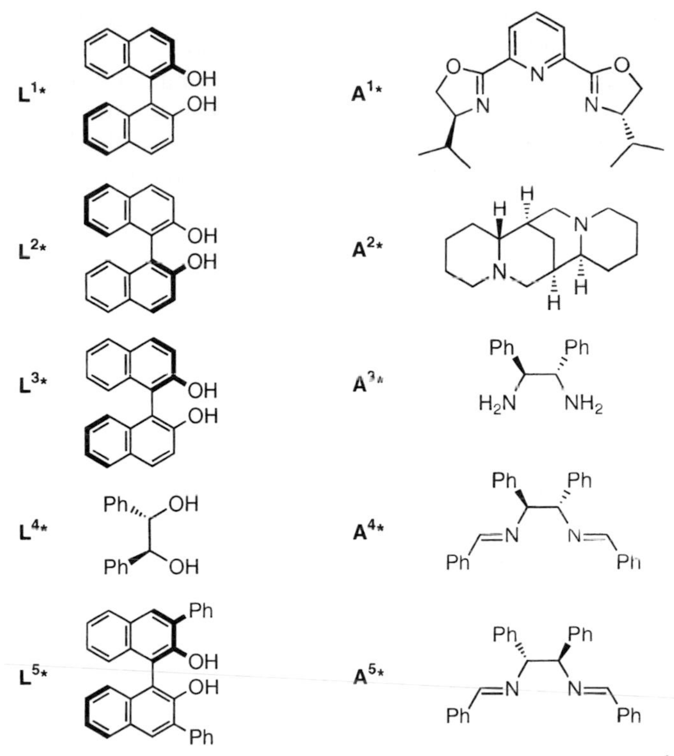

Figure 7-19. Primary combinatorial library of chiral ligands ($L^1* \sim L^5*$) and chiral activators ($A^1* \sim A^5*$).

On the basis of the results collected from the primary combinatorial library, we then create the next generation library of diimines (activators) with 12 members (A^4*–$A^{15}*$) (Fig. 7-20).

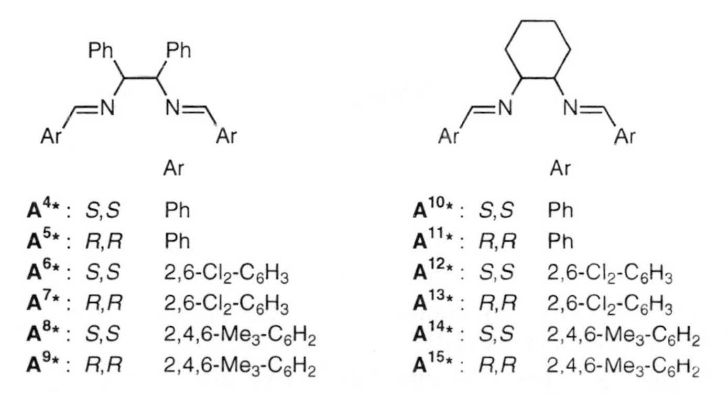

Figure 7-20. Second generation library of diimines as chiral activators ($A^4* \sim A^{15}*$)

All library members significantly activate the **L**^{5}*-Zn complex and produce 1-phenylpropanol in higher yields and enantioselectivities than those obtained by only using the ligands themselves. The steric hindrance of the chiral activators is crucial, and hence the activator **A**^{9}* provides the best results. The reaction catalyzed by the best combination **L**^{5}*/**A**^{9}* is further optimized by tuning the lower reaction temperature. (*S*)-1-phenylpropanol is obtained in 99% ee and quantitative yield (Eq. (7.28)). Even if 2 mol % of **L**^{5}*/**A**^{9}* is used for the reaction, (*S*)-phenylpropanol can be obtained in 97% ee and 100% yield.

PhCHO + Et_2Zn → (**L**^{5}* and **A**^{9}*; CH_2Cl_2, –78 °C to –20 °C) → (*S*)-PhCH(OH)Et (7.28)

L^{5}*/**A**^{9}* (10 mol% each) 99.0% ee (100%)
L^{5}*/**A**^{9}* (2 mol% each) 97.0% ee (100%)

The best combination of chiral ligands and activators can easily be found out in an efficient way by super high throughput screening (SHTS) employing CO-HPLC with achiral stationary phase for affording the most enantioselective activated catalyst to give excellent yields and enantioselectivities.

7.8 Smart Self-Assembly into the Most Enantioselective Activated Catalyst

Sharpless et al. have emphasized the significance of "chiral ligand acceleration" [3] through the construction of an asymmetric catalyst from an achiral pre-catalyst via ligand exchange with a chiral ligand. By contrast, an achiral pre-catalyst combined with several chiral ligand components (**L**^{1}*, **L**^{2}*, – – –) may selectively assemble into the most catalytically active and enantioselective activated catalyst (**ML**^{m}***A**^{n}*) found among the combinatorial library of possible activated catalysts with association of chiral activators (**A**^{1}*, **A**^{2}*, – – –) (Fig. 7-21) [69].

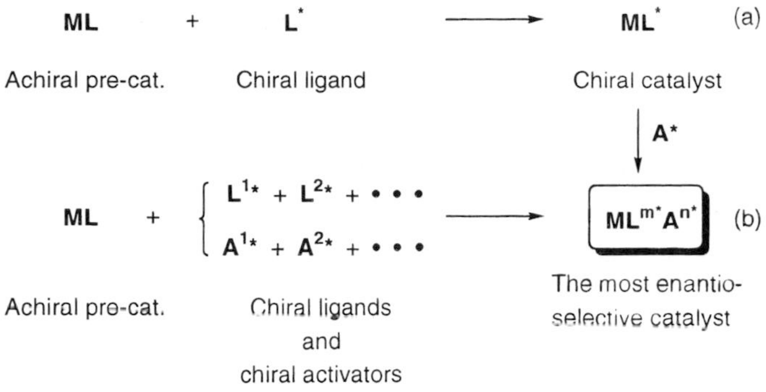

ML + L* ⟶ ML* (a)

Achiral pre-cat. Chiral ligand Chiral catalyst

A*

ML + {L^1* + L^2* + • • • / A^1* + A^2* + • • •} ⟶ ML^m*A^n* (b)

Achiral pre-cat. Chiral ligands and chiral activators The most enantio-selective catalyst

Figure 7-21. General principle of the ligand exchange based on smart self-assembly.

Two patterns of self-assembly are conceivable for an achiral pre-catalyst, Ti(O*i*Pr)$_4$ with couples of chiral diol components into a single chiral titanium complex. In one (Fig. 7-22a), a combination of acidic (*R*)-binol and a relatively basic diol such as taddol [92] in a molar ratio of 1:1:1 suggests a push-pull assembly into a single (*R*)-binolato-Ti-(*R*)-taddolato complex (**6a**); no isopropanol was observed after azeotropic removal with toluene. This (*R*)-binolato-Ti-(*R*)-taddolato complex (**6a**) is obtained from (*R*)-taddolato-Ti(O*i*Pr)$_2$ (**5a**) with (*R*)-binol or from (*R*)-binolato-Ti(O*i*Pr)$_2$ (**2**) with (*R*)-taddol. In the other (Fig. 7-22b), upon addition of (*R*)-binol and a more acidic diol such as (*R*)-5-Cl-bipol to Ti(O*i*Pr)$_4$, (*R*)-binolato-Ti(O*i*Pr)$_2$ (**2**)/(*R*)-bipol complex (**6b**) is obtained. This complex is derived not only from binolato-Ti(O*i*Pr)$_2$ (**2**) [48d, 93] with 5-Cl-bipol [52] but also from 5-Cl-bipolato-Ti(O*i*Pr)$_2$ (**5b**) with binol.

The role of multicomponent ligand assembly into a highly enantioselective catalyst was exemplified in the investigation of enantioselective catalysis of the carbonyl-ene reaction (Eq. (7.29)). The catalyst was prepared by mixing an achiral pre-catalyst, Ti(O*i*Pr)$_4$ and a combination of binol with various chiral diols such as taddol and 5-Cl-bipol in a molar ratio of 1:1:1 (10 mol% equivalent with respect to an olefin and glyoxylate) in toluene (Table 7-6).

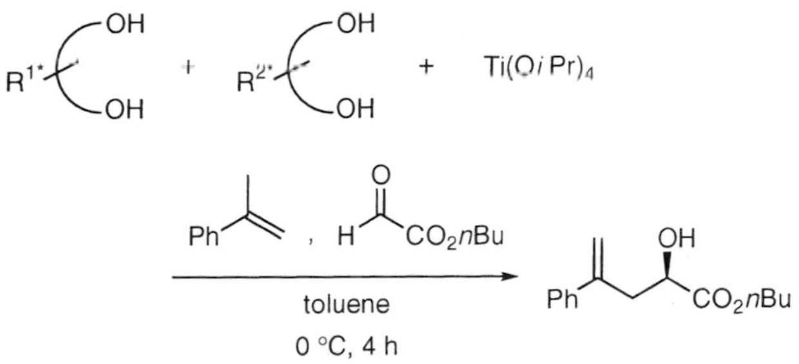

R^1*(OH)(OH) + R^2*(OH)(OH) + Ti(O*i* Pr)$_4$

Ph(C=CH$_2$)CH$_3$, H(C=O)CO$_2$*n*Bu ⟶ (toluene, 0 °C, 4 h) Ph(C=CH$_2$)CH$_2$CH(OH)CO$_2$*n*Bu (7.29)

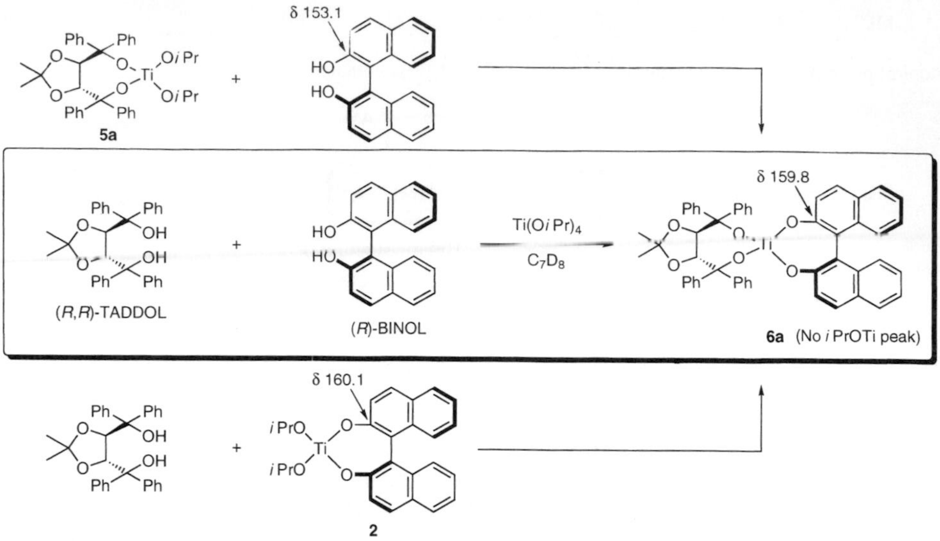

δ 153.1
Ph Ph O O O Ti O*i* Pr O*i* Pr Ph Ph
5a
+
HO HO
Ph Ph O OH OH O Ph Ph
(*R,R*)-TADDOL
+
HO HO
(*R*)-BINOL
Ti(O*i* Pr)$_4$
C_7D_8
δ 159.8
Ph Ph O O O Ti O O Ph Ph
6a (No *i* PrOTi peak)
δ 160.1
Ph Ph O OH OH O Ph Ph
+
i PrO *i* PrO Ti O O
2

Figure 7-22 a. "Smart" self-assembly of the highly activated Ti catalysts (**6a**).

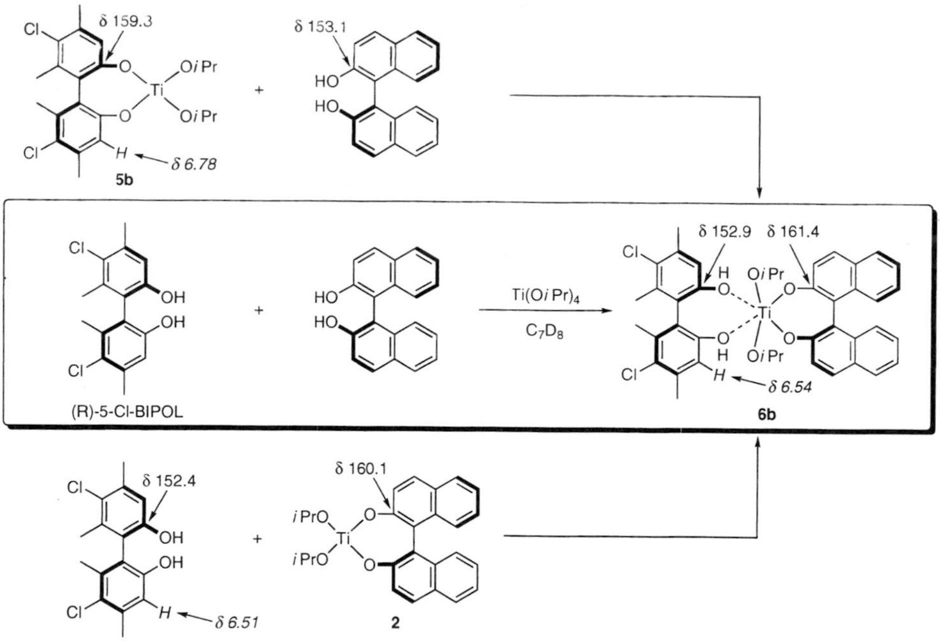

Cl δ 159.3 O Ti O*i* Pr O*i* Pr O Cl H δ 6.78
5b
+
δ 153.1
HO HO
Cl OH OH Cl
(R)-5-Cl-BIPOL
+
HO HO
Ti(O*i* Pr)$_4$
C_7D_8
δ 152.9 δ 161.4
Cl H O O*i* Pr O Ti O O H O*i* Pr Cl H δ 6.54
6b
Cl δ 152.4 OH OH Cl H δ 6.51
+
δ 160.1
i PrO *i* PrO Ti O O
2

Figure 7-22 b. "Smart" self-assembly of the highly activated Ti catalysts (**6b**).

Table 7-6. Asymmetric catalysis by multi-component ligand cooperation.

Run	$R^{1*}(OH)_2$	$R^{2*}(OH)_2$	Yield [%]	Ee [%]
1	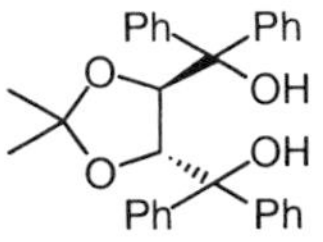 (taddol: Ph, Ph, OH, OH, Ph, Ph)	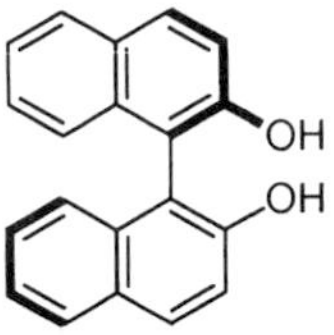 (binol: OH, OH)	50	91
2	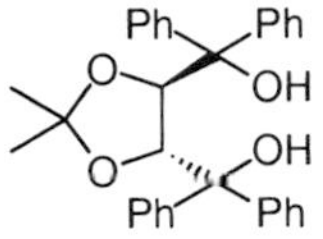 (taddol: Ph, Ph, OH, OH, Ph, Ph)	None	0	–
3	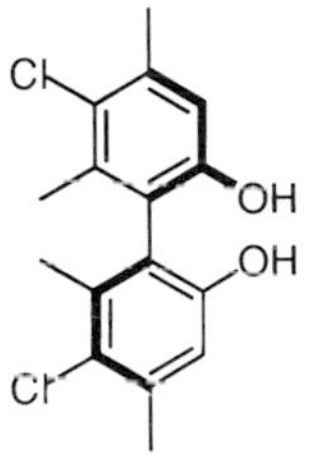 (bipol: Cl, OH, OH, Cl)	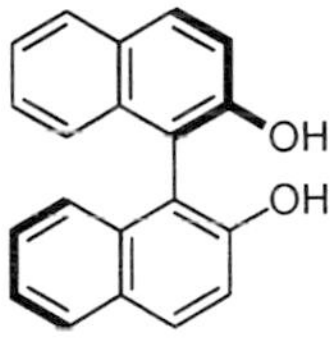 (binol: OH, OH)	66	97
4	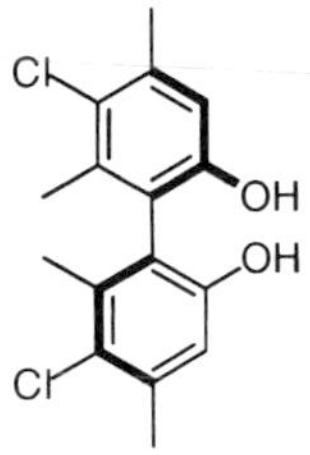 (bipol: Cl, OH, OH, Cl)	None	13	75
5	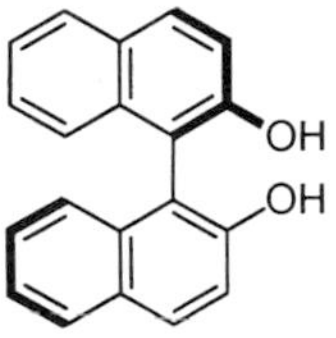 (binol: OH, OH)	None	20	95

Of significance, a quantum jump in chemical yield from 0% to 50% was established in addition to the high enantioselectivity (91.0% ee, *R*) when a combination of (*R*)-taddol and (*R*)-binol was employed (Runs 1 vs 2). Using a combination of (*R*)-bipol and (*R*)-binol, the reaction proceeded quite smoothly to produce the carbonyl-ene product in the highest chemical yield as well as the highest enantioselectivity (Run 3). This finding is in direct contrast to the lower enantiomeric excesses and chemical yields obtained using the (*R*)-bipolato-Ti(IV) catalyst (**5b**) or (*R*)-binolato-Ti(O*i*Pr)$_2$ catalyst (**2**) (Runs 4 and 5).

Hill and Zhang reported the results of an elegant study in which smart self-assembly resulted in the creation of an achiral "immortal" catalyst [94]. We have re-

mura, M.; Suga, S.; Oka, H.; Noyori, R., *J. Am. Chem. Soc.*, **1998**, 9800–9809. (e) Kitamura, M.; Oka, H.; Noyori, R., *Tetrahedron*, **1999**, 3605–3614.

15. Reviews: (a) Mikami, K.; Shimizu, M. *Chem. Rev.,* **1992**, 1021–1050. (b) Mikami, K.; Terada, M.; Shimizu, M.; Nakai, T., *J. Synth. Org. Chem. Jpn*., **1990**, 292–303.
16. Bolm, C., *Tetrahedron Asym*., **1991**, 701–704. Bolm, C.; Ewald, M.; Felder, M., *Chem. Ber.*, **1992**, 1205–1215. Bolm, C.; Felder, M.; Muller, J., *Synlett*, **1992**, 439–441.
17. Kanemasa, S.; Oderaotoshi, Y.; Sakaguchi, S.; Yamamoto, H.; Tanaka, J.; Wada, E.; Curran, D.P., *J. Am. Chem. Soc.*, **1998**, 3074–3088.
18. Diels-Alder: (a) Evans, D.A.; Miller, S.J.; Lectka, T., *J. Am. Chem. Soc.*, **1993**, 6460–6461. (b) Evans, D.A.; Murry, J.A.; von Matt, P.; Norcross, R.D.; Miller, S.J., *Angew. Chem., Int. Ed. Engl.*, **1995**, 798–800. Aldol: (c) Evans, D.A.; Murry, J.A., *J. Am. Chem. Soc.,* **1996**, 5814–5815. (d) Evans, D.A.; Kozlowski, M.C.; Burgey, C.S.; MacMillan, D.W.C., *J. Am. Chem. Soc.*, **1997**, 7893–7894. (e) Evans, D.A.; Burgey, C.S.; Kozlowski, M.C.; Tregay, S.W., *J. Am. Chem. Soc.*, **1999**, 686–699. Hetero Diels-Alder: (f) Evans, D.A.; Olhava, E. J.; Johnson, J.S.; Janey, J. M., *Angew. Chem., Int. Ed.*, **1998**, 3372–3375. (g) Evans, D.A.; Johnson, J.S., *J. Am. Chem. Soc.*, **1998**, 4895–4896. Ene: (h) Evans, D.A.; Burgey, C.S.; Paras, N.A.; Vojkovsky, T.; Tregay, S.W., *J. Am. Chem. Soc.*, **1998**, 5824–5825. Michael: (i) Evans, D.A.; Rovis, T.; Kozlowski, M.C.; Tedrow, J.S., *J. Am. Chem. Soc.*, **1999**, 1994–1995.
19. Evans, D.A.; Kozlowski, M.C.; Murry, J.A.; Burgey, C.S.; Campos, K.R.; Connell, B.T.; Staples, R.J., *J. Am. Chem. Soc.*, **1999**, 669–685. Also see: Evans, D.A.; Lectka, T.; Miller, S.J., *Tetrahedron Lett.*, **1993**, 7027–7030.
20. Brunel, J.-M.; Luukas, T.O.; Kagan, H.B., *Tetrahedron Asymm*., **1998**, 1941–1946.
21. (a) Tanaka, K.; Matsui, J.; Kawabata, Y.; Suzuki, H.; Watanabe, A., *Chem. Commun.,* **1991**, 1632–1634. (b) Tanaka, K.; Matsui, J.; Suzuki, H., *Perkin 1*, **1993**, 153–157.
22. Mikami, K.; Motoyama, Y.; Terada, M., *J. Am. Chem. Soc.*, **1994**, 2812–2820.
23. (a) Keck, G.E.; Krishnamurthy, D.; Grier, M.C., *J. Org. Chem.*, **1993**, 6543–6544. (b) Keck, G.E.; Krishnamurthy, D., *J. Am. Chem. Soc.*, **1995**, 2363–2364.
24. Linear relationship as a probe for stereodetermining step: (a) Schmidt, B.; Seebach, D., *Angew. Chem., Int. Ed. Engl.*, **1991**, 1321–1323. (b) Schwenkreis, T.; Berkessel, *Tetrahedron Lett.*, **1993**, 4785–4788. (c) Giardello, M.A.; Conticello, V.P.; Bard, L.; Gagne, M.R.; Marks, T.J., *J. Am. Chem. Soc.,* **1994**, 10241–10254. (d) Bolm, C.; Bienewald, F., *Angew. Chem., Int. Ed. Engl.*, **1995**, 2640–2642. (e) Ramon, D.J.; Guillena, G.; Seebach, D., *Helv. Chim. Acta,* **1996**, 875–894. (f) Yamaguchi, M.; Shiraishi, T.; Hirama, M., *J. Org. Chem.*, **1996**, 3520–3530. (g) Denmark, S.E.; Christenson, B.L.; O'Connor, S.P., *Tetrahedron Lett.*, **1995**, 2219–2222. (h) Dosa, P.I.; Ruble, J.C.; Fu, G.C., *J. Org. Chem.*, **1997**, 444–445. (i) Mori, M.; Nakai, T., *Tetrahedron Lett.,* **1997**, 6233–6236. (j) Guo, C.; Qiu, J.; Zhang, X.; Verdugo, D.; Larter, M.L.; Christie, R.; Kenney, P.; Walsh, P.J., *Tetrahedron*, **1997**, 4145–4158. (k) Ramon, D.J.; Yus, M., *Tetrahedron*, **1998**, 5651–5666. (l) Pritchett, S.; Woodmansee, D.H.; Gantzel, P.; Walsh, P.J., *J. Am. Chem. Soc.*, **1998**, 6423–6424. (m) Ford, A.; Woodward, S., *Angew. Chem., Int. Ed.,* **1999**, 335–336. n) Simonsen, K.B.; Bayon, P.; Hazell, R. G.; Gothelf, K.V.; Jorgensen, K.A. *J. Am. Chem. Soc.*, **1999**, 3845–3853. o) Evans, D.A.; Johnson, J.S.; Burgey, C.S.; Campos, K.R., *Tetrahedron Lett*., **1999**, 2879–2882.
25. (a) Hansen, K.B.; Leighton, J.L.; Jacobsen, E.N., *J. Am. Chem. Soc.*, **1996**, 10924–10925. (b) Konsler, R.G.; Karl, J.; Jacobsen, E.N., *J. Am. Chem. Soc.,* **1998**, 10780–10781.
26. Denmark, S.E.; Su, X.; Nishigaichi, Y., *J. Am. Chem. Soc.*, **1998**, 12990–12991.
27. (a) Blackmond, D.G., *J. Am. Chem. Soc.*, **1997**, 12934–12939. (b) Blackmond, D.G., *J. Am. Chem. Soc.,* **1998**, 13349–13353.
28. Comprehensive reviews: (a) Bonner, W.A., *Topics in Stereochemistry*, **1998**, Eds: Eliel, E.L.; Wilen, S.H., Wiley, New York, Vol 18, 1–96. (b) Calvin, M., *Chemical Evolution*, **1969**, Oxford University Press, Oxford, 149–152. (c) Decker, P., *Origins of Optical Activity in Nature*, **1997**, Eds: Walker, D.C.; Elsevier, New York, 109–124.
29. (a) Avalos, M.; Babiano, R.; Cintas, P.; Jimenez, J.L.; Palacios, J.C.; Barron, L.D., *Chem. Rev.*, **1998**, 2391–2404. (b) Mason, S. F.; Tranter, G.E., *Chem. Phys. Lett.,* **1983**, *94*, 34–37. (c) Mason, S. F.; Tranter, G.E., *Chem. Commun*., **1983**, 117–119.
30. In quartz, selective adsorption of one enantiomer has been reported: Kavasmaneck, P.R.; Bonner, W.A., *J. Am. Chem. Soc.*, **1997**, 44–50. Also all quartz-promoted highly enantioselective synthe-

sis of a chiral organic compound: Soai, K.; Osanai, S.; Kadowaki, K.; Yonekubo, S.; Shibata, T.; Sato, I., *J. Am. Chem. Soc.*, **1999**, *121*, 11235–11236.
31. Interestingly, the slight natural predominance (50.69%) is favored for the L-crystalline form of quartz: Mason, S.F., *Int. Rev. Phys. Chem.*, **1983**, 217–241.
32. (a) Reviews: Buchardt, O., *Angew. Chem., Int. Ed. Engl.*, **1974**, 179–185. Rau, H., *Chem. Rev.* **1983**, 535–547. (b) Huck, N.P.M.; Jager, W.F.; de Lange, B.; Feringa, B.L., *Science,* **1996**, *273*, 1686–1688. Bailey, J.; Chrysostomou, A.; Hough, J.H.; Gledhill, T.M.; McCall, A.; Clark, S.; Menard, F.; Tamura, M., *Science*, **1998**, *281*, 672–674.
33. (a) Zadel, G.; Eisenbaum, C.; Wolff, G.-J.; Breitmaier, E., *Angew. Chem., Int. Ed. Engl.*, **1994**, 454–456. Also see: (b) Feringa, B.L.; Kellogg, R.M.; Hulst, R.; Zondervan, C.; Kruizinga, W.H., *Angew. Chem., Int. Ed. Engl.*, **1994**, 1458–1459. Kaupp, G.; Marquardt, T., *Angew. Chem., Int. Ed. Engl.,* **1994**, 1459–1461.
34. Wynberg, H., *Chimia*, **1989**, 150–152. Wynberg, H., *J. Macromolec. Sci. Chem.*, **1989**, *A26*, 1033–1041. Bolm, C.; Bienewald, F.; Seger, A., *Angew. Chem., Int. Ed. Engl.*, **1996**, 1657–1659.
35. Frank, F.C., *Biochim. Biophys. Acta*, **1953**, 459–463.
36. Goldanskii, V.I.; Kuz'min, Z.V.V., *Phys. Chem. Leipzig*, **1988**, 269:216.
37. Alberts, A.H.; Wynberg, H., *J. Am. Chem. Soc.,* **1989**, 7265–7266. *Chem. Commun.,* **1990**, 453–454.
38. Danda, H.; Nishikawa, H.; Otaka, K., *J. Org. Chem.*, **1991**, 6740–6741. Also see: Shvo, Y.; Gal, M.; Becker, Y.; Elgavi, A., *Tetrahedron Asymm.*, **1996**, 911–924.
39. ShengJian, L.; Yaozhong, J.; Aiqiao, M.; Guishu, Y., *Perkin 1*, **1993**, 885–886.
40. Soai, K.; Niwa, S.; Hori, H., *Chem. Commun.*, **1990**, 982–983.
41. (a) Soai, K.; Hayase, T.; Shimada, C.; Isobe, K., *Tetrahedron Asymm.*, **1994**, 789–792. (b) Soai, K.; Hayase, T.; Takai, K., *Tetrahedron Asymm.*, **1995**, 637–638. (c) Soai, K.; Inoue, Y.; Takahashi, T.; Shibata, T., *Tetrahedron,* **1996**, 13355–13362. (d) Shibata, T.; Morioka, H.; Tanji, S.; Hayase, T.; Kodaka, Y.; Soai, K., *Tetrahedron Lett.*, **1996**, 8783–8786.
42. (a) Shibata, T.; Choji, K.; Morioa, H.; Hayase, T.; Soai, K., *Chem. Commun.,* **1996**, 751–752. (b) Shibata, T.; Choji, K.; Hayase, T.; Aizu, Y.; Soai, K., *Chem. Commun.*, **1996**, 1235–1236.
43. (a) Soai, K.; Shibata, T.; Morioka, H.; Choji, K., *Nature*, **1995**, 378:767–768. (b) Shibata, T.; Morioka, H.; Hayase, T.; Choji, K.; Soai, K., *J. Am. Chem. Soc.*, **1996**, 471–472. (c) Shibata, T.; Hayase, T.; Yamamoto, J.; Soai, K., *Tetrahedron Asymm.*, **1997**, 1717–1719. (d) Shibata, T.; Yonekubo, S.; Soai, K., *Angew. Chem., Int. Ed.*, **1999**, 659–661.
44. Shibata, T.; Takahashi, T.; Konishi, T.; Soai, K., *Angew. Chem., Int. Ed. Engl.*, **1997**, 2458–2460.
45. Shibata, T.; Yamamoto, J.; Matsumoto, N.; Yonekubo, S.; Osanai, S.; Soai, K., *J. Am. Chem. Soc.*, **1998**, 12157–12158.
46. (a) Alcock, N.W.; Brown, J.M.; Maddox, P.J., *J. Chem. Soc. Chem. Commun.*, **1986,** 1532–1534. (b) Brown, J.M.; Maddox, P.J., *Chirality*, **1991**, 345–354.
47. Maruoka, K.; Yamamoto, H.,. *J. Am. Chem Soc.*, **1989**, 789–790. Also see: Maruoka, K.; Itoh, T.; Shirasaka, T.; Yamamoto, H. *J. Am. Chem. Soc*, **1989**, 310–312.
48. (a) Faller, J.W.; Parr, J., *J. Am. Chem. Soc.*, **1993**, 804–805. (b) Faller, J.W.; Mazzieri, M.R.; Nguyen, J.T.; Parr, J.; Tokunaga, M., *Pure Appl. Chem.*, **1994**, 1463–1469. (c) Faller, J.W.; Tokunaga, M., *Tetrahedron Lett.*, **1993**, 7359–7362. (d) Faller, J.W.; Sams, D.W.I.; Liu, X., *J. Am. Chem. Soc.*, **1996**, 1217–1218. (e) Faller, J.W.; Liu, X., *Tetrahedron Lett.*, **1996**, 3449–3452. (f) Sablong, R.; Osborn, J.A.; Faller, J.W., *J. Organomet. Chem.*, **1997**, *527*, 65–70.
49. An excellent review was quite recently reported on achiral additives as a poison to kill an undesired catalyst species and/or to deoligomerize less active catalysts: Vogl, E.M.; Groger, H.; Shibasaki, M., *Angew. Chem., Int. Ed.* Engl., **1999**, 1570–1577.
50. Reviews: (a) Hoffmann, H.M.R., *Angew. Chem., Int. Ed. Engl.*, **1969**, 556–577. (b) Snider, B.B., *Comprehensive Organic Synthesis,* **1991**, Eds: Trost, B.M.; Fleming, I., Pergamon, London, Vol 2, 527–561; Vol 5, 1–27.
51. Reviews: (a) Weidmann, B.; Seebach, D., *Angew. Chem., Int. Ed. Engl.*, **1983**, 31–45. (b) Yamamoto, Y., *Acc. Chem. Res.*, **1987**, 243–249. (c) Hoffmann, R.W., *Angew. Chem., Int. Ed. Engl.*, **1987**, 489–503. (d) Roush, W.R., *Comprehensive Organic Synthesis*, **1991**, Eds: Trost, B.M.; Fleming, I., Pergamon, London, Vol 2, 1–53. (e) Marshall, J.A., *Chemtracts Org. Chem.*, **1992**, 75–98. (f) Yamamoto, Y.; Asao, N., *Chem. Rev.,* **1993**, 2207–2293.

52. Mikami, K.; Matsukawa, S., *Nature*, **1997**, *385*, 613–615. Also see: Volk, T.; Korenaga, T.; Matsukawa, S.; Terada, M.; Mikami, K., *Chirality*, **1998**, 717–721.
53. Matsukawa, S.; Mikami, K., *Enantiomer*, **1996**, 69–73.
54. Matsukawa, S.; Mikami, K., *Tetrahedron Asymm.*, **1997**, 815–816.
55. Matsukawa, S.; Mikami, K., *Tetrahedron Asymm.*, **1995**, 2571–2574.
56. (a) Ohkuma, T.; Ooka, H.; Hashiguchi, S.; Ikariya, T.; Noyori, R., *J. Am. Chem. Soc.*, **1995**, 2675–2676. (b) Ohkuma, T.; Ooka, H.; Ikariya, T.; Noyori, R., *J. Am. Chem. Soc.*, **1995**, 10417–10418. (c) Ohkuma, T.; Ooka, H.; Yamakawa, M.; Ikariya, T.; Noyori, R., *J. Org. Chem.*, **1996**, 4872–4873. (d) Ohkuma, T.; Ikehira, H.; Ikariya, T.; Noyori, R., *Synlett*, **1997**, 467–468. (e) Doucet, H.; Ohkuma, T.; Murata, K.; Yokozawa, T.; Kozawa, M.; Katayama, E.; England, A.F.; Ikariya, T.; Noyori, R., *Angew. Chem., Int. Ed. Engl.*, **1998**, 1703–1707.
57. Ohkuma, T.; Doucet, H.; Pham, T.; Mikami, K.; Korenaga, T.; Terada, M.; Noyori, R., *J. Am. Chem. Soc.*, **1998**, 1086–1087.
58. Tolbinap=2,2′-bis(di-p-tolylphosphanyl)-1,1′-binaphthyl: (a) Takaya, H.; Mashima, K.; Koyano, K.; Yagi, M.; Kumobayashi, H.; Taketomi, T.; Akutagawa, S.; Noyori, R., *J. Org. Chem.*, **1986**, 629–635. (b) Kitamura, M.; Tokunaga, M.; Ohkuma, T.; Noyori, R., *Org. Synth.*, **1992**, 1–13.
59. dm-binap= 2,2′-bis(di-3,5-xylylphosphanyl)-1,1′-binaphthyl: (a) Mashima, K.; Matsumura, Y.; Kusano, K.; Kumobayashi, H.; Sayo, N.; Hori, Y.; Ishizaki, T.; Akutagawa, S.; Takaya, H., *Chem. Commun.*, **1991**, 609–610. (b) Ohkuma, T.; Koizumi, M.; Doucet, H.; Pham, T.; Kozawa, M.; Murata, K.; Katayama, E.; Yokozawa, T.; Ikariya, T.; Noyori, R., *J. Am. Chem. Soc.*, **1998**, 13529–13530.
60. (a) Mangeney, P.; Tejero, T.; Alexakis, A.; Grosjean, F.; Normant, J., *Synthesis*, **1988**, 255–257. (b) Pikul, S.; Corey, E., *J. Org. Synth.*, **1992**, *71*, 22–29.
61. Review on matched and mismatched pairing in double asymmetric synthesis: Masamune, S.; Choy, W.; Petersen, J.S.; Sita, L.R., *Angew. Chem., Int. Ed. Engl.*, **1985**, 1–30.
62. The active species has been suggested to be a mono- or dihydride species (X=H or Cl): (a) Chowdhury, R.L.; Bäckvall, J.-E., *Chem. Commun.*, **1991**, 1063–1064. (b) Haack, K.-J.; Hashiguchi, S.; Fujii, A.; Ikariya, T.; Noyori, R., *Angew. Chem., Int. Ed. Engl.*, **1997**, 285–288. (c) Noyori, R.; Hashiguchi, S., *Acc. Chem. Res.*, **1997**, 97–102. (d) Aranyos, A.; Csjernyik, G.; Szabo, K.; Bäckvall, J.-E., *Chem. Commun.*, **1999**, 351–352. (e) Persson, B.A.; Larsson, A.L.E.; Ray, M.L.; Bäckvall, J.-E., *J. Am. Chem. Soc.*, **1999**, 1645–1650.
63. Halpern, J., *Asymmetric Synthesis*, **1985**, Ed: Morrison, J.D., Academic, New York, Vol 5, 41–69.
64. Excellent reviews on atropisomerism: (a) Oki, M., Top. *Stereochem.*, **1983**, *14*, 1–81. (b) Oki, M., *The Chemistry of Rotational Isomers*, Springer, New York, **1993**. (c) Eliel, E.L., *Stereochemistry of Carbon Compounds*, McGraw-Hill, New York, **1962**, Chap 6, 156–179. (d) Eliel, E.L.; Wilen, S.H.; Mander, L.N., *Stereochemistry of Organic Compounds*, Wiley, New York, **1994**, Chap 14, 1142–1190.
65. (a) BPBP (2,2′-bis(diphenylphosphanyl)-1,1′-bipheny(l) was also termed for this bisphosphine ligand but synthesized unsuccessfully to give the monophosphine: Uehara, A.; Bailar, Jr. J.C., *J. Organomet. Chem.*, **1982**, *239*, 1–10. (b) Bennett, M.A.; Bhargava, S.K.; Griffiths, K.D.; Robertson, G.B., *Angew. Chem., Int. Ed. Engl.*, **1987**, 260–261. (c) Desponds, O.; Schlosser, M., *J. Organomet. Chem*, **1996**, *507*, 257–261. (d) Desponds, O.; Schlosser, M., *Tetrahedron Lett.*, **1996**, 47–48. (e) Allen, D.W.; Millar, I.T., *J. Chem. Soc. C*, **1968**, 2406–2408. (f) Costa, T.; Schmidbaur, H. *Chem. Ber.*, **1982**, 1367–1373. Also see the 6,6′-substituted analogues: (g) "biphemp" (2,2′-bis(diphenylphosphanyl)-6,6′-dimethyl-1,1′-biphenyl): Svensson, G.; Albertsson, J.; Frejd, T.; Klingstedt, T., *Acta Crystallogr. Sect. C*, **1986**, *42*, 1324–1327. Schmid, R.; Cereghetti, M.; Heiser, B.; Schonholzer, P.; Hansen, H.-J. *Helv. Chim. Acta*, **1988**, 897–929. (h) "bicheps" (2,2′-bis(dicyclohexylphosphanyl)-6,6-dimethyl-1,1′-biphenyl): Chiba, T.; Miyashita, A.; Nohira, H., *Tetrahedron Lett.*, **1991**, 4745–4748. (i) "MeO-biphep" (2,2′-bis(diphenylphosphanyl)-6,6′-dimethoxy-1,1′-biphenyl): Schmid, R., Foricher. J., Cereghetti. M., Schönholzer, P., *Helv. Chim. Acta*, **1991**, 370–389. Schmid, R.; Broger, E.A.; Cereghetti, M.; Crameri, Y.; Foricher, J.; Lalonde, M.; Müller, R.K.; Scalone, M.; Schoettel, G.; Zutter, U., *Pure. Appl. Chem.*, **1996**, 131–138. Trabesinger, G.; Albinati, A.; Feiken, N.; Kunz, R.W.; Pregosin, P. S.; Tschoerner, M.; *J. Am. Chem. Soc.*, **1997**, 6315–6323. (j) (2,2′-bis(diphenylphosphanyl)-6,6′-difluoro-1,1′-biphenyl): Jendralla, H.; Li, C.-H.; Paulus, E., *Tetrahedron Asymm.*, **1994**, 1297–1320.

66. Mikami, K.; Korenaga, T.; Terada, M.; Ohkuma, T.; Pham, T.; Noyori, R., *Angew. Chem., Int. Ed.*, **1999**, 495–497.
67. Brunner, H.; Hammer, B., *Angew. Chem., Int. Ed. Engl.*, **1984**, 312–313.
68. Review on a center of chirality in the central metal of a complex: (a) Brunner, H., *Adv. Organomet. Chem.*, **1980**, *18*, 151–206. (b) von Zelewsky, A., *Stereochemistry of Coordination Compounds*, Wiley, New York, **1996**. (c) Knof, U.; von Zelewsky, A., *Angew. Chem., Int. Ed.* **1999**, 302–322.
69. Mikami, K.; Matsukawa, S.; Volk, T.; Terada, M., *Angew. Chem., Int. Ed. Engl.*, **1997**, 2768–2771.
70. Chavarot, M.; Byrne, J.J.; Chavant, P.Y.; Pardillos-Guindet, J.; Vallee, Y., *Tetrahedron Asymm.*, **1998**, 3889–3894.
71. (a) Hamada, T.; Fukuda, T.; Imanishi, H.; Katsuki, T., *Tetrahedron*, **1996**, 515–530. (b) Noguchi, Y.; Irie, R.; Fukuda, T.; Katsuki, T., *Tetrahedron Lett.*, **1996**, 4533–4536. (c) Ito, Y.N.; Katsuki, T., *Tetrahedron Lett.*, **1998**, 4325–4328. (d) Irie, R.; Hashihayata, T.; Katsuki, T.; Akita, M.; Moro-oka, Y., *Chem. Lett.*, **1998**, 1041–1042.
72. Miura. K.; Katsuki, T., *Synlett*, **1999**, 783–785.
73. Reviews: (a) Okamoto, Y.; Nakano, T., *Chem. Rev.*, **1994**, 349–372. (b) Brintzinger, H.-H.; Fischer, D.; Mulhaupt, R.; Rieger, B.R.; Waymouth, M., *Angew. Chem., Int. Ed. Engl.*, **1995**, 1143–1170. (c) Hoveyda, A.H.; Morken, J.P., *Angew. Chem., Int. Ed. Engl.*, **1996**, 1262–1284. (d) Mikami, K.; Terada, M.; Osawa, A., *Kobunshi/High Polymers Jpn.*, **1997**, *46*, 72–76.
74. Ringwald, M.; Sturmer, R.; Brintzinger, H.H., *J. Am. Chem. Soc.*, **1999**, 1524–1527.
75. Special issue on combinatorial library: (a) *Acc. Chem. Res.*, **1996**, No 3. (b) *Chemical Eng. News*, **1996**, 74, No 4. Reviews: (c) Balkenhohl, F.; Hunnefeld, C.B.; Lansky, A.; Zechel, C., *Angew. Chem., Int. Ed. Engl.*, **1996**, 2288–2337. (d) Gennari, C.; Nestler, H.P.; Piarulli, U.; Salom, B., *Liebigs Ann.*, **1997**, 637–647. (e) *Combinatorial Chemistry: Synthesis and Application*, **1997**, Eds: Wilson, S.R.; Czarink, A.W., Wiley, New York.
76. High Throughput Screening, **1997**, Ed: Devlin, J.P.; Marcel Dekker., New York.
77. (a) Burgess, K.; Lim, H.-J.; Porte, A.M.; Sulikowski, G.A., *Angew. Chem., Int. Ed. Engl.*, **1996**, 220–222. Porte, A.M.; Reibenspies, J.; Burgess, K., *J. Am. Chem. Soc.*, **1998**, 9180–9187. (b) Cole, B.M.; Shimizu, K.D.; Krueger, C.A.; Harrity, J. P.A.; Snapper, M.L.; Hoveyda, A.H., *Angew. Chem., Int. Ed. Engl.*, **1996**, 1668–1671. Shimizu, K.D.; Cole, B.M.; Krueger, C.A.; Kuntz, K.W.; Snapper, M.L.; Hoveyda, A.H., *Angew. Chem., Int. Ed. Engl.*, **1997**, 1703–1707. (c) Sigman, M.S.; Jacobsen, E.N., *J. Am. Chem. Soc.*, **1998**, 4901–4902. Francis, M.B.; Jacobsen, E.N., *Angew. Chem., Int. Ed.*, **1999**, 937–941. (d) Liu, G.; Ellman, J.A., *J. Org. Chem.*, **1995**, 7712–7713.
78. For the application of a CD detection system to measure optical purity by HPLC on nonchiral stationary phase, see: (a) Salvadori, P.; Bertucci, C.; Rosini, C., *Circular Dichroism*, **1994**. Principles and Application, Eds: Nakanishi, K.; Berova, N.; Woody, R.W., VCH, Weinheim, 541–560. (b) Bertucci, C.; Salvadori, P.; Guimaraes, L.F.L., *J. Chromatogr. A.*, **1994**, *666*, 535–539. (c) Mannschreck, A., *Chirality*, **1992**, 163–169. (d) Salvadori, P.; Bertucci, C.; Rosini, C., *Chirality*, **1991**, 376–385. (e) Drake, A.F.; Gould, J.M.; Mason, S.F., *J. Chromatogr*, **1980**, *202*, 239–245.
79. Ding, K.; Ishii, A.; Mikami, K., *Angew. Chem., Int. Ed.* **1999**, 497–501.
80. Soai, K.; Niwa, S., *Chem. Rev.*, **1992**, 833–856.
81. Oguni, N.; Omi, T., *Tetrahedron Lett.*, **1984**, 2823–2824.
82. Reviews: (a) Whitesell, J.K., *Chem. Rev.*, **1989**, 1581–1590. (b) Rosini, C.; Franzini, L.; Raffaelli, A.; Salvadori, P., *Synthesis*, **1992**, 503–517.
83. (a) Kaufmann, D.; Boese, R., *Angew. Chem., Int. Ed. Engl.*, **1990**, 545–546. (b) Hattori, K.; Yamamoto, H., *J. Org. Chem.*, **1992**, 3264–3265. (c) Ishihara, K.; Kurihara, H.; Matsumoto, M.; Yamamoto, H; *J. Am. Chem. Soc.*, **1998**, 6920–6930.
84. (a) Maruoka, K.; Itoh, T.; Shirasaka, T.; Yamamoto, H., *J. Am. Chem. Soc.*, **1988**, 310–312. (b) Bao, J.; Wulff, W.D.; Rheingold, A.L., *J. Am. Chem. Soc.*, **1993**, 3814–3815. (c) Heller, D.P.; Goldberg, D.R.; Wulff, W.D., *J. Am. Chem. Soc.*, **1997**, 10551–10552. (d) Graven, A.; Johannsen, M.; Jorgensen, K.A., *Chem. Commun.*, **1996**, 2373–2374.
85. (a) Reetz, M.T.; Kyung, S.H.; Bolm, C.; Zierke, T., *Chem. Ind.*, London, **1986,** 824–824. (b) Seebach, D.; Beck, A.K.; Imwinkelried, R.; Roggo, S.; Wonnacott, A., *Helv. Chim. Acta*, **1987**, 954–974. (c) Mikami, K.; Terada, M.; Nakai, T., *J. Am. Chem. Soc.*, **1989**, 1940–1941. Mikami, K.; Terada, M.; Nakai, T., *J. Am. Chem. Soc.*, **1990**, 3949–3954. Mikami, K.; Terada, M.; Narisa-

wis acids have been devised. Not only have these led to useful processes for the stereoselective aldol addition reactions, but further study of these systems is leading to increased understanding of the structure and reactivity of coordination complexes and the reactions they catalyze.

Recent developments in the field have also identified novel mechanistic pathways for the development of catalytic, asymmetric aldol processes. Thus in addition to Lewis acid catalysts that mediate the Mukaiyama aldol addition by electrophilic activation of the aldehyde reactant, metal complexes that lead to enolate activation by the formation of a metalloenolate have been documented. Additionally, a new class of Lewis-base-catalyzed addition reactions is now available for the asymmetric aldol addition reaction.

This chapter collects some of the important advances that have taken place over the last five years in asymmetric aldol addition methods. Although a comprehensive review is not possible within the limitations of a chapter, the assortment of methods discussed in this chapter provides a highlight of the important problems that have found innovative solutions in the field. The study of these, while not exhaustive, promises to be of interest to the initiate and expert alike.

8.2 Diastereoselective Aldol Addition Reactions

8.2.1 Acetate Aldol Additions

One of the long-standing problems in asymmetric synthesis has been the development of practical stereoselective, acetate aldol addition reactions [1]. The chiral auxiliaries that perform superbly well in diastereoselective propionate aldol additions with rare exceptions have been unsuccessful in the corresponding additions of unsubstituted acetate-derived enolates [2]. However, recently, Yamamoto and co-workers have reported a novel auxiliary: optically active 2,6-diaryl-3,5-dimethyl-substituted phenol (Eq. (8.1)) [3]. Enolization of the acetate esters **1** or **2** with LDA at –78 °C followed by addition of aldehyde furnishes adducts **3/4** in 94–>99% de and useful yields. The method is easily executed and tolerant of a wide range of aldehyde substrates, including unsaturated, alkynyl, and aliphatic aldehydes. Moreover, auxiliary removal is readily effected at 0 °C (30 min) with Bu_4NOH in THF. The optically active β-hydroxy acids are isolated without loss of the stereochemical integrity of the newly installed stereogenic center.

1) LDA/THF
2) PhCHO

up to >99% ee
(*R*,*R*)-**1**:R = *i*Pr
(*R*,*R*)-**2**:R = Me
(*R*,*R*)-**3**:R = *i*Pr
(*R*,*R*)-**4**:R = Me

(8.1)

Yamamoto has also investigated the use of **1** and **2** with racemic aldehydes such as **5** in a series of addition reactions (Eq. (8.2)). The chiral enolate participates in aldol additions to afford a mixture of *anti-* and *syn*-products **6** and **7** in which the aldehyde facial selectivity has been determined in large part by the overriding bias of the auxiliary. Thus the process offers great promise for the construction of stereochemically complex, densely functionalized fragments common in numerous polypropionate-derived natural products.

(*S*,*S*)-**1** + *rac*-**5** → *anti*-**6** 39% (94% *ee*) + *syn*-**7** 38% (98% *ee*)

(8.2)

8.2.2 *Anti*-Selective Aldol Additions

In addition to the acetate aldol problem, stereoselective aldol additions of substituted enolates to yield 1,2-*anti-* or *threo*-selective adducts has remained as a persistent gap in asymmetric aldol methodology. A number of innovative solutions have been documented recently that provide ready access to such products. The different successful approaches to *anti*-selective propionate aldol adducts stem from the design of novel auxiliaries coupled to the study of metal and base effects on the reaction stereochemistry. The newest class of auxiliaries are derived from *N*-arylsulfonyl amides prepared from readily available optically active vicinal amino alcohols, such as *cis*-1-aminoindan-2-ol and norephedrine.

Masamune has documented the addition of optically active ester enolates that afford *anti*-aldol adducts in superb yields and impressive stereoselectivity (Eq. (8.3)) [4]. The generation of a boryl enolate from **8** follows from groundbreaking studies of ester enolization by Masamune employing dialkyl boryl triflates and amines [5]. Careful selection of di-*n*-alkyl boron triflate (di-*n*-butyl versus dicyclopentyl or dicyclohexyl) and base (triethyl amine versus Hünigs base) leads to the formation of enolates that participate in the *anti*-selective propionate aldol additions. Under optimal conditions, **8** is treated with 1–2 equiv of di-*c*-hexyl boron triflate and triethyl amine at –78 °C followed by addition of aldehyde; the products **9** and **10** are isolated in up to 99 : 1 *anti*:*syn* diastereomeric ratio. The asymmetric aldol process can be successfully carried out with a broad range of substrates including aliphatic, aromatic, unsaturated, and functionalized aldehydes.

An attractive feature of the Masamune process is the subsequent ease of removal of the sulfonamide auxiliary to afford the corresponding acid (LiOH, THF/H_2O) without loss of stereochemical integrity of the products.

(c-Hex)$_2$BOTf, Et$_3$N, then RCHO; 8 → 9 + 10; diastereoselectivity up to >99:1 (8.3)

Gosh has independently reported a second *anti*-selective aldol addition process (Eq. (8.4)) [6]. Amino indanol derived esters such as **11** are enolized with excess $TiCl_4$ (2 equiv) and Hünig's base to furnish a brown solution consisting exclusively of the *Z*-enolate as determined by ^{1}H NMR spectroscopy. Addition of aldehyde (2 equiv) at –78 °C affords the corresponding aldol adducts **12/13** in 44–97% yield and up to 99 : 1 *anti:syn* diastereoselectivity. The optimal substrates in the addition reaction include aliphatic and unsaturated aldehydes. It is interesting to note that the only aromatic aldehyde examined, benzaldehyde, yielded products as a 1 : 1.1 mixture of *anti*:*syn* diastereomers.

$TiCl_4$, *i*Pr$_2$NEt, -78 °C, RCHO; 11 → 12 + 13; diastereoselectivity up to >99:1 (8.4)

Myers has studied the remarkable chemistry of cyclic silyl ketene acetals **14** prepared from optically active (*S*)-prolinol propionamides and dichlorodimethylsilane (Eq. (8.5)) [7]. The reactive species is generated upon deprotonation of the prolinol amide and treatment with the silyl dichloride. The enoxysilane may be purified by distillation under reduced pressure and utilized in aldol additions to afford *anti*-adducts **15** in >99% diastereomeric purity.

RCHO, CH_2Cl_2, 23 °C; 14 → 15 + 16; diastereoselectivity up to >99:1 (8.5)

8.2.3 *Syn*-Selective Aldol Additions

One of the most successful and widely used methods for diastereoselective aldol addition reactions employs Evans' imides **17** and the derived dialkyl borylenolates [8]. The 1,2-*syn* aldol adducts are typically isolated in high diastereoisomeric purity (>250 : 1 dr) and useful yields. More recent investigations of Ti(IV) and Sn(II) enolates by Evans and others have considerably expanded the scope of the aldol process [9]. In 1991, Heathcock documented that diverse stereochemical outcomes could be observed in the aldol process utilizing acyl oxazolidinone imides by variation of the Lewis acid in the reaction mixture [10]. Thus, for example, in contrast to the 1,2-*syn* adduct (**21**) isolated from traditional Evans aldol addition, the presence of excess $TiCl_4$ yields the complementary "non-Evans" 1,2-*syn* aldol diastereomer. This and related observations employing other Lewis acids were suggested to arise from the operation of open transition-state structures wherein a second metal independently activates the aldehyde electrophile.

In recent pioneering studies, Crimmins has reported the use of acyloxazolidinethione auxiliaries (**18**) and $TiCl_4$ for the preparation of either *syn* aldol adducts as a function of the stoichiometry of the amine base and metal (Eq. (8.6)) [11]. The use of 1 : 2 $TiCl_4$ and TMEDA or sparteine yielded the normal 1,2-*syn* "Evans" aldol adduct **21**; however, the use of 2:1 $TiCl_4$: $^iPr_2NEt$ leads to dramatic reversal to give the "non-Evans" 1,2-*syn* diastereomer **20** in remarkable diastereoselectivity (>99 : 1).

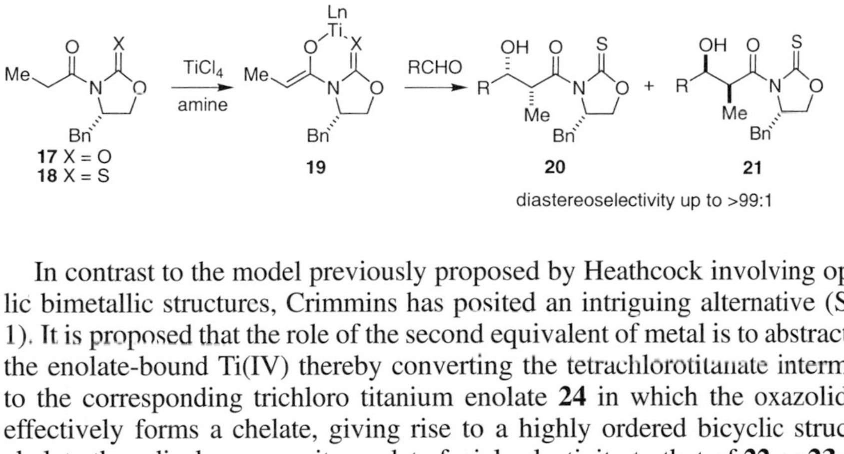

diastereoselectivity up to >99:1 (8.6)

In contrast to the model previously proposed by Heathcock involving open, acyclic bimetallic structures, Crimmins has posited an intriguing alternative (Scheme 8-1). It is proposed that the role of the second equivalent of metal is to abstract Cl^- from the enolate-bound Ti(IV) thereby converting the tetrachlorotitanate intermediate **23** to the corresponding trichloro titanium enolate **24** in which the oxazolidinethione effectively forms a chelate, giving rise to a highly ordered bicyclic structure. The chelate then displays opposite enolate facial selectivity to that of **22** or **23** to furnish the non-Evans *syn* aldol. An important feature of the auxiliary examined by Crimmins is that the thiono-oxazolidinone is proposed to exhibit greater affinity for coordination to Ti(IV). A series of investigations by ^1H NMR spectroscopy have yielded results that are consistent with this proposal. The addition of a second equivalent of $TiCl_4$ produces a new enolate species that is distinct from that of the Ti(IV)-enolate initially observed spectroscopically. In addition, the addition of an equivalent of $AgSbF_6$ in lieu of a second equivalent of $TiCl_4$ furnishes identical results.

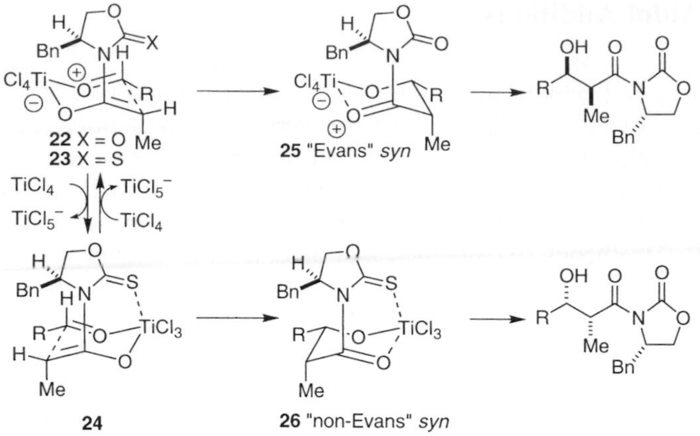

22 X = O
23 X = S
25 "Evans" *syn*
24
26 "non-Evans" *syn*

Scheme 8-1.

8.2.4 Enol Silanes

The pioneering discovery by Mukaiyama in 1974 of the Lewis acid mediated aldol addition reaction of enol silanes and aldehydes paved the way for subsequent explosive development of this innovative method for C–C bond formation. One of the central features of the Mukaiyama aldol process is that the typical enol silane is unreactive at ambient temperatures with typical aldehydes. This reactivity profile allows exquisite control of the reaction stereoselectivity by various Lewis acids; additionally, it has led to the advances in catalytic, enantioselective aldol methodology. Recent observations involving novel enol silanes, such as enoxy silacyclobutanes and *O*-silyl *N*,*O*-ketene acetals have expanded the scope of this process and provided additional insight into the mechanistic manifolds available to this versatile reaction.

Denmark has shown that enoxy silacyclobutanes such as **27** display unusual reactivity; addition to aldehydes takes place rapidly at ambient temperature (Eq. (8.7)) [12]. These enolates are generated following deprotonation of the corresponding ester and treatment with chloro- alkyl- or aryl-, silacyclobutane. The rate at which these enol silanes undergo addition has been shown to be highly dependent on the nature of the substituents on the silane and the geometry of the enolate. In this regard, the *E*-enolate was observed to furnish products possessing 1,2-*syn* relative to configuration **28** in up to 99% diastereoselectivity. In addition, the reaction was shown to be catalyzed by metal alkoxides, allowing for the possibility for asymmetric catalysis by chiral Lewis bases.

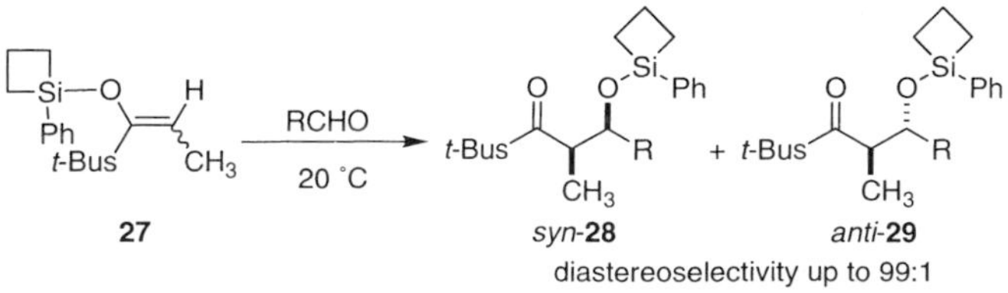

27 —RCHO, 20 °C→ *syn*-**28** + *anti*-**29** (8.7)

diastereoselectivity up to 99:1

8.2.5 Tandem Reactions

A series of innovative investigations by Kiyooka and co-workers have introduced the use of tandem reaction processes that commence with a stereoselective aldol addition reaction and are followed by C=O reduction [13]. A chiral oxazaborolidine complex prepared from BH_3·THF and *N*-*p*-toluenesulfonyl (L)-valine controls the absolute stereochemical outcome of the aldol reaction. In a subsequent reaction, the β-alkoxyboronate effects intramolecular reduction of the ester to furnish the corresponding β-hydroxy aldehyde.

Woerpel has recently reported a tandem double asymmetric aldol/C=O reduction sequence that diastereoselectively affords propionate stereo-triads and -pentads commonly found in polyketide-derived natural products (Scheme 8-2) [14]. When the lithium enolate of propiophenone is treated with excess aldehyde, the expected aldolates **30/31** are formed; however, following warming to ambient temperature a mono-protected diol **34** can be isolated. In a powerful demonstration of the method, treatment of 3-pentanone with 1.3 equiv of LDA and excess benzaldehyde yielded product in corporating five new stereocenters in 81% as an 86:5:5:3 mixture of diastereomers (Eq. (8.8)). A series of elegant experiments have shown that under the condition that the reaction is conducted, the aldol addition reaction is rapidly reversible with an irreversible intramolecular Tischenko reduction serving as the stereochemically determining step (**32** → **34**, Scheme 8-2).

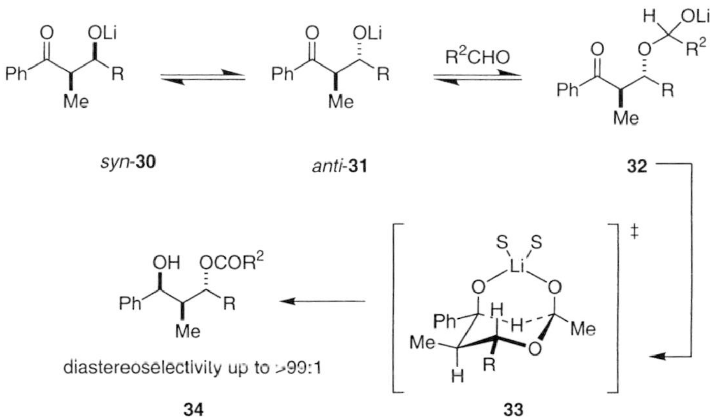

syn-**30** ⇌ anti-**31** ⇌ (R^2CHO) **32** → [**33**]‡ → **34** (diastereoselectivity up to >99:1)

Scheme 8-2.

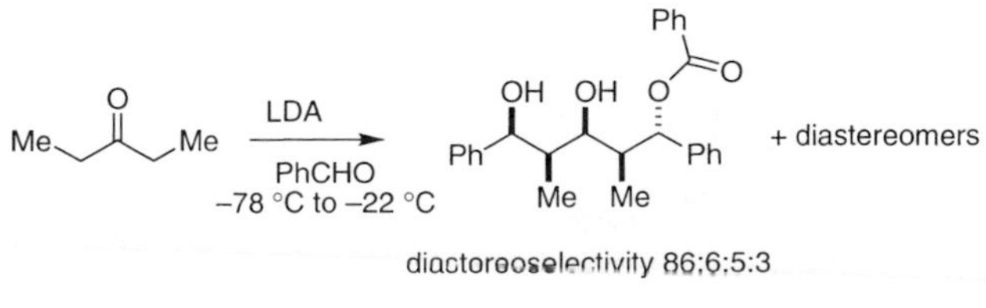

$$\text{3-pentanone} \xrightarrow[\text{PhCHO, } -78\,^\circ\text{C to } -22\,^\circ\text{C}]{\text{LDA}} \text{product} + \text{diastereomers} \qquad (8.8)$$

diastereoselectivity 86:6:5:3

8.2.6 Substrate-Controlled Aldol Additions

Many interesting, powerful applications of inter- and intra-molecular aldol addition reactions have been reported in the context of complex molecule synthesis. These demonstrate the power of aldol bond construction in providing rapid access to stereochemically complex fragments in a stereocontrolled manner. While a comprehensive review of these is well beyond the scope of any one chapter, some recent examples merit examination as they provide insight into interesting and unusual reactivities that may result in the design of novel stereoselective aldol processes.

In an elegant study by Corey culminating in the total synthesis of the medicinally important natural product lactamysin, a critical aldol addition reaction provided an opportunity for the development of double-face-selective Mukaiyama aldols (Eq. (8.9)) [15]. In these investigations, MgI_2 was identified as optimal for effective stereocontrol. Corey has proposed that the unique aspects of MgI_2 are its ability to effect exert chelation control. The unique ability of MgI_2 in this respect stems more facile dissociation of the gegenion (I^-) from coordinated Mg(II), promoting chelate formation between the hindered *β*-nitrogen and the aldehyde C=O.

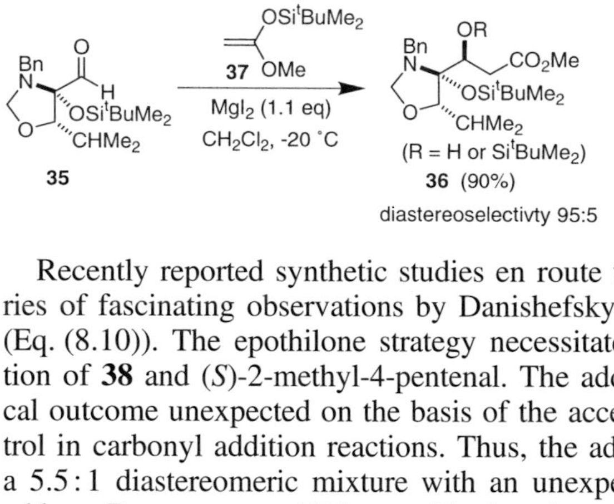

(8.9)

Recently reported synthetic studies en route to the epothilones documents a series of fascinating observations by Danishefsky of a novel aldol addition reaction (Eq. (8.10)). The epothilone strategy necessitated an unusual aldol addition reaction of **38** and (*S*)-2-methyl-4-pentenal. The addition reaction gave a stereochemical outcome unexpected on the basis of the accepted models for acyclic stereocontrol in carbonyl addition reactions. Thus, the addition affords adduct **39** and **40** as a 5.5 : 1 diastereomeric mixture with an unexpected preference for the *anti*-Cram adduct. By contrast, addition to (*S*)-phenyl acetaldehyde affords the Cram adduct as an 11 : 1 mixture of diastereomers. In a series of studies, Danishefsky has noted that the positioning of unsaturation in the substrate in relation to the aldehyde C=O appears to be critical.

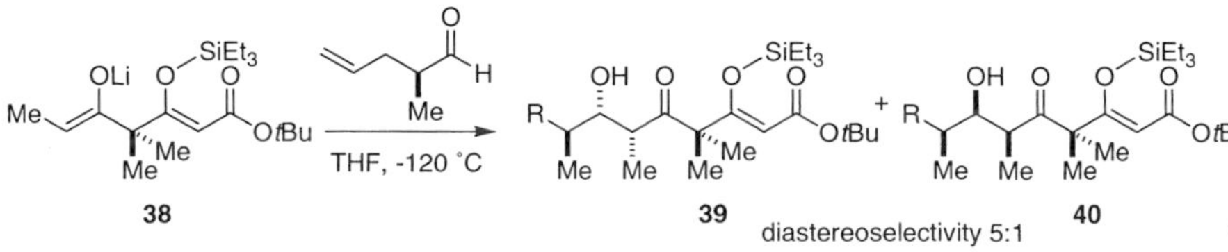

(8.10)

Danishefsky has proposed that the unusual behavior of the unsaturated aldehyde as substrate is accounted for by an energetically stabilizing interaction between polarized aldehyde carbonyl and olefin π-electrons. The seemingly parallel behavior of similarly functionalized aldehydes is consistent with this proposal. This type of electronic stabilization may prove general, offering innovative avenues for the future design of stereoselective aldol addition reactions.

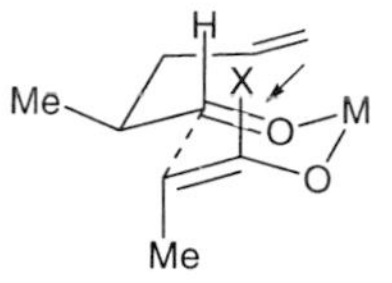

8.3 Enantioselective Catalysis

In addition to advances in diastereoselective aldol addition methods, there have been impressive advances in catalytic, asymmetric aldol addition methodology [16]. The early pioneering work in this area by Mukaiyama and Kobayashi had been focussed primarily on Sn(II)·diamine complexes as asymmetric promoters and catalysts. Over the last five years, the classes of complexes that function competently as catalysts have been expanded considerably to include coordination compounds of B(III), Ag(I), Au(I), Sn(II), La(III), Cu(II), Ti(IV), Ln(III), Si(IV), Pt(II) and Pd(II). The rapid evolution of the field follows from innovative designs and the discovery of ligands based on nitrogen, oxygen, and phosphorus donors. Additionally important discoveries are the documentation of new processes proceeding by activation of the enolate component via metalloenolate intermediates. This contrasts the more traditional methods for catalysis of the Mukaiyama aldol addition reaction involving electrophilic activation of the aldehyde substrate.

8.3.1 Lewis Acids

Oxazaborolidenes. Corey has reported the use of a novel oxazaborolidene complex **41** prepared from borane and *N*-tosyl (*S*)-tryptophan. This complex functions in a catalytic fashion in enantioselective, Mukaiyama aldol addition reactions (Scheme 8-3) [17]. The addition of ketone-derived enol silanes **42–43** gives adducts in 56–100% yields and up to 93% ee. The use of 1-trimethylsilyloxycyclopentene **43** in the addition reactions to benzaldehyde affords adducts **46** as a 94:6 mixture of diastereomers favoring the *syn* diastereomer in 92% ee. Addition reactions with dienol silanes **44** furnishes products **47** in up to 82% ee. Corey also demonstrated the use of these adducts as important building blocks for the synthesis of corresponding dihydropyrones; treatment of **47** with trifluoroacetic acid affords the cyclic product in good yields.

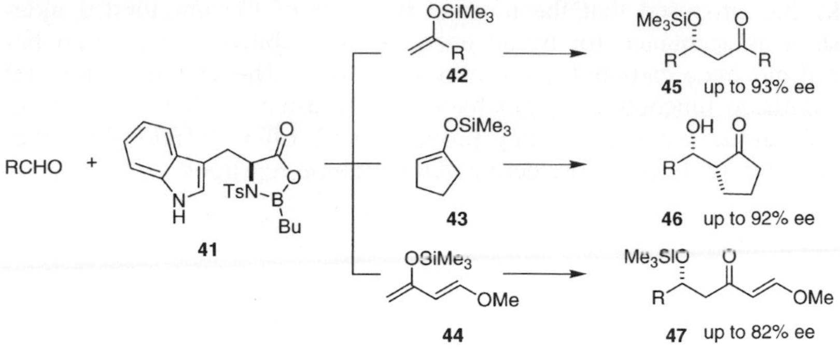

Scheme 8-3.

The unique properties of this complex and its associated ligands have been investigated and discussed. Corey has proposed a model that incorporates two critical features: (1) the electron-rich indole engages the metal-bound polarized aldehyde in an energetically favorable donor-acceptor interaction [18]; (2) the geometry of the bound aldehyde is defined and rigidified by a putative hydrogen bond between the formyl C–H and an oxyanionic ligand [19].

Ti(IV). Carreira has reported a novel class of tridentate ligands whose complexes with Ti(IV) **48** serve as catalysts for a variety of enantioselective aldehyde additions [20]. The reactions that have been examined include acetate and dienolate aldol additions as well as ene-like reactions of 2-methoxy propene [21]. The salient features of these catalytic systems include the fact that a wide range of aldehyde substrates may be utilized, the ability to carry out the reaction employing 0.2–5 mol% catalyst loading, and the experimental ease with which the process is executed. The typical experimental procedure prescribes the use of an in situ generated catalyst, at –10 to 23 °C in a variety of solvents, employing as little as 0.5 mol% catalyst.

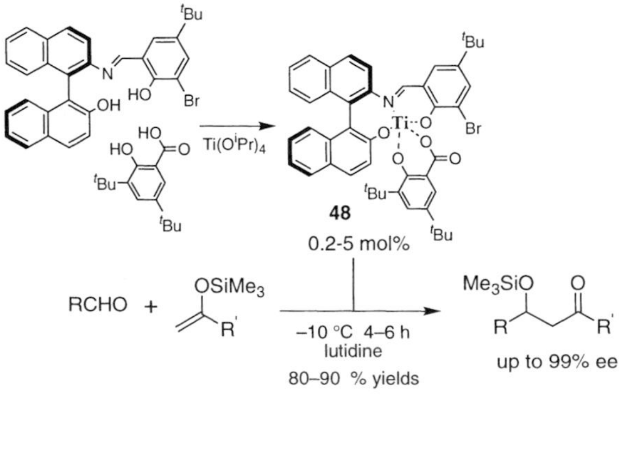

(8.11)

In the initial reports by Carreira and co-workers, the Ti(IV) complex **48** prepared in situ from $Ti(O\textit{i}Pr)_4$, tridentate ligand, and 2,6-di-*tert*-butylsalicylic acid effectively mediates the addition of the methyl acetate-derived silyl ketene acetal to a large range of aldehydes, giving adducts in up to 98.6% ee. The original catalyst preparation protocol prescribed mixing the tridentate ligand, $Ti(O\textit{i}Pr)_4$, salicylic acid, and lutidine in toluene with subsequent removal of the released isopropanol by evaporation of the solvent. A subsequent modification was reported that prescribes mixing the ligands, $Ti(O\textit{i}Pr)_4$, Me_3SiCl, and Et_3N [22]. The released *iso*-propanol forms the corresponding trimethylsilyl ether, thereby obviating any further manipulation of the catalyst. The catalytic acetate aldol process has found application in the preparation of a number of natural product syntheses including Simon's total synthesis of depsipeptide antitumor antibiotic FR-901,228 [23], Rychnovsky's roflamycoin synthesis [24], and the synthesis of macrolactin A by Carreira [25].

The catalytic aldol addition process has been extended to include the addition reactions of dienolsilane **49** to a broad range of aldehydes (Eq. (8.12)) [26]. The addition reactions of **49** are conducted at 23 °C utilizing 5 mol% of catalyst, giving adducts in up to 94% ee. This dienolsilane is easily prepared by enolization of the commercially available acetone-ketene adduct followed by quenching with chlorotrimethyl silane. The resulting dienolsilane is isolated typically in 78% yield as a clear colorless liquid that can be conveniently purified by distillation.

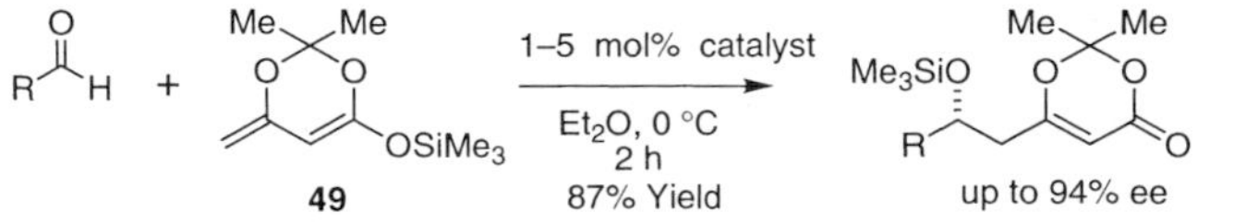

(8.12)

In the study of catalytic, dienolate addition reactions, the use of stannyl propenal **50** as a substrate in aldol methodology has been introduced (Scheme 8-4). The adduct **51** produced from the process is isolated in 92% ee and, importantly, serves as a useful building block for subsequent synthetic elaboration. It is amenable for further manipulations such as Stille cross-coupling reactions to give a diverse family of protected acetoacetate adducts **52**.

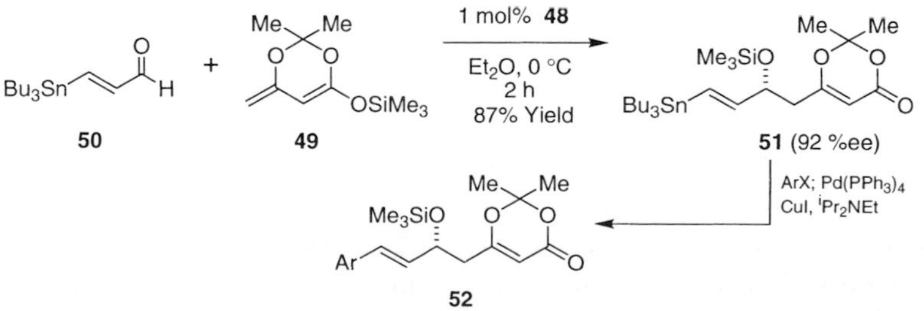

Scheme 8-4.

Sato and co-workers have also investigated the dienolate addition reactions of **49** with benzaldehyde and pentanal (Scheme 8-5). When 20 mol% of Mikami catalyst **53** was employed the aldol adducts were isolated in 38 and 55% yield and 88–92% ee [27]. The use of Yamamoto's oxazaborolidene catalyst **54** afforded the products with diminished optical purity.

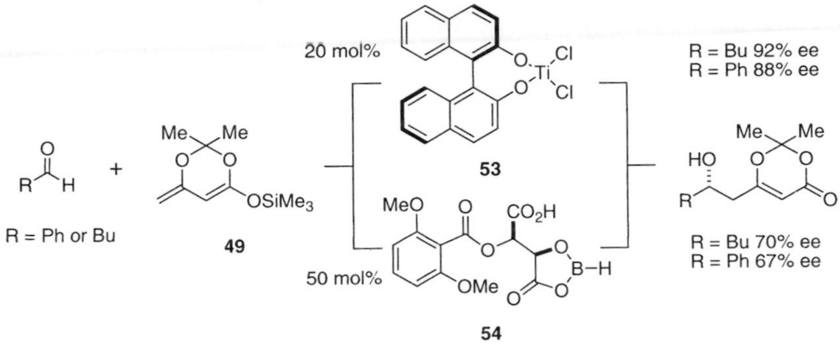

R = Ph or Bu

49

20 mol%

53

50 mol%

54

R = Bu 92% ee
R = Ph 88% ee

R = Bu 70% ee
R = Ph 67% ee

Scheme 8-5.

As part of a series of studies on the use of BINOL·Ti(IV) complex **53** as a catalyst in a number of C–C bond-forming reactions, Mikami has reported the aldol addition reactions of thioacetate-derived silyl ketene acetals **55**, **56** to a collection of highly functionalized aldehydes (Eq. (8.13)) [28]. As little as 5 mol% of the catalyst mediates the addition reaction and furnishes adducts **57** in excellent yields and up to 96% ee. One of the noteworthy features of the Mikami process is the fact that aldehyde substrates containing polar substituents can be successfully employed, a feature exhibited by few other Lewis-acid-catalyzed aldehyde addition reactions.

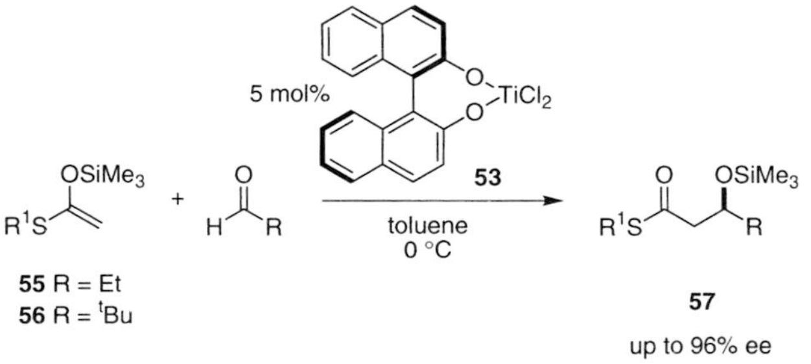

5 mol% **53**, toluene, 0 °C

55 R = Et
56 R = ^{t}Bu

57
up to 96% ee

(8.13)

In addition to processes involving thioacetate aldols, Mikami has studied the aldol addition reaction of thiopropionate-derived enolsilanes **58**, **59** (Eq. (8.14)). The *Z*-enol silane derived from *tert*-butyl thiopropionate undergoes addition to benzyloxyacetaldehyde to give products as a 92 : 8 *anti* : *syn* mixture of diastereomers with the major *anti* stereoisomer **61** isolated in 90% ee. The additions of *E* or *Z*- *S*-ethyl thiopropionate-derived silyl ketene acetals **59** with benzyloxyacetaldehyde afforded adducts in up to 98% ee, albeit with diminished levels of simple diastereoinduction 72:28–48:52, *syn*/*anti*. By contrast, the addition of *E*-**61** to *n*-butyl glyoxylate leads to the formation of *syn* diastereomer **60** (92:8 *syn*/*anti*) in 98% ee.

(8.14)

58 R = ^{t}Bu
59 R = Et

R = Et (77% *E*) R' = CH_2OBn 90% ee 72/28 syn/anti
R = Et (77% *E*) R' = CO_2Bu 98% ee 92/8 syn/anti
R = ^{t}Bu (93% *Z*) R' = CH_2OBn 90% ee 8:92 syn/anti

The catalytic, enantioselective additions of thioacetate-derived enol silanes has also been studied by Keck (Eq. (8.15)) [29]. In these studies, the active catalyst (**62**) is readily generated upon mixing binol, $TiCl_2(OiPr)_2$, and 4 Å molecular sieves in Et_2O at –20 °C followed by an aging period. The addition reactions are best conducted with 10 mol% catalyst in ether at –20 °C; the *tert*-butyl thioacetate adducts are isolated in up to 98% ee and 90% yield.

(8.15)

up to >98% ee

The use of C_a unsubstituted and substituted stannyl enolates has been studied by Yamamoto in a series of elegant reports involving a novel bisphosphine Ag(I) complex **64** as a catalyst for C–C bond formation [30]. The addition of methyl ketone and acetate-derived enolates furnishes adducts in up to 96% ee. The use of E-stannyl enolates yields the 1,2-*anti* diastereomer as the major product in up to 96% ee. The use of acyclic *Z*-enol stannanes provided the complementary *syn*-substituted adducts as the major adduct in equally high diastereoselectivity and enantioselectivity. The observed correlation between enolate geometry and the simple diastereoselectivity of the product (E-enolates yield *anti* adducts while *Z*-enolates yield *syn* adducts) has led Yamamoto to postulate the involvement of a closed, cyclic transition-state structure.

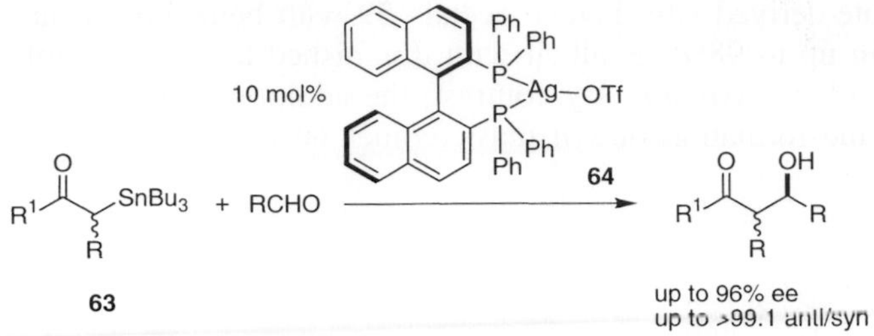

$$\mathbf{63} + \text{RCHO} \xrightarrow{10\ \text{mol\%}\ \mathbf{64}} \text{aldol product (up to 96\% ee, up to >99:1 anti/syn)} \quad (8.16)$$

Cu(II) and Sn(II). In a series of elegant studies, Evans has documented a class of highly enantioselective, catalytic aldol addition reactions involving pyruvate, benzyloxyacetaldehyde, and glyoxylates [31]. The reactions are efficiently mediated by a family of Cu(II) and Sn(II) complexes **68–73** prepared from bisoxazoline ligands and metal salts incorporating poorly coordinating counterions (Chart 8-1). The critical structural feature of the substrates that are ideally suited for the enantioselective addition is their ability to coordinate to the metal center by formation of a five-membered ring chelate [32]. The addition of trimethylsilyl enoxysilane prepared from acetone and methyl pyruvate provides an illustration of this powerful method for ketone-derived enol silanes (Eq. (8.17)). Utilizing as little as 10 mol% Cu(II) catalyst at –78 °C in CH_2Cl_2, the hydroxyketo ester adduct is isolated in superb yields and up to 96% ee.

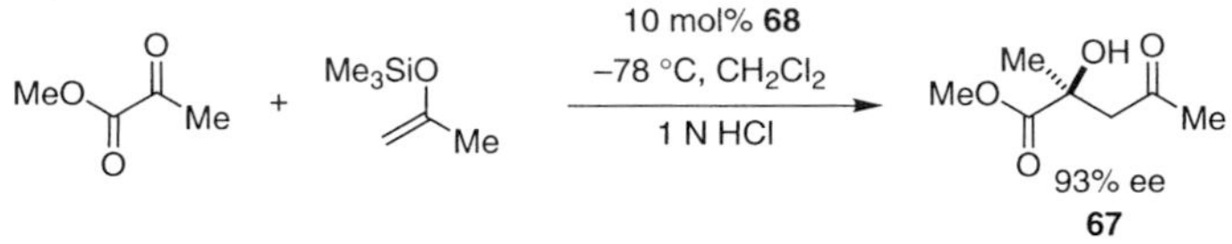

$$\text{MeO}_2\text{CC(O)Me} + \text{Me}_3\text{SiO(C=CH}_2\text{)Me} \xrightarrow[\text{1 N HCl}]{\text{10 mol\% } \mathbf{68},\ -78\,^\circ\text{C, CH}_2\text{Cl}_2} \mathbf{67}\ \text{(93\% ee)} \quad (8.17)$$

68 (Cu, tBu, TfO/OTf); 69 (Sn, Bn, TfO/OTf); 70a: X = OTf; 70b: X = SbF_6; 71 (Sn, Ph, TfO/OTf)

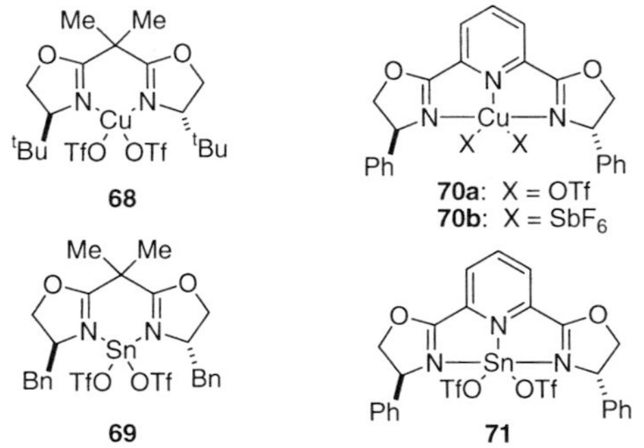

Chart 8-1.

The addition reaction of thioacetate-derived enoxysilanes to the same substrates has also been investigated (Scheme 8-6). Thus, treatment of *tert*-butyl thioacetate-derived silyl thioketene acetal and benzyloxyacetaldehyde, methyl glyoxylate, or pyruvates in the presence of as little as 0.5 mol% **68/69** in CH_2Cl_2 at –78 °C affords aldol adducts in up to 99% ee [33].

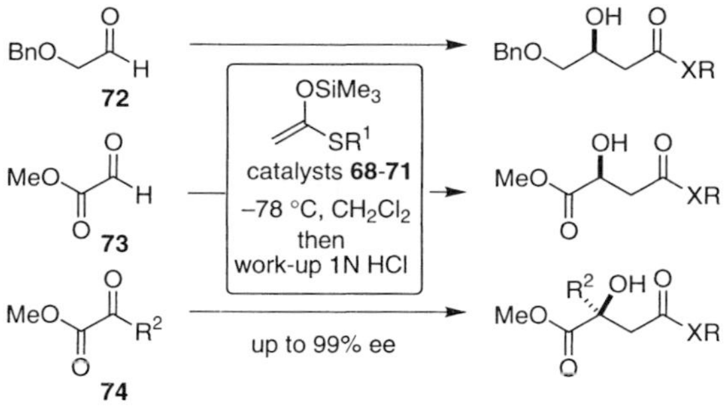

Scheme 8-6.

The same bisoxazoline Cu(II) and Sn(II) complexes have been utilized successfully in the corresponding propionate aldol addition reactions (Scheme 8-7). A remarkable feature of these catalytic processes is that either *syn* or *anti* simple diastereoselectivity may be accessed by appropriate selection of either Sn(II) or Cu(II) complexes. The addition of either *E*- or *Z*-thiopropionate-derived silyl ketene acetals catalyzed by the Cu(II) complexes afford adducts **78**, **80**, and **82** displaying 86:14–97:3 (*syn*/*anti*) simple diastereoselectivity. The optical purity of the major *syn* diastereomer isolated from the additions of both *Z*- and *E*-enol silanes were excellent (85–99% ee). The stereochemical outcome of the aldol addition reactions mediated by Sn(II) are complementary to the Cu(II)-catalyzed process and furnish the corresponding *anti*-stereoisomers **79**, **81**, and **83** as mixtures of 10:90–1:99 *syn*/*anti* diastereomers in 92–99% ee.

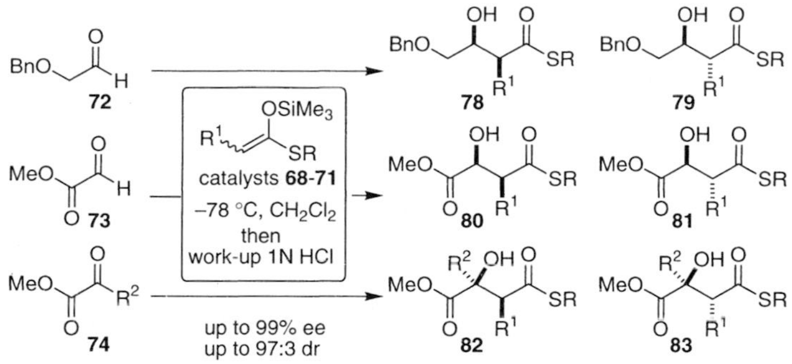

Scheme 8-7.

A remarkable feature of the Evans process is its ability to mediate enantio-, chemo-, and diastereo-selective additions to 1,2-diketones (Eq. (8.18)). The Cu(II) and Sn(II) bisoxazoline complexes display superb group selectivity, differentiating between ethyl and methyl groups in the addition of thiopropionate-derived *Z*-silyl ketene acetal to **84**. As discussed above, the Cu(II) and Sn(II) catalysts elicit complementary simple diastereoselectivity with the Cu(II) catalyst leading to the formation of the 1,2-*syn* dialkyl substituted adduct **85** and the Sn(II) catalysts generating the 1,2-*anti*-substituted diastereomer **86** (Eq. (8.19)).

84 + $OSiMe_3$ / S^tBu / Me — 10 mol% **71**, −78 °C, CH_2Cl_2 then work-up 1N HCl → **85** (Et, Me, OH, XR, Me); 98% ee, anti:syn 99:1, regioselectivity 97:3 (8.18)

84 + $OSiMe_3$ / S^tBu / Me — 10 mol% **68**, −78 °C, CH_2Cl_2 then work-up 1N HCl → **86** (Et, Me, OH, XR, Me); 97% ee, anti:syn 7:93, regioselectivity 98:2 (8.19)

Evans has also reported the addition of a number of other synthetically useful enol silanes. In this regard, the Cu(II)-catalyzed addition of butyrolactone enol silane give the *syn* diastereomer (96 : 4 *syn*/*anti*) in 92% ee. The addition of dienolates **49** and **89** furnishes acetoacetate adducts in 92–94% yields and 92–97% ee (Eqs. (8.20) and (8.21)).

70 (BnO, H) + **49** (Me, Me, O, O, $OSiMe_3$) — 5 mol% **70b**, −78 °C, CH_2Cl_2 → BnO, RO, Me, Me, O, O, O; 92 % ee; **87** R = $SiMe_3$; **88** R = H (8.20)

70 (BnO, H) + **89** (Me_3SiO, $OSiMe_3$, OMe) — 0.5 mol % **70b**, −78 °C, CH_2Cl_2 then $Me_4NBH(OAc)_3$ → **91** (BnO, OH, OH, O, OMe); 97% ee (8.21)

Recently, Chen has synthesized and resolved chiral suberyl carbenium ions and utilized these as catalysts for enantioselective Mukaiyama aldol addition reactions (Eq. (8.22)) [34]. Thus the reaction of the ethyl acetate-derived silyl ketene acetal with benzaldehyde in the presence of 10–20 mol% of catalyst afforded the corresponding adduct in 50% ee. The enantioselectivity of the process proved sensitive to the nature of the cation, consistent with observations previously highlighted by Denmark in related studies [35]. Although at the current level of development the selectivities are modest, the study documents a novel class of metal-free Lewis acidic agents.

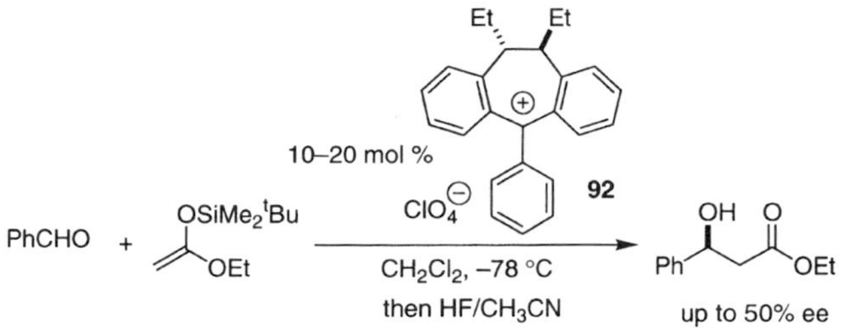

(8.22)

8.3.2 Metallo Enolates

A great majority of the catalytic aldol processes that have been developed over the last two decades involve Lewis acids derived from complexes of titanium, boron, tin, and, more recently, copper as well as silver. A recent, exciting area of rapid development for aldehyde addition reactions is represented by the catalytic aldol methods that utilize soft-metal and lanthanide coordination complexes which mediate addition reactions through metalloenolate intermediates.

Ln. Shibasaki has pioneered the use of alkali-metal/lanthanide aryloxy complexes **93** as catalysts for a wide range of aldehyde addition reactions (Eq. (8.23)). The heterobimetallic alkoxide complexes that have been investigated by Shibasaki are proposed to function by dual nucleophilic and electrophilic activation of the reacting partners [36, 37]. Importantly, the incorporation of basic and Lewis-acidic sites in the active catalyst produces processes that utilize ketones directly in the addition reaction, obviating the need for prior preparation of the corresponding enol silane. In the presence of 20 mol% **93**, ketones undergo addition to aldehydes to give adducts in good yields and up to 94% ee. It is interesting to note that enolizable aldehydes such as cyclohexane carboxaldehyde can be utilized in the addition reactions. Thus, the addition of acetophenone with cyclohexane carboxaldehyde gave the corresponding adduct in 44% ee and 72% yield.

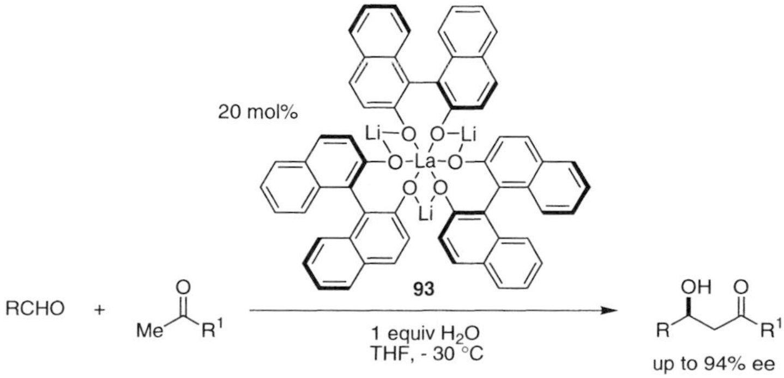

(8.23)

dered chair-like transition-state structure, with the enolate and aldehyde organized around a hexacoordinate siliconate center.

101 + RCHO → (10 mol% 102, -78 °C, CH_2Cl_2) syn 103 + anti 104; up to 93%ee, up to 1:99 syn/anti (8.28)

105 + RCHO → (15% 102, -78 °C, CH_2Cl_2) 106 + 107; up to 96% ee, up to 18:1 syn/anti (8.29)

8.4 Conclusion

The survey of asymmetric aldol addition reactions presented herein attests to the phenomenal advances that have been made in the field recently. In this regard, the range of substrates that can be reliably utilized in catalytic aldol additions has been expanded considerably with the optical activity of the adducts isolated from various procedures being uniformly higher than was attainable with prior art. Moreover, in general, the accessibility of catalysts as well as the experimental execution of the prescribed procedures has improved. The realization of these advances has been possible as a consequence of novel catalysts operating in concert with new mechanistic insight. The standards in asymmetric synthesis are continuously raised, providing the investigators in the field with new challenges. Building upon the discoveries highlighted in this review, the next decade promises to yield further revolutionary advances in asymmetric aldol addition reactions.

References

1. (a) Braun, M.; Sacha, H. *J. Prakt. Chem.* **1993**, *335*, 653. (b) Noyori, R. *Asymmetric Catalysis in Organic Synthesis,* **1994**, Wiley, New York.
2. Gennari, C. In: Trost, B.M.; Fleming, I.; Heathcock C.H. (Eds.) *Comprehensive Organic Synthesis* **1991**, Additions to C–X π-Bonds, Pergamon Press, New York, Chap 2.4, p 629. (i) Yamamoto, H.; Maruoka, K. In: Ojima, I. (Ed.) *Catalytic Asymmetric Synthesis* **1993**, VCH, New York, Chap 9; p 413. (j) Ito, Y.; Sawamura, M. In: Ojima I (Ed) *Catalytic Asymmetric Synthesis* **1993**, VCH, New York, Chap 7; p 367.
3. Saito, S.; Hatanaka, K.; Kano, T.; Yamamoto, H. *Angew. Chem. Int. Ed. Engl.* **1998**, *37*, 3378.
4. Ghosh, A.K.; Onishi, M. *J. Am. Chem. Soc.* **1996**, *118*, 2527.
5. Abiko, A.; Liu, J.-F.; Masamune, S. *J. Org. Chem.* **1996**, *61*, 2590.
6. Ghosh, A.K.; Onishi, M. *J. Am. Chem. Soc.* **1996**, *118*, 2527.
7. (a) Myers, A.G.; Widdowson, K.L. *J. Am. Chem. Soc.* **1990**, *112*, 9672. (b) Myers, A.G.; Widdowson, K.L.; Kukkola, P.J. *J. Am. Chem. Soc.* **1992**, *114*, 2765.
8. (a) Evans, D.A.; Bartroli, J.A.; Shih, T. L. *J. Am. Chem. Soc.* **1981**, *103*, 2127. (b) Evans, D.A. *Aldrichimica Acta* **1982**, *15*, 23.
9. (a) Evans, D.A.; Rieger, D.L.; Bilodeau, M.T.; Urpi, F. *J. Am. Chem. Soc.* **1991**, *113*, 1047.
10. Walker, M.A.; Heathcock, C.H. *J. Org. Chem.* **1991**, *56*, 5747.
11. Crimmins, M.T.; King, B.W.; Tabet, E.A. *J. Am. Chem. Soc.* **1997**, *119*, 7883.
12. Denmark, S.E.; Griedel, B.D.; Coe, D.M.; Schnute, M.E. *J. Am. Chem. Soc.* **1994**, *116*, 7026.
13. Kiyooka, S.; Kaneko, Y.; Kume, K. *Tetrahedron Lett.* **1992**, *33*, 4927.
14. Bodnar, P.M.; Shaw, J.T.; Woerpel, K.A. *J. Org. Chem.* **1997**, *62*, 5674.
15. Corey, E.J.; Li, W.; Reichard, G.A. *J. Am. Chem. Soc.* **1998**, *120*, 2330.
16. For a recent review, see: Nelson, S.G. *Tetrahedron: Asymmetry* **1998**, *9*, 357.
17. Corey, E.J., Cywin, C.L., Roper, T.D. *Tetrahedron Lett.* **1992**, *33*, 6907.
18. Corey, E.J., Loh, T.-P., Roper, T.D., Azimioara, M.D., Noe, M.C. *J. Am. Chem. Soc.* **1992**, *114*, 8290.
19. Corey, E.J.; Rohde, J.J.; Fischer, A.; Azimioara, M.D. *Tetrahedron Lett.* **1997**, *38*, 33. (b) Corey, E.J.; Rohde, J.J. *Tetrahedron Lett.* **1997**, *38*, 37. (c) Corey, E.J.; Barnes-Seeman, D.; Lee, T.W. *Tetrahedron Lett.* **1997**, *38*, 1699. (d) Corey, E.J.; Barnes-Seeman, D.; Lee, T.W. *Tetrahedron Lett.* **1997**, *38*, 4351. (e) Corey, E.J.; Barnes-Seeman, D.; Lee, T.W.; Goodman, S.N. *Tetrahedron Lett.* **1997**, *38*, 6513.
20. (a) Carreira, E.M.; Singer, R.A.; Lee, W. *J. Am. Chem. Soc.* **1994**, *116*, 8837. (b) Carreira, E.M.; Singer, R.A. *DDT* **1996**, *1*, 145.
21. Carreira, E.M.; Lee, W.; Singer, R.A. *J. Am. Chem. Soc.* **1995**, *117*, 3649.
22. Singer, R.A.; Carreira, E.M. *Tetrahedron Lett.* **1997**, *38*, 927.
23. Li, K.W.; Wu, J.; Xing, W.; Simon, J.A. *J. Am. Chem. Soc.* **1996**, *118*, 7237.
24. Rychnovsky, S.D.; Khire, U.R.; Yang, G. *J. Am. Chem. Soc.* **1997**, *119*, 2058.
25. Kim, Y.;. Singer, R.A.; Carreira, E.M. *Angew. Chem., Int. Ed. Engl.* **1998**.
26. Singer, R.A.; Carreira, E.M. *J. Am. Chem. Soc.* **1995**, *117*, 12360.
27. (a) Sato, M.; Sunami, S.; Sugita, Y.; Kaneko, C. *Heterocycles* **1995**, *41*, 1435. (b) Sato, M.; Sunami, S.; Sugita, Y.; Kaneko, C. *Chem. Pharm. Bull.* **1994**, *42*, 839.
28. Mikami, K.; Matsukawa, S. *J. Am. Chem. Soc.* **1994**, *116*, 4077.
29. Keck, G.E.; Krishnamurthy, D. *J. Am. Chem. Soc.* **1995**, *117*, 2363.
30. Yanagisawa, A.; Matsumoto, Y.; Nakashima, H.; Asakawa, K.; Yamamoto, H. *J. Am. Chem. Soc.* **1997**, *119*, 9319.
31. For a thorough discussion of these processes, see: (a) Evans, D.A.; Burgey, C.S.; Kozlowski, M.C.; Tregay, S. W. *J. Am. Chem. Soc.* **1999**, *121*, 686. (b) Evans, D.A.; Kozlowski, M.C.; Murry, J.A.; Burgey, C.S.; Connell, B.T.; *J. Am. Chem. Soc.* **1999**, *121*, 669.
32. Evans, D.A.; Murry, J.A.; Kozlowski, M.C. *J. Am. Chem. Soc.* **1996**, *118*, 5814.
33. (a) Evans, D.A.; MacMillan, D.W.C.; Campos, K.R. *J. Am. Chem. Soc.* **1997**, *119*, 10859. (b) Evans, D.A.; Kozlowski, M.C.; Burgey, C.S.; MacMillan, D.W.C. *J. Am. Chem. Soc.* **1997**, *119*, 7893.

34. Chen, C.-T.; Chao, S.-D.; Yen, K.-C.; Chen, C.-H.; Chou, I.-C.; Hon, S.-W. *J. Am. Chem. Soc.* **1997**, *119*, 11341.
35. Denmark, S.E.; Chen, C.-T. *Tetrahedron Lett.* **1994**, *35*, 4327.
36. (a) Yamada, Y.M.; Yoshikawa, N.; Sasai, H.; Shibasaki, M. *Angew. Chem., Int. Ed. Engl.* **1997**, *36*, 1871. (b) Yamada, Y.M.; Shibasaki, M. *Tetrahedron Lett.* **1998**, *39*, 5561.
37. For a comprehensive review of these bimetallic catalysts, see: Shibasaki, M.; Sasai, H.; Arai, T. *Angew. Chem., Int. Ed. Engl.* **1997**, *36*, 1236.
38. Yamada, Y.M.A.; Shibasaki, M. *Tetrahedron Lett.* **1998**, *39*, 5561.
39. Sodeoka, M.; Ohrai, K.; Shibasaki, M. *J. Org. Chem.* **1995**, *60*, 2648.
40. Krüger, J.; Carreira, E.M. *J. Am. Chem. Soc.* **1998**, *120*, 837.
41. For an application, see: Krüger, J.; Carreira, E.M. *Tetrahedron Lett.* **1998**, *39*, 7013.
42. Pagenkopf, B.; Krüger, J.; Stojanovic, A.; Carreira, E.M. *Angew. Chem., Int. Ed. Engl.* **1998**, *37*, 3124.
43. Fujimura, O. *J. Am. Chem. Soc.* **1998**, *120*, 10032.
44. Anklin, C.; Pregosin, P.S.; Bachechi, F.; Mura, P.; Zambonelli, L. *J. Organomet. Chem.* **1981**, *222*, 175.
45. (a) Denmark, S.E.; Wong, K.-T.; Stavenger, R.A. *J. Am. Chem. Soc.* **1997**, *119*, 2333. (b) Denmark, S.E.; Winter, S.B.D.; Su, X.; Wong, K.-T. *J. Am. Chem. Soc.* **1996**, *118*, 7404.

9 Stereoselective Aldol Reactions in the Synthesis of Polyketide Natural Products

Ian Paterson, Cameron J. Cowden and Debra J. Wallace

9.1 Introduction

The ability to form new carbon-carbon bonds in a regio-, stereo- and enantioselective fashion plays a fundamental role in organic synthesis. In recent years, the aldol reaction has been developed into arguably the most powerful and versatile method in modern carbonyl chemistry for the control of acyclic stereochemistry [1]. This has directly facilitated the efficient assembly of complex polyoxygenated natural products, particularly those of polyketide origin. The directed aldol reaction, in one of its many variants, forms a new carbon-carbon bond between two selected carbonyl compounds, enabling the introduction of one or two stereocenters with predictable configuration. Such a process can be used in the controlled synthesis of a wide variety of β-hydroxy carbonyl compounds, which can serve as functionalized building blocks for further transformations such as stereoselective *syn* or *anti* reduction to provide 1,3-polyols. In a more demanding form, the aldol reaction can be used for the stereodefined coupling of advanced intermediates in natural product synthesis.

In this Chapter, we provide an overview of the various aldol control elements available for achieving synthetically useful stereoselectivity and then analyze some representative syntheses of polyketide natural products (particularly macrolide targets) which are based primarily on the strategic use of aldol chemistry. These examples are chosen to illustrate the variety of aldol processes that have been applied to structurally complex targets and are taken largely from the recent literature (1989–1999). This selection covers some of our own research along with important contributions from other groups.

9.2 Stereochemical Control Elements in Asymmetric Aldol Reactions

In most cases, the relative stereochemistry of an aldol product is determined by the geometry of the enolate component, where (*Z*)-enolates give *syn* aldol adducts and (*E*)-enolates afford *anti* products. Asymmetric induction in the aldol reaction requires that significant π-face selectivity is realized upon addition of the enolate to the aldehyde component. In general (Scheme 9-1), this π-face selection can be imparted from one or more of the following sources: *substrate control* using a chiral ketone or aldehyde; *auxiliary control* using a temporary chiral directing group; *reagent control* using either chiral ligands on the enolate metal or a chiral Lewis acid. Each of these aldol control elements will now be briefly discussed with particular emphasis on reactions used in total synthesis or which, we believe, offer scope for future use.

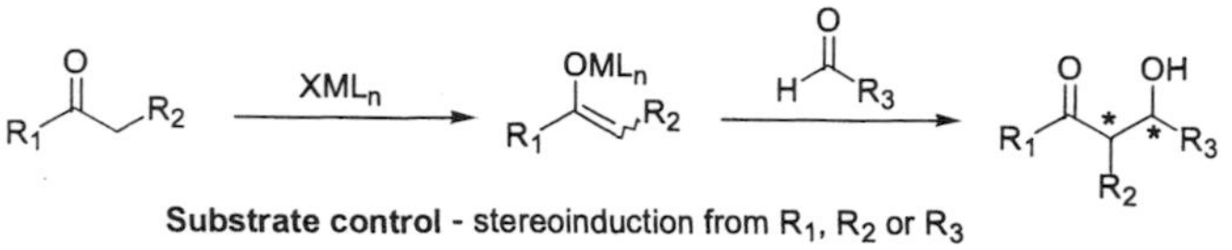

Substrate control - stereoinduction from R_1, R_2 or R_3
Auxiliary control - stereoinduction from R_1 = Auxiliary
Reagent control - stereoinduction from ML_n or added Lewis acid

Scheme 9-1.

9.2.1 Substrate-Controlled Aldol Reactions

A substrate-controlled aldol reaction occurs when either a chiral enolate (usually derived from a ketone) or aldehyde imparts a π-facial bias. This can lead to a concise synthesis of polyketide natural products as steps to incorporate and remove auxiliaries, for example, are not required. This is only possible, however, when the two carbonyl components for aldol coupling favour the desired product. Fortunately, several variables can be tuned to give the desired stereochemical outcome. For example, the enolate geometry can be varied, as can the enolate metal, or changing ligands on the metal itself may alter the π-facial bias. Further variation may also result in changing from a cyclic transition state to an acyclic one. Hence, there is great flexibility when devising a synthetic plan based upon strategically placed aldol disconnections.

9.2.1.1 Stereoinduction from a chiral aldehyde

The combination of an achiral enolate with a chiral aldehyde is a simple method of achieving acyclic stereocontrol. If the enolate geometry can be faithfully translated into predictable product stereochemistry and the aldehyde partner displays a π-facial bias then a synthetically useful result is attained. Unfortunately, the reaction of common metal enolates such as boron, lithium, titanium, etc., with chiral aldehydes usually does not lead to high selectivities as other controlling features play more important roles in reactions proceeding through cyclic transition states. Exceptions to this generalization are aldehydes with a strong Felkin-Anh bias such as those with an α-heteroatom.

In order to maximize the π-facial control from a chiral aldehyde, the Mukaiyama aldol reaction is usually employed. As shown in Scheme 9-2, Felkin-Anh control arising from the α-stereocenter of the aldehyde can have a strong influence on the reaction outcome [2]. When these two reactions were performed using the lithium enolates, the selectivity for **1** and **2** dropped considerably.

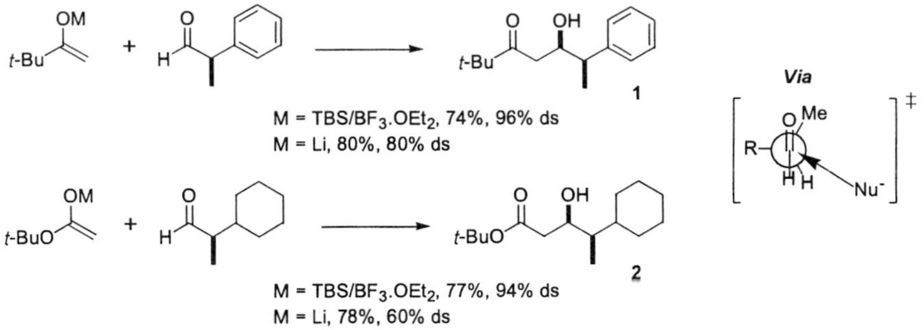

M = TBS/$BF_3.OEt_2$, 74%, 96% ds
M = Li, 80%, 80% ds

M = TBS/$BF_3.OEt_2$, 77%, 94% ds
M = Li, 78%, 60% ds

Scheme 9-2.

As well as the influence of the aldehyde α-stereocenter, a β-alkoxy group can also impart a strong facial bias on the aldehyde partner. As shown in Scheme 9-3, poor selectivity was obtained for lithium, titanium and boron enolates, while the Mukaiyama aldol addition proceeded with 92% ds in favour of adduct **3** [3]. An opposed dipoles argument accounts for the 1,3-*anti* relationship between the new and pre-existing β-oxy-substituent in this product. When the α- and β-stereocenters of an aldehyde are arranged in a mutually reinforcing sense, as in aldehyde **5**, almost complete stereochemical control can be achieved (Scheme 9-3). This simple control element arising from the aldehyde component has great potential in total synthesis.

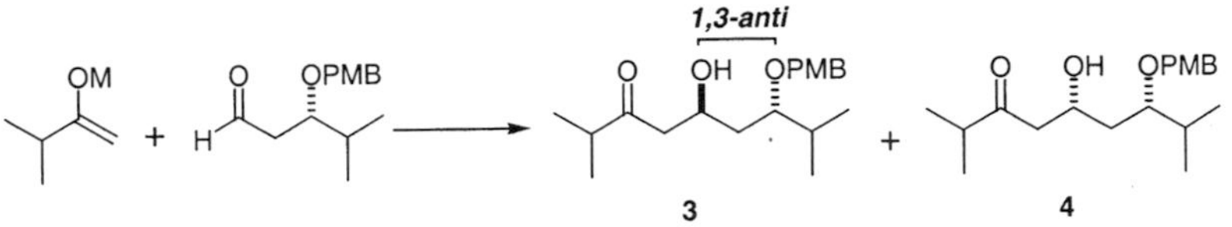

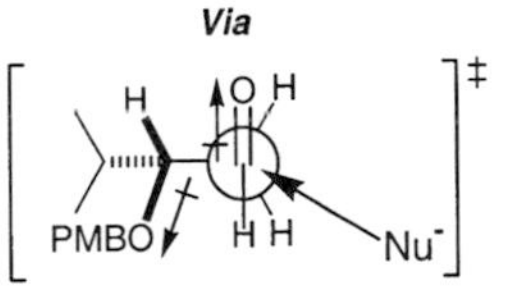

Metal (M)	3:4 (Yield)
Li	71:29 (100%)
$TiCl_n$	60:40 (98%)
B(9-BBN)	42:58 (82%)
TMS/$BF_3.OEt_2$	92:8 (91%)

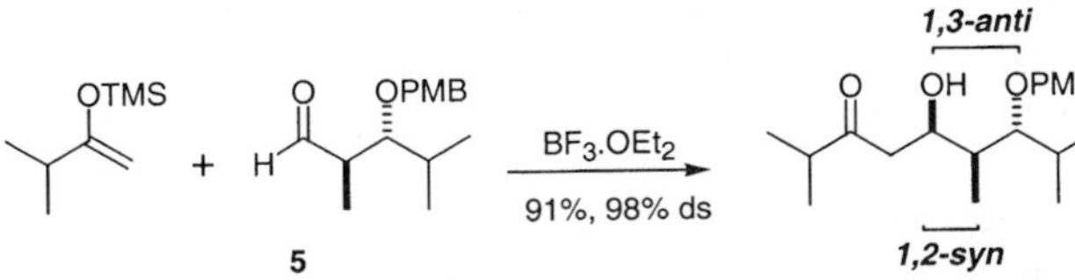

$BF_3.OEt_2$
91%, 98% ds

Scheme 9-3.

The use of a coordinating Lewis acid allows the exploitation of chelation control in Mukaiyama aldol reactions. The aldol coupling shown in Scheme 9-4 led to the *anti*-Felkin adduct **6** as the only observed product and was a key step in the synthesis of tautomycin [4].

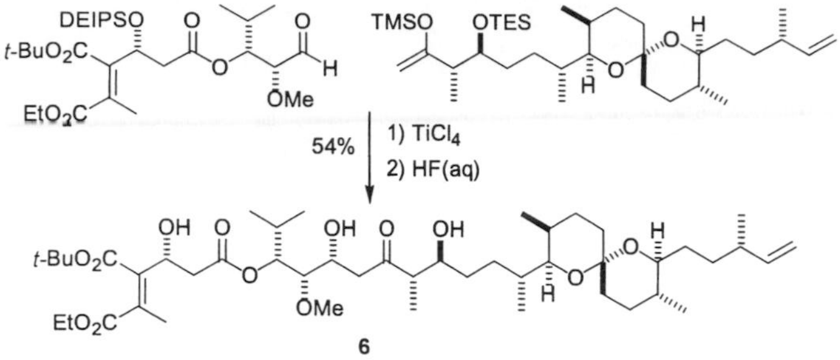

Scheme 9-4.

The Mukaiyama aldol reaction of ethyl ketones can lead to the controlled introduction of two adjacent stereocenters. While enolate geometry may not be transferred faithfully to the relative stereochemistry of the aldol product (*syn* versus *anti*), stereoconvergent reactions are possible. In the example shown in Scheme 9-5, it should be noted that π-facial control from the chiral aldehyde is strong as both products **7** and **8** arise from Felkin selectivity [5].

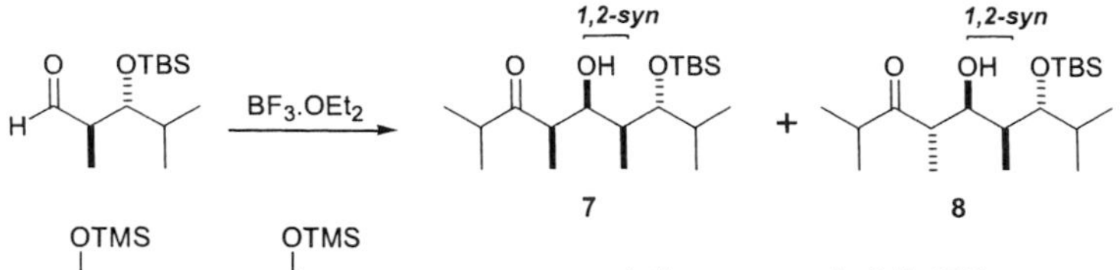

enol ether	selectivity (**7**:**8**)
(*E*)-**9**	95:5
(*Z*)-**9**	87:13

Scheme 9-5.

9.2.1.2 Stereoinduction from a chiral ketone

This is the most common form of substrate control in asymmetric aldol reactions. In general, π-facial discrimination arises from the α-stereocenter of the enolate component; however, there are many cases where a β-oxygen substituent plays an important role.

We have shown that useful levels of 1,4-asymmetric induction can be achieved with the boron enolate **10**, derived from methyl ketone **11** [6], in which a variety of protecting groups can be accommodated (e.g. TBS, TIPS, Bn, PMB)

(Scheme 9-6). The selectivity for adducts such as **12** can also be improved using reinforcing chiral ligands on boron.

Boron-mediated aldol reactions of *α*-oxygenated methyl ketones are normally unselective, and chiral ligands are needed to achieve useful levels of control. However, as shown in Scheme 9-6, a Mukaiyama aldol reaction can be used where induction from silyl enol ether **13** is high, favouring adduct **14** [7, 8].

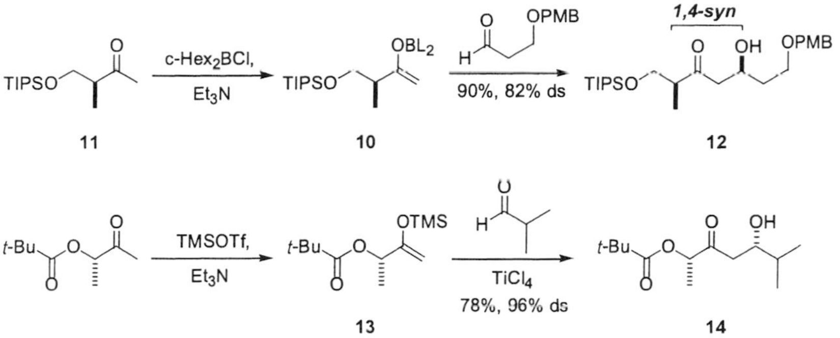

Scheme 9-6.

More surprising was our recent finding that high levels of 1,5-induction were imparted by *β*-oxygenated methyl ketones such as **15**, leading to adduct **16** (Scheme 9-7) [9]. This reaction now opens the way for the stereoselective synthesis of a variety of 1,3-polyols as found in polyacetate natural products. A closely related series of ketones using cyclic protection of the *β*-oxygen protecting group (e.g. ketone **17**), as studied by Evans and co-workers, also led to high levels of 1,5-*anti* induction [10].

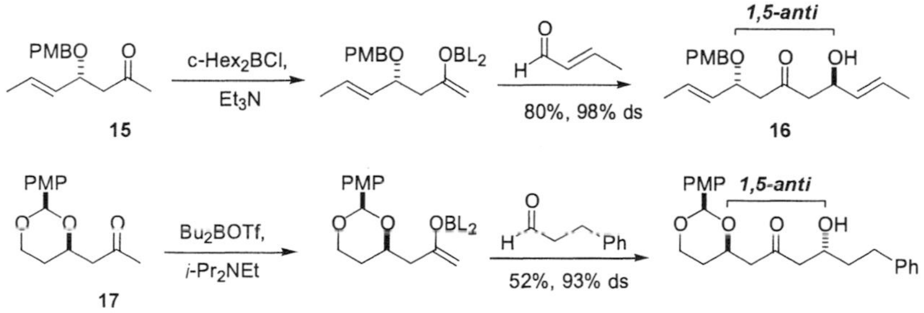

Scheme 9-7.

Ethyl ketones (*R*)- and (*S*)-**18** (Scheme 9-8), as introduced by our group, have been used extensively as versatile dipropionate reagents in polyketide synthesis [11]. Selective formation of the (*E*)-enol borinate **19** is possible using *c*-Hex_2BCl and the resulting *anti* aldol products, e.g. **20**, are formed with ca. 97% ds [6a]. Several different hydroxyl protecting groups can be accommodated, but benzyl or *p*-methoxybenzyl lead to best results. This work has been extended to include *α*-alkoxymethyl ketones with comparable selectivities obtained in aldol reactions. For example, ketone **21** was used in the synthesis of adduct **22**, a C_{24}–C_{32} subunit of rapamycin [12]. We have used these same ketones to access *syn* aldol products, e.g. **23**, this time using tin(II) enolate **24** [13].

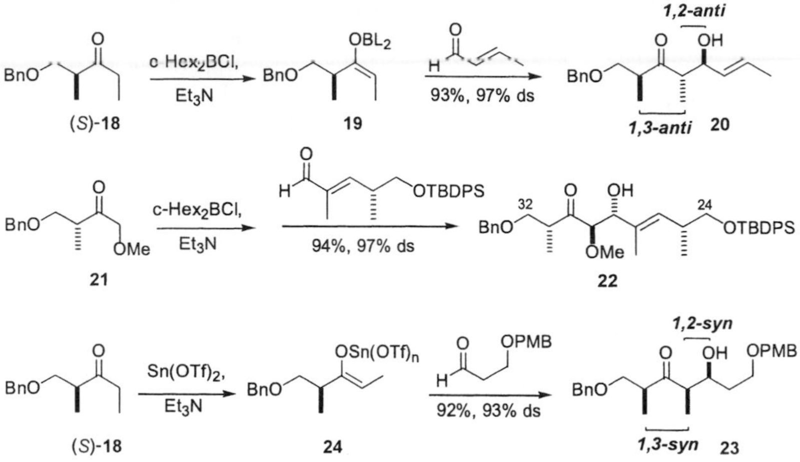

Scheme 9-8.

In a related manner, *β*-keto imide **25** also functions as a versatile dipropionate reagent with three different stereoselective aldol reactions being reported by the Evans group (Scheme 9-9). Both *syn* aldol isomers, **26** and **27**, are available from either the titanium or tin(II) enolates [14] and the *anti* adduct **28** can be accessed using the dicyclohexyl boron enolate [15]. While a chiral auxiliary is present, it is the ketone *α*-stereocenter that controls the *π*-facial selectivity in these aldol reactions.

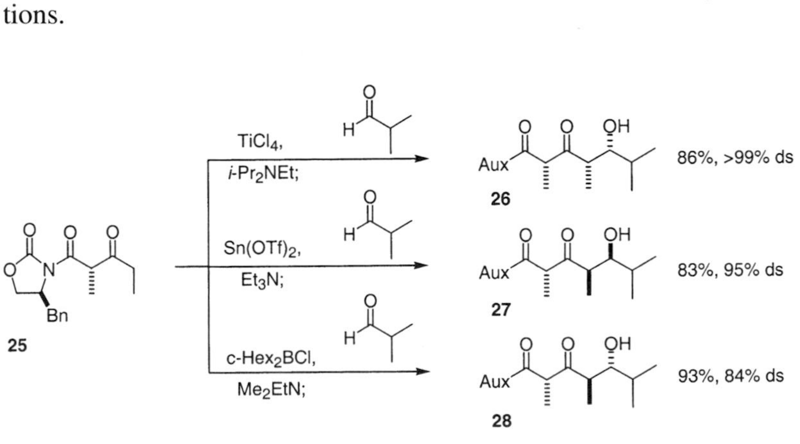

Scheme 9-9.

More elaborate ketones incorporating further stereocenters have also proved synthetically useful. The β-oxygenated ketone **29** gave rise to highly selective, boron-mediated, aldol reactions [16] (Scheme 9-10). This selectivity was only observed when the unusual chlorophenyl boron enolate was used – a system now known to give high *syn* diastereoselectivity with a range of ketones [17].

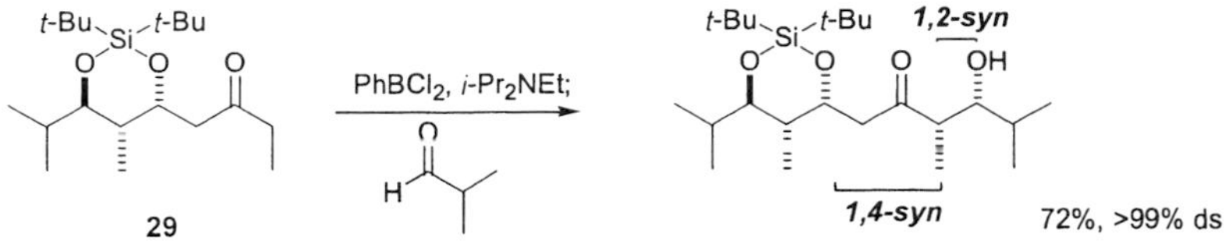

t-Bu Si t-Bu O O O 29 — $PhBCl_2$, i-Pr_2NEt; H(C=O) → t-Bu Si t-Bu O O O OH 1,2-syn 1,4-syn 72%, >99% ds

Scheme 9-10.

The use of α,β-chiral ketones has been studied (Scheme 9-11). In general, the α-stereocenter of the ketone controls the sense of addition and this can be seen in the boron-mediated *anti* aldol reaction of ketone **30** where the configuration of the C_5 stereocenter makes little difference to the selectivity of the reaction [15].

We have shown that the analogous boron-mediated *syn* aldol reactions of these ketones are also highly stereoselective (Scheme 9-11) [18]. A steric model, where the CH(OTBS)R substituent is considered as the large group, accounts for the *syn* relationship between the methyl groups flanking the ketone in adduct **31** [19]. Titanium enolates also give the *syn* aldol adducts preferentially and the stereocontrol from such reactions can be very high, especially if matched in an *anti*-Felkin sense with an α-chiral aldehyde.

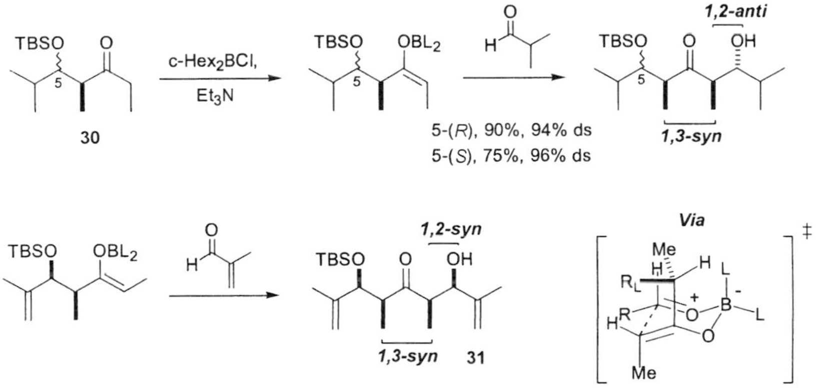

TBSO O 5 30 — c-Hex_2BCl, Et_3N → TBSO OBL_2 5 — H(C=O) → TBSO O OH 5 1,2-anti 1,3-syn

5-(*R*), 90%, 94% ds
5-(*S*), 75%, 96% ds

TBSO OBL_2 — H(C=O) → TBSO O OH 1,2-syn 1,3-syn 31

Via [Me H H L R_L + O B L R H O Me]‡

Scheme 9-11.

As mentioned previously, it can be more difficult to predict *syn*:*anti* diastereoselectivity in Mukaiyama aldol reactions of substituted ketones. However, the reward of high stereocontrol in these reactions is attainable as shown in Scheme 9-12. While the second example shows disappointing aldehyde face selectivity, there is strong enolate facial bias (1,3-*anti* in both **32** and **33**). Therefore,

An *anti* aldol reaction with Felkin control was now needed to couple the two spiroacetal fragments and generate the correct stereochemistry at C_{15} and C_{16} of the spongistatins. A study of the individual fragments indicated that while the enolate showed little facial selectivity, the aldehyde component had a considerable bias for the desired Felkin product. Best results were obtained with the lithium-mediated aldol coupling, which gave adduct **104** in good yield and acceptable selectivity [56c].

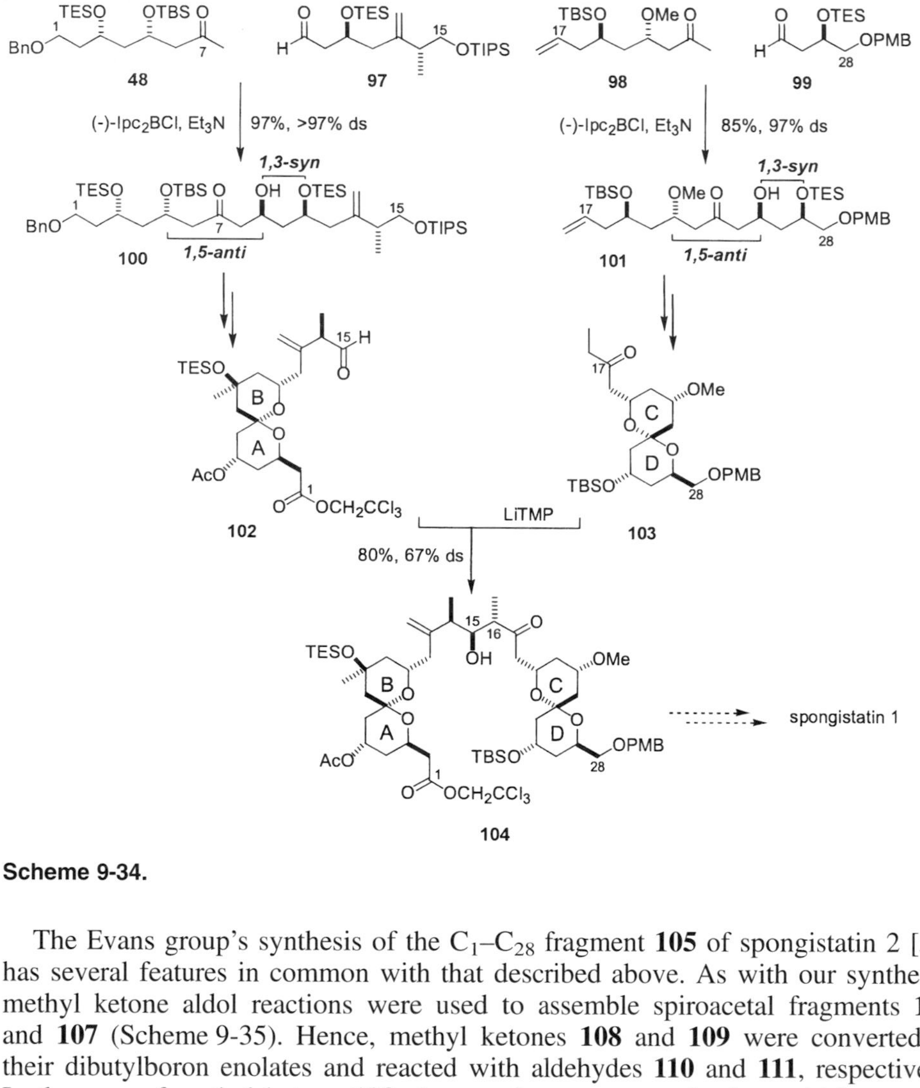

Scheme 9-34.

The Evans group's synthesis of the C_1–C_{28} fragment **105** of spongistatin 2 [55] has several features in common with that described above. As with our synthesis, methyl ketone aldol reactions were used to assemble spiroacetal fragments **106** and **107** (Scheme 9-35). Hence, methyl ketones **108** and **109** were converted to their dibutylboron enolates and reacted with aldehydes **110** and **111**, respectively. In the case of methyl ketone **108**, the reaction was non-selective, which was not detrimental to the synthesis as the C_7 alcohol was subsequently oxidized. As already noted, use of chiral ligands would usually be required for high selectivity when a β-siloxy group is used on the ketone. However, with methyl ketone **109**, product **112** was produced with excellent stereocontrol, even using achiral ligands (96% ds) [10]. Following elaboration to the spiroacetal fragments **106** and **107**, a similar *anti* aldol reaction to that carried out by our group was required. In this case, boron was the metal of choice and the (*E*)-enolate, derived from ketone **107**, reacted with aldehyde **106** to afford the desired adduct **105** (90% ds). Presumably, this excellent selectivity results solely from the aldehyde component which shows a high facial bias for the Felkin product in reaction with (*E*)-enolates.

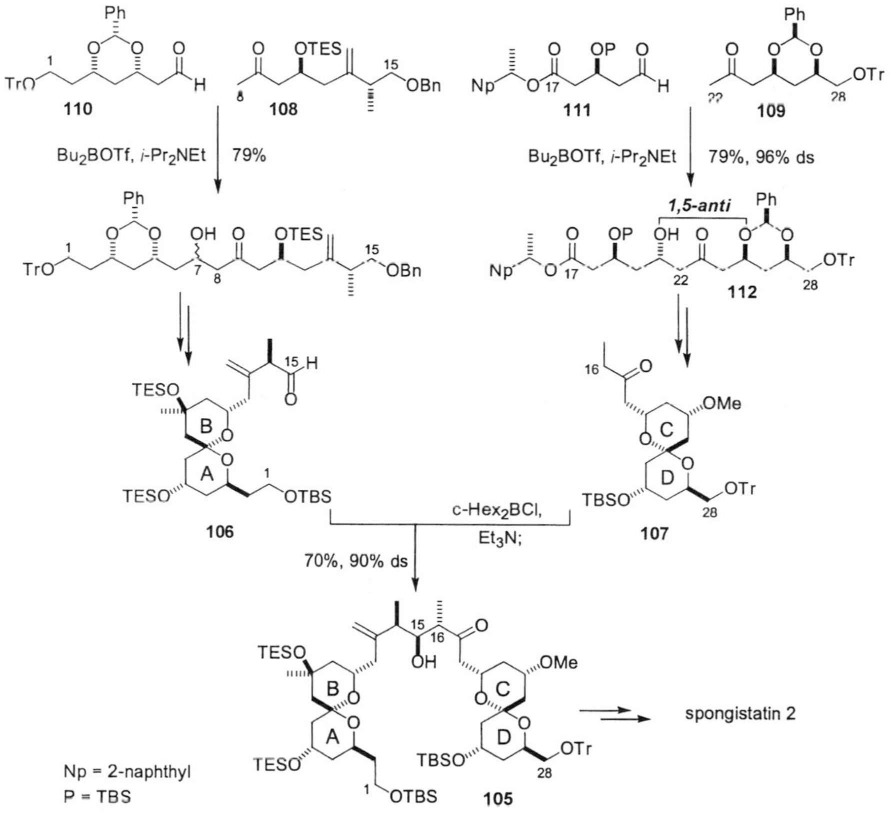

Np = 2-naphthyl
P = TBS

Scheme 9-35.

Bryostatins. The bryostatins are a group of cytotoxic marine macrolides, isolated from invertebrate filter feeders, that exhibit promising anticancer activity (Scheme 9-36). Two completed total syntheses of the bryostatins have demonstrated the versatility and power of aldol reactions in the concise assembly of complex polyketides.

Masamune's synthesis of bryostatin 7 (**114**) [36] contains early examples of double asymmetric induction, where the aldol reaction of chiral ketones could be

In the synthesis of the C_{17}–C_{27} subunit **128**, the aldol reaction of ketone **129** was non-selective under various conditions, and hence asymmetric induction from chiral ligands on boron was required to introduce the C_{23} stereocenter in **130**. Use of the commercially available (–)-Ipc_2BCl reagent gave aldol adduct **130** with 93% ds. Further manipulations afforded the third bryostatin fragment **128**, and the synthesis of bryostatin 2 was completed following the coupling of the above three fragments.

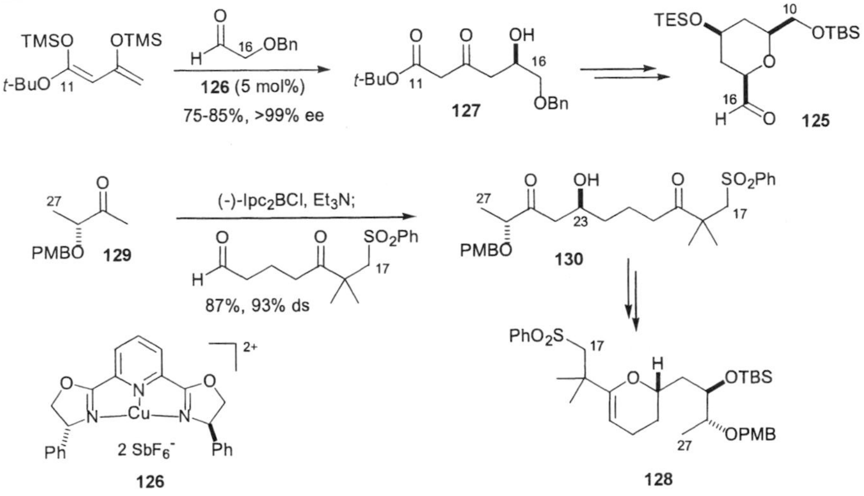

Scheme 9-40.

In the previous synthesis, two asymmetric aldol reactions using dienyl silyl ethers were described, one using a chiral Lewis acid for stereoinduction while the other used substrate control from a chiral aldehyde. This can be compared with the use of chiral dienolate **131** in the synthesis of a C_1–C_{16} fragment of the bryostatins (Scheme 9-41) [59]. Here, the menthyl-derived auxiliary is covalently attached to the enolate, and again an excellent level of asymmetric induction was achieved on addition to aldehyde **132** to give adduct **133**.

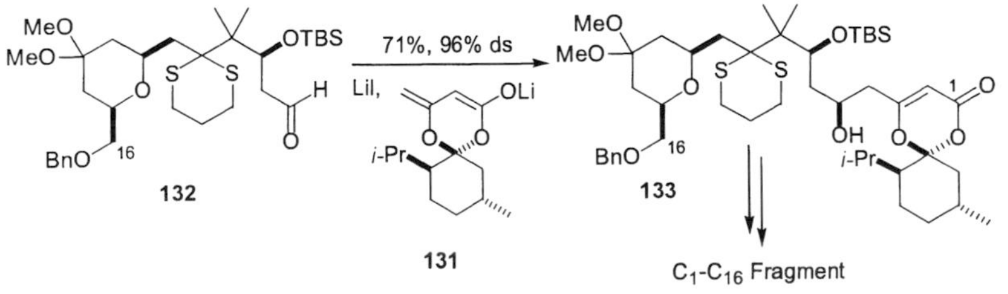

Scheme 9-41.

Other approaches to the bryostatins have also used enantio- or diastereoselective aldol reactions. An interesting iterative strategy for the synthesis of the C_1–C_9 polyacetate region **134** has been disclosed where each aldol addition proceeds with excellent stereocontrol (99:1) under the catalytic influence of oxazaborolidine **135** (Scheme 9-42) [60]. Finally, a moderately selective, auxiliary controlled, acetate aldol reaction has been used for the introduction of the C_3 stereocenter of the bryostatins giving adduct **136** (84% ds) [61].

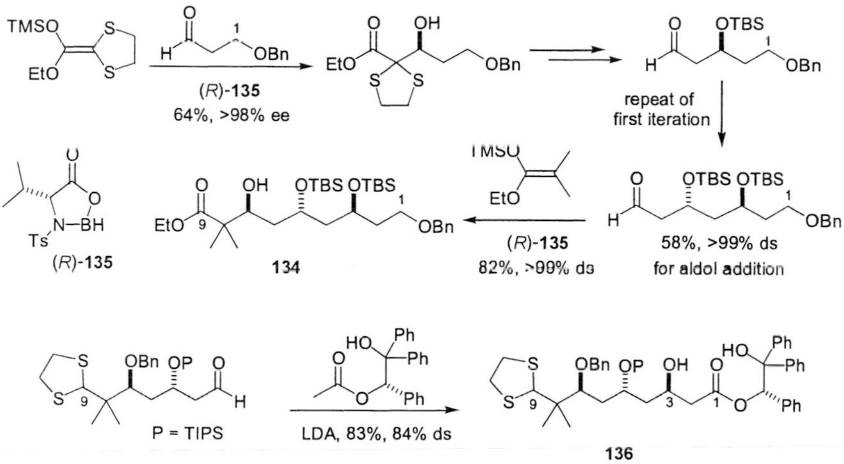

Scheme 9-42.

Rutamycins. Rutamycins A (**137**) and B (**138**) are 26-membered macrolide antibiotics isolated from *Streptomyces* cultures (Scheme 9-43). They are closely related to the oligomycin family of natural products and more distantly to the 22-membered cytovaricin. Each of these macrocycles consists of a highly functionalized spiroacetal unit bridged by a polypropionate-derived chain. In the rutamycins, the asymmetric aldol reaction could be envisaged to play a key role in assembling the spiroacetal portion and the C_4–C_{14} polypropionate sector with various strategic disconnections possible.

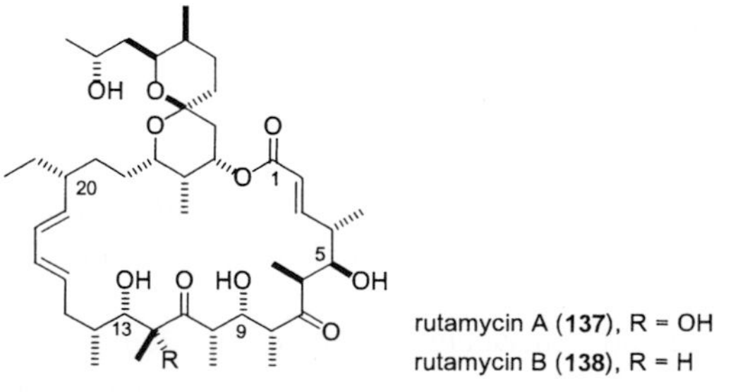

Scheme 9-43.

The first total synthesis of rutamycin B (**138**) was reported by Evans et al. in 1993 [62]. In this route, two auxiliary controlled *syn* aldol reactions were used to introduce the stereocenters at C_{23}/C_{24} and C_{30}/C_{31} of the spiroacetal (Scheme 9-44). In both cases, the same enantiomer of auxiliary **36** was used and, as expected, excellent stereocontrol over the newly formed centers in **139** and **140** was achieved (>99% ds). Elaboration of the aldol adducts to **141** and **144**, subsequent coupling and further functionalization afforded the spiroacetal containing vinyl boronic acid **143**.

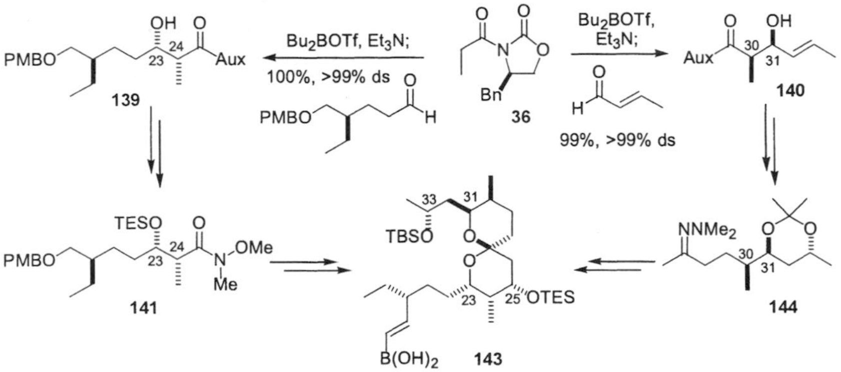

Scheme 9-44.

In the Evans synthesis of the polypropionate region (Scheme 9-45), the boron-mediated *anti* aldol reaction of *β*-ketoimide *ent*-**25** with *α*-chiral aldehyde **145** afforded **146** with 97% ds in what is expected to be a matched addition. Adduct **146** was then converted into aldehyde **147** in readiness for union with the C_1–C_8 ketone. This coupling was achieved using the titanium-mediated *syn* aldol reaction of enolate **148** leading to the formation of **149** with 97% ds.

Considering that (*Z*)-enolates usually have a small bias towards *anti*-Felkin adducts due to an unfavourable *syn*-pentane interaction in the transition state, the high selectivity for aldol product **149** is surprising [52]. One explanation would be that enolate **148** has a particularly strong influence; however, further studies showed that aldehyde **147** was playing an important role, as changing either the aldehyde *β*-stereocenter or protecting group led to an erosion of aldol selectivity. An indication of the stereochemical influences operating in this situation comes from the reaction of the related aldehyde **150** with achiral enolates [63]. This reveals that aldehyde **150** has a small bias for the Felkin adduct **151**, and hence the C_8–C_9 rutamycin coupling to give **149** is probably a matched reaction. Such an observation has also been made by White et al., and in their recent total synthesis of rutamycin they also used an impressive titanium-mediated aldol coupling for C_8–C_9 bond formation [64].

A synthesis of both the polypropionate and spiroacetal fragments of rutamycin B have been described by Panek and Jain [65]. The majority of stereocenters were introduced via asymmetric allylation and crotylation reactions; however, an aldol

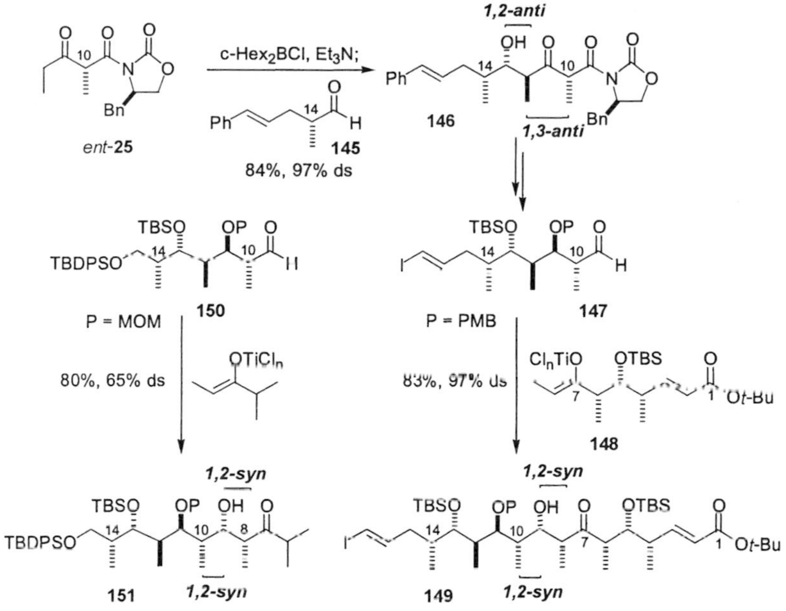

Scheme 9-45.

reaction between ketone **152** and aldehyde **153** was used to form the C_{12}–C_{13} bond in **154** (Scheme 9-46). This disconnection requires an *anti* aldol reaction to proceed with Felkin control – normally a matched reaction with an achiral enolate; however, the 1,3-*anti* relationship of the methyl groups flanking the carbonyl group in **154** would be difficult to control using common enolates (for example see Scheme 9-11). As discussed earlier (Scheme 9-5), the Mukaiyama aldol is best used in this situation, and indeed, reaction between aldehyde **153** and the silyl enol ether derived from ketone **152**, in the presence of $BF_3{\cdot}OEt_2$, generated product **154** with 88% ds.

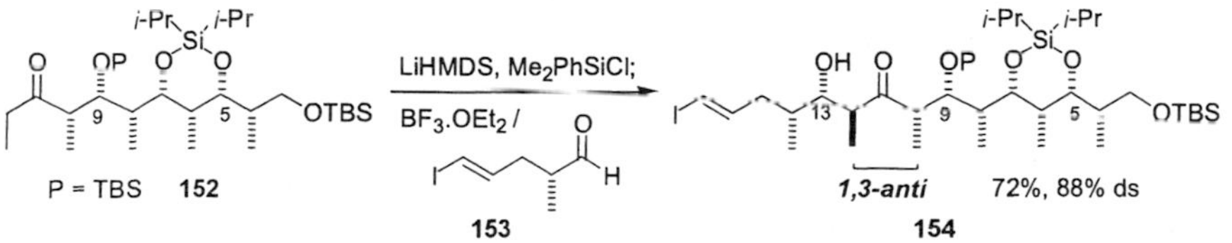

Scheme 9-46.

Aplyronine A. Aplyronine A (**155**) was isolated in 1993 from a Japanese sea hare and is an unusual 24-membered macrolide displaying potent antitumor activity (Scheme 9-47). The three stereotetrads within the molecule, C_7–C_{10}, C_{23}–C_{26} and C_{29}–C_{32}, make it an attractive target to demonstrate aldol methodology. To date, the only total synthesis was completed by Yamada et al., which served to confirm the structural and stereochemical assignment of the natural product which, incidentally, was proposed by the same group [66].

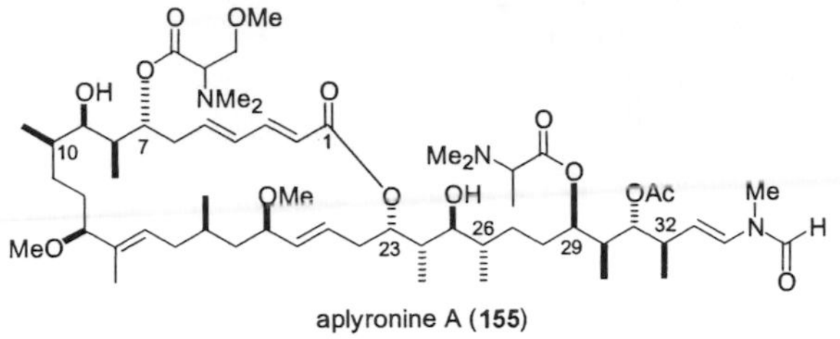

aplyronine A (**155**)

Scheme 9-47.

In the construction of the C_5–C_{11} segment **156**, the boron-mediated *syn* aldol reaction of the Evans imide **157** with α-chiral aldehyde **158** afforded adduct **159** (Scheme 9-48) [66]. Following conversion to an allylic alcohol, the C_7 stereocenter in **156** was introduced by a Sharpless asymmetric epoxidation. The Evans auxiliary was again used to introduce the C_{23}/C_{24} and C_{29}/C_{30} stereocenters leading to *syn* aldol adducts **160** and **161**. These were then converted into allylic alcohols **162** and **163**, and the final two stereocenters were introduced by asymmetric epoxidation and regioselective epoxide opening with methyl cuprate to generate **164** and **165**, respectively. While these reactions proceeded with high yields and selectivities, a significant number of steps are required for the introduction of each stereocenter – outlining one limitation of chiral auxiliary methodology in synthesis.

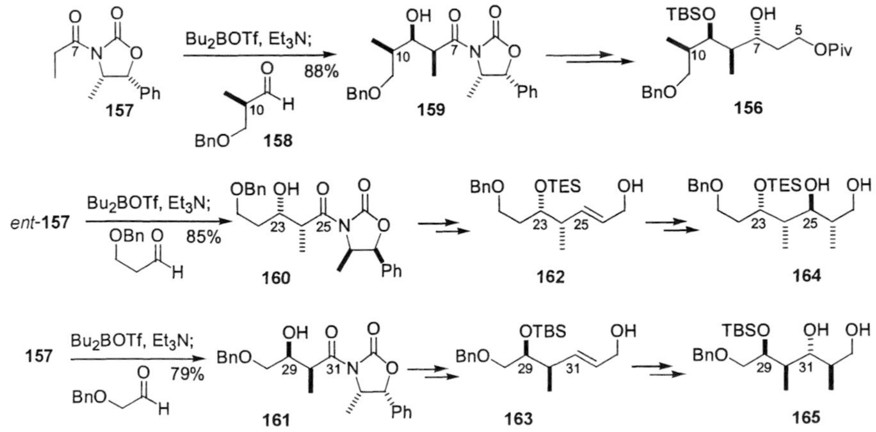

Scheme 9-48.

An alternative strategy has been used by ourselves in the synthesis of the aplyronine macrocycle **166**, whereby chiral ketones **167** and **168** were used in two substrate-controlled aldol reactions (Scheme 9-49) [67]. Following reduction, this installs four contiguous stereocenters in just two steps. Hence, the boron-mediated *anti* aldol reaction between ketone **167** and aldehyde **169** proceeded in 96% yield with 97% ds. An Evans-Tishchenko reduction was then used to install the C_9 stereocenter to give **170**. For the C_{23}–C_{26} fragment, a *syn* aldol reaction between ketone **168** and aldehyde **171** was carried out using the (*Z*)-tin enolate to ensure high selectivity. A subsequent *anti* reduction using a borohydride reagent gave adduct **172** which, along with the C_1–C_{11} fragment **170**, were carried through to the synthesis of macrocycle **166**.

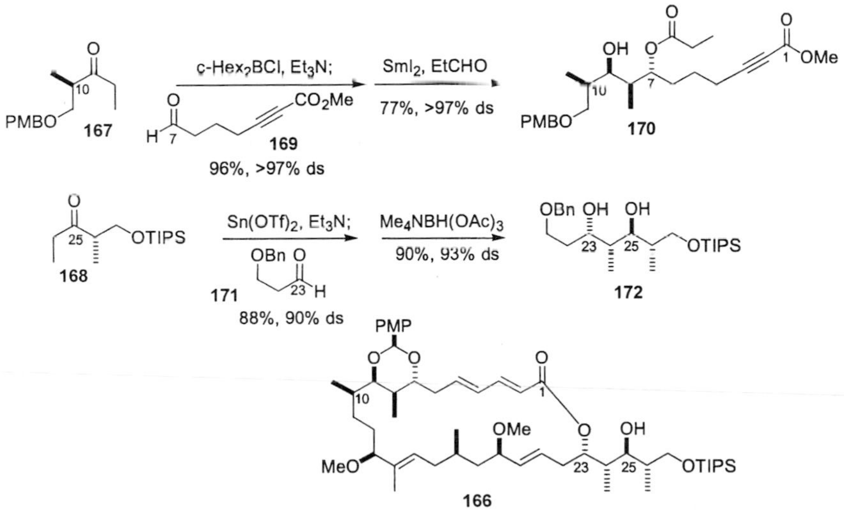

Scheme 9-49.

Bafilomycin A$_1$ and Concanamycin F. Bafilomycin A$_1$ (**173**) and concanamycin A (**174**) are members of the hygrolide family of macrolide antibiotics and act as potent, relatively specific, membrane ATPase inhibitors. As is clear from the structures, there is a strong stereochemical homology between these compounds (Scheme 9-50).

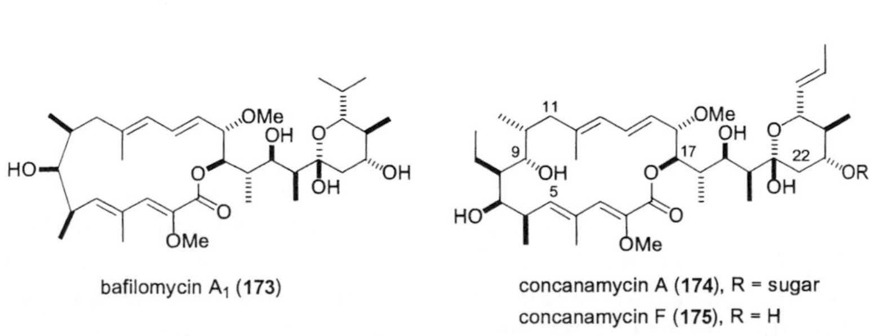

bafilomycin A$_1$ (**173**)

concanamycin A (**174**), R = sugar
concanamycin F (**175**), R = H

Scheme 9-50.

In the first total synthesis of bafilomycin A_1 by Evans and Calter [16], the *syn* aldol reaction between ketone **29** and aldehyde **176** was a pivotal transformation (Scheme 9-51). Using a (*Z*)-enolate, it could be expected that aldehyde **176** would have a small bias for the desired *anti*-Felkin adduct, however, control from the ketone component would be needed for high stereoselectivity. Use of common metal enolates led to poor stereocontrol; however, model studies indicated that the (*Z*)-chlorophenyl boron enolate, in conjunction with cyclic protection of the C_{21}–C_{23} diol, induced high selectivity in the desired sense. In practice, the coupling of the required aldehyde **176** and enolate **77** afforded **178** with >95% ds. Compound **178** was then successfully elaborated to give bafilomycin A_1. In the second reported synthesis of bafilomycin A_1, Toshima et al. carried out the same aldol coupling to form the C_{17}–C_{18} bond [68].

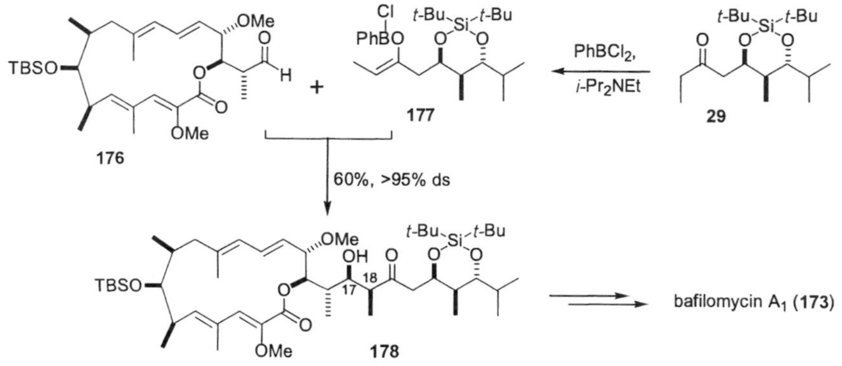

Scheme 9-51.

In our synthesis of the C_{13}–C_{25} fragment **179** of bafilomycin A_1 (Scheme 9-52), a boron-mediated *syn* aldol reaction was employed between an ethyl ketone and, this time, the truncated aldehyde **180** (as opposed to the macrocyclic aldehyde **176**) [69]. This aldehyde contains an α-methylene group, thus removing the effect of an α-stereocenter. To test the aldehyde π-facial bias, it was first reacted with diethylketone and found to favour the required 2,3-*syn*-3,5-*anti* adduct **181** with a range of enolates [70]. The π-facial bias of the ketone was also examined and, as with the Evans study, cyclic protection of the ketone hydroxyl groups was essential for high stereoselectivity. When combined, the (*Z*)-dibutyl boron enolate **182** reacted with aldehyde **180** to afford the desired compound **183** with 82% ds. Hydroxyl-directed hydrogenation then installed the required C_{16}-stereocenter and cyclization gave the C_{13}–C_{25} fragment **179**.

Roush has also completed the synthesis of a C_{13}–C_{25} fragment of bafilomycin A_1, now using a methyl ketone aldol reaction between ketone **184** and aldehyde **5** to form the C_{20}–C_{21} bond (Scheme 9-52) [71]. This reaction was only selective (89% ds) under carefully defined conditions which included choice of metal enolate and, remarkably, the remote C_{15} oxygen protecting group. Replacing the C_{15} MOM ether with a silyl ether, as in **185**, led to a ca. 1:1 mixture of aldol products. Further experiments showed that the C_{17}-protecting group had little effect on the reaction and that the aldehyde had low facial bias [72]. This led to the proposal that a transition state involving secondary chelation of the C_{15} Lewis basic group was operating. For another example of an unexpected result due to remote functionality, see Scheme 9-59 where a secondary orbital interaction was proposed.

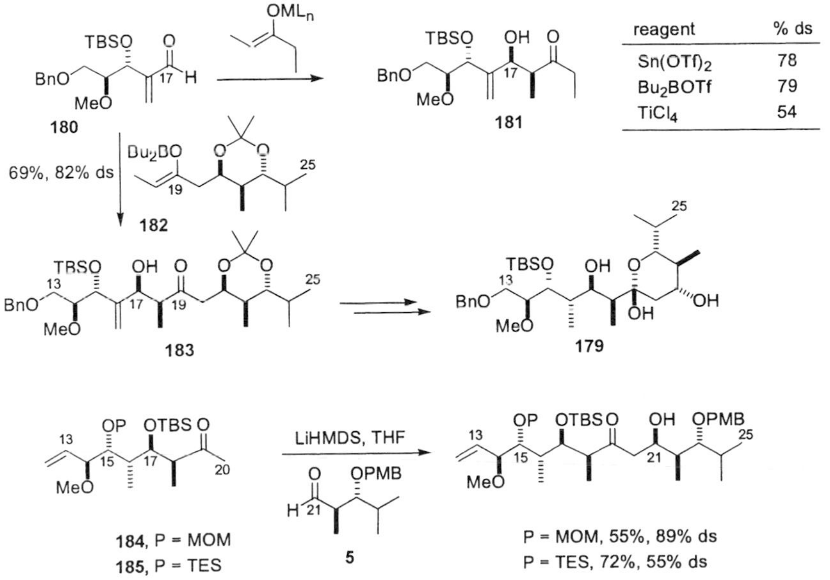

reagent	% ds
$Sn(OTf)_2$	78
Bu_2BOTf	79
$TiCl_4$	54

Scheme 9-52.

In our group's synthesis of concanamycin F (**175**), we exploited several aldol reactions that had been developed in our laboratory [73]. *Anti* aldol reactions of both propyl and ethyl ketones **186** and (*R*)-**18** (Scheme 9-53) proceeded with excellent stereocontrol (≥95% ds). We have also introduced the in situ $LiBH_4$ reduction of a boron aldolate to complement the various *anti* reduction protocols and, in the present case, the aldol/reduction process provided the *syn* diol **187** with 95% ds in a single step. The *anti* reduction of adduct **188** was achieved using the samarium-mediated, Evans-Tishchenko reaction which simultaneously protected the incipient hydroxyl group. Functional group manipulation of adduct **189** then provided aldehyde **190**, which was combined with (*E*)-boron enolate **191** affording *anti* aldol product **192** (>98% ds). The major stereochemical influence in this reaction comes from the ketone, and this constitutes a matched situation. In this scheme, just three aldol reactions and simple manipulations have contributed more than half of concanamycin's stereocenters, highlighting how quickly acyclic stereocontrol can assemble complex, stereochemically rich, polyketide fragments.

The penultimate step in our synthesis of concanamycin F required the coupling of macrocyclic enolate **193** with aldehyde **194** – itself derived from a stereoselec-

The final example highlights the use of an aldol reaction to form macrocycle **223**, which was then used in the synthesis of epothilone A (Scheme 9-60). In earlier work by Danishefsky on a related reaction to close the 35-membered macrocyclic ring of rapamycin [80], only a low yield of the desired product was obtained. In this current example, however, the reaction was cleverly conceived using a non-enolizable aldehyde. Interestingly, higher selectivities were achieved when the reaction was allowed to warm to room temperature, indicating a reversible aldol addition process where adduct **223** is the thermodynamically favoured product [77].

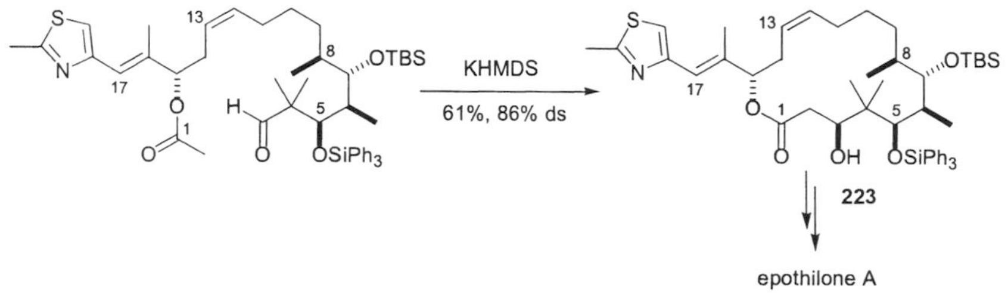

KHMDS

61%, 86% ds

223

epothilone A

Scheme 9-60.

Oleandolide, 6-Deoxyerythronolide B and Erythronolide A. The 14-membered macrolides, oleandolide (**224**), 6-deoxyerythronolide B (**225**) and erythronolide A (**226**) have proved to be popular synthetic targets for the development of new reactions for acyclic stereocontrol. The first two compounds differ only in the presence/absence of the epoxide functionality at C_8 and substitution at C_{14} while erythronolide A has two tertiary alcohols: one at C_6 and the other at C_{12} (Scheme 9-61).

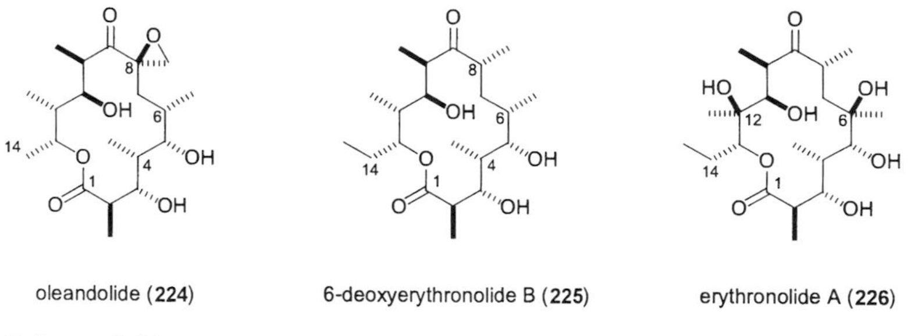

oleandolide (**224**) 6-deoxyerythronolide B (**225**) erythronolide A (**226**)

Scheme 9-61.

The Paterson [81] and Evans [82] syntheses of oleandolide are related in that each use their own dipropionate reagents to synthesize two stereopentad units, which are then connected near the center of the molecule (C_7–C_8 or C_8–C_9) (Scheme 9-62). In each case, the final C_9 carbonyl is "protected" at the lower hydroxyl oxidation state, although the two groups use different C_9-epimers.

Our synthesis of the C_1–C_7 fragment **227** of oleandolide started with a substrate-controlled tin-mediated aldol reaction of *α*-chiral ketone (*S*)-**18** which afforded *syn* adduct **52** with 93% ds. This same transformation could also be achieved using reagent control with $(Ipc)_2BOTf$, albeit with lower selectivity (90% ds). In a key step, treatment of the aldol adduct **52** with (+)-$(Ipc)_2BH$ led to controlled reduction of the C_3 carbonyl together with stereoselective hydroboration of the C_6–C_7 olefin, affording the desired triol **228** with 90% ds.

For the synthesis of the C_8–C_{14} fragment **229**, we again started from (*S*)-**18** and, on this occasion, a substrate-controlled *anti* aldol reaction was used to give adduct **230** with 97% ds. Although the newly generated stereocenter at C_9 was not retained in the final product, earlier work using the C_9 epimer had failed at a late stage of the synthesis, justifying the control of this center. Functionalization of adduct **230** afforded aldehyde **229**, which was then coupled with sulfoxide **227**.

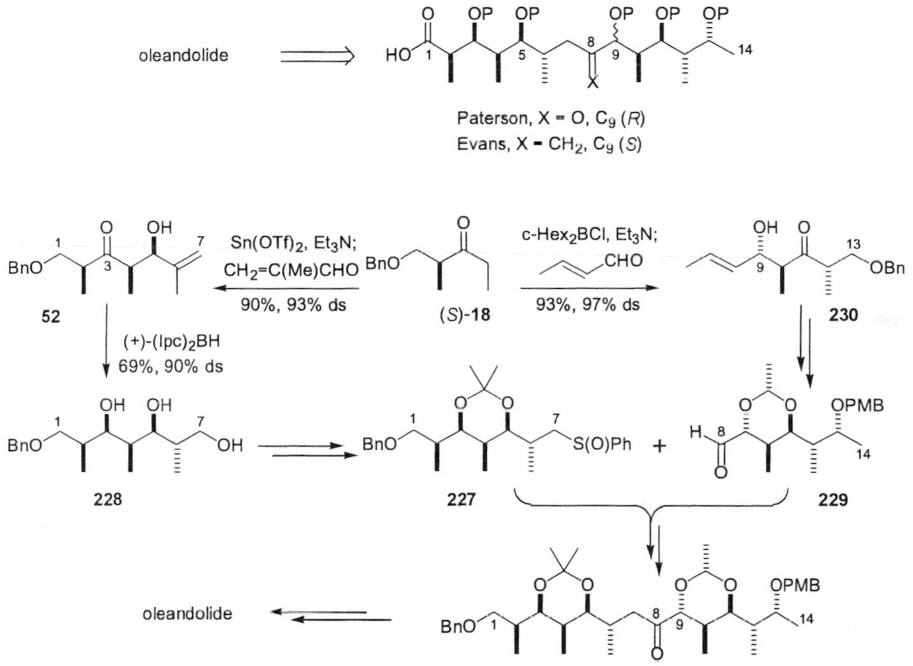

Scheme 9-62.

The Evans synthesis of oleandolide [82] began with substrate-controlled titanium and tin aldol reactions of dipropionate reagent *ent*-**25** (Scheme 9-63). In the first of these reactions, the titanium enolate was generated using $Ti(i\text{-}OPr)Cl_3$ instead of the more common $TiCl_4$, and, when reacted with *α*-chiral aldehyde **231**, the *anti*-Felkin product **232** was formed with >95% ds. Following stereoselective reduction, adduct **233** was then used in the synthesis of the C_1–C_8 fragment **234**, ready for a Pd-catalyzed Stille cross-coupling reaction. The synthesis of the C_9–C_{14} acid chloride **235** again started from imide *ent*-**25**. This time, the tin(II) enolate provided *syn* aldol adduct **236** with moderate selectivity (83% ds). Following *anti* reduction and selective protecting group chemistry, acid chloride **235** was produced. The Stille coupling reaction of **234** and **235** gave enone **237**, an advanced intermediate in the successful synthesis of oleandolide.

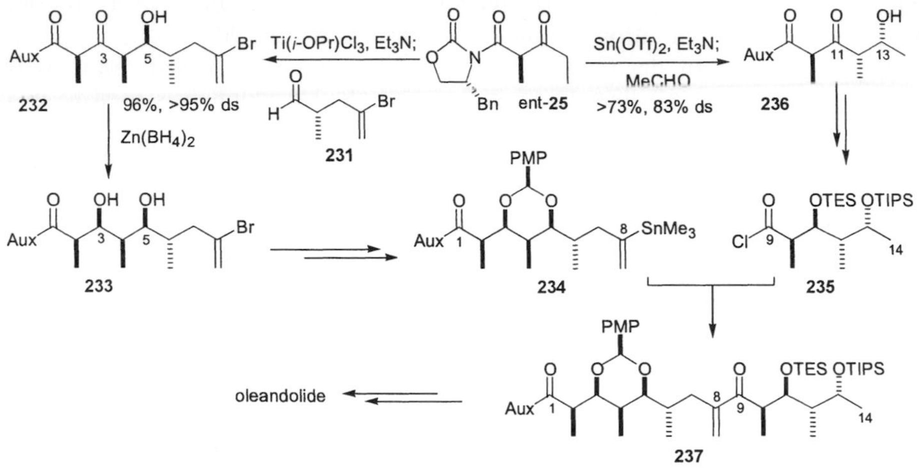

Scheme 9-63.

Both these syntheses of oleandolide relied upon substrate-controlled aldol reactions of dipropionate reagents (*S*)-**18** and *ent*-**25**. Substrate control is also evident in the way both groups incorporated the exocyclic epoxide with greater than 95% ds. While we chose to use macrocyclic control for this transformation, the Evans synthesis used acyclic stereocontrol and the directing influence of a nearby hydroxyl group.

The Evans synthesis of 6-deoxyerythronolide B [82] used similar chemistry to that described above to synthesize two functionalized building blocks that contain nine of the ten required stereocenters (Scheme 9-64). The coupling of aldehyde **238** with the appropriate ketone needed a selective reaction as, although the C_7 hydroxyl was to be removed at a later stage, the C_8 stereocenter of the eventual macrolide still needed to be set correctly. This was achieved using a Lewis acid promoted Mukaiyama aldol reaction of (*Z*)-enol silane **239** with aldehyde **238** proceeding under Felkin control. The reaction is noteworthy in that titanium or most boron-mediated couplings would provide the undesired compound with a *syn* methyl relationship across the carbonyl group, i.e. the incorrect stereochemistry at C_8. This aldol strategy is also interesting as it introduces an extra hydroxyl at C_7, which then requires deoxygenation, a process that also occurs in the biosynthesis of 6-deoxyerythronolide B.

At the time (1981), the Masamune synthesis of 6-deoxyerythronolide B was a landmark achievement in the art of acyclic stereocontrol [83]. Four aldol reactions were used in the synthesis, all proceeding with high selectivity (>93% ds). The first aldol reaction depicted in Scheme 9-65 was used in the synthesis of aldehyde

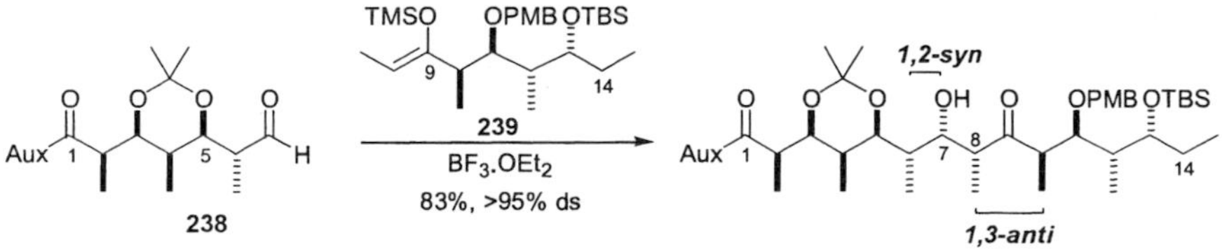

Scheme 9-64.

240 and employed the mandelic acid-derived auxiliary (*R*)-**241**. Transformation of the *syn* aldol adduct provided aldehyde **240** ready for coupling with ketone **242**, itself synthesized using two aldol reactions of substrate (*S*)-**241**. The aldol union of **240** and **242** was best achieved using the lithium enolate and led to adduct **243** with 94% ds. Several factors must be contributing to this matched result, including a small preference for the (*Z*)-enolate to provide the *anti*-Felkin product. The presence of chelation from the *β*-oxygen of the aldehyde is possible, though recent studies indicate that this is unlikely [3]. Even today, understanding the sense of asymmetric induction in complex coupling reactions remains difficult as many factors are competing.

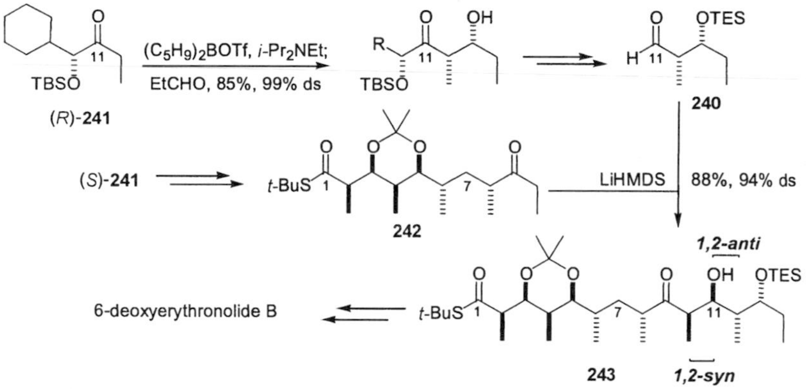

Scheme 9-65.

Our synthesis of (9*S*)-dihydroerythronolide A, which constitutes a formal synthesis of erythronolide A (**226**), depends on a key aldol reaction between the racemic aldehyde **244** and imide auxiliary **245** (Scheme 9-66) [84]. In this reaction, the auxiliary overrides any aldehyde facial bias, thus leading to an equimolar mixture of separable *syn* adducts **246** and **247**. These two compounds were then processed separately and together provide five of the ten necessary stereocenters of erythronolide A (C_9 will be oxidized). This synthesis also features the thioalkylation of silyl enol ether **248** giving ketone **249**, a process which can be compared with the Mukaiyama addition to aldehydes. Presumably, Felkin selectivity controls the C_{11} stereocenter while the mixture of C_{12} epimers was not detrimental as epimerization could be effected in the subsequent elimination step.

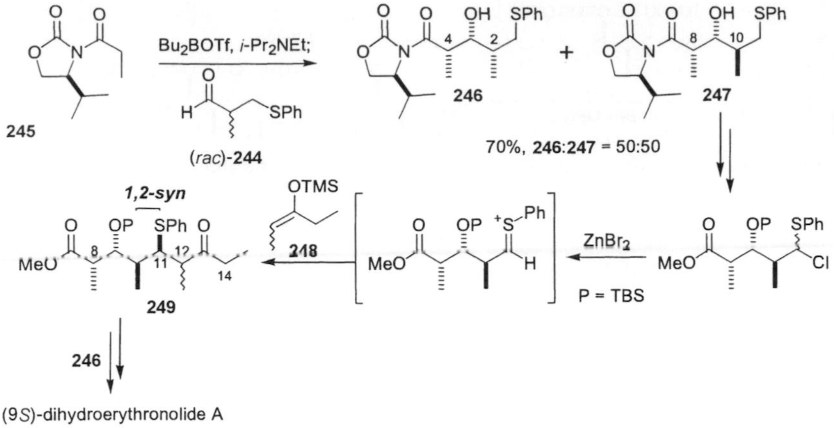

Scheme 9-66.

Denticulatins A and B. The denticulatins A (**250**) and B (**251**) were isolated in 1983 from the marine mollusc *Siphonaria denticulata* and are highly oxygenated structures with seven contiguous stereocenters (Scheme 9-67). Three total syntheses of these molecules rely upon aldol reactions to assemble triketone **252**, a key intermediate, forming either the C_{10}–C_{11} or C_9–C_{10} bond at a late stage.

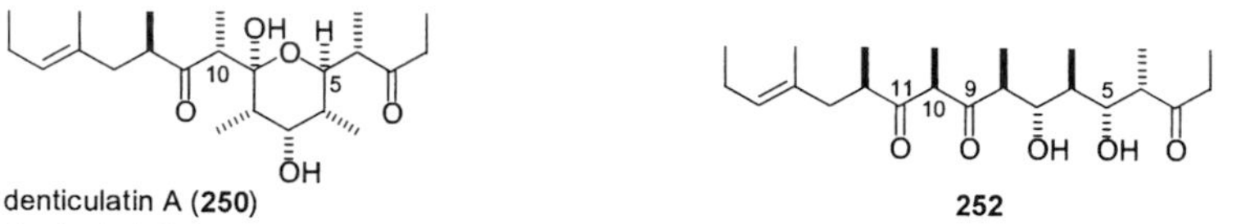

denticulatin A (**250**)
denticulatin B (**251**), C_{10} epimer

252

Scheme 9-67.

The first stereoselective synthesis of denticulatin B was achieved by our group in 1992 [85] and began with a well-documented aldol/reduction/hydroboration strategy (Scheme 9-68). Hence, an *anti* aldol reaction between ketone (*R*)-**18** and aldehyde **253**, with in situ reduction of dicyclohexylboron aldolate **254** afforded diol **255** with high selectivity (96% ds). Interestingly, this reduction proved more successful than the traditional method of aldol adduct isolation followed by reduction as a separate step. After protection of diol **255**, stereoselective hydroboration introduced the C_8 stereocenter, and subsequently ketone **256** was produced. While the lithium aldol reaction between this ketone and aldehyde (*R*)-**257** afforded a mixture of isomers, a titanium-mediated *syn* aldol reaction proceeded in a selective manner giving Felkin adduct **258** (83% ds). Here, the titanium enolate **259** showed a high level of diastereoface selection, as the minor aldol component resulted from reaction with antipodal aldehyde. Reaction of titanium enolate **259** with (*R*)-**257** was shown to be a mismatched reaction, as using racemic aldehyde led to a kinetic resolution giving a mixture of adducts now favouring the *anti*-

Felkin product (69% ds, 1,2-*anti*). A (*Z*)-titanium enolate usually favours the *anti*-Felkin adduct, and the subsequent Oppolzer synthesis of denticulatins A (see Scheme 9-69) highlights this behaviour (see also Scheme 9-30); however, exceptions can be found (Scheme 9-45). Oxidation of the C_3 and C_{11} hydroxyls of **258**, and cyclization, under carefully controlled conditions to preserve the configuration of the C_{10} stereocenter, then allowed the selective synthesis of denticulatin B.

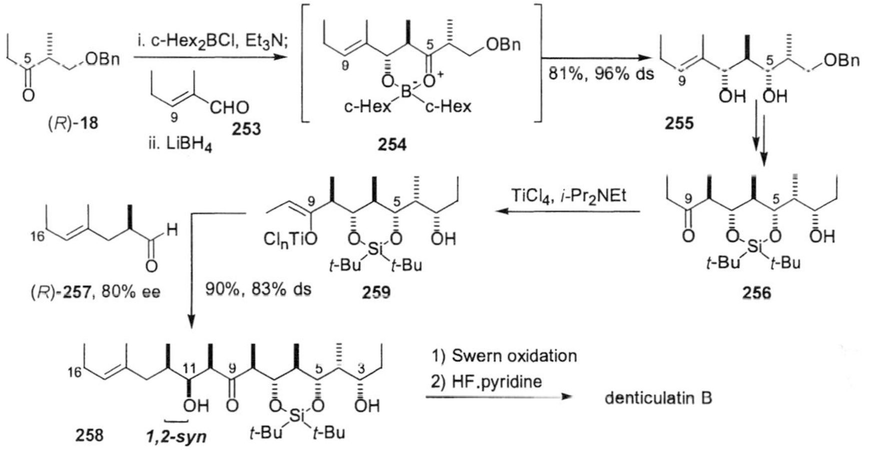

Scheme 9-68.

The first selective synthesis of denticulatin A was completed by the Oppolzer group and a titanium-mediated aldol reaction was now used to form the C_9–C_{10} bond (Scheme 9-69) [86]. However, the most striking feature of this synthesis was the enantiotopic group desymmetrization of a *meso* dialdehyde. The (*Z*)-boron enolate **260** of camphor-derived *ent*-**38** reacted with aldehyde **261** to give a mixture of lactols **262** with a strong preference for the *anti*-Felkin *syn* product. No double addition of the enolate was observed, possibly due to internal protection of the second aldehyde moiety as the hemiacetal. With the C_4–C_8 stereopentad thus installed, simple transformations gave access to keto aldehyde **263**. The *syn* aldol reaction of titanium enolate **264** with aldehyde **263** afforded the desired *anti*-Felkin adduct **265** with 87% ds. It would appear that the aldehyde *α*-stereocenter plays a controlling role in imposing an *anti*-Felkin bias on the (*Z*)-enolate. The C_{10} stereocenter was then retained through the oxidation and cyclization steps, enabling a stereoselective synthesis of denticulatin A.

Hoffmann and co-workers completed the first synthesis of both denticulatins via a C_9–C_{10} aldol bond construction (Scheme 9-70) [87]. In this case, aldehyde **266** was assembled using asymmetric crotylboration reactions to introduce the C_4–C_8 stereocenters. The (*Z*)-boron enolate **267** was then reacted with aldehyde **266** to afford the desired *anti*-Felkin adduct with 80% selectivity where the minor diastereomer resulted from reaction of the enantiomer of the starting ketone. Unfortunately, the C_5-PMB ether protecting group could not be removed without epimerization at C_{10}, and denticulatins A and B were formed in equimolar amounts.

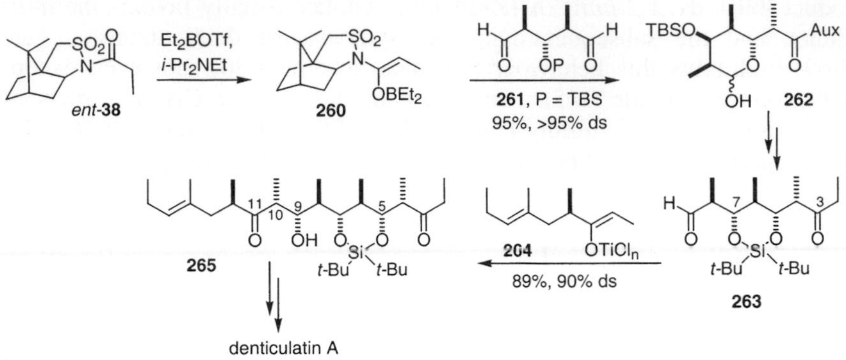

Scheme 9-69.

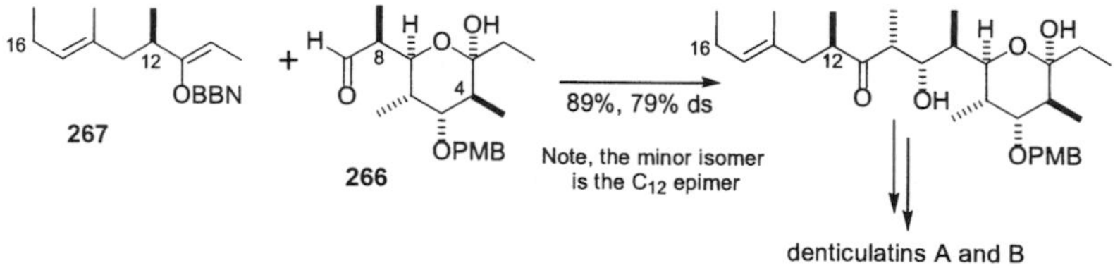

Scheme 9-70.

Muamvatin. Muamvatin (**268**) was isolated from the pulmonate mollusc *Siphonaria normalis*, and, while extensive NMR studies allowed for assignment of the relative configuration at C_4–C_6 and C_8, the side-chain C_{10}–C_{11} stereochemistry and the absolute configuration remained elusive (Scheme 9-71). Independently, Hoffmann [88] and our own group [89] synthesized aldehyde **269**, a degradation product from muamvatin, which then allowed for the assignment of both relative and absolute stereochemistry.

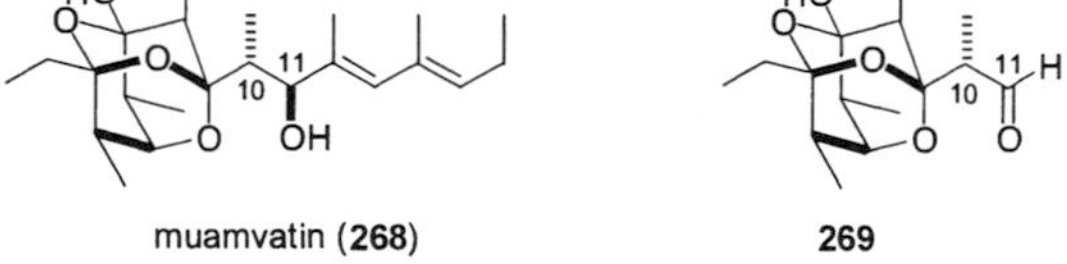

Scheme 9-71.

In our synthesis, iterative aldol reactions of dipropionate reagent (*R*)-**18** allowed for the control of the C_3–C_{10} stereocenters (Scheme 9-72) [89]. Hence, a tin-mediated, *syn* aldol reaction followed by an *anti* reduction of the aldol product afforded **270**. Diol protection, benzyl ether deprotection and subsequent oxidation gave aldehyde **271** which reacted with the (*E*)-boron enolate of ketone (*R*)-**18** to afford *anti* aldol adduct **272**. While the ketone provides the major bias for this reaction, it is an example of a matched reaction based on Felkin induction from the aldehyde and hence proceeded with excellent selectivity (98% ds). Elaboration of **272** then gave aldehyde **269**, which was identical to the previously isolated degradation product. Addition of the dienyl side-chain allowed for the synthesis of muamvatin (**268**) and full assignment of the relative and absolute stereochemistry.

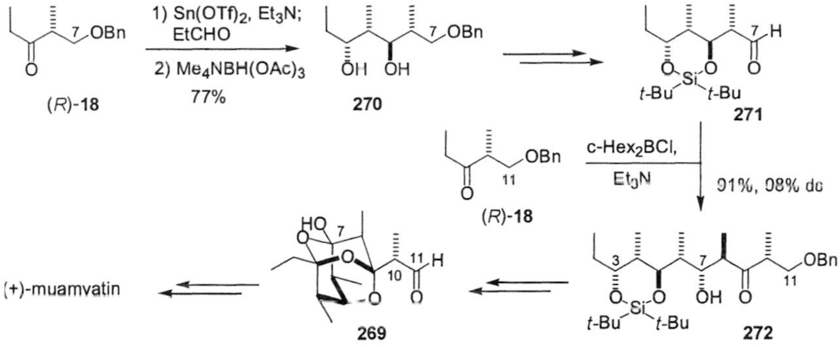

Scheme 9-72.

In order to unambiguously ascertain the C_{10} stereochemistry, the Hoffmann group elected to separately synthesize both C_{10} epimers of aldehyde **269** (Scheme 9-73) via aldol additions of both enantiomers of ketone **18** to aldehyde **273** [88]. Notably, the boron-mediated *syn* aldol reactions of this ketone are *non-selective* in the absence of chiral ligands (see Scheme 9-8 for selective *syn* aldol reactions of **18**). In this case, the C_7 hydroxyl was ultimately oxidized to a ketone, and the C_8 stereocenter epimerized during cyclization so the lack of selectivity was not detrimental to the synthesis.

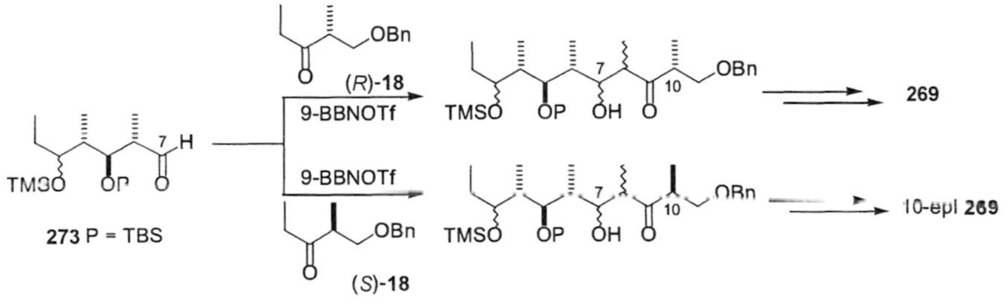

Scheme 9-73.

Ebelactone A and B. The ebelactones are a small group of β-lactone enzyme inhibitors isolated from a cultured strain of soil actinomycetes. Our synthesis of ebelactone A (**274**) and B (**275**) employed three different types of aldol reactions including the enantioselective *syn* aldol reaction of diethylketone and ethacrolein which afforded aldol adduct **276** with 86% ee (Scheme 9-74) [90]. Following

TBS-protection, a second, boron-mediated, *syn* aldol reaction led to the formation of **277** with 95% ds. In this case, ketone **278** controlled the stereochemical outcome of the reaction, and chiral ligands on boron were not required. A simple steric model accounts for this selectivity (see Scheme 9-11), and a titanium-mediated aldol reaction would be expected to give the same product. Following elaboration, including an Ireland-Claisen rearrangement, aldehyde **279** was prepared.

The completion of the synthesis of ebelactone A required an *anti* aldol reaction of a suitable three-carbon unit to proceed with *anti*-Felkin selectivity, i.e. a mismatched reaction. Conversion of thioester **280** into its (*E*)-enol borinate and reaction with aldehyde **279** gave two *anti* aldol adducts, unfortunately with little stereochemical preference. The minor isomer **281** from this reaction was used in the successful synthesis of ebelactone A (**274**), and the same chemistry, now using thioester **282**, was employed to complete the first synthesis of ebelactone B (**275**).

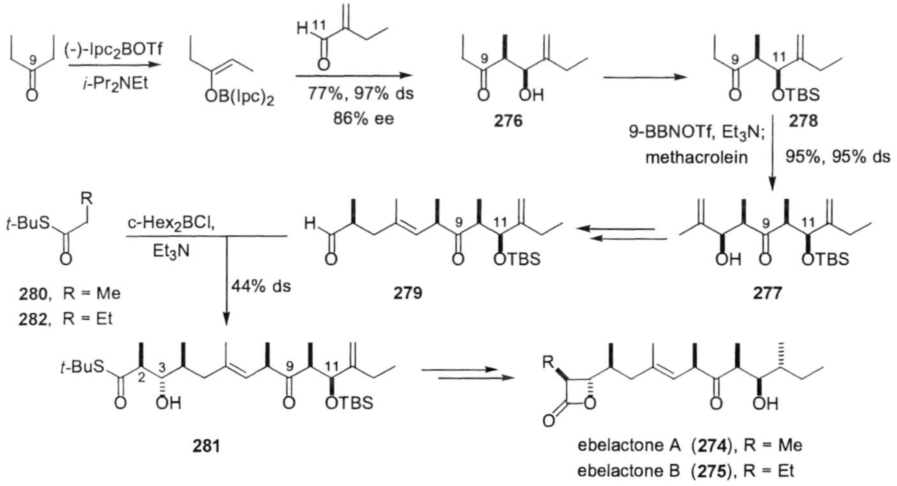

Scheme 9-74.

At the time, many unsuccessful attempts were made to improve the selectivity of the mismatched *anti* aldol reaction mentioned above, outlining the limitations of some chiral ligands or auxiliaries at overcoming inherent substrate bias in *anti* aldol reactions. Since the completion of this work, we have introduced the lactate-derived ketones (*R*)- and (*S*)-**39**, which should now allow the stereoselective synthesis of the ebelactones. As shown in Scheme 9-75, each enantiomer of the parent ketone acts as a propionate equivalent with a covalently attached auxiliary which will overturn the facial bias of most aldehydes [27, 28].

We have used this methodology in a recent synthesis of the anti-obesity drug tetrahydrolipstatin (**283**) (Scheme 9-76) [91]. Hence, the (*E*)-boron enolate of ketone **284** was reacted with aldehyde **285** to afford the desired *anti* aldol adduct

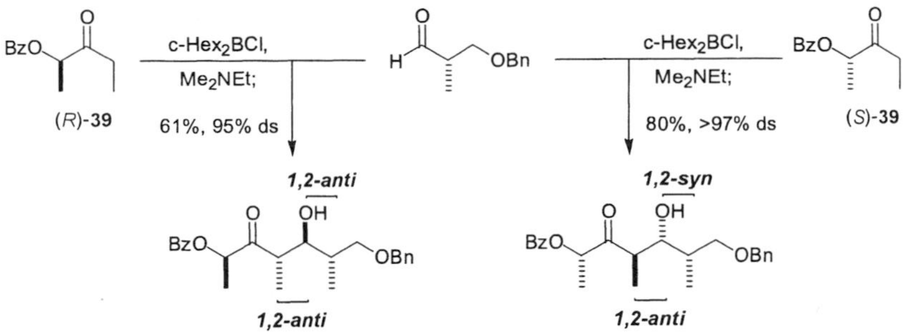

Scheme 9-75.

286 with high stereocontrol (>97% ds). One-step reduction of the ester and ketone, glycol cleavage and oxidation to acid **287** were all carried out with the C_3 hydroxyl unprotected to afford *β*-hydroxyacid **287**, which cyclized to give **288** allowing access to tetrahydrolipstatin (**283**).

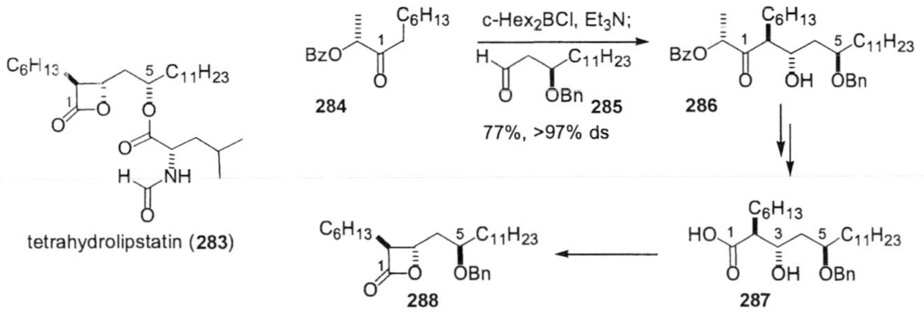

Scheme 9-76.

(–)-ACRL toxin IIIB. (–)-ACRL toxin IIIA (**289**) was isolated from the phytopathogenic fungus *Alternaria citri* and was characterized as its methyl ether, ACRL toxin IIIB (**290**). As already mentioned, lactate-derived ketones (*R*)- and (*S*)-**39** afford *anti* aldol adducts with excellent control (>95% ds), and these adducts can be manipulated in a variety of ways (see Scheme 9-14). Applying this methodology to the synthesis of ACRL toxin IIIB [92] (Scheme 9-77), addition of ketone (*S*)-**39** to tiglic aldehyde afforded *anti* adduct **291** in 86% yield and > 98% ds, thus installing the C_{12}/C_{13} stereocenters. Three-step manipulation gave aldehyde **292**, which was then homologated to enal **293**, and a second *anti* aldol reaction, again using ketone (*S*)-**39**, installed the C_8/C_9 stereocenters with 98% ds. Manipulation of this adduct then gave aldehyde **294**, which was used to complete the synthesis of ACRL toxin IIIB (**290**).

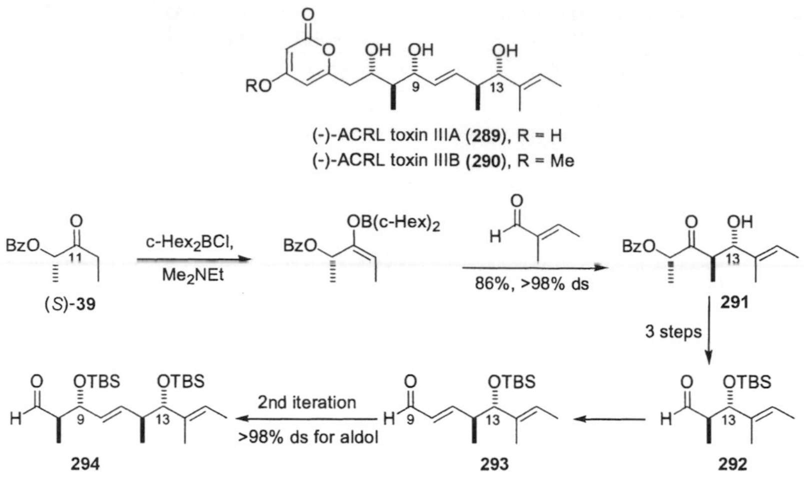

Scheme 9-77.

References

1. For recent reviews of the aldol reaction, see (a) Cowden, C.J.; Paterson, I. *Org. React.* **1997**, *51*, 1. (b) Franklin, A.S.; Paterson, I. *Cont. Org. Synthesis* **1994**, *1*, 317. (c) Heathcock, C.H.; Kim, B.M.; Williams, S.F.; Masamune, S.; Paterson, I.; Gennari, C. in Comprehensive Organic Synthesis, Trost, B.M., Ed. Pergamon, Oxford, **1991**, Vol. 2. (d) Evans, D.A.; Nelson, J.V.; Taber, T.R. *Top. Stereochem.* **1982**, *13*, 1. (e) Heathcock, C.H. In *Asymmetric Synth.*, Ed. Morrison, J.D., Academic Press, New York, **1984**, *3*, 111. (f) Mukaiyama, T.; Kobayashi, S. *Org. React.* **1994**, *46*, 1.
2. Heathcock, C.H.; Flippin, L.A. *J. Am. Chem. Soc.* **1983**, *105*, 1667.
3. Evans, D.A.; Duffy, J.L.; Dart, M.J.; Yang, M.G. *J. Am. Chem. Soc.* **1996**, *118*, 4322.
4. Oikawa, M.; Ueno, T.; Oikawa, H.; Ichihara, A. *J. Org. Chem.* **1995**, *60*, 5048.
5. Evans, D.A.; Yang, M.G.; Dart, M.J.; Duffy, J.L.; Kim, A.S. *J. Am. Chem. Soc.* **1995**, *117*, 9598.
6. (a) Paterson, I.; Goodman, J.M.; Isaka, M. *Tetrahedron Lett.* **1989**, *30*, 7121. (b) Paterson, I.; Oballa, R.M. *Tetrahedron Lett.* **1997**, *38*, 8241.
7. Trost, B.M.; Urabe, H. *J. Org. Chem.* **1990**, *55*, 3982.
8. Denmark, S.E.; Stavenger, R.A. *J. Org. Chem.* **1998**, *63*, 9524.
9. Paterson, I.; Gibson, K.R.; Oballa, R.M. *Tetrahedron Lett.* **1996**, *37*, 8585.
10. Evans, D.A.; Coleman, P.J.; Cote, B. *J. Org. Chem.* **1997**, *62*, 788.
11. (a) Paterson, I.; Lister, M.A. *Tetrahedron Lett.* **1988**, *29*, 585. (b) Paterson, I.; Norcross, R.D.; Ward, R.A.; Romea, P.; Lister, M.A. *J. Am. Chem. Soc.* **1994**, *116*, 11287. (c) Paterson, I. *Pure Appl. Chem.* **1992**, *64*, 1821.
12. Paterson, I.; Tillyer, R.D. *J. Org. Chem.* **1993**, *58*, 4182.
13. Paterson, I.; Tillyer, R.D. *Tetrahedron Lett.* **1992**, *33*, 4233.
14. Evans, D.A.; Clark, J.S.; Metternich, R.; Novack, V.J.; Sheppard, G.S. *J. Am. Chem. Soc.* **1990**, *112*, 866.
15. Evans, D.A.; Ng, H.P.; Clark, J.S.; Rieger, D.L. *Tetrahedron* **1992**, *48*, 2127.
16. Evans, D.A.; Calter, M.A. *Tetrahedron Lett.* **1993**, *34*, 6871.
17. Ramachandran, P.V.; Xu, W.-C.; Brown, H.C. *Tetrahedron Lett.* **1997**, *38*, 769.

18. (a) Paterson, I.; McClure, C.K. *Tetrahedron Lett.* **1987**, *28*, 1229. (b) Paterson, I.; Hulme, A.N.; Wallace, D.J. *Tetrahedron Lett.* **1991**, *32*, 7601.
19. Bernardi, A.; Gennari, C.; Goodman, J.M.; Paterson, I. *Tetrahedron Asymmetry* **1995**, *6*, 2613.
20. (a) Evans, D.A.; Bartroli, J.; Shih, T.L. *J. Am. Chem. Soc.* **1981**, *103*, 2127. (b) Gage, J.R.; Evans, D.A. *Org. Synth.* **1990**, *68*, 83.
21. (a) Jones, T.K.; Reamer, R.A.; Desmond, R.; Mills, S.G. *J. Am. Chem. Soc.* **1990**, *112*, 2998. (b) Evans, D.A.; Sjogren, E.B.; Weber, A.E.; Conn, R.E. *Tetrahedron Lett.* **1987**, *28*, 39. (c) Evans, D.A.; Weber, A.E. *J. Am. Chem. Soc.* **1986**, *108*, 6757. (d) Abdel-Magid, A.; Pridgen, L.N.; Eggleston, D.S.; Lantos, I. *J. Am. Chem. Soc.* **1986**, *108*, 4595.
22. Oppolzer, W.; Blagg, J.; Rodriguez, I.; Walther, E. *J. Am. Chem. Soc.* **1990**, *112*, 2767.
23. (a) Walker, M.A.; Heathcock, C.H. *J. Org. Chem.* **1991**, *56*, 5747. (b) Oppolzer, W.; Lienard, P. *Tetrahedron Lett.* **1993**, *34*, 4321.
24. Dahmann, G.; Hoffmann, R.W. *Liebigs Ann. Chem.* **1994**, 837.
25. Oppolzer, W.; Starkemann, C.; Rodriguez, I.; Bernardinelli, G. *Tetrahedron Lett.* **1991**, *32*, 61.
26. For other auxiliaries introduced for *anti* aldol reactions, see (a) Gennari, C.; Colombo, L.; Bertolini, G.; Schimperna, G. *J. Org. Chem.* **1987**, *52*, 2754. (b) Abiko, A.; Liu, J.-F.; Masamune, S. *J. Am. Chem. Soc.* **1997**, *119*, 2586. (c) Van Draanen, N.A.; Arseniyadis, S.; Crimmins, M.T.; Heathcock, C.H. *J. Org. Chem.* **1991**, *56*, 2499.
27. (a) Paterson, I.; Wallace, D.J.; Velazquez, S.M. *Tetrahedron Lett.* **1994**, *35*, 9083. (b) Paterson, I.; Wallace, D.J.; Cowden, C.J. *Synthesis* **1998**, *639*.
28. Paterson, I.; Wallace, D.J. *Tetrahedron Lett.* **1994**, *35*, 9087.
29. Figueras, S.; Martin, R.; Romea, P.; Urpi, F.; Vilarrasa, J. *Tetrahedron Lett.* **1997**, *38*, 1637.
30. Nagao, Y.; Hagiwara, Y.; Kumagai, T.; Ochiai, M.; Inoue, T.; Hashimoto, K.; Fujita, E. *J. Org. Chem.* **1986**, *51*, 2391.
31. Romo, D.; Rzasa, R.M.; Shea, H.A.; Park, K.; Langenhan, J.M.; Sun, L.; Akhiezer, A.; Liu, J.O. *J. Am. Chem. Soc.* **1998**, *120*, 12237.
32. Brown, H.C.; Ramachandran, P.V. *J. Organomet. Chem.* **1995**, *500*, 1.
33. Paterson, I.; Goodman, J.M. *Tetrahedron Lett.* **1989**, *30*, 997.
34. Paterson, I.; Oballa, R.M.; Norcross, R.D. *Tetrahedron Lett.* **1996**, *37*, 8581.
35. Paterson, I.; Goodman, J.M.; Lister, M.A.; Schumann, R.C.; McClure, C.K.; Norcross, R.D. *Tetrahedron* **1990**, *46*, 4663.
36. (a) Blanchette, M.A.; Malamas, M.S.; Nantz, M.H.; Roberts, J.C.; Somfai, P.; Whritenour, D.C.; Masamune, S.; Kageyama, M.; Tamura, T. *J. Org. Chem.* **1989**, *54*, 2817. (b) Kageyama, M.; Tamura, T.; Nantz, M.H.; Roberts, J.C.; Somfai, P.; Whritenour, D.C.; Masamune, S. *J. Am. Chem. Soc.* **1990**, *112*, 7407.
37. Gennari, C.; Vulpetti, A.; Donghi, M.; Mongelli, N.; Vanotti, E. *Angew. Chem. Int. Ed. Engl.* **1996**, *35*, 1723.
38. Corey, E.J.; Huang, H.-C. *Tetrahedron Lett.* **1989**, *30*, 5235.
39. (a) Oertle, K.; Beyeler, H.; Duthaler, R.O.; Lottenbach, W.; Riediker, M.; Steiner, E. *Helv. Chim. Acta* **1990**, *73*, 353. (b) Duthaler, R.O.; Herold, P.; Wyler-Helfer, S.; Riediker, M. *Helv. Chim. Acta* **1990**, *73*, 659.
40. Sheppeck, J.E. (II); Liu, W.; Chamberlin, A.R. *J. Org. Chem.* **1997**, *62*, 387.
41. Iwasawa, N.; Mukaiyama, T. *Chem. Lett.* **1983**, 297.
42. Mukaiyama, T.; Shiina, I.; Iwadare, H.; Saitoh, M.; Nishimura, T.; Ohkawa, N.; Sakoh, H.; Nishimura, K.; Tani, Y.; Hasegawa, M.; Yamada, K.; Saitoh, K. *Chem. Eur. J.* **1999**, *5*, 121.
43. Kim, Y.; Singer, R.A.; Carreira, E.M. *Angew. Chem. Int. Ed. Engl.* **1998**, *37*, 1261.
44. Evans, D.A.; Kozlowski, M.C.; Murry, J.A.; Burgey, C.S.; Campos, K.R.; Connell, B.T.; Staples, R.J. *J. Am. Chem. Soc.* **1999**, *121*, 669.
45. Evans, D.A.; Burgey, C.S.; Kozlowski, M.C.; Tregay, S.W. *J. Am. Chem. Soc.* **1999**, *121*, 686.
46. Furuta, K.; Maruyama, T.; Yamamoto, H. *J. Am. Chem. Soc.* **1991**, *113*, 1041.
47. (a) Kiyooka, S.; Kaneko, Y.; Komura, M.; Matsuo, H.; Nakano, M. *J. Org. Chem.* **1991**, *56*, 2276. (b) Kiyooka, S.; Kaneko, Y.; Kume, K. *Tetrahedron Lett.* **1992**, *33*, 4927.
48. (a) Kiyooka, S.; Yamaguchi, T.; Maeda, H.; Kira, H.; Hena, M.A.; Horiike, M. *Tetrahedron Lett.* **1997**, *38*, 3553. (b) Kiyooka, S.; Maeda, H. *Tetrahedron Asymmetry* **1997**, *8*, 3371.
49. Hena, M.A.; Kim, C.-S.; Horiike, M.; Kiyooka, S. *Tetrahedron Lett.* **1999**, *40*, 1161.

50. (a) Paterson, I.; Cumming, J.G.; Ward, R.A.; Lamboley, S. *Tetrahedron* **1995**, *51*, 9393. (b) Paterson, I.; Smith, J.D.; Ward, R.A. *Tetrahedron* **1995**, *51*, 9413. (c) Paterson, I.; Ward, R.A.; Smith, J.D.; Cumming, J.G.; Yeung, K-S. *Tetrahedron,* **1995**, *51*, 9437. (d) Paterson, I.; Yeung, K-S.; Ward, R.A.; Smith, J.D.; Cumming, J.G.; Lamboley, S. *Tetrahedron,* **1995**, *51*, 9467.
51. Nicolaou, K.C.; Patron, A.P.; Ajito, K.; Richter, P.K.; Khatuya, H.; Bertinato, P.; Miller, R.A.; Tomaszewski, M.J. *Chem. Eur. J.* **1996**, *2*, 847.
52. (a) Roush, W.R. *J. Org. Chem.* **1991**, *56*, 4151. (b) Gennari, C.; Vieth, S.; Comotti, A.; Vulpetti, A.; Goodman, J.M.; Paterson, I. *Tetrahedron* **1992**, *48*, 4439.
53. Nagasawa, K.; Shimizu, I.; Nakata, T. *Tetrahedron Lett.* **1996**, *37*, 6885.
54. (a) Guo, J.; Duffy, K.J.; Stevens, K.L.; Dalko, P.I.; Roth, R.M.; Hayward, M.M.; Kishi, Y. *Angew. Chem. Int. Ed. Engl.* **1998**, *37*, 187. (b) Hayward, M.M.; Roth, R.M.; Duffy, K.J.; Dalko, P.I. Stevens, K.L.; Guo, J.; Kishi, Y. *Angew. Chem. Int. Ed. Eng.* **1998**, *37*, 192.
55. (a) Evans, D.A.; Coleman, P.J.; Dias, L.C. *Angew. Chem. Int. Ed. Engl.* **1997**, *36*, 2738. (b) Evans, D.A.; Trotter, B.W.; Cote, B.; Coleman, P.J. *Angew. Chem. Int. Ed. Engl.* **1997**, *36*, 2741. (c) Evans, D.A.; Trotter, B. W.; Cote, B.; Coleman, P.J.; Dias, L.C.; Tyler, A.N. *Angew. Chem. Int. Ed. Eng.* **1997**, *36*, 2744.
56. (a) Paterson, I.; Keown, L.E. *Tetrahedron Lett.* **1997**, *38*, 5727. (b) Paterson, I.; Wallace, D.J.; Gibson, K.R. *Tetrahedron Lett.* **1997**, *38*, 8911. (c) Paterson, I.; Wallace, D.J.; Oballa, R.M. *Tetrahedron Lett.* **1998**, *39*, 8545.
57. Duplantier, A.J.; Nantz, M.H.; Roberts, J.C.; Short, R.P.; Somfai, P.; Masamune, S. *Tetrahedron Lett.* **1989**, *30*, 7357.
58. Evans, D.A.; Carter, P.H.; Carreira, E.M.; Prunet, J.A.; Charette, A.B.; Lautens, M. *Angew. Chem. Int. Ed. Engl.* **1998**, *37*, 2354.
59. Ohmori, K.; Suzuki, T.; Miyazawa, K.; Nishiyama, S.; Yamamura, S. *Tetrahedron Lett.* **1993**, *34*, 4981.
60. Kiyooka, S.; Maeda, H. *Tetrahedron Asymmetry* **1997**, *8*, 3371.
61. Weiss, J.M.; Hoffmann, H.M.R. *Tetrahedron Asymmetry* **1997**, *8*, 3913.
62. Evans, D.A.; Ng, H.P.; Rieger, D.L. *J. Am. Chem. Soc.* **1993**, *115*, 11446.
63. Gustin, D.J.; VanNieuwenhze, M.S.; Roush, W.R. *Tetrahedron Lett.* **1995**, *36*, 3447.
64. (a) White, J.D.; Tiller, T.; Ohba, Y.; Porter, W.J.; Jackson, R.W.; Wang, S.; Hanselmann, R. *Chem. Commun.* **1998**, 79. (b) White, J.D.; Porter, W.J.; Tiller, T. *Synlett.* **1993**, *535*.
65. (a) Jain, N. F.; Panek, J.S. *Tetrahedron Lett.* **1997**, *38*, 1345. (b) Jain, N. F.; Panek, J.S. *Tetrahedron Lett.* **1997**, *38*, 1349.
66. Kigoshi, H.; Suenaga, K.; Mutou, Y.; Ishigaki, T.; Atsumi, T.; Ishiwata, H.; Sakakura, A.; Ogawa, T.; Ojika, M.; Yamada, K. *J. Org. Chem.* **1996**, *61*, 5326.
67. (a) Paterson, I.; Cowden, C.J.; Woodrow, M.D. *Tetrahedron Lett.* **1998**, *39*, 6037. (b) Paterson, I.; Woodrow, M.D.; Cowden, C.J. *Tetrahedron Lett.* **1998**, *39*, 6041.
68. Toshima, K.; Jyojima, T.; Yamaguchi, H.; Noguchi, Y.; Yoshida, T.; Murase, H.; Nakata, M.; Matsumura, S. *J. Org. Chem.* **1997**, *62*, 3271.
69. Paterson, I.; Bower, S.; McLeod, M.D. *Tetrahedron Lett.* **1995**, *36*, 175.
70. Paterson, I.; Bower, S.; Tillyer, R.D. *Tetrahedron Lett.* **1993**, *34*, 4393.
71. Roush, W.R.; Bannister, T.D. *Tetrahedron Lett.* **1992**, *33*, 3587.
72. (a) Roush, W.R.; Bannister, T.D.; Wendt, M.D. *Tetrahedron Lett.* **1993**, *34*, 8387. (b) Gustin, D.J.; VanNieuwenhze, M.S.; Roush, W.R. *Tetrahedron Lett.* **1995**, *36*, 3443.
73. (a) Paterson, I.; McLeod, M.D. *Tetrahedron Lett.* **1995**, *36*, 9065. (b) Paterson, I.; McLeod, M.D. *Tetrahedron Lett.* **1997**, *38*, 4183. (c) Paterson, I.; Doughty, V.A.; McLeod, M.D.; Trieselmann, T. unpublished results.
74. (a) Jyojima, T.; Katohno, M.; Miyamoto, N.; Nakata, M.; Matsumura, S.; Toshima, K. *Tetrahedron Lett.* **1998**, *39*, 6003. (b) Jyojima, T.; Miyamoto, N.; Katohno, M.; Nakata, M.; Matsumura, S.; Toshima, K. *Tetrahedron Lett.* **1998**, *39*, 6007.
75. (a) Schinzer, D.; Bauer, A.; Schieber, J. *Synlett* **1998**, 861. (b) Schinzer, D.; Limberg, A.; Bauer, A.; Bohm, O.M.; Cordes, M. *Angew. Chem. Int. Ed. Engl.* **1997**, *36*, 523.
76. Gabriel, T.; Wessjohann, L. *Tetrahedron Lett.* **1997**, *38*, 1363.
77. Meng, D.; Bertinato, P.; Balog, A.; Su, D.-S.; Kamenecka, T.; Sorensen, E.J.; Danishefsky, S.J. *J. Am. Chem. Soc.* **1997**, *119*, 10073.
78. Mulzer, J.; Mantoulidis, A.; Ohler, E. *Tetrahedron Lett.* **1998**, *39*, 8633.

79. Balog, A.; Harris, C.; Savin, K.; Zhang, X.-G.; Chou, T.C.; Danishefsky, S.J. *Angew. Chem. Int. Ed. Engl.* **1998**, *37*, 2675.
80. Hayward, C.H.; Yohannes, D.; Danishefsky, S.J. *J. Am. Chem. Soc.* **1993**, *115*, 9345.
81. Paterson, I.; Norcross, R.D.; Ward, R.A.; Romea, P.; Lister, M.A. *J. Am. Chem. Soc.* **1994**, *116*, 11287.
82. Evans, D.A.; Kim, A.S.; Metternich, R.; Novack, V.J. *J. Am. Chem. Soc.* **1998**, *120*, 5921.
83. Masamune, S.; Hirama, M.; Moris, S.; Ali, S.A.; Garvey, D.S. *J. Am. Chem. Soc.* **1981**, *103*, 1568.
84. (a) Paterson, I.; Laffan, D.D.P.; Rawson, D.J. *Tetrahedron Lett.* **1988**, *29*, 1461. (b) Paterson, I.; Rawson, D.J. *Tetrahedron Lett.* **1989**, *30*, 7463.
85. Paterson, I.; Perkins, M.V. *Tetrahedron* **1996**, *52*, 1811.
86. De Brabander, J.; Oppolzer, W. *Tetrahedron* **1997**, *53*, 9169.
87. Andersen, M. W.; Hildebrandt, B.; Hoffmann, R.W. *Angew. Chem. Int. Ed. Engl.* **1991**, *30*, 97.
88. Hoffmann, R.W.; Dahmann, G. *Tetrahedron Lett.* **1993**, *34*, 1115.
89. Paterson, I.; Perkins, M.V. *J. Am. Chem. Soc.* **1993**, *115*, 1608.
90. Paterson, I.; Hulme, A.N. *J. Org. Chem.* **1995**, *60*, 3288.
91. Paterson, I.; Doughty, V.A. *Tetrahedron Lett.* **1999**, *40*, 393.
92. Paterson, I.; Wallace, D.J.; Cowden, C.J. *Synthesis* **1998**, 639.

10 Allylation of Carbonyls: Methodology and Stereochemistry

Scott E. Denmark and Neil G. Almstead

10.1 Introduction

10.1.1 Allylmetal Additions to Aldehydes and Ketones

The invention and development of new methods for the synthesis of complex molecules of both natural and unnatural origin remains an enduring challenge in organic chemistry. Over the past two decades one of the major efforts in this arena has been directed towards the controlled construction of open-chain systems bearing sequences of stereocenters such as those found in the important class of polyketide natural products. Such structural challenges marked a divergence from the (no less difficult but nonetheless distinct) problems presented by the construction of complex polycyclic compounds that characterized earlier epochs. In response to these challenges, many methods have been developed to synthesize the long sequences of stereocenters present in these molecules. These methods include, inter alia, the aldol addition, the addition of allylmetal reagents to aldehydes, epoxidation, hydroboration, and stereocontrolled reductions. The key features in all of these reactions are the high degree of stereocontrol and a significant level of predictability which assures success in the application of such methods in new scenarios. Both of these critical features evolve from an understanding of the reaction mechanism at least at the level where an operational model for the origin of stereocontrol is available.

The allylmetal-aldehyde addition reaction [1] has proven to be an enormously successful method for the controlled construction of contiguous stereocenters. Some of the reasons for the popularity of the method are: (1) the high degree of both diastereo- and enantioselectivity observed (2) the extreme diversity of reagent reactivity based on metal, (3) the ability to access different stereodyads and triads etc. and (4) the latent functionality in the homoallylic alcohol product, which makes the reaction ideal for synthetic planning. Moreover, the reactions are mechanistically intriguing, and their utility stimulated an important synergy between fundamental studies of stereochemistry and applications in target-oriented synthesis.

One of the more interesting aspects of these reactions that illustrates the attraction to both mechanistically and synthetically oriented chemists is the dramatic dependence of diastereoselectivity observed in the allylmetal-aldehyde additions. This dependence has been classified into the following three groups that relate the stereochemical outcome of the reaction to the geometry of the double bond [2]; (1) reactions wherein the *syn/anti* (see below) ratio reflects the *Z/E* ratio of the starting allylmetal (Type 1); (2) reactions wherein the product is predominantly *syn*, independent of the geometry of the allylmetal (Type 2); and (3) reactions wherein the product is predominantly *anti*, independent of the geometry of the allylmetal (Type 3). Representative examples of the various types of additions are shown in Table 10-1. These few examples illustrate the fascination that stimulated organic chemists around the world to ask: why do these transformations proceed as they do, what is the origin of stereocontrol, how can this reaction be applied to the synthesis of interesting target structures, what can be done to expand the scope and selectivity of the process, and what new variations can be invented? The objective of this Chapter is to provide a summary of the answers to the first two questions. The middle question is addressed in the accompanying Chapter by Chemler and Roush (Chapter 11), and the last two questions are left as a challenge to the reader.

Table 10-1. Examples of allylmetal-aldehyde additions.

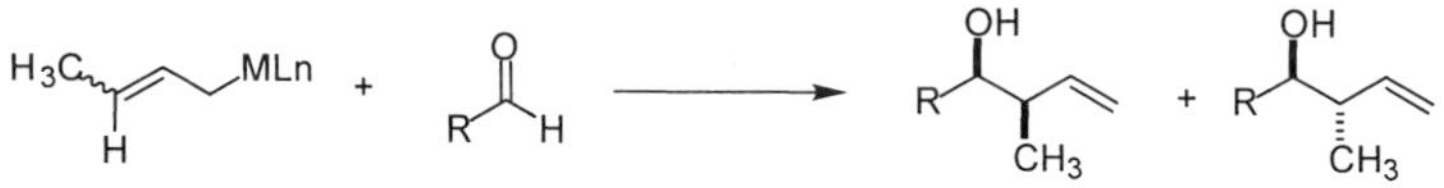

Metal	R	*E/Z* ratio	Conditions	*syn/anti*
$SiMe_3$	*i*-Pr	99/1	$TiCl_4/CH_2Cl_2/-78\,^\circ C$	97/3
$SiMe_3$	*i*-Pr	3/97	$TiCl_4/\ CH_2Cl_2/-78\,^\circ C$	64/36
$SnBu_3$	Ph	100/0	$BF_3{\cdot}OEt_2/CH_2Cl_2/-78\,^\circ C$	98/2
$SnBu_3$	Ph	0/100	$BF_3{\cdot}OEt_2/CH_2Cl_2/-78\,^\circ C$	99/1
$(RO)_2B$	Ph	93/9	hexane/–78 °C	6/94
$(RO)_2B$	Ph	$<5/>95$	hexane/–78 °C	96/4
L_2CrCl_2	Ph	100/0	THF/room temperature	0/100
L_2CrCl_2	Ph	0/100	THF/room temperature	0/100
$TiCp_2Cl$	Ph	*E*	$BF_3{\cdot}OEt_2/THF/-78\,^\circ C$	14/86
$ZrCp_2Cl$	Ph	*E*	THF/–78 °C	19/81

10.1.2 Definition of Stereochemical Issues and Nomenclature

The reaction of a C(3)-substituted allylmetal with an aldehyde will result in the formation of two diastereomeric homoallylic alcohols (Scheme 10-1, Eq. (10.1). The new stereocenters are generated in concert with the formation of a new carbon-carbon bond. The relative configuration of these new centers is a consequence of the simplest form of diastereoselection, i.e. the relative topicity of approach of the two reacting sp^2 centers. In some stereochemistry treatises this is referred to as simple diastereoselectivity [3]. We find this nomenclature to be devoid of intrinsic meaning and propose the use of **relative stereoselection** throughout this chapter. The use of "relative" diastereoselection is easy to understand because the selectivity features pertain uniquely to the *relative* configuration of the stereodyad. To describe the products observed in these reactions the stereochemical descriptors suggested by Masamune [4] will be employed. The *syn* isomer is defined as having both of the two stereocenters comprising the newly formed carbon-carbon bond projecting towards or away from the viewer when the product is drawn in an extended conformation. The *anti* isomer is defined as having one of the stereocenters projecting forward and the other back when the product is again drawn in an extended conformation.

The next level of stereoselection pertains to the existence of stereocenters resident in either of the reactants. In these scenarios, illustrated in Scheme 10-1 (Eqs. (10.2)–(10.5)), the newly formed centers are created under the influence of these covalently bound subunits and will be referred to as arising from **internal stereoselection**. The resident stereocenter can be anywhere on the aldehyde or allylmetal, and this kind of selection process is easy to identify when the stereocenters persist in both the educts and the product (Eqs. (10.2), (10.3)). However there are two important cases where they do not, namely, when the resident stereocenter bears the metal subunit (Eq. (10.4)) or is the metal subunit (Eq. (10.5)). Because these stereocenters are covalently bound in the educts and (to the extent that they influence the stereochemical outcome) in the transition structure, they will be considered under internal stereocontrol [5].

Finally, the allylmetal aldehyde addition can also operate under the influence of stereocontrolling reagents, in particular, chiral Lewis acids and related activators (Eq. (10.6)). In these cases the stereochemical influence on the steric course of the reaction is due to a non-covalently bound agent that is found in neither the educts or products. Thus, the term **external stereoselection** will be used to describe the enantiofacial outcome at the newly formed stereogenic centers.

10.1.3 Organization of Chapter and Logic of Presentation

The most logical organization of the mechanistic and stereochemical features of the allylmetal addition reaction is by metal. These two most important components of the reaction are inexorably bound and critically dependent on the nature of the metal. In turn, the metal also carries with it the types of ligands and activators that are often also integral to a discussion of the mechanism and moreover influential in the stereochemical outcome of the process. Thus, each subsection, defined by metal or metalloid class, will contain a discussion of the current structure of understanding of the reaction mechanism followed by the stereochemical consequences at all levels of stereoselection outlined above.

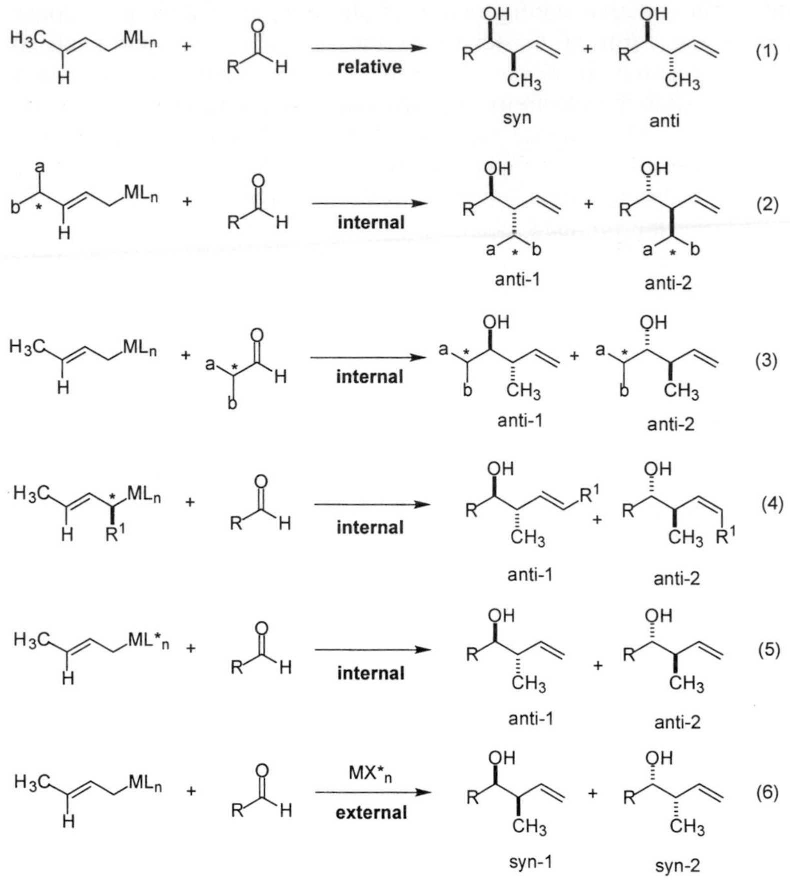

Scheme 10-1.

10.2 Allylic Silicon Reagents

10.2.1 Allylic Trialkylsilanes

The addition of an allylsilane to an electrophile was first documented in 1948 by Sommer et al. [6]. These workers predicted that the allylsilane would react with an electrophile to generate a silicon-stabilized cationic intermediate. In 1956, Calas and co-workers demonstrated that allylsilanes undergo an allylic shift in the protiodesilylation of a cyclohexenylsilane to afford a methylidenecyclohexane [7]. The first report of the reaction of allylsilanes with carbonyl compounds (1974) is also due to Calas [8]. These authors used activated substrates such as perfluoroacetone and chloroacetone and $AlCl_3$, $GaCl_3$ or $InCl_3$ as Lewis acids to promote the reaction. The utility of this reaction was greatly expanded by the discovery that $TiCl_4$ could promote the regioselective addition of allylic silanes to unactivated carbonyl groups in high yield [9]. Since this discovery, the reaction of allylic silanes with various electrophiles has developed into one of the most useful methods of carbon-carbon bond formation [10]. One of the advantages of allylsilanes when compared to other reagents is their stability, relative inertness to water and low toxicity. These reagents are readily handled and can usually be stored for long periods of time without special precautions.

10.2.1.1 Mechanism of addition

The stepwise nature of electrophilic attack on the π-bond of an allylsilane was demonstrated by Fleming in 1981 using the silanes **1** and **2**, which are both simultaneously vinylsilanes and allylsilanes [11]. Protiodesilylation of either silane **1** or **2** provided the same 4/1 ratio of allylsilanes **3** and **4**, presumably through the common intermediate **i** (Scheme 10-2). The carbocation that is initially formed by electrophilic attack is stabilized by hyperconjugative overlap [12] with the silicon atom.

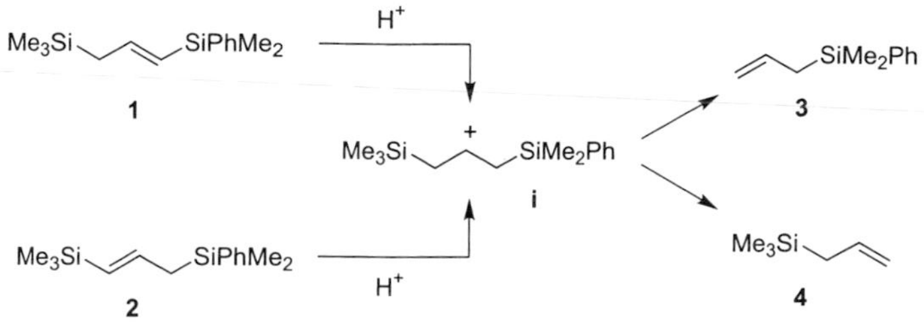

Me3Si SiPhMe2 **1**
Me3Si SiPhMe2 **2**
H+
Me3Si + SiMe2Ph **i**
SiMe2Ph **3**
Me3Si **4**

Scheme 10-2.

Proposals to rationalize the stereochemical course of the addition have been advanced for Type 2 (Si, Sn) reactions, which involve an open-chain arrangement of the reacting species [13]. In any description of this process there are two defining geometric issues: (1) the topicity [14] of the two reacting faces of the π-systems (relative stereoselection) and (2) the orientation of the metal electrofuge with respect to the incoming electrophile (*anti* or *syn* $S_{E'}$). The transition structures developed to explain the selectivities observed in these reactions have in addition identified the torsional angle between the two double bonds in two limiting arrangements, synclinal (60°) and antiperiplanar (180°) (Scheme 10-3).

The Lewis acid-mediated addition of electrophiles to allylsilanes has been extensively studied [10c]. In most cases the addition of an electrophile to an allylsilane proceeds via an *anti* $S_{E'}$ process. In the ground state structure, simple allylsilanes are known to prefer the conformation wherein the allylic hydrogen eclipses the double bond. The electrophile can then approach the double bond from the same side as the allylmetal (*syn* $S_{E'}$) or from the side opposite the allyl-

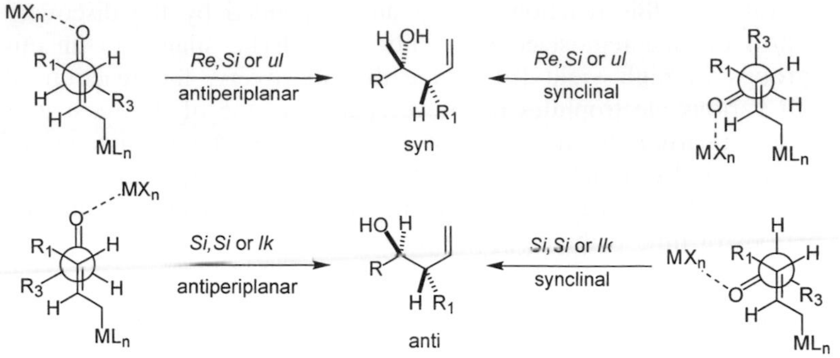

Scheme 10-3.

metal (*anti* $S_{E'}$). The configuration of the newly formed stereogenic center is therefore dependent upon the directionality of attack (Scheme 10-4). After attack of the electrophile on the double bond only a slight rotation of the C–C bond is necessary for the formation of intermediates **ii** and **iii**, which are stabilized by hyperconjugation with the silicon atom. The silyl group is then released, resulting in the stereoselective formation of a *trans*-double bond.

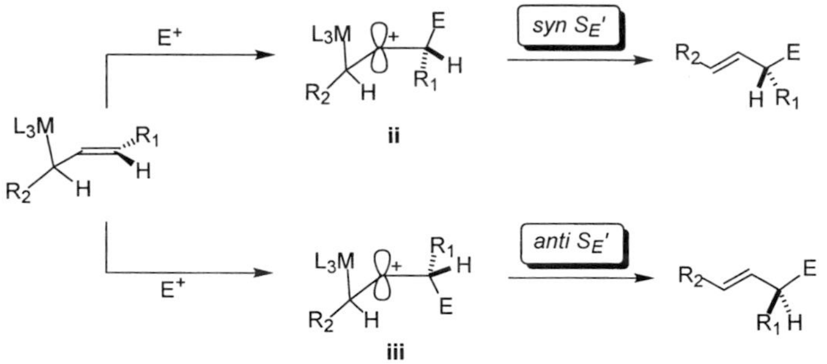

Scheme 10-4.

The regiochemical and the stereochemical course of electrophilic additions to allylsilanes has been modeled computationally by Hehre [15]. In this study the conformational profile of 2-silylbut-3-ene was determined and three energy minima were observed (Chart 10-1). In the two most stable conformers the C–Si bond is perpendicular to the C–C double bond. Experimental evidence has been obtained (microwave [16a, b], electron diffraction [16c], infrared and Raman [16d]) which is in agreement with the computational results.

The interaction of a point charge (a proton) and the allylsilane was next studied in the three low-energy conformers. By using this "test" electrophile an electrostatic potential map can be developed. The electrophilic attack onto the two low-energy conformers of 2-silylbut-3-ene, **iv** and **v**, occurs *anti* to the silyl group. In the high-energy conformer **vi**, attack will occur *anti* to the methyl group. These

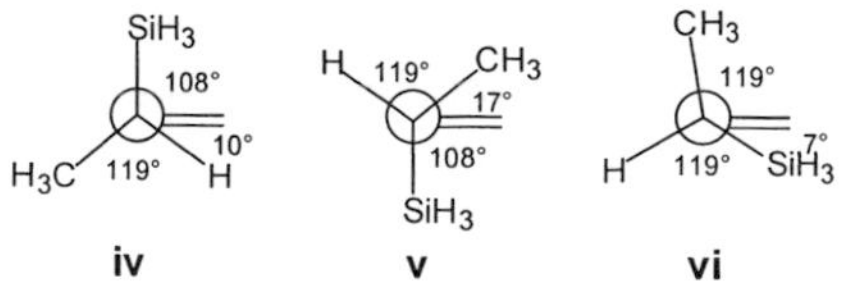

Chart 10-1.

results are interpreted as a tendency of the approaching electrophile to avoid regions of high positive charge (the silyl group) due to electrostatic repulsion.

To provide an unambiguous correlation between product stereochemistry and transition structure geometry, an intramolecular allylsilane-aldehyde condensation has been examined [2, 17]. Cyclization of model system **5** in the presence of Lewis acids leads to the formation of two diastereomeric products **6** and **7**. Cyclization through a synclinal arrangement of the groups will afford the proximal alcohol **6** (hydroxyl group close to the olefin), whereas cyclization through an antiperiplanar arrangement results in the formation of the distal alcohol **7** (hydroxyl group away from the olefin) (Scheme 10-5).

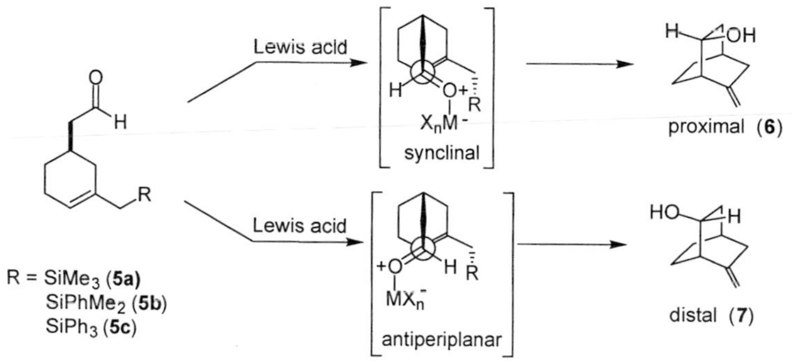

Scheme 10-5.

Table 10-2. Combined results for the cyclization of models **5a**, **5b**, **5c**.

Entry	Reagent	Proximal/distal (**6**/**7**)		
		$SiMe_3$ (**5a**)	$SiPhMe_2$ (**5b**)	$SiPh_3$ (**5c**)
1	$SnCl_4$	47/53	56/44	82/18
2	Et_2AlCl	64/36	67/33	66/34
3	$BF_3 \cdot OEt_2$	79/21	82/18	78/22
4	$SiCl_4$	99/1	98/2	92/8
5	CF_3SO_3H	94/6	94/6	90/10

The results obtained from cyclization of the model systems demonstrate a preference for the proximal product (synclinal arrangement of double bonds in the transition structure) but the observed selectivity is dependent upon the Lewis acid used (Table 10-2). The use of the bulky Lewis acid $SnCl_4$ leads to a non-selective reaction. Brønsted acid-initiated cyclization (entry 5) resulted in a very selective reaction favoring the proximal diastereomer. If *E*-complexation geometry is assumed between the Lewis acid and the aldehyde, then the major steric contribution in the model system would arise from the silylmethylene group [18, 19]. The formation of any of the proximal product with $SnCl_4$, a Lewis acid known to form 2/1 complexes with aldehydes [20], is interpreted as a stereoelectronic advantage for the synclinal transition structure. The results with $BF_3{\cdot}OEt_2$ are not unexpected because $BF_3{\cdot}OEt_2$ is known to form 1/1 complexes with aldehydes [21]. To fully interpret the stereochemical significance of the silicon electrofuge, the complete data on the cyclizations of the trimethylsilyl, phenyldimethylsilyl, and triphenylsilyl models should be considered. The data in Table 10-2 clearly demonstrate that the bulk of the silicon electrofuge does not significantly affect the observed selectivity in these reactions. Instead, the preference for these reactions to proceed via the synclinal transition structure appears to be related to the bulk of the Lewis acid; the larger the agent, the less selective the reaction for the proximal diastereomer. Very little change in selectivity is observed with a change in the steric environment around the silane; therefore the silicon electrofuge is thought to be disposed *anti* to the approaching electrophile as shown in Scheme 10-5. The major unfavorable steric contribution in the synclinal transition structure would then arise from the interaction of the Lewis acid and the (trialkylsilyl)methylene unit. The only departure from this observation is the reaction of model **5c** with $SnCl_4$. It is possible that metathesis of the allylsilane occurs with tin to give a trichlorostannane. Reaction of this species may then occur through a *syn* $S_{E'}$ pathway leading to higher than expected selectivity for the proximal alcohol.

To eliminate any potential diastereomeric bias inherent to model system **5**, a second generation system **8** was designed (Scheme 10-6). The stereochemical analysis required the specific placement of a ^{13}C-label in the exo methylidine group [22]. Intramolecular cyclization of **8** leads to the formation of the pseudoenantiomeric bicyclic alcohols **9a** and **9b**. The results from the cyclization of **8** with various Lewis acids are shown in Table 10-3. The product ratios were determined by integration of the ^{13}C NMR signals for the labeled product alcohols. A preference for the synclinal orientation is observed with the Lewis acids, although this preference is not strong. Unfortunately, it has been shown that transmetallation of the starting allylsilane could occur with the Lewis acids studied. Therefore, the results of these cyclizations do not provide an unambiguous assessment of the synclinal versus antiperiplanar preference in a diastereomerically unbiased system.

The results obtained with model systems **5** and **8** demonstrate a clear preference for the synclinal transition structure, but several questions regarding the ability of this model to predict the stereochemical outcome of the intermolecular allylsilane-aldehyde condensation are still at issue. Model system **10** was designed to remove any steric bias that may be present in model **5** (Scheme 10-7) [17e]. Cyclization of **10** was induced by various Lewis acids and the results are shown

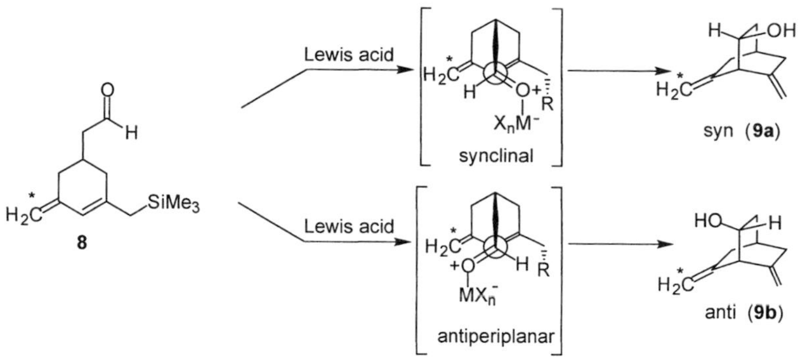

Scheme 10-6.

Table 10-3. Cyclization of model system **8** with various Lewis acids [22].

Entry	Lewis acid	Time, min	temp, °C	*syn*, %	*anti*, %	Yield, %
1	Et_2AlCl	180	–70	73	27	12
2	$BF_3{\cdot}OEt_2$	150	–70	70	30	6
3	$FeCl_3$	150	–70	70	30	20
4	$SnCl_4$	60	–70	67	31	12
5	n-$Bu_4N^+F^-$	60	20	53	47	21

in Table 10-4. All of the Lewis acids examined were selective for the proximal diastereomer. The results obtained with $BF_3{\cdot}OEt_2$ and CF_3SO_3H are almost identical to those obtained with model system **5** (compare entries 3 and 4, Table 10-4, with entries 3 and 5, Table 10-2). The cyclization with $TiCl_4$ and $SnCl_4$ are found to be highly selective for the proximal diastereomer. The cyclizations with $SnCl_4$ (the sterically most demanding Lewis acid) actually afford a 90/10 ratio of diastereomers favoring the *syn* isomer (entry 1, Table 10-4).

The results obtained from the cyclization of model **5** indicated that the size of the Lewis acid-aldehyde complex influences the selectivity of the reaction. For model system **10** it appears that the steric bulk of the Lewis acid does not play a significant role in determining the stereochemical outcome of the reaction. In model system **10** no external methylene unit exists which could interact with the Lewis acid-aldehyde complex. In fact, the silane is fixed in an *anti* orientation with respect to the approaching aldehyde (*anti* $S_{E'}$). The cyclization of model system **10** with fluoride ion affords primarily the distal product resulting from cyclization through an antiperiplanar transition structure. Thus, the antiperiplanar transition structure is accessible, but is not favored in reactions with the Lewis acids.

The high selectivity for the synclinal transition structure in these models may also arise from frontier orbital interactions. Anh and Thanh have suggested that the stereochemical outcome of an aldol reaction may be controlled by the overlap between the frontier molecular orbitals [23a]. This same cycloaddition-like transition structure was first used by Mulzer to explain the high selectivity observed in

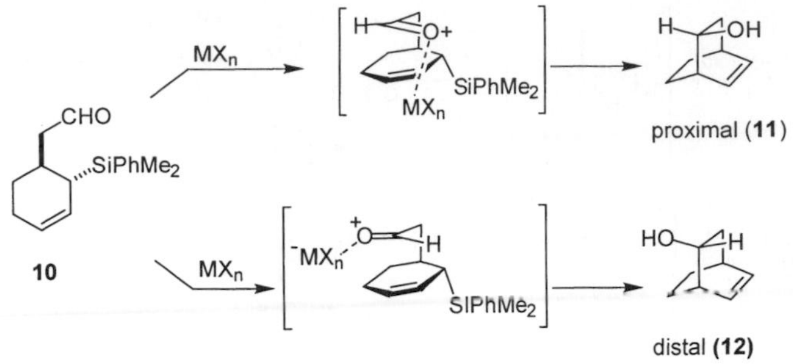

Scheme 10-7.

Table 10-4. Cyclization of model **10.**

Entry	Reagent	Time, min	Proximal/distal, %	Mass recovery, %
1	$SnCl_4$	20	90/10	80
2	$TiCl_4$	3	94/6	88
3	$BF_3{\cdot}OEt_2$	40	80/20	80
4	CF_3SO_3H	1	95/5	80
5	$ZrCl_4$	90	78/22	89
6	n-$Bu_4N^+F^-$	720	16/84	71

an aldol reaction [23b]. In this proposal an out-of-phase overlap would be energetically disfavored. Thus, cyclization of model system **10** would proceed through the synclinal transition structure where in-phase overlap is possible. No overlap would be possible with the antiperiplanar transition structure. The high selectivity observed for the proximal diastereomer in the cyclization of model system **10** must result from a stereoelectronic preference, *not* a steric preference, for the synclinal arrangement of double bonds in the transition structure.

The relative disposition of the silicon electrofuge and the aldehyde (*syn* or *anti* $S_{E'}$) was studied by the cyclization of model system **13** [17d]. The cyclization of *u*-**13** and *l*-**13** with various reagents could afford the four possible diastereomeric alcohols (*E*)- and (*Z*)-**14**-**15** (Scheme 10-8). The alcohols (*E*)-**14** and (*Z*)-**14** would result from reaction through a synclinal arrangement of double bonds in the transition structure. The position of deuterium can then be established to determine the stereochemical course of the reaction. The alcohols (*E*)-**15** and (*Z*)-**15** result from reaction through an antiperiplanar arrangement of double bonds in the transition structure. An *E*-complexation geometry is assumed throughout.

As was seen previously for model system **5**, the ratio of proximal to distal diastereomers **14/15** displays a Lewis acid dependence (Table 10-8). The reactions strongly favor the *anti* $S_{E'}$ pathway (selectivities >95%) regardless of the Lewis acid employed and regardless of the relative stereochemical outcome. Reactions with either diastereomer (*l*)- or (*u*)-**13** gave identical results, therefore only the results obtained with *l*-**13** are presented. The only divergence from this behavior is seen when fluoride ion was used to initiate the reaction. With fluoride, the distal product was dominant as before, but was formed by a combination of both *syn* and *anti* $S_{E'}$ pathways. On the basis of these reactions it is clear that the silicon electrofuge is located away from the approaching electrophile regardless of Lewis acid or double bond orientation. The Lewis acid only influences the synclinal vs antiperiplanar orientation of double bonds, most likely due to differences in effective bulk of the Lewis acid-aldehyde complex.

The most intriguing results are those from the reaction promoted by fluoride. The reaction of fluoride with an allylmetal reagent is thought to proceed through either an allyl anion or an allyl fluorosiliconate intermediate [24]. Allyl anions [25] and pentacoordinate silicon species [26] have been proposed as intermediates in the fluoride-induced allylation of a carbonyl compound. The results obtained with the deuterium model rule out the intermediacy of a free allyl anion since the ratio **14/15** is different for the *syn* compared to the *anti* $S_{E'}$ pathways.

Remarkably, the ratio of proximal to distal products was dependent upon the mode of cyclization (*syn* versus *anti* $S_{E'}$). Cyclization through the *syn* $S_{E'}$ pathway led to the predominant formation of the distal products (**8/1**), whereas cyclization

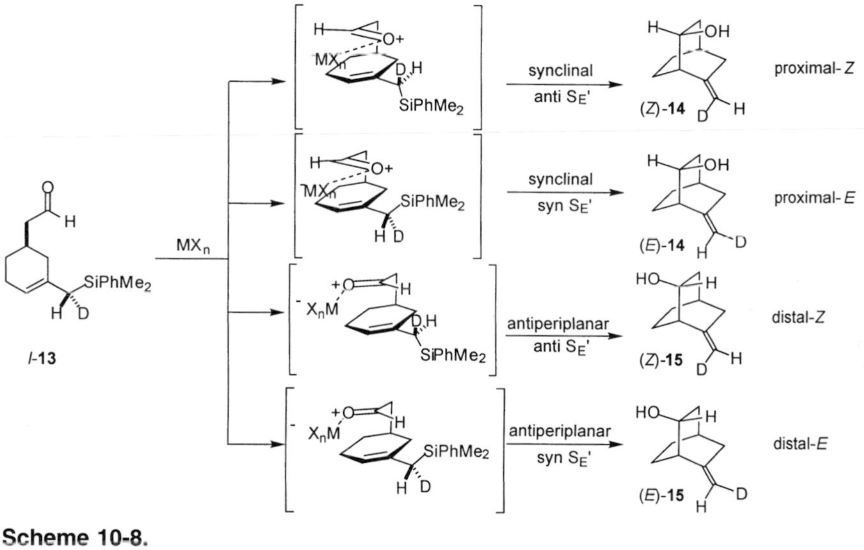

Scheme 10-8.

Table 10-5. Cyclization of model **13**.

Model	Reagent	Proximal/ distal (**14/15**)	**14** *Z/E*	**15** *Z/E*	Proximal % *anti* $S_{E'}$ [a]	Distal % *anti* $S_{E'}$ [a]
l-**13**	$BF_3 \cdot OEt_2$	75/25	94/6	94/6	100	100
l-**13**	$SnCl_4$	60/40	91/9	94/6	97	100
l-**13**	CF_3SO_3H	95/5	93/7	94/6	99	100
l-**13**	$SiCl_4$	98/2	95/5	–	100	–
l-**13**	n-$Bu_4N^+F^-$	20/80	80/20	60/40	85	65

[a]) Percent *anti* $S_{E'}$ based on 94.5% *d*-content in **13**.

through the *anti* $S_{E'}$ pathway was less selective for the distal products (**3/1**). This preference for the distal products with the *syn* $S_{E'}$ pathway suggests that a steric contribution exists for relativ stereoselection in the reactions with fluoride ion. In the distal somer, little preference for cyclization through either the *syn* or *anti* pathways was observed. Possibly, the difference could be due to a weak Coulombic repulsion which favors formation of the distal products and reaction through an *anti* $S_{E'}$ pathway. That such a steric component exists also rules out the intermediacy of closed transition structures. These transition structures were previously proposed by both Corriu [27a] and Sakurai [27b] to explain the selectivities observed with fluoro- and alkoxysiliconates.

The Lewis acid-promoted cyclization of the deuterium-labeled model **13** is found to give products corresponding to an *anti* $S_{E'}$ reaction. All of the cyclizations with Lewis acids are greater than 95% selective for the *anti* $S_{E'}$ reaction. The high selectivity observed demonstrates that *in a sterically unbiased* $S_{E'}$ *reaction an anti orientation of the electrophile with respect to silicon is preferred.* The products from both the synclinal and antiperiplanar transition structures are found to be *anti* selective. The arrangement of double bonds in the transition structure does not affect the relative disposition of the silicon electrofuge, which must be disposed away from the approaching electrophile.

10.2.1.2 Stereochemical course of addition

Relative stereoselection

When an allylsilane containing a C(3) substituent reacts with an aldehyde, two new stereogenic centers are established. The reaction of (*E*)- and (*Z*)-2-butenyltrialkylsilanes with aldehydes was first reported by Hayashi and Kumada in 1983 [28]. In these studies either (*E*)- and (*Z*)-2-butenyltrimethylsilane or (*E*)- and (*Z*)-cinnamyltrimethylsilane combined with various aldehydes in the presence of $TiCl_4$ (Scheme 10-9). The (*E*)-2-butenylsilane and (*E*)-cinnamylsilane both afford >95% of the *syn* diastereomer upon reaction with either propanal, isobutyraldehyde, or pivalaldehyde. When the *Z*-silanes were subjected to the same reaction conditions much lower selectivities were observed (65–72% *syn* selectivity). An acyclic transition structure with an antiperiplanar arrangement of double bonds was proposed to account for the diastereoselectivity observed in these reactions. A transition structure which includes a synclinal arrangement of double bonds may be necessary to explain the lower selectivities observed with the *Z*-allylsilanes.

Internal stereoselection

Achiral allylic silanes and chiral aldehydes. The stereochemical course of reaction for simple allylsilanes is largely governed by Cram (Felkin-Anh) [29] or chelation control [10h]. Two examples which illustrate this phenomenon are shown in Scheme 10-10 [30]. In the first example, the reaction of 2-phenylpropanal **19** with allyltrimethylsilane **20** affords the homoallylic alcohols **22** and **24** with almost no selectivity. However, methallylsilane **21** afforded a much higher *syn* selectivity under the influence of $BF_3{\cdot}OEt_2$. The reaction of the α-alkoxy aldehyde

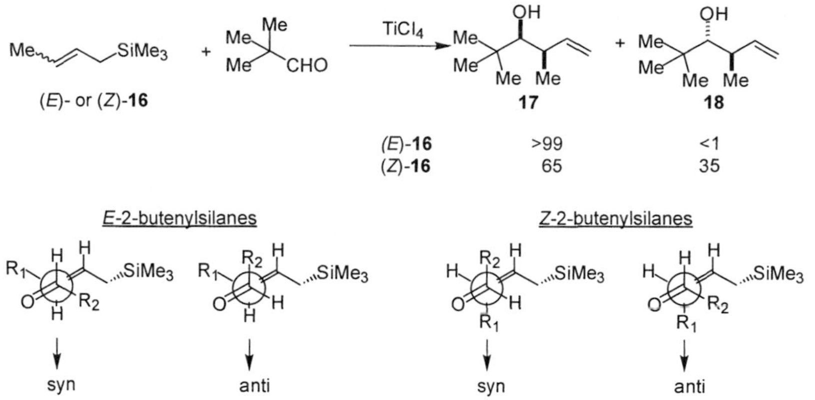

	17	18
(*E*)-**16**	>99	<1
(*Z*)-**16**	65	35

Scheme 10-9.

26 with **20** was next examined utilizing either $SnCl_4$ or $BF_3{\cdot}OEt_2$ as the Lewis acid. With $SnCl_4$, a Lewis acid known to form bidentate chelates, the reaction proceeded with a high degree of stereoselectivity giving primarily the homoallylic alcohol **22**. With $BF_3{\cdot}OEt_2$, the reaction proceeded via Cram control, and in this example provided almost a 1/1 mixture of **22** and **24**.

The effect of concentration in the chelation-controlled reactions of allylsilanes and α- and β-alkoxy aldehydes has been studied (Scheme 10-11) [31]. The α-alkoxy aldehyde **27** was allowed to react with varying amounts of $TiCl_4$ and allyltrimethylsilane to produce the homoallylic alcohol **28**. With less than 0.5 equivalents of $TiCl_4$, the reaction affords a mixture of products. When 0.5 equivalents or more of $TiCl_4$ are used, the reaction gives only the product of chelation control. Intervention of chelation control with the α-alkoxy aldehyde was independent of substrate concentration. The reaction of the β-alkoxy aldehyde **29** is found to be highly sensitive to both the substrate concentration and the stoichiometry of $TiCl_4$ employed. The reaction gave primarily the product of chelation control **30** when

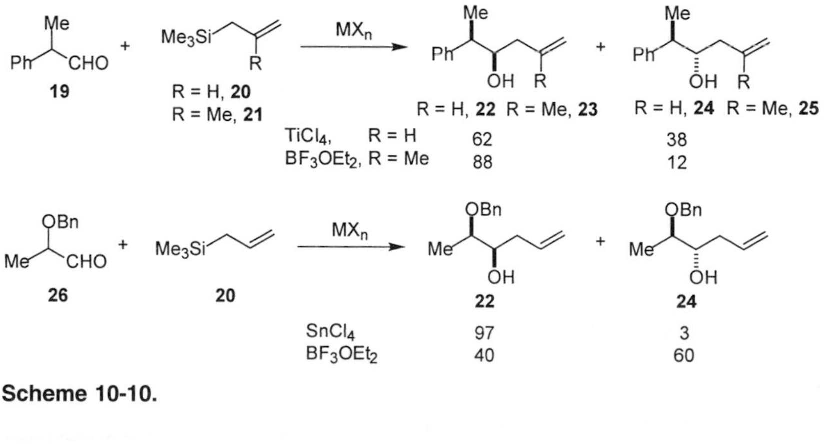

	R = H, **22** R = Me, **23**	R = H, **24** R = Me, **25**
$TiCl_4$, R = H	62	38
BF_3OEt_2, R = Me	88	12

	22	**24**
$SnCl_4$	97	3
BF_3OEt_2	40	60

Scheme 10-10.

between 0.5 and 1 equiv of $TiCl_4$ were employed at substrate concentrations lower than 0.1 M. The curve of diastereoselectivity versus $TiCl_4$ loading appears as a plateau with extremely steep sides. According to the authors, the observed differences in diastereoselectivities observed with the two aldehydes when greater than one equivalent of $TiCl_4$ was employed can be attributed to the difference in stability of the five- vs six-membered ring chelates of titanium. The reaction of the *β*-alkoxyaldehyde may occur via an open-chain non-chelated 2/1 complex ($TiCl_4$/aldehyde) when more than one equivalent of titanium is used.

Scheme 10-11.

The stereochemical complexity of the reaction can be further increased when an (*E*)- or (*Z*)-2-butenylsilane reacts with a chiral aldehyde. Herein both diastereoselection processes are operative, relative (between the reacting faces) and internal with respect to the original stereogenic center in the aldehyde. Thus, the reaction of *β*-benzyloxy aldehyde **32** and silane (*E*)-**31** with bivalent Lewis acids ($SnCl_4$, $TiCl_4$) was examined in the presence of an additive, e.g. $MgBr_2$, $ZrCp_2Cl_2$, $TiCp_2Cl_2$ (Scheme 10-12) [32]. The reactions all afford mixtures of the four possible diastereomeric products, favoring the *syn* homoallylic alcohol. When the com-

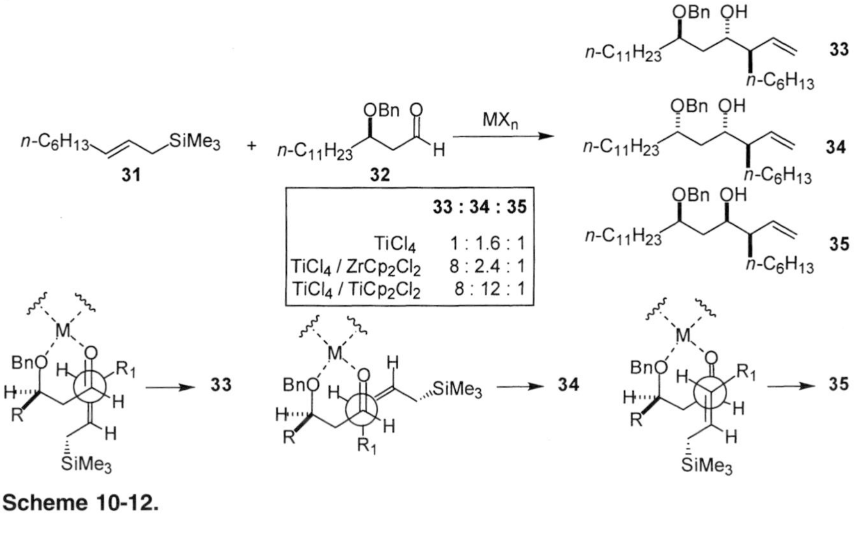

	33 : 34 : 35
$TiCl_4$	1 : 1.6 : 1
$TiCl_4$ / $ZrCp_2Cl_2$	8 : 2.4 : 1
$TiCl_4$ / $TiCp_2Cl_2$	8 : 12 : 1

Scheme 10-12.

bination of $TiCl_4$ and $TiCp_2Cl_2$ is used in the reaction, an 8/12/1 mixture of diastereomers is formed. Possible transition structures for the reaction are formulated below. The synclinal transition structure is proposed to be favored, with the $TiCp_2Cl_2$ species chelated between the aldehyde carbonyl and the benzyloxy ether. These pictures are extremely speculative as the nature of the actual titanium species has not been established.

The reaction of β-methyl-2-butenylsilanes **36** and stannanes with chiral α-alkoxyaldehydes has also been reported [33]. Surprisingly, the *anti* homoallylic alcohols were predominantly observed (94/6, *anti/syn*) when a bivalent Lewis acid such as $SnCl_4$ was used (Scheme 10-13). A synclinal transition structure is proposed to account for the observed selectivity. In the chelation-controlled reactions the synclinal transition structure is favored over the corresponding antiperiplanar transition structure because there exists an open space wherein the complexed Lewis acid can reside. The monovalent Lewis acid $BF_3 \cdot OEt_2$ provides the expected *syn* homoallylic alcohol, presumably through the antiperiplanar transition structure shown (66% of the product was the *syn* alcohol **37**).

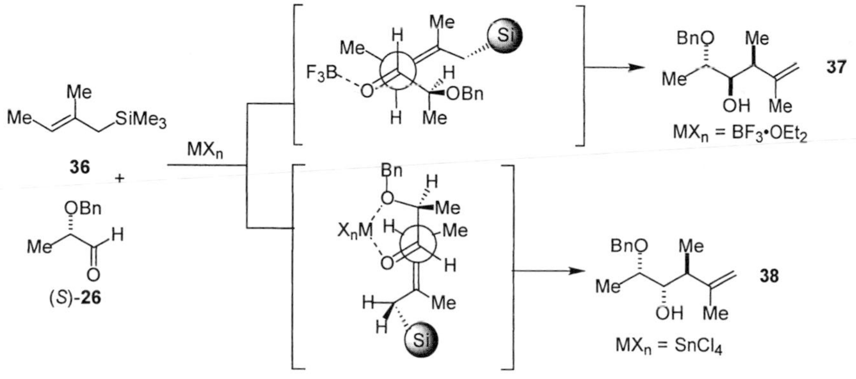

Scheme 10-13.

The addition of chiral *E*-2-butenylsilanes to an α- or β-alkoxy aldehyde can be used in the synthesis of substituted furans (Scheme 10-14) [10h, 34]. An antiperiplanar transition structure is proposed to account for the observed selectivity in this *anti* $S_{E'}$ reaction. A 1,2-silyl migration follows electrophilic addition of the aldehyde. Usually, this 1,2-silyl migration is not competitive with elimination, but in this instance it competes favorably. Cationic rearrangements of silicon are well known in the absence of nucleophilic reagents and have been extensively studied [35]. The intermediate carbocation which is produced by this silyl migration undergoes bond rotation followed by intramolecular cyclization to generate the substituted tetrahydrofuran **40**. This reaction proceeds in high yield with greater than 95% diastereomeric excess of the desired tetrahydrofuran. The products of this electrophilic addition correspond to reaction through the *anti* $S_{E'}$ pathway.

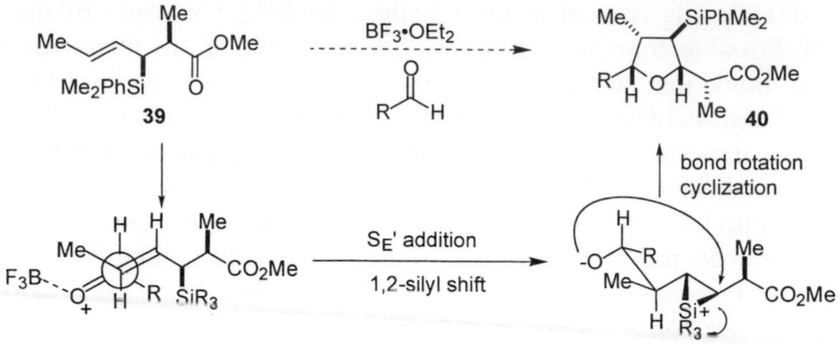

Scheme 10-14.

Chiral allylic silanes and achiral aldehydes. To clarify the various stereochemical features of the reaction, addition of enantiomerically enriched allyl- and 2-butenylsilanes to aldehydes and $TiCl_4$ has been examined (Scheme 10-15) [28b,c]. Here again, both diastereoselection processes are operative, relative (between the reacting faces) and internal with respect to the original stereogenic centers in the allylsilanes **41** and **42**. In these studies: (1) the enantiomeric excess of the products was essentially the same as the starting materials; (2) the *E*-allylsilanes reacted with high diastereoselectivity (*syn/anti*, 92/8 → 99/1); (3) the *Z*-allylsilanes were less selective with the resulting *syn/anti* ratio of products dependent upon the structure of the aldehydes (*syn/anti*, 50/50 → 99/1). The configuration of the products obtained for all of the reactions studied is interpreted in terms of an *anti* $S_{E'}$ reaction with the aldehyde approaching the double bond from the side opposite the trimethylsilyl group. It is interesting to note that in the reaction of (*Z*)-2-butenyltrimethylsilane [(*Z*)-**16**] and pivalaldehyde (Scheme 10-9), the relative induction was significantly reduced from that observed with the chiral *α*-phenyl-2-butenylsilane (**42**) and pivalaldehyde (*syn/anti*, 65/35 → 99/1), illustrating that these two types of induction are not necessarily operating independently.

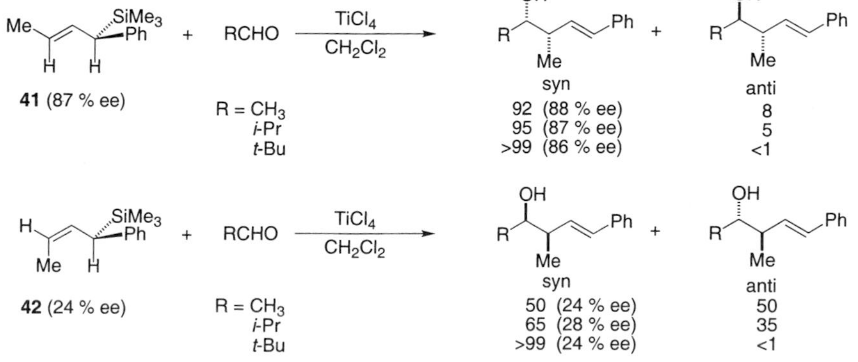

41 (87 % ee) + RCHO → ($TiCl_4$, CH_2Cl_2) syn + anti

R	syn	anti
CH_3	92 (88 % ee)	8
i-Pr	95 (87 % ee)	5
t-Bu	>99 (86 % ee)	<1

42 (24 % ee) + RCHO → ($TiCl_4$, CH_2Cl_2) syn + anti

R	syn	anti
CH_3	50 (24 % ee)	50
i-Pr	65 (28 % ee)	35
t-Bu	>99 (24 % ee)	<1

Scheme 10-15.

To explain the observed selectivities, acyclic linear transition structures are invoked. In these transition structures the double bonds are placed in an antiperiplanar relationship. The observed diastereoselectivities are proposed to result from a minimization of steric interactions in the transition structures (Scheme 10-16). Because the enantiomeric excess of the product matches that of the starting material, this reaction is also selective for the *anti* $S_{E'}$ pathway [28c].

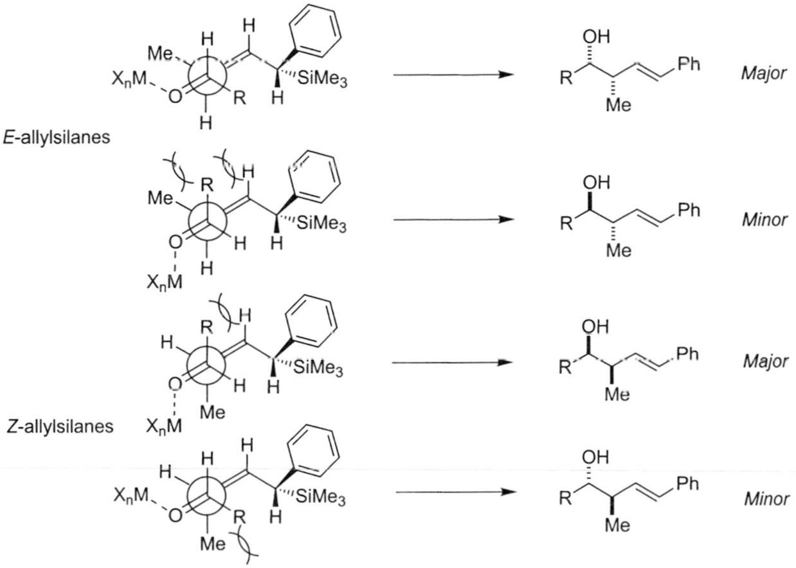

E-allylsilanes

Z-allylsilanes

Major

Minor

Major

Minor

Scheme 10-16.

The Lewis acid-promoted reaction of chiral *β*-methyl (*E*)-2-butenylsilane (*S*)-**43** with *α*-(benzyloxy)propanal affords homoallylic alcohols favoring the *anti* or *syn* diastereomers depending on the Lewis acid employed (Scheme 10-17) [36]. The $BF_3 \cdot OEt_2$-promoted reaction provided a 6.5/1 (**44/45**) ratio of the *syn* to *anti* diastereomers presumably through the antiperiplanar transition structure shown below. When $MgBr_2$ was employed, the *anti* homoallylic alcohol was favored by a 12/1 (**45/44**) ratio. A synclinal transition structure accounts for the reversal in selectivity observed with the bivalent Lewis acid.

Chiral allylic silanes and chiral aldehydes. This combination of agents provides fascinating opportunities for double asymmetric induction and allows the magnitude of the various controlling features to be expressed. The sense and level of 1,2-asymmetric induction in the Lewis acid-promoted addition of chiral *E*-2-butenylsilanes to chiral *α*-alkoxy aldehydes has been examined as well (Scheme 10-18) [37].

The chiral 2-butenylsilanes (*S*)-**43** and (*R*)-**43** react with either the benzyl protected *α*-alkoxyaldehyde (*S*)-**26** or the *t*-butyldiphenylsilyl protected *α*-alkoxyalde-

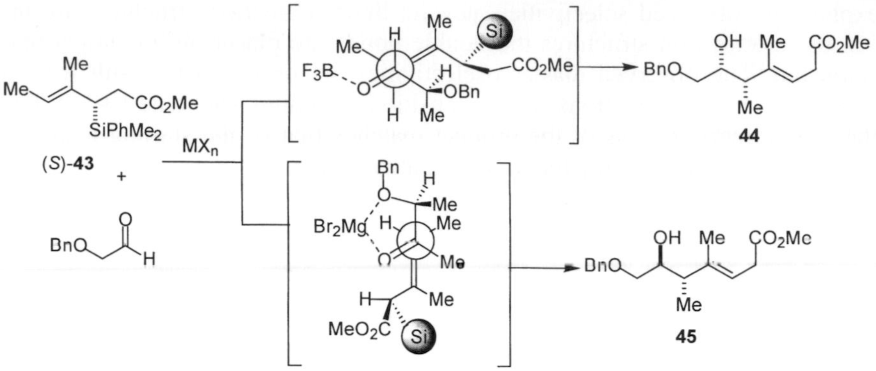

Scheme 10-17.

hyde (*S*)-**46** using $TiCl_4$ or $BF_3{\cdot}OEt_2$ as the Lewis acid promoter. The $BF_3{\cdot}OEt_2$-promoted reaction of the benzyl- or silyl-protected *α*-alkoxy aldehydes with the allylsilane (*S*)-**43** afford the *syn* homoallylic alcohols **47** or **48** presumably through the antiperiplanar transition structure shown. The $TiCl_4$-promoted reaction of (*S*)-**26** and the allylsilane (*S*)-**43** affords the *anti* homoallylic alcohol **49** through a chelated transition structure. The configuration of the emerging hydroxyl group appears to be influenced by the chirality of the aldehydes while the absolute stereochemical relationships are dictated by the configuration of the C–SiR_3 bond.

The Lewis acid-promoted reaction of the allylsilane (*R*)-**43** with either of the (*S*)-*α*-alkoxy aldehydes provides some surprising insights (Scheme 10-19) [34]. The $BF_3{\cdot}OEt_2$-promoted reaction of the silane (*R*)-**43** and either (*S*)-**26** or (*S*)-**46** afforded predominantly the *syn* homoallylic alcohols **50** and **51**, even though this is presumed to be a mismatched combination of reagents. The $TiCl_4$-promoted reaction of (*R*)-**43** with (*S*)-**26** or (*S*)-**46** also produces the *syn* homoallylic alcohols, presumably through a Cram chelate transition structure model (albeit with lesser selectivity for **46**). These experiments indicate that the chirality of the *R*-silane re-

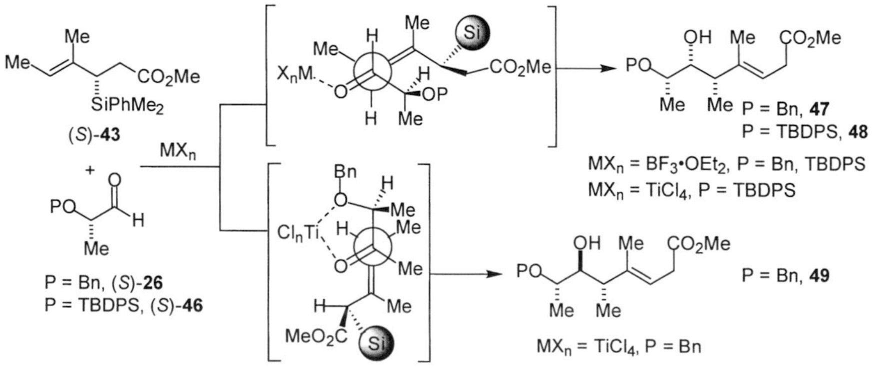

Scheme 10-18.

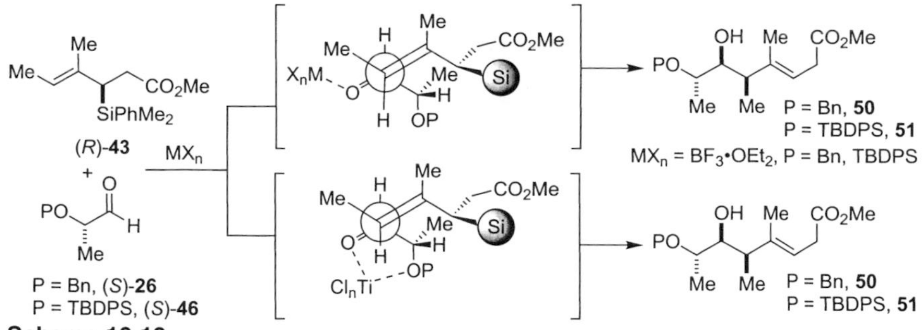

Scheme 10-19.

agents is capable of overriding the inherent preference of the aldehyde to afford the *syn* homoallylic alcohols. The utility of these double-stereodifferentiating reactions has been demonstrated in the total synthesis of Macbecin [38].

The reaction of β-alkoxy aldehydes [e.g. (*S*)-**52**] with β-alkyl-substituted silane reagents **43** has been studied [39]. The configuration of the C-SiR_3 center was again found to determine the configuration of the center bearing the methyl group, while the chirality of the aldehyde controlled the configuration of the oxygen-bearing stereocenter. The stereochemical models shown in Scheme 10-20 help illustrate the unique features of these reactions.

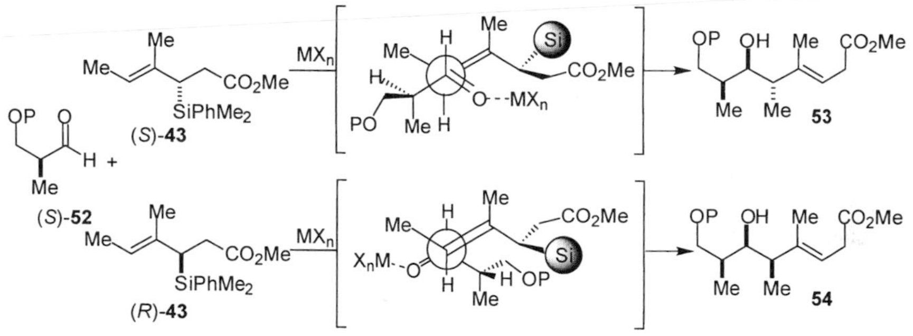

Scheme 10-20

The Lewis acid-promoted reaction of α-amino aldehydes with chiral *E*-2-butenylsilanes **55** has also been examined [40]. In the reaction of (*S*)-**56**, it was determined that the stereochemical outcome of the reaction is largely determined by the absolute configuration of the C–SiR_3 center. The reaction of both the (*R*)- and (*S*)-2-butenylsilanes proceeded in high yield, but the (*S*)-2-butenylsilane gave poor levels of diastereoselectivity (Scheme 10-21).

Chiral-auxiliary modified aldehydes. A well-established strategy for the asymmetric allylation of aldehydes is the conversion of the aldehyde carbonyl group to an acetal with a chiral diol [41]. Despite the excellent selectivities achieved, this

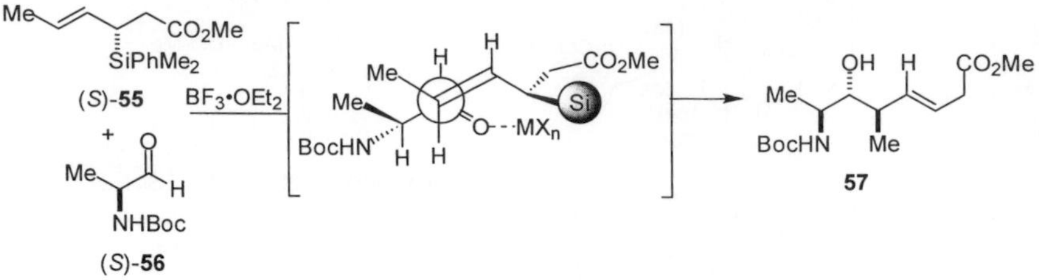

Scheme 10-21.

method is hampered by the difficulty of removing the modifier auxiliary after opening the acetal. To address this shortcoming, a new method has been developed for the asymmetric allylation of achiral aldehydes with allyltrimethylsilane that utilizes a pseudoephedrine derivative (Scheme 10-22) [42]. The aldehyde is treated with the norpseudoephedrine derivative **58** and 10 mol% TMSOTf for 1 h and then allyltrimethylsilane was added to afford the homoallylic ether **60**. The chiral inductor could then be removed by a sodium-in-ammonia reduction to give the homoallylic alcohol **61**. The allylation proceeds at extremely diastereoselectivity when aliphatic aldehydes are employed. For example, the allylation of ethanal or pivalaldehyde provides the homoallylic ether in ~55% yield and >99% de. When the method is extended to aromatic aldehydes lower diastereoselectivity is observed. A possible mechanism has been proposed which involves the formation of an adduct (**59**) that suffers ionization to an oxocarbenium ion (**vii**) which is suggested to exist in a closed form as oxazolidinium ion (**viii**) which can then undergo nucleophilic attack to generate the homoallylic ether **60** [48c]. The mechanism assumes that at low temperature the S_N^2 displacement of **viii** by allyltrimethylsilane is faster than proton transfer. It was proposed that aromatic aldehydes react preferentially through the oxocarbenium ion **vii** which may be lower in energy.

The asymmetric allylation of unfunctionalized aliphatic ketones has also been described (Scheme 10-23) [43]. Simple aliphatic ketones are treated with a mixture of the trimethylsilyl ether of norpseudoephedrine (**58**), two equivalents of allyltrimethylsilane, and a catalytic amount of triflic acid. The homoallylic ethers

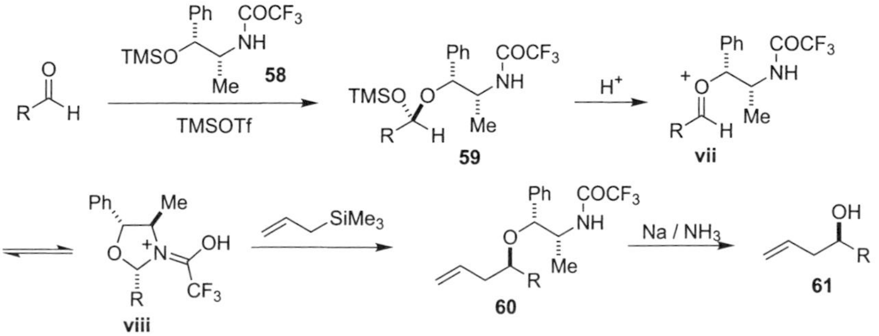

Scheme 10-22.

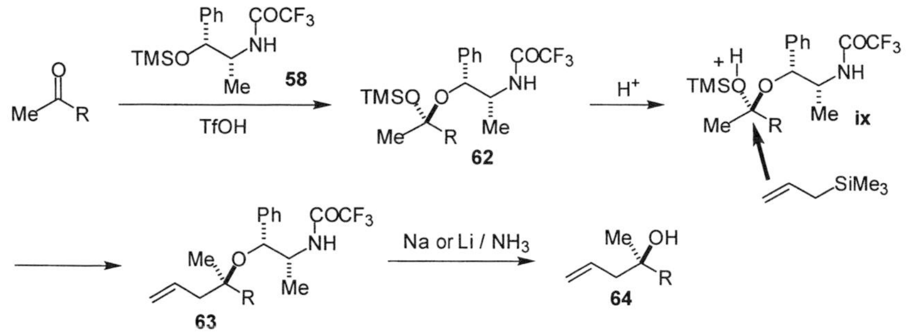

Scheme 10-23.

63 were obtained in good yield and diastereoselectivity. The cleavage of the ethers to give the homoallylic alcohols **64** is performed by reductive removal with lithium or sodium in ammonia. A mechanism has been proposed for the reaction of ketones which involves the initial formation of the mixed acetal **62**. Allylation of the protonated mixed acetal **ix** by allyltrimethylsilane in an S_N^2 type fashion can then occur to produce the observed homoallylic ether.

External stereoselection

The importance and advantages of catalytic enantioselective variants of synthetic organic reactions cannot be overstated [44]. In view of the importance of Lewis acidic activation of allylsilane-aldehyde addition reactions, it is not surprising that chiral Lewis acids developed for other reactions were also applied here. Indeed, the first example of the catalytic enantioselective allylation of aldehydes with allylsilanes was reported in 1991 by Yamamoto [45]. The chiral (acyloxy)borane (CAB) catalyst (originally developed for enantioselective Diels-Alder reactions) was used (20 mol%) to produce the desired homoallylic alcohols in moderate to good yields when substituted allylsilanes were employed. The reaction gives the best yield and selectivity when *β*-alkyl-substituted allylsilanes were used in conjunction with aromatic aldehydes. For example, the CAB-promoted reaction of benzaldehyde and **36** affords the desired *syn* homoallylic alcohol **65** (97/3, *syn/anti*) in 74% yield and 96% ee (Scheme 10-24). An acyclic, antiperiplanar transition structure is proposed to account for the observed selectivity. Unfortunately, due to the uncertain structure of the CAB catalyst it is difficult at best to provide a rationale for the observed enantioselectivity. The high *syn* selectivity in the reaction is especially noteworthy, as the $BF_3{\cdot}OEt_2$-promoted reaction produced a 53/47 *syn/anti* mixture of homoallylic alcohols. In general, the reaction of aromatic aldehydes with the allylsilane produces homoallylic alcohols with enantioselectivities ranging from 80 to 96%. The observed selectivities are independent of the allylsilane geometry. Aliphatic aldehydes afford the homoallylic alcohol in 20–36% yield, albeit with good enantioselectivity (85–90%). The reaction is highly solvent dependent, giving the highest diastereo- and enantioselectivity with the polar solvent propionitrile. Modification of the CAB reagent by the use of aromatic boronic acids produces a more Lewis acidic reagent which resulted in higher yields and enantioselectivities [45 b].

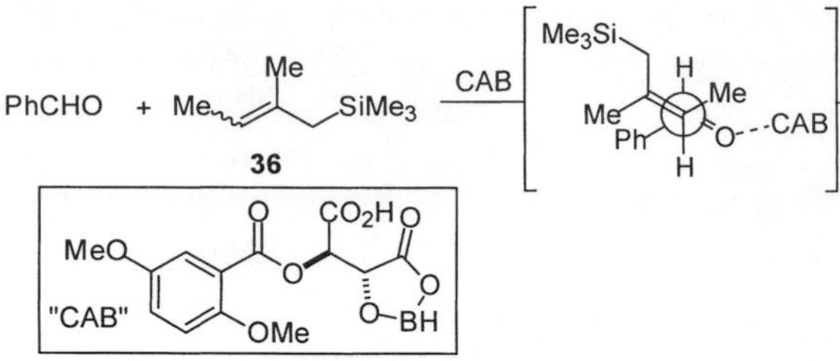

Scheme 10-24.

Another enantioselective variant of this reaction has been developed with a binol-titanium complex **66** [46]. This catalyst affords homoallylic alcohols in moderate diastereo- and enantioselectivity with 2-butenylsilane (*E*)-**16** and methyl glyoxylate (Scheme 10-25). Reaction of (*Z*)-**16** was much less selective, providing homoallylic alcohols with low enantioselectivity. An antiperiplanar transition structure accounts for the formation of the *syn* homoallylic alcohol **67**. A more reactive complex formed from BINOL and TiF_4 has also found utility [46c].

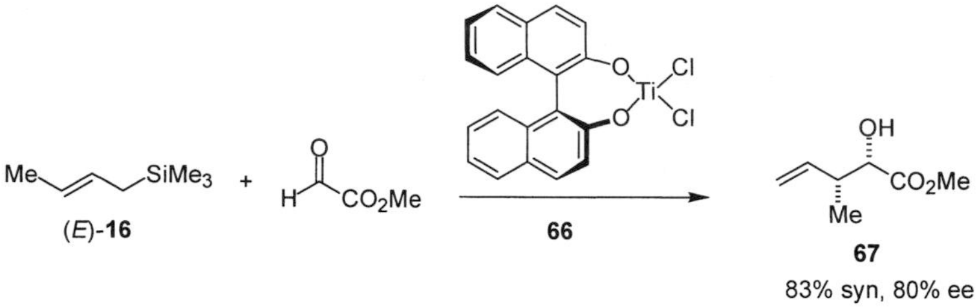

Scheme 10-25.

10.2.2 Allylic Trihalosilanes

Allylic trihalosilanes are electronically complementary to their trialkyl relatives. These species are rather electrophilic at the silicon atom and accordingly are activated by the addition of Lewis bases rather than Lewis acids. The appeal of these reagents stems from the control of both internal and absolute stereochemistry [47]. When the electron-deficient metal center is present (Type I reaction) a high degree of internal stereocontrol (*syn*/*anti* diastereocontrol) is possible through a change in double-bond geometry by reaction through a chair-like transition structure.

10.2.2.1 Mechanism of addition

The fluoride-promoted reaction of allyltrifluorosilanes with aldehydes to generate homoallylic alcohols was first reported by Sakurai in 1987 [48]. The reactions are found to be highly regioselective (carbon-carbon bond formation exclusively at the γ-carbon) and proceed with a high degree of stereoselectivity. The mechanism proposed for the reaction involves the formation of a pentacoordinate allylsiliconate which can then react with the aldehyde in a six-membered cyclic transition structure to afford the homoallylic alcohols (Scheme 10-26). The significant Lewis acidity of the tetrafluorosiliconate and the enhanced nucleophilicity of the γ-carbon of the allylsiliconate may stabilize a cyclic transition structure. An ab initio study of the reaction of pentacoordinate allylsilicates with aldehydes indicates that the cyclic arrangement is a more favorable transition structure than the open-chain form [49].

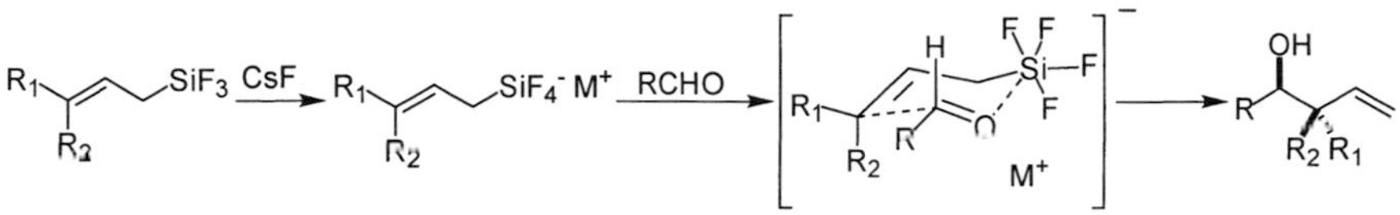

Scheme 10-26.

The addition of allyl- and 2-butenyltrichlorosilanes to aldehydes has recently emerged as a useful synthetic method [50]. The reaction of these species proceeds under neutral conditions using DMF as solvent at 0 °C and affords homoallylic alcohols in high yield. The coordination of DMF to the silicon atom is suggested (and supported by ^{29}Si NMR studies) to play a key role in the reaction forming the penta- or hexacoordinate siliconate complex (Scheme 10-27). The hypervalent silicon atom is then sufficiently Lewis acidic from the electron-withdrawing chlorine groups and sufficiently nucleophilic from electron donation from the hypervalent silicon atom for the reaction to proceed smoothly and stereoselectively.

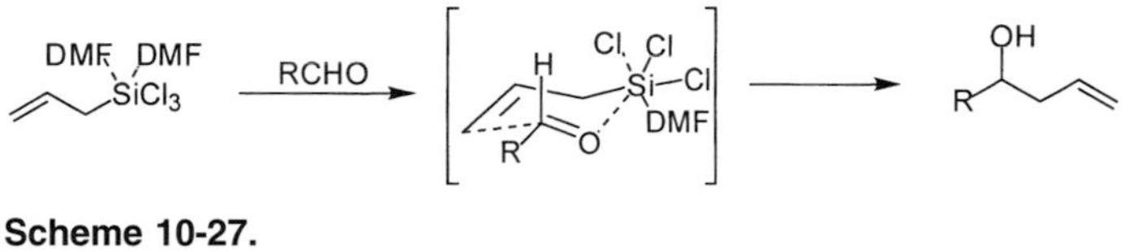

Scheme 10-27.

The rate of allylation of aldehydes with allyltrichlorosilane is greatly influenced by O-donor ligands [51]. The reaction of benzaldehyde with allyltrichlorosilane was promoted by a stoichiometric quantity of DMF in combination with various salts. When the salts are added to the reaction, a dramatic increase in the rate of allylation was observed. This effect is interpreted in terms of an ionization event at a key step in the reaction pathway. The dissociation of a chloride ion is proposed to create a reactive complex in solution. Conductivity experiments indicate that ion-paired complexes may play a role in the ligand-activated allyl- and 2-butenylation of aldehydes. The conductivity of allyltrichlorosilane in dichloromethane (a solvent which does not promote allylation) is 1000 times less than at the same concentration of allyltrichlorosilane in DMF.

10.2.2.2 Stereochemical course of addition

Relative stereoselection

The addition of 2-butenyltrifluorosilanes to aldehydes promoted by fluoride or triethylamine has proven to be a highly stereoselective method of carbon-carbon bond formation [50, 52]. The stereochemical outcome of the reaction is dependent on the geometry of the starting 2-butenylsilane. The *E*-silane (*E*)-**68** leads to the almost exclusive formation of the *anti* homoallylic alcohol, while the *Z*-silane (*Z*)-**68** produces the *syn* homoallylic alcohol upon reaction with the aldehyde [52a]. The unprotected *α*-hydroxy ketone also reacts with allyltrifluorosilane in the presence of triethylamine to produce the corresponding tertiary homoallylic alcohol in extremely high regio- and diastereoselectivity [52b,d]. For example, reaction of (*E*)-**68** (97/3 *E*/*Z*) with *α*-hydroxyacetone affords the *syn* homoallylic alcohol **69** (97/3 *syn*/*anti*) (Scheme 10-28). The high regio- and diastereoselectivities observed suggest that the reaction proceeds via a 1,3-bridged cyclohexane-like transition structure (Scheme 10-28) [52b].

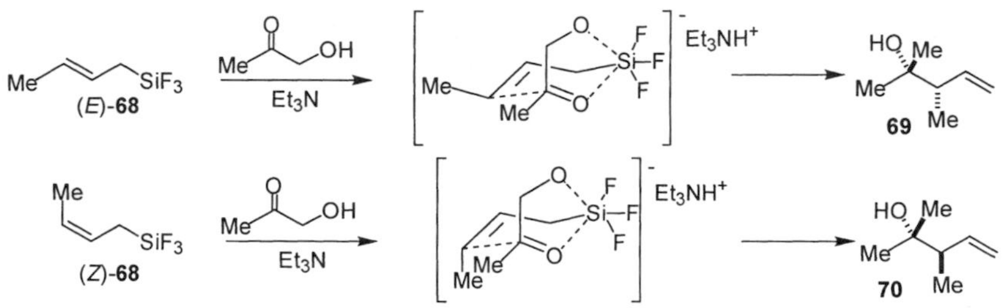

Scheme 10-28.

The reaction of (*E*)- and (*Z*)-**71** with aldehydes has been demonstrated to proceed smoothly with high regio- and diastereoselectivity [50]. Reaction of the (*E*)-**71** provides almost exclusively the *syn* homoallylic alcohols, while (*Z*)-**71** provides the corresponding *anti* alcohols. The stereochemical course of the reaction has been attributed to the intermediacy of a chairlike, six-membered transition structure assembly which incorporates all three elements and places the C(3) substituent in pseudoequatorial or pseudoaxial orientations according to olefin geometry (Scheme 10-29).

The asymmetric allylation of aldehydes with allyltrichlorosilane modified by a chiral diol has recently been demonstrated by Wang [53a,b] and Kira [53c]. In these reactions, diisopropyl tartrate is added to the appropriate allylchlorosilane in the presence of triethylamine to generate the modified allylsilane. In the system developed by Wang, the allylsilane was allowed to react with the aldehydes in the presence of either DMF or triethylamine at room temperature to give the corresponding homoallylic alcohol in moderate yield and enantioselectivity. The Kira group did not use any additional promoter, suggesting instead that the carbonyl group on the tartrate ester is able to coordinate to silicon. Both authors suggest that the reaction occurs through the intermediacy of a six-membered cyclic transition structure.

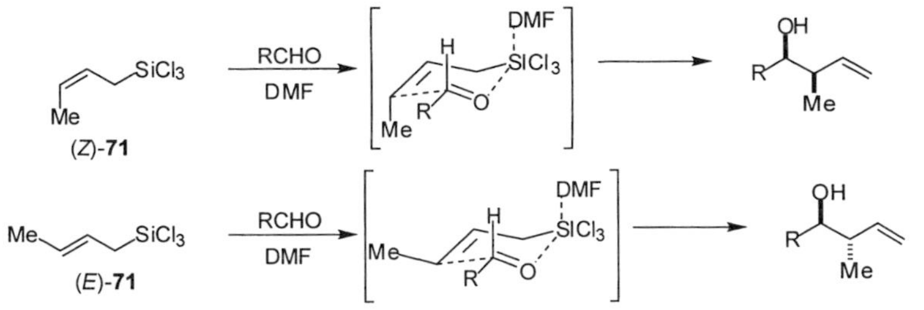

Scheme 10-29.

Internal stereoselection

The addition of 2-butenyltrifluorosilane **68** to chiral aldehydes has been examined by Roush in the synthesis of the *anti,anti*-dipropionate stereotriad, a common but difficult-to-synthesize subunit in polyketide-derived natural products [54]. The synthetic approach involves the 2-butenylation reaction of α-methyl-β-hydroxy aldehydes **72** with (*Z*)-**68** (Scheme 10-30). Using this approach, the *anti,anti*-dipropionate **73** could be obtained in excellent selectivity. The best selectivity is observed when *anti*-β-hydroxy aldehydes are used. When the *syn* aldehyde is used, a mixture of homoallylic alcohols is produced which may arise from a nonchelated Zimmerman-Traxler transition structure.

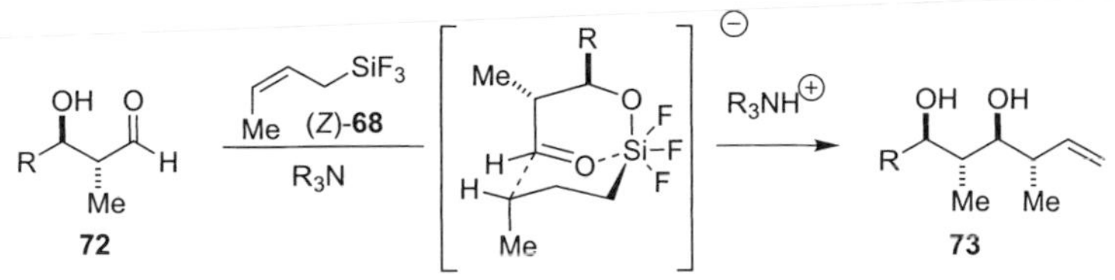

Scheme 10-30.

External stereoselection

The observed activation of allyltrihalosilanes with fluoride ion and DMF and the proposition that these agents are bound to the silicon in the stereochemistry-determining transition structures clearly suggested the use of chiral Lewis bases for asymmetric catalysis. The use of chiral Lewis bases as promoters for the asymmetric allylation and 2-butenylation of aldehydes was first demonstrated by Denmark in 1994 (Scheme 10-31) [55]. In these reactions, the use of a chiral phosphoramide promoter **74** provides the homoallylic alcohols in high yield, albeit modest enantioselectivity. For example, the (*E*)-**71** and benzaldehyde affords the *anti* homoallylic alcohol **75** (98/2 *anti*/*syn*) in 66% ee. The sense of relative stereoinduction clearly supports the intermediacy of a hexacoordinate silicon species. The stereochemical outcome at the hydroxy center is also consistent with a cyclic transition structure.

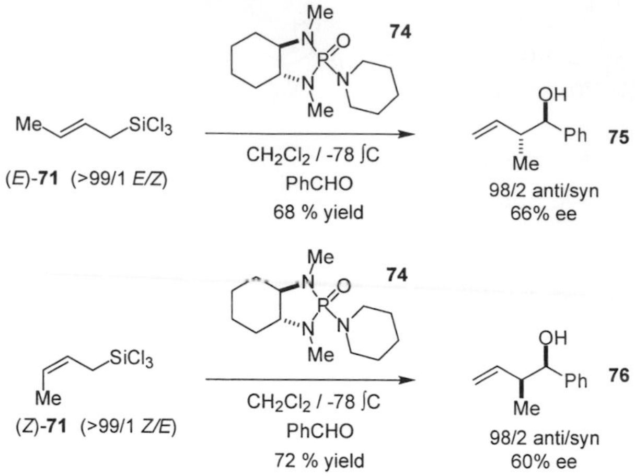

Scheme 10-31.

Since this initial report, several research groups have developed chiral promoters for this reaction such as the proline-derived phosphoramides **77** [56a, b], formamide **78** [56c, d], pyridinyl oxazoline **79**, [57] and a biisoquinoline *N,N'*-dioxide **80** [58] (Chart 10-2). Enantioselectivities as high as 92% have been achieved.

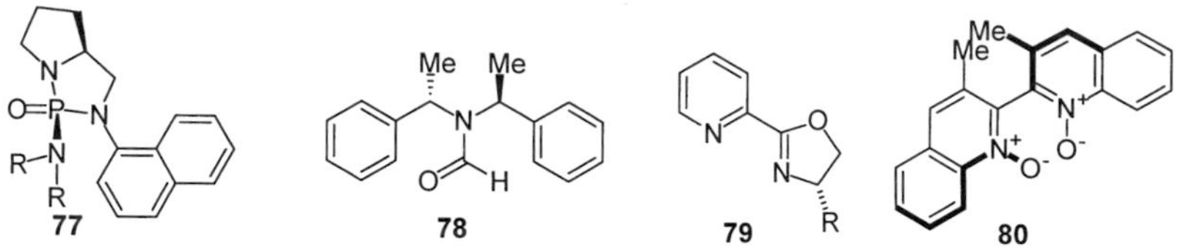

Chart 10-2.

A proposed transition structure for the chiral formamide-promoted reaction of 2-butenyltrichlorosilanes with aldehydes has been put forward by Iseki (Fig. 10-1) [56d]. The hexacoordinate silicon atom is bound to the oxygen atoms of the aldehyde, the catalyst, and HMPA. The chiral formamide-promoted reaction of allyltrichlorosilane with aliphatic aldehydes proceeds in high yield and excellent enantioselectivity to produce homoallylic alcohols. When either (*E*)-**71** or (*Z*)-**71** was used, the reactions with the same aldehydes were extremely sluggish (21 days). The diastereo- and enantioselectivity were quite high for (*E*)-**71** but less so for (*Z*)-**71**. This transition structure is not consistent with the author's observation of a nonlinear dependence on formamide enantiopurity.

The asymmetric allylation of achiral aldehydes with allyltrichlorosilane has also been promoted by biisoquinoline *N,N'*-dioxide and derivatives (Scheme 10-32) [58]. The reaction gave the highest yield and enantioselectivities when aromatic aldehydes were employed. For example, the biisoquinoline *N,N'*-dioxide (**80**) pro-

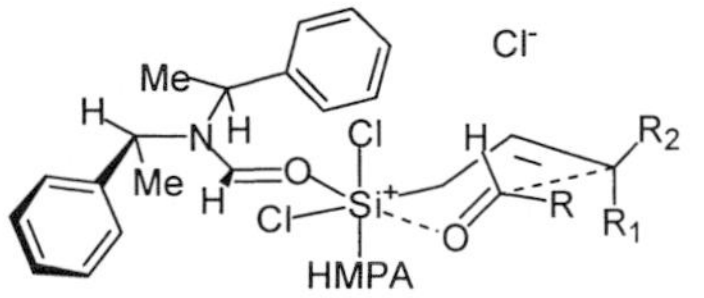

Figure 10-1. Cyclic chair-like transition structure.

moted reaction of benzaldehyde with allyltrichlorosilane afforded an 85% yield of the homoallylic alcohol in 88% ee (*R* configuration at the newly formed stereogenic center). The mechanistic profile of the reaction was examined by studying the allylations of benzaldehyde and (*E*)- and (*Z*)-**71**. The (*E*)-**71** provides exclusively the *anti* homoallylic alcohol, while (*Z*)-**71** affords the *syn* homoallylic alcohol. In each case, the same sense and magnitude of asymmetric induction at the hydroxy center is observed. The results suggest that the reaction proceeds through a six-membered, chair-like transition structure (Scheme 10-32).

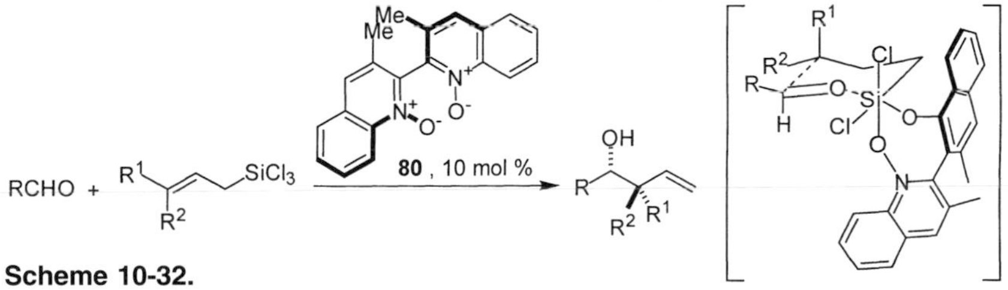

Scheme 10-32.

10.2.3 Pentacoordinated Siliconates

Stable pentacoordinated allylsiliconates have been employed in aldehyde addition reactions. These reagents require no activation by Lewis acids or Lewis bases, but have found only limited applications in synthesis to date. The use of these agents in addition to aldehydes was first described in 1987 by Corriu [59] and Hosomi [60] and by Kira and Sakurai [61] in 1988. In these reactions, the addition of a catechol or 2,2′-biphenol-derived allylsiliconate to an achiral aldehyde led to the highly regio- and stereoselective formation of homoallylic alcohols. For example, the addition of the catechol-derived 2-butenylsiliconate **81** (90/10 *E*/*Z*) provided a diastereomeric mixture of homoallylic alcohols **74** and **75** in a 90/10 ratio (Scheme 10-33) [60c].

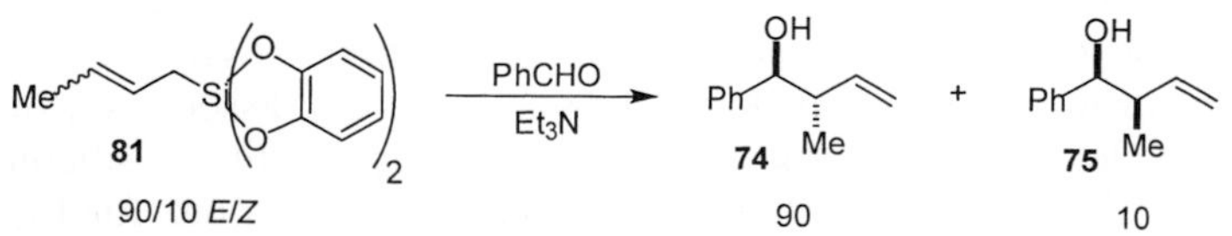

Scheme 10-33.

The high *anti* selectivity observed in the reaction of the *E*-2-butenylsiliconate suggests that the reaction proceeds through a cyclic six-membered transition structure (Scheme 10-34). The reaction of the allylsiliconate is highly sensitive to the solvent employed. The reaction proceeded smoothly in either CH_2Cl_2, $CHCl_3$, or EtOH. However, when a dipolar aprotic solvent such as DMF is used, the reaction does not give a high yield of the homoallylic alcohol. The dipolar aprotic solvents might interfere in the reaction by occupying a coordination site on the silicon atom before complexation with the carbonyl group of the aldehyde can occur, thus slowing down the reaction [60c].

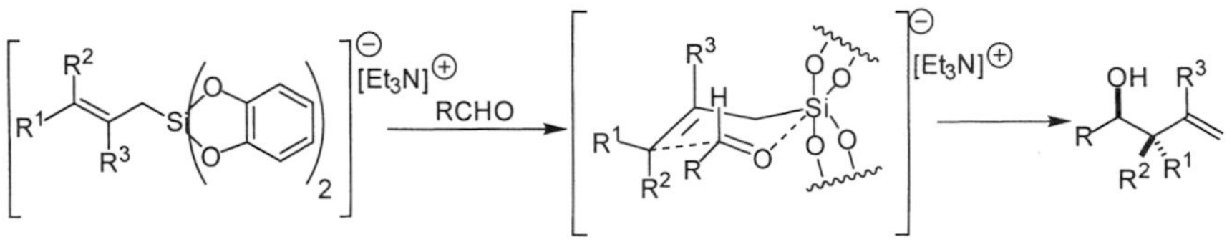

Scheme 10-34.

Enantiomerically pure allyl(triethoxy)silanes (**82**) react readily with aldehydes to provide homoallylic alcohols (Scheme 10-35) [62]. The γ-carbon of the allylsiliconate attacks the aldehyde on the same side of the allyl group (*syn* S_E'). The observation of high *syn*-relative diastereoselectivity and high *syn* S_E' (internal diastereoselectivity) is readily explained by a cyclic six-membered transition structure.

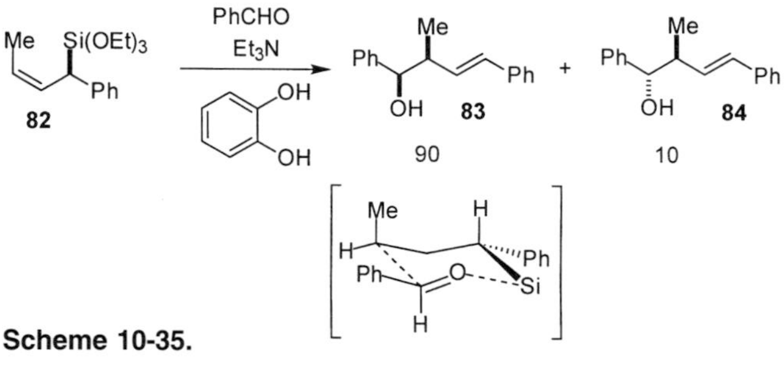

Scheme 10-35.

The reaction of highly strained allylsilacyclobutanes with aldehydes has recently been developed to produce homoallylic alcohols with a high degree of regio- and stereoselectivity (Scheme 10-36) [63]. These species are structurally akin to the allyltrialkylsilanes, but are more mechanistically aligned with the allyltrihalosilanes. The *E*-2-butenylsilacyclobutane upon reaction with an aldehyde at elevated temperature will produce almost exclusively the *anti* homoallylic alcohol. When the *Z*-2-butenylsilacyclobutane is used instead, the *syn* homoallylic alcohol is obtained. The mechanism proposed for the reaction involves the association of aldehyde and allylsilacyclobutane to form an activated pentacoordinate silicon complex. A closed, chair-like transition structure is proposed to account for the observed stereoselectivity in the reaction (Scheme 10-36). A theoretical examination has found that the reaction of the allylsilacyclobutane has a lower activation barrier than the reaction of allylsilane or methyl-substituted allylsilanes [64]. The study also found that the reaction will take place via a pentacoordinated silicon species where the oxygen of the aldehyde is located at the apical site of the silicon center, while the allyl group departs from the silicon center in the equatorial plane without causing any pseudorotation.

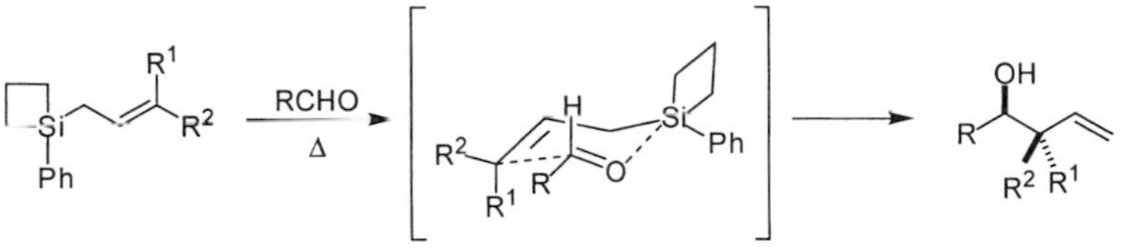

Scheme 10-36.

10.3 Allylic Tin Reagents

The allylation of aldehydes with an allyltin reagent was first reported in 1967 by König and Neumann in the thermally promoted addition of allyltrialkyltin reagents to generate homoallylic alcohols [65]. Servens and Pereyre determined that the reaction of activated aldehydes such as chloral with allyltin reagents could take place at much lower temperature [66]. The reaction of allyltin chlorides with aldehydes was demonstrated to proceed under mild reaction conditions by Tagliavini in 1977 [67]. The Lewis acid-promoted addition of allyltrialkyltin reagents to aldehydes was reported independently in 1979 by Naruta [68a] and by Sakurai and Hosomi [68b]. Since these initial reports, the synthetic utility of allyl- and 2-butenyltin reagents has increased dramatically [1]. The following section describes the mechanism and stereochemistry of the reaction of allyltin reagents with aldehydes and ketones.

10.3.1 Mechanism of Addition

10.3.1.1 Thermally promoted addition

The stereochemical course of the thermally promoted addition of allylic trialkylstannanes to aldehydes is dependent upon the geometry of the 2-butenyl unit [67, 69]. The reaction is believed to proceed via a cyclic, six-membered, chair-like transition structure. Reaction of an *E*-2-butenylstannane provides the *anti* homoallylic alcohol, while an *Z*-2-butenylstannane affords the corresponding *syn* homoallylic alcohol (Scheme 10-37). The allylation of aldehydes with allylic stannanes has also been performed under high pressure and neutral conditions [70]. The stereochemical outcome of the reaction of *E*- and *Z*-2-butenylstannanes with aldehydes under high pressure was almost identical to the results obtained thermally.

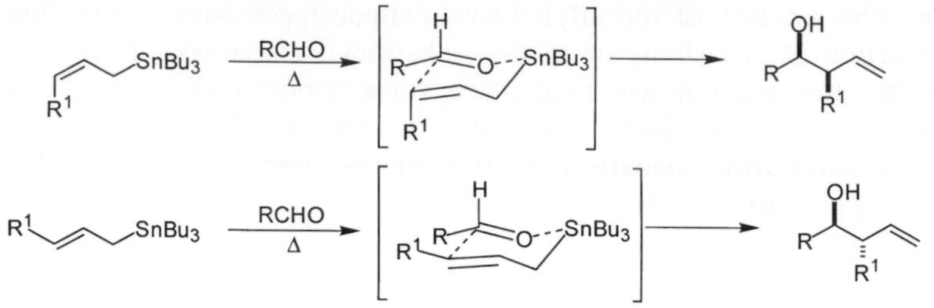

Scheme 10-37.

Essentially the same cyclic transition structures were proposed for the reaction performed under high pressure.

10.3.1.2 Lewis acid-promoted addition

The Lewis acid-promoted addition of 2-butenyltin reagents to aldehydes was reported by Yamamoto [71] in 1980 and has been extensively reviewed [1, 72]. In the initial studies, it was observed that the $BF_3{\cdot}OEt_2$-promoted reaction of 2-butenylstannanes (**85**) with achiral aldehydes afforded >90% of the *syn* homoallylic alcohol regardless of the geometry of the 2-butenyl unit. To explain the observed high diastereoselectivity, an acyclic transition structure was invoked wherein the double bonds are in an antiperiplanar arrangement (Scheme 10-38). Since the boron trifluoride is coordinated to the carbonyl oxygen to activate the addition, further association of the tin center to the oxygen is precluded. The structure of the Lewis acid-aldehyde complex is therefore believed to play a significant role in determining the stereochemical outcome of the reaction.

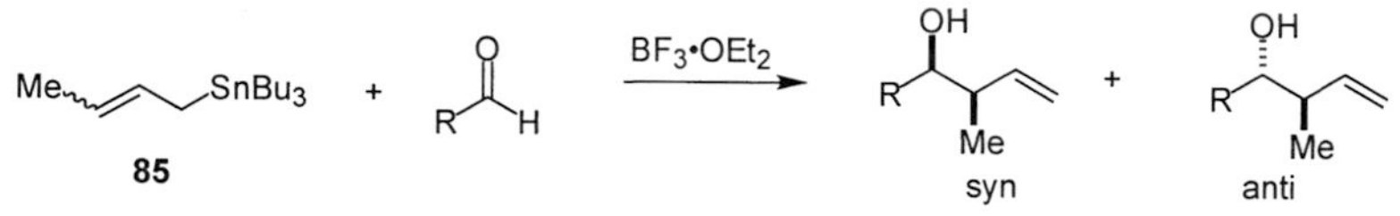

R		syn	anti
R = Ph	(*E*)-**85**	98	2
	(*Z*)-**85**	99	1
R = *i*-Pr	(*E*)-**85**	95	5

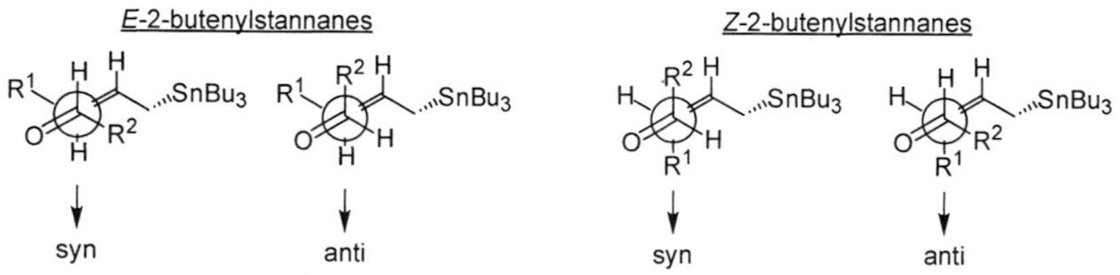

Scheme 10-38.

Model system **86** was designed to evaluate the relative importance of the synclinal vs antiperiplanar geometries (Scheme 10-39 and Table 10-6) [17a,b,c]. Reaction of **86** with various Lewis acids resulted in the predominant formation of the proximal alcohol (**6**). The selectivities obtained with model **86** do not correlate with the size of the Lewis acid employed, as previously observed with the allylsilane model **5a**. An early transition structure for these reactions is proposed to explain the insensitivity of the cyclization to the size of the Lewis acid-aldehyde complex. The following proposal for Type 2 reactions which proceed by direct addition was advanced: (1) there exists a preference for the synclinal orientation of double bonds and (2) the bulk of the Lewis acid-aldehyde complex and the stoichiometry of complexation are stereochemically significant [73].

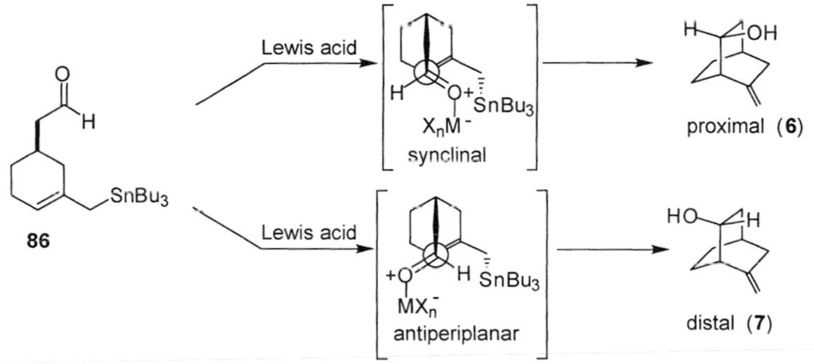

Scheme 10-39.

Table 10-6. Cyclization of allylstannane **86** with various Lewis acids [24a].

Lewis acid	Time, min	Proximal, %	Distal, %
$BF_3 \cdot OEt_2$	15	87	13
$ZrCl_4$	10	90	10
$SnCl_4$	5	93	7
$SiCl_4$	20	99	1
CF_3COOH	10	99	1

The relative disposition of the tin electrofuge (S_E' stereochemistry) has been established in the study of model system **87** [74]. The cyclization of compounds *u*-**87** and *l*-**87** with various reagents could afford the four diastereomeric alcohols (*E*)- and (*Z*)-**14-15** (Scheme 10-40). The alcohols (*E*)-**14** and (*Z*)-**14** result from reaction through a synclinal arrangement of double bonds in the transition structure. The alcohols (*E*)-**15** and (*Z*)-**15** result from reaction through an antiperiplanar arrangement of double bonds in the transition structure. The position of the deuterium can then be established in order to determine if the reaction proceeds through a *syn* or *anti* S_E' pathway.

Cyclization of *l*-**87** with the Lewis and Brønsted acids proceeds rapidly and with high selectivity for the proximal diastereomer (Table 10-7). The reactions also proceed via an *anti* S_E' pathway except when the cyclization was performed under thermal conditions. The thermolysis of *l*-**87** affords exclusively **14** which is indicative of a synclinal/*syn* S_E' pathway. The lower *anti* S_E' preference for $SnCl_4$ is suggested to arise from partial metathesis to an allyltrichlorostannyl species which might react via a closed transition structure. However, when 5 equivalents of $SnCl_4$ are employed, the *anti* selectivity does not change, indicating that metathesis is not a major pathway for the cyclization.

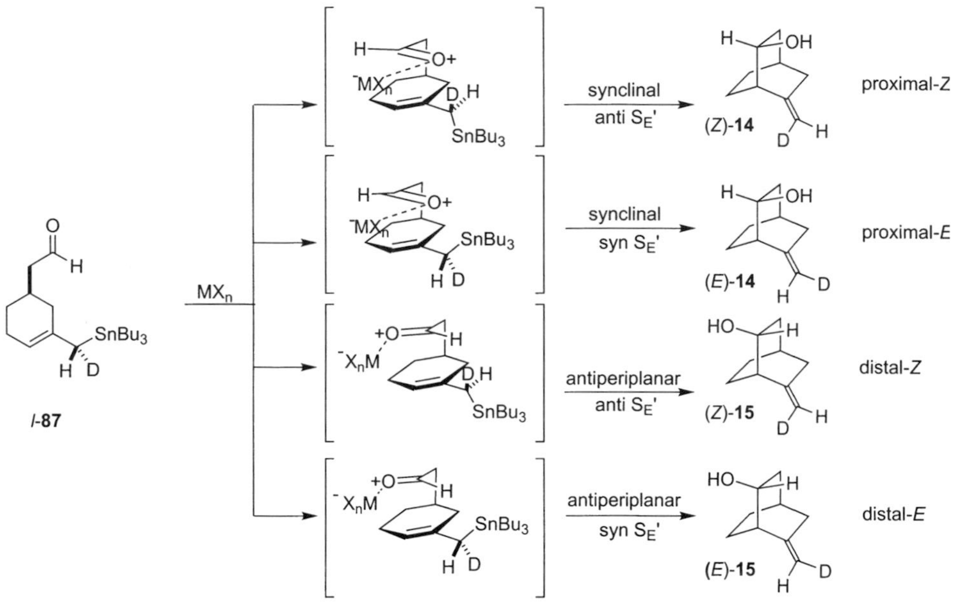

Scheme 10-40.

Table 10-7. Cyclization of model **87** [74].

reagent	Proximal/ distal (**14**/**15**)	**14** *Z/E*	**15** *Z/E*	Proximal *anti/syn* S_E' [a]	Distal % *anti* S_E' [a]
$TiCl_4$	88/12	89/11	95/5	94/6	>99/1
$SnCl_4$	94/6	86/14	95/5	91/9	>99/1
$BF_3 \cdot OEt_2$	86/14	92/8	95/5	97/3	>99/1
CF_3SO_3H	97/3	93/7		98/1	
CF_3CO_2H	>99/1	93/7		98/2	
CCl_3CO_2H	99/1	93/7		98/2	
n-$Bu_4N^+F^-$	>99/1	5/95		<1/99	

[a]) Percent *anti* S_E' based on 94.5% *d*-content in **87**.

Keck has also examined the Lewis acid-promoted addition of allylstannanes to aldehydes to elucidate the origin of stereocontrol in these reactions [75]. The diastereoselectivity observed in the reaction of 2-butenylstannanes with aldehydes is found to be dependent upon the aldehyde structure, stannane configuration and Lewis acid employed [75 a]. The results obtained from the $BF_3{\cdot}OEt_2$-promoted reaction of simple aldehydes with 2-butenylstannanes enriched in either the *Z* or *E* isomer are shown in Table 10-8. A significant correlation between the levels of diastereoselectivity observed in the addition reactions and the *E*/*Z* configuration of the 2-butenylstannane was established. The most dramatic change in diastereoselectivity is observed with cyclohexanecarboxaldehyde (entries 1–3, Table 10-8). This aldehyde was highly sensitive to 2-butenylstannane geometry, affording almost a 1/1 mixture of diastereomers when the *Z*-enriched 2-butenylstannane was employed.

Table 10-8. Stereoselectivity in reactions of *E*/*Z* 2-butenylstannanes with simple aldehydes [75 a].

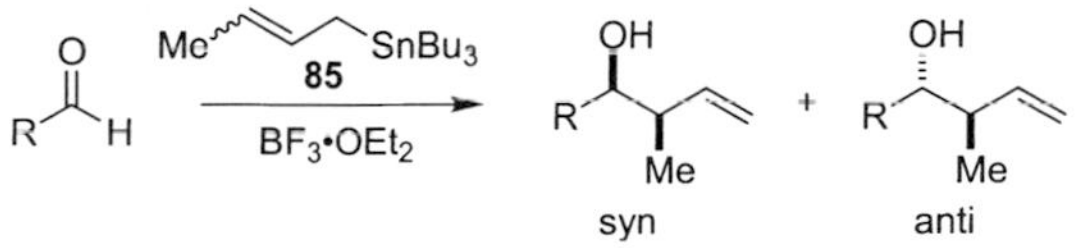

Entry	R	(*E*) /(*Z*)-**85**	*syn*	*anti*	Yield, %
1	*c*-Hex	90/10	94	6	88
2	*c*-Hex	74/26	86	14	80
3	*c*-Hex	12/88	58	42	82
4	Ph	90/10	98	2	85
5	Ph	74/26	95	5	86
6	Ph	12/88	81	19	80
7	PhCH=CH	90/10	98	2	81
8	PhCH=CH	74/26	92	8	86
9	PhCH=CH	12/88	82	18	82

Six potential transition structure arrays for the (*E*)- and (*Z*)-2-butenylstannanes were considered to explain the observed selectivity (Fig. 10-2). For (*E*)-**85**, transition structures E_2, E_3, and E_6 would appear to be favored based on steric arguments. The high *syn* selectivity observed with (*E*)-**85** is believed to result from a preference for the sterically unencumbered *syn*-synclinal arrangement E_2 over the other synclinal and antiperiplanar arrangements. No particular transition structure is favored upon inspection of the (*Z*)-**85**/aldehyde pairs. This is indeed observed empirically, as the addition reactions with the *Z*-2-butenylstannanes are much less selective than the corresponding *E*-stannanes.

The reaction of α-alkoxy and β-alkoxy aldehydes with 2-butenylstannanes enriched in either the *E*- or *Z*-isomer is very informative [75 a]. The results for the reaction of the α-alkoxy aldehyde (*R*)-**26** are shown in Table 10-9. Only the internal diastereoselectivity of the reaction is revealed by these results, as both **90** and

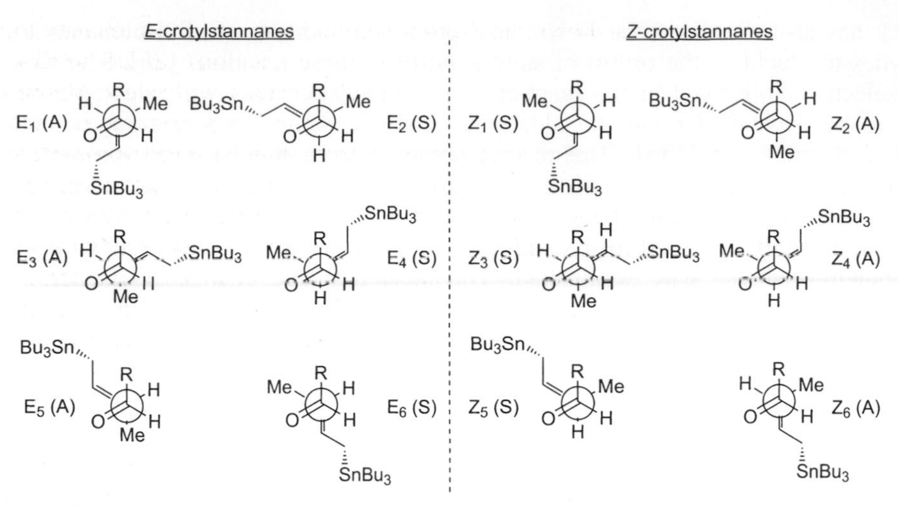

Figure 10-2. Staggered transition structures for reaction of *E*/*Z* 2-butenylstannanes with simple aldehydes.

91 were not formed in the reaction. In reactions of chelated structures, the energies of the synclinal and antiperiplanar transition structures must be quite close, as the preference for one transition structure over another can be changed by subtle changes in stannane substitution or changes in the Lewis acid employed. The *syn*-synclinal transition structure again appeared to be favored in the $MgBr_2$-promoted reactions of (*E*)-**85** with aldehydes, while weaker Lewis acids appeared to prefer the antiperiplanar arrays.

A thorough mechanistic study of the intramolecular allylstannane-aldehyde cyclization has recently been reported [75b]. The intramolecular cyclization substrates described allow the double bond geometry of the allylstannane to be varied. The Lewis acid-promoted cyclization of either the *E*- or *Z*-stannane can lead to the four limiting stereoisomeric products (Scheme 10-41).

Table 10-9. Stereoselectivity in reactions of *E*/*Z* 2-butenylstannanes with α-alkoxy aldehydes [75a].

(*R*)-**26** + **85** ($MgBr_2 \cdot OEt_2$) → **88** + **89** + **90** + **91**

(*E*)/(*Z*)-**85**	**88**	**89**	Yield, %
90/10	92.5	7.5	84
76/24	91.2	8.8	79
12/88	78.6	21.4	82

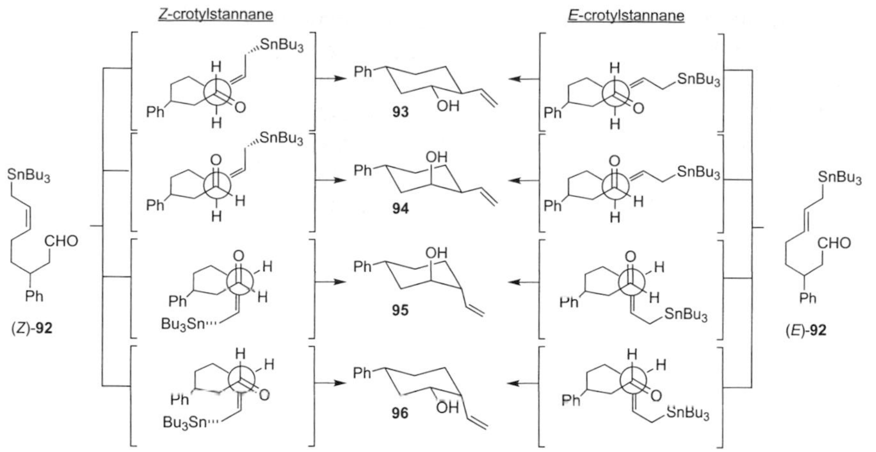

Scheme 10-41.

The cyclization results for the (*E*)- and (*Z*)-**92** are shown in Table 10-10. The cyclization of (*Z*)-**92** proceeds largely through the *syn*-synclinal transition structure providing product **94**. The predominance of **94** can be rationalized using FMO theory. When the allylstannane is oriented in the *syn*-synclinal geometry a possibility for secondary orbital overlap exists between the oxygen of the carbonyl and the stannyl methylene carbon. The antiperiplanar and the remaining synclinal transition structures would not benefit from any orbital stabilization. The cyclization of (*E*)-**92** can also afford the same four products through the transition structures shown in Scheme 10-41. In this reaction, two different transition structures (those leading to products **93** and **94**) could benefit from secondary orbital overlap. Steric considerations would favor the formation of **93** with exclusively equatorial substitution. This is indeed the case with the Brønsted Lewis acids; however, as the size of the Lewis acid increases, the selectivity for **93** decreases.

As part of this study, the intramolecular cyclization of a β-benzyloxy-substituted aldehyde was also examined [75b]. Similar results are obtained with the reaction of *E*- and *Z*-2-butenylstannanes with that acceptor. A rationale for the observed stereoselectivity was put forward based on secondary orbital overlap.

Table 10-10. Cyclization results for the (*E*)- or (*Z*)-**92** [75b].

(*Z*)-**92**					(*E*)-**92**				
Reagent	**93**	**94**	**95**	**96**	Reagent	**93**	**94**	**95**	**96**
CF_3CO_2H	4	96	0	0	CF_3CO_2H	81	6	1	8
$BF_3{\cdot}OEt_2$	15	85	<1	<1	$BF_3{\cdot}OEt_2$	60	24	6	6
$MgBr_2{\cdot}OEt_2$	15	85	0	0	$MgBr_2{\cdot}OEt_2$	55	32	2	11
$TiCl_4$	5	91	~1	<1	$TiCl_4$	35	39	18	8
$SnCl_4$	10	90	<1	<1	$SnCl_4$	29	6	51	14
Thermal	5	95	0	0	Thermal	9	80	2	9

These results do not provide suitable models or predictions for the intermolecular addition of 2-butenylstannanes to aldehydes. The *E*-2-butenylstannanes generally provide higher levels of stereoselectivity than the corresponding *Z*-2-butenylstannanes in intermolecular reactions. However, the hypothesis of secondary orbital overlap influencing stereoselectivity could also be applied to the intermolecular reactions.

10.3.1.3 Lewis base-promoted addition

There are very few examples of Lewis base-promoted allylations of aldehydes with allylstannanes. In 1992 Baba disclosed an intriguing method for allylation of aldehydes with allyl- and 2-butenyltributylstannanes in the presence of catalytic amounts of dibutyltin dichloride and certain coactivators such as tetrabutylammonium iodide, tributylphosphine oxide, HMPA or tetraphenylphosphonium iodide [76]. No definitive mechanistic information is available on the role of the co-activators; the authors speculate that the ligands accelerate the metathesis to form allyldibutyltin chloride which is the actual nucleophile. The same group has recently reported the use of a lead(II) iodide/HMPA catalyst for the allylation of α,β-epoxyketones [76b].

In their initial disclosure on the chiral phosphoramide-catalyzed allylation of aldehydes with allyltrichlorosilanes, Denmark also reported that allyltrichlorostannanes were effective, albeit in significantly lower yield and selectivity (cf. Scheme 10-31; 51% yield, 16% ee) [55].

10.3.1.4 Transmetallation

The transmetallation or the metal-metal exchange reaction of an allylic tin species with an electrophile was first observed in 1970 [77]. The possibility that transmetallation may play a role in the Lewis acid-promoted reaction of allylstannanes with aldehydes was initially discussed by Tagliavini [78], Keck [79], Yamamoto [71b, 80], and Maruyama [81]. It is believed that upon transmetallation with either $SnCl_4$ or $TiCl_4$, the addition of an allylstannane and aldehyde will occur via a cyclic, six-membered transition structure. The reaction occurs by coordination of the aldehyde carbonyl with the Lewis acidic trichlorotin or trichlorotitanium reagent, thus affording the *anti* homoallylic alcohol (Scheme 10-42).

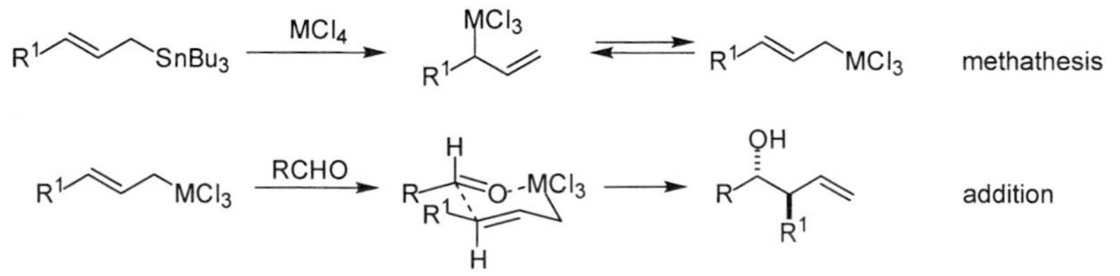

Scheme 10-42.

Denmark has spectroscopically examined the reaction of both allyl- and 2-butenylstannanes with aldehydes using the Lewis acids $SnCl_4$ and $BF_3 \cdot OEt_2$ [73, 82]. First, the metathesis of both allyltributylstannane and tetraallyltin with $SnCl_4$ was determined (by ^{13}C NMR spectroscopy) to be instantaneous at –80 °C. The reaction of allyltributylstannane with a complexed aldehyde was determined to be significantly more complicated. When a molar equivalent of $SnCl_4$ per aldehyde was employed, metathesis was determined to be the preferred pathway for aldehydes. When one half a molar equivalent of $SnCl_4$ per aldehyde is used, the reaction pathways and product distribution become very sensitive to both the aldehyde structure and addition order. A spectrum of mechanistic pathways was documented ranging from direct addition (acetaldehyde) to complete metathesis (pivalaldehyde) to a competitive addition and metathesis (4-*t*-butylbenzaldehyde). The results obtained with a molar equivalent of $SnCl_4$ are most relevant, as this reagent stoichiometry is most commonly used in the addition reactions.

Keck has also spectroscopically examined the mechanism of the $SnCl_4$-promoted additions of allylstannanes to aldehydes [83]. In this report, the $SnCl_4$-promoted reaction of two β-benzyloxyaldehydes (**52** and **97**) and a β-siloxyaldehyde (**98**) with allyltributylstannane was studied (Chart 10-3). These authors also determined that the pathways are extremely sensitive to aldehyde structure and the stoichiometry of $SnCl_4$ employed in the reaction. When one equivalent of $SnCl_4$ is used, the reactions did not proceed via a transmetallation pathway except for the β-benzyloxyaldehyde **97**. A 1/1 mixture of $SnCl_4$ and aldehyde **97** gives three signals by ^{119}Sn NMR corresponding to free $SnCl_4$, a 2/1 aldehyde/$SnCl_4$ complex, and a bidentate chelate. In the reaction of aldehyde **97**, transmetallation with $SnCl_4$ is virtually instantaneous, thus the reaction of aldehyde **97** likely occurs via the transmetallated species. According to the authors, transmetallation is a phenomenon that depends critically upon the stability of the complexed or chelated intermediates.

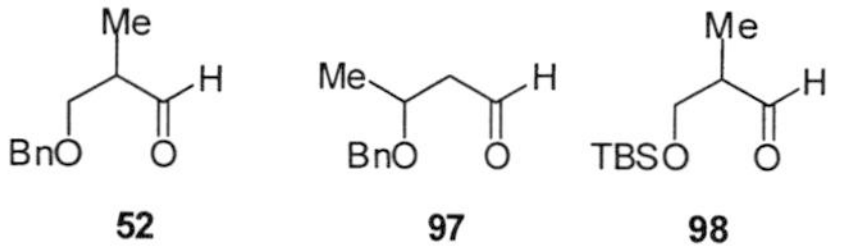

Me H BnO O **52** Me H BnO O **97** Me H TBSO O **98**

Chart 10-3.

10.3.2 Stereochemical Course of Addition

10.3.2.1 Allylic trialkylstannanes

Internal stereoselection

The Lewis acid-promoted addition of allylic trialkylstannanes to achiral aldehydes has been demonstrated to provide *syn* homoallylic alcohols in high yield [71 a, b]. The relative stereochemical outcome of the reaction of simple aldehydes with 2-butenylstannanes is discussed earlier in this Chapter. Studies on the addition of al-

allylstannanes to produce homoallylic alcohols in good yield and moderate to high regio- and enantioselectivity (Scheme 10-47).

This group has recently reported the CAB-promoted additions of 2-butenylstannanes to branched and chiral aldehydes [88]. The reactions were optimized using cyclohexanecarboxaldehyde and (*E*)-**85** in the presence of the CAB catalyst and trifluoroacetic anhydride. Under optimized conditions, the reaction affords a 92/8 mixture of *syn*/*anti* homoallylic alcohols in 71% yield and the *syn* diastereomer is obtained in 93% ee.

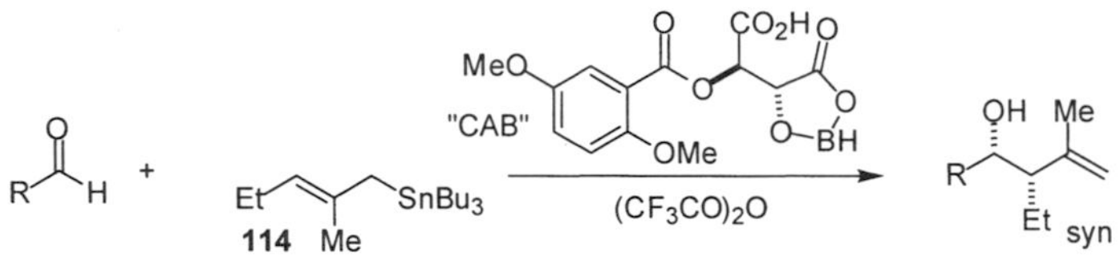

R	yield, %	syn : anti	ee, %
Ph	99	90 : 10	85
BuC≡C	92	71 : 29	70
n-C_3H_7	61	97 : 3	81
i-C_3H_7	44	94 : 6	85

Scheme 10-47.

In a direct comparison, the CAB catalyst proved to be superior to the titanium binol catalyst in the reaction of branched aldehydes in both reaction times and selectivity. Finally, the CAB-promoted reaction of a chiral aldehyde with (*E*)-**85** was examined (Scheme 10-48). When the aldehyde (*R*)-**106** reacts with the 2-butenylstannane in the presence of the CAB catalyst, a 98/2 mixture of diastereomers is obtained. The CAB-promoted reaction of the aldehyde (*S*)-**106** with the (*E*)-**85** affords a 90/10 mixture of diastereomers. This adduct is the minor isomer from the BF_3-promoted addition of (*E*)-**85** to aldehyde (*S*)-**106**. Thus, the CAB-promoted additions are strongly reagent controlled, essentially overriding the intrinsic facial preference of the aldehyde substrate.

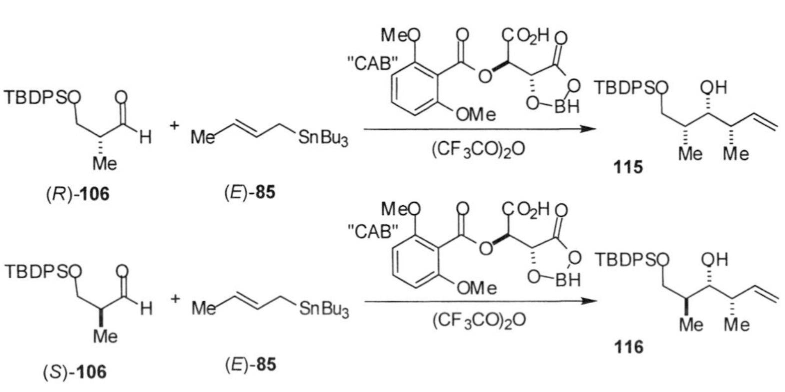

Scheme 10-48.

Keck [89a–c], Tagliavini [89d,e], and Yu [89f] have extensively studied the BINOL-Ti- or binol-Zr promoted reactions of achiral aldehydes with allylstannanes. The initial studies employed BINOL and either $Ti(Oi\text{-}Pr)_4$ or $TiCl_2(Oi\text{-}Pr)_2$ as the Lewis acid promoter in the reaction of achiral aldehydes with allyltributylstannane. The reaction affords good yields of the desired homoallylic alcohol with a high degree of enantioselectivity even with as little as 10 mol% of the chiral catalyst (Scheme 10-49) [89a]. The rate and turnover of the catalytic, asymmetric allylation reaction have also been optimized. It was found that when *i*-$PrSSiMe_3$ is added to the reaction, a rate acceleration occurs, allowing as little as 1–2% of the catalyst to be used [89f].

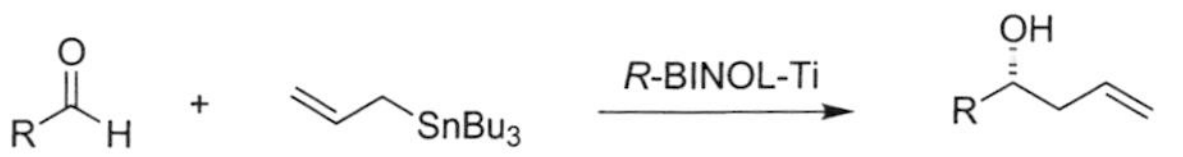

RCHO + allyl-$SnBu_3$ → (*R*-BINOL-Ti) → R–CH(OH)–$CH_2CH=CH_2$

R	yield, %	ee, %
Ph	88	95
c-Hex	66	94
furyl	89	96
i-Pr	73	96

Scheme 10-49.

The reaction of methallyltri-*n*-butylstannane **117** with achiral aldehydes is also effectively promoted by the binol-Ti complex [89c]. In all but one case (cyclohexanecarboxaldehyde), the yields and enantioselectivities observed with the methallylstannane are identical or higher than those obtained in the reactions with allyltributylstannane with only 10 mol% of the binol-Ti complex (Scheme 10-50). Insight into the nature of the titanium catalyst is provided by the observation of asymmetric amplification [89b] and chiral poisoning [89g]. An intruiging hypothesis on the origin of enantioselection in allylation and related reactions [89h].

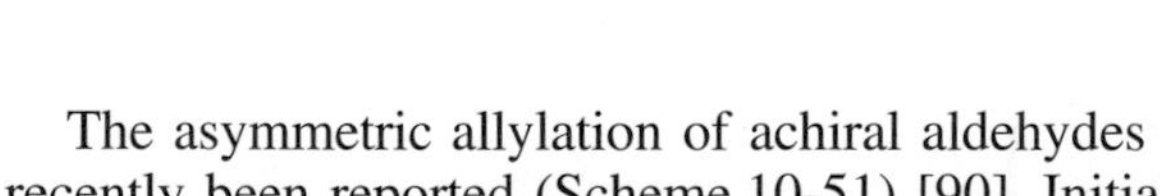

RCHO + methallyl-$SnBu_3$ (**117**) → (*R*-BINOL-Ti) → R–CH(OH)–$CH_2C(Me)=CH_2$

R	yield, %	ee, %
Ph	95	96
c-Hex	50	84
furyl	99	88
$PhCH_2CH_2$	97	98

Scheme 10-50.

The asymmetric allylation of achiral aldehydes with a novel silver complex has recently been reported (Scheme 10-51) [90]. Initially, it was shown that the silver-promoted reaction of allyltributylstannane with benzaldehyde could be accelerated by triphenylphosphine. A survey of various chiral phosphine reagents and silver salts identified the combination of binap and AgOTf as optimal. The reaction of benzaldehyde and allyltributylstannane promoted by 5 mol% of the binap·AgOTf catalyst affords an 88% yield of the homoallylic alcohol in 96% ee. Although the mechanism has not yet been elucidated, it is believed that the binap·Ag complex acts as a chiral Lewis acid catalyst rather than an allylsilver reagent. The aliphatic aldehyde afforded the lowest yield and enantioselectivity.

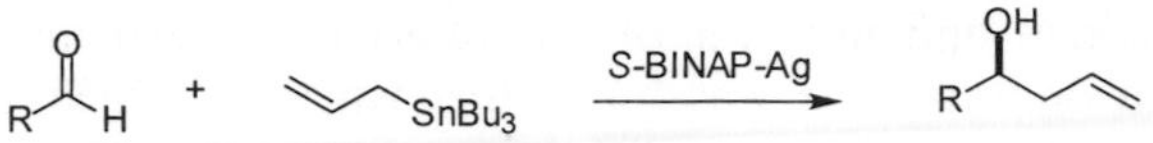

S-BINAP-Ag

R	yield, %	ee, %
Ph	88	96
furyl	94	93
$PhCH_2CH_2$	47	88

Scheme 10-51.

The reaction of 2-butenyltributylstannane with benzaldehyde using the binap·Ag complex has also been reported [90b]. When 20 mol% of the binap·Ag complex was employed, a 45% yield of the homoallylic alcohols is obtained (85/15, *anti*/*syn*). The *syn*/*anti* ratio of homoallylic alcohols is found to be independent of the *E*/*Z* ratio of the starting 2-butenylstannane. These results are especially surprising as they are contrary to the general behavior of 2-butenylstannanes, whose reactions are classified as Type 2, i.e. to afford mainly the *syn* homoallylic alcohols independently of olefin geometry. Several transition structure models were considered in order to help explain the *anti* selectivity observed with the binap·Ag complex (Scheme 10-52). It is possible that the antiperiplanar transition structure shown leads to the observed *anti* homoallylic alcohol. Reaction through this transition structure may be favored because of the steric requirements of the large binap·Ag complex. Cyclic transition structures are also proposed which may arise via transmetallation of the 2-butenylstannane to a 2-butenylsilver reagent. No transmetallation of the stannane is observed before aldehyde addition; therefore the Lewis acid mechanism appears to be preferred, although the recovered 2-butenylstannane had slightly isomerized during the reaction.

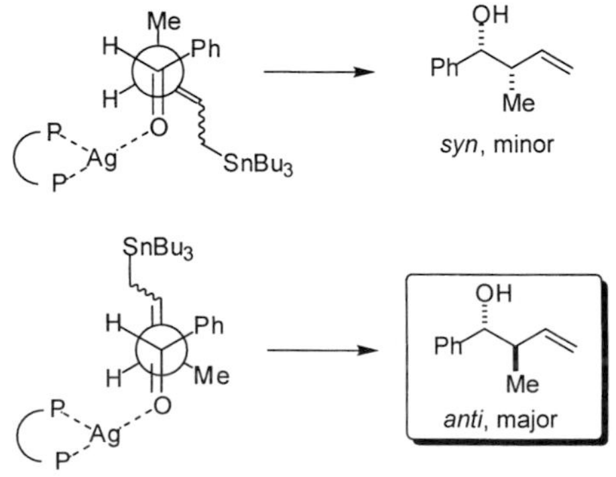

syn, minor

anti, major

Scheme 10-52.

Chiral rhodium catalyst **118**, pioneered by Nishiyama, has been put to use in the addition of allyltributylstannane to achiral aldehydes [91]. This catalyst is relatively insensitive to water and can even be purified by silica gel chromatography. The optimized allylation conditions employ 1 equiv of the aldehyde, 1.5 equiv of allyltributylstannane, and 5 mol% of **118** (Scheme 10-53). The reactions with many different aldehydes can all be performed at room temperature to provide good yields of the desired homoallylic alcohols albeit in moderate to poor enantioselectivity.

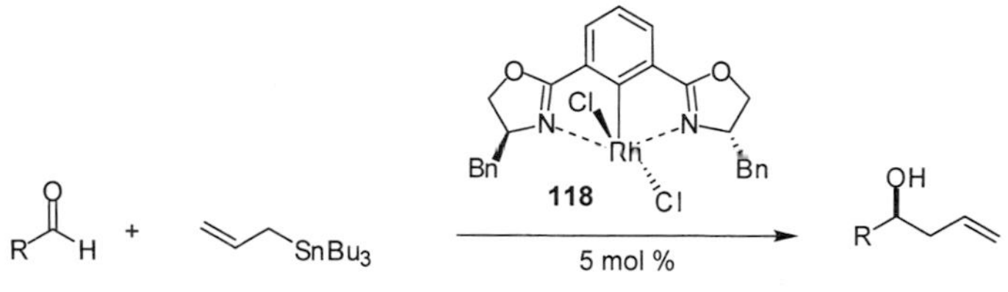

R	yield, %	ee, %
Ph	88	61
furyl	94	58
$PhCH_2CH_2$	84	63

Scheme 10-53.

10.3.2.2 Allylic trihalostannanes

Relative stereoselection

The diastereoselectivity associated with the reaction of (*E*)- and (*Z*)-2-butenyltrichlorostannanes with simple aldehydes was discussed in Section 3.1.4 on metathesis. This transformation is complicated by the generation of both stereo- and regioisomeric reagents by combination of $SnCl_4$ with the 2-butenylstannane. Much greater success has been reported with allyltrichlorostannane itself as well as with substituted trichlorostannanes bearing internal coordinating groups.

Internal stereoselection

In all the following studies, the allylic trihalostannanes are generated in situ by transmetallation with $SnCl_4$. Accordingly, for all but allyltrichlorostannane (**119**) itself, the intermediacy of the trihalostannane is (reasonably) implied but the actual structure of the reactive species is rarely established.

Achiral allylic stannanes and chiral aldehydes. Thomas has pioneered the use of allylic trichlorostannanes in various combinations of chiral substrates and reagents [92]. The simplest illustration is the reaction of in-situ generated **119** with α- and β-alkoxy aldehydes (Scheme 10-54). Not surprisingly the reactions of (*S*)-**26** and (*S*)-**52** afford products primarily of chelation control [93].

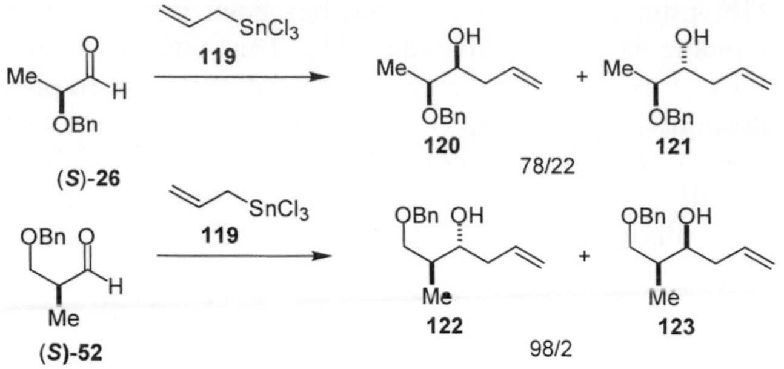

(*S*)-**26** + **119** ($SnCl_3$) → **120** + **121** (78/22)

(*S*)-**52** + **119** ($SnCl_3$) → **122** + **123** (98/2)

Scheme 10-54.

Chiral allylstannanes and achiral aldehydes. To address the problem of rapid isomerization and configurational inhomogeneity of the allylic trichlorostannane unit, the use of remote heteroatom substituents to facilitate the transmetallation and to stabilize the resulting transmetallated intermediate has been extensively investigated. Thus, treatment of enantiomerically pure alkoxystannane (*S*)-**124** with $SnCl_4$ followed by a variety of different aldehydes affords the addition products in good yield and high stereoselectivity [94]. In general, the *Z*-homoallylic alcohol is formed exclusively with a high preference for the 1,5-*syn* configuration (Scheme 10-55). A number of other Lewis acids have been investigated among which only $SnBr_4$ proves comparable in yield and stereoselectivity. In addition, other protective groups such as the PMB and MOM groups show similar stability and directing effects as compared with the benzyl group.

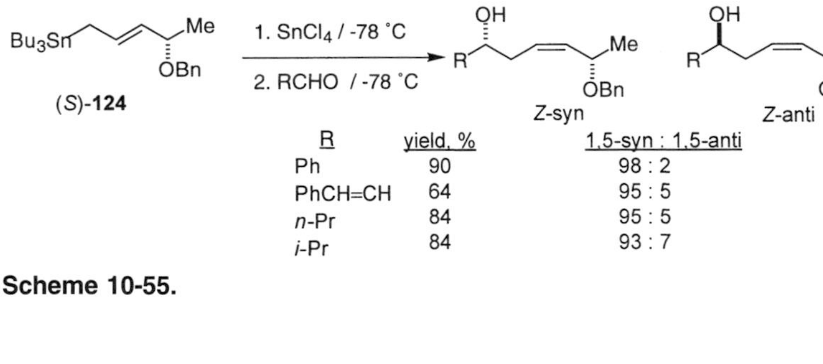

(*S*)-**124** → 1. $SnCl_4$ / -78 °C; 2. RCHO / -78 °C → *Z*-syn + *Z*-anti

R	yield, %	1,5-syn : 1,5-anti
Ph	90	98 : 2
PhCH=CH	64	95 : 5
n-Pr	84	95 : 5
i-Pr	84	93 : 7

Scheme 10-55.

The preferential formation of *Z*-homoallylic alcohols and the significant 1,5-stereoinduction have been rationalized in terms of a directed, stereoselective transmetallation to form oxastannacyclobutane intermediate **x** followed by reaction via a closed, six-membered ring transition structure to form trichlorostannyl ether **xi**, which after hydrolysis affords the observed homoallylic alcohol (Scheme 10-56). While neither the intermediacy of **x** nor the origin of its selective formation have been established, similar four-membered ring structures have been reported [95].

This work has been extended to include substrates wherein the coordinating heteroatom is further removed. Thus, allylic stannane (*R*)-**125**, when subjected to

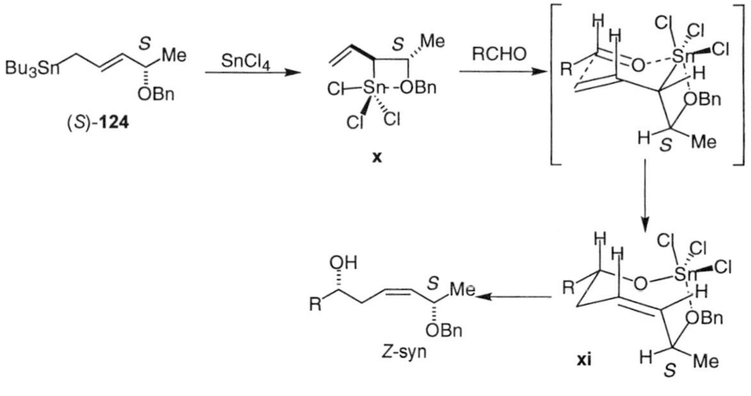

Scheme 10-56.

transmetallation with $SnCl_4$ followed by reaction with a series of aldehydes, gives rise to *Z*-homoallylic alcohols with excellent 1,5 *anti* induction (Scheme 10-57) [96]. The origin of the *anti* diastereoselection is rationalized in terms of the metathesis of **125** to form allylic trichlorostannane (**xii**) which reacts with aldehydes via a closed, chairlike transition structure.

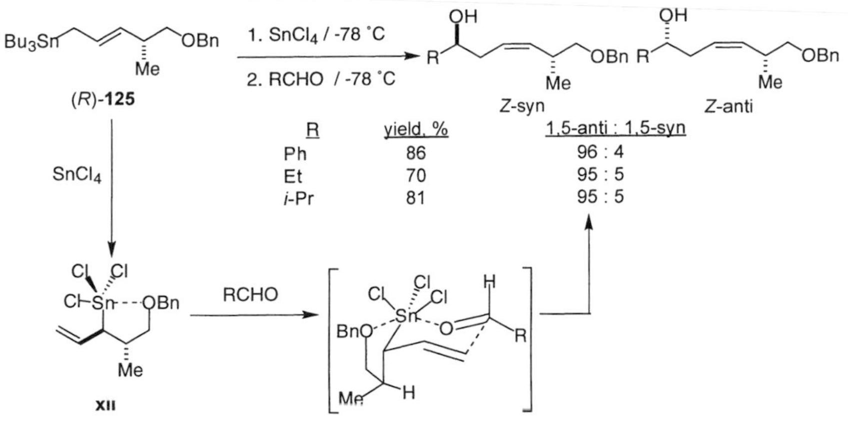

R	yield, %	1,5-anti : 1,5-syn
Ph	86	96 : 4
Et	70	95 : 5
i-Pr	81	95 : 5

Scheme 10-57.

Extension to higher homologs **126** [97a] and **129** [97b] has been detailed recently as well. The selectivities observed with these reagents (1,6-*syn* and 1,7-*syn*) are also impressive and can be reasonably well rationalized by conformational analysis of the intermediate allylic trichlorostannyl species (Scheme 10-58).

Chiral allylic stannanes and chiral aldehydes. Pairwise combination of these chiral allylstannane reagents with chiral *α*- and *β*-alkoxy aldehydes revealed a

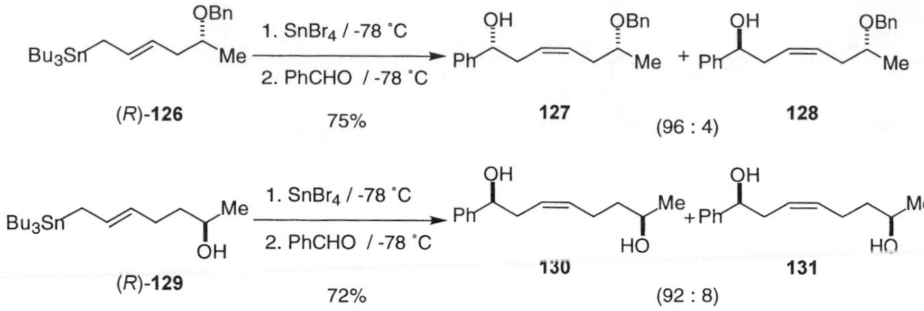

Scheme 10-58.

strong reagent-controlled selectivity for virtually all additions (Scheme 10-59). Thus, either (*R*)- or (*S*)-**52**, (*R*)- or (*S*)-**97**, and (*S*)-**26** reacted selectively at the *Re* face of the aldehyde with [(*S*)-**124**/$SnCl_4$]. Only (*R*)-**26** failed to react selectively [93 a]. Other reagents, (*R*)-**125** and (*R*)-**126**, also reacted with high facial selectivity with (*R*)- or (*S*)-**26** [96].

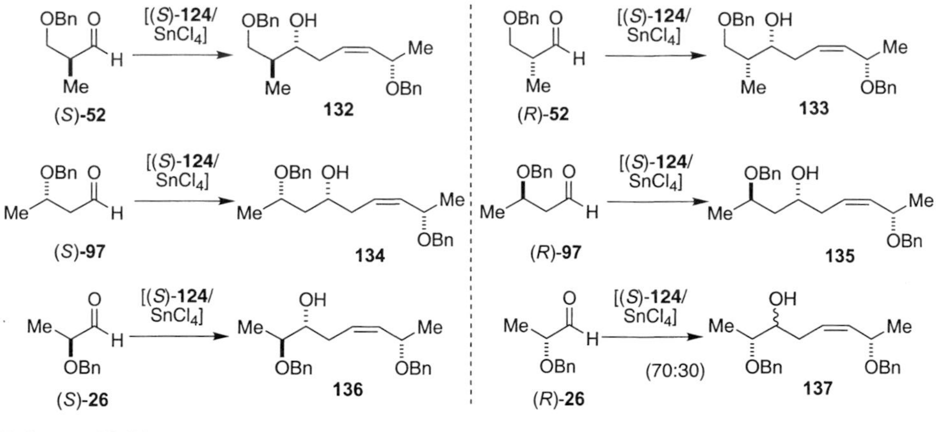

Scheme 10-59.

10.3.2.3 Heteroatom-substituted allylic stannanes

Relative stereoselection

The preparation of vicinal polyol triads requires the placement of oxygen functionality at the γ-position of the allylic stannane. The Lewis acid-promoted reaction of γ-alkoxyallylstannanes with achiral aldehydes was first reported by Koreeda (Scheme 10-60) [98]. The reactions proceed in moderate to high yield and with good diastereoselectivity to produce the homoallylic alcohols. As in the case of simple (*E*)- and (*Z*)-**138**, the reactions are stereoconvergent giving rise to predominantly the *syn* diastereomer independently of olefin geometry. It was speculated that the reaction proceeds via an acyclic transition structure.

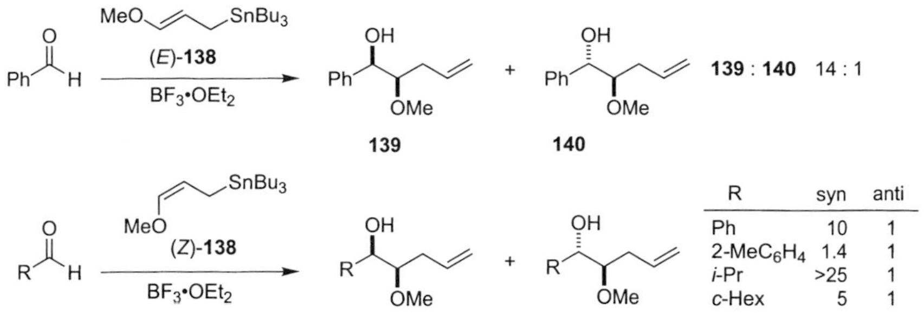

139 : 140 14 : 1

R	syn	anti
Ph	10	1
2-MeC_6H_4	1.4	1
i-Pr	>25	1
c-Hex	5	1

Scheme 10-60.

Internal stereoselection

Keck [99] first disclosed the Lewis acid-promoted reaction of a *γ*-siloxy or alkoxyallylstannane with either *α*- or *β*-alkoxy aldehydes (Scheme 10-61). Reaction of the *α*-alkoxy aldehyde **141** with the allylstannane **142** affords the homoallylic alcohol **143** as the only product. Reaction of the *β*-alkoxy aldehyde **52** with the allylstannane **142** also proceeds in high diastereoselectivity (~50/1) to produce the homoallylic alcohol **144**. The stereochemical outcome of these reactions is consistent with a chelation-control mechanism.

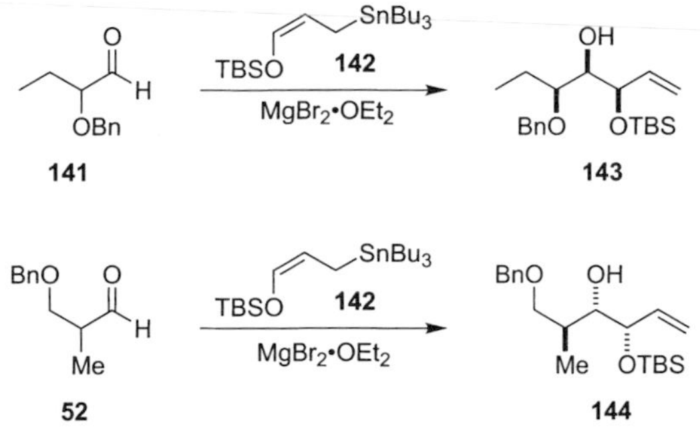

Scheme 10-61.

The thermally promoted reaction of an enantiomerically pure *α*-alkoxyallylstannane with achiral aldehydes was first reported by Thomas in 1984 [100]. The *α*-alkoxyallylstannane **145** (prepared from menthol and a racemic stannol) is heated with the aldehyde at 130 °C to produce the homoallylic alcohol **146** as a single diastereomer in good yield (Scheme 10-62). Chairlike, six-membered transition structures can account for the observed diastereoselectivity in these reactions. The *α*-alkoxy group prefers to adopt an axial position in the transition structure, ensuring that the diastereomers **145** and **147** react selectively with the aldehyde from only one face of the carbonyl group.

Following this report, Marshall disclosed the $BF_3 \cdot OEt_2$-promoted addition of enantiomerically enriched *α*-alkoxyallylstannanes to achiral aldehydes in 1989 [101].

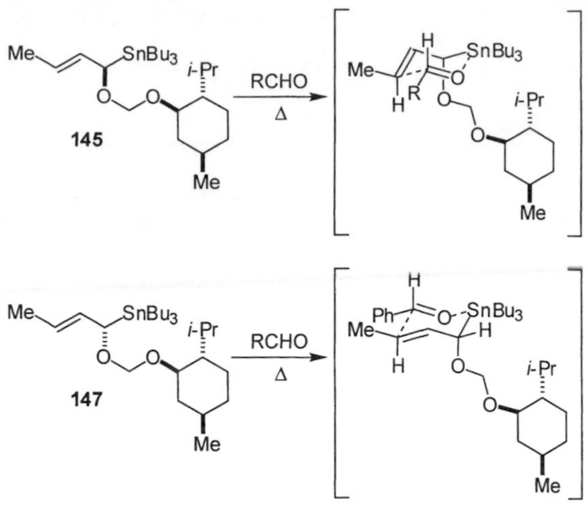

Scheme 10-62.

The reaction afforded the *syn* homoallylic alcohol highly selectively, although both the *E*- and *Z*-olefins were produced in the reaction (Scheme 10-63).

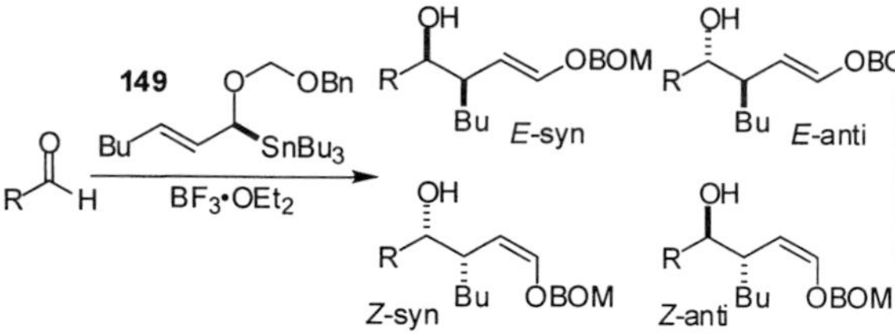

R	*E*-syn	*E*-anti	*Z*-syn	*Z*-anti
n-Hex	70	<1	27	3
E-BuCH=CH	80	1	17	2
BuC≡C	51	7	25	17

Scheme 10-63.

The stereochemical outcome of the addition of α-alkoxyallylstannanes to achiral aldehydes is consistent with an acyclic transition structure [101]. The antiperiplanar arrangements **xiii** and **xiv** shown in Fig. 10-3 are proposed to account for the observed selectivity. Synclinal transition structures are dismissed on the belief that minimization of the steric interactions between the stannane and aldehyde substituents Bu and R could best be accommodated by the antiperiplanar transition structures shown.

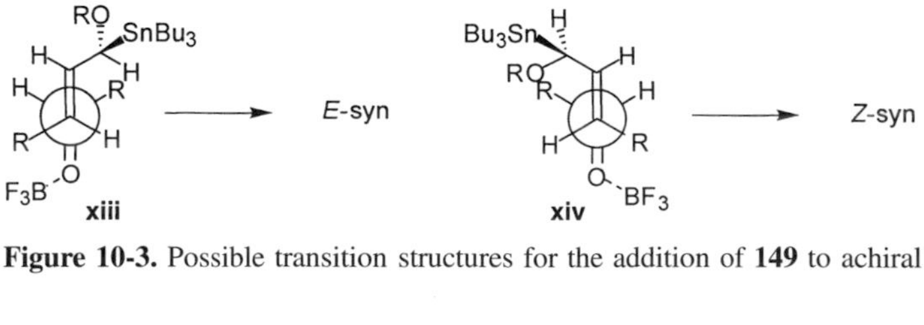

Figure 10-3. Possible transition structures for the addition of **149** to achiral aldehydes.

The Lewis acid-promoted addition of γ-alkoxyallylstannanes to achiral aldehydes was shown to proceed in high diastereoselectivity by an *anti* S_E' pathway (Scheme 10-64) [102]. An acyclic antiperiplanar transition structure (very similar to that shown in Fig. 10-3) was proposed to rationalize the stereoselectivity. When enantiomerically enriched allylstannanes are employed, the reaction proceeds in high enantio- and diastereoselectivity to give the homoallylic alcohols [102b]. The stereochemical outcome of the reaction is consistent with an *anti* S_E' pathway.

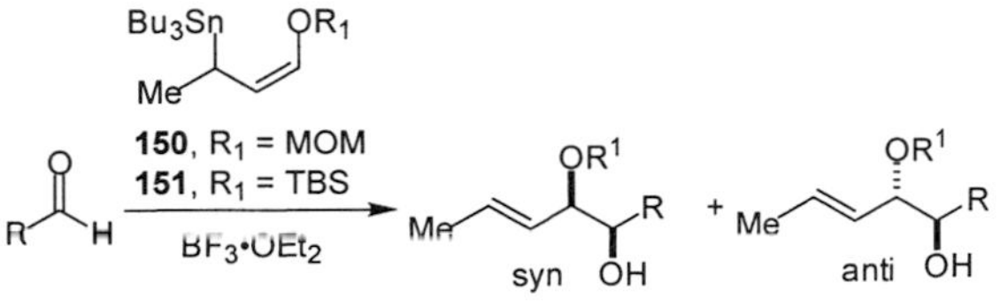

R	R_1	syn	anti
n-Hex	TBS	97	3
BuC≡C	TBS	93	7
c-Hex	MOM	95	5

Scheme 10-64.

The $BF_3{\cdot}OEt_2$-promoted reaction of the enantiomerically enriched MOM-protected alkoxystannane (*S*)-**150** with the aldehyde (*S*)-**26** (mismatched series) affords a mixture of *syn* and *anti* homoallylic alcohols **152** and **153** (Scheme 10-65) [103]. The matched series gives much higher selectivity for the *syn* homoallylic alcohol. Reaction of the TBS-protected alkoxystannane (*S*)-**151** with the aldehyde (*S*)-**26** provided the *syn* homoallylic alcohol **154** and a cyclopropane derivative (not shown). When the bivalent Lewis acid $MgBr_2$ was employed, the stereochemical outcome of the reaction was significantly altered. The *anti* homoallylic alcohol **156** was favored with the MOM-protected alkoxystannane, while the *syn* homoallylic alcohol **159** was the only product observed with the TBS-protected alkoxystannane. Antiperiplanar transition structures are proposed to account for the observed selectivity in these reactions.

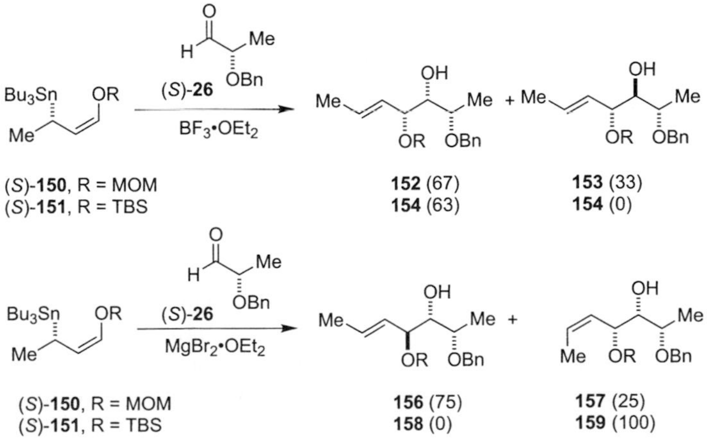

Scheme 10-65.

10.3.2.4 Allenylstannanes

The addition of allenylstannanes to aldehydes bears many similarities to the allylstannane reactions. All of the same stereochemical issues are relevant and have been addressed. Mechanistically, there are also strong resemblances and finally these reactions have been used successfully as a key transformation in the total synthesis of several natural products [104]. The reaction proceeds in high yield and excellent regio- and stereoselectivity to afford the desired homopropargylic alcohols.

Relative and internal stereoselection

Chiral allenylstannanes and achiral aldehydes. A systematic investigation on the stereochemical course of addition of allenylstannanes to aldehydes has been carried out [105]. In preliminary experiments, the additions of an enantiomerically enriched allenylstannane **160** to achiral aldehydes promoted by either $BF_3{\cdot}OEt_2$ or $MgBr_2{\cdot}OEt_2$ have been studied (Scheme 10-66). The selectivity observed with these reactions is found to be dependent upon both the aldehyde and the Lewis acid used. The sterically congested isobutyraldehyde and pivalaldehyde afford almost exclusively the *syn* diastereomer with $BF_3{\cdot}OEt_2$. Unbranched aldehydes such as heptanal favor the *anti* homoallylic alcohol. The Lewis acid $MgBr_2{\cdot}OEt_2$ is less selective for the *syn* diastereomer, possibly because of the formation of the larger 2/1 complex with the aldehyde. An antiperiplanar transition structure (Scheme 10-66) has been proposed to rationalize the observed sense of relative stereoselectivity, while the internal selectivity is controlled by an *anti* S_E' pathway to afford the homopropargylic alcohol.

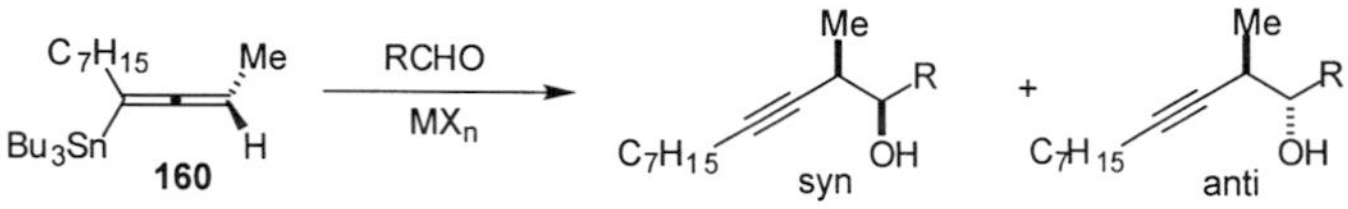

RCHO	MX_n	syn : anti
n-HexCHO	$BF_3{\cdot}OEt_2$	39 : 61
n-HexCHO	$MgBr_2{\cdot}OEt_2$	66 : 34
i-PrCHO	$BF_3{\cdot}OEt_2$	99 : 1
i-PrCHO	$MgBr_2{\cdot}OEt_2$	88 : 12
t-BuCHO	$BF_3{\cdot}OEt_2$	99 : 1

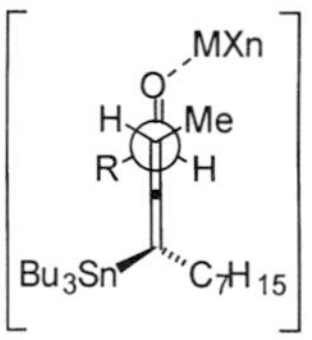

Scheme 10-66.

The addition of a **160** to achiral aldehydes in the presence of $SnCl_4$ produces only the *anti* homopropargylic alcohols (Scheme 10-67) [106]. The reaction is believed to proceed via transmetallation of the tributylstannyl moiety to the trichlorostannyl group. The formation of the *anti* product then occurs by a *syn* S_N2' pathway through a cyclic six-membered transition structure (product ee was identical to the ee of **160**).

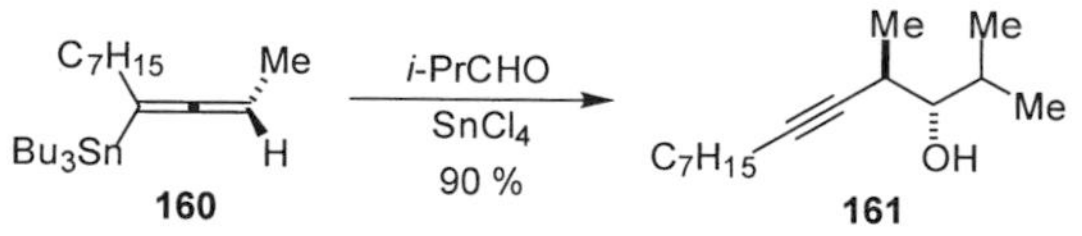

Scheme 10-67.

Chiral allenylstannanes and chiral aldehydes. The reaction of chiral aldehydes with chiral allenylstannanes gives rise to interesting double diastereoselection which is dependent on the Lewis acid employed [105, 106]. For example, the reaction of the allenylstannane (*R*)-**162** with the aldehyde (*S*)-**26** (Scheme 10-68) proceeds in high diastereoselectivity to form either the *syn* or *anti* homopropargylic alcohols. With $BF_3 \cdot OEt_2$, reaction of (*R*)-**162** provides the *syn* isomer **163**. An acyclic transition structure with an antiperiplanar arrangement of double bonds is proposed to rationalize the stereochemical outcome. In the $MgBr_2$-promoted reaction, attack of the allenylstannane on the chelated aldehyde leads selectively to the *anti* product **164** (99/1, *anti/syn*). The combination of (*S*)-**162** and (*S*)-**26** is mismatched as the selectivity is only 87/13, *syn/anti*.

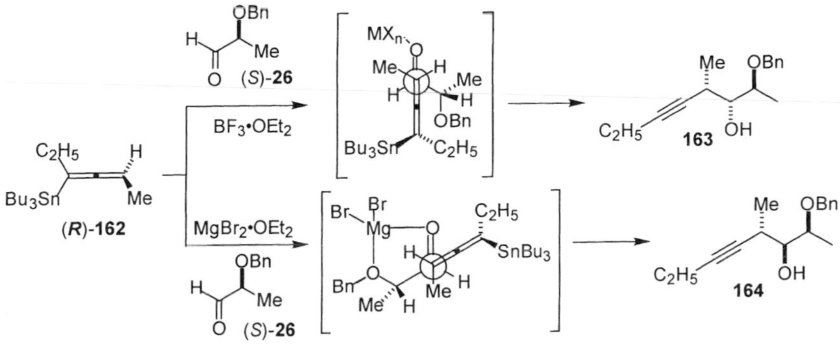

Scheme 10-68.

The synthesis of four stereotriads [1d] from a single pair of enantiomeric reagents has been accomplished with allenylstannane (*S*)-**165** (Scheme 10-69) [107]. The four different stereotriads are obtained by simple modulation of the reaction conditions (Lewis acid and solvent) and by using the appropriate configuration of starting material. Four different transition structures are proposed to account for the observed selectivity. The $BF_3 \cdot OEt_2$-promoted reaction with (*S*)-**52** is believed to proceed through an open transition structure under Felkin-Ahn control. When $MgBr_2$ is employed with this aldehyde, a reaction under chelation control occurs, providing the *anti* homopropargylic alcohol **166**. The reactions promoted by $SnCl_4$ are believed to proceed via transmetallation of the tributylstannyl group and subsequent reaction with (*S*)- or (*R*)-**52** in a cyclic transition structure or by chelation control.

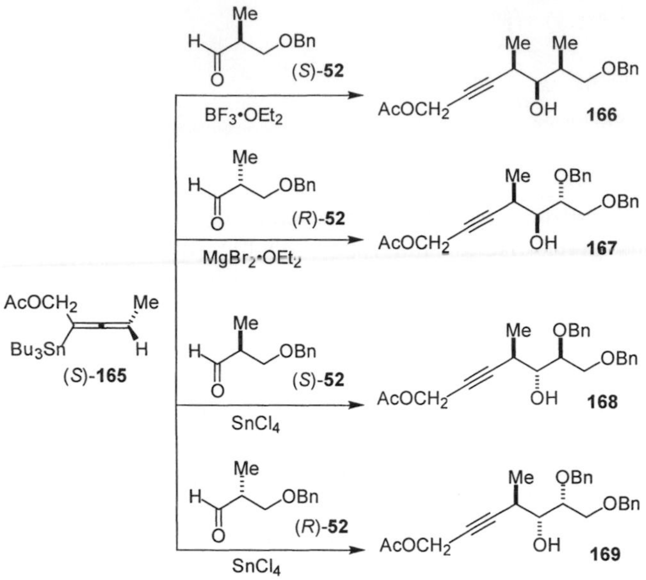

Scheme 10-69.

External stereoselection

The catalytic asymmetric propargylation [108] and allenylation [109] of achiral aldehydes has been performed with high levels of enantioselection. The asymmetric propargylation promoted by the chiral Lewis acid derived from binol and Ti(O*i*-Pr)$_4$ are representative. Between 50 and 100 mol% of titanium is required for these reactions to go to completion (Scheme 10-70). The reaction of benzaldehyde with allenyltributylstannane **170** and the chiral promoter produced the homopropargylic alcohol **171** in >99% ee and 48% yield (7% of the undesired allenyl alcohol was also obtained).

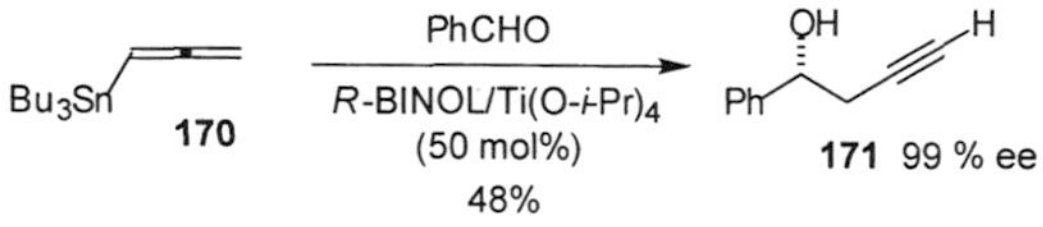

Scheme 10-70.

10.4 Allylic Boron Reagents

10.4.1 Mechanism of Addition

The chemistry of allylic boron reagents has a rich history owing to the extensive studies of the Mikhailov and Bubnov research group [110]. The first documented addition of allylic boranes to carbonyl compounds dates back to 1964 as does the demonstration of allylic inversion in the reaction of 2-butenylboranes [111]. This same group is responsible for the first demonstration of the addition of allylboronic ester to carbonyl compounds in 1968 [112]. Allylic boron reagents are the prototypical examples of Type 1 reactions; their high stereospecificity has been explained by a closed, chair-like transition structure where the boron is coordinated to the carbonyl oxygen (Scheme 10-71). The aldehyde is oriented in such a manner that the R group is placed in an equatorial position of the chair to minimize steric interactions between the ligands on boron and the allyl unit. This proposal explains the high degree of stereoselection observed when isomerically pure *E*- or *Z*-2-butenylboronates react with aldehydes. Thus, the *E*-2-butenyl isomer leads to the *anti* homoallylic alcohol while the *Z*-2-butenylboronate gives the *syn* product.

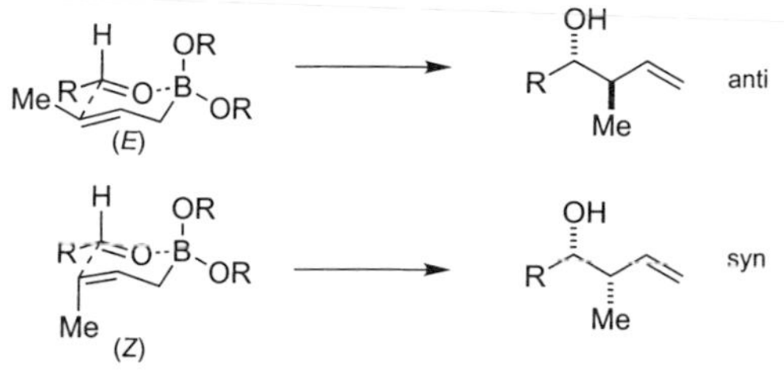

Scheme 10-71.

The transition structure geometries for the allylboration of carbonyl compounds have also been modeled computationally. Ab initio calculations identified a strong preference for the chair-like arrangement of the two components in the addition of allylboranes or allylboronic acids to aldehydes [113]. Force field calculation also provide insights into the transition structures of these reactions. The origins of selectivity in these additions have been elucidated by studying the reactions of both allylboranes [114a] and allylboronates [114b] with aldehydes. The force field generated in these studies is able to reproduce the ab initio geometries and energies of the relevant transition structures of the reaction. The force field is then used to correctly reproduce the experimental *syn/anti* stereoselectivity for the addition of *E*- and *Z*-2-butenylboranes to acetaldehyde. Accurate predictions are obtained for the stereoselectivity observed in the intermolecular reactions of *E*- and *Z*-2-butenylboronates, but the enantiofacial selectivity in the reaction of chiral allylboronates to aldehydes could not be modeled accurately.

The nature of the solvent has a significant effect on the rates of allylboration reactions [115]. Polar solvents, including $CHCl_3$, CH_2Cl_2, and Et_2O, which are either poorly coordinating or non-coordinating, enhance the rate of allylboration, while solvents capable of stronger coordination with boron, such as THF, retard the rate. Highly substituted aldehydes such as pivalaldehyde react significantly more slowly than less substituted aldehydes. Acyclic allylboronates react much more rapidly than the corresponding cyclic allylboronates. Electron-withdrawing groups on the cyclic allylboronate improve reactivity.

A theoretical study of the effects of structure and substituents on reactivity in allylboration has recently been completed [116]. Electron delocalization from the oxygen of an attacking aldehyde to the boron p-type atomic orbital is crucial in the allylboration reaction. The ab initio molecular orbital study indicates that the complex between the allylic borane and the aldehyde is weak. The relative electrophilicity of the boron atom in allylboron reagents is estimated by projecting out the unoccupied reactive orbital having the maximum amplitude onto the boron p-type atomic orbital. Two factors which are considered to be of importance in the reaction include the electron accepting level of the reactive unoccupied orbital and the efficiency of localization of the orbital in the unoccupied MO space.

10.4.2 Stereochemical Course of Addition

10.4.2.1 Allylic boron reagents

Relative stereoselection

The configurational instability of 2-butenylboranes precludes their use in stereoselective synthesis [110]. Fortunately, the corresponding boronic esters are configurationally stable and have contributed significantly to the popularity of this method [117]. The reaction of substituted *E*- and *Z*-2-butenylboronates with achiral aldehydes has been used to prepare *syn* or *anti* homoallylic alcohols in high diastereoselectivity. In general, the *syn* homoallylic alcohol is produced by reaction of the *Z*-2-butenylborane, while the *anti* homoallylic alcohol is obtained from the corresponding *E*-2-butenylborane (Scheme 10-71). Among the first examples of this reaction are the additions of the pinacol-derived 2-butenylboronate to aldehydes to afford the homoallylic alcohols as demonstrated by Hoffmann (Table 10-11) [118]. The only reactions where the observed diastereoselectivity deviates from the *E*/*Z* isomeric ratio was with the sterically demanding aldehyde isobutyraldehyde and the γ-alkoxyallylboronates.

The allylboration of α-oxocarboxylic acids, α-amino ketones and α- and β-hydroxy aldehydes and ketones has also been examined [119]. A representative example of the reaction of an α-oxocarboxylic acid with *E*- or *Z*-2-butenylboronates is shown in Scheme 10-72. Clearly, the carboxylic acid α-substituents exert a significant effect on the regio- and diastereoselectivity of these reactions. The reactions are proposed to occur via the bicyclic transition structure shown. The reac-

Table 10-11. Diastereoselectivity in the reactions of achiral aldehydes with 2-butenylboronates.

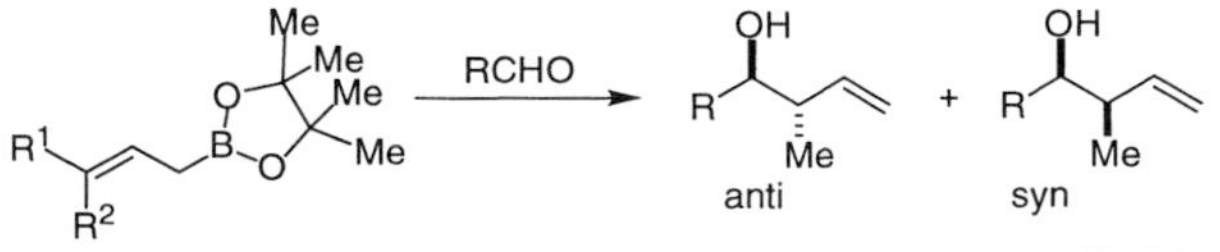

RCHO	R^1	R^2	*E/Z* ratio	Yield, %	*syn/anti*
PhCHO	Me	H	93/7	80	94/6
MeCHO	Me	H	93/7	40	93/7
i-PrCHO	Me	H	93/7	59	93/7
PhCHO	H	Me	5/95	80	5/95
MeCHO	H	Me	5/95	20	7/93
i-PrCHO	H	Me	5/95	51	7/93
PhCHO	MeO	H	90/10	87	95/5
PhCHO	$TMSCH_2CH_2O$	H	90/10	92	95/5
i-PrCHO	MeO	H	90/10	77	>98/2
PhCHO	H	MeO	5/95	86	5/95
PhCHO	H	$TMSCH_2CH_2O$	5/95	98	5/95
i-PrCHO	H	MeO	5/95	94	11/89

tion of *β*-hydroxy ketones or aldehydes with the allylboronates are less selective, producing a mixture of diastereomers. The reaction of the *β*-hydroxy ketones are significantly more diastereoselective than the corresponding aldehydes.

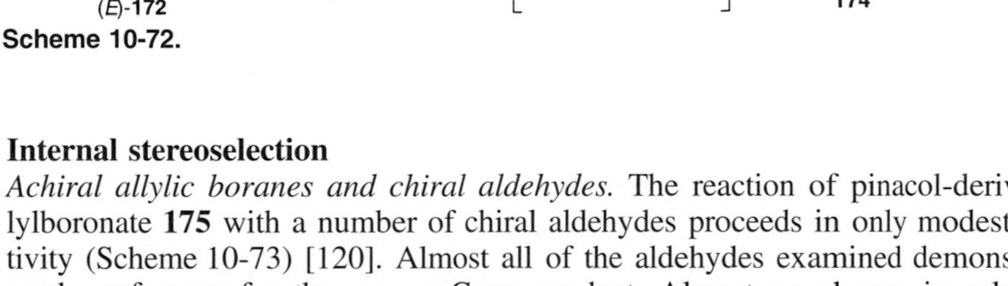

Scheme 10-72.

Internal stereoselection

Achiral allylic boranes and chiral aldehydes. The reaction of pinacol-derived allylboronate **175** with a number of chiral aldehydes proceeds in only modest selectivity (Scheme 10-73) [120]. Almost all of the aldehydes examined demonstrate a weak preference for the *syn* or Cram product. Almost no change in selectivity was observed when the protecting group on the alcohol is changed from a *t*-butylsiloxy to a methoxymethyl group.

The internal diastereoselectivity in the reaction of achiral 2-butenylboronates with chiral aldehydes has also been extensively examined [121]. The stereochemical

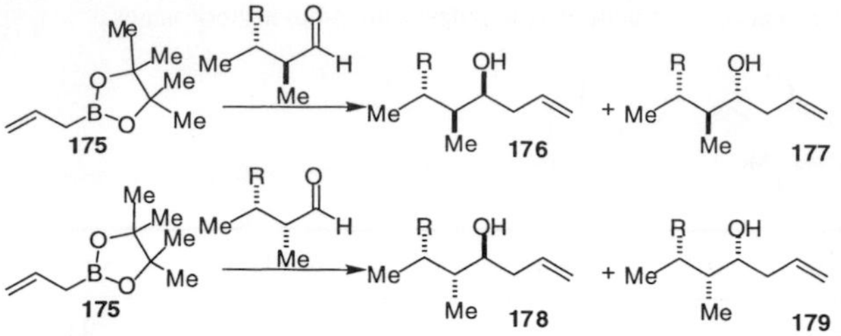

R	176	177
OH	49	51
OTBS	61	39
OMOM	62	38

R	178	179
OTBS	79	21
OMOM	79	21

Scheme 10-73.

course of reaction of pinacol-derived 2-butenylboronates (*E*)- and (*Z*)-**180** with chiral aldehydes is highly dependent upon the nature of the aldehyde and the 2-butenylboronate employed (Table 10-12). A 2-butenylboronate, (*E*)-**180**, provides the *anti* homoallylic alcohols **184** and **186**. The homoallylic alcohol **184** is formed preferentially, indicating a preference for *syn* bond construction with this reagent. When the butenylboronate (*Z*)-**180** is employed the reaction produced the *syn* homoallylic alcohols **183** and **185**, but in this case the "anti-Felkin" bond construction product **185** is favored. These results indicated that the internal (induced) diastereoselectivity observed in the reaction of a 2-butenylboronate with a chiral aldehyde is dependent upon the geometry and substitution pattern of the boronate. The magnitude of diastereoselectivity in these reactions is dependent upon the aldehyde. Unexpectedly, the products observed in reaction of the *Z*-2-butenylboronates correspond to reaction through an *anti*-Felkin pathway. The rationalization of the stereochemical outcome of these reactions has been discussed in considerable detail by Roush [1f, 117]. This pathway is favored because of minimization of gauche pentane interactions present between the 2-butenyl unit and the substituents on the chiral aldehyde substrate as shown in Fig. 10-4. Such interactions are apparently less important in the corresponding *E*-2-butenylboronates which do follow the Felkin paradigm. Computational modeling has provided support for this analysis [122].

Table 10-12. Diastereoselectivity observed in 2-butenylboration of chiral aldehydes.

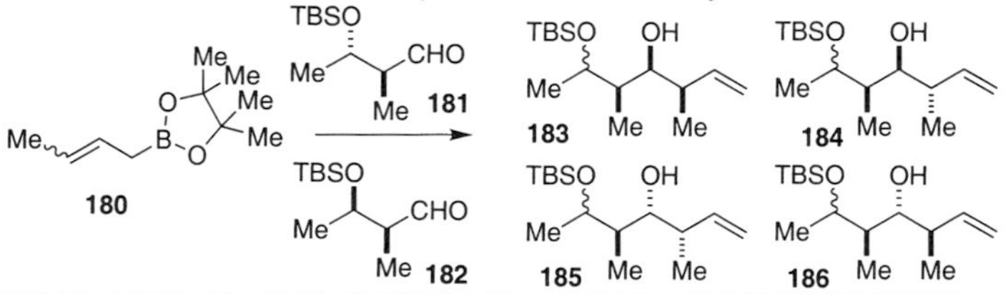

Borolane	Aldehyde	Product ratio, %			
		183	**184**	**185**	**186**
(*E*)-**180**	**181**	–	95		5
(*Z*)-**180**	**181**	9	–	91	–
(*E*)-**180**	**182**	–	98	–	2
(*Z*)-**180**	**182**	40	–	60	–

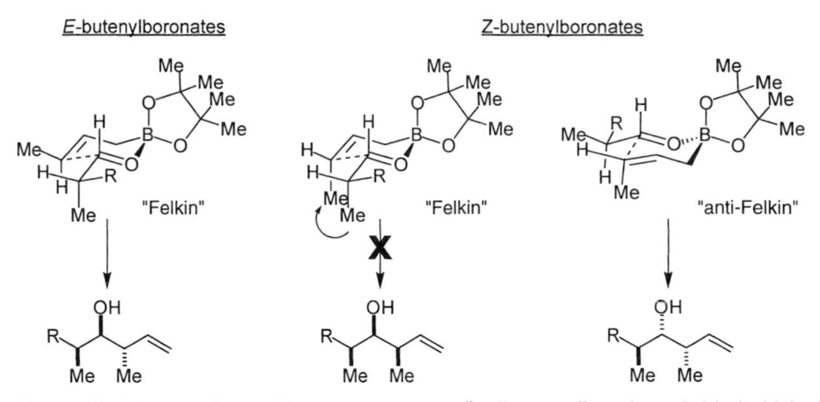

Figure 10-4. Proposed transition state structures for 2-butenylboration of chiral aldehydes.

Chiral allylic boranes and achiral aldehydes. The high degree of organization that characterizes the putative transition structures for these Type 1 reactions has stimulated the development of a myriad of chirally modified allylic boron reagents for asymmetric allylation of carbonyl compounds. Pioneering studies by Hoffmann demonstrated the use of camphor-derived allylboron reagents (**187**) in the preparation of enantiomerically enriched homoallylic alcohols in 1978 [123]. The reaction proceeds in high yield to produce the desired homoallylic alcohols (Scheme 10-74). Unfortunately, the highest selectivity is observed with simple aliphatic aldehydes (65–76%). With benzaldehyde, the homoallylic alcohol is obtained in only 40% ee.

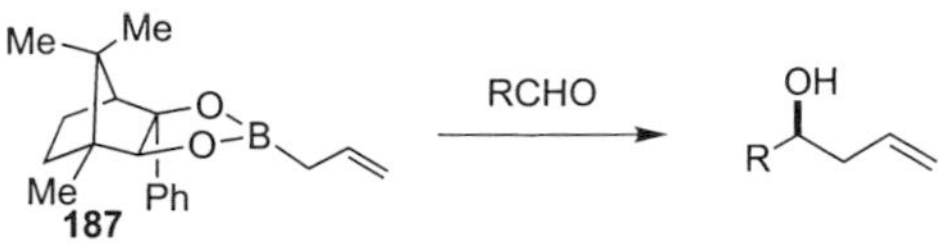

187 RCHO

Scheme 10-74.

Since that initial report, many research groups including those of Roush [124], Brown [125], Corey [126], and others [127] have designed chiral auxiliaries for attachment to the boron unit. High levels of enantioselection have been realized by the use of commercially available tartrate esters and simple cyclic monoterpenes. Chiral allylboranes derived from (+) or (–)-pinene, (+)-3-carene and (+)-2-carene (Chart 10-4) have been extensively investigated in reactions with achiral aldehydes [125]. The reaction of these chirally modified allylboranes with aldehydes proceeds almost instantaneously when no magnesium salts are present even at –100 °C. The reactions performed in the presence of magnesium salts are slower and elevated reaction temperatures were required (–78 °C). Extremely high levels of enantioselection can be obtained with these reagents (Table 10-13). The allylation of heterocyclic aldehydes affords heterocyclic substituted homoallylic alcohols in extremely high enantioselectivity [125f].

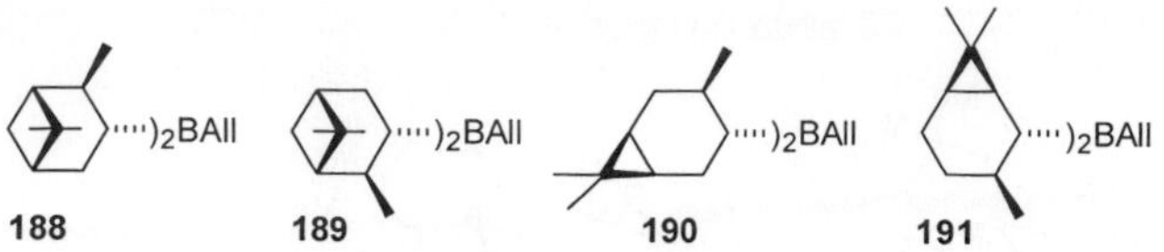

188 189 190 191

Chart 10-4. Terpene-derived chiral allylborane reagents.

Table 10-13. Reaction of achiral aldehydes with chiral allylborane reagents [125e].

RCHO	Allylborane reagent, % ee		
	188	**190**	**191**
MeCHO	>99 (*R*)	>99 (*R*)	>99 (*S*)
n-PrCHO	96 (*R*)	98 (*R*)	>99 (*S*)
i-PrCHO	96 (*S*)	98 (*S*)	>99 (*R*)
t-BuCHO	>99 (*S*)	>99 (*S*)	>99 (*R*)
CH_2=CHCHO	96 (*S*)	98 (*S*)	>99 (*R*)
PhCHO	96 (*S*)	98 (*S*)	>99 (*R*)

The use of the terpene-derived auxiliaries has been extended to the reaction of chiral 2-butenylboranes with achiral aldehydes [125d,h]. The relative diastereoselectivity observed in the reactions of the 2-butenyldiisopinocamphenylboranes **192** is greater than 98%. The major diastereomer obtained is also produced with high enantioselectivity for all of the aldehydes examined. No change in stereoselectivity is observed in reactions of aliphatic versus the unsaturated or aromatic aldehydes. All aldehydes provide products with uniformly high selectivity (Table 10-14). The requisite cyclic transition structure is proposed to account for the

Table 10-14. Stereoselective reaction of achiral aldehydes with chiral 2-butenylborane **192** [125d].

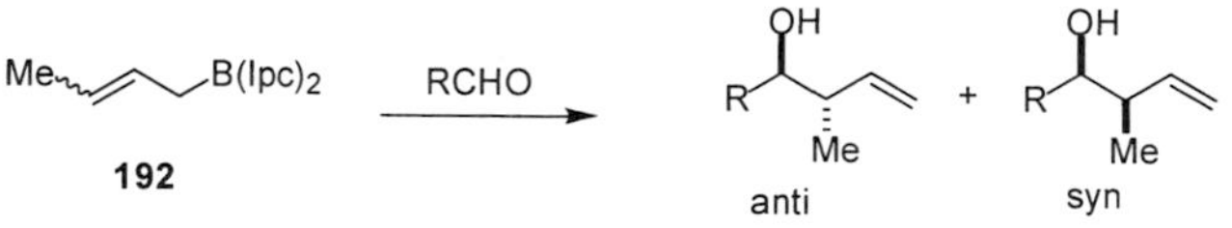

Me~~B(Ipc)$_2$ **192** —RCHO→ anti + syn

Aldehyde	Borane	Product [a])	ee, %
MeCHO	*d*-(*E*)-**192**	*anti*	90
MeCHO	*d*-(*Z*)-**192**	*syn*	90
EtCHO	*d*-(*E*)-**192**	*anti*	90
EtCHO	*d*-(*Z*)-**192**	*syn*	90
H_2C=CHCHO	*d*-(*E*)-**192**	*anti*	90
H_2C=CHCHO	*d*-(*Z*)-**192**	*syn*	90
PhCHO	*d*-(*E*)-**192**	*anti*	88
PhCHO	*d*-(*Z*)-**192**	*syn*	88

[a]) Diastereoselectivity observed in all reactions ≥98%.

high observed stereoselectivity in these reactions. The facial selectivity is obviously governed by the bulky isopinocampheyl group, although a compelling model for the exact origin of selectivity is lacking.

The asymmetric allylboration of achiral aldehydes with a substituted chiral allylborolane **193** and (*E*)- or (*Z*)-**194** has been reported [128]. The enantioselectivity observed with (*S*)-**193** at –100 °C and aldehydes is uniformly high with all of the achiral aldehydes examined (Scheme 10-75). The enantioselection observed with the borolane **193** is proposed to be primarily steric in origin and not from any stereoelectronic component. The reaction likely proceeds via a closed, six-membered transition structure in which the aldehyde is coordinated such that the trimethylsilyl group is oriented *anti* to the developing B–O bond.

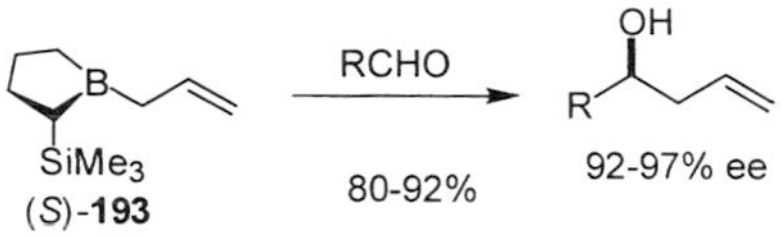

SiMe3 (*S*)-**193** —RCHO, 80-92%→ OH, R, 92-97% ee

Scheme 10-75.

The reaction of chiral 2-butenylborolanes (*E*)- or (*Z*)-(*R*,*R*)-**194** with achiral aldehydes proceeds in high yield and the homoallylic alcohol was obtained with a high degree of diastereo- and enantioselectivity (Table 10-15) [128]. The selectivities are not dependent upon the steric environment around the aldehyde. Despite the excellent selectivities, these reagents have enjoyed much less use than other chiral boron reagents owing to the difficulty of synthesis of the boron precursor and the impossibility of recovering the auxiliary after addition.

Table 10-15. Reaction of 2-butenylborolane (*R*,*R*)-**194** with achiral aldehydes [128].

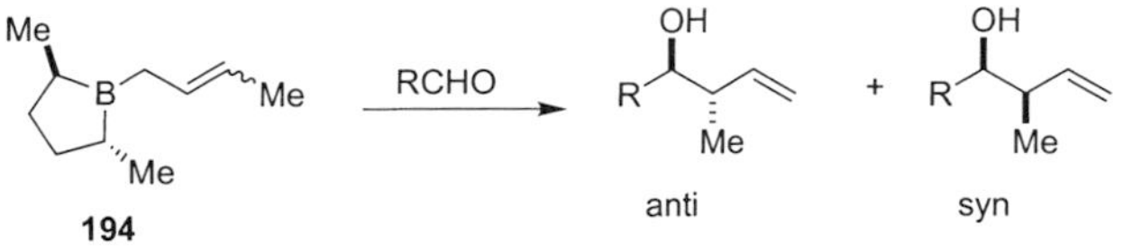

194 —RCHO→ anti + syn

Borane	RCHO	Yield, %	*anti/syn*	ee, %
(*E*)-(*R*,*R*)-**194**	EtCHO	81	93/7	96
(*E*)-(*R*,*R*)-**194**	*t*-BuCHO	72	96/4	95
(*Z*)-(*R*,*R*)-**194**	EtCHO	73	7/93	86
(*Z*)-(*R*,*R*)-**194**	*t*-BuCHO	75	5/95	97

By far the most extensively investigated and most useful of the chiral allyl- and 2-butenylboron reagents are those developed in the Roush laboratories (Chart 10-5) [124]. The tartrate ester-modified allyl- and 2-butenylboronates are attractive alternatives to the allylborane reagents because of their ease of prepara-

Table 10-17. Reaction of achiral aldehydes with chiral allylborane reagents [126].

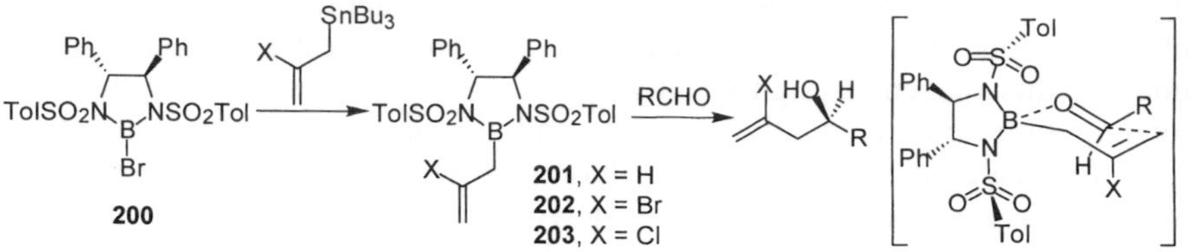

Borane	RCHO	X	Yield, %	ee, %
201	PhCHO	H	>90	95
201	(*E*)-PhCHCHCHO	H	>90	97
201	*c*-HexCHO	H	>90	97
201	*n*-PentCHO	H	>90	95
202	PhCHO	Br	73	79
203	PhCHO	Cl	79	84
202	*c*-HexCHO	Br	75	94
203	*c*-HexCHO	Cl	81	99
203	*n*-PentCHO	Cl	77	99

the enantiomerically enriched boronate (*E*)-**204** with the chiral aldehyde (*S*)-**205** provides a 92/8 mixture of diastereomeric homoallylic alcohols **206** and **207** with the Cram product predominating (Scheme 10-76). Reaction of the corresponding (*Z*)-**204** was not diastereoselective, providing almost a 1/1 mixture of homoallylic alcohols.

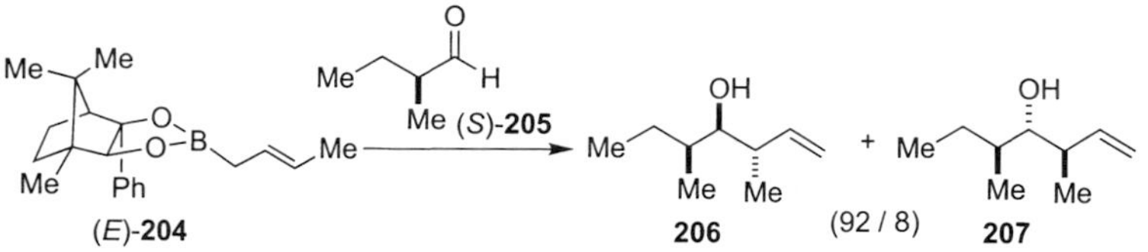

Scheme 10-76.

The 2-butenyldiisopinocamphenyl boranes (*E*)- and (*Z*)-**192** have also been employed in double diastereoselective additions with chiral aldehydes [130]. The reactions proceed in high diastereoselectivity for both the (*E*)- and (*Z*)-**192**, and both reagents give consistently high levels of reagent-controlled selectivity in the addition to chiral aldehydes (Table 10-18). The observed diastereoselectivity in these reactions is dependent upon the configuration of the 2-butenylborane employed.

The chiral borolanes (*E*)- and (*Z*)-**194** (of both *R,R* and *S,S* configuration) also show high reagent-controlled diastereoselection in the addition to chiral aldehydes [127b]. The results obtained from reaction of the substituted borolanes with (*R*)-glyceraldehyde acetonide (*R*)-**212** are summarized in Table 10-19. The matched

Table 10-18. Double diastereoselection addition with chiral 2-butenylboranes **192**.

Borane	Yield, %	Product ratio, %			
		208	**209**	**210**	**211**
d-(*E*)-**192**	75		96		4
l-(*E*)-**192**	70	–	9	–	91
d-(*Z*)-**192**	79	4	–	96	–
l-(*Z*)-**192**	73	82	–	18	–

series of reagents (entries 1 and 2) provide the homoallylic alcohol in high diastereoselectivity. The mismatched series of reagents (entries 3 and 4) are also quite selective, with the chiral borolane determining the stereochemical outcome of the reaction.

Table 10-19. Double diastereoselective addition with chiral 2-butenylboranes **194**.

Entry	Borane	Yield, %	Product ratio, %			
			213	**214**	**215**	**216**
1	*E*-(*R*,*R*)-**194**	71	96.1	2.8	0.9	0.2
2	*Z*-(*R*,*R*)-**194**	66	4.5	2.3	91.6	1.6
3	*E*-(*S*,*S*)-**194**	74	12.4	85.6	0.4	1.6
4	*Z*-(*S*,*S*)-**194**	65	0.5	2.4	15.4	81.7

The chiral allyl- and 2-butenylboronates derived from tartrate esters (Chart 10-5) have been used in combination with a wide variety of chiral aldehydes to produce homoallylic alcohols in high yield and moderate to high enantioselectivity [124]. The results obtained from reaction of selected chiral aldehydes (Chart 10-6) with the tartrate-modified allylboronates **195** and **197** (Chart 10-5) are shown in Table 10-20. As with the achiral aldehydes, the highest enantioselectivities are obtained when the chiral aldehydes are combined with allylboronate **197**. A strong reagent-induced selectivity is apparent, but is nevertheless dependent on the intrinsic bias of the aldehyde.

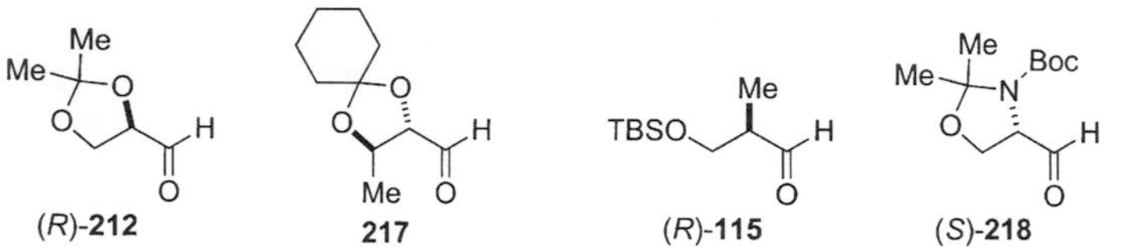

(*R*)-**212** **217** (*R*)-**115** (*S*)-**218**

Chart 10-6. A selection of chiral aldehydes.

Table 10-20. Double diastereoselective addition with tartrate-derived 2-butenylboranes [124].

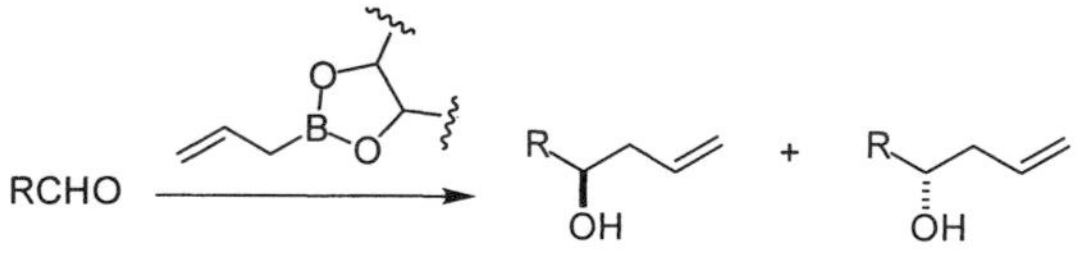

RCHO → **219** + **220**

Boronate	Aldehyde	Yield, %	**219/220**
(*S,S*)-**195**	(*R*)-**212**	91	96/4
(*R,R*)-**195**	(*R*)-**212**	79	10/90
(*S,S*)-**195**	**217**	94	98/2
(*R,R*)-**195**	**217**	–	36/64
(*S,S*)-**195**	(*S*)-**218**	77	11/89
(*R,R*)-**195**	(*S*)-**218**	84	10/90
(*S,S*)-**197**	(*R*)-**212**	84	98/2
(*R,R*)-**197**	(*R*)-**212**	81	<1/>99
(*S,S*)-**197**	(*R*)-**98**	46	97/3
(*R,R*)-**197**	(*R*)-**98**	43	3/97

The demonstration of effective double diastereoselection with reagent control has been extended by Roush to include the family of tartrate-modified 2-butenylboronates [124d]. The results obtained from reaction of a representative aldehyde with the chiral boronates (*E*)- and (*Z*)-**196** are shown in Table 10-21. The reactions proceed with high diastereoselectivity in the matched series, whereas in the mismatched series the observed diastereoselectivity is significantly reduced. As the steric requirements of the aldehyde increases relative to the methyl group on the 2-butenylboronate it becomes increasing difficult to synthesize the *syn,syn* diastereomer **224**. The transition structures for the double diastereoselective reactions of the chiral 2-butenylboronates with the chiral aldehydes can be readily constructed by combination of the salient features of the individual transition structures for the reference reactions as shown in Figs. 10-4 and 10-5.

Table 10-21. Reaction of chiral 2-butenylboronates with chiral aldehyde (*S*)-**98** [124d].

Boronate	Yield, %	**221**	**222**	**223**	**224**
(*R,R*)-(*E*)-**196**	80	97	3	–	–
(*S,S*)-(*E*)-**196**		16	81	3	–
(*R,R*)-(*E*)-**196**	71	–	4	95	1
(*S,S*)-(*E*)-**196**		12	2	45	41

Generally the reaction of unsaturated aldehydes (aromatic, olefinic and acetylenic) with chiral boronates has provided homoallylic alcohols in low to moderate enantioselectivity [124]. However, the enantioselectivity of the allyl- and 2-butenylborations of benzaldehyde and unsaturated aldehydes is significantly improved when a metal carbonyl complex is utilized as the substrate [131]. For example, the reaction of iron carbonyl-complexed diene **225**, chromium carbonyl-complexed benzaldehyde **226** and dicobalt hexacarbonyl complexed acetylene **227** all give significantly increased allyl and 2-butenylboration selectivities compared to the parent aldehydes (Fig. 10-6). In the case of chiral substrates **225** and **226**, these species can be obtained in enantioenriched form by kinetic resolution by use of the asymmetric allylboration reaction.

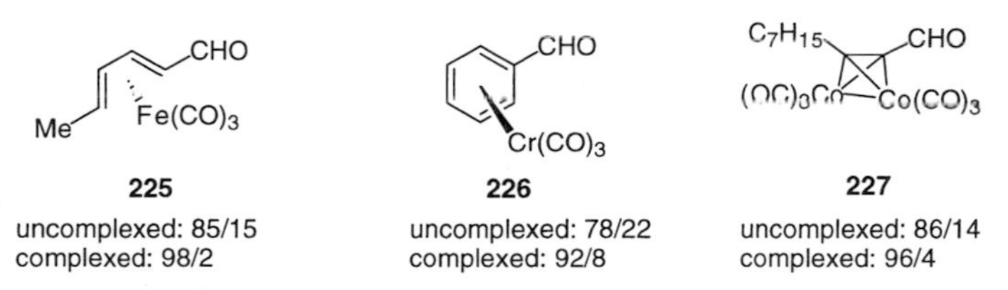

Figure 10-6. Asymmetric allylborations with (*R,R*)-**195.**

10.4.2.2 Allenylboron reagents

The observation that allenylboronic esters react with carbonyl compounds to afford homopropargylic compounds was first made by Favre and Gaudemar [132].

This transformation is important by virtue of the high regioselectivity obtained in the preparation of homopropargylic alcohols, which is not possible with other organometallic reagents. 9-Allenyl-9-BBN, prepared by the addition of allenylmagnesium bromide to a solution of 9-chloro-9-BBN, reacts with achiral aldehydes or ketones to afford the homopropargylic alcohol in high yield (Scheme 10-77) [133]. The reaction works equally well with aliphatic aldehydes, ketones, and aromatic aldehydes. From a comparison of the reaction of allenylmagnesium bromide, di(*n*-butyl)allenylboronate and 9-allenyl-9-BBN (**228**) it is evident that the allenylation with **228** is preferred. Reagent **228** produced only the homopropargylic alcohols, while the other reagents gave mixtures of the homopropargylic alcohol and some allenyl alcohol.

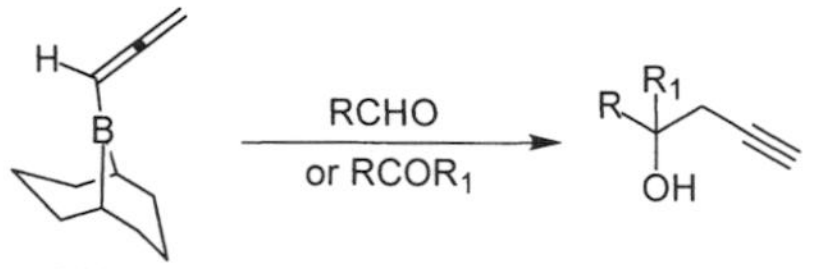

H B RCHO or $RCOR_1$ R R_1 OH

228

Scheme 10-77.

Internal stereoselection

Chiral modification of allenylboronic esters with tartrate-derived auxiliaries has given rise to a useful class of reagents. Addition of (*R,R*)- or (*S,S*)-**229** to aldehydes affords homopropargylic alcohols in high enantioselectivity and yield [134]. The chiral allenylboronic ester is prepared from a dialkyl tartrate and allenylboronic acid. A survey of various tartrate esters indicates that a bulkier tartrate ester leads to higher enantioselectivity [134a]. The results obtained using the tartrate derived from 2,4-dimethyl-3-pentanol are summarized in Table 10-22. The reaction proceeds with a high level of enantioselectivity for all of the saturated aldehydes examined. The reaction of the allenylboronate with aromatic or unsaturated aldehydes affords the homopropargyl alcohols in low yield and relatively poor enantioselectivity.

Table 10-22. Reaction of chiral allenylboronate **229** with aldehydes [134b].

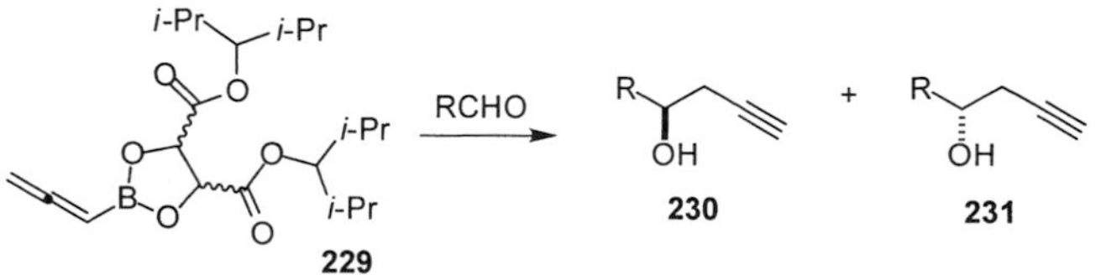

i-Pr *i*-Pr O O O B O O *i*-Pr *i*-Pr O RCHO R OH + R OH

229 **230** **231**

Aldehyde	Tartrate	Yield, %	Product	ee, %
n-PentCHO	L-(+)	81	**231**	94
c-HexCHO	L-(+)	88	**231**	98
c-HexCHO	D-(–)	82	**230**	99
(*S*)-citronellal	L-(+)	74	**231**	92
(*R*)-citronellal	L-(+)	67	**231**	98

By making use of the clean allylic inversion observed in the transmetallation of allenyl- and propargylstannanes with boron reagents, Corey developed a versatile and highly enantioselective method for allenylation and propargylation of aldehydes [135]. Thus, treatment of the chiral bromoborane **200** with allyltributylstannane afforded the propargyl borane **232**, which underwent extremely enantioselective addition to a variety of aldehydes to produce the allenyl alcohols **233** in excellent yields (Table 10-23). Alternatively, treatment of **200** with propargyltriphenylstannane affords the allenylborane **234**, which reacts with aldehydes to selectively produce the homopropargylic alcohols **235** again with excellent enantioselectivity.

Table 10-23. Selective allenylation and propargylation of aldehydes [135].

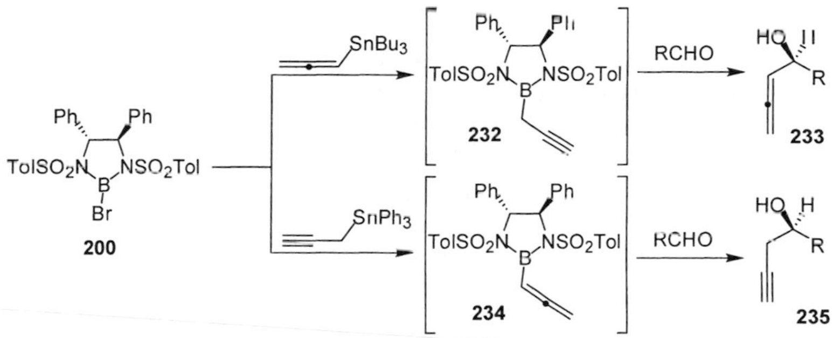

Borane	RCHO	Product	Yield, %	ee, %
232	PhCHO	**233**	72	99
232	(*E*)-PhCHCHCHO	**233**	74	99
232	*c*-HexCHO	**233**	78	99
232	*n*-PentCHO	**233**	82	99
234	PhCHO	**235**	76	96
234	(*E*)-PhCHCHCHO	**235**	79	98
234	*c*-HexCHO	**235**	82	92
234	*n*-PentCHO	**235**	81	91

Insight into the mechanism of the reaction of allenylboranes with aldehydes is available from the reaction of enantiomerically enriched allenylborane; (*S*)-**236** with benzaldehyde produced the homopropargylic alcohols **237** and **238** (Scheme 10-78) [136]. Although the relative diastereoselectivity was poor (*syn*/*anti*, 3/1), the induced selectivity at the methyl-bearing center is conserved. Moreover, the configuration of the methyl-bearing center indicates that the reaction proceeded via a *syn* S_E' pathway through a cyclic transition structure.

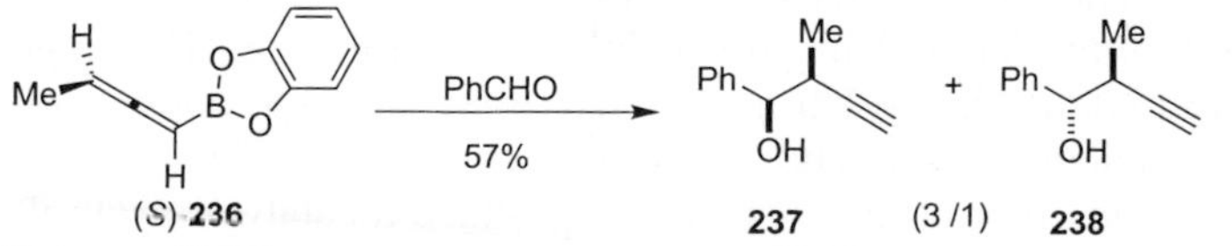

Scheme 10-78.

10.5 Allylic Chromium Reagents

10.5.1 Mechanism of Addition

The chromium(II)-mediated reaction of allylic halides with aldehydes or ketones was first demonstrated by Hiyama in 1977 [137]. In this example, 1-bromo-2-butene and benzaldehyde affords a single homoallylic alcohol, which was later determined to be the *anti* product by Heathcock [138]. The reaction of a crotylchromium reagent with an aldehyde (Nozaki-Hiyama-Kishi Reaction) has since been determined to afford the *anti* homoallylic alcohol in high diastereoselectivity regardless of the geometry of the starting allylic halide (Type 3) [139]. The high diastereoselectivity observed in the reaction of crotylchromium reagents with aldehydes has been explained by assuming that the intermediate crotylchromium reagents (*E*)-**xv** and (*Z*)-**xv** equilibrate rapidly via the chromium intermediate **xvi** before reaction with the aldehyde (Scheme 10-79). The diastereoselective formation of the *anti* homoallylic alcohol results by reaction through a closed, six-membered transition structure. This generalization breaks down, however, with bulky aldehydes or highly substituted allylic substrates for which isomerization is slower than addition (see below).

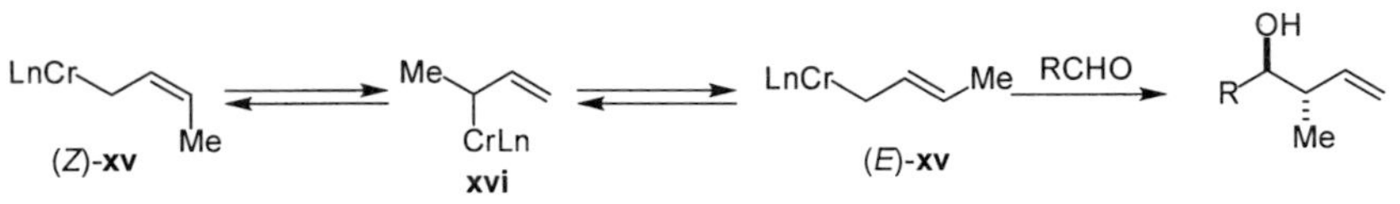

LnCr (*Z*)-**xv** Me ⇄ Me CrLn **xvi** ⇄ LnCr Me (*E*)-**xv** —RCHO→ R OH Me

Scheme 10-79.

A mechanistic proposal has been put forward for the formation and reaction of allylic chromium species [139f]. It is believed that a dinuclear ketyl complex reversibly forms by single electron transfer (SET) from $CrCl_2$ and the carbonyl compound. Oxidative addition of the allyl halide followed by a second, reversible (SET) affords a diradical which can undergo coupling, reduction or dimerization of the allylic units (Fig. 10-7). The low rotational barrier of the allylic radical explains the stereoconvergence of the addition products independently of the configuration of the allylic precursor. The arrangement of the groups around the chromium center loosely resembles the features of a chair-like transition structure, which explains the predominance of the *anti* stereoisomer.

A catalytic variant of the Nozaki-Hiyama-Kishi reaction was recently introduced by Fürstner [140]. The stoichiometric reaction generally requires at least three equivalents of chromium for the transformation to be complete. The large excess of $CrCl_2$ and the toxicity of the chromium salts precludes the application of this reaction in industrial processes. The reaction developed by Fürstner employs manganese powder and chlorotrimethylsilane to produce a catalytic cycle illustrated in Fig. 10-8 for the addition of vinyl iodides to aldehydes. The stereo-

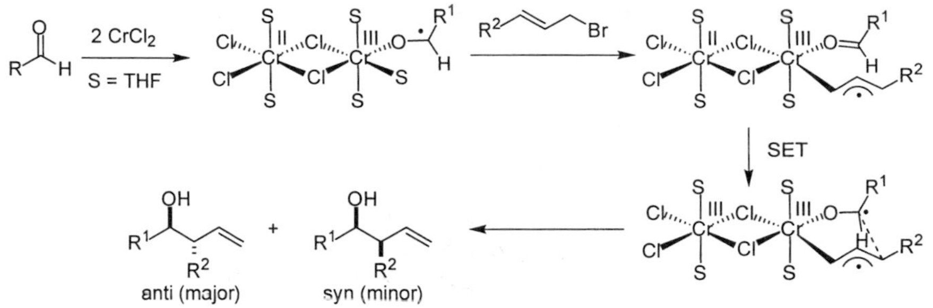

Figure 10-7. Proposed mechanism for the formation and addition of allylic chromium reagents.

chemical outcome with 2 butenyl bromide is comparable with reactions that employ an (over) stoichiometric amount of $CrCl_2$.

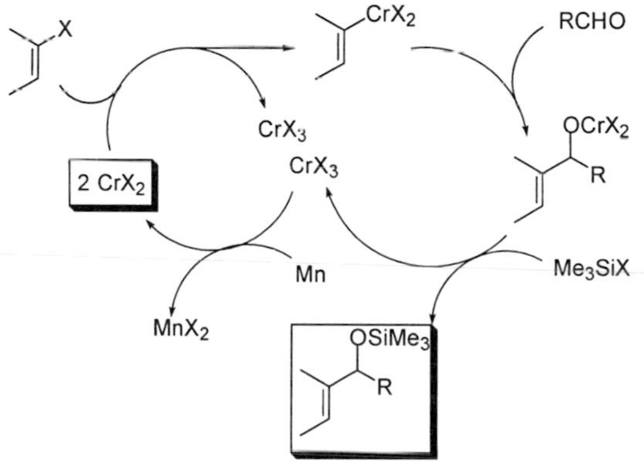

Figure 10-8. Proposed catalytic cycle for the Nozaki-Hiyama-Kishi reaction.

The allylic chromium reagents have also been generated by treatment of a 1,3-diene with $CrCl_2$ and B_{12} in methodology developed by Takai [141]. The crotylchromium species thus generated is able to undergo reaction with a variety of aldehydes to produce homoallylic alcohols in high yield. The cobalt(I) species, B_{12s}, derived by the reduction of B_{12} with 2 equiv of chromium(II) chloride affords hydridocobalamin by reaction with water. Hydrocobaltation of the cobalt hydride species to a 1,3-diene produces an allylcobalt(III) species. Homolytic cleavage of allylcobalt(III) to an allyl radical followed by trapping with chromium(II) gives allylchromium(III), which adds to the carbonyl compounds in a selective manner. Reduction of cobalt(II) with chromium(II) regenerates the cobalt(I) species. Takai has used this method to produce a variety of substituted homoallylic alcohols in moderate to high selectivity. The ratio of diastereomers obtained in these reactions is almost identical to the ratio observed in reactions of 2-butenyl bromide with $CrCl_2$. It is assumed that rapid equilibration between the (*E*)- and (*Z*)-2-butenylchromium intermediates occurs in the solvent, giving rise to the observed products.

10.5.2 Stereochemical Course of Addition

10.5.2.1 Relative stereoselection

The chromium-mediated reaction of 1-bromo-2-butene with achiral aldehydes shows good generality and high selectivity [142]. The reaction affords a high yield of the desired *anti* homoallylic alcohol with a high degree of diastereoselectivity (Table 10-24). The preference for the *anti* homoallylic alcohol is observed with all aldehydes except when the more sterically demanding pivalaldehyde is employed.

Table 10-24. Reaction of 1-bromo-2-butene with aldehydes using chromium(II) reagents.

RCHO + Br⌒⌒Me $\xrightarrow{CrCl_2}$ anti + syn

R	Yield, %	*anti*/*syn*
Ph	96	100/0
n-Pr	59	93/7
i-Pr	55	95/5
n-Pent	70	97/3
t-Bu	64	35/65

The chromium-mediated reaction of allylic phosphates with aldehydes constitutes an important advance in stereoselection [143]. The reaction of γ-disubstituted and β,γ-disubstituted allylic phosphates with aldehydes proceeds with good to excellent diastereoselectivity. All of the reactions examined were stereodivergent, with the stereochemical outcome dependent upon the configuration of the starting allyl phosphate (Table 10-25). The difference in behavior of the allylic phosphates is attributed to a very slow isomerization of the intermediate γ-disub-

Table 10-25. Reaction of allylic phosphates with aldehydes using chromium(II) reagents.

$R^1R^2C{=}C(R^3)CH_2OP(O)(OEt)_2$ $\xrightarrow[R^4CHO]{CrCl_2}$ **239** + **240**

R^1	R^2	R^3	R^4	Yield	**239/240**
Me	*n*-Bu	H	Ph	95	97/3
Me	*n*-Bu	H	*c*-Hex	89	93/7
n-Bu	*n*-Pr	H	*n*-Hex	64	97/3
n-Pr	*n*-Bu	H	*n*-Hex	90	1/99
H	*n*-Bu	$SiMe_3$	Ph	83	87/13
n-Bu	H	$SiMe_3$	Ph	91	16/84

stituted allylic chromium species compared to the addition rate of the aldehyde. In the case of the γ-monosubstituted allylic chromium reagents this isomerization is fast. Clearly, the geometry of the starting allylic phosphate reagent determines the stereochemical outcome of the reaction.

10.5.2.2 Internal stereoselection

Chiral aldehydes

One of the advantages of the allylic chromium reagents is the high degree of functional group compatibility they display. Accordingly, they have been employed in many reactions of functionalized aldehydes, and in this section the stereochemical consequences of α-substituents are discussed. The reaction of chiral aldehydes with allyl and 2-butenylchromium reagents has been examined for an extensive selection of aldehydes bearing heteroatom substituents at the α- and β-carbons. In general it is seen that chelation control is not involved in reactions of β-alkoxy aldehydes with the chromium reagents [144]. Instead, it is apparent that the nature of the larger substituent on the α-carbon of the aldehyde determines the stereochemical outcome of the reaction (Scheme 10-80). The highest selectivity is observed with the ketal-derived aldehyde, which is significantly more bulky than the corresponding benzyl ether. Simple α-alkoxy aldehydes also displays only modest selectivity in reaction with these reagents [145].

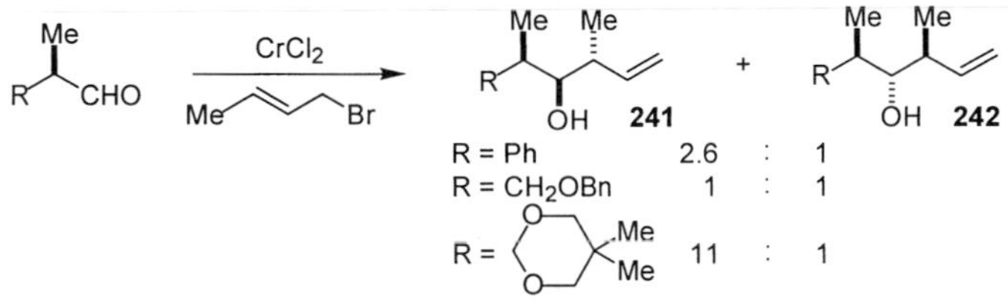

Me, R, CHO; $CrCl_2$, Me–CH=CH–CH$_2$–Br → Me Me, R, OH **241** + Me Me, R, OH **242**

R	241	:	242
R = Ph	2.6	:	1
R = CH_2OBn	1	:	1
R = 2,2-dimethyl-1,3-dioxan-4-yl (O, O, Me, Me)	11	:	1

Scheme 10-80.

N-Protected amino aldehydes have been employed in the chromium-mediated reaction of allylic halides [146]. A series of amino aldehydes bearing various *N*-protective groups are treated with allyl bromide and $CrCl_2$ to produce the homoallylic alcohols **243** and **244** (Table 10-26). The reactions are very unselective with the simple allylchromium reagents unless a disubstituted amino group is employed.

The addition of the 2-butenylchromium reagents to the α-amino aldehydes give predominantly the "*syn*" isomer **245** independently of the nature of the group bonded to the nitrogen (Table 10-27) [146b]. Ohta has recently examined the addition of allyl chromium reagents to Garner's aldehyde [147]. Poor to moderate diastereoselectivity is observed in these reactions.

Table 10-26. Allylation of *N*-boc- or *N*-Cbz-protected α-amino aldehydes.

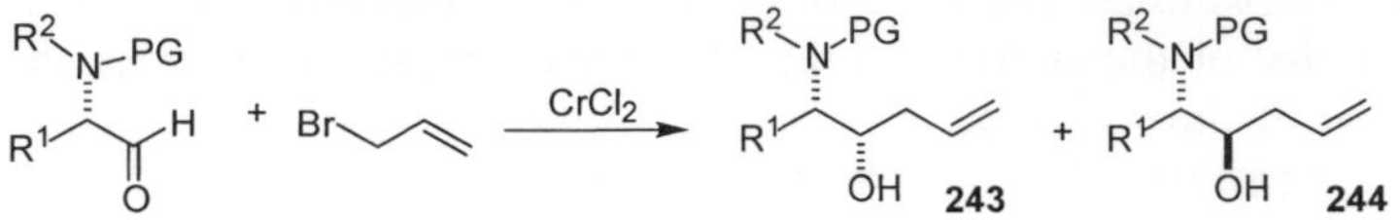

R^1	R^2	PG	Yield, %	**243/244**
Me	H	Boc	72	60/40
i-Pr	H	Boc	65	65/35
$PhCH_2$	H	Boc	77	60/40
$CbzNH(CH_2)_3$	H	Cbz	48	60/40
$PhCH_2$	$PhCH_2$	Tosyl	n. a.	90/10

Table 10-27. Reaction of 2-butenylchromium reagents with *N*-protected α-amino aldehydes.

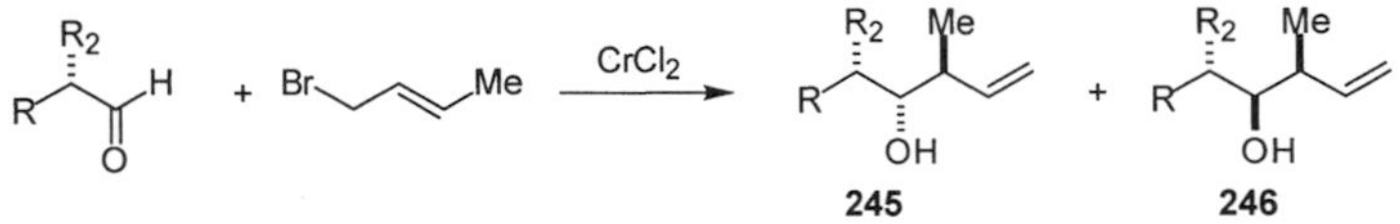

R	R_2	Yield, %	**245/246**
$PhCH_2$	NHBoc	73	9/1
3′-Indolyl-CH_2	NHBoc	71	8/1
Me	Bn_2N	63	15/1

Chiral allylic bromides

The addition of chiral allylic bromides of the general type **247** to achiral aldehydes mediated by $CrCl_2$ proceeds with a high degree of stereocontrol in which the bromide acts as the stereodominant component (Scheme 10-81) [148]. In all cases examined, major diastereomer **248** has an all-*syn* arrangement of the β'-OH, γ-vinyl and δ-methyl substituents. On the basis of this stereochemical outcome the following conclusions can be drawn: (1) the δ-center determines the stereochemical outcome of the newly created stereocenters (γ and β'), (2) the relative diastereoselection of the reaction is not affected by the presence of stereogenic centers in the allylic bromide, and (3) additional stereocenters in the ε- and ζ-positions of the bromide increase the diastereofacial selectivity but have no influence on the sense of the asymmetric induction.

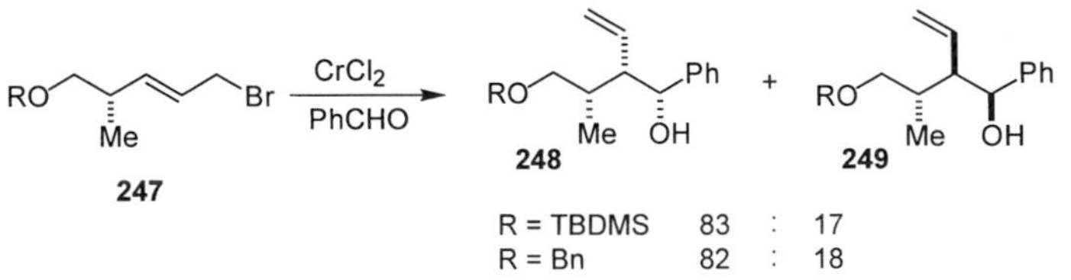

R = TBDMS 83 : 17
R = Bn 82 : 18

Scheme 10-81.

The level of 1,4-asymmetric induction in the chromium-mediated coupling of allyl bromides bearing α'-stereogenic centers to aldehydes is generally good to excellent (Table 10-28) [149]. The origin of stereoselection has been rationalized on the basis of the reactive conformation of the allylic chromium reagent in which the oxygen function is perpendicular to the allylic plane (minimizing $A^{1,3}$ interactions). Approach of the aldehyde antiperiplanar to the oxygen in this conformation gives rise to the observed *syn* diastereomers (Scheme 10-82).

Table 10-28. Coupling of allyl bromides to aldehydes.

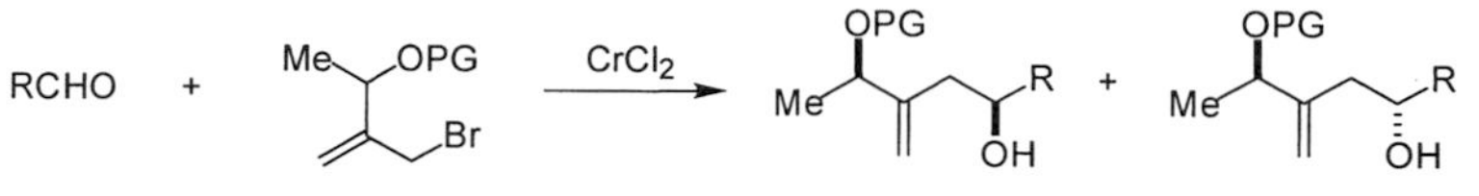

RCHO + 250 —$CrCl_2$→ 251 + 252

Aldehyde	PG	Yield, %	251/252
PhCHO	Bn	64	87/13
MeCHO	Bn	75	83/17
t-BuCHO	Bn	68	85/15
PhCHO	TBS	34	83/17
PhCHO	MOM	61	81/19

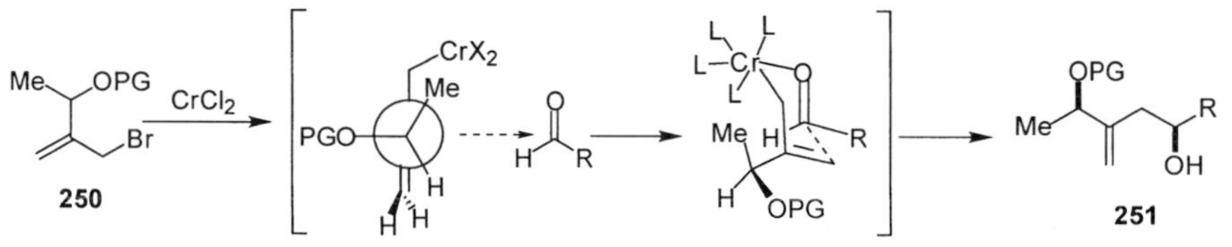

Scheme 10-82.

10.5.2.3 External stereoselection

Despite the obvious advantages of a reagent-controlled enantioselective addition of allylic chromium reagents, no such modification has yet to be reported. The enantioselective allylation of benzaldehyde with allyl bromide and $NiCl_2/CrCl_2$ in the presence of chiral bipyridyl ligands (employed stoichiometrically) affords modest enantioselectivities [150]. The highest selectivity was observed in the reaction of benzaldehyde with allyl bromide using only $CrCl_2$ and the bipyridyl ligand **253**. In this example, the reaction afforded a 6.7/1 enantiomeric ratio of homoallylic alcohols (Scheme 10-83).

The most preparatively useful enantioselective allylation of aldehydes in this domain involves dialkoxyallylchromium (III) complexes. These stoichiometric reagents are prepared by treatment of $CrCl_2$ with lithium alkoxides of chiral alcohols [151]. From a variety of chiral modifiers, the *N*-benzoyl-L-prolinol derivatives are found to give the best yields and enantioselectivities (Scheme 10-84).

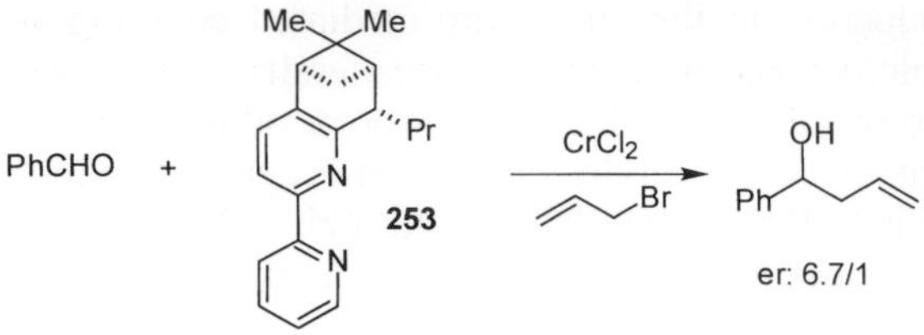

PhCHO + 253 → ($CrCl_2$, allyl bromide) → Ph–CH(OH)–$CH_2CH=CH_2$

er: 6.7/1

Scheme 10-83.

The highest selectivity is obtained when the chromium(II) dialkoxide **254** is prepared at room temperature and then allowed to react with the aldehyde at –30 °C.

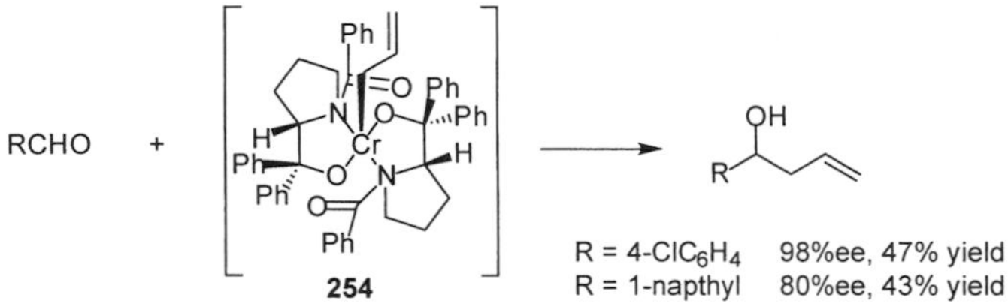

RCHO + 254 → R–CH(OH)–$CH_2CH=CH_2$

R = 4-ClC_6H_4 98%ee, 47% yield
R = 1-napthyl 80%ee, 43% yield

Scheme 10-84.

The development of a viable catalytic enantioselective variant of this transformation is clearly a worthy objective. While it is clear that donor solvents are crucial for the success of the reaction, the actual role of the solvent is not clear. In addition, the use of chiral agents to modulate the stereochemical course of reaction is complicated by the weak affinity of Cr(II) for common chiral ligands. Identifying that class of ligand that can compete with solvent for the Cr(II) center, accelerate the addition, and not irreversibly bind to Cr(II) constitutes a significant challenge.

10.6 Allylic Lithium, Magnesium and Zinc Reagents

10.6.1 Mechanism of Addition

The reaction of allylic lithium, magnesium and zinc reagents with ketones or aldehydes has been used extensively to prepare homoallylic alcohols [2]. The corresponding 2-butenylmetal reagents are configurationally unstable, existing as mixtures of rapidly equilibrating *E*- and *Z*-isomers (Scheme 10-85) [152]. The barrier to topomerization for allyllithium itself has been measured to be 10.7 kcal/mol [153]. 2-Decenyllithium is configurationally stable only below –90 °C [154]. In addition to being configurationally unstable [154], allylic magnesium halides also undergo rapid 1,3-migrations and engage in a Schlenk equilibrium with the diallylmagnesium species [155]. Diallylzinc reagents are (like their organomagnesium counterparts) known to be dynamic [156].

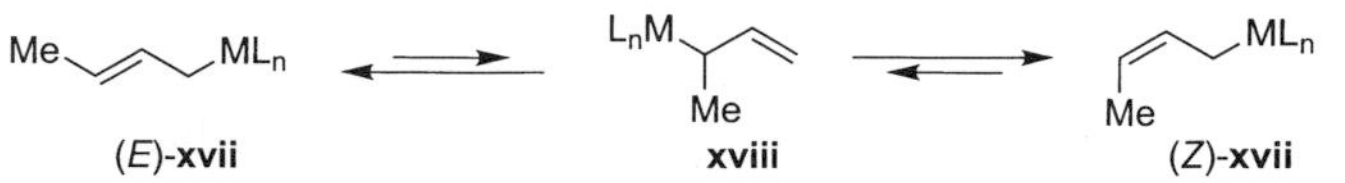

(*E*)-**xvii** ⇌ **xviii** ⇌ (*Z*)-**xvii**

Scheme 10-85.

Gajewski has examined the secondary deuterium isotope effects in the addition of allyllithium and allyl Grignard additions to benzaldehyde [157]. With allyllithium and allylmagnesium halides a normal secondary deuterium isotope effect was observed. The results indicate that rate-determining single-electron transfer occurs with the allyl reagents.

10.6.2 Stereochemical Course of Addition

10.6.2.1 Lithium reagents

2-Butenyllithium is configurationally unstable and demonstrates poor regio- and stereoselectivity in reactions with aldehydes [158]. The utility of the reactions increases when a heteroatom or heterocycle-stabilized allylic anion is employed [159]. The control of α- vs γ-substitution in allyl anions depends upon the complex interplay between the nature of the stabilizing group, charge delocalization, steric effects, solvation, the type of electrophile, and the counterion [160]. A significant body of literature exists for heteroatom-substituted allylic lithium reagents (oxygen, nitrogen, sulfur, selenium, silicon) [161]. For the most part, these functionalized reagents show their greatest utility after transmetallation with a titanium alkoxide and will be discussed in the following section on titanium reagents.

10.6.2.2 Magnesium reagents

The 2-butenyl and 3-methyl-2-butenyl Grignard reagents react with aromatic and aliphatic aldehydes in the γ-position [152]. The stereoselectivity observed in reactions of crotyl Grignard reagents with aldehydes and ketones is generally quite low, with the direction and magnitude of the selectivity depending upon the structure of the reagent and the nature of the aldehyde [162]. The preparative interest of allyl Grignard reagents is largely based on the relative insensitivity of the reagent towards steric crowding around the carbonyl group [163].

10.6.2.3 Zinc reagents

The allylic zinc reagents are generally prepared by mixing zinc metal with the requisite allyl halide using either sonication or a saturated aqueous solution of

NH_4Cl/THF to initiate the reaction. The allylation of aldehydes and ketones with allylic zinc reagents proceeds in moderate to high yield and high regioselectivity to afford homoallylic alcohols. The 2-butenylzinc reagents are configurationally unstable, providing a mixture of *syn* and *anti* homoallylic alcohols upon reaction with aldehydes [158 a]. For example, treatment of (*E*)-1-chloro-2-butene with zinc in saturated aqueous NH_4Cl followed by the addition of *n*-heptanal affords a 1/1 mixture of *syn*/*anti* homoallylic alcohols **255** in 85% yield. Increasing the steric requirements of the reaction leads to an increase in the stereoselectivity of the reaction. The reaction of cinnamylzinc with *n*-hexanal affords a 75/25 mixture of *syn*/*anti* homoallylic alcohols. When benzaldehyde is employed in the reaction the selectivity increases to provide an 87/13 mixture of homoallylic alcohols **256** (Scheme 10-86).

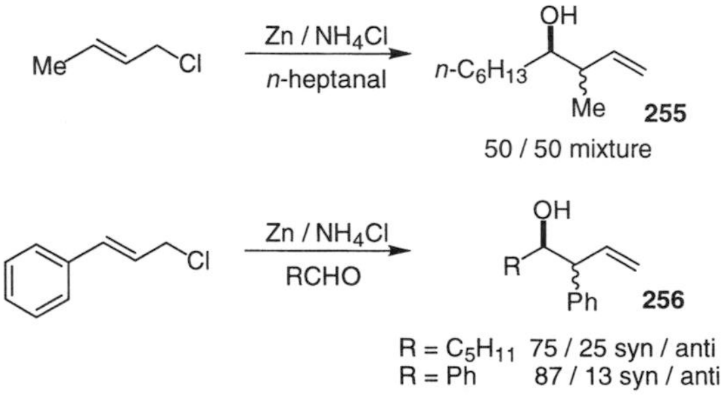

Me, Cl; Zn / NH_4Cl, *n*-heptanal; OH, *n*-C_6H_{13}, Me **255**
50 / 50 mixture

Cl; Zn / NH_4Cl, RCHO; OH, R, Ph **256**
R = C_5H_{11} 75 / 25 syn / anti
R = Ph 87 / 13 syn / anti

Scheme 10-86.

A clever application of masked allylic zinc reagents for highly diastereoselective allylation reactions has recently been developed [164]. The addition of allylic organozinc reagents to electrophiles has been known for some time to proceed in a reversible fashion [165]. Knochel hypothesized that, because of the reversibility of these reactions, a sterically hindered tertiary homoallylic alcohol **257** could, upon generation of a zinc alkoxide **xix**, undergo fragmentation to generate the allylic zinc reagent **xx** which could then undergo reaction with a suitable electrophile (Scheme 10-87). The reaction of the bis(*t*-butyl)alcohol **258** with benzaldehyde proceeds smoothly, affording an 89% yield of the homoallylic alcohol **259**.

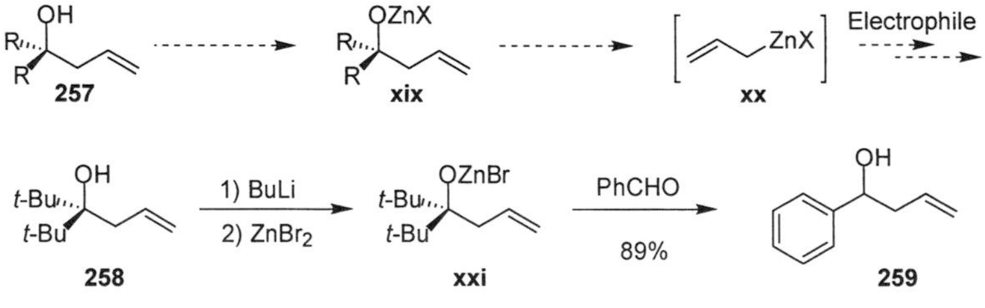

OH, R, R **257** → OZnX, R, R **xix** → [ZnX] **xx** → Electrophile

OH, *t*-Bu, *t*-Bu **258**; 1) BuLi, 2) $ZnBr_2$; OZnBr, *t*-Bu, *t*-Bu **xxi**; PhCHO, 89%; OH **259**

Scheme 10-87.

A range of aldehydes were treated with the homoallylic alcohol **260** and $ZnCl_2$ to afford the desired homoallylic alcohols in high yield and excellent diastereoselectivity (Fig. 10-9).

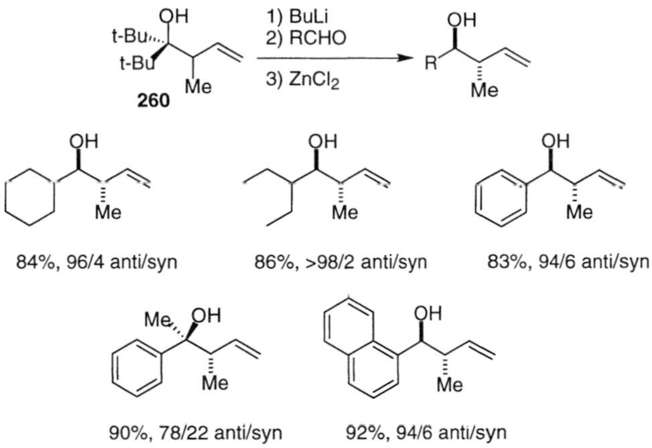

t-Bu, t-Bu, OH, Me, **260**; 1) BuLi 2) RCHO 3) $ZnCl_2$; R, OH, Me

84%, 96/4 anti/syn

86%, >98/2 anti/syn

83%, 94/6 anti/syn

90%, 78/22 anti/syn

92%, 94/6 anti/syn

Figure 10-9. Generation and addition of 2-butenylzinc from **260**.

All of the reactions proceed in high yield and diastereoselectivity except for the reaction of acetophenone. None of the undesired γ-substituted isomer is detected. The high diastereoselectivity observed in these reactions is especially surprising, as the addition of 2-butenylzinc to an aldehyde has already been observed to proceed with almost no diastereoselectivity [166].

The stereochemical outcome of these reactions is well accommodated by a mechanistic proposal involving a double allylic transposition (Scheme 10-88). Generation of the zinc alkoxide complexed by the aldehyde provides intermediate **xxii**. Allylic transposition in a cyclic transition structure affords the 2-butenyl reagent **xxiii** solely of *E*-configuration (presumably also complexed to the parent ketone arising from fragmentation). At –78 °C this species is configurationally stable and undergoes no isomerization. The second allylic transposition in the addition

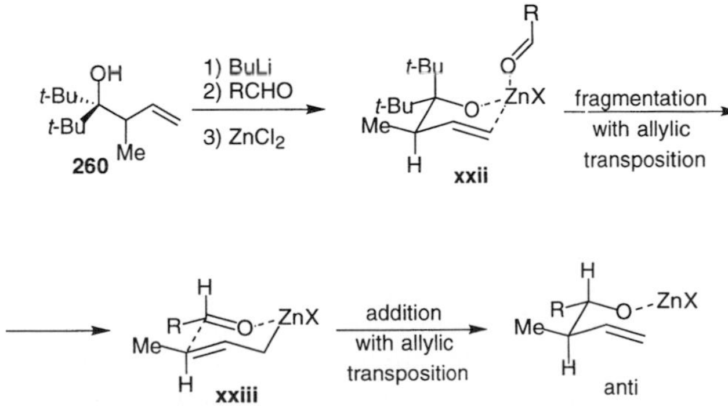

t-Bu, *t*-Bu, OH, Me, **260**; 1) BuLi 2) RCHO 3) $ZnCl_2$; **xxii**; fragmentation with allylic transposition; **xxiii**; addition with allylic transposition; anti

Scheme 10-88.

step provides the homoallylic alcohol, predominantly as the *anti* isomer. The high diastereoselectivity is attributed to the generation of pure (*E*)-2-butenylzinc in the presence of the electrophile. These reactions can proceed in high yield even with only a catalytic (10 mol%) amount of zinc.

Addition of diallylzinc and allylzinc halides to α- and β-alkoxy aldehydes proceeds in high yields and good selectivities. The major diastereomers arise from steric approach control [167].

The zinc-promoted reaction of allylic bromides **262** with chiral *N*-protected α-amino aldehydes **261** affords the *anti* diastereomer **263** with generally high selectivity (Table 10-29) [168]. The choice of protecting group is critical to the stereochemical outcome of the reaction. When the amino acids are protected by an N-Boc group, poor diastereoselectivity is observed.

Table 10-29. Allylation of N-protected α-amino aldehydes.

$R^1CH(NBn_2)CHO$ (**261**) + $R^2_2C{=}C(R^3)CH_2Br$ (**262**) $\xrightarrow[NH_4Cl]{Zn\,/\,THF}$ **263** + **264** (Bzl$_2$N, OH, R^2, R^2, R^3)

R^1	R^2	R^3	Yield, %	**263/264**
Me	H	H	95	6/1
Me	Me	H	90	20/1
Me	H	CO_2Me	95	5.5/1
i-Pr	H	H	84	8/1
i-Pr	Me	H	81	7.5/1
i-Pr	H	CO_2Me	92	7.5/1
$PhCH_2$	H	H	90	9/1
$PhCH_2$	Me	H	87	8/1
$PhCH_2$	H	CO_2Me	95	6/1

10.7 Allylic Titanium, Zirconium and Indium Reagents

10.7.1 Allylic Titanium Reagents

10.7.1.1 Mechanism and stereochemical course of addition

There exist three distinct classes of allylic titanium reagents which have been used in carbonyl addition reactions (Chart 10-7). The structures of these species have not been established, but can be inferred from the reaction products and related titanium species. The η^1-triheterosubstituted reagents are preparatively the most important and are easily prepared from allylic magnesium or lithium reagents. 2-Alkenyltitanium(IV) derivatives **265** are monomeric and fluxional at temperatures as low as –100 °C [169]. Thus all configurations are labile, except in the important class of α-heterosubstituted reagents. Chirally modified η^1-monocyclopentadienyl reagents **266** are readily prepared from inexpensive chiral alcohols and diols and show excellent enantioselectivity. The η^3-dicyclopentadienyl reagents **267** have found only limited application, but are mild, selective reagents. In almost all cases, closed, chair-like transition structures have been invoked to rationalize stereoselectivity.

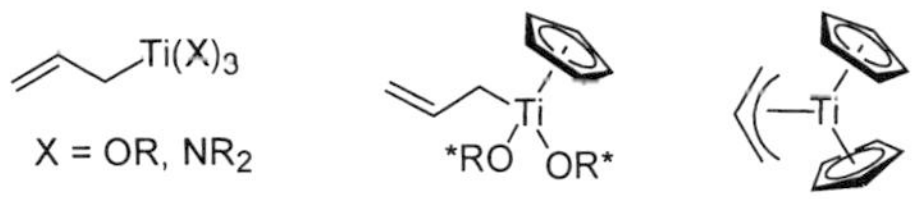

$Ti(X)_3$
X = OR, NR_2
265

Ti
RO OR
266

Ti
267

Chart 10-7. Allylic titanium reagents.

Nonheteroatom-substituted allylic reagents

The addition of η^1-2-butenyltitanium reagents to aldehydes and ketones has been demonstrated to give *anti* homoallylic alcohols preferentially [170]. These reagents (Chart 10-8) are prepared by treatment of Cp_2TiX_2, $(RO)_3TiCl$ or $(Et_2N)_3TiCl$ with 2-butenylmagnesium halides. In general, the relative diastereoselectivity for the different reagents increases in the order **268**>**269**>**270**.

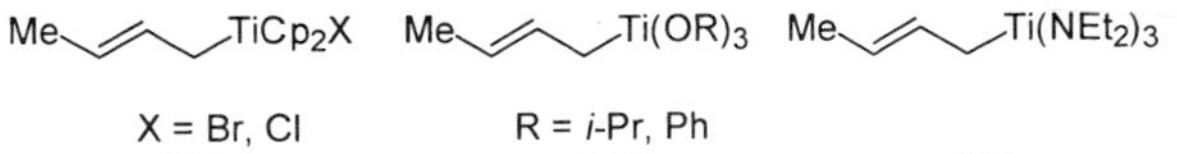

Me $TiCp_2X$
X = Br, Cl
268

Me $Ti(OR)_3$
R = *i*-Pr, Ph
269

Me $Ti(NEt_2)_3$
270

Chart 10-8. Selected η^1-2-butenyltitanium reagents.

η^1-Allyltitanium and allenyltitanium complexes can also be prepared by the combination of a reduced titanium species with an allylic electrophile [171]. The titanium reagents are prepared by the reaction of an allylic bromide or carbonate with $Ti(Oi\text{-}Pr)_4$ and *i*-PrMgBr. To effect addition, the aldehyde is then added to the allyltitanium reagent and the homoallylic alcohol is obtained in high yield and excellent diastereoselectivity. The method also provides a highly efficient method for the preparation of substituted allyltitanium compounds including those with functional groups. The induced diastereoselection with chiral η^1-2-alkenyltitanium reagents **27** with aldehydes has been reported (Scheme 10-89) [172]. The highest selectivity is observed when simple aliphatic aldehydes are employed. The reactions are believed to proceed via a closed, chair-like transition structure which minimizes interaction between the amino group and the aldehyde.

The addition of η^3-2-butenyltitanium reagents **274** to aldehydes produces *anti* homoallylic alcohols in good yield and excellent diastereoselectivity [173]. The size of the substituents on the cyclopentadienyl (Cp) ring has a marked effect on the yield and diastereoselectivity in the addition of 2-butenyltitanocenes to aldehydes (Table 10-30) [174]. Increasing the steric hindrance at the metal center favorably affects the diastereoselectivity of the addition process. The stereochemical

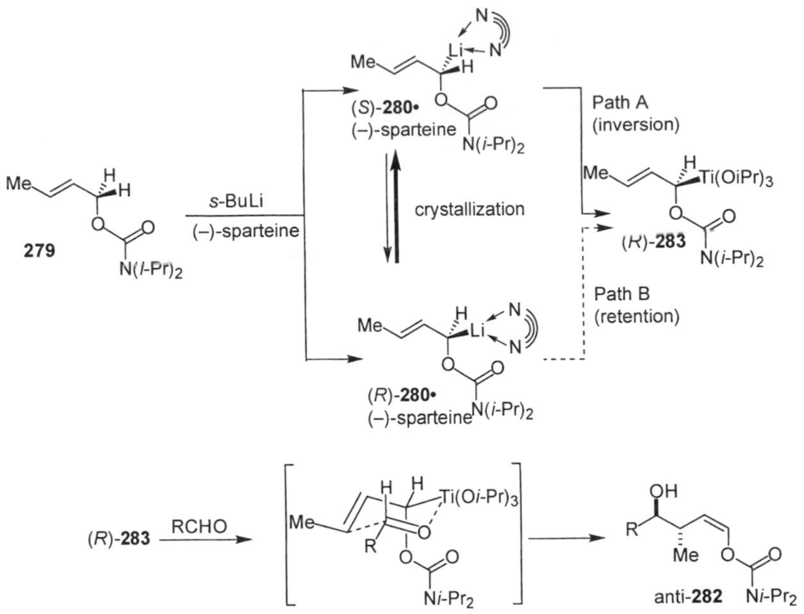

Scheme 10-95.

The X-ray crystal structure of *N*-Boc-*N*-*p*-methoxyphenyl-3-phenylallyl-lithium·(–)-sparteine complex has been reported [184]. This structure differs from the previous structure in that the lithium is associated in an η^3-fashion. The lithium·(–)-sparteine complex resides on the *Re* face of the allyl unit. Stannylation of the lithium complex was established to occur with inversion of configuration.

In an interesting stereochemical cycle, a new configurationally stable allylic trichlorotitanium species has been generated and combined with aldehydes. The new titanium reagent is generated by transmetallation of allylstannane with $TiCl_4$ (Scheme 10-96). Stannylation of the tri-*iso*propoxytitanium intermediate (*R*)-**283** leads with high efficiency to the *R*-configured stannane (*R*)-**284** which is obtained in about 95% ee [185]. The crystalline suspension of (*S*)-**280**·(–)-sparteine has also been trapped with preservation of enantiomeric purity using the trialkyltin chlorides although low regioselectivity and only moderate enantioselectivity are obtained in this process. The stannane (*R*)-**284** is then treated with $TiCl_4$ followed by the requisite aldehyde or ketone to afford the homoallylic alcohol **282** with high enantioselectivity. The configuration and the enantiomeric purities of the compounds obtained in this sequence are identical to those obtained by direct use of the tri-*iso*-propoxytitanium derivative (*R*)-**283**. It is thus established that the exchange of trialkylstannyl for trichlorotitanyl moieties [(*R*)-**284** → (*R*)-**285**] proceeds in an *anti* S_E' sense.

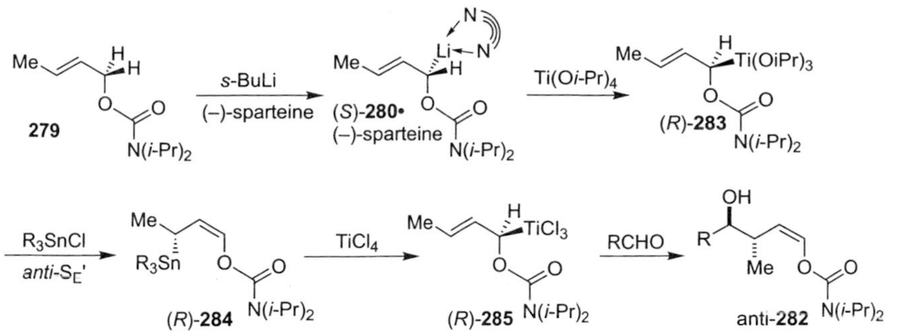

Scheme 10-96.

10.7.2 Allylic Zirconium Reagents

10.7.2.1 Mechanism and stereochemical course of addition

2-Butenylzirconium reagents have been prepared in situ by the addition of one, two or three equivalents of 2-butenyllithium or 2-butenylmagnesium bromide to Cp_2ZrCl_2 in THF. The addition of these reagents to aldehydes affords homoallylic alcohols in high yield and moderate diastereoselectivity (favoring the *anti* homoallylic alcohol) [186]. Low temperature NMR examination of these reagents indicates that they exist as mixtures of *E* and *Z* isomers. The selectivity observed in reaction closely parallels the isomeric purity of the reagent, indicating that the reaction proceeds via a closed, six-membered transition structure.

The preparation of allylic zirconium reagents from allylic ethers and a low valent zirconium source has been described [187]. The subsequent reaction with aldehydes proceeds in good to excellent yield to afford homoallylic alcohols. The diastereoselectivity observed is highly dependent upon the substitution of the start-

Table 10-31. Addition of allylic zirconium reagents to aldehydes.

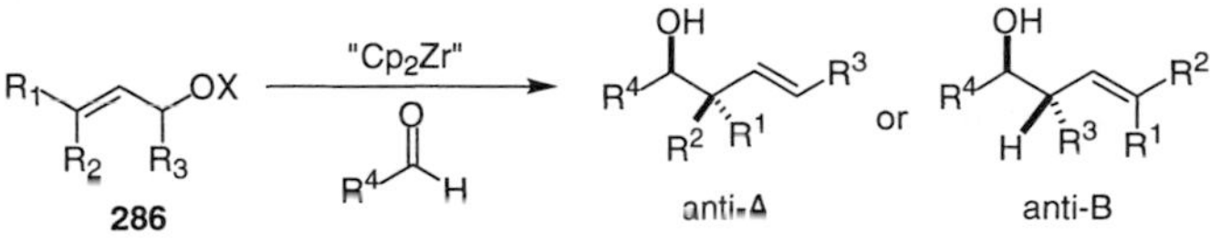

X	R^1	R^2	R^3	R^4	Product	Yield, %	*anti/syn*
Me	Ph	H	H	Ph	**A**	79	10/1
$PhCH_2$	Ph	H	H	Ph	**A**	89	15/1
TBS	Ph	H	H	Ph	**A**	86	23/1
TBS	H	H	Ph	Ph	**B**	93	19/1
TBS	Ph	H	H	*i*-Pr	**A**	96	49/1
$PhCH_2$	Me	H	H	Ph	**A**	76	1.8/1
$PhCH_2$	H	Me	H	Ph	**A**	67	1.9/1
TBS	*c*-Hex	H	H	Ph	**A**	90	10/1
TBS	*i*-Pr	Me	*c*-Hex	Ph	**B**	84	16/1

ing allylic ether. The *anti* homoallylic alcohol is always favored, but higher selectivity is observed when sterically more demanding substituents are present (Table 10-31).

The generation and subsequent reaction of oxy-functionalized allylic zirconium reagents to a wide range of aldehydes proceeds with excellent *anti* selectivity [188]. Allylic zirconium reagents can also be prepared from the hydrozirconation of allenes [189]. Very high levels of diastereoselectivity for both simple aliphatic and aromatic aldehydes are observed in these reactions for the production of the *anti* homoallylic alcohol (Scheme 10-97).

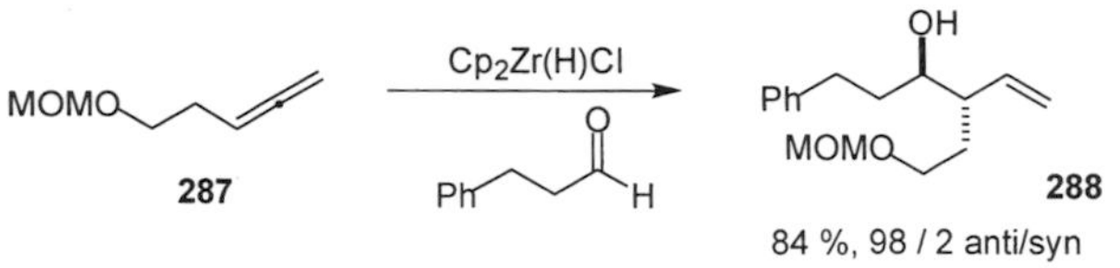

Scheme 10-97.

10.7.3 Allylic Indium Reagents

10.7.3.1 Mechanism of addition

The preparation and subsequent reaction of allylic indium reagents with aldehydes or ketones was first reported by Butsugan in 1988 [190]. The reaction of the 2-butenylindium reagents with aldehydes produced homoallylic alcohols in high yield albeit low selectivity.

Since this initial report, several reviews have detailed the synthetic utility of the reaction, and the degree of stereocontrol that can be obtained with the allylic indium reagents [191]. These species are formed by two general methods, (1) oxidative metallation of allylic halides or phosphates with indium metal or (2) transmetallation of allylic stannanes with indium trichloride. The synthetic utility of these reactions is due, in part, to the relatively low toxicity of the reagents and the stability of allylindium reagent to aqueous conditions. Very little mechanistic information is available for these reactions, though single electron transfer mechanisms have been suggested [192].

10.7.3.2 Stereochemical course of addition: oxidative metallation

Relative stereoselection

To determine the origin of regio- and diastereoselectivity in the indium-mediated coupling of aldehydes with allyl bromides, a series of aldehydes and allylic bromides **289** was examined [193]. From examination of the results collected in Table 10-32, the following conclusions can be drawn: (1) the regioselectivity appears to be governed by the size of the γ-substituent, but not the degree of substitution; (2) in γ-regioselective reactions leading to a mixture of *syn* and *anti* diastereomers, the diastereoselectivity is governed by the size of the substituent on the aldehyde to give mainly the *anti* isomer; (3) the diastereoselectivity appears to be independent of the configuration of the double bond (compare entries 2 and 3, Table 10-32).

Table 10-32. Indium-mediated allylation of carbonyl compounds in water.

RCHO + $R_1(R_2)C{=}CH{-}CH_2Br$ **289**, In / H_2O → anti + syn homoallylic alcohols ($R_{1(2)}$)

Entry	R	R^1	R^2	Yield, %	*anti/syn*
1	Ph	Ph	H	88	96/4
2	*i*-Pr	Ph	H	88	96/4
3	*i*-Pr	H	Ph	79	90/10
4	*c*-Hex	Ph	H	75	90/10
5	*n*-Oct	Ph	H	80	69/31
6	Ph	CO_2Me	H	75	84/16
7	*i*-Pr	CO_2Me	H	81	92/8
8	Ph	Me	H	92	50/50
9	*i*-Pr	Me	H	88	84/16
10	*t*-Bu	Me	H	87	80/20

A convergent mechanism proceeding through closed, chair-like transition structures has been proposed to account for the above observations (Scheme 10-98). The initially formed allylindium species **xxiv** can exist in equilibrium with its regioisomer **xxv**. Because of this equilibrium, the configuration of the double bond is permutable to favor the more stable *E*-isomer over the *Z*-isomer. The coupling reaction with the aldehyde can proceed through a number of plausible transition structures. The stereochemical outcome of the reaction ultimately depends upon the relative energies of the different transition structures; representative possibilities are shown.

The diastereoselective production of homoallylic indium alkoxides can be accomplished by a kinetic resolution process [194]. The indium-mediated reaction of benzaldehyde with 2-butenyl bromide has always been observed to be unselective. The use of alkoxide or halide modifiers in the reactions of allylindium reagents has previously been shown to provide synthetically useful reagents [195]. Upon addition of 2-butenylindium sesquibromide to benzaldehyde it was determined that newly formed *syn* and *anti* homoallylic alcohols undergo decomposition at a similar rate, but as the concentration of the *anti* homoallylic alcohol reaches zero, the rate of decomposition of the *syn* alcohol slows dramatically. Thus, the *syn* homoallylic alcohol can be obtained in high diastereoselectivity, albeit low yield.

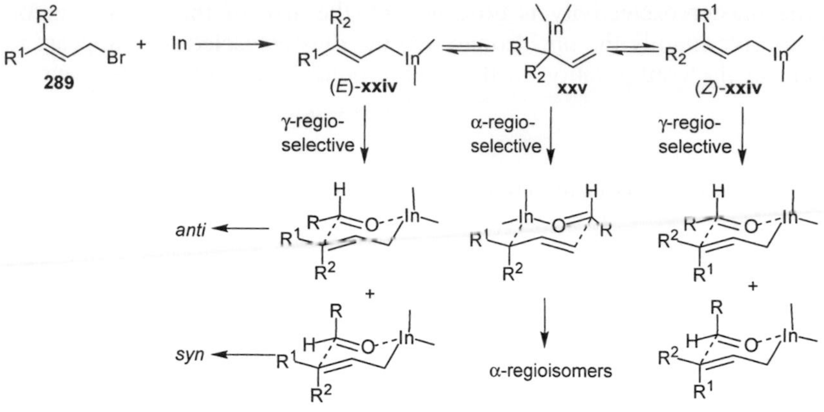

Scheme 10-98.

Internal stereoselection

Achiral allylic indium reagents and chiral aldehydes. Allylindium reagents generated in water react smoothly with aldehydes and ketones (Scheme 10-99) [196]. The reaction of achiral aldehydes and α-oxygenated aldehydes **290** with the allyl indium reagents proceeds smoothly to homoallylic alcohols without the need for external promoters. It is interesting to note that the α-hydroxyl aldehyde was selective for the *syn* (chelation-controlled) product even in water.

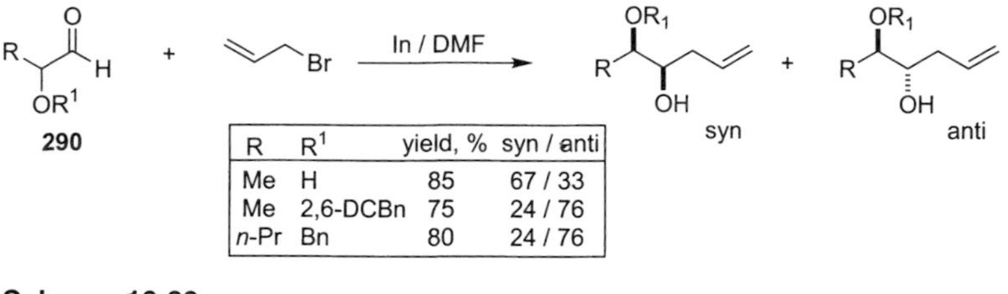

R	R^1	yield, %	syn / anti
Me	H	85	67 / 33
Me	2,6-DCBn	75	24 / 76
n-Pr	Bn	80	24 / 76

Scheme 10-99.

From an extensive examination of the addition of allylindium reagents to α-oxygenated aldehydes **291** it has been established that the stereochemical outcome is dependent on both the α-alkoxy substituent and the solvent employed (Table 10-33) [197]. The silyl and benzyl ethers favor the formation of the *anti* homoallylic alcohol, whereas the MOM and the hydroxyl aldehydes favor the *syn* alcohol. The rate of the reaction is dependent upon the solvent (faster in a mixed THF/water solvent) and the pH (faster at lower pH).

The reaction of β-oxygenated aldehydes **292** are in general less selective (Table 10-34) [197]. The reaction produces a mixture of *syn* and *anti* homoallylic alcohols in high overall yield. The stereochemical outcome of the reaction is believed to be determined by the possible chelation of the β-alkoxy aldehyde. The

Table 10-33. Addition of allylindium bromide to α-oxygenated aldehydes.

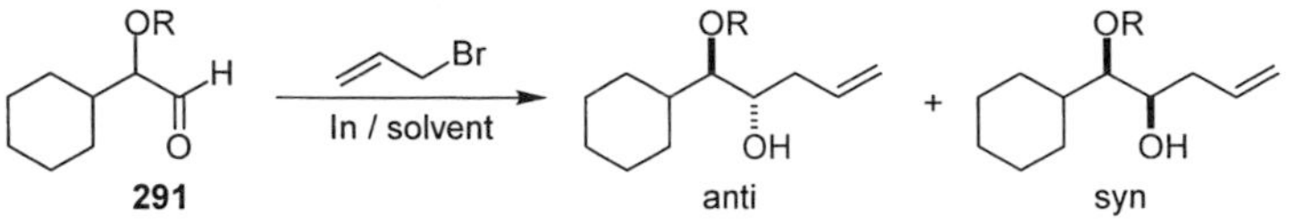

R	Solvent	Yield, %	*anti/ syn*
TBS	H_2O	90	80/20
TBS	THF	88	81/19
$PhCH_2$	H_2O	92	55/45
$PhCH_2$	THF	81	76/24
MOM	H_2O	83	32/68
H	H_2O	87	9/91

Table 10-34. Addition of allylindium bromide to β-oxygenated aldehydes.

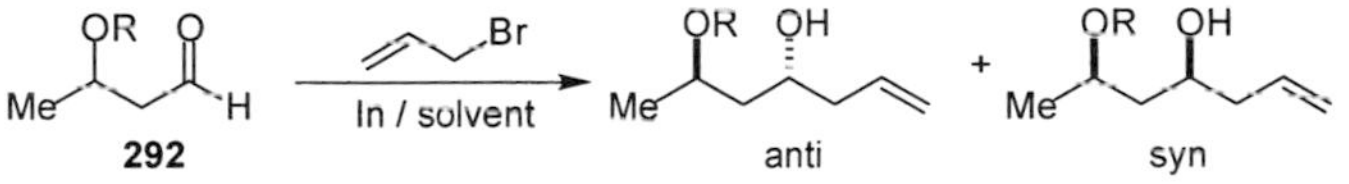

R	Solvent	Yield, %	*anti/syn*
H	H_2O	90	91/9
H	H_2O-THF	87	90/10
$PhCH_2$	H_2O	80	50/50
$PhCH_2$	THF	82	50/50
TBS	H_2O	84	50/50
TBS	THF	77	37/63
Me	H_2O	78	80/20

highest selectivity is observed with the β-hydroxy or β-methoxy aldehydes. The silyloxy or benzyloxy aldehydes afford homoallylic alcohols unselectively.

The reaction of both (*E*)- and (*Z*)-butenylindium bromides with α-alkoxy aldehydes has been examined to assess the direction and sense of both relative and internal stereoinduction. (Scheme 10-100). High selectivities for the 3,4-*syn*/4,5-*anti* diastereomers are observed in the reaction of the *Z*-2-(bromomethyl)-2-butenoate (**293**) with chiral aldehydes **294**. The stereochemical outcome of the reaction is believed to result from reaction through transition structure **xxvi**. As the size of the R substituent increases only modest erosion of the coupling diastereoselectivities is observed. This is reminiscent of the "anti-Felkin" selectivity observed with (*Z*)-2-butenylboronates with chiral aldehydes (Section 10.4.2.1).

The indium-promoted reaction of *E*-cinnamyl bromide with α-alkoxy aldehyde **295** affords the expected 3,4-*anti*/4,5-*anti* diastereomer as the major product (Scheme 10-101). Normal Felkin-type selectivity through transition structure **xxviii** is restored here in view of the *E* configuration of the indium reagent.

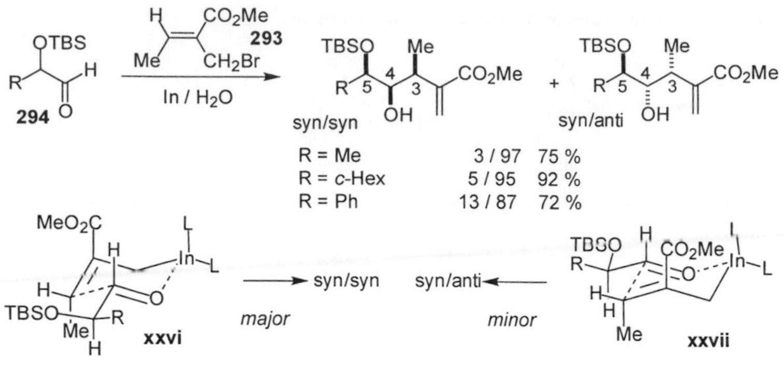

R = Me 3 / 97 75 %
R = *c*-Hex 5 / 95 92 %
R = Ph 13 / 87 72 %

Scheme 10-100.

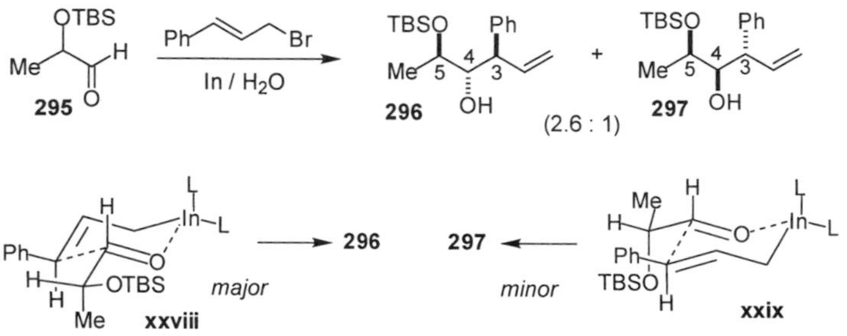

(2.6 : 1)

Scheme 10-101.

The reaction of allylic indium reagents with α-thio aldehydes displays minimal diastereoselectivity, indicating that the allylic indium reagent is not thiophilic [198]. Chelation is not observed, and the π-facial discrimination is achieved through Felkin-Anh transition structures. With the α-amino aldehydes, the stereoselectivity of the aqueous allylation can be "tuned" by proper choice of amino protecting group [198]. The highest selectivities are observed with the *N,N*-dibenzyl and *N,N*-dimethyl analogs which produce the *syn* homoallylic alcohols in up to 99% diastereoselectivity.

Chiral allylic indium reagents and achiral aldehydes. Chiral indium reagents have been generated by the oxidative metallation of allylic bromides, which bear remote stereogenic subunits. For example, the 1,4-asymmetric induction in the indium-mediated coupling of allyl bromides **298** with benzaldehyde affords the *syn*-adducts as the major products (Table 10-35) [199]. The stereoselectivity of the reaction improves with more sterically encumbered allyl bromides.

The 1,4-asymmetric induction in the indium-promoted coupling reaction of aldehydes to protected and unprotected 3-substituted 3-oxy-bromomethylidenepropanes has been investigated [200]. The highest selectivity was observed in reactions of the silyl-protected bromomethylvinyl alkanols. A representative series is shown below in Scheme 10-102. The free hydroxyl and the methyl ether containing allylic bromides were less selective than the corresponding silyl derivative [201].

Table 10-35. Indium-mediated addition of substituted allylic bromides.

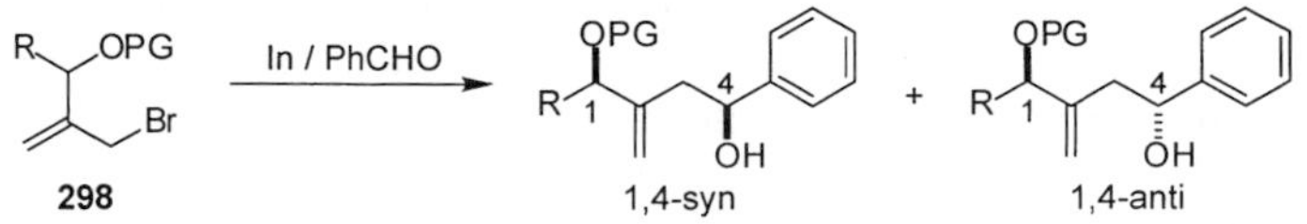

R	PG	Yield, %	*syn*/*anti*
Me	Bn	72	86/14
Ph	Bn	79	88/12
i-Pr	Bn	59	96/4
Me	TBS	67	86/14
Ph	TBS	75	90/10
i-Pr	TBS	71	97/3
Me	MOM	89	73/27
Ph	MOM	83	79/21
i-Pr	MOM	85	82/18

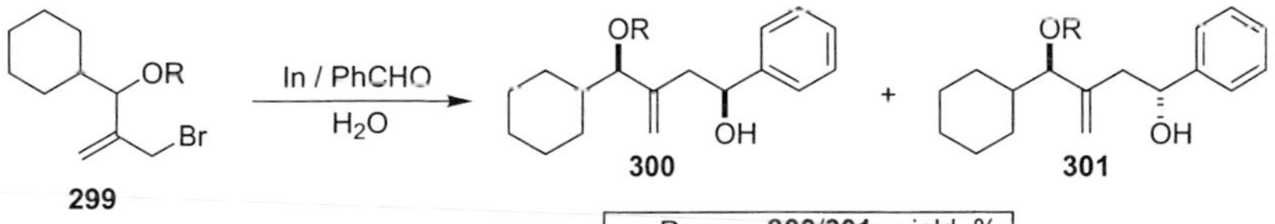

R	300/301	yield, %
TBS	99 / 1	90
Me	69 / 31	87
H	50 / 50	76

Scheme 10-102.

The 1,4-*syn* product is preferred only when the steric bulk of the oxygenated substituent on the allylic bromide is appreciable, as in the silyl ether. The transition structures shown in Fig. 10-10 rationalize the stereochemical outcome in this reaction. For the hydroxy and methoxy bromides, it is possible that the *syn*-selectivity of the reaction is eroded because the O-inside conformation **xxx** is now populated due to the balanced steric demands of the R and OR^2 substituents.

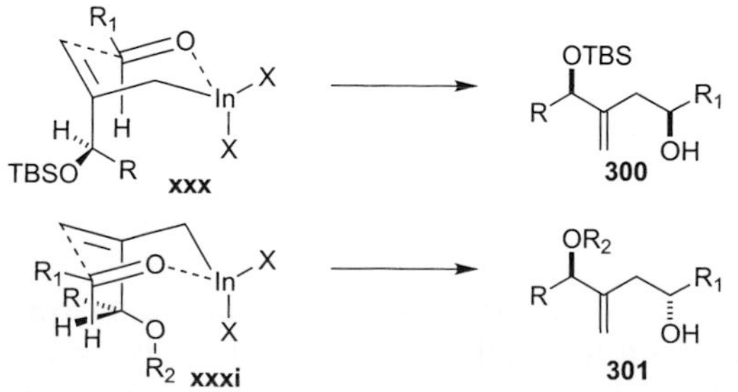

Figure 10-10. Transition structure models for substituted allylic indium reagents.

1,4-Stereoinduction with an alkoxymethyl substituent has been demonstrated recently in the indium-promoted coupling reaction of protected and unprotected bromomethylvinyl alkanols **302** (Scheme 10-103) [202]. A significant rate difference is observed among the hydroxy, methoxy and siloxy allylic bromides. The silyl-protected reagent is much less reactive in competitive allylations than either the hydroxy or the methoxy allylic bromides. The most reactive species, the hydroxy allylic bromide, is proposed to react via a bicyclic or chelated transition structure **xxxii** (Scheme 10-103). The reaction of the hydroxy allyl bromide **302** with various aldehydes affords the 1,4-*syn* homoallylic alcohol in high yield and good diastereoselectivity (~9/1 *syn*/*anti* for PhCHO).

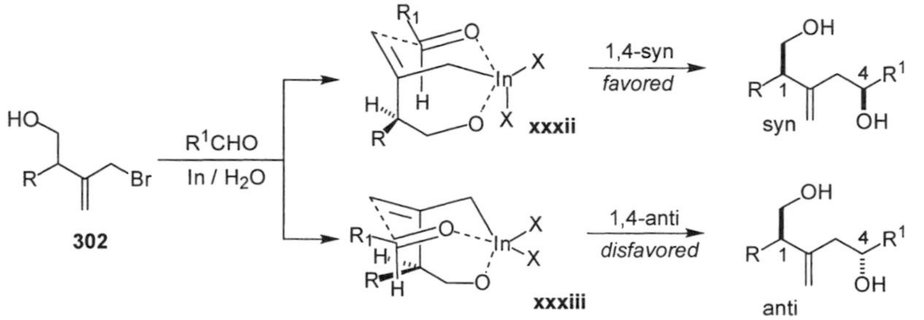

Scheme 10-103.

10.7.3.3 Stereochemical course of addition: transmetallation

Relative stereoselection

The generation of allylic indium reagents by transmetallation of allylic stannanes with $InCl_3$ and subsequent reaction of these with aldehydes represents an important advance for diastereoselective synthesis of homoallylic alcohols [203]. In these reactions, the stannane is added to a premixed solution of the aldehyde and $InCl_3$ in acetone. In this way, the reaction of cyclohexanecarboxaldehyde with the 2-butenylindium reagent affords a 98/2 *anti*/*syn* mixture of homoallylic alcohols (Scheme 10-104).

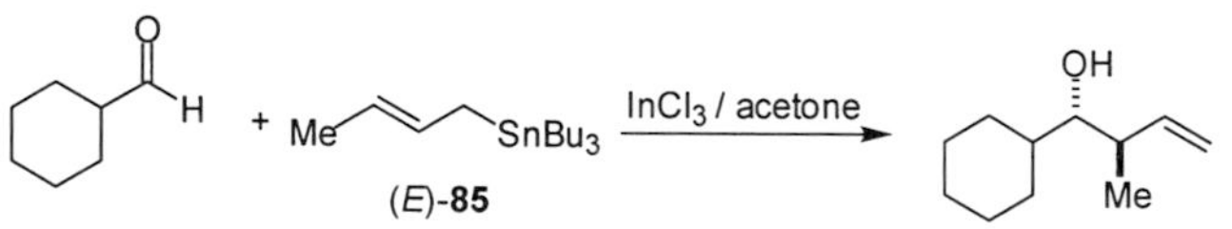

Scheme 10-104.

The relative diastereoselectivity in the indium trichloride-mediated allylation with allylic stannanes in water is for the *anti* homoallylic alcohol [204]. The reactions are *anti* selective regardless of the geometry of the starting 2-butenylbromide. The in-situ-generated allylic indium species undergoes reaction with the corresponding aldehyde via a familiar six-membered transition structure also proposed for the reductive metallation pathway.

A highly stereoselective synthesis of trifluoromethylated homoallylic alcohols is possible using the transmetallation [Sn(II) to In(III)] pathway [205]. The indium trichloride/tin-promoted reaction of trifluorobromobutene with various aldehydes afforded the homoallylic alcohols in extremely high yield and diastereoselectivity (Scheme 10-105). The strong preference for the *anti* products with the simple aldehydes is expected on the basis of previous observations with 2-butenylindium reagents explained by the cyclic transition structure **xxxiv** shown. The *syn* homoallylic alcohol was obtained upon reaction of glyoxylic acid and 2-pyridinyl carboxaldehyde. The *syn* products were proposed to arise by reaction via the 5-membered chelate transition structures **xxxv** and **xxxvi**.

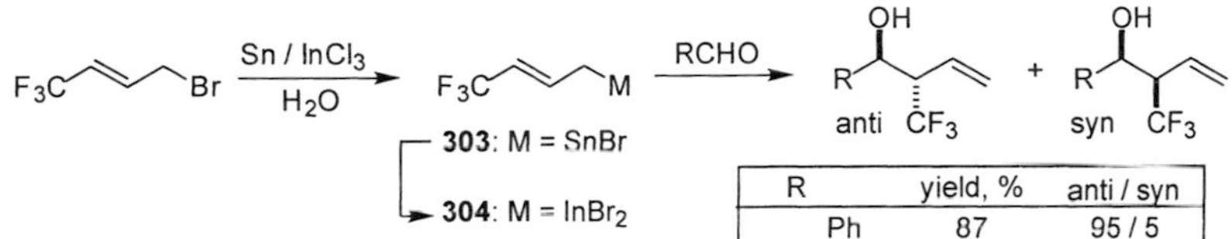

R	yield, %	anti / syn
Ph	87	95 / 5
c-C_6H_{11}	92	100 / 0
3-pyridyl	95	92 / 8
2-pyridyl	96	0 / 100
CO_2H	83	0 / 100

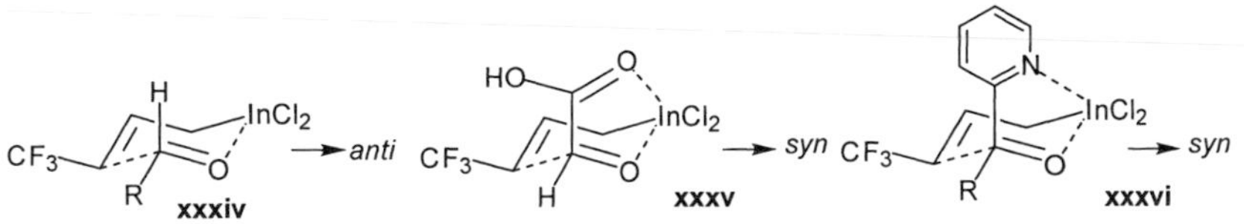

Scheme 10-105.

Internal stereoselection

The $InCl_3$-promoted reaction of enantiomerically enriched α-alkoxystannanes **305** with achiral aldehydes produces a mixture of homoallylic alcohols with a high degree of relative diastereoselectivity and excellent enantioselectivity (Scheme 10-106). A plausible explanation first invokes the formation of indium reagent **306** produced via *anti* S_E2' attack of $InCl_3$ on the α-alkoxystannane. Addition to the aldehyde can then occur via the chair-like transition structure **xxxvii** to afford the *anti* homoallylic alcohol preferentially.

The scope of this transformation has been extended by employing enantiomerically pure aldehydes for the synthesis of the eight stereoisomeric hexoses [206]. The stereochemical outcome of these reactions is controlled by the configuration of the chiral α-alkoxystannane.

The preparation of allenylindium reagents and their reaction with aldehydes has further expanded the utility of indium in synthesis. By making use of the transmetallation protocol, enantiomerically enriched allenyltin reagents **308** (themselves generated by an *anti* S_N2' substitution of propargylic mesylate **307** with

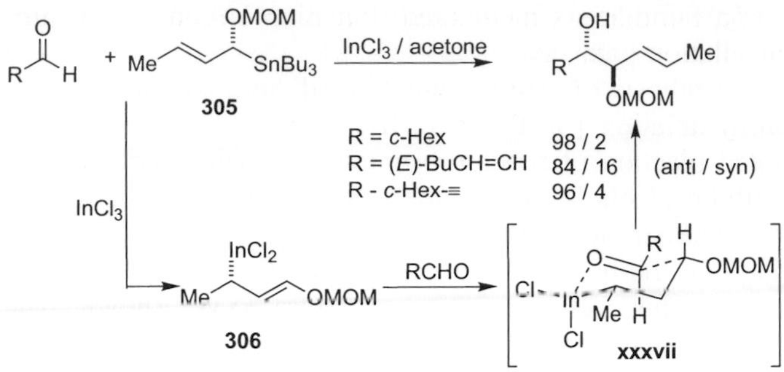

R	anti / syn
R = *c*-Hex	98 / 2
R = (*E*)-BuCH=CH	84 / 16
R - *c*-Hex-≡	96 / 4

Scheme 10-106.

Bu_3SnLi) undergo smooth exchange with $InCl_3$ to form the allenylindium species **309** (Scheme 10-107) [207]. This transmetallation takes place primarily via the unexpected *syn* S_E' pathway. These reagents afford homopropargylic alcohols **310** in high yield but modest enantio- and diastereoselectivity, especially in additions involving *α*-branched aliphatic aldehydes. Unfortunately, the formation of Bu_3SnX by-products prevents the widespread usage of these reagents.

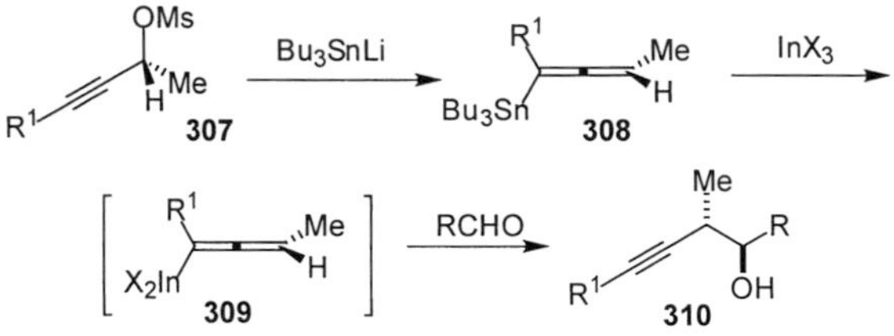

Scheme 10-107.

A very recent improvement in the transmetallation pathway involves the transient formation of chiral allenylindium reagents from propargylic mesylate (*R*)-**311** through an oxidative transmetallation pathway (Scheme 10-108) [208]. Optimized conditions for the reaction require the addition of 5 mol% of the catalyst $Pd(dppf)Cl_2$ and indium iodide to the mesylate at room temperature. The reaction of achiral aldehydes with the newly generated allenylindium reagent affords homopropargylic alcohols in high yield and moderate to good diastereoselectivity.

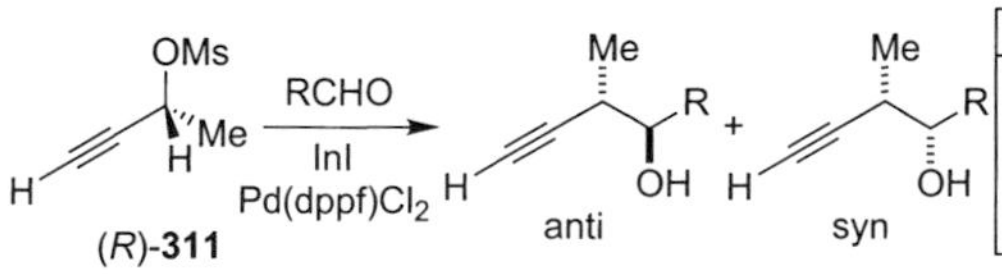

R	yield, %	anti / syn	ee, %
c-Hex	76	95 / 5	95
n-Hex	73	82 / 18	96
1-heptynyl	62	72 / 28	95
Ph	85	45 / 55	92

Scheme 10-108.

The scope of the reaction is further demonstrated by the reaction of matched and mismatched reagent pairs with aldehyde (*S*)-**106** (Scheme 10-109).The *anti,anti* adduct **313** is obtained in 87% yield from (*R*)-**312** and the *anti,syn* adduct **314** is formed in 88% yield from (*S*)-**312**. Only a trace of the diastereomeric product was detected by ^{1}H-NMR analysis. These reactions indicate that the addition is strongly reagent controlled.

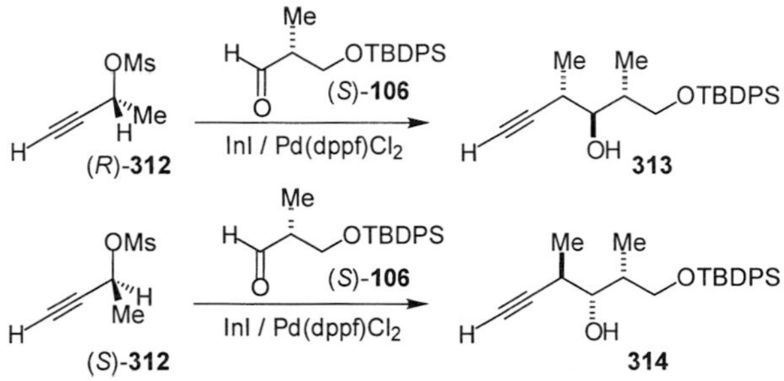

Scheme 10-109.

10.8 Conclusions and Future Outlook

The nearly twenty years of intensive research in the development, application and understanding of allylation reactions has provided spectacular successes. Many useful methods are now available for installation of homoallylic and homopropargylic subunits with excellent levels of stereocontrol in a myriad of structural settings. Good functional group compatibility has been demonstrated and factors that influence the various dimensions of stereocontrol have been identified in many cases. There remain, nevertheless, important challenges in the area of introducing newer methods, particularly catalytic enantioselective (and diastereoselective) variants. These modifications are most visibly lacking in the important subclass of allylic chromium and indium reagents which have proven otherwise so useful in recent years. Likewise, the advent of chiral Lewis base catalysis, which has shown respectable promise in the reactions of allylic silanes, is certainly applicable with other allylic element reagents. In view of the accelerated development of new asymmetric transformations, it is easy to predict that these problems will be addressed and that creative and practical solutions will be found. It is our hope that this review will serve to stimulate both the experienced practitioner and the newcomer to the field to undertake these endeavors.

References

1. Reviews: (a) Yamamoto, Y.; Asao, N. *Chem. Rev.* **1993**, *93*, 2207. (b) Yamamoto, Y. *Acc. Chem. Res.* **1987**, *20*, 243. (c) Yamamoto, Y. *Aldrichim. Acta* **1987**, *30*, 45. (d) Hoffmann, R.W. *Angew. Chem. Int. Ed. Engl.* **1987**, *26*, 489. (e) Nishigaichi, Y.; Takuwa, A.; Naruta, Y.; Maruyama, K. *Tetrahedron* **1993**, *49*, 7395. (f) Roush, W.R. In *Comprehensive Organic Synthesis, Vol. 2, Additions to C–X Bonds, Part 2*; Heathcock, C.H., Ed.; Pergamon Press: Oxford, 1991, pp 1. (g) *Stereoselective Synthesis, Methods of Organic Chemistry (Houben-Weyl)*; Edition E21; Helmchen, G.; Hoffmann, R.; Mulzer, J.; Schaumann, E. Eds.; Thieme, Stuttgart, 1996; Vol. 3; pp 1357–1602.
2. Denmark, S.E.; Weber, E.J. *Helv. Chim. Acta* **1983**, *66*, 1655.
3. Helmchen, G. In *Stereoselective Synthesis, Methods of Organic Chemistry (Houben-Weyl)*; Edition E21; Helmchen, G.; Hoffmann, R.; Mulzer, J.; Schaumann, E. Eds.; Thieme, Stuttgart, 1996; Vol. 1; pp 1–74.
4. Masamune, S.; Kaiho, T.; Garvey, D.S. *J. Am. Chem. Soc.* **1982**, *104*, 5521.
5. There is good reason to adapt the Helmchen[3] "substrate-induced" and "auxiliary induced" stereoselection terms to differentiate the two different scenarios here.
6. Sommer, L.H.; Tyler, L.J.; Whitmore, F.C. *J. Am. Chem. Soc.* **1948**, *70*, 2872.
7. Frainnet, E.; Calas, R. *C. R. Hebd. Séances Acad. Sci.* **1955**, *240*, 203.
8. (a) Calas, R.; Dunogues, J.; Deleris, G.; Pisciotti, F. *J. Organomet. Chem.* **1974**, *69*, C15. (b) Deleris, G.; Dunogues, J.; Calas, R. *J. Organomet. Chem.* **1975**, *93*, 43.
9. Hosomi, A.; Sakurai, H. *Tetrahedron Lett.* **1976**, 1295.
10. Reviews: (a) Sakurai, H. *Pure Appl. Chem.* **1982**, *54*, 1. (b) Majetich, G. In *Organic Synthesis: Theory and Applications*; Hudlicky, T., Ed.; JAI Press, Greenwich, 1989; Vol. 1, p. 173. (c) Fleming, I.; Dunogués, J.; Smithers, R. In *Organic Reactions*, Wiley, New York, 1989; Vol. 37, pp 57–575. (d) Hosomi, A. *Acc. Chem. Res.* **1988**, *21*, 200. (e) The intramolecular reactions of allylsilanes have also been reviewed: Schinzer, D. *Synthesis* **1988**, 263. (f) Fleming, I. In *Comprehensive Organic Synthesis, Vol. 2, Additions to C–X Bonds, Part 2;* Heathcock, C.H., Ed.; Pergamon Press, Oxford, 1991, pp 563–593. (g) Fleming, I.; Barbero, A.; Walter, D. *Chem. Rev.* **1997**, *97*, 2063. (h) Masse, C.; Panek, J. *Chem. Rev.* **1995**, *95*, 1293.
11. Fleming, I.; Langley, J.A. *J. Chem. Soc. Perkin I* **1981**, 1421.
12. Review: Jarvie, A.W.P. *Organomet. Chem. Rev. A.* **1970**, *6*, 153. The effect of hyperconjugation in chemical reactions has recently been investigated: (a) Lambert, J.B.; Finzel, R.B.; *J. Am. Chem. Soc.* **1982**, *104*, 2020. (b) Wierschke, S.G.; Chandrasekhar, J.; Jorgensen, W.L. *J. Am. Chem. Soc.* **1985**, *107*, 1496.
13. (a) Yamamoto, Y.; Yatagai, H.; Maruyama, K. *J. Org. Chem.* **1980**, *45*, 195. (b) Koreeda, M.; Tanaka, Y. *J. Chem Soc., Chem. Comm.* **1982**, 845.
14. Seebach, D.; Prelog, V. *Angew. Chem., Int. Ed. Engl.* **1982**, *21*, 652.
15. Kahn, S.D.; Pau, C.F.; Chamberlin, A.R.; Hehre, W.J. *J. Am. Chem. Soc.* **1987**, *109*, 650.
16. (a) Imachi, M.; Nakagawa, J.; Hayashi, M. *J. Mol. Struct.* **1983**, *102*, 403. (b) Hayashi, M.; Imachi, M. *Chem. Lett.* **1977**, 121. (c) Beagley, B.; Foord, A.; Moutran, R.; Roszondai, B. *J. Mol. Struc.* **1977**, 117. (d) Ohno, K.; Taga, K.; Murata, M. *Bull. Chem. Soc. Jpn.* **1977**, *50*, 2870.
17. (a) Denmark, S.E.; Weber, E.J. *J. Am. Chem. Soc.* **1984**, *106*, 7970. (b) Denmark, S.E.; Henke, B.R.; Weber, E.J. *J. Am. Chem. Soc.* **1987**, *109*, 2512. (c) Denmark, S.E.; Weber, E.J.; Wilson, T.M. *Tetrahedron* **1989**, *45*, 1053. (d) Denmark, S.E.; Almstead, N.G. *J. Org. Chem.* **1994**, *59*, 5130. (e) Denmark, S.E.; Almstead, N.G. unpublished results.
18. Denmark, S.E.; Almstead, N.G. *J. Am. Chem. Soc.* **1993**, *115*, 3133.
19. For excellent review on Lewis acid carbonyl structures see: (a) Shambayati, S.; Crowe, W.E.; Schreiber, S.L. *Angew. Chem., Int. Ed. Engl.* **1990**, *29*, 256. (b) Shambayati, S.; Schreiber, S.L. In *Comprehensive Organic Synthesis, Vol. 1, Additions to C–X p Bonds, Part 1*; Schreiber, S.L., Ed.; Pergamon Press, Oxford, 1991; pp 283–324.
20. (a) Keck, G.E.; Castellino, S. *J. Am. Chem. Soc.* **1986**, *108*, 3847. (b) Keck, G.E.; Castellino, S.; Wiley, M.R. *J. Org. Chem.* **1986**, *51*, 5478. (c) Filippini, F.; Susz, B.-P. *Helv. Chim. Acta* **1971**, *54*, 835. (d) Beattie, I.R. *Quart. Rev.* **1963**, *17*, 382.

21. Reetz, M.T.; Hüllmann, M.; Massa, W.; Berger, S.; Rademacher, P.; Heymanns, P. *J. Am. Chem. Soc.* **1986**, *108*, 2405.
22. Weber, E.J. Ph.D. Thesis, University of Illinois, 1985.
23. (a) Anh, N.T.; Thanh, B.T. *Nouv. J. Chim.* **1986**, *10*, 681. (b) Mulzer, J.; Büntrup, G.; Finke, J.; Zippel, M. *J. Am. Chem. Soc.* **1979**, *101*, 7723.
24. For a discussion of the fluoride-induced reactions of allylsilanes see: Majetich, G. In *Organic Synthesis, Theory and Applications;* Hudlicky, T. Ed.; JAI Press, Inc., Greenwich, CT, 1989; pp 173–240.
25. (a) Hosomi, A.; Shirahata, A.; Sakurai, H. *Tetrahedron Lett.* **1978**, 3043. (b) Hosomi, A.; Shirahata, A.; Sakurai, H. *Chem. Lett.* **1978**, 901.
26. (a) Majetich, G.; Casares, A.; Chapman, D.; Behnke, M. *J. Org. Chem.* **1986**, *51*, 1745. (b) Majetich, G.; Desmond, R.W. Jr.; Soria, J.J. *J. Org. Chem.* **1986**, *51*, 1753.
27. (a) Corriu, R. *Pure Appl. Chem.* **1988**, *60*, 99. (b) Sakurai, H. *Synlett.* **1989**, *1*, 1.
28. (a) Hayashi, T.; Kabeta, K.; Hamachi, I.; Kumada, M. *Tetrahedron Lett.* **1983**, *24*, 2865. (b) Hayashi, T.; Konishi, M.; Ito, H.; Kumada, M. *J. Am. Chem. Soc.* **1982**, *104*, 4963. (c) Hayashi, T.; Konishi, M.; Kumada, M. *J. Org. Chem.* **1983**, *48*, 281.
29. For a recent review on Cram or Felkin-Anh control in addition reactions see: Mikami, K.; Shimizu, M. In *Advances in Detailed Reaction Mechanisms, Vol. 3;* Coxon, J.M. Ed.; JAI Press, Inc.: Greenwich, CT, 1994; pp 45–77.
30. Heathcock, C.H.; Kiyooka, S.; Blumenkopf, T.A. *J. Org. Chem.* **1984**, *49*, 4214.
31. Springer, J.B.; DeBoard, J.; Corcoran, R.C. *Tetrahedron Lett.* **1995**, *36*, 8733.
32. Hanessian, S.; Tehim, A.; Chen, P. *J. Org. Chem.* **1993**, *58*, 7768.
33. Mikami, K.; Kawamoto, K.; Loh, T.-P.; Nakai, T. *J. Chem. Soc., Chem. Commun.* **1990**, 1161.
34. (a) Panek, J.S.; Yang, M. *J. Am. Chem. Soc.* **1991**, *113*, 9868. (b) Panek, J.S.; Beresis, R. *J. Org. Chem.* **1993**, *58*, 809.
35. Brook, A.G.; Bassindale, A.R. In "Rearrangements in the Ground and Excited States"; de Mayo, P. Ed.; Academic Press; New York, 1980, Vol. II, pp 190–192.
36. Panek, J.S.; Cirillo, P.F. *J. Org. Chem.* **1993**, *58*, 999.
37. Jain, N.F.; Cirillo, P.F.; Pelletier, R.; Panek, J.S. *Tetrahedron Lett.* **1995**, *36*, 8727.
38. Panek, J.S.; Xu, F.; Rondón, A.C. *J. Am. Chem. Soc.* **1998**, *120*, 4113.
39. Jain, N.F.; Takenaka, N.; Panek, J.S. *J. Am. Chem. Soc.* **1996**, *118*, 12475.
40. Panek, J.S.; Liu, P. *Tetrahedron Lett.* **1997**, *38*, 5127.
41. Reviews: (a) Alexakis, A.; Mangeney, P. *Tetrahedron: Asymmetry* **1990**, *1*, 477. (b) Seebach, D.; Imwinkelreid, R.; Weber, T. In *Modern Synthetic Methods*; Scheffold, R. Ed.; Springer Verlag: Berlin, **1986**, Vol. 4; p 125.
42. (a) Tietze, L.F.; Dölle, A.; Schiemann, K. *Angew. Chem., Int. Ed. Engl.* **1992**, *31*, 1372. (b) Tietze, L.F.; Schiemann, K.; Wegner, C.; Wulff, C. *Chem. Eur. J.* **1996**, *2*, 1164. (c) Tietze, L.F.; Wulff, C.; Wegner, C.; Schuffenhauer, A.; Schiemann, K. *J. Am. Chem. Soc.* **1998**, *120*, 4276.
43. (a) Tietze, L.F.; Schiemann, K.; Wegner, C. *J. Am. Chem. Soc.* **1995**, *117*, 5851. (b) Tietze, L.F.; Schiemann, K.; Wegner, C.; Wulff, C. *Chem. Eur. J.* **1998**, *4*, 1862.
44. (a) Noyori, R. *Asymmetric Catalysis in Organic Synthesis,* Wiley-Interscience, 1994. (b) *Catalytic Asymmetric Synthesis*, Ojima, I. Ed., VCH, 1993. (c) Gates, B.C. *Catalytic Chemistry,* Wiley, 1992. (d) Parshall, G.W.; Ittel S.D. *Homogeneous Catalysis,* Wiley, Second Edition, 1992. (e) *Comprehensive Asymmetric Catalysis,* Jacobsen, E.N.; Pfaltz, A.; Yamamoto, H., Eds.; Springer Verlag, Heidelberg, 1999.
45. (a) Fururta, K.; Mouri, M.; Yamamoto, H. *Synlett* **1991**, 561. (b) Ishihara, K.; Mouri, M.; Gao, Q.; Maruyama, T.; Furata, K.; Yamamoto, H. *J. Am. Chem. Soc.* **1993**, *115*, 11490.
46. (a) Aoki, S.; Mikami, K.; Terada, M.; Nakai, T. *Tetrahedron* **1993**, *49*, 1783. (b) Mikami, K.; Matsukawa, S. *Tetrahedron Lett.* **1994**, *35*, 3133. (c) Gauthier, D.R.; Carreira, E.M. *Angew. Chem., Int. Ed. Engl.* **1996**, *35*, 2363.
47. Marshall, J.A. *Chemtracts-Org. Chem.* **1998**, *11*, 697.
48. Kira, M.; Kobayashi, M.; Sakurai, H. *Tetrahedron Lett.* **1987**, *28*, 4081.
49. Kira, M.; Sato, K.; Sakurai, H.; Hada, M.; Izawa, M.; Ushiro, J. *Chem. Lett.* **1991**, 387.
50. (a) Kobayashi, S.; Nishio, K. *Tetrahedron Lett.* **1993**, *34*, 3453. (b) Kobayashi, S.; Nishio, K. *J. Org. Chem.* **1994**, *59*, 6620.

51. Short, J.D.; Attenoux, S.; Berrisford, D.J. *Tetrahedron Lett.* **1997**, *38*, 2351.
52. (a) Kira, M.; Hino, T.; Sakurai, H. *Tetrahedron Lett.* **1989**, *30*, 1099. (b) Sato, K.; Kira, M.; Sakurai, H. *J. Am. Chem. Soc.* **1989**, *111*, 6429. (c) Kira, M.; Sato, K.; Sakurai, H. *J. Am. Chem. Soc.* **1990**, *112*, 257. (d) Kira, M.; Sato, K.; Sekimoto, K.; Gewald, R.; Sakurai, H. *Chem. Lett.* **1995**, 281. (e) Gewald, R.; Kira, M.; Sakurai, H. *Synthesis* **1996**, 111.
53. (a) Wang, Z.; Wang, D.; Sui, X. *Chem. Commun.* **1996**, 2261. (b) Wang, D.; Wang, Z.G.; Wang, M.W.; Chen, Y.J.; Liu, L.; Zhu, Y. *Tetrahedron Asym.* **1999**, *10*, 327. (c) Zhang, L.C.; Sakurai, H.; Kira, M. *Chem. Lett.* **1997**, 129.
54. Chemler, S.R.; Roush, W.R. *J. Org. Chem.* **1998**, *63*, 3800.
55. Denmark, S.E.; Coe, D.M.; Pratt, N. E.; Griedel, B.D. *J. Org. Chem.* **1994**, *59*, 6161.
56. (a) Iseki, K.; Kuroki, Y.; Takahashi, M.; Kobayashi, Y. *Tetrahedron Lett.* **1996**, *37*, 5149. (b) Iseki, K.; Kuroki, Y.; Takahashi, M.; Kishimoto, S.; Kobayashi, Y. *Tetrahedron* **1997**, *53*, 3513. (c) Iseki, K.; Mizuno, S.; Kuroki, Y.; Kobayashi, Y. *Tetrahedron Lett.* **1998**, *39*, 2767. (d) Iseki, K.; Mizuno, S.; Kuroki, Y.; Kobayashi, Y. *Tetrahedron* **1999**, *55*, 977.
57. Angell, R.M.; Barrett, A.G.M.; Braddock, D.C.; Swallow, S.; Vickery, B.D. *Chem. Commun.* **1997**, 919.
58. Nakajima, M.; Saito, M.; Shiro, M.; Hashimoto, S. *J. Am. Chem. Soc.* **1998**, *120*, 6419.
59. Cerveau, G.; Chuit, C.; Corriu, R.J. P.; Reye, C. *J. Organomet. Chem.* **1987**, *328*, C17.
60. (a) Hosomi, A.; Kohra, S.; Tominaga, Y. *J. Chem. Soc., Chem. Commun.* **1987**, 1517. (b) Hosomi, A.; Kohra, S.; Tominaga, Y. *Chem. Pharm. Bull.* **1987**, *35*, 2155. (c) Hosomi, A.; Kohra, S.; Ogata, K.; Yanagi, T.; Tominaga, Y. *J. Org. Chem.* **1990**, *55*, 2415.
61. Kira, M.; Sato, K.; Sakurai, H. *J. Am. Chem. Soc.* **1988**, *110*, 4599.
62. Hayashi, T.; Matsumoto, Y.; Kiyoi, T.; Ito, Y. *Tetrahedron Lett.* **1988**, *29*, 5667.
63. Matsumoto, K.; Oshima, K.; Utimoto, K. *J. Org. Chem.* **1994**, *59*, 7152.
64. Omoto, K.; Sawada, Y.; Fujimoto, H. *J. Am. Chem. Soc.* **1996**, *118*, 1750.
65. Konig, K.; Neumann, W.P. *Tetrahedron Lett.* **1967**, 495.
66. Servens, C.; Pereyre, M. *J. Organomet. Chem.* **1972**, *35*, C20.
67. Tagliavini, G.; Peruzzo, V.; Plazzogna, G.; Marton, D. *Inorg. Chim. Acta* **1977**, *24*, L47.
68. (a) Naruta, Y.; Ushida, S.; Maruyama, K. *Chem. Lett.* **1979**, 919. (b) Hosomi, A.; Iguchi, H.; Endo, M.; Sakurai, H. *Chem. Lett.* **1979**, 977.
69. Pratt, A.J.; Thomas, E.J. *J. Chem. Soc., Chem. Commun.* **1982**, 1115.
70. (a) Yamamoto, Y.; Maruyama, K.; Matsumoto, K. *J. Chem. Soc., Chem. Commun.* **1983**, 489. (b) Isaacs, N.S.; Maksimovic, L.; Rintoul, G.B.; Young, D.J. *J. Chem. Soc., Chem. Commun.* **1992**, 1749. (c) Isaacs, N.S.; Marshall, R.L.; Young, D.J. *Tetrahedron Lett.* **1992**, *33*, 3023.
71. (a) Yamamoto, Y.; Yatagai, H.; Naruta, Y.; Maruyama, K. *J. Am. Chem. Soc.* **1980**, *102*, 7107. (b) Yamamoto, Y.; Yatagai, H.; Ishihara, Y.; Maeda, N.; Maruyama, K. *Tetrahedron* **1984**, *40*, 2239.
72. (a) Yamamoto, Y.; Shida, N. In *Advances in Detailed Reaction Mechanisms;* JAI Press, Inc., London, 1994; Vol. 3, pp 1–44. (b) Marshall, J.A. *Chem. Rev.* **1996**, *96*, 31.
73. Denmark, S.E.; Weber, E.J.; Wilson, T.M.; Willson, T.M. *Tetrahedron* **1989**, *45*, 1053.
74. Denmark, S.E.; Hosoi, S. *J. Org. Chem.* **1994**, *59*, 5133.
75. (a) Keck, G.E.; Savin, K.A.; Cressman, E.N.K.; Abbott, D.E. *J. Org. Chem.* **1994**, *59*, 7889. (b) Keck, G.E.; Dougherty, S.M.; Savin, K.A. *J. Am. Chem. Soc.* **1995**, *117*, 6210.
76. (a) Yano, K.;Baba, A.; Matsuda, H. *Bull. Chem. Soc. Jpn.* **1992**, *65*, 66. (b) Shibata, I.; Fukuoka, S.; Yoshimura, N.; Matsuda, H.; Baba, A. *J. Org. Chem.* **1997**, *62*, 3790.
77. Fishwick, M.; Wallbridge, M.G.H. *J. Organomet. Chem.* **1970**, *25*, 69.
78. (a) Tagliavini, G. *Reviews Silicon, Germanium, Tin Lead Cmpds.* **1985**, *8*, 237. (b) Boaretto, A.; Marton, D.; Tagliavini, G.; Ganis, P. *J. Organomet. Chem.* **1987**, *321*, 199. (c) Boaretto, A.; Furlani, D.; Marton, D.; Tagliavini, G.; Gambaro, A. *J. Organomet. Chem.* **1986**, *299*, 157.
79. (a(Keck, G.E.; Abbott, D.E., *Tetrahedron Lett.* **1984**, *25*, 1883. (b) Keck, G.E.; Abbott, D.E.; Boden, E.P.; Enholm, E.J. *Tetrahedron Lett.* **1984**, *25*, 3927.
80. Yamamoto, Y.; Maeda, N.; Maruyama, K. *J. Chem. Soc., Chem. Commun.* **1983**, 742.
81. Naruta, Y.; Nishigaichi, Y.; Maruyama, K. *J. Org. Chem.* **1988**, *53*, 1192.
82. Denmark, S.E.; Wilson, T.; Willson, T.M. *J. Am. Chem. Soc.* **1988**, *110*, 984.
83. Keck, G.E.; Andrus, M.B.; Castellino, S. *J. Am. Chem. Soc.* **1989**, *111*, 8136.

84. (a) Keck, G.E.; Boden, E.P. *Tetrahedron Lett.* **1984**, *25*, 265. (b) Keck, G.E.; Boden, E.P. *Tetrahedron Lett.* **1984**, *25*, 1879. (c) Keck, G.E.; Castellino, S.; Wiley, M.R. *J. Org. Chem.* **1986**, *51*, 5478. (d) Keck, G.E.; Murry, J.A. *J. Org. Chem.* **1991**, *56*, 6606.
85. Mikami, K.; Kawamoto, K.; Loh, T.-P.; Nakai, T. *J. Chem. Soc. Chem. Commun.* **1990**, 1161.
86. Yoshida, T.; Chika, J.-I.; Takei, H. *Tetrahedron Lett.* **1998**, *39*, 4305.
87. Marshall, J.A.; Tang, Y. *Synlett* **1992**, 653.
88. Marshall, J.A.; Palovich, M.R. *J. Org. Chem.* **1998**, *63*, 4381.
89. (a) Keck, G.E.; Tarbet, K.H.; Geraci, L.S. *J. Am. Chem. Soc.* **1993**, *115*, 8467. (b) Keck, G.E.; Krishnamurthy, D.; Grier, M.C. *J. Org. Chem.* **1993**, *58*, 6543. (c) Keck, G.E.; Geraci, L.S. *Tetrahedron Lett.* **1993**, *34*, 7827. (d) Costa, A.L.; Piazza, M.G.; Tagliavini, E.; Trombini, C.; Umani-Rochi, A. *J. Am. Chem. Soc.* **1993**, *115*, 7001. (e) Bedeschi, P.; Casolari, S.; Costa, A.L.; Tagliavini, E.; Umani-Ronchi, A. *Tetrahedron Lett.* **1995**, *36*, 7897. (f) Yu, C. M.; Choi, H.-S.; Jung, W.-H.; Lee, S.-S. *Tetrahedron Lett.* **1996**, *37*, 7095. (g) Faller, J.W.; Sams, D.W.T.; Liu, X. *J. Am. Chem. Soc.* **1996**, *118*, 1217. (h) Corey, E.J.; Barres-Seeman, D.; Lee, T.W. *Tetrahedron Lett.* **1997**, *38*, 1699.
90. (a) Yanagisawa, A.; Nakashima, H., Ishiba, A.; Yamamoto, H. *J. Am. Chem. Soc.* **1996**, *118*, 4723. (b) Yanagisawa, A.; Nakashima, H.; Ishiba, A.; Yamamoto, H. *Synlett* **1997**, 88.
91. Motoyama, Y.; Narusawa, H.; Nishiyama, H. *Chem. Commun.* **1999**, 131.
92. Review: Thomas, E.J. *Chemtracts-Org. Chem.* **1994**, *7*, 207.
93. (a) McNeill, A.H.; Thomas, E.J. *Tetrahedron Lett.* **1992**, *33*, 1369. (b) Keck, G.E.; Park, M.; Krishnamurthy, D.; Grier, M.C. *J. Org. Chem.*. **1993**, *58*, 3787.
94. (a) McNeill, A.H.; Thomas, E.J. *Tetrahedron Lett.* **1990**, *31*, 6239. (a) McNeill, A.H.; Thomas, E.J. *Synthesis* **1994**, 322.
95. Kawashima, T.; Iwama, N.; Okazaki, R. *J. Am. Chem. Soc.* **1993**, *115*, 2507.
96. Carey, J.S.; Coulter, T.S.; Thomas, E.J. *Synlett* **1992**, 585.
97. (a) Carey, J.S.; Thomas, E.J. *Tetrahedron Lett.* **1993**, *34*, 3933. (b) Carey, J.S.; Thomas, E.J. *J. Chem. Soc., Chem. Commun.* **1994**, 283.
98. Koreeda, M.; Tanaka, Y. *Tetrahedron Lett.* **1987**, *28*, 143.
99. Keck, G.E.; Abbott, D.E.; Wiley, M.R. *Tetrahedron Lett.* **1987**, *28*, 139.
100. (a) Jephcote, V.J.; Pratt, A.J.; Thomas, E.J. *J. Chem. Soc., Chem. Commun.* **1984**, 800. (b) Jephcote, V.J.; Pratt, A.J.; Thomas, E.J. *J. Chem. Soc. Perkin Trans. I* **1989**, 1529.
101. Marshall, J.A.; Gung, W.Y. *Tetrahedron* **1989**, *45*, 1043.
102. (a) Marshall, J.A.; Welmaker, G.S. *J. Org. Chem.* **1992**, *57*, 7158. (b) Marshall, J.A.; Jablonowski, J.A.; Elliott, L.M. *J. Org. Chem.* **1995**, *60*, 2662.
103. (a) Marshall, J.A.; Luke, G.P. *J. Org. Chem.* **1991**, *56*, 483. (b) Marshall, J.A.; Jablonowski, J.A.; Luke, G.P. *J. Org. Chem.* **1994**, *59*, 7825.
104. (a) Marshall, J.A.; Johns, B.A. *J. Org. Chem.* **1998**, *63*, 7885. (b) Marshall, J.A.; Liao, J. *J. Org. Chem.* **1998**, *63*, 5962. (c) Marshall, J.A.; Hinckle, K.W. *J. Org. Chem.* **1997**, *62*, 5989.
105. (a) Marshall, J.A.; Wang, X.-J. *J. Org. Chem.* **1990**, *55*, 6246. (b) Marshall, J.A.; Wang, X.-J. *J. Org. Chem.* **1991**, *56*, 3211. (c) Marshall, J.A.; Wang, X.-J. *J. Org. Chem.* **1992**, *57*, 1242.
106. Marshall, J.A.; Perkins, J.F. *J. Org. Chem.* **1994**, *59*, 3509.
107. Marshall, J.A.; Perkins, J.F.; Wolf, M.A. *J. Org. Chem.* **1995**, *60*, 5556.
108. (a) Keck, G.E.; Krishnamurthy, D.; Chen, X. *Tetrahedron Lett.* **1994**, *35*, 8323. (b) Yu, C.-M.; Choi, H.-S.; Yoon, S. -K.; Jung, W.-H. *Synlett* **1997**, 889.
109. Yu, C.-M.; Yon, S.-K.; Baek, K.; Lee, J.-Y. *Angew. Chem. Int. Ed. Engl.* **1998**, *37*, 2392.
110. Mikhailov, B.M.; Bubnov, Yu.N. *Organoboron Compounds in Organic Synthesis*; Harwood Academic Science, Chur, 1984; pp 571–670.
111. (a) Mikhailov, B.M.; Bubnov, Yu.N. *Izv. Akad. Nauk SSSR, Ser. Khim.* **1964**, 1874. (b) Mikhailov, B.M.; Pozdnev, V.F. *Izv. Akad. Nauk SSSR, Ser. Khim.* **1964**, 1477.
112. Mikhailov, B.M.; Ter-Sarkisyan, G.S.; Nikolaeva, N.A. *Izv. Akad. Nauk SSSR, Ser. Khim.* **1968**, 1655.
113. Li, Y.; Houk, K. N. *J. Am. Chem. Soc.* **1989**, *111*, 1236.
114. (a) Vulpetti, A.; Gardner, M.; Gennari, C.; Bernardi, A.; Goodman, J.M.; Paterson, I. *J. Org. Chem.* **1993**, *58*, 1711. (b) Gennari, C.; Fioravanzo, E.; Bernardi, A.; Vulpetti, A. *Tetrahedron* **1994**, *50*, 8815.

115. Brown, H.C.; Racherla, U.S.; Pellechia, P.J. *J. Org. Chem.* **1990**, *55*, 1868.
116. Omoto, K.; Fujimoto, H. *J. Org. Chem.* **1998**, *63*, 8331.
117. Review: Roush, W.R. In *Stereoselective Synthesis, Methods of Organic Chemistry (Houben-Weyl)*; Edition E21; Helmchen, G.; Hoffmann, R.; Mulzer, J.; Schaumann, E. Eds.; Thieme, Stuttgart, 1996; Vol. 3; pp 1410–1486.
118. (a) Hoffmann, R.W.; Zeiss, H.-J. *Angew. Chem., Int. Ed. Engl.* **1979**, *18*, 306. (b) Hoffmann, R.W.; Zeiss, H.-J. *J. Org. Chem.* **1981**, *46*, 1309. (c) Hoffmann, R.W.; Kemper, B.; Matternich, R.; Lehmeier, T. *Liebigs Ann. Chem.* **1985**, 2246.
119. (a) Wang, Z.; Meng, X.-J.; Kabalka, G.W. *Tetrahedron Lett.* **1991**, *32*, 1945. (b) Wang, Z.; Meng, X.-J.; Kabalka, G.W. *Tetrahedron Lett.* **1991**, *32*, 5677. (c) Pace, R.D.; Kabalka, G.W. *J. Org. Chem.* **1995**, *60*, 4838. (b) Kabalka, G.W.; Narayana, C.; Reddy, N.K. *Tetrahedron Lett.* **1996**, *37*, 2181.
120. (a) Hoffmann, R.W.; Weidmann, U. *Chem. Ber.* **1985**, *118*, 3966. (b) Hoffmann, R.W.; Endesfelder, A.; Zeiss, H.-J. *Carbohydr. Res.* **1983**, *123*, 320.
121. Hoffmann, R.W.; Zeiss, H.-J.; Ladner, W.; Tabche, S. *Chem. Ber.* **1982**, *115*, 2357.
122. Hoffmann, R.W.; Brinkmann, H.; Frenking, G. *Chem. Ber.* **1990**, *123*, 2387.
123. (a) Hoffmann, R.W.; Herold, T. *Angew. Chem. Int. Ed. Engl.* **1978**, *18*, 768. (b) Hoffmann, R.W.; Zeiss, H.-J. *Angew. Chem. Int. Ed. Engl.* **1980**, *19*, 218. (b) Herold, T.; Schrott, U.; Hoffmann, R.W.; *Chem. Ber.* **1981**, *114*, 359. (c) Hoffmann, R.W.; Herold, T. *Chem. Ber.* **1981**, *111*, 375. (d) Hoffmann, R.W.; Herold, T. *Chem. Ber.* **1981**, *111*, 5495.
124. (a) Roush, W.R.; Walts, A.E.; Hoong, L.K. *J. Am. Chem. Soc.* **1985**, *107*, 8186. (b) Roush, W.R.; Banfi, L. *J. Am. Chem. Soc.* **1988**, *110*, 3979. (c) Roush, W.R.; Ando, K.; Powers, D.B.; Palkowitz, A.D.; Halterman, R.L. *J. Am. Chem. Soc.* **1990**, *112*, 6339. (d) Roush, W.R.; Ando, K.; Palkowitz, A.D. *J. Am. Chem. Soc.* **1990**, *112*, 6348. (e) Roush, W.R.; Hunt, J.A. *J. Org. Chem.* **1995**, *60*, 798. (f) Roush, W.R.; Palkowitz, A.D.; Palmser, M.J. *J. Org. Chem.* **1987**, *52*, 316.
125. (a) Brown, H.C.; Jadhav, P.K. *J. Am. Chem. Soc.* **1983**, *105*, 2092. (b) Brown, H.C.; Randad, R.S.; Bhat, K.S.; Zaidlewicz, M.; Racherla, U.S. *J. Am. Chem. Soc.* **1990**, *112*, 2389. (c) Brown, H.C.; Bhat, K.S. *J. Am. Chem. Soc.* **1986**, *108*, 5919. (d) Brown, H.C.; Bhat, K.S. *J. Am. Chem. Soc.* **1986**, *108*, 293. (e) Racherla, U.S.; Brown, H.C. *J. Org. Chem.* **1991**, *56*, 401. (f) Racherla, U.S.; Liao, Y.; Brown, H.C. *J. Org. Chem.* **1992**, *57*, 6614. (g) Brown, H.C.; Narla, G. *J. Org. Chem.* **1995**, *60*, 4686. (h) Brown, H.C.; Jodhar, P.K.; Bhat, K.S. *J. Am. Chem. Soc.* **1985**, *107*, 2564.
126. Corey, E.J.; Yu, C.-M.; Kim, S.S. *J. Am. Chem. Soc.* **1989**, *111*, 5495.
127. (a) Reetz, M.T.; Zierka, T. *Chem. Ind. (London)* **1988**, 663. (b) Short, R.P.; Masamune, S. *J. Am. Chem. Soc.* **1989**, *111*, 1892. (c) Mears, R.J.; De Silva, H.; Whiting, A. *Tetrahedron* **1997**, *53*, 17395.
128. Garcia, J.; Kim, B.; Masamune, S. *J. Org. Chem.* **1987**, *52*, 4831.
129. Roush, W.R.; Grover, P.T. *J. Org. Chem.* **1995**, *60*, 3806.
130. (a) Brown, H.C.; Bhat, K.S.; Randad, R.S. *J. Org. Chem.* **1987**, *52*, 319. (b) Brown, H.C.; Bhat, K.S.; Randad, R.S. *J. Org. Chem.* **1987**, *52*, 3701. (c) Brown, H.C.; Bhat, K.S.; Randad, R.S. *J. Org. Chem.* **1989**, *54*, 1570. (d) Ramachandran, P.V.; Chen, G.-M.; Brown, H.C. *Tetrahedron Lett.* **1997**, *38*, 2417.
131. (a) Roush, W.R.; Park, J.C. *J. Org. Chem.* **1990**, *55*, 1143. (b) Roush, W.R.; Park, J.C. *Tetrahedron Lett.* **1990**, *31*, 4707.
132. (a) Favre, E.; Gaudemar, M. *Comp. Rend. C* **1966**, *263*, 1543. (b) Favre, E.; Gaudemar, M. *J. Organomet. Chem.* **1974**, *76*, 297 and 305.
133. (a) Brown, H.C.; Khire, U.R.; Racherla, U.S. *Tetrahedron Lett.* **1993**, *34*, 15. (b) Brown, H.C.; Khire, U.R.; Narla, G.; Racherla, U.S. *J. Org. Chem.* **1995**, *60*, 544.
134. (a) Haruta, R.; Ishiguro, M.; Ikeda, N.; Yamamoto, H. *J. Am. Chem. Soc.* **1982**, *104*, 7667. (b) Ikeda, N.; Arai, I.; Yamamoto, H. *J. Am. Chem. Soc.* **1986**, *108*, 483.
135. Corey, E.J.; Yu, C.-M.; Lee, D.-H. *J. Am. Chem. Soc.* **1990**. *112*, 878.
136. Matsumoto, Y.; Naito, M.; Uozumi, Y.; Hayashi, T. *J. Chem. Soc., Chem. Commun.* **1993**, 1468.
137. Okude, Y.; Hirana, S.; Hiyama, T.; Nozaki, H. *J. Am. Chem. Soc.* **1977**, *99*, 3179.
138. Buse, C.T.; Heathcock, C.H. *Tetrahedron Lett.* **1978**, 1685.

139. For recent reviews see: (a) Cintas, P. *Synthesis* **1992**, 248. (b) Fürstner, A. *Pure Appl. Chem.* **1998**, *70*, 1071. (c) Wessjohann, L. A.; Scheid, G. *Synthesis* **1999**, 1. (d) Fürstner, A. *Chem. Rev.* **1999**, *99*, 991. (e) Saccomano, N. A. In *Comprehensive Organic Synthesis, Vol. 1, Additions to C–X Bonds, Part 1;* Schreiber, S. L. Ed.; Pergamon Press, Oxford, 1991, pp 173. (f) Hoppe, D. In *Stereoselective Synthesis, Methods of Organic Chemistry (Houben-Weyl)*; Edition E21; Helmchen, G.; Hoffmann, R.; Mulzer, J.; Schaumann, E. Eds.; Thieme, Stuttgart, 1996; Vol. 3; pp 1584.
140. (a) Fürstner, A.; Shi, N. *J. Am. Chem. Soc.* **1996**, *118*, 12349. (b) Fürstner, A.; Brunner, H. *Tetrahedron Lett.* **1996**, *37*, 7009.
141. Takai, K.; Toratsu, C. *J. Org. Chem.* **1998**, *63*, 6450.
142. (a) Hiyama, T.; Kimura, K.; Nozaki, H. *Tetrahedron Lett.* **1981**, *22*, 1037. (b) Hiyama, T.; Okude, Y.; Kimura, K.; Nozaki, H. *Bull. Chem. Soc. Jpn.* **1982**, *55*, 561.
143. (a) Jubert, C.; Nowotny, S.; Kornemann, D.; Antes, I.; Tucker, C. E.; Knochel, P. *J. Org. Chem.* **1992**, *57*, 6384. (b) Nowotny, S.; Tucker, Jubert, C.; C. E.; Knochel, P. *J. Org. Chem.* **1995**, *60*, 2762.
144. (a) Nagaoka, H.; Kishi, Y. *Tetrahedron* **1981**, *37*, 3873. (b) Lewis, M. D.; Kishi, Y. *Tetrahedron Lett.* **1982**, *23*, 2343. (c) Jin, H.; Uenishi, J.-I.; Christ, W. J.; Kishi, Y. *J. Am Chem. Soc.* **1986**, *108*, 5644.
145. (a) Martin, S. F.; Li, W. *J. Org. Chem.* **1989**, *54*, 6129. (b) Mulzer, J.; Schulze, T.; Strecker, A.; Denzer, W.; *J. Org. Chem.* **1988**, *53*, 4098.
146. (a) Ciapetti, P.; Taddei, M.; Ulivi, P. *Tetrahedron Lett.* **1994**, *35*, 3183. (b) Ciapetti, P.; Falorni, M.; Taddei, M. *Tetrahedron* **1996**, *52*, 7379.
147. Aoyagi, Y.; Inaba, H.; Hiraiwa, Y.; Kuroda, A.; Ohta, A. *J. Chem. Soc., Perkin Trans. I* **1998**, 3975.
148. Mulzer, J.; Kattner, L.; Strecker, A. R.; Schröder, C.; Buschmann, J.; Lehmann, C.; Luger, P. *J. Am. Chem. Soc.* **1991**, *113*, 4218.
149. Maguire, R. J.; Mulzer, J.; Bats, J. W. *J. Org. Chem.* **1996**, *61*, 6936.
150. Chen, C.; Tagami, K.; Kishi, Y. *J. Org. Chem.* **1995**, *60*, 5386.
151. Sugimoto, K.; Aoyagi, S.; Kibayashi, C. *J. Org. Chem.* **1997**, *62*, 2322.
152. (a) Schlosser, M.; Hartmann, J. *J. Am. Chem. Soc.* **1976**, *98*, 4674. (b) Hutchison, D. A.; Beck, K. R.; Benkeser, R. A.; Grutzner, J. B. *J. Am. Chem. Soc.* **1973**, *95*, 7075. (c) West, P.; Purmort, J. I.; McKinley, S. V. *J. Am. Chem. Soc.* **1968**, *90*, 797. (d) Reich, H. J.; Holladay, J. E.; Mason, J. D.; Sikorski, W. H. *J. Am. Chem. Soc.* **1995**, *117*, 12137.
153. Fraenkel, G.; Winchester, W. R. *J. Am. Chem. Soc.* **1989**, *111*, 3794.
154. Yanagisawa, A.; Habaue, S.; Yamamoto, H. *J. Am. Chem. Soc.* **1991**, *113*, 5893.
155. Courtois, G.; Miginiac, L. *J. Organomet. Chem.* **1974**, *69*, 1.
156. Boersma, J. *Zinc and Cadmium;* In *Comprehensive Organometallic Chemistry;* Wilkinson, G.; Stone, F. G. A.; Abel, E. Eds. Pergamon, Cambridge, 1982; Vol. 2; pp 823–865.
157. Gajewski, J. J.; Bocian, W.; Harris, N. J.; Olson, L. P.; Gajewski, J. P. *J. Am. Chem. Soc.* **1999**, *121*, 326.
158. (a) Courtois, G.; Miginiac, L. *J. Organomet. Chem.* **1974**, *69*, 1. (b) Rautenstrauch, V. *Helv. Chim. Acta* **1974**, *57*, 496.
159. (a) Yamamoto, Y. In *Comprehensive Organic Synthesis, Vol. 2, Additions to C–X Bonds, Part 2;* Heathcock, C. H. Ed.; Pergamon Press, Oxford, 1991, pp 55. (b) Hoppe, D. *Angew. Chem., Int. Ed. Engl.* **1984**, *23*, 932.
160. Epifani, E.; Florio, S.; Ingrosso, G. *Tetrahedron Lett.* **1987**, *28*, 6385.
161. Hoppe, D. In *Stereoselective Synthesis, Methods of Organic Chemistry (Houben-Weyl)*; Edition E21; Helmchen, G.; Hoffmann, R.; Mulzer, J.; Schaumann, E. Eds.; Thieme, Stuttgart, 1996; Vol. 3; pp 1379–1400.
162. Felkin, H.; Gault, Y.; Roussi, G. *Tetrahedron* **1970**, *26*, 3761.
163. El Idrissi, M.; Santelli, M. *J. Org. Chem.* **1988**, *53*, 1010.
164. (a) Jones, P.; Millot, N.; Knochel, P. *Chem. Commun.* **1998**, 2405. (b) Jones, P.; Knochel, P. *Chem. Commun.* **1998**, 2407. (c) Jones, P.; Knochel, P. *J. Org. Chem.* **1999**, *64*, 186.
165. (a) Barbot, F.; Miginiac, P. *Tetrahedron Lett.* **1975**, 3829. (b) Miginiac, P.; Bouchoule, C. *Bull. Chem. Soc. Fr.* **1968**, 4675. (c) Bocoum, A.; Savoia, D.; Umani-Ronchi, A. *J. Chem. Soc., Chem. Commun.* **1993**, 1542.

166. Wilson, S.R.; Guazzaroni, M.E. *J. Org. Chem.* **1989**, *54*, 3087.
167. (a) Mulzer, J.; Angermann, A. *Tetrahedron Lett.* **1983**, *24*, 2843. (b) Fronza, G.; Fuganti, C.; Grasselli, P.; Pedrocchi-Fantoni, G.; Zirotti, C. *Tetrahedron Lett.* **1982**, *23*, 4143.
168. Hanessian, S.; Park, H.; Yang, R.-Y. *Synlett* **1997**, 353.
169. Duthaler, R.O.; Hafner, A. *Chem. Rev.* **1992**, *92*, 807.
170. (a) Sato, F.; Iida, K.; Iijima, S.; Moriya, H.; Sato, M. *J. Chem. Soc., Chem. Commun.* **1981**, 1140. (b) Seebach, D.; Widler, L. *Helv. Chim. Acta*, **1982**, *65*, 1972. (c) Widler, L.; Seebach, D. *Helv. Chim. Acta* **1982**, *65*, 1085. (d) Weidmann, B.; Seebach, D. *Angew. Chem., Int. Ed. Engl.* **1983**, *22*, 31. (e) Reetz, M.T. *Top. Curr. Chem.* **1982**, *106*, 1. (f) Reetz, M.T.; Steinbach, R.; Westermann, J.; Peter, R.; Wenderoth, B. *Chem. Ber.* **1985**, *118*, 1441. (g) Hoppe, D. In *Stereoselective Synthesis, Methods of Organic Chemistry (Houben-Weyl)*; Edition E21; Helmchen, G.; Hoffmann, R.; Mulzer, J.; Schaumann, E. Eds.; Thieme, Stuttgart, 1996; Vol. 3; pp 1551–1583.
171. Kasatkin, A.; Nakagawa, T.; Okamoto, S.; Sato, F. *J. Am. Chem. Soc.* **1995**, *117*, 3881. (b) Nakagawa, T.; Kasatkin, A.; Sato, F. *Tetrahedron Lett.* **1995**, *36*, 3207.
172. Teng, X.; Kasatkin, A.; Kawanaka, Y.; Okamoto, S.; Sato, F. *Tetrahedron Lett.* **1997**, *38*, 8977.
173. (a) Sato, F.; Iijima, S.; Sato, M. *Tetrahedron Lett.* **1981**, *22*, 243. (b) Sato, F.; Suzuki, Y.; Sato, M. *Tetrahedron Lett.* **1982**, *23*, 4589. (c) Sato, F.; Uchiyama, H.; Iida, K.; Kobayashi, Y.; Sato, M. *J. Chem. Soc., Chem. Commun.* **1983**, 921. (d) Kobayashi, Y.; Umeyama, K.; Sato, F. *J. Chem. Soc., Chem. Commun.* **1984**, 621.
174. Collins, S.; Dean, W.P.; Ward, D.G. *Organometallics* **1988**, *7*, 2289.
175. Collins, S.; Kuntz, B.A.; Hong, Y. *J. Org. Chem.* **1989**, *54*, 4154.
176. (a) Szymoniak, J.; Pagneux, S.; Felix, D.; Moïse, C. *Synlett* **1996**, 46. (b) Szymoniak, J.; Thery, N.; Moïse, C. *Synlett* **1997**, 1239.
177. (a) Seebach, D.; Beck, A.K.; Imwinkelried, R.; Roggo, S.; Wonnacott, A. *Helv. Chim. Acta* **1987**, *70*, 954. (b) Takahashi, H.; Kawabata, A.; Niwa, H.; Higashiyama, K. *Chem. Pharm. Bull.* **1988**, *36*, 803. (c) Schmidt, B.; Seebach, D. *Angew. Chem., Int. Ed. Engl.* **1991**, *30*, 99.
178. (a) Hafner, A.; Duthaler, R.O.; Marti, R.; Rihs, G.; Rothe-Streit, P.; Schwarzenbach, F. *J. Am. Chem. Soc.* **1992**, *114*, 2321. (b) Duthaler, R.O.; Hafner, A.; Riediker, M. *Pure Appl. Chem.* **1990**, *62*, 631.
179. Hanko, R.; Hoppe, D. *Angew. Chem., Int. Ed. Engl.* **1982**, *21*, 372.
180. (a) Hoppe, D. *Angew. Chem., Int. Ed. Engl.* **1984**, *23*, 932. (b) Hoppe, D.; Zschage, O. *Angew. Chem., Int. Ed. Engl.* **1989**, *28*, 69. (c) Hoppe, D.; Hense, T. *Angew. Chem., Int. Ed. Engl.* **1997**, *36*, 2283.
181. Zschage, O.; Hoppe, D. *Tetrahedron* **1992**, *48*, 5657.
182. Hoppe, D.; Tarara, G.; Wilckens, M. *Synthesis* **1989**, 83.
183. Marsch, M.; Harms, K.; Zschage, O.; Hoppe, D.; Boche, G. *Angew. Chem., Int. Ed. Engl.* **1991**, *30*, 321.
184. Pippel, D.J.; Weisenburger, G.A.; Wilson, S.R.; Beak, P. *Angew. Chem., Int. Ed. Engl.* **1998**, *37*, 2522.
185. Paulsen, H.; Graeve, C.; Hoppe, D. *Synthesis* **1996**, 141.
186. (a) Yamamoto, Y.; Maruyama, K. *Tetrahedron Lett.* **1981**, *22*, 2895. (b) Mashima, K.; Yasuda, H.; Asami, K.; Nakamura, A. *Chem. Lett.* **1983**, 219.
187. (a) Ito, H.; Taguchi, T.; Hanazawa, Y. *Tetrahedron Lett.* **1992**, *33*, 1295. (b) Ito, H.; Taguchi, T.; Hanzawa, Y. *Tetrahedron Lett.* **1992**, *33*, 7873. (c) Ito, H.; Nakamura, T.; Taguchi, T.; Hanzawa, Y. *Tetrahedron* **1995**, *51*, 4507.
188. Clark, A.J.; Kasumee, I.; Peacock, J.L. *Tetrahedron Lett.* **1995**, *36*, 7137.
189. Chino, M.; Matsumoto, T.; Suzuki, K. *Synlett* **1994**, 359.
190. (a) Araki, S.; Ito, H.; Butsugan, Y. *J. Org. Chem.* **1988**, *53*, 1833. (b) Araki, S.; Shimizu, T.; Johar, P.S.; Jin, S.-J.; Butsugan, Y. *J. Org. Chem.* **1991**, *56*, 2538.
191. (a) Marshall, J.A. *Chemtracts-Organic Chemistry* **1997**, *10*, 481. (b) Cintas, P. *Synlett* **1995**, 1087.
192. Li, C.J.; Chan, T.H. *Tetrahedron Lett.* **1991**, *32*, 7017.
193. Isaac, M.B.; Chan, T.H. *Tetrahedron Lett.* **1995**, *36*, 8957.
194. Lloyd-Jones, G.C.; Russell, T. *Synlett* **1998**, 903.
195. (a) Capps, S.M.; Lloyd-Jones, G.C.; Murray, M.; Peakman, T.M.; Walsh, K.E. *Tetrahedron Lett.* **1998**, *39*, 2853. (b) Höppe, H.A.F.; Lloyd-Jones, G.C.; Murray, M.; Peakman, T.M.;

Walsh, K.E. *Angew. Chem., Int. Ed. Engl.* **1998**, *37*, 1545. (c) Reetz, M.T.; Haning, H. *J. Organomet. Chem.* **1997**, *541*, 117.

196. Li, C.J.; Chan, T.H. *Tetrahedron Lett.* **1991**, *32*, 7017.
197. (a) Paquette, L.A.; Mitzel, T.M. *Tetrahedron Lett.* **1995**, *36*, 6863. (b) Paquette, L.A.; Mitzel, T.M. *J. Am. Chem. Soc.* **1996**, *118*, 1931. (c) Paquette, L.A.; Mitzel, T.M. *J. Org. Chem.* **1996**, *61*, 8799. (d) Isaac, M.B.; Paquette, L.A. *J. Org. Chem.* **1997**, *62*, 5333.
198. Paquette, L.A.; Mitzel, T.M.; Isaac, M.B.; Crasto, C.F.; Schomer, W.W. *J. Org. Chem.* **1997**, *62*, 4293.
199. Maguire, R.J.; Mulzer, J.; Bats, J.W. *J. Org. Chem.* **1996**, *61*, 6936.
200. Paquette, L.A.; Bennett, G.D.; Chhatriwalla, A.; Isaac, M.B. *J. Org. Chem.* **1997**, *62*, 3370.
201. Morikawa and Taguchi have observed a similar phenomenon in the reaction of α-oxygenated allyl halide derivatives with aldehydes. Morikawa, T.; Narasaka, T.; Sakuma, C.; Taguchi, T. *Chem. Pharm. Bull.* **1997**, *45*, 1877.
202. Paquette, L.A.; Bennett, G.D.; Isaac, M.B.; Chhatriwalla, A. *J. Org. Chem.* **1998**, *63*, 1836.
203. Marshall, J.A.; Hinkle, K.W. *J. Org. Chem.* **1995**, *60*, 1920.
204. Li, X.-R.; Loh, T.-P. *Tetrahedron: Asymmetry* **1996**, *7*, 1535.
205. Loh, T.-P.; Li, X.-R. *Angew. Chem. Int. Ed. Engl.* **1997**, *36*, 980.
206. Marshall, J.A.; Hinkle, K.W. *J. Org. Chem.* **1996**, *61*, 105.
207. (a) Marshall, J.A.; Perkins, J.F.; Wolf, M.A. *J. Org. Chem.* **1995**, *60*, 5556. (b) Marshall, J.A.; Pavlovich, M.R. *J. Org. Chem.* **1997**, *62*, 6001.
208. Marshall, J.A.; Grant, C.M. *J. Org. Chem.* **1999**, *64*, 696.

11 Recent Applications of the Allylation Reaction to the Synthesis of Natural Products

Sherry R. Chemler and William R. Roush

11.1 Introduction

The reaction of carbonyl compounds and allylmetal reagents is an important transformation in organic synthesis. Advances in stereoselective carbonyl allylation reactions have been spurred by interest in the synthesis of polypropionate-derived natural products, carbohydrates and other polyhydroxylated compounds. These reactions are ideally suited for the construction of stereochemically rich acyclic skeletons. Additionally, cyclic polyether-containing natural products, among others, have inspired chemists to investigate ring-closing allylation reactions. This review will focus on recent developments in the allylation reaction, with special emphasis on its application towards the synthesis of natural products.

Several reviews of allylmetal chemistry, including applications in natural product synthesis, have appeared. Hoffmann, Yamamoto, Roush, Fleming and Thomas have each written general reviews on allylmetal chemistry [1–6], and Yamamoto has written a general review on the chemistry of allenyl organometallics [7]. Sakurai, Fleming, Majetich, Chan, Panek, Lankopf and Thomas have reviewed the chemistry of allylsilanes [8–15], Marshall and Thomas have reviewed the chemistry of allylstannanes [16, 17], Cintas, Wessjohann and Hoppe have reviewed allylchromium reagents [18–20], Duthaler and Hoppe have reviewed the reactions of allyltitanium reagents [21–23], and Hoffmann and Roush have reviewed allylboron reagents [24–26].

We begin by providing a brief overview of the stereoselectivity of the reactions of achiral allylmetal reagents with achiral aldehydes ("simple diastereoselection") and chiral aldehydes ("relative diastereoselection"). We then examine the most important classes of chiral allylmetal reagents, particularly those involving boron, tin and silicon, which have found the greatest number of applications in natural product synthesis. Recent applications of chiral Lewis acids to the enantioselective allylation of carbonyl compounds are also briefly reviewed. Our intention throughout this Chapter is to highlight the opportunities for stereochemical control using allylmetal reagents.

11.1.1 Simple Diastereoselection Using Type I and Type III Allylmetal Reagents

Allylmetal reagents have been classified into three mechanistically distinct groups [27]. Type I and Type III allylmetal reagents react through cyclic, six-membered transition states, and Type II reagents react through acyclic transition states (see Chapter 10 for a mechanistic discussion) [3, 27]. Type I and Type III crotylmetal reagents differ in their propensity to undergo metallotropic rearrangement, which leads to interconversion of the (*E*)- and (*Z*)-olefin isomers. Type I reagents, including crotylboronate [28], crotyltrihalo- and trialkoxysilane [29–32], trialkylstannane (thermally promoted) [33], allylaluminum [34] and some crotylchromium [35] reagents, undergo allylation reactions with aldehydes faster than they undergo metallotropic rearrangement. Indeed, several of these allylmetal reagents are configurationally stable and can be stored at room temperature for extended periods of time.

Simple diastereoselection in the allylation reaction of aldehydes with Type I crotylmetal reagents is dictated by the double-bond geometry of the reagent. Thus, as first noted by Hoffmann [28, 36], the (*E*)-crotylboronate **1** reacts preferentially through the Zimmerman-Traxler transition state **2** to generate the *anti* α-methyl homoallylic alcohol **3**, and the (*Z*)-crotylboronate **4** reacts preferentially through the Zimmerman-Traxler transition state **5** to generate the *syn* β-methyl homoallylic alcohol adduct **6** (Fig. 11-1). When these reactions are performed with reagents **1** and **4** of >97% isomeric purity, the stereoselectivity of these reactions is remarkably high (>97%) [37].

Type III crotylmetal reagents, including most crotylchromium [38–40], crotyltitanium [23, 41], and crotylzirconium reagents [42, 43], undergo metallotropic rearrangement faster than they undergo the allylation reaction with aldehydes [3]. Thus, the Nozaki-Hiyama (*E*)-crotylchromium reagent **8** can be formed from either (*E*)- or (*Z*)-crotylbromide, where the (*Z*)-crotylchromium reagent **7** is converted to the (*E*)-crotylchromium reagent **8** via metallotropic rearrangement (Fig. 11-2) [39]. The *anti* homoallylic alcohol is usually the major product of re-

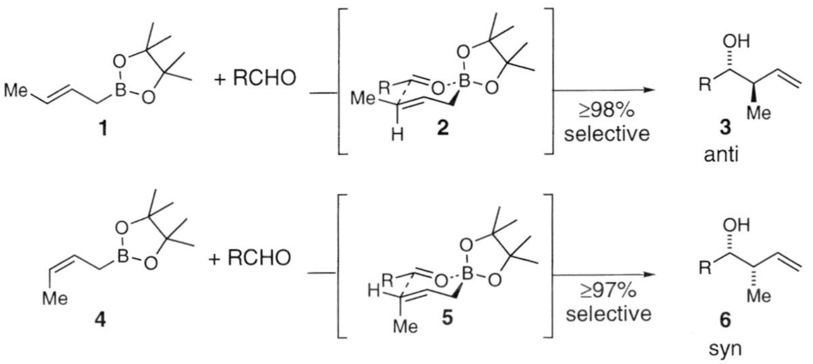

Figure 11-1. Reaction mechanism of type I allylmetal reagents.

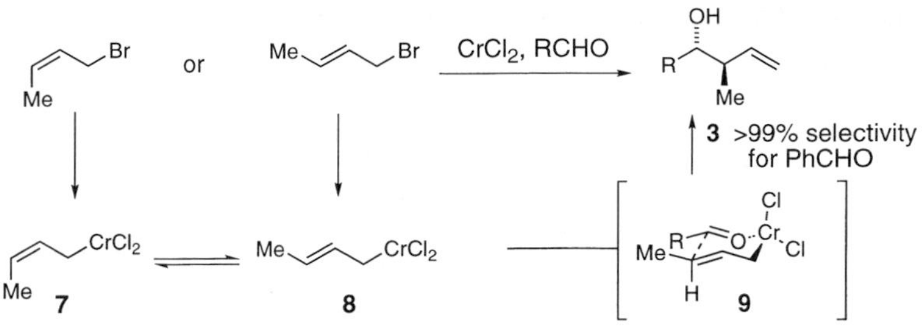

Figure 11-2. Reaction mechanism of type III allylmetal reagent.

actions of Type III crotylmetal reagents; in the case of the crotylchromium reagent, the selectivity for the *anti* alcohol **3** is usually very high [39, 44].

11.1.2 Simple Diastereoselection Using Type II Allylmetal Reagents

Type II allylmetal reagents, including the allyltrialkylsilanes [45, 46] and allyltrialkylstannanes [47], among others [48, 49], react with aldehydes through open transition states where an external Lewis acid is used to activate the aldehyde toward nucleophilic attack. These reactions provide the 3,4-*syn* homoallylic alcohol **6** as the major product via an S_E' substitution [9] (Fig. 11-3). As a general rule, the 3,4-*syn* isomer **6** is the major product using either the (*E*)- or (*Z*)-crotylstannanes or silanes. For example, in 1980, Yamamoto reported that the $BF_3{\cdot}OEt_2$-catalyzed reactions of achiral (*E*)- and (*Z*)-crotyltri-*n*-butylstannanes **10** with achiral aldehydes lead to the stereoselective generation of the *syn* α-methyl homoallylic alcohols **6**, with *syn*:*anti* diastereoselection ranging from 90:10 to 98:2, depending on the size of the aldehyde R group (α-branched aldehydes gave better selectivity than unbranched aldehydes) [47]. The *syn* stereochemistry of adduct **6** can be rationalized by C–C bond formation occurring through either the synclinal transition state **11** or the antiperiplanar transition state **12**. Yamamoto argued that the antiperiplanar transition state **12** should be favored because of minimization of steric interactions between the aldehyde R group and the γ-methyl of the crotyl-

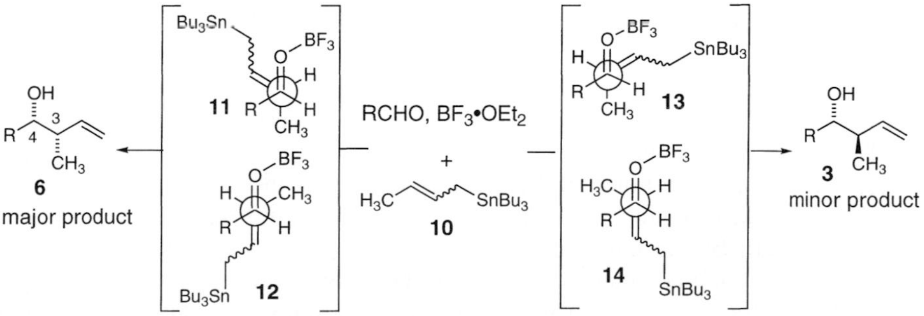

Figure 11-3. Possible mechanisms of the type II allylation reaction.

stannane [47]. Denmark [50–52] subsequently provided evidence that the synclinal transition state **11** may be favored, and, more recently, Keck [53, 54] has provided additional evidence in support of transition state **11**.

Denmark studied the intramolecular allylation reaction of allylstannane **15** in order to differentiate between the *syn* and *anti* S_E' transition states (only *anti* S_E' is shown below) as well as to differentiate between the synclinal and antiperiplanar transition states **19** and **20**, which are analogous to transition states **11** and **14**, respectively (Eq. (11.1)) [51]. Denmark found that the major product of the $BF_3{\cdot}OEt_2$-promoted reaction of **15** was adduct **16**, which must arise from the synclinal, *anti* S_E' transition state **19**. The minor adduct **17** must arise through the antiperiplanar transition state **20**.

$BF_3{\bullet}OEt_2$, CH_2Cl_2, -70 °C

15 **16** **17** **18**

selectivity = 79 : 14 : 7

19 major pathway

20 minor pathway

(11.1)

Denmark argued that the synclinal transition state **19** may be favored due to stabilization by stereoelectronic effects such as secondary orbital overlap or minimization of charge separation. The allylstannane HOMO and the aldehyde LUMO could participate in secondary orbital overlap in transition state **19**, with specific interactions between the allylstannane α-carbon and the aldehyde oxygen [50, 55]. Alternatively, the preference for the synclinal transition state **19** can also be attributed to minimization of charge separation in the transition state, compared to the situation in the antiperiplanar transition state **20** [50, 56].

Keck subsequently studied the allylation reactions of crotyltri-*n*-butylstannane (**10**) and a wide range of achiral aldehydes with $BF_3{\cdot}OEt_2$ as the catalyst. He found that (*E*)-crotyltri-*n*-butylstannane, (*E*)-**10**, is more selective for the formation of the *syn* homoallylic alcohols **6** (Fig. 11-3) than is the (*Z*)-crotyltri-*n*-butylstannane, (*Z*)-**10** [53]. Additionally, Hayashi and co-workers had previously noted that in $TiCl_4$-catalyzed allylation reactions, the (*E*)-crotyltrimethylsilane was much more selective for the *syn* homoallylic alcohols than was the (*Z*)-crotyltrimethylsilane [46]. Keck rationalized his results (his rationale should also apply to Hayashi's results) by attributing the formation of the *syn* adduct **6** to C–C bond formation through the synclinal transition state **11**, and argued that the (*E*)-crotylstannane, (*E*)-**10**, prefers to react through the synclinal transition state (*E*)-**11** to a greater extent than does the (*Z*)-crotyltri-*n*-butylstannane via (*Z*)-**11** (Fig. 11-4).

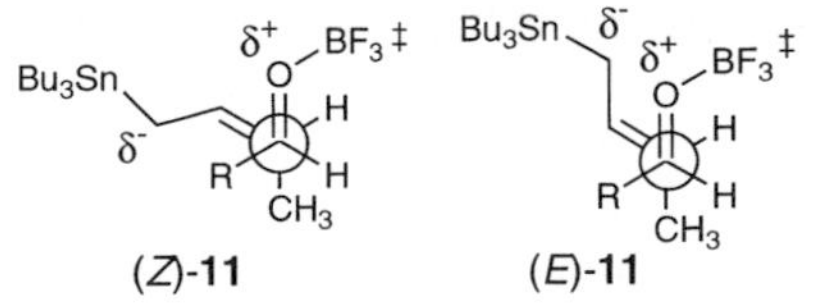

Figure 11-4. Synclinal transition states with (*E*)- and (*Z*)-crotyltri-*n*-butylstannanes.

In his analysis of transition states (*Z*)-**11** and (*E*)-**11**, Keck pointed out that in (*Z*)-**11**, the α-carbon of the (*Z*)-crotyltri-*n*-butylstannane is farther from the aldehyde oxygen in transition state (*Z*)-**11** than is the α-carbon of the (*E*)-crotyltri-*n*-butylstannane in transition state (*E*)-**11**, and thus is less able to participate in secondary orbital overlap interactions. The decrease in stereoelectronic stabilization of transition state (*Z*)-**11**, compared to (*E*)-**11**, allows access to other competing transition states which can lead to the diastereomeric *anti* homoallylic alcohol **3** (e.g. transition states **13** or **14**, Fig. 11-3, see above) in the reactions of the (*Z*)-crotylstannane reagent. Also, the (*Z*)-crotyltri-*n*-butylstannane in transition state (*Z*)-**11** probably experiences increased steric interactions with the aldehyde R group relative to (*E*)-crotyltri-*n*-butylstannane in transition state (*E*)-**11**.

Throughout this review, we will generally invoke the Denmark-Keck synclinal transition state **11** rather than the Yamamoto antiperiplanar transition state **12** in our analysis of Type II allylation reactions, in spite of the fact that many research groups favor the antiperiplanar transition state **12** when rationalizing their results (see below). As previously noted by Keck, subtle changes in electronics, steric interactions of the reacting partners, and the Lewis acid that promotes the reaction ultimately determine whether one of the transition states, **11** or **12**, is favored over the other [53].

The analysis of open transition states in the chelate-controlled allylation reactions of α- and β-alkoxy aldehydes with Type II allylmetal reagents is much simpler (Fig. 11-5). In these cases, only the synclinal transition state **21** and the antiperiplanar transition state **22** are considered as viable possibilities. Other possible transition states have been eliminated because of the perceived requirement that

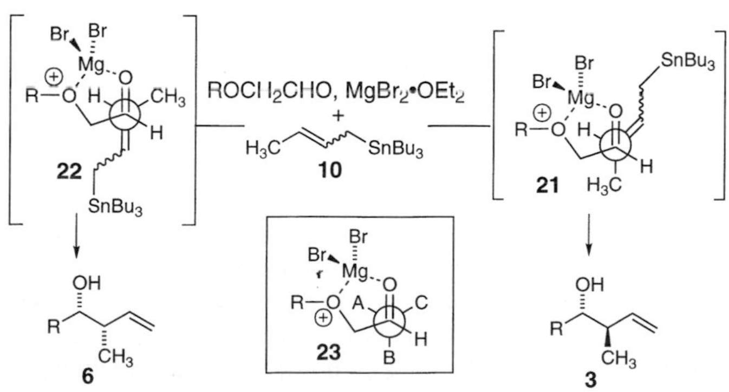

Figure 11-5. Mechanisms of allylation reactions of type II crotylmetal reagents with internally chelated aldehydes.

the C(3)–H of crotylstannane reagent **10** must occupy a position over the chelate ring, e.g. position A in the generalized transition state **23**, which is generally acknowledged as the most sterically demanding position [53, 57, 58]. Positions B and C in transition state **23** are then occupied by the methyl and the $=CHCH_2SnBu_3$ groups of reagent **10**. After analyzing numerous examples for this review, we conclude that, in general, position B is the second most sterically demanding position, which leaves position C as the least sterically demanding position. Thus, product formation through the competing transition states **21** and **22** is then determined by the relative steric demands of the B and C substituents of the allylmetal reagent.

11.2 Relative Diastereoselection in Allylation Reactions of Achiral Type I and Type III Allylmetal Reagents with Chiral Aldehydes; Selected Application Towards the Synthesis of Natural Products

11.2.1 Reactions of Achiral Type I and Type III Allylmetal Reagents with Chiral Aldehydes

Hoffmann and co-workers have investigated the reactions of pinacol allylboronate **24** and crotylboronates (*E*)-**1** and (*Z*)-**4** with the *anti*- and *syn*-α-methyl-β-alkoxy aldehydes **25** and **32** (Table 11-1) [59]. While the reaction of allylboronate **24** does not give useful levels of diastereoselectivity with either aldehyde, the (*E*)-crotylboronate reagent **1** gives high levels of diastereoselectivity for the Felkin (4,5-*syn*) products **28** and **35**. Reactions of the (*Z*)-crotylboronate reagent **4** give good diastereoselectivity for the anti-Felkin adduct **31** with the 2,3-*anti*-aldehyde **25**, but poorer diastereoselection for the anti-Felkin product **38** with the 2,3-*syn* aldehyde **32**. Studies by Kishi and co-workers have shown that the reactions of the Type III (*E*)-crotylchromium reagent (**8**) with α-methyl chiral aldehydes parallel the trend of Felkin selectivity observed in the reactions of (*E*)-crotylboronate reagents [60, 61].

The high selectivity for the 4,5-*syn* adducts **28** and **35**, the major products from the allylation reactions with the (*E*)-crotylboronate **1**, can be rationalized by the Felkin-Anh [62–64] transition state **39** (Fig. 11-6) where the large R group on the α-carbon of the aldehyde occupies the least sterically demanding position *anti* to the forming C–C bond, while the aldehyde α-C–H occupies the position *syn* to the (*E*)-crotylboronate γ-Me substituent [3]. In the competing transition state **41**, which leads to the anti-Felkin adduct **42**, the aldehyde α-methyl group occupies the position *anti* to the forming C–C bond. In transition state **41**, the aldehyde R substituent experiences a gauche interaction with the crotyl γ-Me substituent. Thus, as the size of the aldehyde R substituent increases relative to Me, transition

Table 11-1.

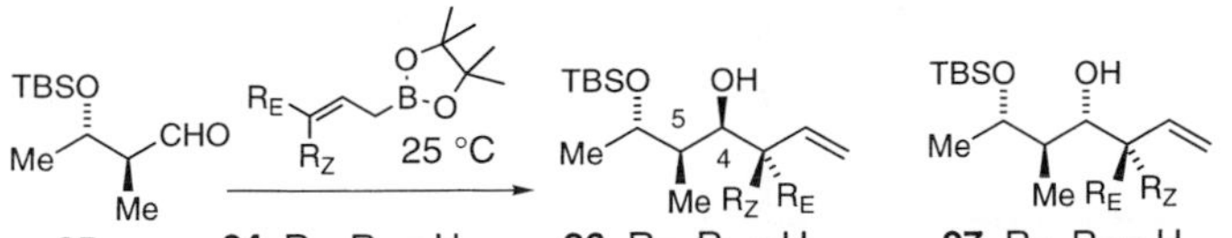

25 + **24**, R_E, R_Z = H; **1**, R_E = Me, R_Z = H; **4**, R_E = H, R_Z = Me (25 °C) → **26**, R_E, R_Z = H; **28**, R_E = Me, R_Z = H; **30**, R_E = H, R_Z = Me + **27**, R_E, R_Z = H; **29**, R_E = Me, R_Z = H; **31**, R_E = H, R_Z = Me

reagent	products	selectivity
24	**26** : **27**	49 : 51
1	**28** : **29**	95 : 5
4	**30** : **31**	9 : 91

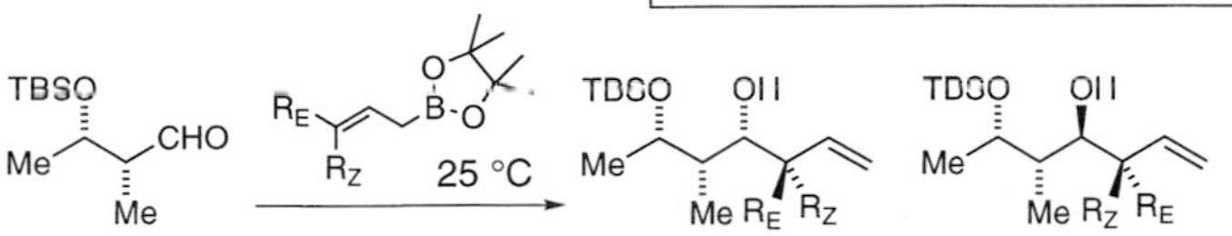

32 + **24**, R_E, R_Z = H; **1**, R_E = Me, R_Z = H; **4**, R_E = H, R_Z = Me (25 °C) → **33**, R_E, R_Z = H; **35**, R_E = Me, R_Z = H; **37**, R_E = H, R_Z = Me + **34**, R_E, R_Z = H; **36**, R_E = Me, R_Z = H; **38**, R_E = H, R_Z = Me

reagent	products	selectivity
24	**33** : **34**	79 : 21
1	**35** : **36**	98 : 2
4	**37** : **38**	40 : 60

state **39**, which minimizes the steric interactions involving R, becomes increasingly favorable. This analysis explains the outstanding selectivity for **28** and **35** in the reactions of **25** and **32** with **1**.

The preference for the (*Z*)-crotylboronate reagent **4** to generate the anti-Felkin homoallylic alcohols **31** and **38** in reaction with α-methyl chiral aldehydes **25** and **32** (Table 11-1) is rationalized by transition state **43**, where the R substituent of the aldehyde occupies the least sterically demanding α-carbon position, *anti* to the forming C–C bond, while the hydrogen occupies the most sterically demanding

39 —*favored* / *sterically (R > Me)*→ **40** Felkin

41 (gauche pentane Me-R) → **42** Anti-Felkin

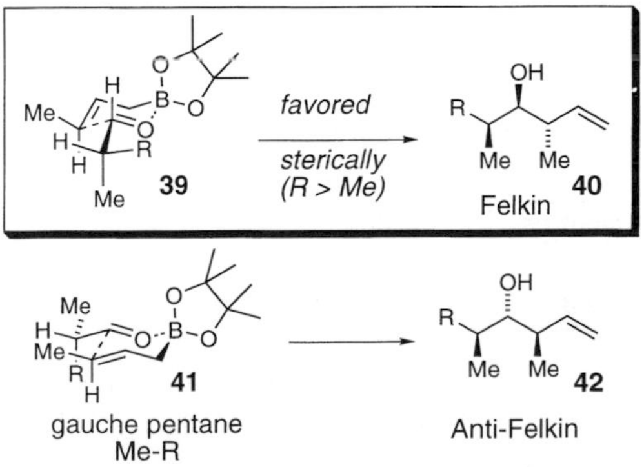

Figure 11-6. Transition states of (*E*)-crotylboronate with α-methyl chiral aldehydes.

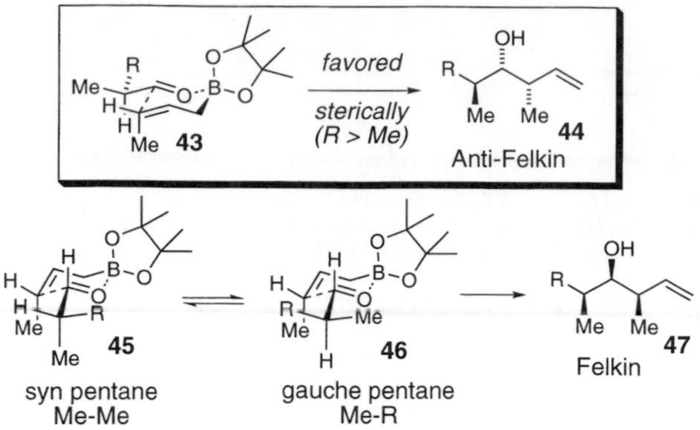

Figure 11-7. Transition states of (*Z*)-crotylboronate with *α*-methyl chiral aldehydes.

position, *syn* to the (*Z*)-crotylboronate *γ*-Me substituent (Fig. 11-7) [3, 65]. The competing transition states **45** and **46**, which lead to the Felkin adduct **47**, are disfavored due to *syn*-pentane and gauche pentane interactions between the crotylboronate *γ*-Me substituent and the aldehyde *α*-methyl and R groups, respectively.

Shortly after Hoffmann's study of the crotylboration reaction of *α*-methyl chiral aldehydes appeared, Roush and co-workers reported a study of the crotylboration reactions of chiral *α*-alkoxy aldehydes [66, 67]. They found that the diastereoselectivity in these reactions was dependent on the geometry of the crotylboronate reagent as well as the steric and stereoelectronic influence of the aldehyde *α*-stereocenter. Thus, the allylboration reaction of glyceraldehyde acetonide **48** with allylboronate reagent **24** resulted in the formation of the 4,5-*anti* adduct **49** as the major isomer with 80:20 selectivity (Table 11-2). The (*E*)-crotylboronate reagent **1** reacted in a considerably less selective manner with **48** to generate adducts **51** and **52** (ratio=55:45) while the (*Z*)-crotylboronate reagent **4** reacted in a highly

Table 11-2.

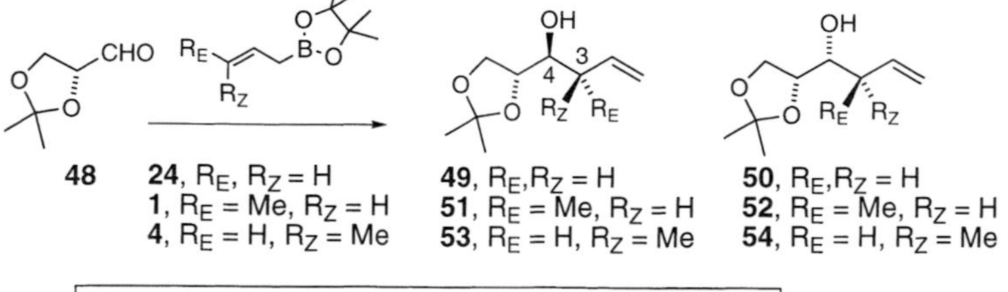

48 — **24**, R_E, R_Z = H; **1**, R_E = Me, R_Z = H; **4**, R_E = H, R_Z = Me

49, R_E,R_Z = H; **51**, R_E = Me, R_Z = H; **53**, R_E = H, R_Z = Me

50, R_E,R_Z = H; **52**, R_E = Me, R_Z = H; **54**, R_E = H, R_Z = Me

reagents	products	selectivity	major t.s.
1	**49 : 50**	80 : 20	
2	**51 : 52**	55 : 45	**63**[a]
3	**53 : 54**	97 : 3	**65**[b]

Reaction conditions: CH_2Cl_2, 23 °C.
[a]) Refer to transition states shown in Fig. 11-8.
[b]) Refer to transition states shown in Fig. 11-9.

stereoselective manner to generate the 3,4-*syn*-4,5-*anti* adduct **53** predominantly (selectivity = 97 : 3). A gauche pentane transition state model, analogous to that summarized in Figs. 11-6 and 11-7, was developed to explain these results (see below) [67].

Hoffmann (with aldehydes **55a,c** and **d**), Mulzner (with aldehyde **55b**) and Wuts (with aldehyde **55e**) subsequently reported similar findings in the crotylation reactions of α-heteroatom-substituted aldehydes with the (*E*)- and (*Z*)-crotylmetal reagents (Tables 11-3 and 11-4) [68–70]. These data, together with the results summarized in Table 11-2, clearly demonstrate that steric effects play a larger role in determining reaction diastereoselectivity than do stereoelectronic effects.

The 4,5-*anti* diastereomer (formally the Felkin product) predominates when the aldehyde α-heteroatom substituent is larger than the aldehyde R group with both the (*E*)- and (*Z*)-crotylmetal reagents (see aldehydes **55b** and **55c**, Tables 11-3 and 11-4). However, when the R substituent is larger than the heteroatom, X, as is the case with aldehyde **55e**, the (*E*)-crotylboronate reagent strongly favors formation of the 4,5-*syn* adduct **57e**, formally the anti-Felkin product (Table 11-3).

Table 11-3. Reactions of α-heterosubstituted aldehydes with (*E*)-crotylmetal reagents.

R–CHO / X **55a-e**; Me–CH=CH–CH$_2$–B(pinacol) **1** or Me–CH=CH–CH$_2$–CrCl$_2$ **8** → **56a-e** + **57a-e**

aldehyde		reagent	products	selectivity	major t.s.*
55a, R = Me,	X = OBn	1	**56a** : **57a**	47 : 53	**64**
55b, R = Me,	X = OTBS	8	**56b** : **57b**	99 : 1	**63**
55c, R = Me,	X = N(CH$_2$Ph)$_2$	1	**56c** : **57c**	90 : 10	**63**
55d, R = Et,	X = Cl	1	**56d** : **57d**	47 : 53	**64**
55e, R = TBSO/BnO (structure)	X = OMe	1	**56e** : **57e**	1 : 99	**64**

*) Refer to transition states shown in Fig. 11-8.

Table 11-4. Reactions of α-hetersubstituted aldehydes with (*Z*)-crotylboronates.

R–CHO / X **55a-d** → **4**, R' = Me; **58**, R' = OMe → **59a-d**, R' = Me; **61**, R' = OMe + **60a-d**, R' = Me; **62**, R' = OMe

aldehyde		reagent	products	selectivity	major t.s.*
55a, R = Me,	X = OBn	4	**59a** : **60a**	88 : 12	65
55a, R = Me,	X = OBn	58	**61** : **62**	82 : 18	
55c, R = Me,	X = N(CH$_2$Ph)$_2$	4	**59c** : **60c**	49 : 51	66
55d, R = Et,	X = Cl	4	**59d** : **60d**	90 : 10	65

*) Refer to transition states shown in Fig. 11-9.

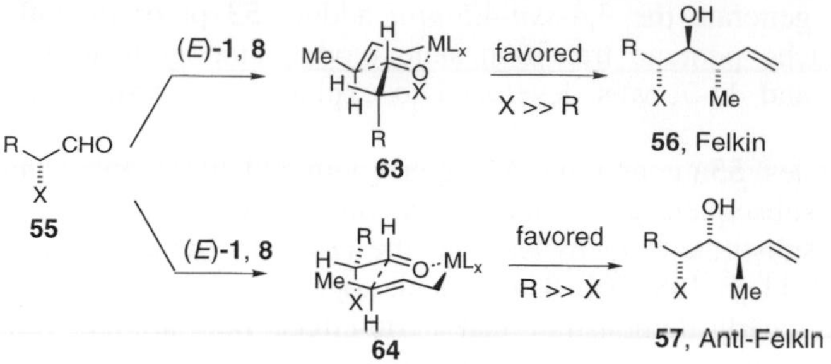

Figure 11-8. Reactions of α-heteroatom aldehydes with (E)-crotylmetal reagents.

The stereochemical outcome of the reactions of α-heteroatom-substituted aldehydes with Type I and Type III (E)-crotylmetal reagents can be rationalized through the competition between transition states **63** and **64**, which lead to the 4,5-*anti* (Felkin) and 4,5-*syn* (anti-Felkin) adducts, **56** and **57**, respectively (Fig. 11-8) [67]. Thus, when X is the largest group, it occupies the least sterically demanding position *anti* to the forming C–C bond in transition state **63**. However, when R is larger than X, transition state **64** becomes favored.

The stereochemical outcome of the reactions of α-alkoxy aldehydes with Type I (Z)-crotylmetal reagents can be rationalized through the competition between transition states **65** and **66**, which lead to the 4,5-*anti* and 4,5-*syn* adducts, **59** and **60**, respectively (Fig. 11-9) [67]. The Cornforth [71] transition state **65** is favored when R is sterically larger than X since, in this transition state, R occupies the least sterically demanding position, *anti* to the forming C–C bond. Conversely, transition state **66** should be favored when X is much larger than R.

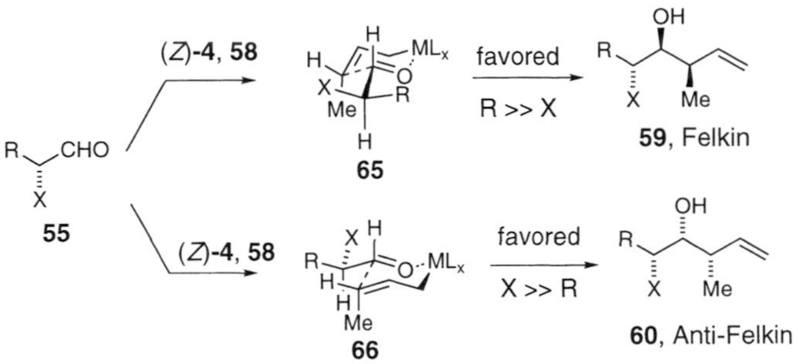

Figure 11-9. Reactions of α-heteroatom aldehydes with (Z)-crotylmetal reagents.

11.2.2 Chelate-Controlled Reactions of Type I Allylmetal Reagents

In certain cases, Type I allylmetal reagents have been shown to give products consistent with α- or β-heteroatom chelation [72–78]. In 1989, Sakurai reported that the allylation reactions of α-hydroxy ketones with allyl- and crotyltrifluorosilanes

Table 11-5.

67 → (reagent, conditions) → 74, R_E, R_Z = H; 75, R_E = Me, R_Z = H; 76, R_E = H, R_Z = Me

77

68, R_E, R_Z = H; 69, R_E = Me, R_Z = H; 70, R_E = H, R_Z = Me

71, R_E, R_Z = H; 72, R_E = Me, R_Z = H; 73, R_E = H, R_Z = Me

reagent	conditions*	product	selectivity	% yield
68	a	**74**	>99 : 1	69
(*E*)-**69**	a	**75**	97 : 3	71
(*Z*)-**70**	a	**76**	95 : 5	75
71	b	**74**	>99 : 1	94
(*E*)-**72**	b	**75**	98 : 2	83
(*Z*)-**73**	b	**76**	95 : 5	80

*) Conditions:
[a]) Et_3N, THF, reflux.
[b]) Et_3N, CH_2Cl_2, $-10\,°C \rightarrow$ reflux.

Table 11-6.

78 → (i-Pr_2NEt, CH_2Cl_2, 4 Å MS, 0 °C) → 79, R_E, R_Z = H; 80, R_E = Me, R_Z = H; 81, R_E = H, R_Z = Me

82

reagent	product	selectivity	% yield
68, R_E, R_Z = H	**79**	90 : 10	74
69, R_E = Me, R_Z = H	**80**	95 : 5	75
70, R_E = H, R_Z = Me	**81**	93 : 7	76

68–70 in the presence of triethylamine produced products **74–76** with high selectivity (Table 11-5) [73]. The stereochemistry of adducts **74–76** is consistent with their emergence through the bicyclic chelate transition state **77**, which involves a hypervalent silicon atom as the ML_x component. Subsequently, Kabalka and coworkers reported similar results using diisopropyl allyl- and crotylboronates **71–73** (Table 11-5) [74].

Roush and Chemler have further expanded this methodology to include chiral *β*-hydroxy aldehydes as substrates (Table 11-6) [77, 78]. They demonstrated that the allyl- and crotyltrifluorosilanes **68–70** react with the 2,3-*anti*-*β*-hydroxy aldehyde **78** to form the 4,5-*anti* adducts **79–81** predominantly, the stereochemistry of which is consistent with formation through the bicyclic chelate transition state **82** (Table 11-6). However, under the same reaction conditions, 2,3-*syn*-*β*-hydroxy aldehydes do not react selectively with **68–70** [78].

11.2.3 Selected Applications of Achiral Type I and Type III Reagents to Natural Product Synthesis

Kishi and co-workers elegantly demonstrated the power of substrate-directed crotylation reactions in their synthesis of the C(15)–C(29) segment **83** of rifamycin S (Fig. 11-10) [61]. This synthesis utilizes two crotylation reactions with the Nozaki-Hiyama crotylchromium reagent **8** [18, 19] and one allylation reaction using allyldichloroiodostannane [79].

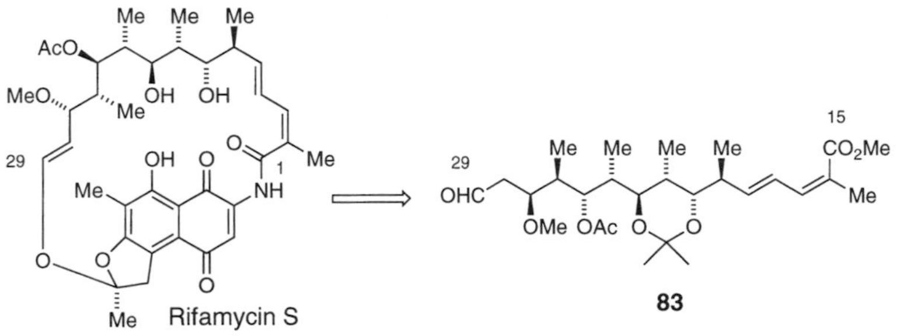

Figure 11-10.

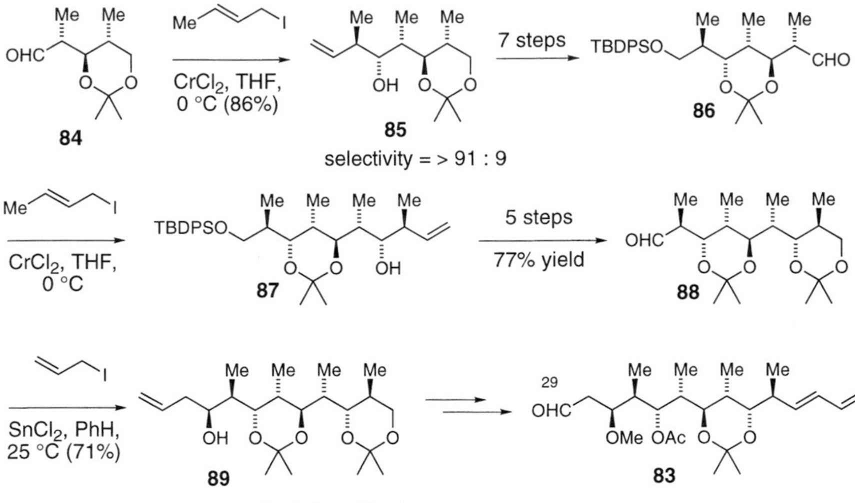

Scheme 11-1.

The synthesis of **83** began with aldehyde **84** (Scheme 11-1) [61]. Substrate-directed reaction of **84** with (*E*)-crotylchromium **8**, generated in situ from (*E*)-crotyliodide, occurred with >91:9 diastereoselectivity for the Felkin adduct **85**. Adduct **85** was subsequently elaborated to aldehyde **86**, which underwent a Felkin-selective reaction with the in situ generated crotylchromium reagent to afford homoallylic alcohol **87** as the only observed adduct. Homoallylic alcohol **87** was converted to aldehyde **88** in five steps (77% overall yield from **86**). Aldehyde **88** underwent a highly diastereoselective allylation reaction with the in situ generated allyldichloroiodostannane reagent derived from allyl iodide and $SnCl_2$ [79], thereby generating homoallylic alcohol **89** as the major adduct, which was subsequently elaborated to the C(15)–C(29) segment **83** of rifamycin S.

The excellent diastereoselection of all three of these substrate-directed allylation reactions is consistent with reaction occurring through Felkin transition states analogous to **39** (Fig. 11-6). These examples illustrate the excellent stereochemical control opportunities that exist in (*E*)-crotylation reactions of α-methyl chiral aldehydes, especially when the β-position is branched (as in the (*E*)-crotylation of **25** and **32**, see above).

In their synthesis of olivin, the aglycon segment of olivomycin A, Roush and coworkers used a highly diastereoselective substrate-directed γ-alkoxy allylation reaction to set the C(1′) stereocenter [80]. Thus, reaction of the aldehyde **90**, derived from L-threonine, with the [(*Z*)-γ-methoxyallyl]boronate **91** resulted in the highly diastereoselective formation of adduct **92**. The stereochemistry of **92** is consistent

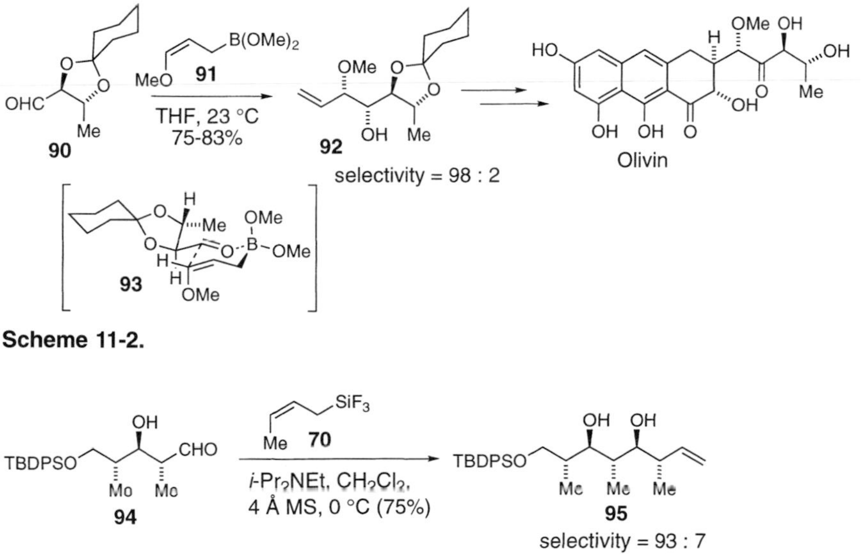

Scheme 11-2.

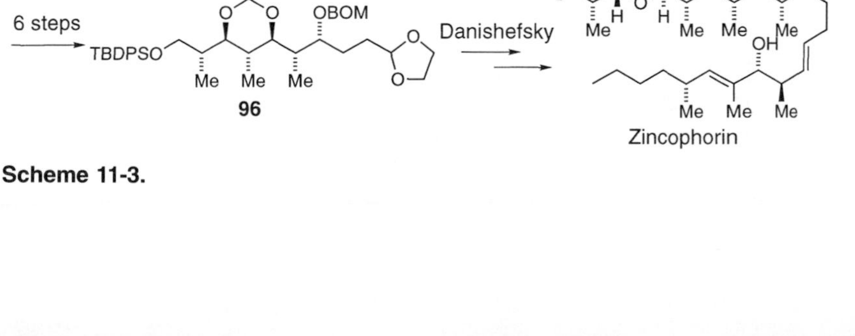

Scheme 11-3.

with its formation through the Cornforth transition state **93** (analogous to **65**, Fig. 11-9). This example illustrates the stereochemical control possibilities in the allylboration of α-alkoxy aldehydes with [(*Z*)-γ-alkoxy]allyl- and crotylboronates.

In their synthesis of the C(7)–C(16) segment **96** of the ionophore antibiotic zincophorin, Roush and Chemler used the (*Z*)-crotyltrifluorosilane reagent **70** in a highly stereoselective synthesis of the all-*anti* stereopentad **95** from the *anti,anti*-β-hydroxy aldehyde **94** (Scheme 11-3) [77]. The stereoselective formation of adduct **95** (75%, selectivity=93:7) is rationalized by C–C bond formation through the bicyclic chelate transition state **82** (see Table 11-6). Adduct **95** was converted in six steps to the C(7)–C(16) segment **96**, whose conversion to zincophorin had previously been performed by Danishefsky and co-workers [81]. It should be noted that intermediate **95** cannot be prepared with acceptable selectivity using crotylboronate technology. This example, therefore, provides a very nice illustration of the use of the (*Z*)-crotyltrifluorosilane methodology to synthesize *anti,anti*-dipropionates, which are difficult to access by using other crotylmetal or aldol methods [82].

11.3 Reactions of Type II Allylmetal Reagents with Chiral Aldehydes; Selected Applications in the Synthesis of Natural Products

11.3.1 Stereochemical Control via 1,2-Asymmetric Induction

Reactions of the α-methyl-β-benzyloxy aldehyde **97** with allyltri-*n*-butylstannane **98** are summarized in Table 11-7 [83]. While little stereocontrol is observed in the $BF_3{\cdot}OEt_2$-promoted allylation reaction, the chelate-controlled reaction catalyzed by either $TiCl_4$ or $SnCl_4$ is much more selective, favoring formation of the Cram-chelate adduct **100** with up to 98:2 selectivity. The chelate transition state **101**, where C–C bond formation occurs *anti* to the aldehyde α-methyl group, rationalizes the observed stereoselective formation of **100**. Although the $BF_3{\cdot}OEt_2$-cata-

Table 11-7.

BnO CHO Me **97** — SnBu3 **98**, CH_2Cl_2, -78 °C → **99** + **100**

Lewis acid	99 : 100	yield
$BF_3{\cdot}OEt_2$	52 : 48	--
$TiCl_4$	18 : 82	--
$SnCl_4$	2 : 98	88%

101

lyzed reaction of **97** and allylstannane **98** is not stereoselective, more sterically demanding β-branched α-methyl-substituted aldehydes display excellent stereoselectively under these reaction conditions (illustrated below in Eqs. (11.8) and (11.9)). The chelate-controlled allylation of α-methyl-β-alkoxy-substituted aldehydes has been used several times in natural product synthesis [84–88].

Although silyloxy groups do not participate in chelates under most conditions, Evans and co-workers have recently reported that the *tert*-butyldimethylsilyl-protected β-alkoxy aldehyde **102** undergoes highly diastereoselective chelate-controlled allylation reactions with allyl and β-methyl allyltri-*n*-butylstannanes **98** using Me_2AlCl (2.5 equiv) as the Lewis acid promoter (Eq. (11.2)) [89].

OHC, OTBS, *i*-Pr, Me, **102**; **98** (R, $SnBu_3$); Me_2AlCl, CH_2Cl_2, -78 °C; **103**; **104** (11.2)

R	yield	103 : 104
Me	63	90:1
H	84	99:10

Table 11-8.

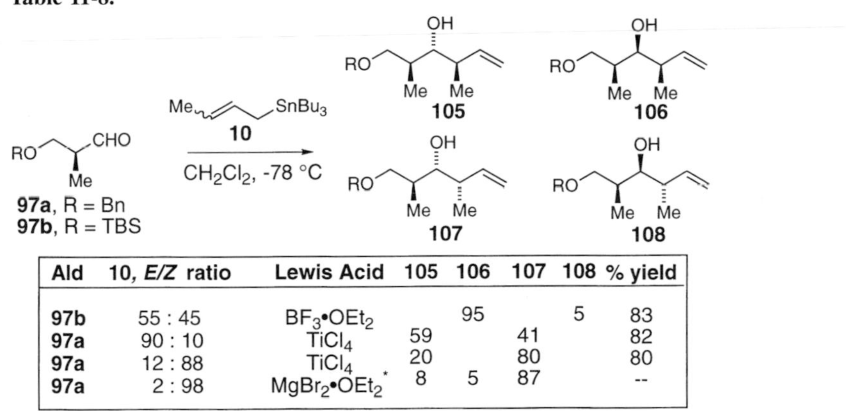

Ald	10, *E/Z* ratio	Lewis Acid	105	106	107	108	% yield
97b	55 : 45	$BF_3{\cdot}OEt_2$		95		5	83
97a	90 : 10	$TiCl_4$	59		41		82
97a	12 : 88	$TiCl_4$	20		80		80
97a	2 : 98	$MgBr_2{\cdot}OEt_2$*	8	5	87		--

*) Reaction run at –23 °C.

The crotylation reactions of α-methyl chiral aldehydes (e.g. **97**) with Type II crotylmetal reagents can give up to four products (e.g., **105–108**). The *syn,syn*-adduct **106** and the 3,4-*syn*-4,5-*anti* diastereomer **107** can be obtained with useful levels of diastereoselectivity via the reaction of **97** with the achiral (*E*)- and (*Z*)-crotyltri-*n*-butylstannanes **10** under appropriate conditions (Table 11-8) [53].

The $BF_3{\cdot}OEt_2$-catalyzed addition of crotyltri-*n*-butylstannane **10** (a 55:45 mixture of *E*- and *Z*-isomers was used in this experiment) to the β-silyloxy aldehyde **97b** furnishes the *syn,syn*-dipropionate adduct **106** with 95:5 diastereoselectivity.

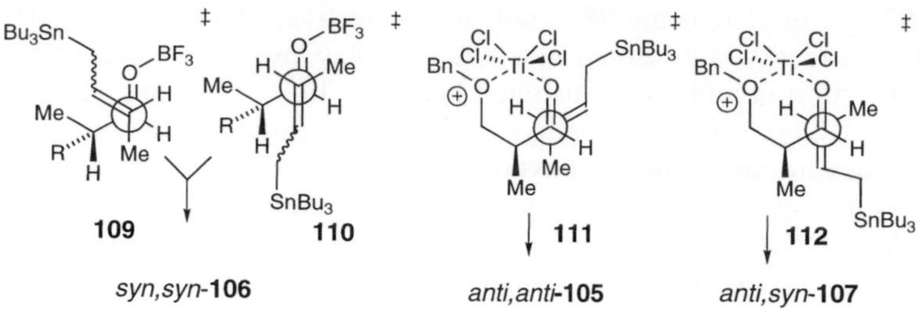

Figure 11-11.

The stereochemistry of adduct **106** is consistent with formation through either the Felkin, synclinal transition state **109** or the Felkin, antiperiplanar transition state **110** (Fig. 11-11). A discussion of the merits of these two transition states appears earlier in this chapter (Section 11.1.2).

Under chelation-controlled conditions ($TiCl_4$ promotion), the *anti-anti-* and *anti,syn*-dipropionate adducts **105** and **107** are the predominant products of the crotylation of aldehyde **97a** using the (*E*)- and (*Z*)-crotyltri-*n*-butylstannanes, respectively. While the (*E*)-tri-*n*-butylcrotylstannane favors formation of the *anti,anti*-adduct **105** through transition state **111**, the *anti,syn*-adduct **107**, which evidently arises through transition state **112**, is obtained preferentially when the (*Z*)-crotyl-tri-*n*-butylstannane is used (Table 11-8 and Fig. 11-11). Use of $MgBr_2{\cdot}OEt_2$ and highly isomerically enriched crotylstannane (*Z*)-**10** improves the selectivity for the *anti,syn* diastereomer **107** to a synthetically useful level (Table 11-8).

Keck further demonstrated that the *anti,syn*-adduct **114** can be formed with high selectivity from the chelate-controlled reaction of aldehyde **97a** with the (γ-silyloxyallyl)tri-*n*-butylstannane **113**, presumably through transition state **115** (Eq. (11.3)) [90].

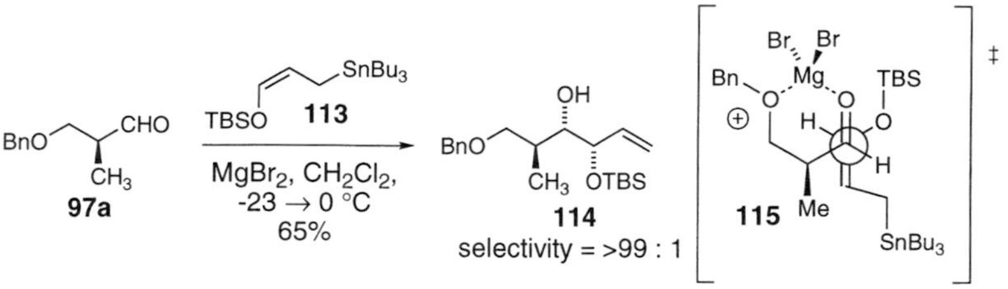

(11.3)

α-Alkoxy aldehydes can undergo stereoselective reaction with Type II allyl and crotylmetal reagents under both non-chelation and chelation conditions. As illustrated in Table 11-9, the $BF_3{\cdot}OEt_2$-catalyzed reaction of allyltri-*n*-butylstannane **98** and α-alkoxy aldehydes **116** provides the Felkin adduct **117**, with up to 95:5 selectivity, depending on the protecting groups. These reactions proceed preferentially by way of the Felkin transition state **119** [91], where the level of selectivity for **117** increases as the size of the alkoxy substituent increases (OTBS>OBn). In contrast, when $MgBr_2$ is used as the Lewis acid promoter, adduct **118** can be ob-

Table 11-9.

aldehyde	Lewis acid	temp.	117 : 118	% yield
116a	$BF_3{\cdot}OEt_2$	-78 °C	61 : 39	--
116b	$BF_3{\cdot}OEt_2$	-78 °C	95 : 5	83
116a	$MgBr_2$	-23 °C	1 : 99	85
116b	$MgBr_2$	-23 °C	79 : 21	--

tained with high selectivity (>99:1) from aldehyde **116a**. Adduct **118** arises from the chelate transition state **120**. However, diastereoselectivity in the $MgBr_2$-promoted reaction of aldehyde **116b** is much lower, and, in fact, formation of the Felkin isomer **117** is favored since the silyloxy group does not readily participate in the formation of a chelate.

Mikami and Keck have each demonstrated that *α*-alkoxy aldehydes can also undergo highly selective chelate-controlled additions with crotyltri-*n*-butylstannane [53, 57]. The $MgBr_2{\cdot}OEt_2$-promoted reaction of the *α*-benzyloxy aldehyde **55** with crotyltri-*n*-butylstannane **10** gave predominantly the *syn,syn*-adduct **121**, whose stereochemistry is consistent with the antiperiplanar, chelate transition state **122** (Eq. (11.4)).

(11.4)

Interestingly, in the analogous reaction of the enantiomeric aldehyde *ent*-**55** with the *β*-methylcrotyltributylstannane **123a**, Mikami found that the *syn,anti*-adduct **124** was the major product (selectivity=75:25, Eq. (11.5)) [57]. This reaction was even more selective (97:3) for adduct **124** in the $SnCl_4$-promoted reaction of aldehyde *ent*-**55** with crotylsilane **123b**. The stereochemistry of adduct **124** is consistent with formation through transition state **126**, where the position between the aldehyde oxygen and the aldehyde hydrogen is occupied by the now more sterically demanding β-carbon of the crotylstannane/silane reagent.

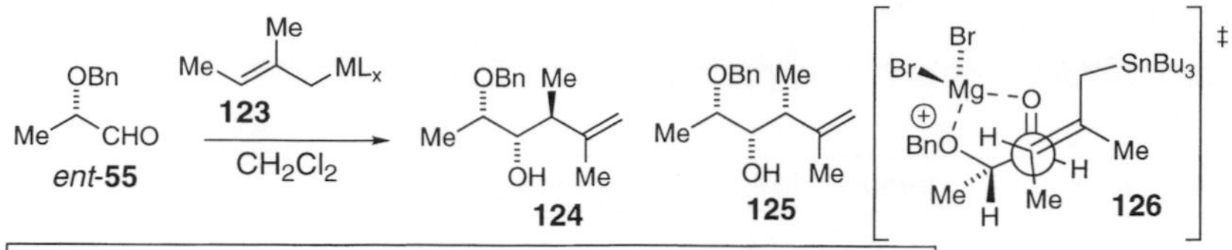

reagent	Lewis acid	temp.	124 : 125	% yield
123a, $ML_x = SnBu_3$	$MgBr_2{\bullet}OEt_2$	-25 °C	75 : 25	85
123b, $ML_x = SiMe_3$	$SnCl_4$	-78 °C	97 : 3	90

(11.5)

In the complementary chelate-controlled reaction of the α-benzyloxy aldehyde **127** with the (γ-silyloxyallyl)stannane **113**, the *syn,syn*-adduct **128** arises as the major adduct, presumably through transition state **129** (Eq. (11.6)) [90].

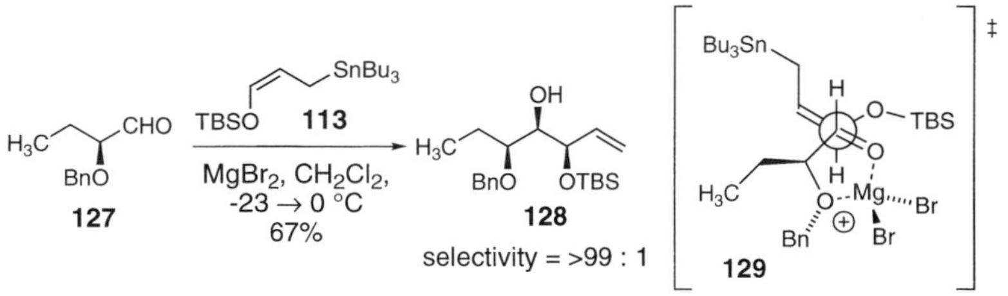

(11.6)

11.3.2 Stereochemical Control via 1,3-Asymmetric Induction

In the reactions of Type II allylmetal reagents with chiral aldehydes, β-alkoxy substituents on the aldehyde can exert a strong influence on the reaction, much more so than in reactions of Type I allylmetal reagents. Reetz and co-workers reported that the $BF_3{\cdot}OEt_2$-catalyzed allylation reaction of the β-benzyloxy aldehyde **130** with allyltrimethylsilane **131** is selective for the 1,3-*anti* diol **132** (Eq. (11.7)) [92]. Evans and co-workers subsequently rationalized this result by invoking tran-

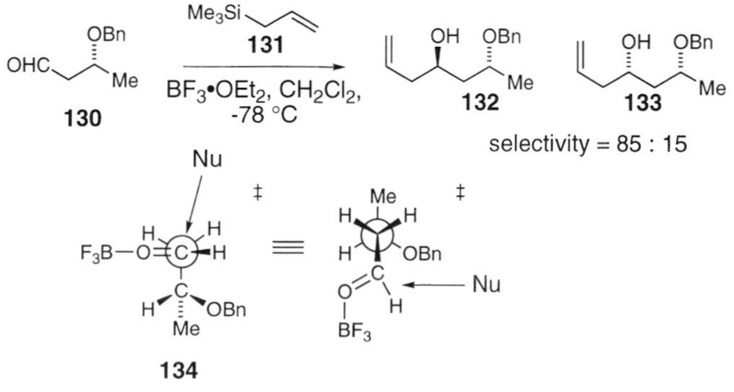

(11.7)

sition state **134** in which preferential addition of the allylsilane to the aldehyde occurs via the conformation shown, that best minimizes dipole interactions between the aldehyde carbonyl and the β-alkoxy group [93].

Merged 1,2 and 1,3-asymmetric induction

In the Type II allylation reactions of α-methyl-β-alkoxy aldehydes, the principles of 1,2- and 1,3-asymmetric induction both contribute to the reaction diastereoselectivity. Evans and co-workers have explained the stereochemical outcome of these reactions in terms of a "merged" 1,2- and 1,3-asymmetric induction model [93]. For example, the 2,3-*anti* aldehyde **135** reacts with allyl- and methallyltri-*n*-butylstannanes **98**, generating the Felkin homoallylic alcohols **136** with >99:1 diastereoselectivity (Eq. (11.8)) [93].

135 + **98** (Bu_3Sn-CH$_2$C(R)=CH$_2$), $BF_3\cdot OEt_2$, CH_2Cl_2, -78 °C → **136a**, R = Me; **136b**, R = H (Felkin); transition state **137** (11.8)

R	yield	selectivity*
Me	86%	>99 : 1
H	95%	>99 : 1

*Felkin : anti-Felkin

The diastereoselectivity of these reactions is consistent with product formation occurring through transition state **137**, where the reactive conformation of the aldehyde in the transition state (corresponding to the normal Felkin-Anh model) minimizes steric interactions with the allylstannane as well as the 1,3-dipole interactions of the aldehyde and the β-alkoxy group. The allylation reaction of the 2,3-*syn* aldehyde **138**, however, with allyltri-*n*-butylstannanes **98**, generates the anti-Felkin adducts **139** preferentially (Eq. (11.9)) [93]. The stereochemistry of these reactions is consistent with product formation occurring preferentially through transition state **140**, in which 1,3-dipole interactions of the aldehyde and the β-

138 + **98** (Bu_3Sn-CH$_2$C(R)=CH$_2$), $BF_3\cdot OEt_2$, CH_2Cl_2, -78 °C → **139a**, R = Me; **139b**, R = H (Anti-Felkin); transition state **140** (11.9)

R	yield	selectivity*
Me	86%	80 : 20
H	85%	87 : 13

*anti-Felkin : Felkin

product **160** of the thermal cyclization of (*E*)-**159** potentially arises through the boat-boat transition state **164**, while the minor adduct **161** could arise through the chair-chair transition state **165**. However, the great difference in thermal reactivity of (*Z*)-**158** and (*E*)-**159** led Keck to postulate that transition states **164** and **165** must be greatly disfavored compared to **162** and **163**, perhaps due to a greater difficulty for the *trans*-olefin of (*E*)-**159** to react through a concerted, pericyclic process. Keck proposed that the formation of the major *syn* adduct **160** from (*E*)-**159** likely arises through isomerization of **159** to **158**, followed by reaction through transition state **162**.

In the $BF_3 \cdot OEt_2$-catalyzed reaction of (*Z*)-**158**, the formation of **160** as the major adduct is rationalized by preferential bond formation through the synclinal transition state **166** (Fig. 11-15). The minor adduct **161** arises through the other synclinal transition state, **167**. The (*E*)-stannane **159**, on the other hand, forms **161** preferentially through the synclinal transition state **168**. In both of the preferred transition states, **166** from (*Z*)-**158** and **168** from (*E*)-**159**, the tin-bearing carbon is in close proximity to the aldehyde oxygen. As noted previously, Keck proposed that this situation is preferable because of secondary orbital overlap be-

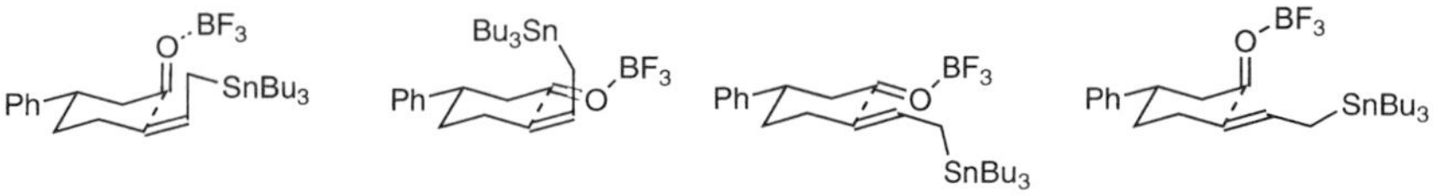

(*Z*)-stannane major pathway, **166** | (*Z*)-stannane minor pathway, **167** | (*E*)-stannane major pathway, **168** | (*E*)-stannane minor pathway, **169**

Figure 11-15. Lewis acid catalyzed carbocyclizations.

Table 11-11.

(*Z*)-**170**
(*E*)-**171**

172a, n = 1 **173a**, n = 1
172b, n = 2 **173b**, n = 2
172c, n = 3 **173c**, n = 3

stannane	conditions	172 : 173	% yield
(*Z*)-**170a**, n = 1	$BF_3 \cdot OEt_2$ [a]	90 : 10	>95
(*Z*)-**170a**, n = 1	100 °C [b]	2 : 98	>95
(*E*)-**171a**, n = 1	$BF_3 \cdot OEt_2$ [a]	97 : 3	79
(*E*)-**171a**, n = 1	100 °C [b]	30 : 70	>95
(*Z*)-**170b**, n = 2	$BF_3 \cdot OEt_2$ [a]	68 : 32	80
(*Z*)-**170b**, n = 2	100 °C [b]	2 : 98	78
(*E*)-**171b**, n = 2	$BF_3 \cdot OEt_2$ [a]	87 : 13	>95
(*E*)-**171b**, n = 2	100 °C [b]	98 : 2	>95
(*Z*)-**170c**, n = 3	$BF_3 \cdot OEt_2$ [a]	>98 : 2	45
(*Z*)-**170c**, n = 3	150 °C [b]	--	trace
(*E*)-**171c**, n = 3	$BF_3 \cdot OEt_2$ [a]	>98 : 2	79
(*E*)-**171c**, n = 3	150 °C [b]	--	trace

[a]) Reaction run in CH_2Cl_2 at –78 °C.
[b]) Reaction run in benzene or toluene.

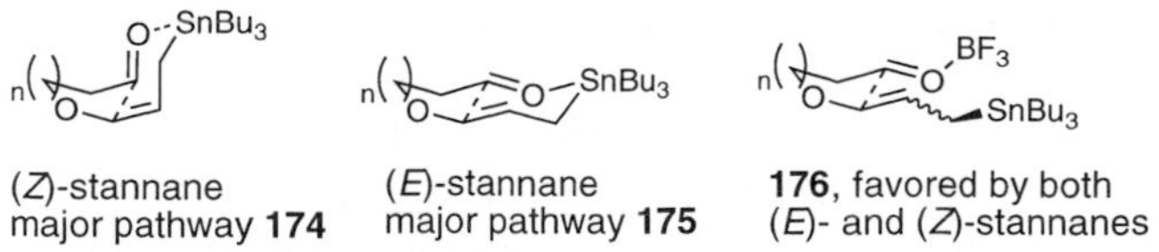

(Z)-stannane major pathway **174** (E)-stannane major pathway **175** **176**, favored by both (E)- and (Z)-stannanes

Figure 11-16. Thermal ether cyclizations Lewis acid catalyzed ether cyclizations.

tween the LUMO of the carbonyl oxygen and the HOMO of the carbon bearing the stannyl group [50, 54]. Minimization of charge separation would also help explain the preference for these synclinal transition states.

Yamamoto [106] and Martín [107] have independently investigated the *exo* cyclizations of (γ-alkoxyallyl)stannanes, potential precursors to polyether natural products. Yamamoto systematically studied the thermal and Lewis acid promoted cyclizations of allylstannanes (Z)-**170** and (E)-**171** (Table 11-11).

Yamamoto found that, in general, the (Z)-stannanes **170** favored formation of the *syn* adducts **173** under thermal conditions, presumably through transition state **174** (Fig. 11-16). Under thermal conditions, however, the (E)-stannanes **171** generally favor formation of the *anti* adducts **172**, where the *anti* stereochemistry is rationalized by reaction occurring through transition state **175**. Also different from Keck's observations, Yamamoto found that the *anti*-adducts **172** were formed preferentially (via transition state **176**) in the Lewis acid-promoted reactions of both the (Z)-**170** and the (E)-**171** substrates. Thus, as noted by Yamamoto, the secondary orbital interactions or the dipole effects experienced in the ring-closing transition states with the (γ-alkoxyallyl)stannanes **170** and **171** are apparently not as important in the reactions of **158** and **159** [106].

11.4.1 Selected Applications of the Ring-Closing Allylation Reaction in Natural Product Synthesis

Yamamoto found that the seven-membered cyclic ether **178** could be formed stereoselectively via $BF_3 \cdot OEt_2$-promoted intramolecular allylation of the [(Z)-γ-alkoxyallyl]stannane **177** (Eq. (11.12)) [108]. This methodology was applied to the synthesis of hemibrevetoxin B [105].

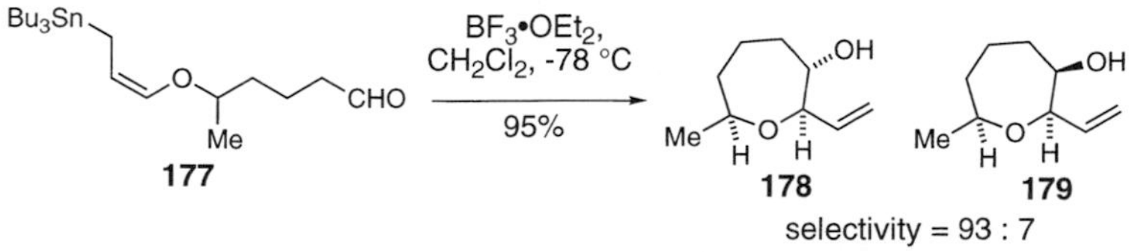

Bu_3Sn ... CHO, Me **177** → $BF_3 \bullet OEt_2$, CH_2Cl_2, -78 °C, 95% → **178** + **179**, selectivity = 93 : 7 (11.12)

The two seven-membered ether rings of hemibrevetoxin B were installed via stereoselective intramolecular *exo* cyclizations of the appropriately advanced (γ-alkoxyallyl)stannane intermediates, **182** and **184** (Scheme 11-6). The synthesis of hemibrevetoxin B began with **180**, which was converted in thirteen steps to the *trans*-fused bis-cyclic ether intermediate **181**. The allyl ether functionality of **181** was converted to the required (γ-alkoxyallyl)stannane unit of **182** via deprotonation with *sec*-butyllithium followed by addition of Bu_3SnCl. Subsequent oxidation of the primary alcohol provided aldehyde **182**. Upon treatment with $BF_3{\cdot}OEt_2$, **182** underwent intramolecular allylation, generating adduct **183** stereoselectively. This adduct was converted in four steps to allylstannane **184**, which also underwent intramolecular allylation to generate the *anti* adduct **185** stereoselectively. The synthesis of hemibrevetoxin B was completed in seven additional steps.

Hoffmann and Krüger have demonstrated that in situ generated (*Z*)-allylboronate intermediates can also undergo intramolecular ring-closing allylation [109] and have applied this methodology to the synthesis of the (+)-laurencin precursor **190** (Scheme 11-7) [104]. Hoffmann's stereoselective synthesis of oxocane **190** began with (*R*)-malic acid (**186**), which was converted in nine steps to vinyl ether **187**, where the Weinreb's amide substituent serves as an aldehyde precursor. Thus, **187** was reduced with diisobutylaluminum hydride (DIBAL), and then the allyl unit was metallated by treatment with *sec*-butyllithium. Addition of methoxypinacol borane then provided the requisite allylboronate in a one-pot sequence. Quenching of this reaction with pH 7 buffer generates the allylboronate **188**, which at room temperature undergoes intramolecular allylation, forming the *cis*-fused *syn* oxocane **189** in 38% overall yield from amide **187** (no other diastereomers were observed). Regioselective reduction of **189** generated intermediate **190**, which has been converted to (+)-laurencin by Holmes et al. [110].

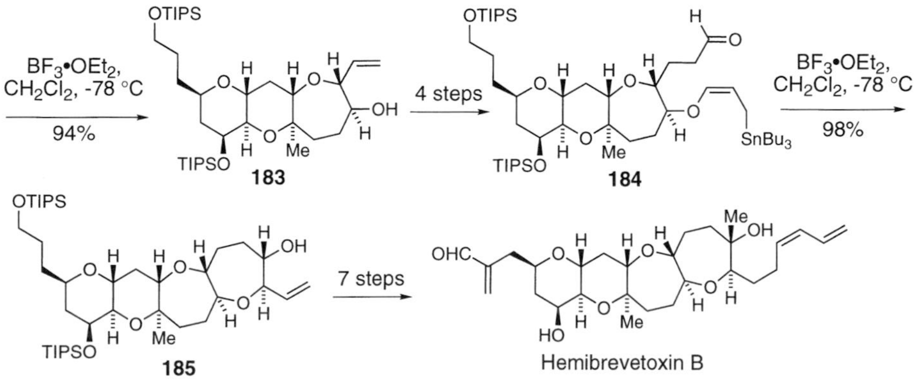

180 → (13 steps) → 181 → (1) s-BuLi, TMEDA, THF, -78 °C, then Bu_3SnCl, -78 °C → rt, 69%; 2) SO_3•py, DMSO, Et_3N, CH_2Cl_2, rt, 90%) → 182 → (BF_3•OEt_2, CH_2Cl_2, -78 °C, 94%) → 183 → (4 steps) → 184 → (BF_3•OEt_2, CH_2Cl_2, -78 °C, 98%) → 185 → (7 steps) → Hemibrevetoxin B

Scheme 11-6.

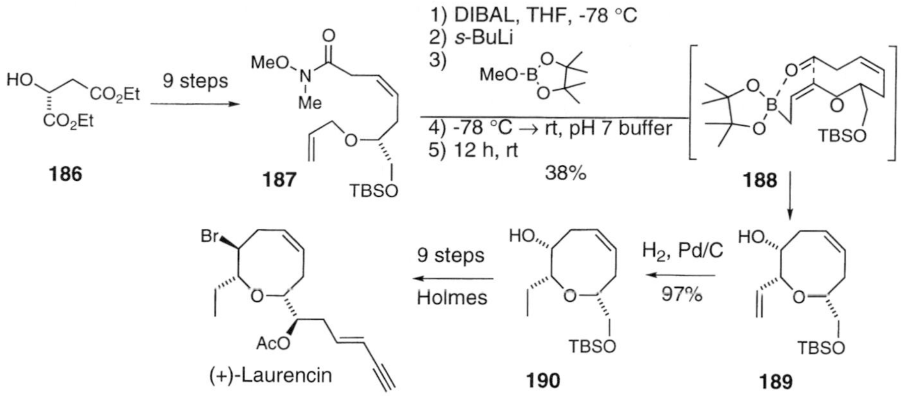

Scheme 11-7.

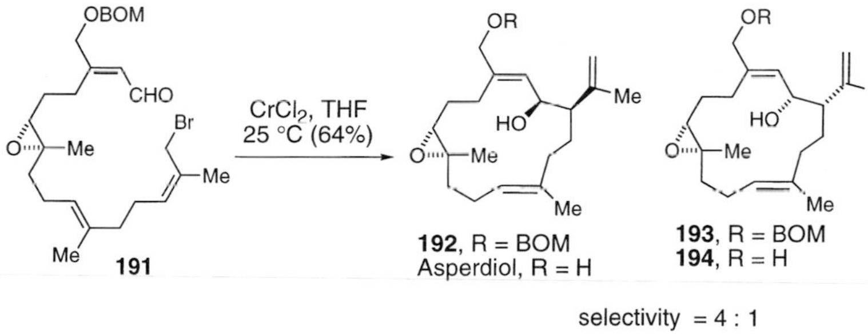

Scheme 11-8.

As a final example of ring-closing allylation in natural product synthesis, Still and co-workers demonstrated that the crotylchromium species derived in situ from the allylbromide **191** undergoes intramolecular allylation to give a 4:1 ratio of adducts **192** and **193**, where the major adduct **192** was subsequently converted to asperdiol (Scheme 11-8) [102].

These selected synthetic applications of the ring-closing allylation reaction illustrate the opportunities for obtaining with high stereoselectivity functionalized heterocycles and carbocycles, common units found in natural products. Other applications of intramolecular allylation reactions have been reported [95–101, 103].

11.5 Overview of Chiral Allylmetal and Allenylmetal Reagents

In the following Sections we review the reactions of chiral allylmetal and allenylmetal reagents and their application to the synthesis of complex natural products. These reagents are useful for the enantioselective allylation of achiral aldehydes as well as for applications in double asymmetric reactions [111] with chiral aldehydes. In the previous sections of this Chapter we highlighted those situations where substrate-directed allylation or crotylation reactions with achiral allylmetal or crotylmetal reagents are synthetically useful. Use of highly enantioselective chiral reagents further expands the utility of carbonyl allylation reactions by enhancing the diastereoselectivity of reactions with otherwise marginally selective substrates (matched double asymmetric diastereoselection), and also by overriding the intrinsic diastereofacial selectivity of chiral aldehydes (mismatched double diastereoselection), thereby giving access to the diastereomers that otherwise are the minor products of substrate-directed allylation reactions.

Over the past twenty years, many chiral allylmetal and allenylmetal reagents have been developed for the enantioselective synthesis of homoallylic and homopropargylic alcohols. Several researchers have developed chiral allylmetal reagents based on boron [26], e.g., allylboron reagents **195–202** [112–127] and allenylboron reagents **203** [128, 129] and **204** [130] (Fig. 11-17). Other chiral allylmetal reagents based on silicon [11, 15] (e.g., reagents **205** [131] and **210** [132]), titanium [21] (e.g. reagents **207** [133] and **208** [134]), molybdenum (e.g., reagent **209**

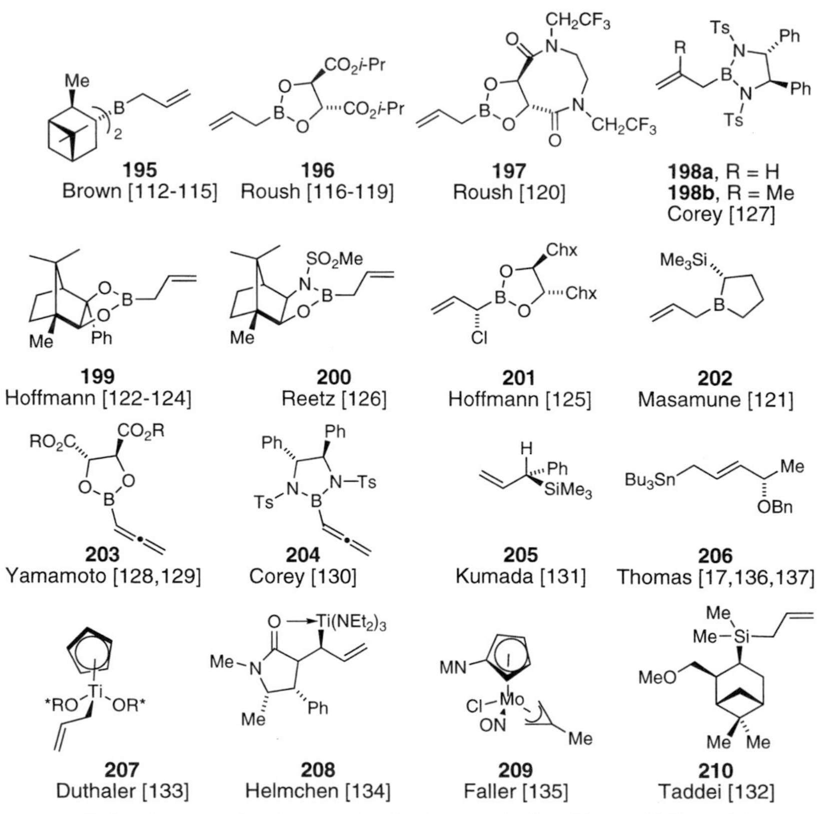

Figure 11-17. Reagents for the enantioselective synthesis of homoallylic and homopropargylic alcohols.

[135]) and tin (e.g., reagent **206** [17, 136, 137]) have also been developed. Of these chiral allylmetal reagents, the Brown reagent **195** has been used most extensively in natural product synthesis. The Roush reagent **196**, the Brown reagent **195**, the Hoffmann reagent **201** and the Corey reagent **198** are discussed further in Sections 11.6–11.9.

Chiral crotylmetal reagents based on boron (e.g., the (*Z*)-crotylboronate reagents **211** [138], **213** [65, 139] and **214** [120], and the (*Z*)-crotylborane reagents **212** [115, 140–142] and **215** [143]), silicon (e.g. **216** [144, 145] and **217** [146–151]) and tin (e.g. **218** [152–154]) have also shown excellent selectivity in the enantioselective synthesis of *syn* α-methylhomoallylic, and, in the case of **218**, homopropargylic alcohols (Fig. 11-18). Use of the Hoffmann reagent **211** (Section 11.8), the Brown reagent **212** (Section 11.7), the Roush reagent **213** (Section 11.6), the Marshall reagent **218** (Section 11.11) and the Panek reagents **217** (Section 11.10) in natural product synthesis is reviewed in this Chapter.

The chiral (*E*)-crotylmetal reagents **217-227** exhibit high selectivity for the formation of *anti* α-methyl homoallylic and, in the case of reagents **218** and **226**, homopropargylic alcohols (Fig. 11-19). Several of these reagents are based on boron, e.g. the (*E*)-crotylboronates **219** [65, 139], **221** [155, 156] and **223** [120], and the (*E*)-crotylboranes **220** [115, 140–142] and **222** [143]. The tin-based reagents **225** [157] and **226** [158–160] are selective for the formation of *anti* adducts under thermal conditions, and, in certain cases, the allenylstannane **218** [152, 158, 161] and the (*E*)-crotylsilanes **217** [58, 150, 151] have also been shown to give the *anti* adducts stereoselectively. The molybdenum and lithium-based reagents **224** [162] and **227** [163] have also exhibited selective formation of *anti* α-methyl homoallylic alcohols. Use of the Hoffmann reagents **221** (Section 11.8), the Brown reagent **220** (Section 11.7), the Roush reagent **219** (Section 11.6), the Marshall reagents **226** and **218** (Section 11.11) and the Panek reagents **217** (Section 11.10) in natural product synthesis is reviewed in this Chapter.

Methods for the preparation of 1,2-*syn* diol adducts using chiral γ-(alkoxyallyl)boron reagents **228** [164] and **229** [165], and (γ-alkoxyallyl)stannane reagents **230** [166–168], **231** [169] and **232** [170], have been reported (Fig. 11-20).

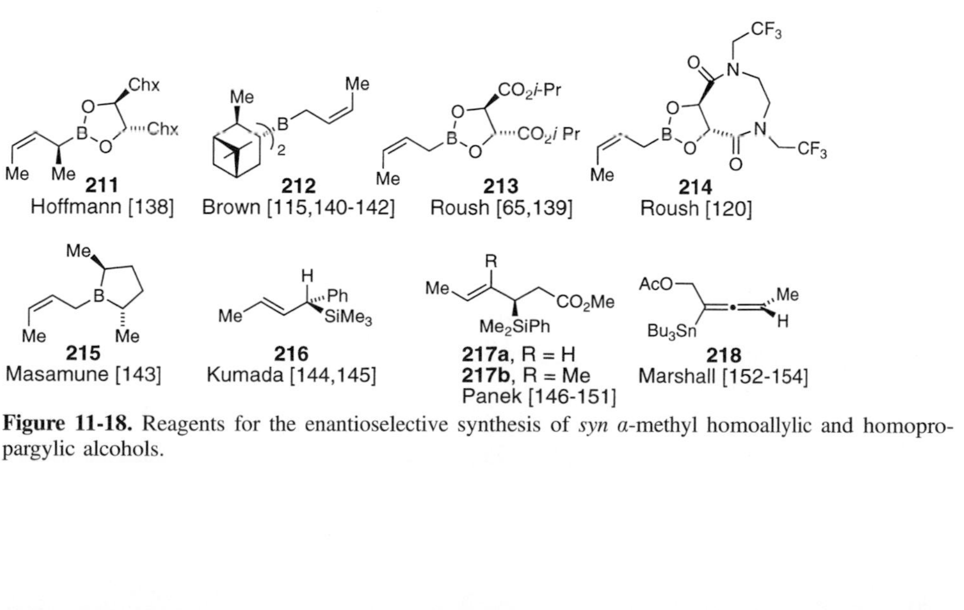

211 Hoffmann [138] **212** Brown [115,140-142] **213** Roush [65,139] **214** Roush [120]

215 Masamune [143] **216** Kumada [144,145] **217a**, R = H **217b**, R = Me Panek [146-151] **218** Marshall [152-154]

Figure 11-18. Reagents for the enantioselective synthesis of *syn* α-methyl homoallylic and homopropargylic alcohols.

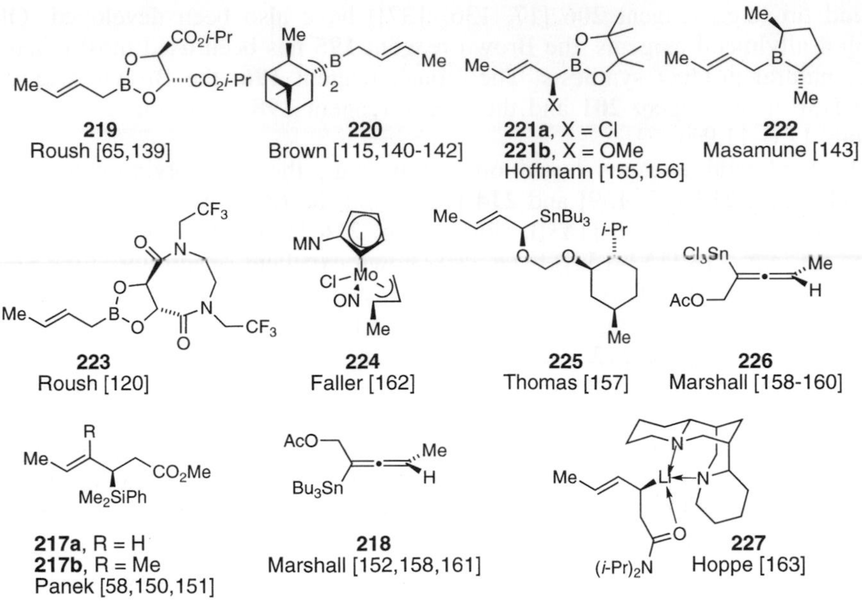

219 Roush [65,139]; **220** Brown [115,140-142]; **221a**, X = Cl; **221b**, X = OMe Hoffmann [155,156]; **222** Masamune [143]

223 Roush [120]; **224** Faller [162]; **225** Thomas [157]; **226** Marshall [158-160]

217a, R = H; **217b**, R = Me Panek [58,150,151]; **218** Marshall [152,158,161]; **227** Hoppe [163]

Figure 11-19. Reagents for the enantioselective synthesis of *anti* α-methyl homoallylic and homopropargylic alcohols.

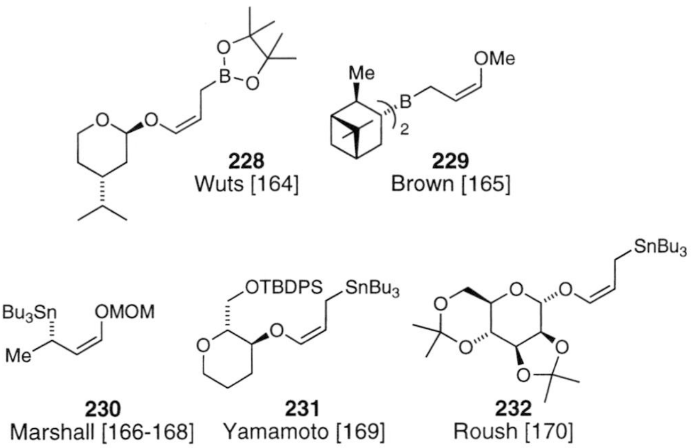

228 Wuts [164]; **229** Brown [165]

230 Marshall [166-168]; **231** Yamamoto [169]; **232** Roush [170]

Figure 11-20. Reagents for the enantioselective synthesis of *syn* 1,2-diols.

Reagents developed for the synthesis of 1,2-*anti* diol adducts include the chiral [(*E*)-γ-alkoxyallyl]indium and [(*E*)-γ-alkoxyallyl]boronate reagents **233** [171] and **234** (Fig. 11-21) [172]. Alternatively, the (*E*)-allylboron reagents **235–237**, which included silicon and boron substituents as hydroxy surrogates, have been independently developed [173–177].

Of these reagents, Brown's reagent **229** and Marshall's reagents **230** and **233** have been used most extensively for appending diol units in natural product syn-

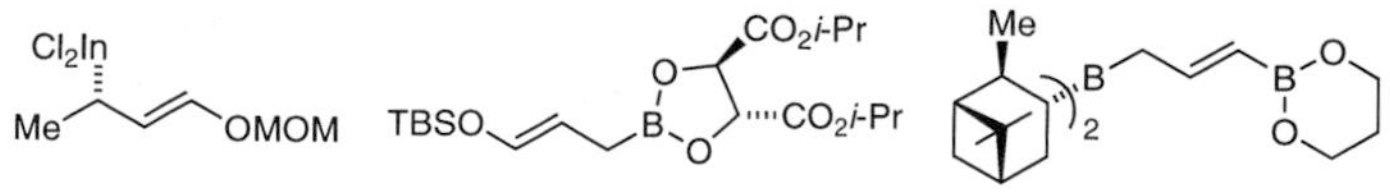

233 Marshall [171] **234** Miyaura [172] **235** Brown [174]

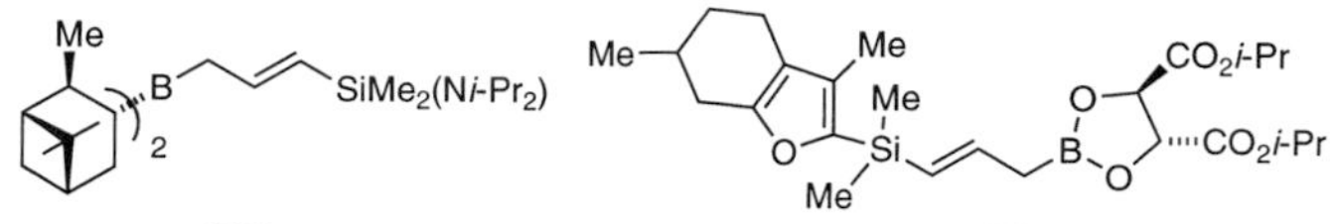

236 Barrett [173] **237** Roush [175-177]

Figure 11-21. Reagents for the enantioselective synthesis of *anti* 1,2-diols.

thesis. The use of Brown's chiral [(*Z*)-γ-alkoxyallyl]borane reagent **229** in natural product synthesis is illustrated in Section 11.7, and the use of Marshall's [(*Z*)-γ-alkoxyallyl]stannane and [(*E*)-γ-alkoxyallyl]indium reagents **230** and **233** is illustrated in Section 11.12.

11.6 Tartrate-derived Chiral Allyl- and Crotylboronate Reagents

Roush and co-workers have developed a family of chiral allyl- and crotylboronate reagents (Fig. 11-22) using tartaric acid as the source of asymmetry. These reagents have been used in the synthesis of complex natural products by the Roush group [175, 178–187] as well as by other researchers [84, 188–197].

Allylboronate **196** is prepared from the reaction of allyl magnesium bromide with trimethylborate followed by esterification with diisopropyl tartrate (DIPT) in the presence of $MgSO_4$ (Eq. (11.13)) [116]. In analogous fashion, the (*E*)- and (*Z*)-crotylboronates **219** and **213** are prepared [139] in high isomeric purity (>98%) from (*E*)- and (*Z*)-2-butene by way of the (*E*)- and (*Z*)-crotylpotassium

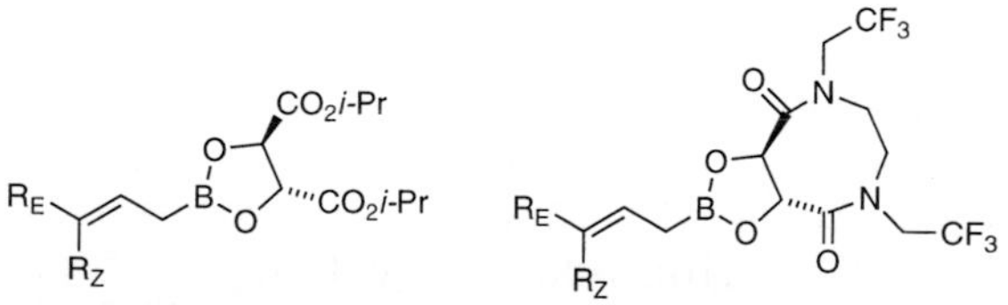

(*R,R*)-**196**, $R_E = R_Z = H$
(*R,R*)-**219**, $R_E = Me$, $R_Z = H$
(*R,R*)-**213**, $R_E = H$, $R_Z = Me$
(*R,R*)-**238**, R_E, $R_Z = Me$
(*R,R*)-**239**, $R_E = SiMe_3$, $R_Z = H$

(*R,R*)-**197**, $R_E = R_Z = H$
(*R,R*)-**223**, $R_E = Me$, $R_Z = H$
(*R,R*)-**214**, $R_E = H$, $R_Z = Me$

Figure 11-22.

anions [198], which are known to be configurationally stable at low temperature (Eqs. (11.14) and (11.15)). Reagents **196**, **219** and **213** are often used without purification and can be stored without decomposition at low temperature for several months. Stock solutions of these reagents in toluene are often prepared, and a titration protocol is used to determine reagent concentration [139].

MgBr — 1) $(MeO)_3B$, -78 °C; 2) H_3O^+, Et_2O; 3) (*R*,*R*)-DIPT, $MgSO_4$; 78% → (*R*,*R*)-**196** (CO_2i-Pr, CO_2i-Pr) (11.13)

1) *n*-BuLi, KO*t*-Bu, THF, -78 → -50 °C, 15 min; 2) $(i\text{-PrO})_3B$, -78 °C; 3) H_3O^+, Et_2O; 4) (*R*,*R*)-DIPT, $MgSO_4$; 70-75% → Me … (*R*,*R*)-**219** (CO_2i-Pr, CO_2i-Pr) (11.14)

1) *n*-BuLi, KO*t*-Bu, THF, -78 → -25 °C, 45 min; 2) $(i\text{-PrO})_3B$, -78 °C; 3) H_3O^+, Et_2O; 4) (*R*,*R*)-DIPT, $MgSO_4$; 70-75% → Me … (*R*,*R*)-**213** (CO_2i-Pr, CO_2i-Pr) (11.15)

Reagents **196**, **219** and **213** give excellent levels of diastereoselectivity and moderate to good levels of enantioselectivity in reactions with achiral aldehydes (Table 11-12) [116, 118, 139]. The (*E*)-crotylboronate **219** is generally the most enantioselective while the (*Z*)-crotylboronate **213** is generally the least. Aliphatic aldehydes undergo allylboration with higher enantioselectivity than unsaturated or α- or β-alkoxy-substituted aldehydes. However, these reagents give very high levels of asymmetric induction (83–98% ee) with metal carbonyl-complexed unsaturated aldehydes [199–201]. The *N,N′*-bis-trifluoroethyl-*N,N′*-ethylene tartramides **197**, **223** and **214** (Fig. 11-22) are much more highly enantioselective reagents: they provide homoallylic alcohols of 94–97% ee in reactions with aliphatic aldehydes [120]. However, the current synthetic route to these reagents is too cumbersome to make them more attractive than other highly enantioselective allylboron reagents (e.g. Brown's reagents).

The asymmetric induction in the reaction of achiral aldehydes with the tartrate-derived crotylboronate reagents is thought to originate from the energy differences between cyclic transition states **243** and **244**, where **244** is disfavored over **243** because of unfavorable Coulombic interactions between the aldehyde oxygen and the tartrate carbonyl, as shown in Fig. 11-23 [116]. Additionally, Corey [202] has proposed that transition state **243** may be favored over **244** due to the possibility in **243** for the formyl hydrogen to participate in a bifurcated hydrogen bond with the axial oxygen of the dioxaborolane ring and the carbonyl oxygen of the -CO_2i-Pr substituent. In transition state **244**, only one such hydrogen bond, between the formyl hydrogen and the axial oxygen of the dioxaborolane ring, is possible.

Table 11-12.

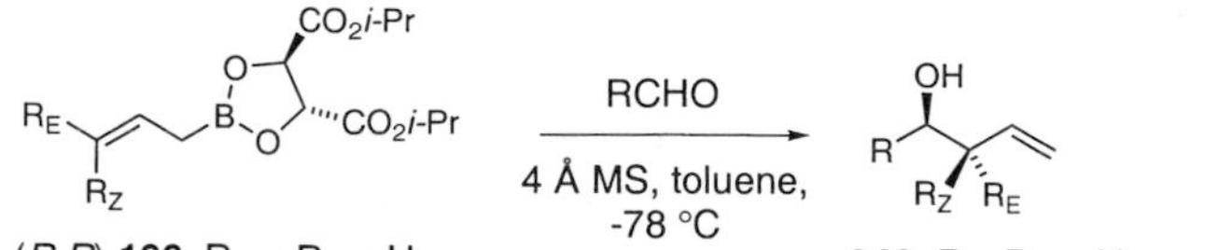

RCHO, 4 Å MS, toluene, -78 °C

(*R,R*)-**196**, $R_E = R_Z = H$
(*R,R*)-**219**, $R_E = Me$, $R_Z = H$
(*R,R*)-**213**, $R_E = H$, $R_Z = Me$

240, R_Z, $R_E = H$
241, $R_E = Me$, $R_Z = H$
242, $R_E = H$, $R_Z = Me$

RCHO	reagent	major product[b]	% yield	% ee
n-$C_9H_{19}CHO$	**196**	**240a**, R = n-C_9H_{19}	86	79
n-$C_9H_{19}CHO$	**219**	**241a**, R = n-C_9H_{19}	87	88
n-$C_9H_{19}CHO$	**213**	**242a**, R = n-C_9H_{19}	80	82
$TBSO(CH_2)_2CHO$	**196**	**240b**, R = $(CH_2)_2OTBS$	ND[c]	66
$TBSO(CH_2)_2CHO$	**219**	**241b**, R = $(CH_2)_2OTBS$	71	85
$TBSO(CH_2)_2CHO$	**213**	**242b**, R = $(CH_2)_2OTBS$	68	72
$C_6H_{11}CHO$	**196**	**240c**, R = C_6H_{11}	72	87
$C_6H_{11}CHO$	**219**	**241c**, R = C_6H_{11}	85	87
$C_6H_{11}CHO$	**213**	**242c**, R = C_6H_{11}	90	83
C_6H_5CHO	**196**	**240d**, R = C_6H_5	78	71
C_6H_5CHO[a]	**219**	**241d**, R = C_6H_5	91	66
C_6H_5CHO[a]	**213**	**242d**, R = C_6H_5	94	55

[a]) Reaction run in THF.
[b]) Diastereoselectivity >95% : 5.
[c]) Yield not determined.

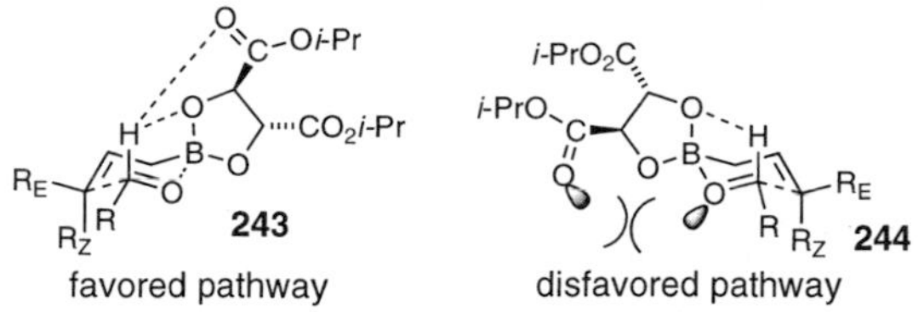

243 favored pathway

244 disfavored pathway

Figure 11-23.

The tartrate-derived crotylboronate reagents are most useful in the context of double asymmetric reactions with chiral aldehydes [118, 203]. Equations (11.16)–(11.19) demonstrate the utility of (*E*)-**219** and (*Z*)-**213** in the synthesis of dipropionate adducts **105–108**.

The TBS-protected (*S*)-*α*-methyl-*β*-alkoxy aldehyde **97b** reacts with the (*R,R*)-(*E*)-crotylboronate **219** to give the *syn,anti*-dipropionate **108b** as the major adduct

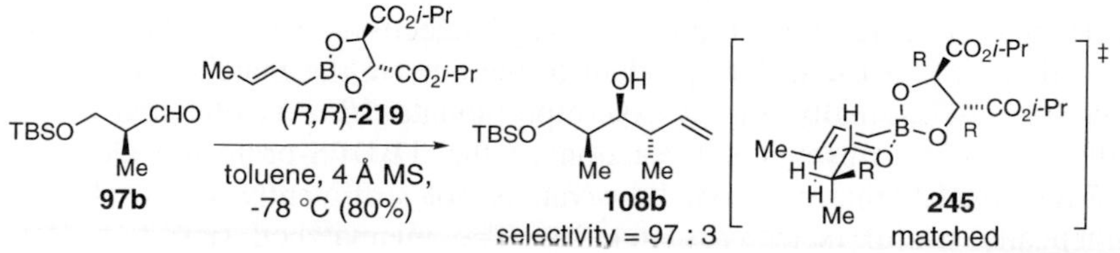

97b + (*R,R*)-**219** → toluene, 4 Å MS, -78 °C (80%) → **108b**, selectivity = 97 : 3 via [**245** matched]‡ (11.16)

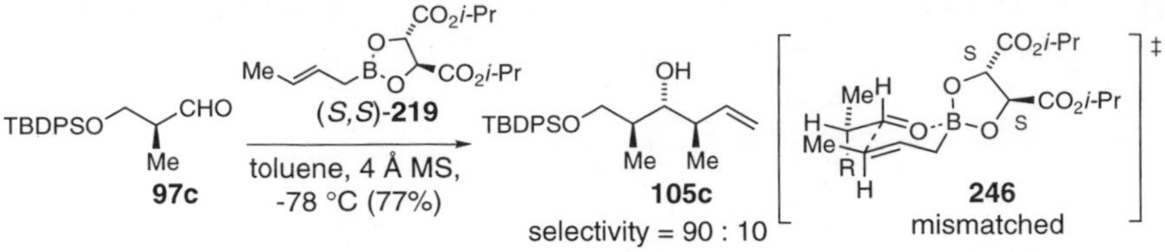

(11.17)

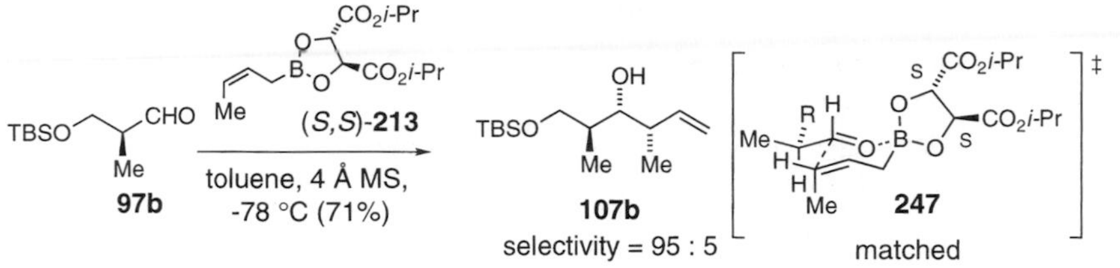

(11.18)

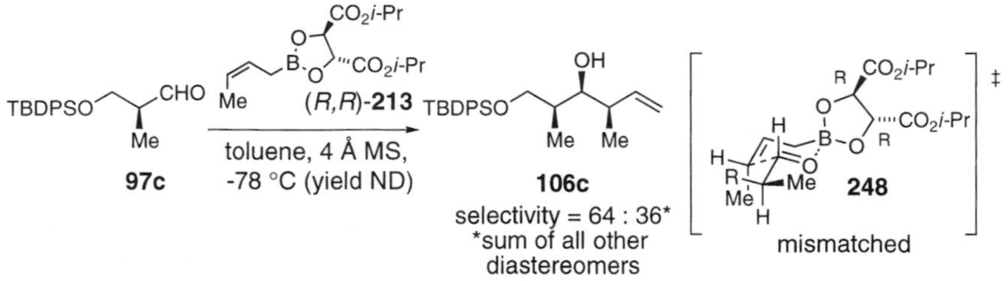

(11.19)

with high diastereoselectivity (selectivity = 97 : 3). The stereochemical outcome of this reaction is rationalized by the matched transition state **245**, where C–C bond formation occurs by addition of the crotylboronate *anti* to the $TBSOCH_2$-substituent of the Felkin rotamer of the aldehyde. This pathway is intrinsically favored by both the crotylboronate reagent and the *α*-chiral aldehyde, which has been shown to favor Felkin addition in reactions with achiral (*E*)-crotylmetal reagents (see Section 11.2). The *anti,anti*-dipropionate **105c** is obtained with useful selectivity (selectivity = 90 : 10) from the reaction of the TBDPS-protected *α*-methyl-*β*-alkoxy aldehyde **97c** with (*S,S*)-(*E*)-**219**. Transition state **246** best rationalizes the stereochemical outcome of this mismatched double asymmetric reaction; C–C bond formation occurs with the crotylboronate adding to the anti-Felkin rotamer of aldehyde **97c**, where the principle destabilizing interaction between the R substituent of the aldehyde and the methyl of the crotyl unit is overcome by the enantioselectivity of the tartrate auxiliary.

In reactions of *α*-methyl chiral aldehydes with achiral (*Z*)-crotylboronates, the anti-Felkin adduct (cf. **107b**) is favored (for further discussion see Section 11.2) [3, 65]. In the double asymmetric reaction of **97b** and (*S,S*)-**213**, the *anti,syn*-dipropionate **107b** is obtained with high selectivity (selectivity = 95 : 5). The stereochemistry of **107b** is consistent with product formation via the matched anti-Felkin transition state **247**. Finally, the *syn,syn*-dipropionate **106c** is obtained as the major product from the mismatched reaction of the TBDPS-protected aldehyde **97c** with (*R,R*)-(*Z*)-**213**; this reaction, however, is not sufficiently stereoselective to be synthetically useful (selectivity = 64 : 36). The mismatched transition state from which adduct **106c** likely arises is destabilized by the gauche interaction between the aldehyde R group and the crotylboronate Me group.

In general, as the aldehyde α-substituents become more sterically demanding, it becomes more difficult to obtain useful levels of diastereoselection for the product expected from reagent control in mismatched double asymmetric reactions between chiral aldehydes and chiral allyl- and crotylboronates [203]. For this reason, in natural product synthesis, mismatched double asymmetric reactions should be designed to occur early rather than late in a synthetic sequence.

In their synthesis of (+)-damavaricin D (Fig. 11-24), Roush and co-workers used crotylboronate methodology three times in the assembly of the C(1)–C(13) polypropionate segment **250** [178, 204, 205]. The synthesis of **250** was designed so that chain growth occurs from C(13) to C(1) and such that the mismatched reaction necessary to install the C(10)–C(12) *anti,anti*-dipropionate stereotriad could be dealt with early in the synthetic sequence, when the aldehyde substrate had a relatively modest diastereofacial bias (Scheme 11-9).

Treatment of aldehyde *ent*-**97c** with (*R,R*)-**219** resulted in the stereoselective (selectivity = 9 : 1) formation of adduct *ent*-**105c**, via the mismatched double asymmetric reaction discussed previously (Eq. (11.17)). Aldehyde **251**, derived in two steps from olefin *ent*-**105c**, underwent an asymmetric aldol reaction [206] with the boron enolate of **252**, generating adduct **253** stereoselectively. Adduct **253** was converted in five steps to aldehyde **254**, which underwent a matched double asymmetric reaction with (*S,S*)-**219**, affording stereoheptad **255** in 90% yield (selectivity = >98 : 2). Adduct **255** was then elaborated to aldehyde **256**, which was directly submitted to the matched double asymmetric reaction with (*R,R*)-**219**, affording the advanced adduct **257** (selectivity = >98 : 2), which was converted in seven steps to the C(1)–C(13) fragment **250** of (+)-damavaracin D.

Chiral crotylboronate technology was used three times in the synthesis of the fully functionalized trioxadecalin portion **258** of mycalamide A by Roush and Marron (Fig. 11-25) [207, 208].

The synthesis commenced with the matched double asymmetric reaction of the (*R,R*)-prenylboronate **238** (synthesized from 3-methyl-1-butene in a manner analogous to the preparation of crotylboronates **219** and **213**) with D-glyceraldehyde pentylidene ketal **259**, forming adduct **260** with >99 : 1 diastereoselectivity

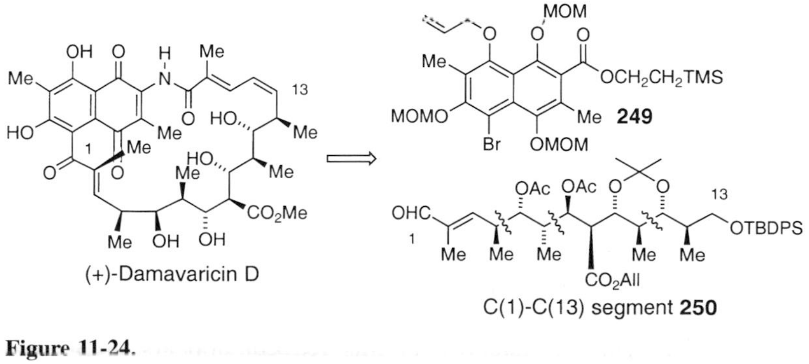

(+)-Damavaricin D

C(1)-C(13) segment **250**

Figure 11-24.

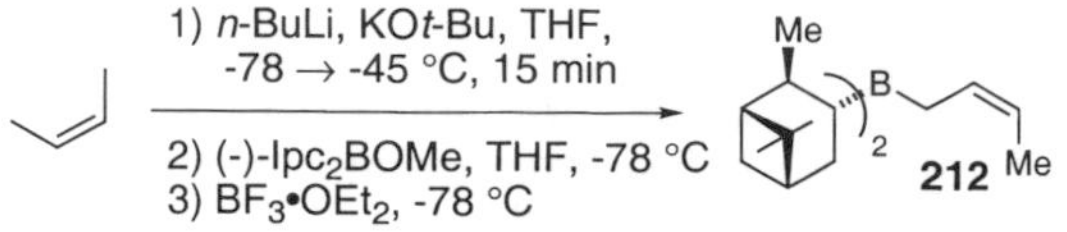

1) n-BuLi, KOt-Bu, THF, $-78 \rightarrow -45$ °C, 15 min
2) (-)-Ipc_2BOMe, THF, -78 °C
3) $BF_3 \cdot OEt_2$, -78 °C
212 (11.22)

The (^{d}Ipc)$_2$BAllOMe reagent **229** is prepared from allyl methyl ether via deprotonation with *sec*-butyllithium and subsequent treatment with (–)-Ipc_2BOMe and then $BF_3 \cdot OEt_2$ (Eq. (11.23)) [165]. For best results, these reagents should be prepared just prior to use because **220**, **212** and **229** are configurationally unstable at temperatures above –78 °C (the olefin isomers interconvert easily through a facile, reversible 1,3-boratropic rearrangement).

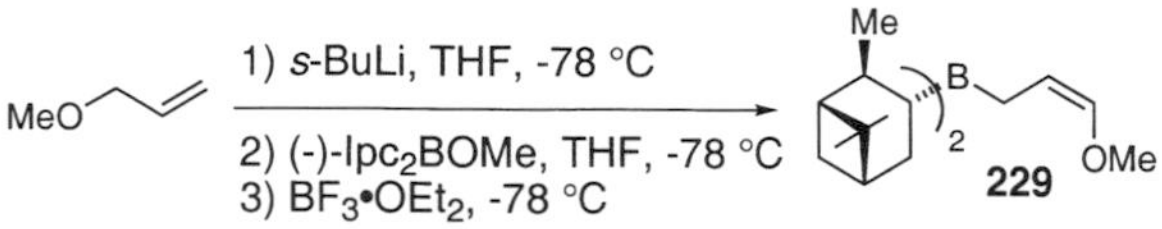

1) s-BuLi, THF, -78 °C
2) (-)-Ipc_2BOMe, THF, -78 °C
3) $BF_3 \cdot OEt_2$, -78 °C
229 (11.23)

Brown and co-workers have demonstrated that these reagents react in a highly diastereo- and enantioselective manner with achiral aldehydes (Table 11-13) [112, 113, 140, 142, 165].

The transfer of chirality to homoallylic alcohols **240–242** and **272** in the reactions of reagents **195**, **220**, **212** and **229** with aldehyde substrates is rationalized by C–C bond formation occurring preferentially through transition state **273** (Fig. 11-27). Transition state **274**, which leads to the enantiomeric products, is disfavored due to destabilizing steric interactions between the α-methylene and the chiral ligands of the allylborane reagent.

The double asymmetric reactions of reagents **195**, **220** and **212** with chiral aldehydes generally result in selective formation of the product predicted from reagent control of asymmetric induction. Results of the reactions of α-methyl chiral aldehyde **97a** and reagents **195**, **220** and **212** are summarized in Table 11-14 [114, 115, 141]. However, as with all chiral allylmetal reagents, as the diastereofacial bias of the aldehyde increases, the degree of reagent-controlled asymmetric induction diminishes in mismatched double asymmetric reactions with reagents **195**, **220** and **212**. An example of a very challenging mismatched double asymmetric reaction is found in the synthesis of calyculin C by Armstrong and co-workers (Scheme 11-13, see below) [232].

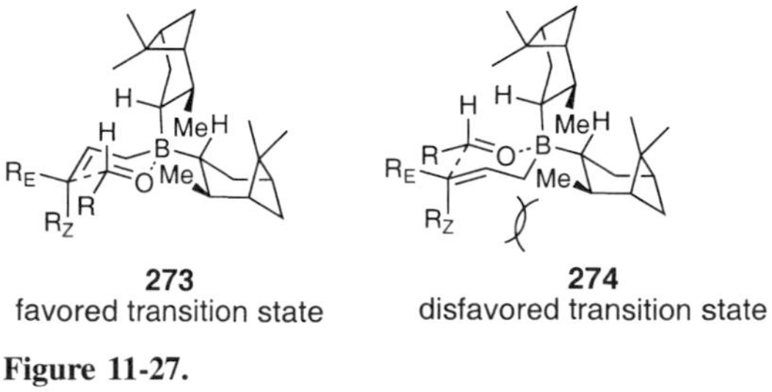

273
favored transition state

274
disfavored transition state

Figure 11-27.

Table 11-14.

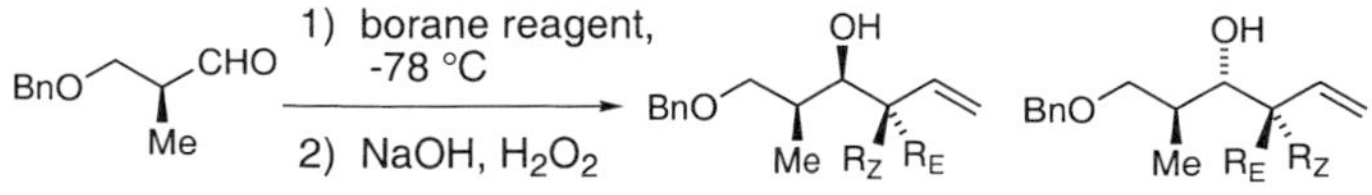

reagent	solvent	% yield	products	selectivity
${}^{d}Ipc_2BAll$ (**195**)	Et_2O	80	**99** : **100**	96 : 4
${}^{l}Ipc_2BAll$ (*ent*-**195**)	Et_2O	NH[b]	**99** : **100**	2 : 98
${}^{d}Ipc_2BCrt^E$ (**220**)	THF, Et_2O[a]	87	**108a** : **105a**	98 : 2
${}^{l}Ipc_2BCrt^E$ (*ent*-**220**)	THF, Et_2O[a]	84	**108a** : **105a**	5 : 95
${}^{d}Ipc_2BCrt^Z$ (**212**)	THF, Et_2O[a]	83	**106a** : **107a**	92 : 8
${}^{l}Ipc_2BCrt^Z$ (*ent*-**212**)	THF, Et_2O[a]	78	**106a** : **107a**	5 : 95

[a]) $BF_3{\cdot}OEt_2$ was used as an additive in this reaction.
[b]) Yield not reported.

Armstrong et al. used pinene-derived allylborane reagents four times in their synthesis of the C(1)–C(25) segment **275** of calyculin C (Fig. 11-28) [231, 232].

The C(22) and C(23) stereocenters of calyculin C were set by the reaction of aldehyde **276** with the $({}^{d}Ipc)_2BCrt^Z$ reagent **212** (Scheme 11-12) [231, 235]. Boeckman and co-workers had previously determined that the diastereoselectivity and enantioselectivity of this conversion are extremely high [235]. The resulting *syn* homoallylic alcohol adduct **277** was converted in two steps to aldehyde **278** in preparation for its upcoming condensation with methyl ketone **281** (Scheme 11-13, see below).

The synthesis of methyl ketone **281** began with the reaction between the tetra-substituted allylborane **279** and 2,3-*O*-isopropylidene-D-glyceraldehyde **48**. The resulting homoallylic alcohol **280**, obtained in 73% yield and excellent selectivity (exact ratio not defined) [231], was converted in two steps to the methyl ketone **281**. Aldol condensation between the lithium enolate of **281** and aldehyde **278** (structure shown in Scheme 11-12) gave, after protection of the initial adduct, the Felkin diastereomer **282** as the only reported product in 54% yield. This adduct

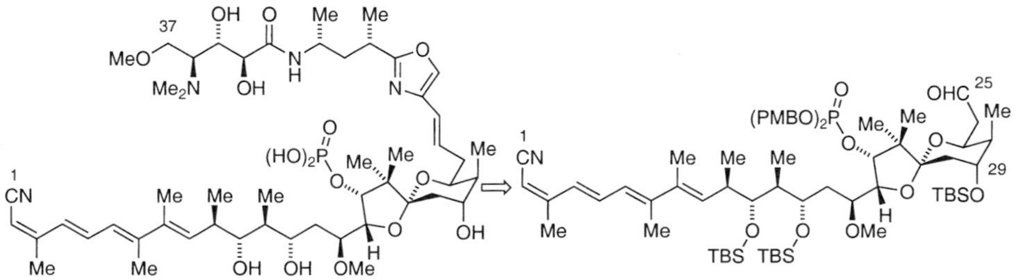

Figure 11-28.

11.8 α-Chiral Allylboronate Reagents

Hoffmann and co-workers have developed a family of chiral α-substituted allylboronate reagents represented by **201**, **211**, **221a** and **221b** (Fig. 11-29) [125, 138, 155, 156], and have demonstrated that these reagents react in a highly stereoselective manner with both achiral and chiral aldehydes to generate homoallylic alcohols with high enantioselectivity. Several applications of these reagents in the synthesis of complex polypropionate-derived natural products have appeared [237, 245].

The synthesis of the α-substituted chiral allyl- and (*Z*)-crotylboronate reagents, **201** [125] and **211** [138], begins with commercially available dihydrobenzoin **293** (Scheme 11-15). Thus, hydrogenation of **293** followed by transesterification with diisopropyl (dichloromethyl)boronate provides the chiral dichloromethylboronate **294**, a common intermediate in the synthesis of **201** and **211**. Treatment of (*R,R*)-**294** with vinylmagnesium chloride followed by $ZnCl_2$, using a modification of a procedure developed by Mattesson [246–249], leads to the (*S,R,R*)-α-chloroallylboronate **201** (Scheme 11-15), which is used directly in the allylation reaction with aldehydes. (This reagent is used immediately after preparation since it racemizes slowly upon storage [125]). Treatment of (*S,S*)-**294** with MeLi followed by $ZnCl_2$ leads to the corresponding chiral α-chloroethylboronate, which, without isolation, is converted to the (*S,S,S*)-(*Z*)-α-methylcrotylboronate **211** (99% ee) via treatment with (*Z*)-propenyllithium.

The (*E*)-α-chloro- and (*E*)-α-methoxycrotylboronates **221a** and **221b** can be synthesized from either antipode of commercially available 3-butyn-2-ol **295** (Scheme 11-16) [155, 156]. Thus, protection of **295** as the TMS ether, hydroboration of the alkyne with dicyclohexylborane and subsequent oxidation of the borane to a vinylboronate followed by transesterification with pinacol affords the intermediate **296**. Allylic rearrangement of **296** with thionyl chloride ($SOCl_2$) generates the chiral (*E*)-α-chlorocrotylboronate **221a** with >98% ee. The (*E*)-α-methoxycrotylboronate **221b** is synthesized from **221a** via S_N^2 displacement with LiOMe. This reaction, however, occurs with some loss of enantiopurity as reagent **221b** is obtained with an estimated 90% ee [156].

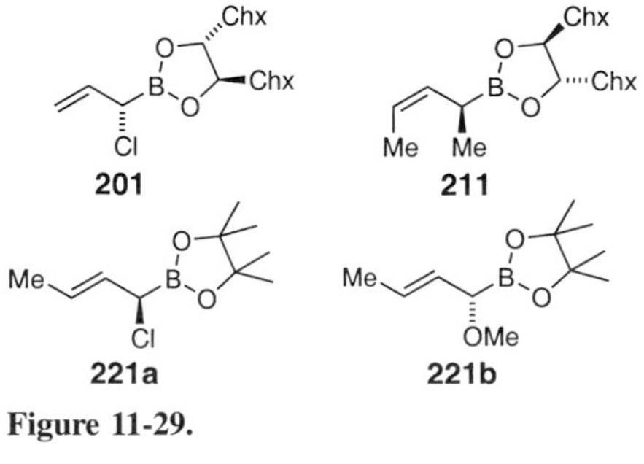

Figure 11-29.

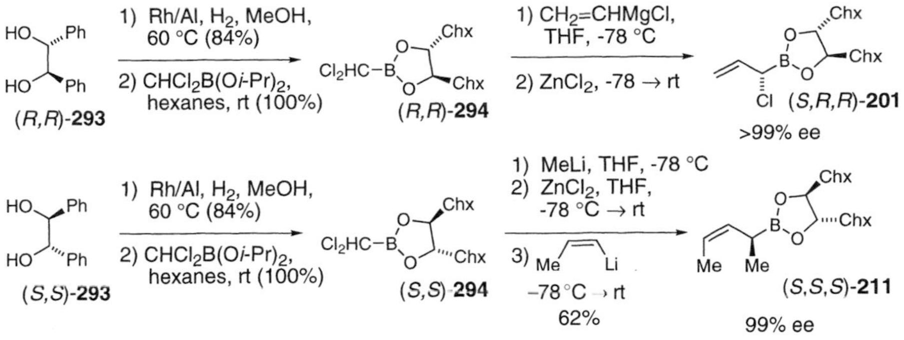

Scheme 11-15.

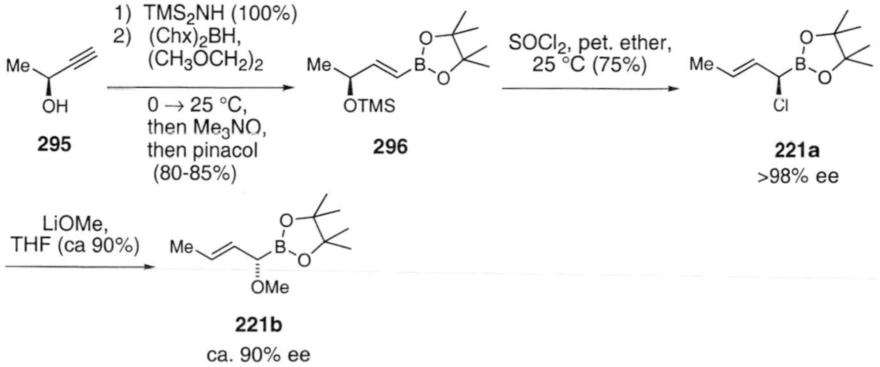

Scheme 11-16.

Freshly prepared (*S,R,R*)-α-chloroallylboronate **201** reacts with achiral aldehydes to form preferentially the (*Z*)-chlorohomoallylic alcohols **297** (Table 11-15) [125]. The preponderance of alcohols **297** indicates that transition state **299** is favored over **300** which leads to the minor diastereomeric alcohol **298**, due to the unfavorable steric interactions between the equatorially-placed α-chlorine substituent and the cyclohexyl substituent of the dioxaborolane unit in **300**. Polar α-substituents, e.g. –Cl, also increase the preference for reaction through transition state **299**, presumably due to minimization of dipole and Coulombic repulsion [24].

The (*S,S,S*)-α-methylcrotylboronate **211** reacts in a highly selective manner with a variety of aldehydes to form preferentially the (*E*)-*syn* homoallylic alcohols **301** (Table 11-16) [250]. In this reaction, homoallylic alcohols **301** are produced preferentially via transition state **303** since the competing transition state **304** is destabilized by unfavorable 1,3-interactions between the α-methyl and the γ-methyl groups of the crotylboronate **211**.

The (*E*)-α-chloro- and α-methoxycrotylboronates **221a** and **221b** react with a variety of aldehydes to give the *anti* (*Z*)-chloro- and (*Z*)-methoxyhomoallylic alcohols **305** as major products (Table 11-17) [155, 156]. The α-chlorocrotylboronate

Table 11-15.

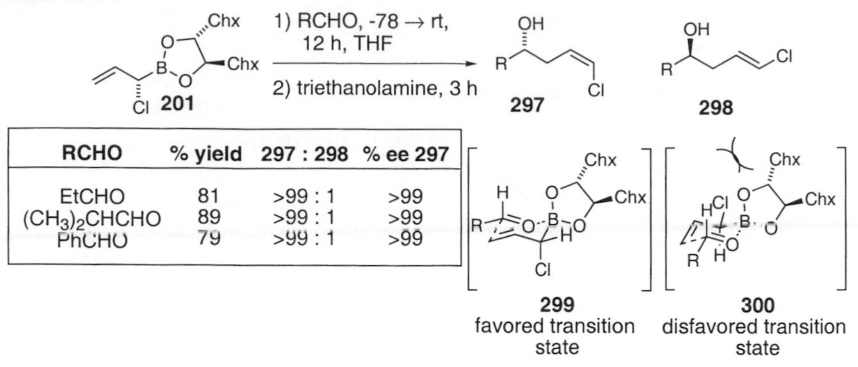

RCHO	% yield	297 : 298	% ee 297
EtCHO	81	>99 : 1	>99
$(CH_3)_2CHCHO$	89	>99 : 1	>99
PhCHO	79	>99 : 1	>99

Table 11-16.

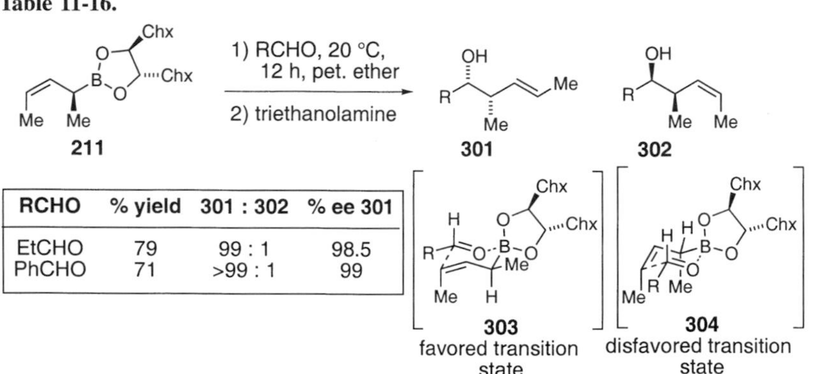

RCHO	% yield	301 : 302	% ee 301
EtCHO	79	99 : 1	98.5
PhCHO	71	>99 : 1	99

221a is more enantioselective, but less diastereoselective than the α-methoxycrotylboronate **221b**. The decrease in enantioselectivity of reagent **221b** is attributed to its lower enantiomeric purity (ca. 90% ee). The formation of **305** as the major adduct of these reactions reflects the preference for reagents **221** to react with aldehydes through transition state **307** rather than transition state **308**, which leads to adduct **306**.

The matched double asymmetric reactions of the α-methyl chiral aldehyde **309** with the (*E*)-α-chloro- and (*E*)-α-methoxycrotylboronates **221a** and **221b** occur with equally high selectivity (ratio = 98 : 2) to generate the *syn,anti*-homoallylic alcohol **310** as the major adduct (Table 11-18). In contrast, the α-methoxycrotylboronate (*S*)-**221b** exerts a higher degree of reagent control (no other diastereomers were observed) than does the α-chlorocrotylboronate (*S*)-**221a** (selectivity = 90 : 10, 50–60% yield) in the mismatched double asymmetric reactions with aldehyde **309** which generate the *anti,anti*-adducts **311** as the major reaction products [155, 156].

Double asymmetric reactions with the α-chloroallylboronate **221a** and the (*Z*)-α-methylcrotylboronate **221b** have also been performed with chiral aldehydes, and, in general, a high degree of reagent control is observed [125, 138].

Table 11-17.

1) RCHO, 20 °C, 12 h, pet. ether
2) triethanolamine

221a, X = Cl
221b, X = OMe

305a, X = Cl
305b, X = OMe

306a, X = Cl
306b, X = OMe

RCHO	reagent	% yield	305 : 306	% ee 305
EtCHO	**221a**	47-65	95 : 5	96
EtCHO	**221b**	81	>99 : 1	90
$(CH_3)_2CHCHO$	**221a**	55-84	95 : 5	95
$(CH_3)_2CHCHO$	**221b**	82	>99 : 1	90
PhCHO	**221a**	53-68	95 : 5	96
PhCHO	**221b**	79	>99 : 1	90

307
favored transition state

308
disfavored transition state

Table 11-18.

309

(*R*)-**221a**, X = Cl
(*R*)-**221b**, X = OMe

310a, X = Cl
310b, X = OMe

311a, X = Cl
311b, X = OMe

reagent	major product*	selectivity
(*R*)-**221a**	**310a**	98 : 2
(*R*)-**221b**	**310b**	only **310b** observed
(*S*)-**221a**	**311a**	10 : 90
(*S*)-**221b**	**311b**	only **311b** observed

Conditions: 1) Reagent+**309**, pet. ether, –78 °C → rt. 2) Triethanol amine.

The utility of reagents (*Z*)-**211** and (*E*)-**221b** is illustrated in Hoffmann's synthesis of denticulatins A and B (Scheme 11-17) [239, 240]. The synthesis was initiated by the crotylation of propionaldehyde with (*S,S,S*)-(*Z*)-**211**, which gave the *syn* homoallylic alcohol **301a** with a high level of enantiomeric purity. Aldehyde **312**, derived from adduct **301a** in two steps, underwent a highly diastereoselective (selectivity = 95 : 5) substrate-directed crotylation reaction with the achiral (*E*)-pinacol crotylboronate **1**, generating the Felkin adduct **313** (see Section 11.2 for a discussion of the diastereoselectivity of this reaction). Aldehyde **314**, generated from **313** in four steps, underwent a mismatched double asymmetric reaction with

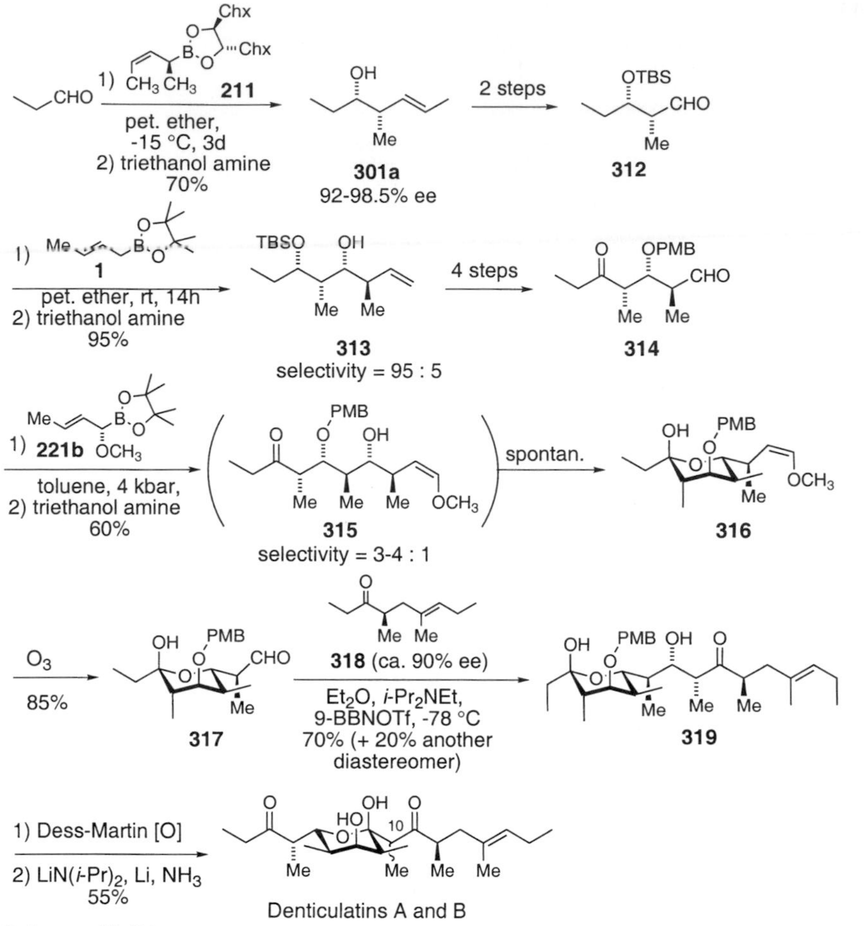

Scheme 11-17.

the (*E*)-*α*-methoxycrotylboronate **221b**, giving the *anti*-Felkin 3,4-*anti*-4,5-*anti* adduct **315** with 75–80% diastereoselectivity. The level of diastereoselection in this reaction is admirable considering that *β*-branched *α*-methyl chiral aldehydes such as **314** are highly disposed to form the Felkin product in reactions with (*E*)-crotylboronate reagents (see Section 11.2 for additional discussion). Adduct **315** spontaneously cyclized to the hemiketal **316**. Ozonolysis of **316** gave aldehyde **317**, which underwent a diastereoselective aldol reaction with the boron enolate of the ethyl ketone **318** [239, 240], generating the anti-Felkin aldol **319** as the major adduct. The anti-Felkin stereochemistry is favored in the reaction of chiral aldehydes with (*Z*)-enolates for the same reasons that the anti-Felkin adduct is favored with (*Z*)-crotylboronates (refer to Section 11.2) [65]. Oxidation of the secondary alcohol of **319** followed by deprotection of the *para*-methoxybenzyl ether led to thermodynamic hemiketal formation with concomitant racemization of C(10), thus completing the total syntheses of denticulatins A and B.

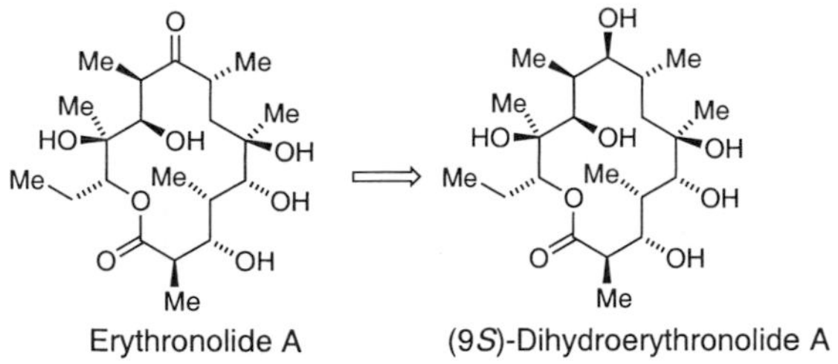

Erythronolide A ⟹ (9*S*)-Dihydroerythronolide A

Figure 11-30.

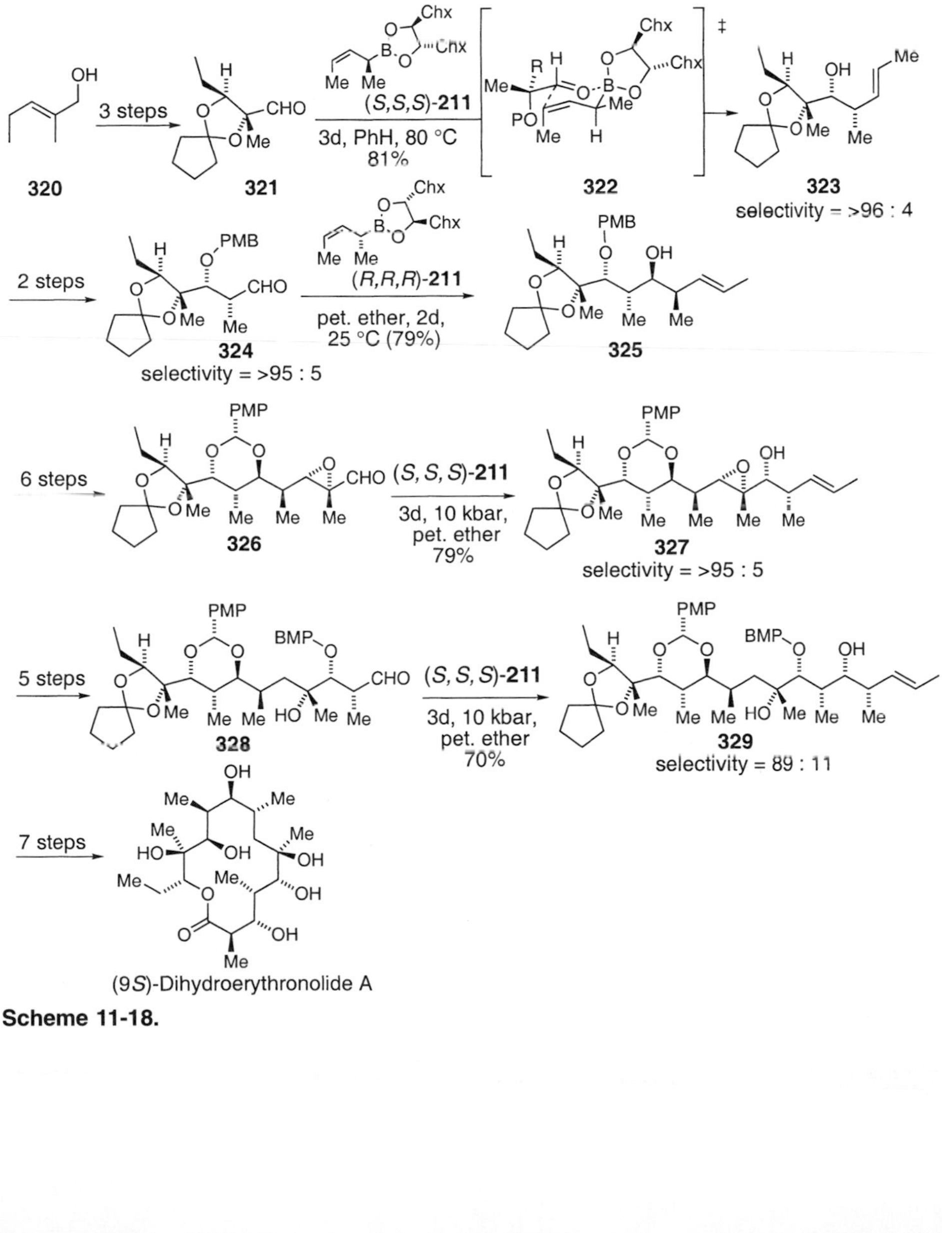

320 —3 steps→ 321 —(*S*,*S*,*S*)-**211**, 3d, PhH, 80 °C, 81%→ [322]‡ → 323 (selectivity = >96 : 4)

—2 steps→ 324 (selectivity = >95 : 5) —(*R*,*R*,*R*)-**211**, pet. ether, 2d, 25 °C (79%)→ 325

—6 steps→ 326 —(*S*, *S*, *S*)-**211**, 3d, 10 kbar, pet. ether, 79%→ 327 (selectivity = >95 : 5)

—5 steps→ 328 —(*S*, *S*, *S*)-**211**, 3d, 10 kbar, pet. ether, 70%→ 329 (selectivity = 89 : 11)

—7 steps→ (9*S*)-Dihydroerythronolide A

Scheme 11-18.

Hoffmann and co-workers demonstrated [242, 243] the utility of (*Z*)-**211** several times in their synthesis of (9*S*)-dihydroerythronolide A, a known precursor of erythronolide A (Fig. 11-30) [251, 252].

The synthesis of (9*S*)-dihydroerythronolide A began with allyl alcohol **320**, which was converted to aldehyde **321** via a three-step sequence which included a Sharpless asymmetric epoxidation reaction (Scheme 11-18) [253]. Aldehyde **321** underwent a highly diastereoselective crotylboration reaction with (*S,S,S*)-(*Z*)-**211**. The resulting homoallylic alcohol **323** is probably the result of a matched double asymmetric pairing of reagent **211** with aldehyde **321**, where the oxygen substituent of the aldehyde occupies the most sterically demanding position, *syn* to the γ-Me of the crotylboronate reagent in transition state **322**. Adduct **323** was converted to aldehyde **324**, which underwent a matched double asymmetric reaction with (*R,R,R*)-(*Z*)-**211**, affording homoallylic alcohol **325** with >95% diastereoselection. Aldehyde **326**, generated from **325**, underwent a highly selective crotylboration reaction with (*S,S,S*)-(*Z*)-**211**, generating the homoallylic alcohol **327**, the stereochemistry of which is consistent with formation through a matched transition state analogous to **322**. After elaboration of **327** to aldehyde **328**, a mismatched double asymmetric reaction with (*S,S,S*)-(*Z*)-**211** provided the Felkin adduct **329**, the product of reagent control, as the major adduct (selectivity = 89 : 11). Adduct **329** was subsequently converted to (9*S*)-dihydroerythronolide A.

11.9 Stein-Based Allylboron Reagents

Corey and co-workers developed the highly enantioselective allylboron reagent **198** [127], whose chiral 1,2-diamino-1,2-diphenylethane (stein) auxiliary [254] serves as the source of asymmetry. In an extension of this methodology, Williams et al. have demonstrated the utility of the bromoborane **332** for the preparation of synthetically complex allylborane reagents [255] and have applied this methodology in two natural product syntheses [256, 257] (see below).

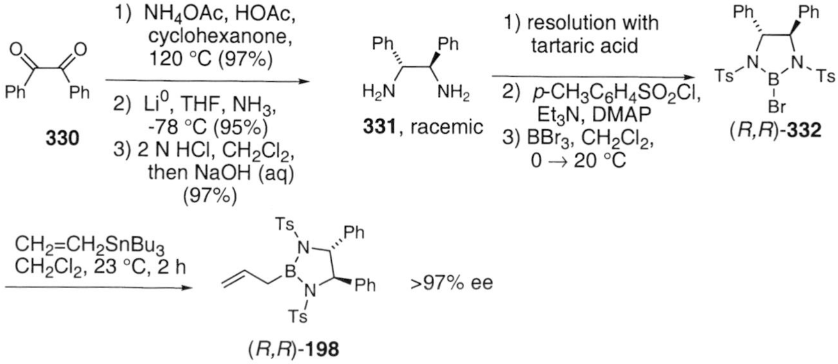

330 → 1) NH_4OAc, HOAc, cyclohexanone, 120 °C (97%); 2) Li^0, THF, NH_3, -78 °C (95%); 3) 2 N HCl, CH_2Cl_2, then NaOH (aq) (97%) → **331**, racemic → 1) resolution with tartaric acid; 2) p-$CH_3C_6H_4SO_2Cl$, Et_3N, DMAP; 3) BBr_3, CH_2Cl_2, 0 → 20 °C → (*R,R*)-**332** → $CH_2{=}CH_2SnBu_3$, CH_2Cl_2, 23 °C, 2 h → (*R,R*)-**198**, >97% ee

Scheme 11-19.

Either antipode of bromoborane **332** can be prepared in a six-step sequence from benzil (**330**, Scheme 11-19) [127, 254, 258]. Reaction of benzil with cyclohexanone in the presence of ammonium acetate and acetic acid generates a cyclic bis-imine which is subsequently reduced with lithium in ammonia. The resulting racemic *trans*-imidazolidine is subsequently hydrolyzed to the diamine **331**. Resolution of **331** is accomplished by crystallization with either antipode of tartaric acid. The enantiomerically enriched stein ligand **331** is then sulfonylated and condensed with boron tribromide, giving the chiral bromoborane **332**. Transmetallation of allyltri-*n*-butylstannane with bromoborane (*R,R*)-**332** then affords the allylboron reagent (*R,R*)-**198**.

Allylboron reagent (*R,R*)-**198** reacts with a variety of achiral aldehydes, affording homoallylic alcohols **240** in excellent yields and enantiomeric excess (Table 11-19). Corey rationalized that adducts **240** should arise through transition state **333**. The alternative transition state, **334**, is disfavored due to unfavorable steric interactions between the *α*-methylene group of the reagent and the adjacent sulfonamide ligand. Note that the toluene substituents of the sulfonamides in transition states **333** and **334** are spatially oriented so as to avoid steric interactions with the phenyl substituents of the chiral auxiliary.

Table 11-19.

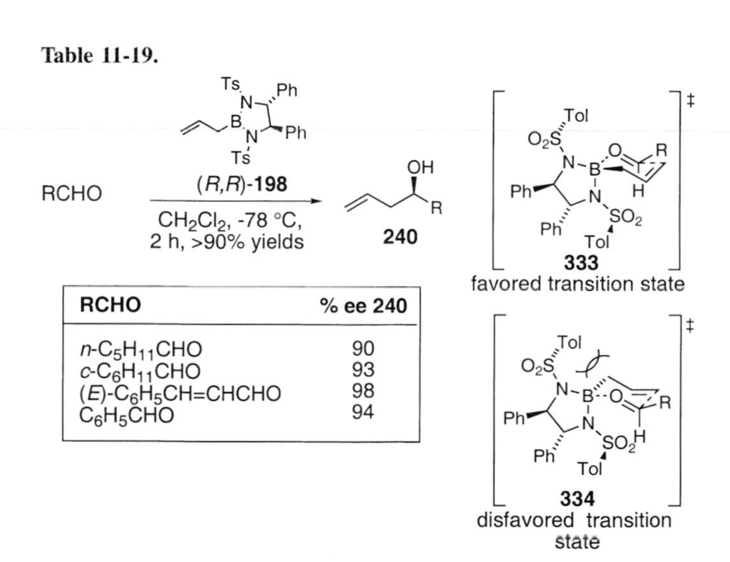

RCHO	% ee 240
n-$C_5H_{11}CHO$	90
c-$C_6H_{11}CHO$	93
(E)-$C_6H_5CH{=}CHCHO$	98
C_6H_5CHO	94

Williams and co-workers have applied this methodology to synthetically advanced allylstannane intermediates [255]. They found that with the allylboron reagents generated in situ from either enantiomer of allylstannane **335**, the chiral auxiliary dominated the asymmetric induction (rather than the C(4)-OTBS stereocenter) in the formation of homoallylic alcohols **337** and **338** from aldehyde **336a** (Eqs. (11.24) and (11.25)). Thus, the allylboron reagent derived from (*S*)-**335** and (*R,R*)-**332** reacts with aldehyde **336a** to generate the homoallylic alcohol **337a** with >20:1 stereoselectivity, where the newly formed alcohol stereocenter can be rationalized by a transition state analogous to **333**. Likewise, the allylboron reagent derived from (*R*)-**335** and (*R*,*R*)-**332** reacts with aldehyde **336a** to generate the homoallylic alcohol **338** (ratio >20:1).

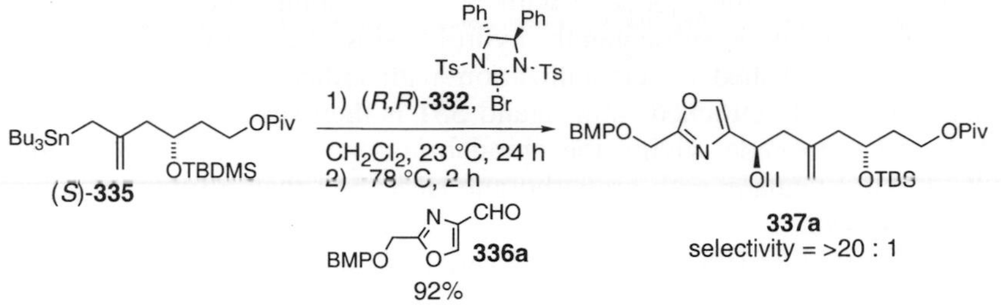

Bu_3Sn … OPiv, OTBDMS (*S*)-**335** → 1) (*R*,*R*)-**332** (Ph, Ph, Ts-N, B, N-Ts, Br), CH_2Cl_2, 23 °C, 24 h; 2) 78 °C, 2 h; BMPO–oxazole–CHO **336a**; 92% → BMPO … OH … OTBS … OPiv **337a** selectivity = >20 : 1 (11.24)

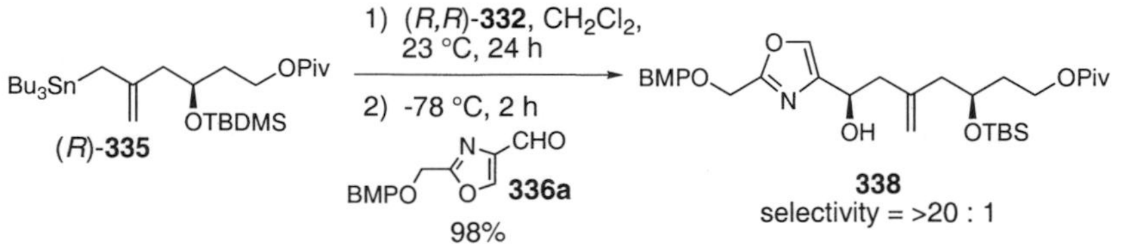

Bu_3Sn … OPiv, OTBDMS (*R*)-**335** → 1) (*R*,*R*)-**332**, CH_2Cl_2, 23 °C, 24 h; 2) -78 °C, 2 h; BMPO–oxazole–CHO **336a**; 98% → BMPO … OH … OTBS … OPiv **338** selectivity = >20 : 1 (11.25)

Williams and co-workers applied this methodology to the synthesis of (–)-hennoxazole A and in the C(3)–C(19) segment **340** of phorboxazole A (Fig. 11-31) [256, 257]. The allylstannane (*S*)-**335** is used in the synthesis of both the *cis* and the *trans* pyran units of **340** (Scheme 11-20).

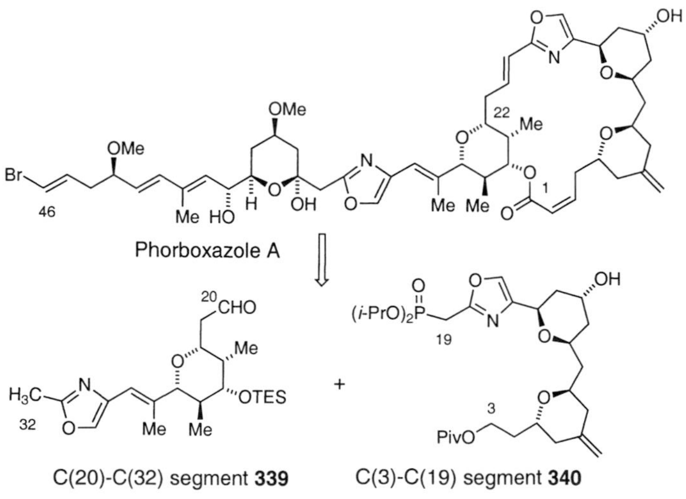

Phorboxazole A (Br, 46, OMe, Me, HO, H, OH, 22, Me, 1, OH)

⇓

H_3C (32) … OTES, Me, Me, ^{20}CHO — C(20)-C(32) segment **339**

\+

$(i\text{-PrO})_2P(O)$ (19) … OH … PivO (3) — C(3)-C(19) segment **340**

Figure 11-31.

The synthesis of bis-pyran **340** was initiated by the allylation reaction of aldehyde **336b** with the allylboron reagent derived from allylstannane (*S*)-**335** and

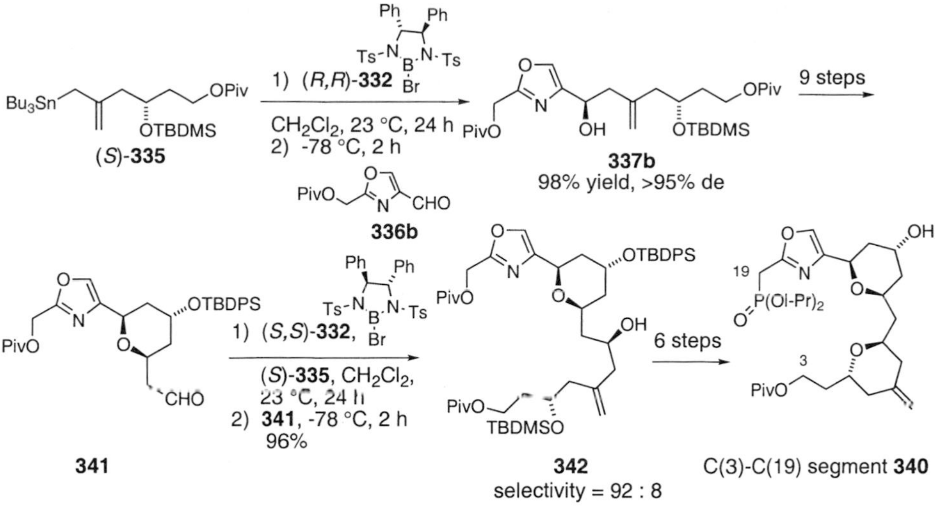

Scheme 11-20.

bromoborane (*R,R*)-**332**. The resulting 1,5-diol derivative **337b** was converted in nine steps to the *cis* pyran aldehyde **341**. Reaction of **341** with the allylboron reagent prepared from the transmetallation of allylstannane (*S*)-**335** with the (*S,S*)-bromoborane **332** generated stereoselectively the homoallylic alcohol **342** in 96% yield and with a diastereomeric ratio of 92:8. Adduct **342** was subsequently converted to segment **340** through a six-step sequence including formation of the *trans* pyran unit.

11.10 Chiral Crotylsilane Reagents

Several research groups have developed chiral allyl- and crotylsilane reagents and studied the enantioselectivity of their reactions with aldehydes [12, 15]. Of these, the chiral crotylsilane reagents developed by Panek and co-workers (Fig. 11-32) have been most extensively applied to the synthesis of natural products.

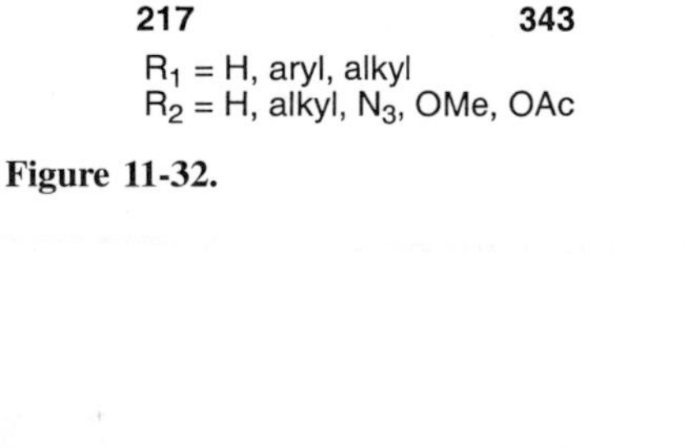

R_1 = H, aryl, alkyl
R_2 = H, alkyl, N_3, OMe, OAc

Figure 11-32.

Chiral crotylsilane reagents **217** and **343** (Fig. 11-32) can be synthesized in 3–6 steps, depending on their degree of substitution [146–148, 259]. The syntheses of crotylsilanes **217a** ($R_1=R_2=H$) and **217b** ($R_1=Me$, $R_2=H$) are outlined in Schemes 11-21 and 11-22. The synthesis of **217a** begins with the hydrosilylation of racemic 3-butyne-2-ol with phenyldimethylsilane in the presence of a catalytic amount of the Pt(0) catalyst **344**, thereby generating the vinylsilane **345** (Scheme 11-21). Resolution of **345** with AK Lipase (purchased from Amano Enzyme Co.) generates a 1:1 mixture of the (*S*)-alcohol **345** and the (*R*)-acetate **345b**. The acetate and alcohol **345b** and **345** are easily separated chromatographically, and the acetate is converted to (*R*)-**345** via treatment with $LiAlH_4$. Either antipode of alcohol **345** can then be submitted to a Johnson orthoester Claisen rearrangement, thereby generating either enantiomer of the chiral crotylsilane **217a**.

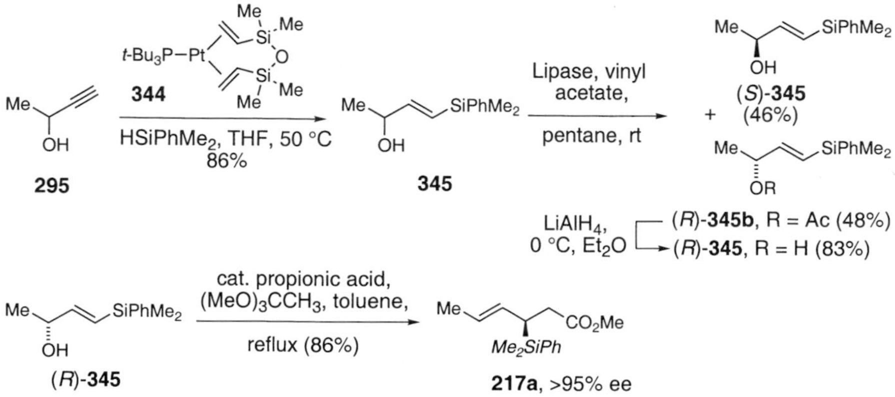

Scheme 11-21.

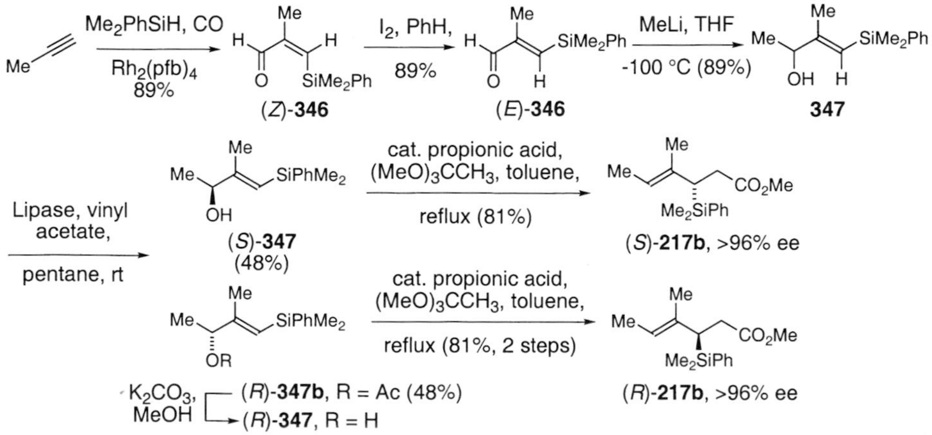

Scheme 11-22.

β-Substituted chiral crotylsilanes, e.g. **217b** (Scheme 11-22), are synthesized from the appropriately substituted terminal alkyne [147]. Thus, propyne undergoes a dirhodium(II)-catalyzed silylformylation [260] to generate the vinylsilane (*Z*)-**346**. Subsequent iodine-catalyzed olefin isomerization affords vinylsilane (*E*)-**346**, which is then treated with methyllithium, thereby generating a racemic mixture of alcohol **347**. Lipase resolution and subsequent Claisen rearrangement generates either antipode of the β-methylcrotylsilane **217b**. These crotylsilane reagents are quite stable and can be stored for long periods of time without decomposition.

Panek and co-workers have demonstrated that crotylsilanes **217** and **343** react with a variety of electrophiles including aldehydes, α,β-unsaturated ketones, acetals and imines under appropriate activation conditions (usually Lewis acidic) to form homoallylic ethers [149, 261], homoallylic alcohols [58, 150, 151], tetrahydrofurans [262, 263], cyclopentanes [264], pyrrolidines and homoallylic amines [265] with high levels of enantio- and diastereoselectivity [12]. This review will focus on the reactions of crotylsilanes **217** with Lewis acid-activated acetals and aldehydes, and the application of these reactions to the synthesis of polypropionate natural products [266–271].

Panek has reported the reactions of chiral crotylsilanes, e.g. (*S*)-**217c**, with a variety of achiral acetals, resulting in the formation of homoallylic ethers **348** with high enantio- (>95% ee) and variable diastereoselectivities (Table 11-20) [149, 261]. The acetals can be formed in situ from the corresponding aldehydes via treatment with TMSOBn or TMSOMe in the presence of catalytic TMSOTf.

Table 11-20.

RCHO → [(*S*)-**217c**, TMSOTf, TMSOBn, -78 C, CH_2Cl_2] → **348** + **349**

RCHO	348 : 349	% yield
CH_3CHO	2 : 1	97
n-BuCHO	3 : 1	51
i-PrCHO	19 : 1	60
$BnOCH_2CHO$	20 : 1	53
2,5-dimethoxy-3-nitrobenzaldehyde (OMe, O_2N, CHO, OMe)	30 : 1	92

The observed selectivity for the 5,6-*syn* adduct **348** can be explained by either the *anti* S_E' synclinal transition state **350** or the *anti* S_E' antiperiplanar transition state **351** (Fig. 11-33). It is generally accepted that the silicon substituent adopts a position *anti* to the incoming electrophile (*anti* S_E') so as to minimize steric interactions and to maximize the stereoelectronic effects (3d → 2p donation by the silicon substituent) [9, 52]. When explaining the selectivity for products **348**, Panek

Figure 11-33.

and co-workers tend to favor the Yamamoto-type [47] antiperiplanar transition state **351**, where the sterically demanding R group of the aldehyde and vinyl methyl group of the crotylsilane are farthest from one another. However, in this transition state, the benzyl ether of the oxonium ion, assumed to take a position *anti* to the aldehyde R group [272], experiences destabilizing non-bonded interactions with the γ-methyl group of the crotylsilane reagent.

In the synclinal transition state **350**, which also leads to **348**, the γ-hydrogen of the crotylsilane reagent occupies a position gauche to the sterically demanding benzyl group of the oxonium ion, thus minimizing steric interactions. This transition state also benefits from electronic effects, such as minimization of charge separation, or maximization of secondary orbital overlap between the oxonium oxygen and the silicon-bearing carbon of the crotylsilane (see Section 11.3 for further discussion) [53].

As shown in Table 11-20, unbranched aldehydes (e.g. CH_3CHO) react much less selectively with chiral crotylsilanes, where the *anti* adduct **349** is the minor product. This observation can be explained by an increase in product formation through the antiperiplanar transition state **352**, which places the γ-methyl group of the crotylsilane in a position gauche to the aldehyde R group. Thus, as the size of the aldehyde R group decreases, transition state **352** becomes more viable.

Panek and Cirillo demonstrated that the β-methyl chiral crotylsilane (*S*)-**217b** favors formation of the 5,6-*anti* diastereomer in the chelate-controlled reaction with the achiral α-benzyloxy aldehyde **353** (Eq. (11.26)) [57, 58]. Here, the synclinal transition state **355** best explains the stereochemistry of the major adduct.

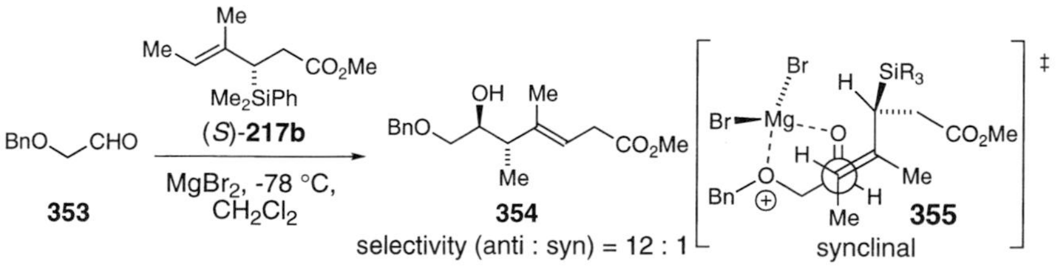

(11.26)

Using chelation and non-chelation strategies in the context of double asymmetric reactions, Panek and Jain demonstrated that by judicious choice of protecting groups all four of the dipropionate stereotriad fragments can be prepared [151]. Reactions of the TBDPS-protected β-alkoxy aldehyde **97c** with (*R*)-crotylsilane reagents **217** (R=H, Me, Et) with $TiCl_4$ catalysis leads with high diastereoselectivity to the *syn,syn* dipropionates **356** (Eq. (11.27)). Both the Felkin synclinal and the Felkin antiperiplanar transition states **357** and **358** potentially can give rise to the observed adduct **356**.

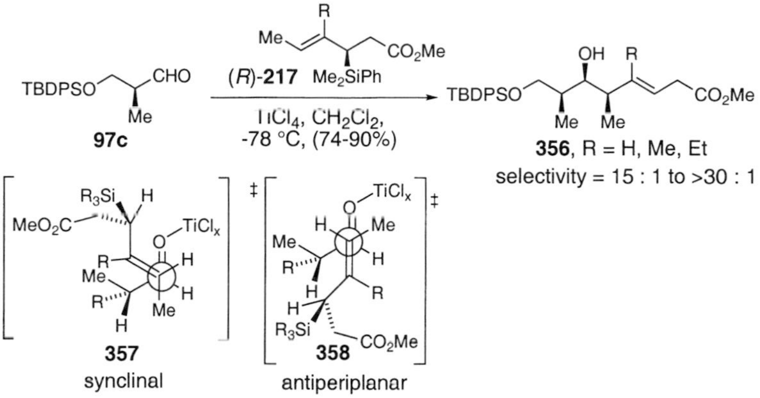

(11.27)

Reaction of the same aldehyde **97c** with enantiomeric crotylsilanes (*S*)-**217** (R=Me, Et) results in preferential formation of the *syn,anti*-dipropionates **359**. These adducts can arise either through the Felkin synclinal transition state **360** or the Felkin antiperiplanar transition state **361** (Eq. (11.28)). In the reactions of aldehyde **97c** with both the (*R*)- and (*S*)-crotylsilane reagent **217**, the major products result from crotylsilane addition to the aldehyde via the normally favored Felkin orientation in the transition state. The chirality of the crotylsilane and the stereoelectronic preference for *anti* S_E' addition then dictate the facial selectivity of the crotylsilane reagent, which is translated into the stereochemistry of the C(5) methyl substituent of the product.

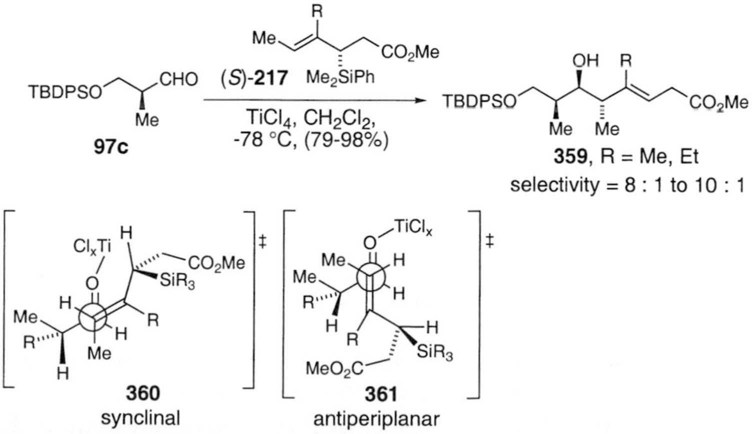

(11.28)

97a → 362, R = H, Et; (R)-217, $TiCl_4$, CH_2Cl_2, -78 °C, (35-64%); selectivity = 10 : 1 to >30 : 1; 363 (11.29)

97a → 364, R = Me, Et; (S)-217, $TiCl_4$, CH_2Cl_2, -78 °C, (35-64%); selectivity = 10 : 1 to 15 : 1; 365 antiperiplanar (11.30)

Reaction of the β-benzyloxy-α-methyl chiral aldehyde **97a** with (*R*)-crotylsilanes **217** (R=H, Et) under catalysis by $TiCl_4$ affords the *anti,anti*-dipropionate adduct **362** (Eq. (11.29)). The diastereoselectivity in this reaction is best explained by *anti* S'_E addition of the chiral crotylsilane to the least hindered face of the β-alkoxy aldehyde chelate, as shown in the synclinal transition state **363**. Finally, the *anti,syn*-dipropionate **364** may be obtained as the major adduct when aldehyde **97a** is treated under the same conditions with the enantiomeric crotylsilane reagents (*S*)-**217** (Eq. (11.30), R=Me, Et). This adduct should arise from the antiperiplanar transition state **365**, where the *anti* S'_E facial selectivity of the crotylsilane reagent and the facial bias of the chiral aldehyde are maintained. In these cases, the factors that dictate the utilization of the synclinal vs the antiperiplanar transition states are: (1) the requirement that a small substituent (H) occupy the position over the chelate ring, (2) that C–C bond formation occurs *anti* to the sterically demanding α-methyl group of the aldehyde and (3) the requirement for an *anti* S'_E mechanism, which dictates the stereochemistry of C(5) of the adducts **362** and **364**.

The synthetic potential of the crotylation reaction of acetals and aldehydes with chiral crotylsilanes **217** was demonstrated by Panek and Jain in the total synthesis of oligomycin C [267, 273, 274]. The natural product was partitioned into the C(1)–C(17) fragment **367** and the C(18)–C(34) fragment **366**, which were coupled in the final stages of the synthesis through a Stille coupling and macrocyclization (Fig. 11-34). The spiroketal fragment **366** derives from the advanced C(19)–C(34) polypropionate fragment **368**. Six of the stereogenic centers of **367** and six of the stereogenic centers of **368** were set using chiral crotylsilane reagents.

The synthesis of the **368** began with the asymmetric crotylation reaction of (*S*)-**217a** with aldehyde **369** in the presence of TMSOBn and TMSOTf (Scheme 11-23). Adduct **370** was converted to the *syn* α-methyl-β-benzyloxy aldehyde **371**,

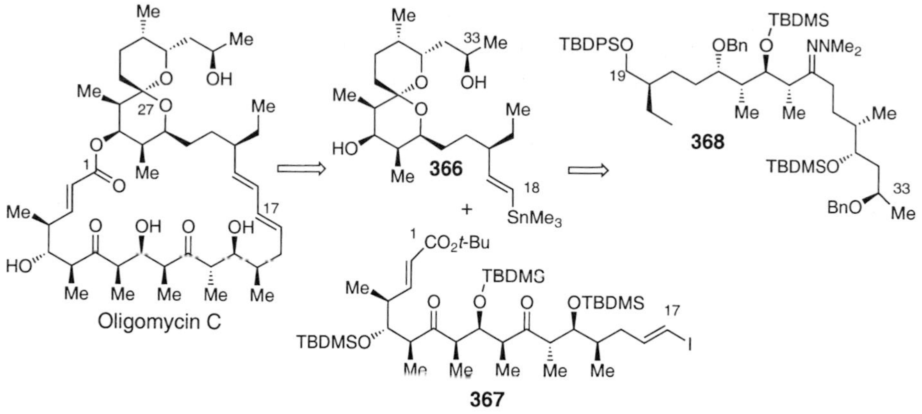

Figure 11-34.

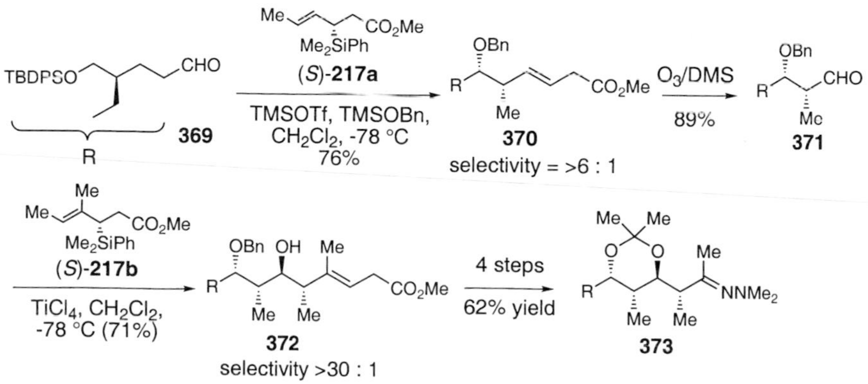

Scheme 11-23.

which when treated with the (*S*)-*β*-methyl crotylsilane **217b** in the presence of $TiCl_4$ afforded exclusively the *syn,anti,anti*-stereotetrad **372**, where the stereochemistry of **372** is consistent with reaction through a transition state analogous to **363** (Eq. (11.29)). This adduct was converted in four steps to the *N,N*-dimethyl hydrazone **373**, which was used subsequently in a condensation reaction with iodide **375** (Scheme 11-24, see below).

Homoallylic alcohol **374** was synthesized from aldehyde **130** via chelate-controlled asymmetric crotylation with crotylsilane (*S*)-**217a**, promoted by $TiCl_4$ (Scheme 11-24). Adduct **374** was transformed into iodide **375**, which underwent an alkylation reaction with the lithium anion of hydrazone **373**, affording the C(19)–C(33) oligomycin segment **368**.

The synthesis of the C(1)–C(17) segment of oligomycin C was initiated with the $TiCl_4$-catalyzed crotylation reaction of aldehyde **97c** with (*R*)-**217a** (Scheme 11-25). The resulting homoallylic alcohol **356a** was converted to aldehyde **376**, which subsequently underwent stereoselective $TiCl_4$-promoted crotylation with (*R*)-**217a**, af-

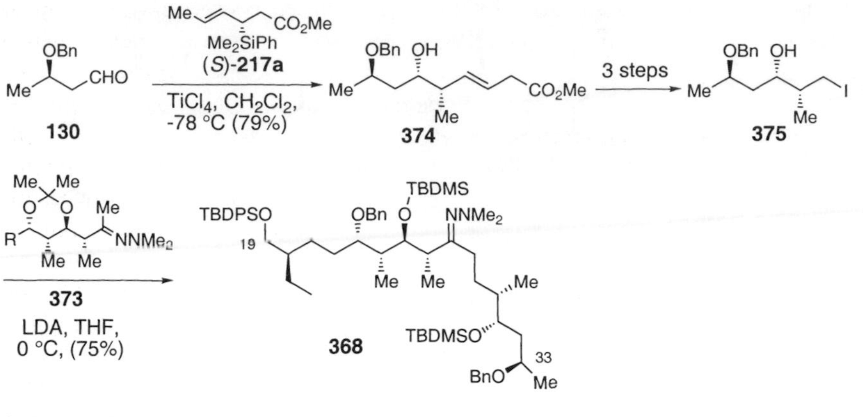

Scheme 11-24.

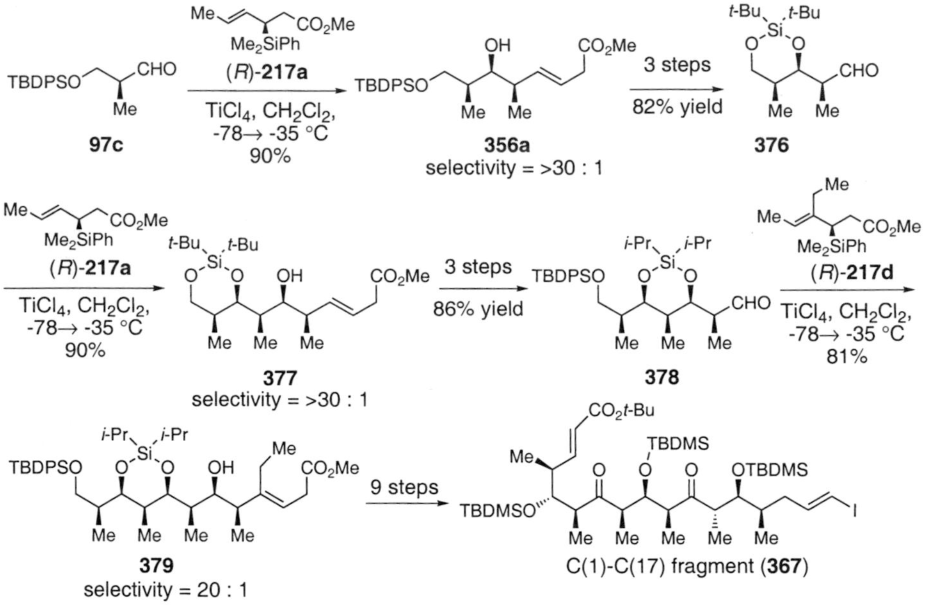

Scheme 11-25.

fording the all *syn* stereopentad **377**. Conversion of **377** to aldehyde **378** and subsequent reaction with the (*R*)-*β*-ethylcrotylsilane **217d**, catalyzed by $TiCl_4$, afforded the all *syn* stereoheptad **379**. The stereochemistry of all three crotylation adducts can be rationalized by their formation through the synclinal transition state **357** (Eq. (11.27)). The synthesis of the C(1)–C(17) oligomycin fragment was completed from **379** in nine additional steps.

The extensive use of chiral (*E*)-crotylsilane reagents in the synthesis of oligomycin clearly illustrates the utility of these reagents in solving a wide range of stereochemical problems associated with the synthesis of polypropionate-derived natural products.

11.11 Chiral Allenylstannane Reagents

Marshall and co-workers have developed methodology for the synthesis of homopropargylic alcohols using chiral allenylmetal reagents [16]. This methodology involves use of chiral allenylstannane [152, 153, 158, 159, 161], allenylindium [160, 171, 275] and allenylzinc [276] reagents. The chiral allenylstannanes can be prepared in eight steps with very high levels of enantiomeric purity from commercially available (*R*)- or (*S*)-methyl lactate (**380**), as shown for the preparation of allenylstannane (*S*)-**218b** in Scheme 11-26 [152, 154].

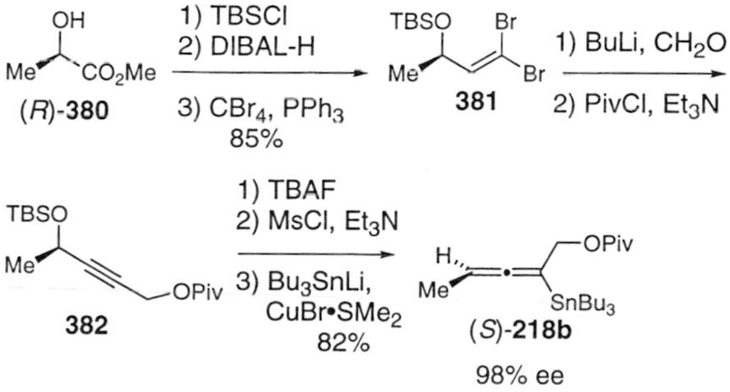

(*R*)-**380** → 1) TBSCl, 2) DIBAL-H, 3) CBr_4, PPh_3, 85% → **381** → 1) BuLi, CH_2O, 2) PivCl, Et_3N → **382** → 1) TBAF, 2) MsCl, Et_3N, 3) Bu_3SnLi, $CuBr{\bullet}SMe_2$, 82% → (*S*)-**218b**, 98% ee

Scheme 11-26.

Marshall et al. have demonstrated that high levels of diastereoselectivity favoring the *syn* homopropargylic alcohols **383** are obtained in the $BF_3{\cdot}OEt_2$-catalyzed reaction of chiral allenylstannanes (*S*)-**218** with branched aldehydes (e.g. *i*-PrCHO, Table 11-21) [153]. However, unbranched and unsaturated aldehydes give lower levels of diastereoselectivity in $BF_3{\cdot}OEt_2$-catalyzed reactions [153, 154]. Alternatively, when $SnCl_4$ is used as the Lewis acid catalyst, the *anti* homopropargylic alcohol is obtained with excellent selectivity [160, 161].

Table 11-21.

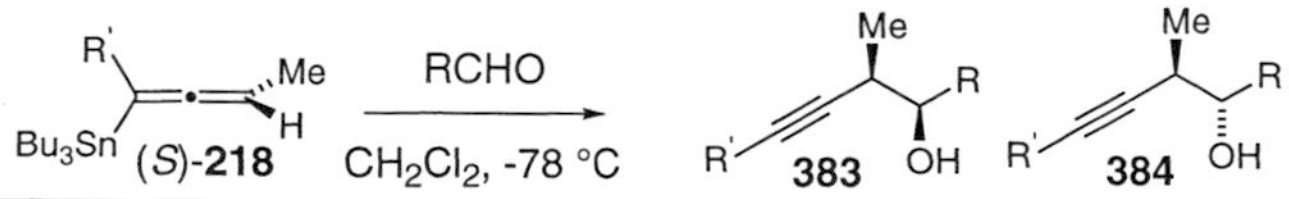

(*S*)-**218** + RCHO → (CH_2Cl_2, -78 °C) → **383** + **384**

RCHO	R'	Lewis Acid	yield	383 : 384	ee%
n-hexCHO	*n*-hept	$BF_3{\bullet}OEt_2$	83	39 : 61	ND
i-PrCHO	*n*-hept	$BF_3{\bullet}OEt_2$	95	99 : 1	92*
c-$C_6H_{11}CH{=}C(CH_3)CHO$	CH_2OPiv	$BF_3{\bullet}OEt_2$	58	81 : 19	ND
i-PrCHO	*n*-hept	$SnCl_4$	90	1 : 99	ND
$TBDPSO(CH_2)_2CHO$	CH_2OAc	$SnCl_4$	81	1 : 99	98

*Starting allenylstannane was ca. 90% ee.

Using the enantiomeric allenylstannane (*S*)-**218a** with $MgBr_2$ as the promoter, the *anti,syn*-dipropionate **398** is obtained in high yield and diastereoselectivity, where the diastereoselection is consistent with reaction occurring through the chelate antiperiplanar transition state **399** (Eq. (11.32)).

OHC–CH(Me)–CH₂OBn **97a** → [(*S*)-**218a**; $MgBr_2$, CH_2Cl_2, -23 °C, (95%); selectivity = >99 : 1] → **398**; transition state **399** (antiperiplanar) (11.32)

The *anti,anti*-dipropionate **400** can be obtained selectively when the allenylstannane (*R*)-**218a** is pre-treated with $SnCl_4$, to give (*S*)-**226a** as discussed above, followed by addition of the aldehyde **97a**. There is some evidence that this reaction occurs through a cyclic transition state where the *β*-alkoxy aldehyde is engaged in a chelate, perhaps through transition state **401** (Eq. (11.33)) [158]. However, Marshall has also demonstrated that the *anti,anti*-dipropionate can also be obtained through reaction of the appropriate allenylindium reagent **392** with the TBDPS-protected aldehyde **97c**. This indicates that chelation involving the aldehyde OR group is not necessary to obtain the *anti,anti* stereochemistry since TBDPS-protected alcohols do not participate in chelates [277].

97a → [(*R*)-**218a**; $SnCl_4$, CH_2Cl_2, -78 °C, (99%); via (*S*)-**226a**] → **400** (selectivity > 95 : 5); transition state **401** (chelated, cyclic ts) (11.33)

Finally, the *syn,anti*-dipropionate **402** is best obtained through the reactions of the TBDPS-protected aldehyde **97c** with the allenylindium (or allenylzinc) reagent (*R*)-**392**, formed in situ from mesylate (*S*)-**389**. The diastereoselectivity of this reaction is best rationalized through the cyclic Felkin transition state **403**, where the aldehyde alkyl group and the Me group of the allene adopt an *anti* relationship in the transition state (Eq. (11.34)).

Marshall has applied the chiral allenylmetal reagents in a number of natural product syntheses [154, 278–280]. The synthetic utility of these reagents is maximized when the alkyne functionality of Marshall's products is used to further elaborate the carbon skeleton of the ultimate synthetic target. This methodology is il-

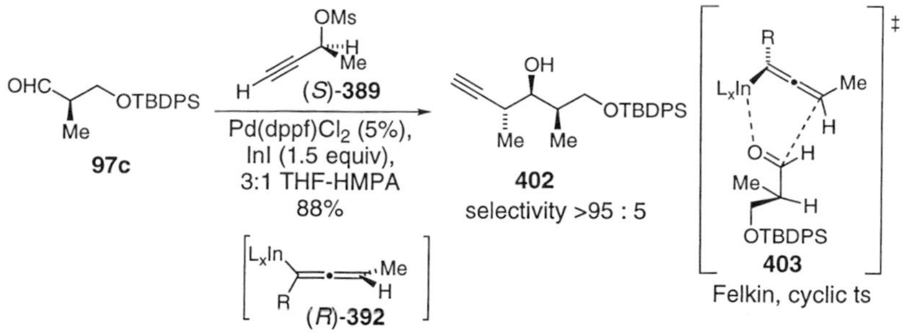

(11.34)

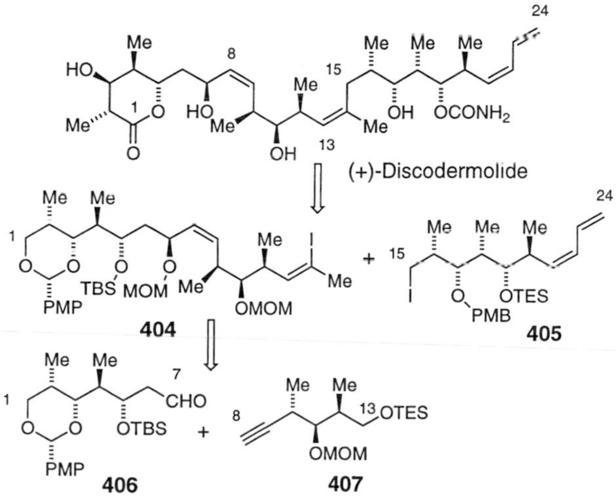

Figure 11-35.

lustrated in Marshall's total synthesis of (+)-discodermolide [278, 279]. The molecule was partitioned into the C(1)–C(14) and C(15)–C(24) segments **404** and **405**, and the C(1)–C(14) segment was further partitioned into the C(1)–C(7) and C(8)–C(13) segments **406** and **407** (Fig. 11-35).

Synthesis of the C(1)–C(7) segment **406** began with aldehyde **97d**, which was converted to the *syn,anti*-dipropionate **402b** via reaction with the chiral allenylzinc reagent prepared in situ from mesylate (*R*)-**389** (Scheme 11-29) [276]. The stereochemistry of the major adduct is consistent with reaction occurring through a transition state analogous to **403** (Eq. (11.34)). This homopropargylic alcohol was converted in four steps to alkyne **408**. The alkyne functionality of **408** was reduced to give the *trans* allylic alcohol, which was subsequently epoxidized using the Sharpless asymmetric epoxidation protocol [253]. The resulting epoxide was regioselectively reduced to afford diol **409**, which was converted to the C(1)–C(7) aldehyde **406**.

The C(8)–C(13) discodermolide subunit **410** was obtained from the previously obtained intermediate **402b** (Scheme 11-29) via etherification of the alcohol as the

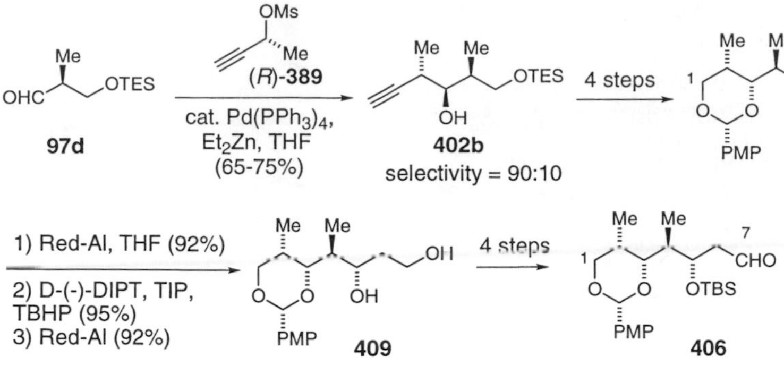

Scheme 11-29.

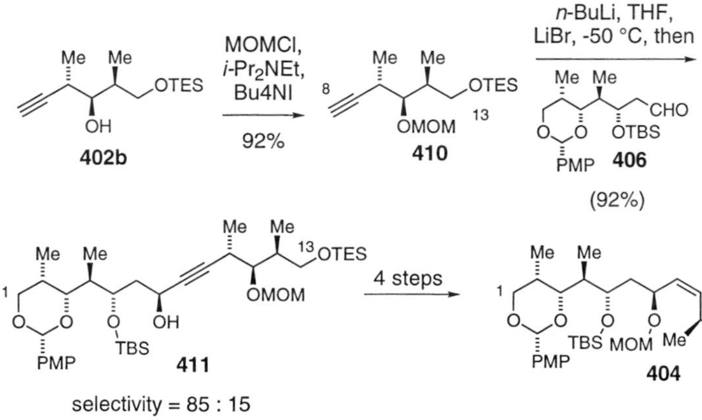

Scheme 11-30.

methoxymethyl ether (Scheme 11-30). Coupling of alkyne **410** and aldehyde **406** involved lithiation of **410** and addition of the resulting anion to aldehyde **406**, thereby providing the intermediate **411** with 85 : 15 diastereoselectivity. The C(5)–C(7) 1,3-*anti* diol stereochemistry of adduct **411** is explained by Evans' dipole model for 1,3-induction [93]. Adduct **411** was converted in four steps to the necessary vinyl iodide **404**.

The C(15)-C(24) segment **405** was constructed in a manner similar to the construction of aldehyde **406** (see above). The $BF_3 \cdot OEt_2$-catalyzed addition of allenylstannane (*S*)-**218a** to aldehyde **97b** provided the *syn,syn*-dipropionate **412** (Scheme 11-31). Adduct **412** was converted in three steps to the *trans* allylic alcohol **413**, which underwent Sharpless asymmetric epoxidation followed by regioselective epoxide opening with $Me_2CuCNLi_2$ to yield diol **414**. Adduct **414** was converted in eight steps to the primary iodide **405** which was then converted to the corresponding organoborane and coupled with the C(1)–C(14) vinyl iodide **404** under Suzuki conditions, thereby generating the fully protected discodermolide precursor **415**. This adduct was converted to (+)-discodermolide in eight additional steps.

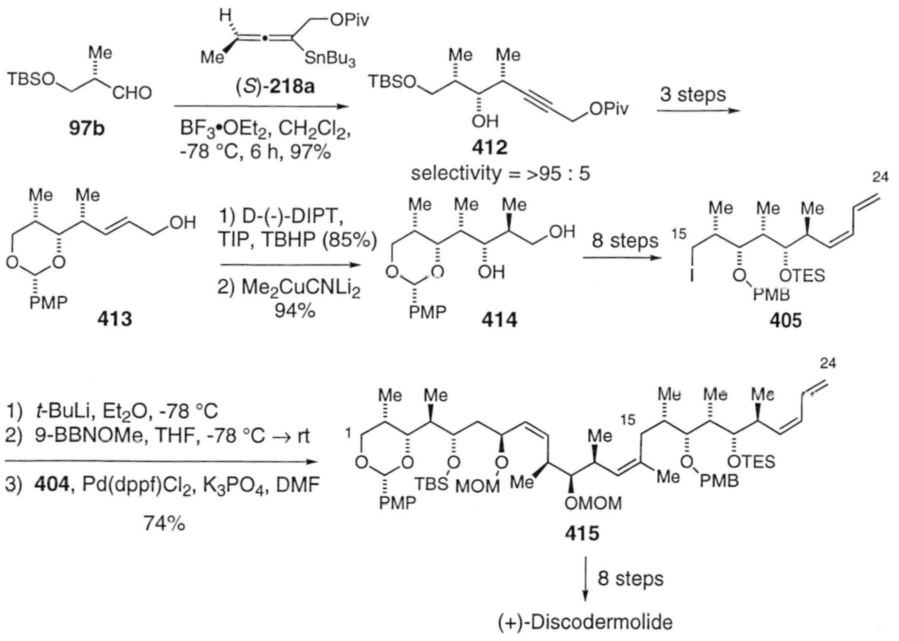

Scheme 11-31.

11.12 Chiral [(*Z*)-γ-Alkoxyallyl]stannane and [(*E*)-γ-Alkoxyallyl]indium Reagents

Marshall and co-workers have demonstrated that the [(*Z*)-γ-alkoxyallyl]stannane and [(*E*)-γ-alkoxyallyl]indium reagents **230** and **233** are useful reagents for the synthesis of *syn* and *anti* diol derivatives, respectively (Fig. 11-36). The use of these reagents in natural product synthesis has recently been reviewed [16].

Reagents **230** and **233** are synthesized from a common intermediate, [(*E*)-γ-(alkoxy)allyl]stannane **417**, which is prepared from crotonaldehyde (Eq. (11.35)) [167, 281]. 1,2-Addition of tri-*n*-butylstannyllithium to crotonaldehyde followed by in situ oxidation of the resulting lithium alkoxide with 1,1′-azodicarbonyldipiperidine (ADD) gives the intermediate acylstannane **416**. Asymmetric reduction of **416** with Noyori's BINAL-H reagent [282] followed by protection of the propargyl alcohol as either a MOM or TBS ether generates the chiral α-(alkoxy)allyl]

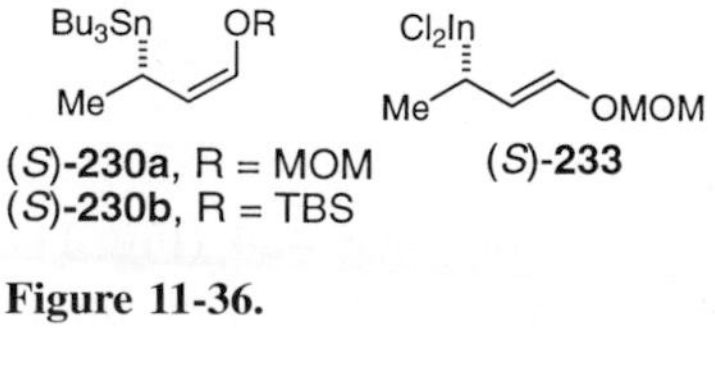

(*S*)-**230a**, R = MOM (*S*)-**233**
(*S*)-**230b**, R = TBS

Figure 11-36.

Me–CH=CH–CHO → [1) Bu_3SnLi, THF, -78 °C; 2) 1,1'-(azodicarbonyl)dipiperidine, 0 °C] → **416** → [1) (*R*)-BINAL-H, THF, -78 °C; 2) MOMCl, or TBSCl, *i*-Pr_2NEt, CH_2Cl_2 (45% overall)] → (*S*)-**417a**, R = MOM; (*S*)-**417b**, R = TBS (11.35)

(*S*)-**417a**, R = MOM; (*S*)-**417b**, R = TBS → [BF_3•OEt_2, CH_2Cl_2, -78 °C] → (*S*)-**230a**, R = MOM (80% yield); (*S*)-**230b**, R = TBS (80% yield); 95% ee (11.36)

(*R*)-**417a** → [$InCl_3$, EtOAc] → ((*S*)-**233**) (11.37)

stannane intermediates **417**. 1,3-Allylic isomerization of the -$SnBu_3$ group, catalyzed by $BF_3 \cdot OEt_2$, gives exclusively the [(*Z*)-γ-(alkoxy)allyl]stannanes **230** with 95% enantiomeric excess (Eq. (11.36)) [167, 281]. Through crossover experiments, Marshall and co-workers provided evidence that indicates that this isomerization reaction occurs in an intermolecular manner, presumbly via an *anti* S_E' mechanism [167]. The [(*Z*)-γ-(alkoxy)allyl]indium reagent **233** is prepared directly in the allylation reaction from [α-(alkoxy)allyl]stannane **417a** by an *anti* S_E' transmetallation with $InCl_3$ (Eq. (11.37)) [171].

[γ-(Alkoxy)allyl]stannane **230a** reacts with achiral aldehydes using $BF_3 \cdot OEt_2$ to generate predominantly the *syn* 1,2-diol adducts **418** in good yield and excellent enantiomeric excess (95% ee, Table 11-23) [167].

Table 11-23.

(*S*)-**230a** → [RCHO, BF_3•OEt_2, CH_2Cl_2, -78 °C] → **418** + **419**

RCHO	yield	418 : 419	% ee 418
n-C_6H_{13}CHO	75	96 : 4	95
(*E*)-BuCH=CHCHO	84	94 : 6	95
BuC≡CCHO	70	90 : 10	95

Marshall has put forth the antiperiplanar *anti* S_E' transition state **420** (Fig. 11-37) to rationalize the formation of the *syn* adducts **418**. This transition state mini-

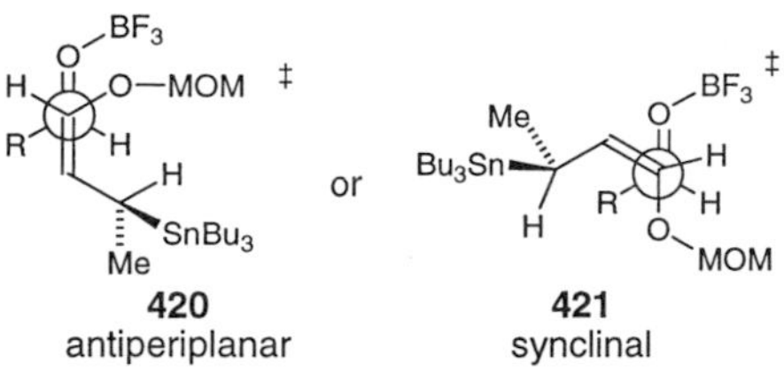

Figure 11-37.

mizes steric interactions between the aldehyde R group and the bulky OMOM substituent of the [(*Z*)-γ-(alkoxy)allyl]stannane **230a**. However, this transition state suffers from destabilizing interactions between the Lewis acid (assumed to adopt a *trans* orientation with respect to the aldehyde R group [272]) and the MOM alkoxy group of **230a**, which should favor the *S-trans* vinyl ether orientation in order to relieve steric interacts with the allylstannane α-carbon [170, 283, 284]. Coulombic repulsion between the aldehyde oxygen and the reagent alkoxy group may also disfavor transition state **420** [93]. The alternative synclinal transition state **421**, which also leads to the *syn* products **418**, relieves the Lewis acid-MOM ether interactions and Coulombic repulsions, but does introduce steric interactions between the α-carbon of reagent **230a** and the aldehyde R group.

The diastereoselectivity of the allylation reaction increases when the [γ-(silyloxy)allyl]stannane reagent **230b** is used [285]. This trend is consistent with a synclinal transition state analogous to **421** being dominant as the steric interactions between the BF_3-complexed aldehyde and the TBS group would be increased relative to the MOM group in transition state **420**.

Complementary reactions of the [(*E*)-γ-(alkoxy)allyl]indium reagent **233**, prepared in situ from [α-(alkoxy)allyl]stannane **417a** (see Eq. (11.37)), with achiral aldehydes provides the *anti* diol derivatives **419**, generally with high levels of diastereoselectivity and enantiomeric purity (Table 11-24) [171, 275]. Analogous reactions performed using the [γ-(silyloxy)allyl]stannane **417b** as precursor to an allylindium reagent were much less diastereoselective [286].

Marshall has proposed transition state **422** to rationalize the formation of the major *anti* diol derivatives (Eq. (11.38)). The minor *syn* diol adduct **418** is thought to arise via the (*Z*)-γ-alkoxyindium intermediate **423** by way of transition state **424** (Eq. (11.39)) [171].

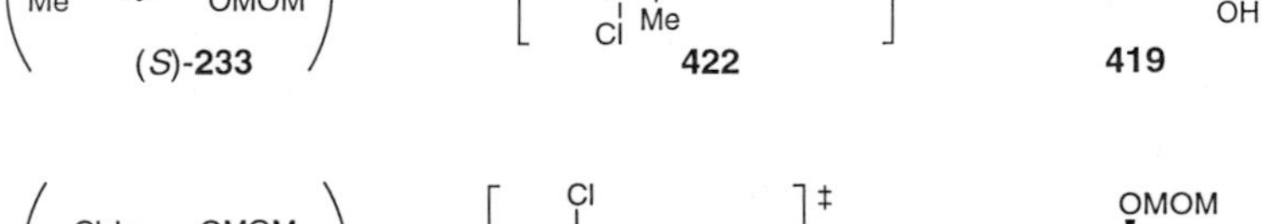

(11.38)

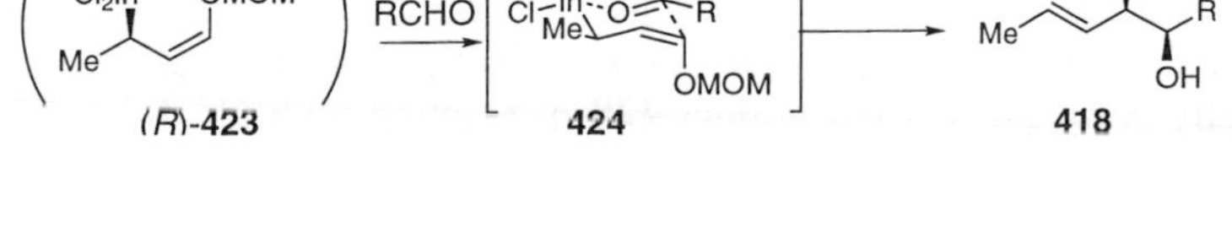

(11.39)

Table 11-24.

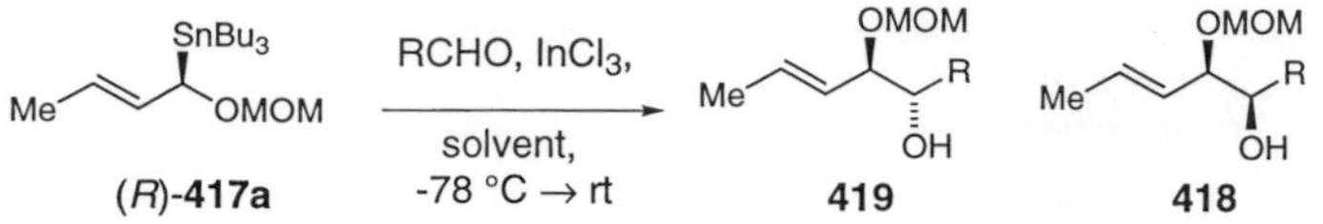

RCHO	solvent	yield	419 : 418	% ee 419
n-C_6H_{13}CHO	acetone	88	98 : 2	>95
(*E*)-BuCH=CHCHO	MeCN	87	83 : 17	>95
c-C_6H_{11}C≡CCHO	MeCN	70	96 : 4	nd

Double asymmetric reactions between [γ-(alkoxy)allyl]stannanes **230** and the α-benzyloxy aldehyde **55** exhibited clear matched and mismatched behavior [168]. With $BF_3{\cdot}OEt_2$ catalysis, the matched double asymmetric reaction between (*R*)-**230a** and aldehyde (*S*)-**55** generates exclusively the *syn,anti* adduct **425** (Eq. (11.40)). Formation of **425** can be rationalized through either the antiperiplanar, Felkin transition state **426** (as proposed by Marshall) or the synclinal Felkin transition state **427**.

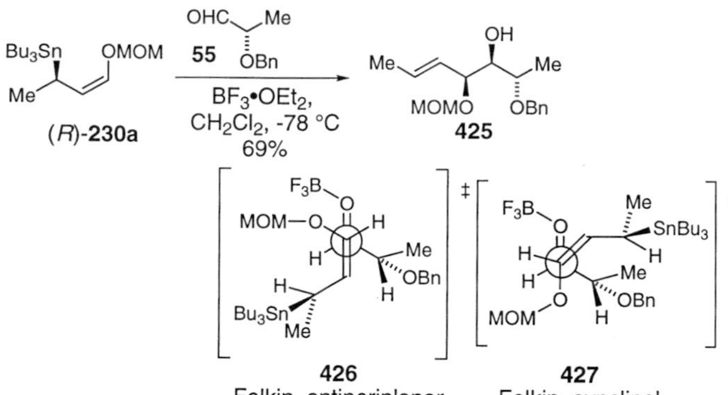

426 Felkin, antiperiplanar   427 Felkin, synclinal   (11.40)

The $BF_3{\cdot}OEt_2$-promoted mismatched reaction between (*S*)-**230a** and aldehyde **55** generates an approximate 2:1 mixture of the anti-Felkin product **428**, the result of reagent control, and the Felkin product **429**, which contains an *anti* relationship between the two new diol stereocenters (Eq. (11.41)). The major anti-Felkin adduct **428** can emerge through either the synclinal transition state **430** or the antiperiplanar transition state **431** (as proposed by Marshall). The *anti,anti*-adduct **429** can arise through the Felkin synclinal transition state **432**, where the steric interactions between the MOM group and the Lewis acid can potentially be relieved by the *S-trans* orientation of the MOM group.

The $MgBr_2{\cdot}OEt_2$-promoted reactions of reagents **230** and aldehyde **55** also display matched/mismatched characteristics (Eq. (11.42)). Both reagent (*S*)-**230a** and (*S*)-**230b** preferentially form the *syn,syn* adduct **428** in reaction with aldehyde **55**.

Bu3Sn OMOM
Me
(*S*)-**230**
OHC Me
55 OBn
$BF_3{\cdot}OEt_2$,
CH_2Cl_2, -78 °C
97%
OH
Me Me
MOMO OBn
428
OH
Me
Me O OBn
MOM **429**
selectivity = 67 : 33

F3B O, H, H, Me, SnBu3, H, OBn, MOM O, H, Me ‡
430
anti-Felkin, synclinal

F3B O, MOM—O, H, OBn, H, Me, H, H, Bu3Sn, Me ‡
431
anti-Felkin, antiperiplanar

Me SnBu3, H, F3B O, MOM O, H, Me, H, OBn, H ‡
432
Felkin, synclinal

(11.41)

Bu3Sn OR
Me
(*S*)-**230a**, R = MOM
(*S*)-**230b**, R = TBS
OHC Me
55 OBn
$MgBr_2{\cdot}OEt_2$,
CH_2Cl_2, -23 °C
OH
Me Me
RO OBn
428a, R = MOM,
(83% yield)
428b, R = TBS,
(71% yield)

Br, R, O- -Mg Br, O, H, O, Bn, H, H, Bu3Sn, Me, Me ‡
433
antiperiplanar, chelate

(11.42)

The stereochemistry of this adduct is best rationalized through transition state **433**.

Interestingly, the $MgBr_2{\cdot}OEt_2$-promoted reactions between aldehyde **55** and reagents (*R*)-**230a** and (*R*)-**230b** occur in a stereodivergent manner: reagent (*R*)-**230a** leads preferentially to the *anti,syn* adduct **434** (selectivity = 75 : 25) while reagent (*R*)-**230b** generates exclusively the *syn,syn* adduct **435** (Eq. (11.43)). These results suggest that the synclinal transition state **436**, which leads to adduct **434**, becomes more disfavored as the size of the alkoxy group of **230** increases

Bu3Sn OR
Me
(*R*)-**230a**, R = MOM
(*R*)-**230b**, R = TBS
OHC Me
55 OBn
$MgBr_2{\cdot}OEt_2$,
CH_2Cl_2, -23 °C
OH
Me Me
RO OBn
434a, R = MOM
434b, R = TBS
OH
Me
Me O OBn
R
435a, R = MOM
435b, R = TBS

434a : **435a** = 75 : 25
(74% yield)
434b : **435b** = <1 : 99
(66% yield)

Br, Me, O- -Mg Br, H, Bu3Sn, H, H, O, Bn, O, R, Me ‡
436
synclinal, chelate

Br, R, O- -Mg Br, O, H, O, Bn, H, Me, Me, Bu3Sn, H ‡
437
antiperiplanar, chelate

(11.43)

(TBS≫MOM). Barring some difference in electronics, these data seem to be a direct indication that the -OR group interacts disfavorably with the aldehyde *α*-carbon in transition state **436**. Relief of this steric interaction, by placement of the R group in an *S-cis* orientation, causes interactions between the R group and the *α*-carbon of the reagent. Thus, the formation of adduct **435**, which most probably arises through transition state **437**, becomes competitive, and is the major pathway in the reaction of (*R*)-**230b**.

In a demonstration of the synthetic utility of chiral *γ*-alkoxy stannane reagents, Marshall and co-workers applied this methodology to the synthesis of the gypsy moth pheromones (+)- and (–)-disparlure [287]. The synthesis required the production of the [*γ*-(alkoxy)allyl]stannane reagent **440** (Scheme 11-32).

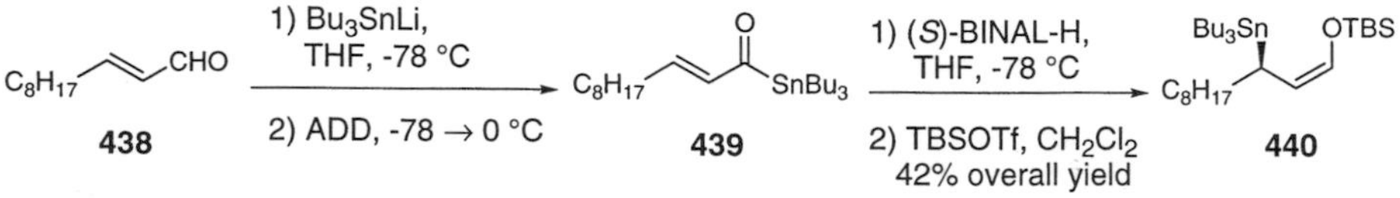

C_8H_{17} CHO **438** → 1) Bu_3SnLi, THF, -78 °C; 2) ADD, -78 → 0 °C → C_8H_{17} O SnBu$_3$ **439** → 1) (*S*)-BINAL-H, THF, -78 °C; 2) TBSOTf, CH_2Cl_2, 42% overall yield → Bu_3Sn OTBS C_8H_{17} **440**

Scheme 11-32.

Marshall found that reagents like **440** exhibit greater diastereoselectivity with *α*,*β*-unsaturated aldehydes than with saturated aldehydes [167]. With this in mind, they began the synthesis of disparlure with the allylation of *α*,*β*-unsaturated aldehyde **441** using $BF_3{\cdot}OEt_2$ to promote the reaction. The *syn* diol adduct **442** (90% ee) was obtained in 73% yield with >95:5 *syn*:*anti* diastereoselectivity. This adduct was converted to (+)-disparlure via tosylation of the free alcohol and subsequent cyclization upon removal of the TBS protecting group with TBAF. Reversing the nucleophilic and electrophilic centers in **442** led in a complementary manner, via intermediate **443**, to the (–)-antipode of disparlure (Scheme 11-33).

OHC **441** → Bu_3Sn OTBS C_8H_{17} **440**, $BF_3{\bullet}OEt_2$, CH_2Cl_2, -78 °C (73%), selectivity > 95 : 5, 90%ee → C_8H_{17} OTBS OH **442** → 1) TsCl, pyr 95%; 2) TBAF, THF (89%) → (+)-Disparlure

442 → 1) Ac_2O, pyr (96%); 2) $BF_3{\bullet}OEt_2$, CH_2Cl_2 (95%); 3) TsCl, pyr (92%) → OTs OAc **443** → Triton-B 95% → (-)-Disparlure

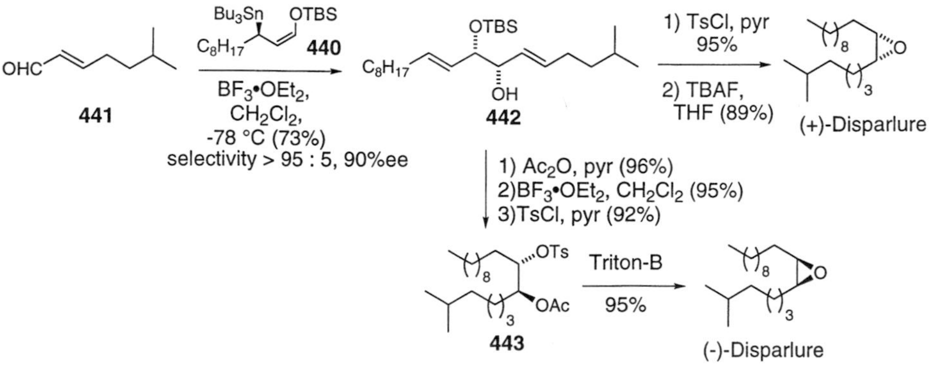

Scheme 11-33.

Using insights deriving from the studies of double asymmetric reactions summarized in Eqs. (11.40)–(11.43), Marshall demonstrated that with judicious choice of reaction conditions, one can obtain all four possible diastereomers selectively

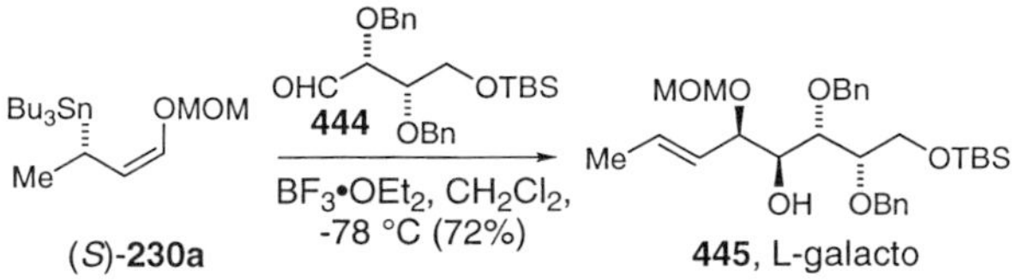

(11.44)

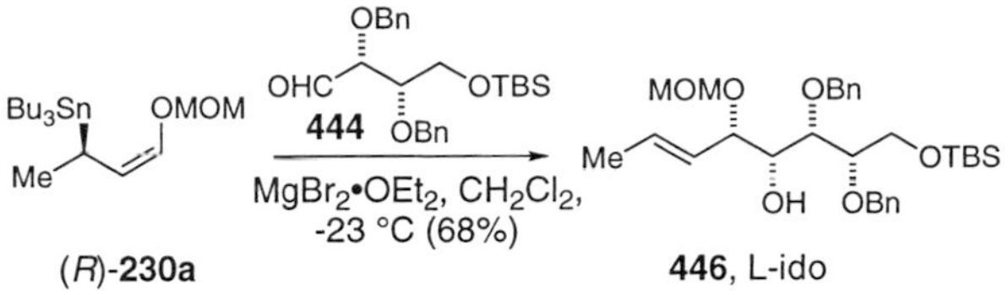

(11.45)

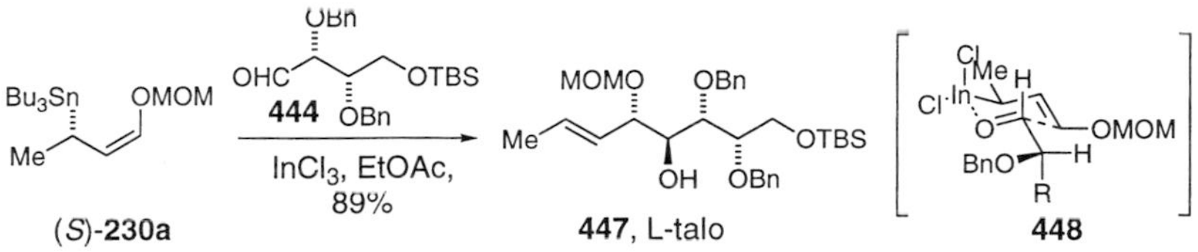

(11.46)

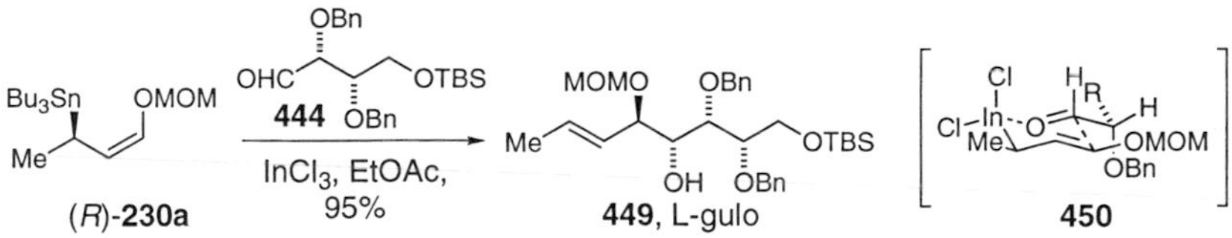

(11.47)

in the reaction of the threose derivative **444** with the chiral [γ-(alkoxy)allyl]stannane and [γ-(alkoxy)allyl]indium reagents **230a** and **233** (prepared in situ from **230a**) [275]. The resulting adducts **445-449** (Eqs. (11.44)–(11.47)) are potentially useful in the context of carbohydrate synthesis. Thus, the $BF_3{\cdot}OEt_2$-promoted matched double asymmetric reaction between (*S*)-**230a** and aldehyde **444** leads stereoselectively (no other diastereomer reported) to the L-galacto adduct **445**, presumably by way of transition states **426** or **427** (Eq. (11.40)). The $MgBr_2{\cdot}OEt_2$-promoted reaction between (*R*)-**230a** and aldehyde **345** leads stereoselectively (no other diastereomer reported) to the L-ido adduct **446**, where the stereochemistry of adduct **446** is consistent with its formation through a transition state analogous to **433** (Eq. (11.42)).

The L-talo and L-gulo adducts **447** and **449** were obtained with very high stereoselectivity (no other diastereomers reported) from the reaction of aldehyde **444** with the [γ-(alkoxy)allyl]indium reagents generated from (*S*)-**230a** and (*R*)-**230a**, respectively. In these double asymmetric reactions, reagent control is clearly dominant. The stereochemistry of adduct **447** is rationalized by the Felkin transition state **448** while the stereochemistry of adduct **449** is rationalized by the *anti*-Felkin transition state **450** [275].

Table 11-26.

RCHO + 131 (allyl-$SiMe_3$) → [(S)-455, conditions] → 240

RCHO	% yield	% ee
$(CH_3)_3CCHO$	91	94
PhCHO	85	80
c-$C_6H_{11}CHO$	72	60
$PhCH_2CH_2CHO$	69	61

Conditions: (1) 10 mol% (*S*)-**455**, CH_2Cl_2, MeCN, 0 °C, 4 h (2) TBAF, THF

Table 11-27.

RCHO + 456 ($SnBu_3$) → [(R)-451, conditions] → 457 + 458

RCHO	Lewis acid	cat. loading	additive	rxn time	% yield	457:458	% ee 457
PhCHO	(*R*)-**451**	50 mol%	–	100 h	48	93:7	>99
PhCHO	(*S*)-**451***	10 mol%	*i*-$PrSBEt_2$	15 h	52	>98:2	92
$PhCH_2CH_2CHO$	(*R*)-**451**	50 mol%	–	72 h	76	96:4	95
$PhCH_2CH_2CHO$	(*S*)-**451***	10 mol%	*i*-$PrSBEt_2$	9 h	86	>98:2	94
c-$C_6H_{11}CHO$	(*R*)-**451**	50 mol%	–	100 h	64	80:20	89
c-$C_6H_{11}CHO$	(*S*)-**451***	10 mol%	*i*-$PrSBEt_2$	15 h	73	>98:2	91

* *Ent*-**457** and *ent*-**458** were the products of these reactions.

enriched homopropargylic alcohols **457** along with a minor amount of allenes **458** in fair yields and good enantioselectivities under Keck's conditions, which utilize high catalyst loading (Table 11-27) [303]. Using a stoichiometric amount of the synergetic reagent *i*-$PrSBEt_2$, Yu was able to decrease the reaction time and catalyst loading while increasing the product yields (Table 11-27) [304]. Yu has extended this methodology to the selective synthesis of allenes **458** as well [305].

Several researchers have observed the "asymmetric amplification" phenomenon with BINOL-Ti(IV) catalyst **451**. Less expensive chiral alcohols added to a racemic BINOL-Ti(IV) catalyst complex can serve to activate or to poison one BINOL-Ti(IV) enantiomer over the other; the more active enantiomer then catalyzes the allylation of aldehydes with high levels of enantioselectivity. These methods have the potential to make large-scale use of catalyst **352** in asymmetric transformations economically attractive [306, 307].

11.13.2 Catalytic Asymmetric Allylation with the CAB Catalyst

The chiral (alkyloxy)borane (CAB) catalysts **459** are prepared from tartaric acid as shown in Scheme 11-34 [288, 308, 309].

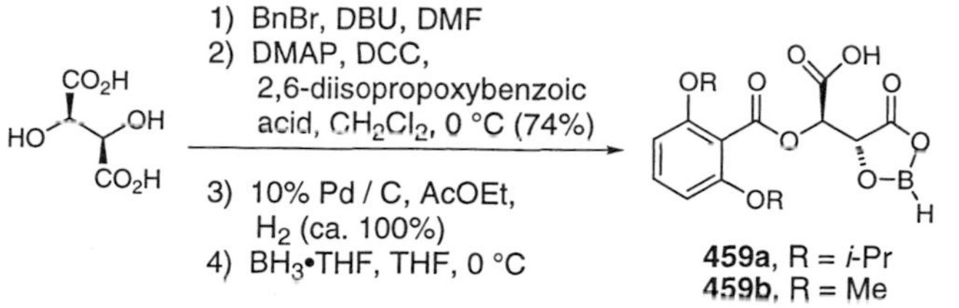

1) BnBr, DBU, DMF
2) DMAP, DCC, 2,6-diisopropoxybenzoic acid, CH_2Cl_2, 0 °C (74%)
3) 10% Pd / C, AcOEt, H_2 (ca. 100%)
4) BH_3•THF, THF, 0 °C

459a, R = *i*-Pr
459b, R = Me

Scheme 11-34.

The CAB-catalyzed reactions of aldehydes with allyl- and crotylsilanes are summarized in Table 11-28 [288, 309]. These data show that the CAB-catalyzed crotylation reactions with crotylsilanes **123b** and **460** are highly diastereo- and enantioselective, while the corresponding allylation reactions with allylsilanes **131** occur with somewhat lower enantioselectivity.

Table 11-28.

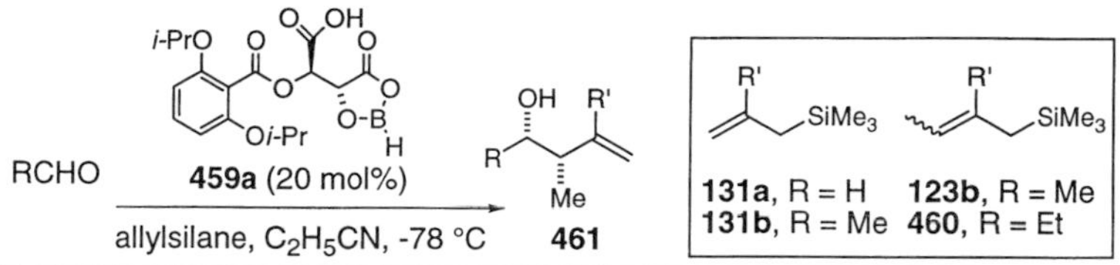

RCHO → (**459a** (20 mol%), allylsilane, C_2H_5CN, -78 °C) → **461**

131a, R = H; **131b**, R = Me; **123b**, R = Me; **460**, R = Et

RCHO	allylsilane	yield	selectivity (*syn*:*anti*)	% ee 461
PhCHO	**131a**	46	–	55
PhCHO	**131b**	68	–	82
PhCHO	**123b**	63	96:4	90
C_4H_9CHO	**123b**	30	94:6	85
PhCHO	**460**	74	97:3	96
trans-CH_3CH=CHCHO	**460**	21	95:5	89
C_3H_7CHO	**460**	36	95:5	86

Marshall subsequently demonstrated that the efficiency of the crotylation reactions catalyzed by CAB catalyst **459b** could be improved by using the more reactive crotylstannanes and employing two equivalents of $(CF_3CO)_2O$ to aid catalyst turnover [310]. In a comparative study of the catalytic asymmetric allylation and crotylation reactions of cyclohexane carboxaldehyde with allyl- and crotylstannanes **98** and **10**, Marshall demonstrated the complementarity between the BINOL-Ti(O-*i*-Pr)$_4$ catalyst **451** and the CAB catalyst **459b** (Table 11-29).

In a demonstration of the synthetic utility of the CAB-catalyzed crotylation reaction, Marshall synthesized the commonly used polypropionate adducts **106** and

Table 11-29.

$R\diagup\!\!\diagdown SnBu_3$ — **98**, R = H; **10**, R = Me; Lewis acid → **462** + **463**

stannane	Lewis acid	conditions	% yield	462 (% ee) : 463 (% ee)
10	BINOL-Ti(IV) (**451**)	a	18	65 (95): 35 (49)
10	CAB (**459b**)	b	71	93 (93): 7 (80)
98	BINOL-Ti(IV) (**451**)	a	53	87
98	CAB (**459b**)	b	42	55

Conditions: (a) 20 mol% (*R*)-BINOL, 10 mol% Ti(O*i*-Pr)$_4$, CH_2Cl_2, 4 Å MS, -20 °C, 70 h. (b) 0.5 equiv tartrate, 0.75 equiv BH_3, 1 equiv stannane, 1 equiv aldehyde, 2 equiv $(CF_3CO)_2O$, 78 °C, 10 h

ent-**97c** + $Me\diagup\!\!\diagdown SnBu_3$ (**10**) → *ent*-**106c** + *ent*-**107c** (11.48)

Lewis acid	conditions	% yield	106 : 107
BF_3•OEt_2	a	71	90 : 10
CAB (**459b**)	b	68	98 : 2

Conditions: (a) CH_2Cl_2, -78 °C, 6 h; (b) **459b** (1 equiv), $(CF_3CO)_2O$ (2 equiv), CH_3CH_2CN, -78 (10 h) → -10 °C (12 h)

97c + $Me\diagup\!\!\diagdown SnBu_3$ (**10**) → **106c** + **107c** (11.49)

Lewis acid	conditions	% yield	106 : 107
BF_3•OEt_2	a	75	90 : 10
CAB (**459b**)	b	65	10 : 90

Conditions: (a) CH_2Cl_2, -78 °C, 3 h; (b) **459b** (1 equiv), $(CF_3CO)_2O$ (2 equiv), CH_3CH_2CN, -78 (10 h) → -10 °C (12 h)

107 (Eqs. (11.48) and (11.49)) [310]. While adduct *ent*-**106c**, obtained with 98 : 2 diastereoselectivity, results from the matched crotylation of *ent*-**97c** with crotylstannane **10**, the *anti,syn* adduct **107c**, obtained in 90 : 10 selectivity, is the result of a mismatched double asymmetric crotylation reaction. However, with these aldehydes, a stoichiometric amount of the CAB catalyst is necessary for synthetic efficiency.

11.13.3 Selected Applications of the Catalytic Enantioselective Allylation Reaction in Natural Product Synthesis

In their synthesis of kallolide A, Marshall and co-workers compared the BINOL-Ti(IV) and CAB catalyst systems **451** and **459b** in the fragment-coupling allylation reaction between the unsaturated aldehyde **464** and the synthetically advanced allylstannane **465**. Consistent with previous findings with crotylstannane reagents (Table 11-28), they found the CAB catalyst **459b** to be more enantio- and diastereoselective than the BINOL-Ti(IV) catalyst **451** in the formation of the *syn* adduct **466** (Scheme 11-35) [311]. However, the efficiencies of both reactions were poor. Marshall and co-workers converted adduct **466** to kallolide A in a sequence which involved the necessary inversion of the alcohol stereocenter of **466**.

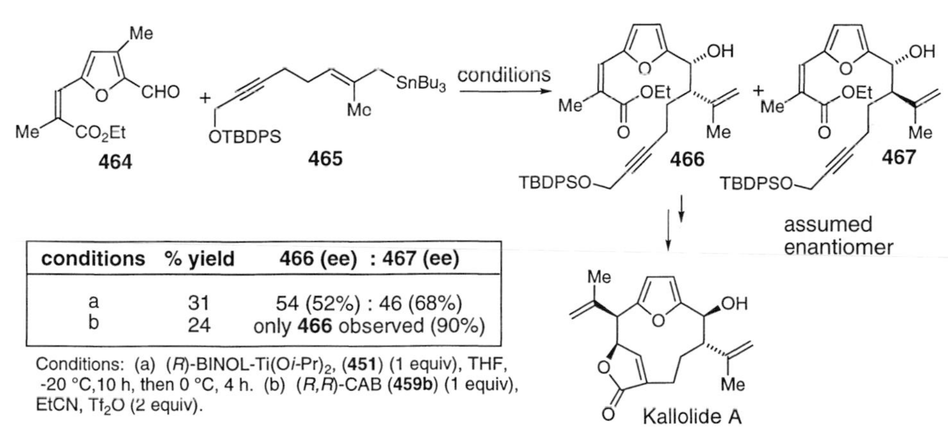

conditions	% yield	466 (ee) : 467 (ee)
a	31	54 (52%) : 46 (68%)
b	24	only **466** observed (90%)

Conditions: (a) (*R*)-BINOL-Ti(O*i*-Pr)$_2$, (**451**) (1 equiv), THF, -20 °C, 10 h, then 0 °C, 4 h. (b) (*R,R*)-CAB (**459b**) (1 equiv), EtCN, Tf_2O (2 equiv).

Scheme 11-35.

Of the BINOL/BINAP-metal catalyst complexes, only the allylation procedure described by Keck using the BINOL-Ti(IV) catalyst **451** has been applied in the synthesis of natural products, presumably because it has the most substrate generality and the field is so new. In a preliminary report, Evans disclosed the synthesis of the 4-hydroxy buteneolide terminus **470** of mucocin, where he uses Keck's original catalytic asymmetric allylation procedure to effect conversion of aldehyde **469** to the homoallylic alcohol **470** in good yield and high diastereoselectivity (Scheme 11-36) [312].

In their synthesis of the *cis*-octahydronaphthalene nucleus **471** of superstolide A (Fig. 11-39), Roush and co-workers demonstrated the use of Keck's original catalytic allylation procedure to effect the diastereoselective conversion of aldehyde **472** to the 1,3-*syn* diol **473** (79% yield, selectivity = 94:6) (Scheme 11-37) [313]. This transformation constitutes a mismatched reaction since the 1,3-*anti* diol is favored under substrate-controlled allylation (see Section 11.3 for a discussion of 1,3-stereoinduction) [93].

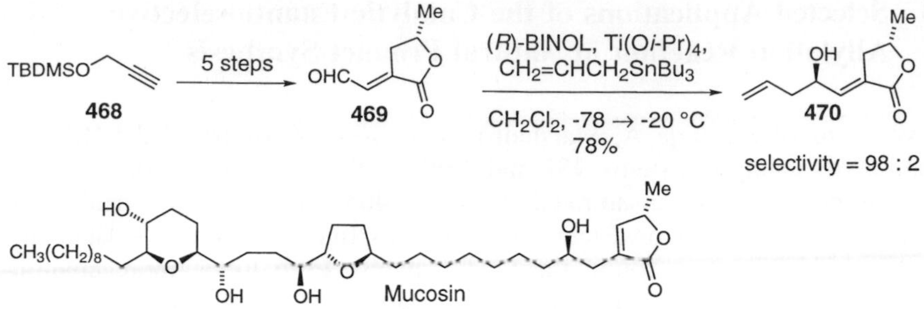

TBDMSO 468 → 5 steps → OHC 469 → (*R*)-BINOL, Ti(O*i*-Pr)$_4$, $CH_2{=}CHCH_2SnBu_3$, CH_2Cl_2, -78 → -20 °C, 78% → 470, selectivity = 98 : 2

Mucosin

Scheme 11-36.

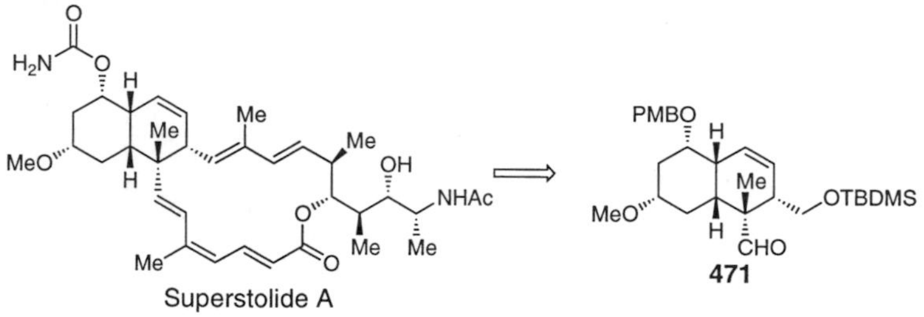

Superstolide A ⟹ 471

Figure 11-39.

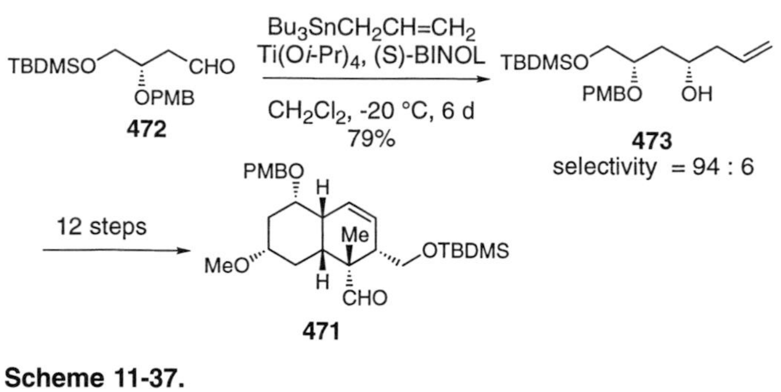

TBDMSO, CHO, OPMB 472 → $Bu_3SnCH_2CH{=}CH_2$, Ti(O*i*-Pr)$_4$, (S)-BINOL, CH_2Cl_2, -20 °C, 6 d, 79% → TBDMSO, PMBO, OH 473, selectivity = 94 : 6

12 steps → 471

Scheme 11-37.

R = H, Epothilone A
R = CH_3, Epothilone B

474 → A or B → 475

A. (*S*)-BINOL, Ti(O*i*-Pr)$_4$, $CH_2{=}CHCH_2SnBu_3$, CH_2Cl_2, -20 °C, 70 h, 60%, >95% ee
B. [(-)-Ipc]$_2BCH_2CH{=}CH_2$ (**195**), Et_2O, -100 °C, then 3 N NaOH, 30% H_2O_2, 83%, >95% ee

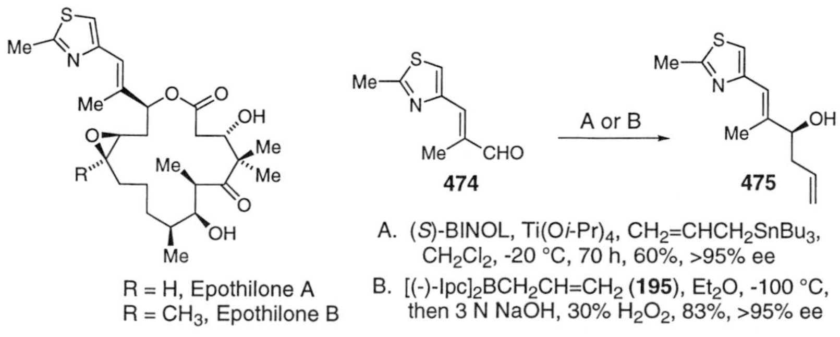

Scheme 11-38.

Using Keck's original catalytic allylation procedure, Danishefsky and co-workers converted aldehyde **474** to the homoallylic alcohol **475** (conditions A, Scheme 11-38, 60% yield, >95% ee) used in their total synthesis of epothilones A and B [314]. Asymmetric allylation with a stoichiometric amount of Brown's reagent, $[(-)\text{-Ipc}]_2$BAll (**195**), however, was higher yielding and required a shorter reaction time (conditions B, Scheme 11-38, 83% yield, >95% ee).

The fact that several laboratories have already applied the enantioselective catalytic allylation reaction to the synthesis of complex natural products illustrates the eagerness with which the synthetic community has welcomed this methodology. It is hoped that further efforts to find conditions that promote high enantio- and diastereoselectivity and low catalyst loading for a variety of aldehyde substrates will continue in this promising new direction of the allylation reaction.

References

1. Hoffmann, R.W. *Angew. Chem. Int. Ed. Engl.* **1982**, *21*, 555–642.
2. Yamamoto, Y.; Maruyama, K. *Heterocycles* **1982**, *18*, 357–386.
3. Roush, W.R. in *Comprehensive Organic Synthesis*, Trost, B.A.; Fleming, I. Eds.; Pergamon Press, Oxford, 1991; Vol. 2, pp 1–53.
4. Yamamoto, Y.; Asao, N. *Chem. Rev.* **1993**, *93*, 2207–2293.
5. Fleming, I. In *Comprehensive Organic Synthesis*, Trost, B.M.; Fleming, I. Eds.; Pergamon Press, Oxford, 1991; Vol. 2, pp 563–593.
6. Thomas, E.J. In *Stereocontrolled Organic Synthesis*, Trost, B.M. Ed.; Blackwell Scientific Publications: Cambridge, 1994, pp 235–258.
7. Yamamoto, H. In *Comprehensive Organic Synthesis*, Trost, B.M.; Fleming, I. Eds.; Pergamon Press, Oxford, 1991; Vol. 2, pp 81–98.
8. Sakurai, H. *Pure & Appl. Chem.* **1982**, *54*, 1–22.
9. Fleming, I.; Dunoguès, J.; Smithers, R. *Org. React.* **1989**, *37*, 57–193.
10. Majetich, G. *Organic Synthesis: Theory and Application* **1989**, *1*, 173.
11. Chan, T.H.; Wang, D. *Chem. Rev.* **1992**, *92*, 995–1006.
12. Masse, C.E.; Panek, J.S. *Chem. Rev.* **1995**, *95*, 1293–1316.
13. Langkopf, E.; Schinzer, D. *Chem. Rev.* **1995**, *95*, 1375–1408.
14. Thomas, E.J. In *Stereoselective Synthesis*, Helmchen, G., Hoffmann, R.W., Mulzer, J.; Schaumann, E. Eds.; Thieme, Stuttgart New York, 1995; Vol. E 21b, pp 1491–1507.
15. Fleming, I.; Barbero, A.; Walter, D. *Chem. Rev.* **1997**, *97*, 2063–2192.
16. Marshall, J.A. *Chem. Rev.* **1996**, *96*, 31–47.
17. Thomas, E.J. *J. Chem. Soc., Chem. Comm.* **1997**, 411–418.
18. Cintas, P. *Synthesis* **1992**, 248–257.
19. Wessjohann, L.A.; Scheid, G. *Synthesis* **1999**, 1–36.
20. Hoppe, D. In *Stereoselective Synthesis*, Helmchen, G., Hoffmann, R.W., Mulzer, J.; Schaumann, E. Eds.; Thieme, Stuttgart New York, 1995; Vol. E 21b, pp 1584–1595.
21. Duthaler, R.O.; Hafner, A. *Chem. Rev.* **1992**, *92*, 807–832.
22. Duthaler, R.O.; Hafner, A.; Alsters, P.L.; Rothe-Streit, P.; Rihs, G. *Pure & Appl. Chem.* **1992**, *64*, 1897–1910.
23. Hoppe, D. in *Stereoselective Synthesis*, Helmchen, G., Hoffmann, R.W., Mulzer, J.; Schaumann, E. Eds.; Thieme, Stuttgart New York, 1995; Vol. E 21b, pp 1551–1583.
24. Hoffmann, R.W. In *Stereocontrolled Organic Synthesis*, Trost, B. M. Ed.; Blackwell Scientific Publications, Cambridge, 1994, pp 259–274.

25. Roush, W.R. In *Trends in Synthetic Carbohydrate Chemistry*, Horton, D., Hawkins, L.D.; McGarvey, G. Ed.; ACS:,Washington, 1989; Vol. American Chemical Society Symposium Series Vol. 386, pp 242–277.
26. Roush, W.R. in *Stereoselective Synthesis*, Helmchen, G., Hoffmann, R.W., Mulzer, J. and and Schaumann, E. Ed.; Thieme Stuttgart: New York, 1995; Vol. E 21b, pp 1410–1486.
27. Denmark, S.E.; Weber, E.J. *Helv. Chim. Acta* **1983**, *66*, 1655–1660.
28. Hoffmann, R.W.; Zeiss, H.-J. *Angew. Chem. Int. Ed. Engl.* **1979**, *18*, 306–307.
29. Kira, M.; Kobayashi, M.; Sakurai, H. *Tetrahedron Lett.* **1987**, *28*, 4081–4084.
30. Kobayashi, S.; Nishio, K. *J. Org. Chem.* **1994**, *59*, 6620–6628.
31. Kira, M.; Sato, K.; Sakurai, H. *J. Am. Chem. Soc.* **1988**, *110*, 4599–4602.
32. Hosomi, A.; Kohra, S.; Ogata, K.; Yanagi, T.; Tominaga, Y. *J. Org. Chem.* **1990**, *55*, 2415–2420.
33. Servens, C.; Pereyre, M. *J. Organomet. Chem.* **1972**, *35*, C20–C22.
34. Roush, W.R. In *Stereoselective Synthesis*, Helmchen, G., Hoffmann, R.W., Mulzer, J.; Schaumann, E. Eds.; Thieme, Stuttgart New York, 1995; Vol. E 21b, pp 1487–1490.
35. Nowotny, S.; Tucker, C.E.; Jubert, C.; Knochel, P. *J. Org. Chem.* **1995**, *60*, 2762–2772.
36. Hoffmann, R.W.; Zeiss, H.-J. *J. Org. Chem.* **1981**, *46*, 1309–1314.
37. Roush, W.R.; Ratz, A.M. *Unpublished results* **1989–90**.
38. Okude, Y.; Hirano, S.; Hiyama, T.; Nozaki, H. *J. Am. Chem. Soc.* **1977**, *99*, 3179–3181.
39. Hiyama, T.; Kimura, K.; Nozaki, H. *Tetrahedron Lett.* **1981**, *22*, 1037–1040.
40. Hiyama, T.; Okuda, Y.; Kimura, K.; Nozaki, H. *Bull. Chem. Soc. Jpn.* **1982**, *55*, 561–568.
41. Sato, F.; Iijima, S.; Sato, M. *Tetrahedron Lett.* **1981**, *22*, 243–246.
42. Yamamoto, Y.; Maruyama, K. *Tetrahedron Lett.* **1981**, *30*, 2895–2898.
43. Mashima, K.; Yasuda, H.; Asami, K.; Nakamura, A. *Chem. Lett.* **1983**, 219–222.
44. Buse, C.T.; Heathcock, C.H. *Tetrahedron Lett.* **1978**, *22*, 1685–1687.
45. Hosomi, A.; Sakurai, H. *Tetrahedron Lett.* **1976**, *16*, 1295–1298.
46. Hayashi, T.; Kabeta, K.; Hamachi, I.; Kumada, M. *Tetrahedron Lett.* **1983**, *24*, 2865–2868.
47. Yamamoto, Y.; Yatagai, H.; Naruta, Y.; Maruyama, K. *J. Am. Chem. Soc.* **1980**, *102*, 7107–7109.
48. Yamamoto, Y.; Maruyama, K. *J. Organomet. Chem.* **1985**, *284*, C45–C48.
49. Reetz, M.T.; Sauerwald, M. *J. Org. Chem.* **1984**, *49*, 2292–2293.
50. Denmark, S.E.; Weber, E.J.; Wilson, T.M.; Willson, T.M. *Tetrahedron* **1989**, *45*, 1053–1065.
51. Denmark, S.E.; Hosoi, S. *J. Org. Chem.* **1994**, *59*, 5133–5135.
52. Denmark, S.E.; Almstead, N.G. *J. Org. Chem.* **1994**, *59*, 5130–5132.
53. Keck, G.E.; Savin, K.A.; Cressman, E.N.K.; Abbott, D.E. *J. Org. Chem.* **1994**, *59*, 7889–7896.
54. Keck, G.E.; Dougherty, S.M.; Savin, K.A. *J. Am. Chem. Soc.* **1995**, *117*, 6210–6223.
55. Mulzner, J.; Bruntrup, G.; Finke, J.; Zippel, M. *J. Am. Chem. Soc.* **1979**, *101*, 7723–7725.
56. Seebach, D.; Golinski, J. *Helv. Chim. Acta* **1981**, *64*, 1413–1423.
57. Mikami, K.; Kawamoto, K.; Loh, T.-P.; Nakai, T. *J. Chem. Soc. Chem. Comm.* **1990**, 1161–1163.
58. Panek, J.S.; Cirillo, P.F. *J. Org. Chem.* **1993**, *58*, 999–1002.
59. Hoffmann, R.W.; Weidmann, U. *Chem. Ber.* **1985**, *118*, 3966–3979.
60. Lewis, M.D.; Kishi, Y. *Tetrahedron Lett.* **1982**, *23*, 2343–2346.
61. Nagaoka, H.; Kishi, Y. *Tetrahedron* **1981**, *37*, 3873–3888.
62. Chérest, M.; Felkin, H.; Prudent, N. *Tetrahedron Lett.* **1968**, 2199–2208.
63. Anh, N.T.; Eisenstein, O. *Nouv. J. Chim.* **1977**, *1*, 61–70.
64. Anh, N.T. *Top. Curr. Chem.* **1980**, *88*, 145–162.
65. Roush, W.R. *J. Org. Chem.* **1991**, *56*, 4151–4157.
66. Roush, W.R.; Adam, M.A.; Harris, D.J. *J. Org. Chem.* **1985**, *50*, 2000–2003.
67. Roush, W.R.; Adam, M.A.; Walts, A.E.; Harris, D.J. *J. Am. Chem. Soc.* **1986**, *108*, 3422–3434.
68. Mulzer, J.; Schulze, T.; Streker, A.; Denzer, W. *J. Org. Chem.* **1988**, *53*, 4098–4103.
69. Brinkmann, H.; Hoffmann, R.W. *Chem. Ber.* **1990**, *123*, 2395–2401.
70. Wuts, P.G.M.; Bigelow, S.S. *J. Org. Chem.* **1988**, *53*, 5023–5034.
71. Cornforth, J.W.; Cornforth, R.H.; Mathew, K.K. *J. Chem. Soc.* **1959**, 112–127.
72. Reetz, M.T.; Jung, A. *J. Am. Chem. Soc.* **1983**, *105*, 4833–4835.
73. Sato, K.; Kira, M.; Sakurai, H. *J. Am. Chem. Soc.* **1989**, *111*, 6429–6431.
74. Wang, Z.; Meng, X.-J.; Kabalka, G.W. *Tetrahedron Lett.* **1991**, *32*, 1945–1948.
75. Kabalka, G.W.; Narayana, C.; Reddy, N.K. *Tetrahedron Lett.* **1996**, *37*, 2181–2184.
76. Brzezinski, L.J.; Leahy, J.W. *Tetrahedron Lett.* **1998**, *39*, 2039–2042.

77. Chemler, S.R.; Roush, W.R. *J. Org. Chem.* **1998**, *63*, 3800–3801.
78. Chemler, S.R.; Roush, W.R. *Tetrahedron Lett.* **1999**, *40*, 4643–4647.
79. Mukaiyama, T.; Harada, T.; Shoda, S.-i. *Chem. Lett.* **1980**, 1507–1510.
80. Roush, W.R.; Michaelides, M.R.; Tai, D.F.; Lesur, B.M.; Chong, W.K.M.; Harris, D.J. *J. Am. Chem. Soc.* **1989**, *111*, 2984–2995.
81. Danishefsky, S.J.; Selnick, H.G.; Zelle, R.E.; DeNinno, M.P. *J. Am. Chem. Soc.* **1988**, *110*, 4368–4378.
82. Hoffmann, R.W.; Dahmann, G.; Andersen, M.W. *Synthesis* **1994**, 629–638.
83. Keck, G.E.; Abbott, D.E. *Tetrahedron Lett.* **1984**, *25*, 1883–1886.
84. Fisher, M.J.; Myers, C.D.; Joglar, J.; Chen, S.-H.; Danishefsky, S. J. *J. Org. Chem.* **1991**, *56*, 5826–5834.
85. Keck, G.E.; Savin, K.A.; Weglarz, M.A.; Cressman, E.N.K. *Tetrahedron Lett.* **1996**, *37*, 3291–3294.
86. Keck, G.E.; Park, M.; Krishnamurthy, D. *J. Org. Chem.* **1993**, *58*, 3787–3788.
87. Panek, J.S.; Yang, M.; Solomon, J. *Tetrahedron Lett.* **1995**, *36*, 1003–1006.
88. Roush, W.R.; Marron, T.G.; Pfeifer, L.A. *J. Org. Chem.* **1997**, *62*, 474–478.
89. Evans, D.A.; Allison, B.D.; Yang, M.G. *Tetrahedron Lett.* **1999**, *40*, 4457–4460.
90. Keck, G.E.; Abbott, D.E.; Wiley, M.R. *Tetrahedron Lett.* **1987**, *28*, 139–142.
91. Keck, G.E.; Boden, E.P. *Tetrahedron Lett.* **1984**, *25*, 265–268.
92. Reetz, M.T.; Kesseler, K.; Jung, A. *Tetrahedron Lett.* **1984**, *25*, 729–732.
93. Evans, D.A.; Dart, M.J.; Duffy, J.L.; Yang, M.G. *J. Am. Chem. Soc.* **1996**, *118*, 4322–4343.
94. Micalizio, G.C.; Roush, W.R. *Tetrahedron Lett.* **1999**, *40*, 3351–3354.
95. Semmelhack, M.F.; Wu, E.S.C. *J. Am. Chem. Soc.* **1976**, *98*, 3384–3386.
96. Nishitani, K.; Yamakawa, K. *Tetrahedron Lett.* **1987**, *28*, 655–658.
97. Kuroda, C.; Shimizu, S.; Satoh, J.Y. *J. Chem. Soc., Chem. Commun.* **1987**, 286–288.
98. Kuroda, C.; Shimizu, S.; Satoh, J.Y. *J. Chem. Soc. Perkin Trans. 1* **1990**, 519–524.
99. Semmelhack, M.F.; Yamashita, A.; Tomesch, J.C.; Hirotsu, K. *J. Am. Chem. Soc.* **1978**, *100*, 5565–5567.
100. Marshall, J.A.; Crooks, S.L.; DeHoff, B.S. *J. Org. Chem.* **1988**, *53*, 1616–1623.
101. Marshall, J.A.; Gung, W.Y. *Tetrahedron Lett.* **1988**, *29*, 1657–1660.
102. Still, W.C.; Mobilio, D. *J. Org. Chem.* **1983**, *48*, 4785–4786.
103. Paquette, L.A.; Astles, P.C. *J. Org. Chem.* **1993**, *58*, 165–169.
104. Krüger, J.; Hoffmann, R.W. *J. Am. Chem. Soc.* **1997**, *119*, 7499–7504.
105. Kadota, I.; Jung-Youl, P.; Koumura, N.; Pollaud, G.; Matsukawa, Y.; Yamamoto, Y. *Tetrahedron Lett.* **1995**, *36*, 5777–5780.
106. Kadota, I.; Kawada, M.; Gevorgyan, V.; Yamamoto, Y. *J. Org. Chem.* **1997**, *62*, 7439–7446.
107. Alvarez, E.; Díaz, M.T.; Pérez, R.; Ravelo, J.L.; Regueiro, A.; Vera, J.A.; Zurita, D.; Martín, J.D. *J. Org. Chem.* **1994**, *59*, 2848–2876.
108. Yamamoto, Y.; Yamada, J.-I.; Kadota, I. *Tetrahedron Lett.* **1991**, *32*, 7069–7072.
109. Hoffmann, R.W.; Münster, I. *Tetrahedron Lett.* **1995**, *36*, 1431–1434.
110. Burton, J.W.; Clark, J.S.; Derrer, S.; Stork, T.C.; Bendall, J.G.; Holmes, A.B. *J. Am. Chem. Soc.* **1997**, *119*, 7483–7498.
111. Masamune, S.; Choy, W.; Petersen, J.S.; Sita, L.R. *Angew. Chem. Int. Ed. Engl.* **1985**, *24*, 1–30.
112. Brown, H.C.; Jadhav, P.K. *J. Am. Chem. Soc.* **1983**, *105*, 2092–2093.
113. Racherla, U.S.; Brown, H.C. *J. Org. Chem.* **1991**, *56*, 401–404.
114. Brown, H.C.; Bhat, K.S.; Randad, R.S. *J. Org. Chem.* **1987**, *52*, 319–320.
115. Brown, H.C.; Bhat, K.S.; Randad, R.S. *J. Org. Chem.* **1989**, *54*, 1570–1576.
116. Roush, W.R.; Walts, A.E.; Hoong, L.K. *J. Am. Chem. Soc.* **1985**, *107*, 8186–8190.
117. Roush, W.R.; Palkowitz, A. D.; Palmer, M.A.J. *J. Org. Chem.* **1987**, *52*, 316–318.
118. Roush, W.R.; Hoong, L.K.; Palmer, M.A.J.; Straub, J.A.; Palkowitz, A.D. *J. Org. Chem.* **1990**, *55*, 4117–4126.
119. Roush, W.R.; Hoong, L.K.; Palmer, M.A.J.; Park, J.C. *J. Org. Chem.* **1990**, *55*, 4110–4117.
120. Roush, W.R.; Grover, P.T. *J. Org. Chem.* **1995**, *60*, 3806–3813.
121. Short, R.P.; Masamune, S. *J. Am. Chem. Soc.* **1989**, *111*, 1892–1894.
122. Hoffmann, R.W.; Herold, T. *Chem. Ber.* **1981**, *114*, 375–383.

123. Hoffmann, R.W.; Helbig, W. *Chem. Ber.* **1981**, *114*, 2802–2807.
124. Hoffmann, R.W.; Zeiss, H.J.; Ladner, W.; Tabche, S. *Chem. Ber.* **1982**, *115*, 2357–2370.
125. Stürmer, R.; Hoffmann, R.W. *Synlett* **1990**, 759–760.
126. Reetz, M.T.; Zierke, T. *Chem. Industry (UK)* **1988**, *20*, 663–664.
127. Corey, E.J.; Yu, C.-M.; Kim, S.-S. *J. Am. Chem. Soc.* **1989**, *111*, 5495–5496.
128. Haruta, R.; Ishiguro, M.; Ikeda, N.; Yamamoto, H. *J. Am. Chem. Soc.* **1982**, *104*, 7667–7668.
129. Ikeda, N.; Arai, I.; Yamamoto, H. *J. Am. Chem. Soc.* **1986**, *108*, 483–486.
130. Corey, E.J.; Yu, C.-M.; Lee, D.-H. *J. Am. Chem. Soc.* **1990**, *112*, 878–879.
131. Hayashi, T.; Konishi, M.; Kumada, M. *J. Org. Chem.* **1983**, *48*, 281–282.
132. Coppi, L.; Mordini, A.; Taddei, M. *Tetrahedron Lett.* **1987**, *28*, 969–972.
133. Riediker, M.; Duthaler, R.O. *Angew. Chem. Int. Ed. Engl.* **1989**, *28*, 494–495.
134. Roder, H.; Helmchen, G.; Peters, E.-M.; Peters, K.; Schnering, H.-G. v. *Angew. Chem. Int. Ed. Engl.* **1984**, *23*, 898–899.
135. Faller, J.W.; Linebarrier, D.L. *J. Am. Chem. Soc.* **1989**, *111*, 1937–1939.
136. Mortlock, S.V.; Thomas, E.J. *Tetrahedron Lett.* **1988**, *29*, 2479–2482.
137. McNeill, A.H.; Thomas, E.J. *Tetrahedron Lett.* **1990**, *31*, 6239–6242.
138. Hoffmann, R.W.; Ditrich, K.; Koster, G.; Sturmer, R. *Chem. Ber.* **1989**, *122*, 1783–1789.
139. Roush, W.R.; Ando, K.; Powers, D.B.; Palkowitz, A.D.; Halterman, R.L. *J. Am. Chem. Soc.* **1990**, *112*, 6339–6348.
140. Brown, H.C.; Bhat, K.S. *J. Am. Chem. Soc.* **1986**, *108*, 293–294.
141. Brown, H.C.; Bhat, K.S.; Randad, R.S. *J. Am. Chem. Soc.* **1987**, *52*, 3702–3704.
142. Brown, H.C.; Bhat, K.S. *J. Am. Chem. Soc.* **1986**, *108*, 5919–5923.
143. Garcia, J.; Kim, B.-M.; Masamune, S. *J. Org. Chem.* **1987**, *52*, 4831–4832.
144. Hayashi, T.; Konishi, M.; Ito, H.; Kumada, M. *J. Am. Chem. Soc.* **1982**, *104*, 4962–4963.
145. Hayashi, T.; Konishi, M.; Kumada, M. *J. Am. Chem. Soc.* **1982**, *104*, 4963–4965.
146. Beresis, R.T.; Solomon, J.S.; Yang, M.G.; Jain, N.F.; Panek, J.S. *Organic Synthesis* **1998**, *75*, 78–88.
147. Jain, N.F.; Cirillo, P.F.; Schaus, J.V.; Panek, J.S. *Tetratrahedron Lett.* **1995**, *36*, 8723–8726.
148. Panek, J.S.; Beresis, R.; Xu, F.; Yang, M. *J. Org. Chem.* **1991**, *56*, 7341–7344.
149. Panek, J.S.; Yang, M.; Xu, F. *J. Org. Chem.* **1992**, *57*, 5790–5792.
150. Jain, N.F.; Cirillo, P.F.; Pelletier, R.; Panek, J.S. *Tetrahedron Lett.* **1995**, *36*, 8727–8730.
151. Jain, N.F.; Takenaka, N.; Panek, J.S. *J. Am. Chem. Soc.* **1996**, *118*, 12475–12476.
152. Marshall, J.A.; Wang, X.-J. *J. Org. Chem.* **1991**, *56*, 3211–3213.
153. Marshall, J.A.; Wang, X.-J. *J. Org. Chem.* **1992**, *57*, 1242–1252.
154. Marshall, J.A.; Xie, S. *J. Org. Chem.* **1995**, *60*, 7230–7237.
155. Hoffmann, R.W.; Dresely, S. *Angew. Chem. Int. Ed. Engl.* **1986**, *25*, 189–190.
156. Hoffmann, R.W.; Dresely, S. *Chem. Ber.* **1989**, *122*, 903–909.
157. Jephcote, V.J.; Pratt, A.J.; Thomas, E.J. *J. Chem. Soc., Chem. Commun.* **1984**, 800–802.
158. Marshall, J.A.; Perkins, J.F.; Wolf, M.A. *J. Org. Chem.* **1995**, *60*, 5556–5559.
159. Marshall, J.A.; Yu, R.H.; Perkins, J.F. *J. Org. Chem.* **1995**, *60*, 5550–5555.
160. Marshall, J.A.; Palovich, M.R. *J. Org. Chem.* **1997**, *62*, 6001–6005.
161. Marshall, J.A.; Perkins, J.F. *J. Org. Chem.* **1994**, *59*, 3509–3511.
162. Faller, J.W.; John, J.A.; Mazzieri, M.R. *Tetrahedron Lett.* **1989**, *30*, 1769–1772.
163. Hoppe, D.; Zschage, O. *Angew. Chem. Int. Ed. Engl.* **1989**, *28*, 69–71.
164. Wuts, P.G.M.; Bigelow, S.S. *J. Chem. Soc., Chem. Comm.* **1984**, 736–737.
165. Brown, H.C.; Jadhav, P.K.; Bhat, K.S. *J. Am. Chem. Soc.* **1988**, *110*, 1535–1538.
166. Marshall, J.A.; Gung, W.Y. *Tetrahedron Lett.* **1989**, *30*, 2183–2186.
167. Marshall, J.A.; Welmaker, G.S.; Gung, B.W. *J. Am. Chem. Soc.* **1991**, *113*, 647–656.
168. Marshall, J.A.; Jablonowski, J.A.; Luke, G.P. *J. Org. Chem.* **1994**, *59*, 7825–7832.
169. Yamamoto, Y.; Kobayashi, K.; Okano, H.; Kadota, I. *J. Org. Chem.* **1992**, *57*, 7003–7005.
170. Roush, W.R.; VanNieuwenhze, M.S. *J. Am. Chem. Soc.* **1994**, *116*, 8536–8543.
171. Marshall, J.A.; Hinkle, K.W. *J. Org. Chem.* **1995**, *60*, 1920–1921.
172. Yamamoto, Y.; Miyairi, T.; Ohmura, T.; Miyaura, N. *J. Org. Chem.* **1999**, *64*, 296–298.
173. Barrett, A.G.M.; Malecha, J.W. *J. Org. Chem.* **1991**, *56*, 5243–5245.
174. Brown, H.C.; Narla, G. *J. Org. Chem.* **1995**, *60*, 4686–4687.
175. Hunt, J.A.; Roush, W.R. *J. Org. Chem.* **1997**, *62*, 1112–1124.

176. Roush, W.R.; Grover, P.T.; Lin, X. *Tetrahedron Lett.* **1990**, *31*, 7563–7566.
177. Roush, W.R.; Grover, P.T. *Tetrahedron* **1992**, *48*, 1981–1998.
178. Roush, W.R.; Coffey, D.S.; Madar, D.J. *J. Am. Chem. Soc.* **1997**, *119*, 11331–11332.
179. Roush, W.R.; Wada, C.K. *J. Am. Chem. Soc.* **1994**, *116*, 2151–2152.
180. Roush, W.R.; Bannister, T.D. *Tetrahedron Lett.* **1992**, *33*, 3587–3590.
181. Roush, W.R.; Koyama, K.; Curtin, M.L.; Moriarty, K.J. *J. Am. Chem. Soc.* **1996**, *118*, 7502–7512.
182. Roush, W.R.; Brown, B.B. *J. Am. Chem. Soc.* **1993**, *115*, 2268–2278.
183. Roush, W.R.; Straub, J.A.; VanNieuwenhze, M.S. *J. Org. Chem.* **1991**, *56*, 1636–1648.
184. Roush, W.R.; Hunt, J.A. *J. Org. Chem.* **1995**, *60*, 798–806.
185. Roush, W.R.; Reilly, M.L.; Koyama, K.; Brown, B.B. *J. Org. Chem.* **1997**, *62*, 8708–8721.
186. Scheidt, K.A.; Tasaka, A.; Bannister, T.D.; Wendt, M.D.; Roush, W.R. *Angew. Chem. Int. Ed. Engl.* **1999**, *38*, 1652–1655.
187. Roush, W.R.; Lane, G.C. *Org. Lett.* **1999**, *1*, 95–98.
188. Lin, G.-Q.; Xu, W.-C. *Tetrahedron* **1996**, *52*, 5907–5912.
189. May, S.A.; Grieco, P.A. *Chem. Commun.* **1998**, 1597–1598.
190. White, J.D.; Tiller, T.; Ohba, Y.; Porter, W.J.; Jackson, R.W.; Wang, S.; Hanselmann, R. *Chem. Commun.* **1998**, 79–80.
191. Smith, A.B.; Friestad, G.K.; Duan, J.J.-W.; Barbosa, J.; Hull, K.G.; Iwashima, M.; Qiu, Y.; Spoors, P.G.; Bertounesque, E.; Salvatore, B.A. *J. Org. Chem.* **1998**, *63*, 7596–7597.
192. Danishefsky, S.J.; Armistead, D.M.; Wincott, F.E.; Selnick, H.G.; Hungate, R. *J. Am. Chem. Soc.* **1987**, *109*, 8117–8119.
193. Kang, S.H.; Kim, C.M. *Synlett* **1996**, 515–516.
194. Zheng, W.; DeMattei, J.A.; Wu, J.-P.; Duan, J.J.-W.; Cook, L.R.; Oinuma, H.; Kishi, Y. *J. Am. Chem. Soc.* **1996**, *118*, 7946–7968.
195. Burke, S.D.; Hong, J.; Lennox, J.R.; Mongin, A.P. *J. Org. Chem.* **1998**, *63*, 6952–6967.
196. Williams, D.R.; Rojas, C.M.; Bogen, S.L. *J. Org. Chem.* **1999**, *64*, 736–746.
197. Makino, K.; Kimura, K.-I.; Nakajima, N.; Hashimoto, S.-I.; Yonemitsu, O. *Tetrahedron Lett.* **1996**, *37*, 9073–9076.
198. Fujita, K.; Schlosser, M. *Helv. Chim. Acta* **1982**, *65*, 1258–1263.
199. Roush, W.R.; Park, J.C. *J. Org. Chem.* **1990**, *55*, 1143–1144.
200. Roush, W.R.; Park, J.C. *Tetrahedron Lett.* **1991**, *32*, 6285–6288.
201. Roush, W.R.; Wada, C.K. *Tetrahedron Lett.* **1994**, *35*, 7347–7350.
202. Corey, E.J.; Rohde, J.J. *Tetrahedron Lett.* **1997**, *38*, 37–40.
203. Roush, W.R.; Palkowitz, A.D.; Ando, K. *J. Am. Chem. Soc.* **1990**, *112*, 6348–6359.
204. Roush, W.R.; Palkowitz, A.D. *J. Org. Chem.* **1989**, *54*, 3009–3011.
205. Roush, W.R.; Coffey, D.S.; Madar, D.; Palkowitz, A.D. *J. Braz. Chem. Soc.* **1996**, *7*, 327–334.
206. Evans, D.A.; Sjogren, E.B.; Bartroli, J.; Dow, R.L. *Tetrahedron Lett.* **1986,** *27*, 4957–4960.
207. Marron, T.G.; Roush, W.R. *Tetrahedron Lett.* **1995**, *36*, 1581–1584.
208. Roush, W.R.; Marron, T.G. *Tetrahedron Lett.* **1993**, *34*, 5421–5424.
209. Marron, T.G. *Ph.D. Thesis, Indiana University* **1995**.
210. Brown, H.C.; Ramachandran, P.V. *J. Organomet. Chem.* **1995**, *500*, 1–19.
211. Brown, H.C.; Ramachandran, P.V. In *Advances in Asymmetric Synthesis*, Hassner, A. Ed.; JAI Press Inc., Greenwich, London, 1995; Vol. 1.
212. Hanessian, S.; Tehim, A.; Chen, P. *J. Org. Chem.* **1993**, *58*, 7768–7781.
213. Kende, A.S.; Koch, K.; Dorey, G.; Kaldor, I.; Liu, K. *J. Am. Chem. Soc.* **1993**, *115*, 9842–9843.
214. Rychnovsky, S.D.; Hoye, R.C. *J. Am. Chem. Soc.* **1994**, *116*, 1753–1765.
215. Patron, A.P.; Richter, P.K.; Tomaszewski, M.J.; Miller, R.A.; Nicolaou, K.C. *J. Chem. Soc., Chem. Commun.* **1994**, 1147–1150.
216. Roush, W.R.; Follows, B.C. *Tetrahedron Lett.* **1994**, *35*, 4935–4938.
217. Bratz, M.; Bullock, W.H.; Overman, L.E.; Takemoto, T. *J. Am. Chem. Soc.* **1995**, *117*, 5958–5966.
218. Dinh, T.Q.; Armstrong, R.W. *J. Org. Chem.* **1995**, *60*, 8118–8119.
219. Steel, P.G.; Thomas, E.J. *J. Chem. Soc., Perkin Trans. 1* **1997**, 371–380.

220. Nagamitsu, T.; Sunazuka, T.; Tanaka, H.; Omura, S.; Sprengeler, P.A.; Smith, A.B. *J. Am. Chem. Soc.* **1996**, *118*, 3584–3590.
221. Maurer, K.W.; Armstrong, R.W. *J. Org. Chem.* **1996**, *61*, 3106–3116.
222. White, J.D.; Kim, T.-S.; Nambu, M. *J. Am. Chem. Soc.* **1995**, *117*, 5612–5613.
223. Andrus, M.B.; Argade, A.B. *Tetrahedron Lett.* **1996**, *37*, 5049–5052.
224. Jadhav, P.K.; Man, H.-W. *Tetrahedron Lett.* **1996**, *37*, 1153–1156.
225. Ramachandran, P.V.; Chen, G.-M.; Brown, H.C. *Tetrahedron Lett.* **1997**, *38*, 2417–2420.
226. Menager, E.; Merifield, E.; Smallridge, M.; Thomas, E.J. *Tetrahedron* **1997**, *53*, 9377–9392.
227. Andrus, M.B.; Lepore, S.D.; Turner, T.M. *J. Am. Chem. Soc.* **1997**, *119*, 12159–12169.
228. McRae, K.J.; Rizzacasa, M.A. *J. Org. Chem.* **1997**, *62*, 1196–1197.
229. Jyojima, T.; Katohno, M.; Miyamoto, N.; Nakata, M.; Matsumura, S.; Toshima, K. *Tetrahedron Lett.* **1998**, *39*, 6003–6006.
230. Paterson, I.; Yeung, K.-S.; Watson, C.; Ward, R.A.; Wallace, P.A. *Tetrahedron* **1998**, *54*, 11935–11954.
231. Scarlato, G.R.; DeMattei, J.A.; Chong, L.S.; Ogawa, A.K.; Lin, M.R.; Armstrong, R.W. *J. Org. Chem.* **1996**, *61*, 6139–6152.
232. Ogawa, A.K.; Armstrong, R.W. *J. Am. Chem. Soc.* **1998**, *120*, 12435–12442.
233. Coleman, R.S.; Kong, J.-S. *J. Am. Chem. Soc.* **1998**, *120*, 3538–3539.
234. Barrett, A.G.M.; Bennett, A.J.; Menzer, S.; Smith, M.L.; White, A.J.P.; Williams, D.J. *J. Org. Chem.* **1999**, *64*, 162–171.
235. Boekman, R.K.; Charette, A.B.; Asberom, T.; Johnston, B.H. *J. Am. Chem. Soc.* **1987**, *109*, 7553–7555.
236. Mancuso, A.J.; Swern, D. *Synthesis* **1981**, 165–185.
237. Hoffmann, R.W.; Ladner, W.; Ditrich, K. *Liebigs Ann. Chem.* **1989**, 883–889.
238. Hoffmann, R.W.; Ditrich, K. *Liebigs Ann. Chem.* **1990**, 23–29.
239. Andersen, M.W.; Hildebrandt, B.; Hoffmann, R.W. *Angew. Chem. Int. Ed. Engl.* **1991**, *30*, 97–99.
240. Andersen, M.W.; Hildebrandt, B.; Dahmann, G.; Hoffmann, R.W. *Chem. Ber.* **1991**, *124*, 2127–2139.
241. Hoffmann, R.W.; Schlapbach, A. *Tetrahedron* **1992**, *48*, 1959–1968.
242. Stürmer, R.; Ritter, K.; Hoffmann, R.W. *Angew. Chem. Int. Ed. Engl.* **1993**, *32*, 101–103.
243. Stürmer, R.; Hoffmann, R.W. *Chem. Ber.* **1994**, *127*, 2519–2526.
244. Hoffmann, R.W.; Rolle, U. *Tetrahedron Lett.* **1994**, *35*, 4751–4754.
245. Hoffmann, R.W.; Rolle, U.; Göttlich, R. *Liebigs Ann.* **1996**, 1717–1724.
246. Matteson, D.S.; Majumdar, D. *J. Am. Chem. Soc.* **1980**, *102*, 7588–7590.
247. Matteson, D.S.; Kandil, A.A. *Tetrahedron Lett.* **1986**, *27*, 3831–3834.
248. Matteson, D.S. *Acc. Chem. Res.* **1988**, *21*, 294–300.
249. Matteson, D.S. *Synthesis* **1986**, 973–985.
250. Aldersen, M.; Hildebrandt, B.; Koster, G.; Hoffmann, R.W. *Chem. Ber.* **1989**, *122*, 1777–1782.
251. Kochetkov, N.K.; Sviridov, A.F.; Ermolenko, M.S.; Yashunsky, D.V.; Borodkin, V.S. *Tetrahedron* **1989**, *45*, 5109–5136.
252. Nakata, M.; Arai, M.; Tamooka, K.; Ohsawa, N.; Kinoshita, M. *Bull. Chem. Soc. Jpn.* **1989**, *62*, 2618–2635.
253. Gao, Y.; Hanson, R.M.; Klunder, J.M.; Ko, S.Y.; Masamune, H.; Sharpless, K.B. *J. Am. Chem. Soc.* **1987**, *109*, 5765–5780.
254. Corey, E.J.; Imwinkelried, R.; Pikul, S.; Xiang, Y.B. *J. Am. Chem. Soc.* **1989**, *111*, 5493–5495.
255. Williams, D.R.; Brooks, D.A.; Meyer, K.G.; Clark, M.P. *Tetrahedron Lett.* **1998**, *39*, 7251–7254.
256. Williams, D.R.; Clark, M.P.; Berliner, M.A. *Tetrahedron Lett.* **1999**, *40*, 2287–2290.
257. Williams, D.R.; Brooks, D.A.; Berliner, M.A. *J. Am. Chem. Soc.* **1999**, *121*, 4924–4925.
258. Pikul, S.; Corey, E.J. *Org. Syn.* **1992**, *71*, 22–29.
259. Sparks, M.A.; Panek, J.S. *J. Org. Chem.* **1991**, *56*, 3431–3438.
260. Doyle, M.P.; Shanklin, M.S. *Organometallics* **1993**, *12*, 11–12.
261. Panek, J.S.; Yang, M. *J. Am. Chem. Soc.* **1991**, *113*, 6594–6600.
262. Panek, J.S.; Yang, M. *J. Am. Chem. Soc.* **1991**, *113*, 9868–9870.
263. Panek, J.S.; Beresis, R. *J. Org. Chem.* **1993**, *58*, 809–811.

264. Panek, J. S.; Jain, N. F. *J. Org. Chem.* **1993**, *58*, 2345–2348.
265. Panek, J. S.; Jain, N. F. *J. Org. Chem.* **1994**, *59*, 2674–2675.
266. Panek, J. S.; Xu, F.; Rodón, A. C. *J. Am. Chem. Soc.* **1998**, *120*, 4113–4122.
267. Panek, J. S.; Jain, N. F. *J. Org. Chem.* **1998**, *63*, 4572–4573.
268. Masse, C. E.; Yang, M.; Solomon, J.; Panek, J. S. *J. Am. Chem. Soc.* **1998**, *120*, 4123–4134.
269. Panek, J. S.; Beresis, R. T.; Celatka, C. A. *J. Org. Chem.* **1996**, *61*, 6494–6495.
270. Panek, J. S.; Beresis, R. T. *J. Am. Chem. Soc.* **1996**, *61*, 6496–6497.
271. Panek, J. S.; Masse, C. E. *Angew. Chem. Int. Ed. Engl.* **1999**, *38*, 1093–1095.
272. Reetz, M. T.; Hullmann, M.; Massa, W.; Berger, S.; Rademacher, P.; Heymanns, P. *J. Am. Chem. Soc.* **1986**, *108*, 2405–2408.
273. Jain, N. F.; Panek, J. S. *Tetrahedron Lett.* **1997**, *38*, 1345–1348.
274. Jain, N. F.; Panek, J. S. *Tetrahedron Lett.* **1997**, *38*, 1349–1352.
275. Marshall, J. A.; Hinkle, K. W. *J. Org. Chem.* **1996**, *61*, 105–108.
276. Marshall, J. A.; Adams, N. D. *J. Org. Chem.* **1998**, *63*, 3312–3313.
277. Marshall, J. A.; Grant, C. M. *J. Org. Chem.* **1999**, *64*, 696–697.
278. Marshall, J. A.; Johns, B. A. *J. Org. Chem.* **1998**, *63*, 7885–7892.
279. Marshall, J. A.; Lu, Z.-H.; Johns, B. A. *J. Org. Chem.* **1998**, *63*, 817–823.
280. Marshall, J. A.; Palovich, M. R. *J. Org. Chem.* **1998**, *63*, 3701–3705.
281. Marshall, J. A.; Welmaker, G. S. *Tetrahedron Lett.* **1991**, *32*, 2101–2104.
282. Chan, P. C.-M.; Chong, J. M. *J. Org. Chem.* **1988**, *53*, 5584–5586.
283. Bernardi, F.; Epiotis, N. D.; Yates, R. L.; Schlegel, H. B. *J. Am. Chem. Soc.* **1976**, *98*, 2385–2390.
284. Bond, D.; Schleyer, P. v. R. *J. Org. Chem.* **1990**, *55*, 1003–1013.
285. Marshall, J. A.; Welmaker, G. S. *J. Org. Chem.* **1992**, *57*, 7158–7163.
286. Marshall, J. A.; Garofalo, A. W. *J. Org. Chem.* **1996**, *61*, 8732–8738.
287. Marshall, J. A.; Jablonowski, J. A.; Jiang, H. *J. Org. Chem.* **1999**, *64*, 2152–2154.
288. Furuta, K.; Mouri, M.; Yamamoto, H. *Synlett* **1991**, 561–562.
289. Keck, G. E.; Tarbet, K. H.; Geraci, L. S. *J. Am. Chem. Soc.* **1993**, *115*, 8467–8468.
290. Costa, A. L.; Piazza, M. G.; Tagliavini, E.; Trombini, C.; Umani-Ronchi, A. *J. Am. Chem. Soc.* **1993**, *115*, 7001–7002.
291. Bedeschi, P.; Casolari, S.; Costa, A. L.; Tagliavini, E.; Umani-Ronchi, A. *Tetrahedron Lett.* **1995**, *36*, 7897–7900.
292. Gauthier, D. R.; Carreira, E. M. *Angew. Chem. Int. Ed. Engl.* **1996**, *35*, 2363–2365.
293. Yanagisawa, A.; Nakashima, H.; Ishiba, A.; Yamamoto, H. *J. Am. Chem. Soc.* **1996**, *118*, 4723–4724.
294. Marshall, J. A. *Chemtracts-Organic Chemistry* **1996**, *9*, 280–285.
295. Cozzi, P. G.; Tagliavini, E.; Umani-Ronchi, A. *Gazzetta Chimica Italiana* **1997**, *127*, 247–254.
296. Keck, G. E.; Krishnamurthy, D.; Grier, M. C. *J. Org. Chem.* **1993**, *58*, 6543–6544.
297. Keck, G. E.; Krishnamurthy, D. *Org. Syn.* **1998**, *75*, 12–18.
298. Keck, G. E.; Geraci, L. S. *Tetrahedron Lett.* **1993**, *34*, 7827–7828.
299. Yu, C.-M.; Choi, H.-S.; Jung, W.-H.; Lee, S.-S. *Tetrahedron Lett.* **1996**, *37*, 7095–7098.
300. Yu, C.-M.; Choi, H.-S.; Yoon, S.-K.; Jung, W.-H. *Synlett.* **1997**, 889–890.
301. Yu, C.-M.; Choi, H.-S.; Jung, W.-H.; Kim, H.-J.; Shin, J. *Chem. Commun.* **1997**, 761–762.
302. Yanagisawa, A.; Ishiba, A.; Nakashima, H.; Yamamoto, H. *Synlett.* **1997**, 88–90.
303. Keck, G. E.; Krishnamurthy, D.; Chen, X. *Tetrahedron Lett.* **1994**, *35*, 8323–8324.
304. Yu, C.-M.; Yoon, S.-K.; Choi, H.-S.; Baek, K. *Chem. Commun.* **1997**, 763–764.
305. Yu, C.-M.; Yoon, S.-K.; Baek, K.; Lee, J.-Y. *Angew. Chem. Int. Ed. Engl.* **1998**, *37*, 2392–2395.
306. Faller, J. W.; Sams, D. W. I.; Liu, X. *J. Am. Chem. Soc.* **1996**, *118*, 1217–1218.
307. Mikami, K.; Matsukawa, S. *Nature* **1997**, *385*, 613–615.
308. Sato, M.; Sunami, S.; Sugita, Y.; Kaneko, C. *Chem. Pharm. Bull.* **1994**, *42*, 839–845.
309. Ishihara, K.; Mouri, M.; Gao, Q.; Maruyama, T.; Furuta, K.; Yamamoto, H. *J. Am. Chem. Soc.* **1993**, *115*, 11490–11495.
310. Marshall, J. A.; Palovich, M. R. *J. Org. Chem.* **1998**, *63*, 4381–4384.
311. Marshall, J. A.; Liao, J. *J. Org. Chem.* **1998**, *63*, 5962–5970.
312. Evans, P. A.; Murthy, V. S. *Tetrahedron Lett.* **1998**, *39*, 9627–9628.

313. Roush, W.R.; Champoux, J.A.; Peterson, B.C. *Tetrahedron Lett.* **1996**, *37*, 8989–8992.
314. Meng, D.; Bertinato, P.; Balog, A.; Su, D.-S.; Kamenecka, T.; Sorensen, E.J.; Danishefsky, S.J. *J. Am. Chem. Soc.* **1997**, *119*, 10073–10092.

12 Asymmetric Michael-Type Addition Reaction

Kiyoshi Tomioka

12.1 Introduction

The Michael-type addition reaction of a carbonucleophile with an activated olefin constitutes one of the most versatile methodologies for carbon-carbon bond formation [1]. Because of the usefulness of the reaction as well as the product, many approaches to the asymmetric Michael-type addition reactions have been reported, especially using chirally modified olefins [2–8]. However, the approach directed towards the enantioselective Michael-type addition reaction is a developing area. In this Chapter, the recent progress of the enantioselective Michael-type addition reaction of active methylene compounds and also organometallic reagents with achiral activated olefins under the control of an external chiral ligand or chiral catalysts will be summarized [9].

The reaction is between two components, nucleophilic and electrophilic reagents. Asymmetric reaction, therefore, becomes possible when these reagents are chiral. The chiral Michael acceptor has been used as the substrate, which usually has an ester group bearing a chiral auxiliary. The substrate sublimed to a chiral Lewis acid-coordinated carbonyl group, which became the chiral target of ketene silyl acetal. Another modification has been to use the chiral Lewis base which generates a chiral Michael donor. A multifunctional catalyst that is expected to work as a chiral Lewis base as well as simultaneously a chiral Lewis acid is the recent growing focus.

12.2 Reaction of an Active Methylene Compound – Typical and Classic Asymmetric Michael Reaction

The Michael reaction is an addition of an active methylene compound to an activated olefin, indicating that both Lewis base and Lewis acid are necessary for the activation of the Michael donor as well as the Michael acceptor. The recent focus is centered on the multifunctional catalyst which works as both Lewis base and Lewis acid.

The recent progress in the Michael addition reaction of active methylene compounds with α,β-unsaturated carbonyl compounds has been characterized by the use of a catalyst generated from chiral binaphthol (binol). The optically active lanthanum-sodium-binol complex (LSB) **1**, prepared from La(O-*i*-Pr)$_3$, (*R*)-binol, and NaO-*t*-Bu, was effective as an asymmetric catalyst for various Michael reactions of malonate derivatives to give the adducts in high ee (Eq. (12.1)). The basic LSB complex is proposed to act as a Lewis acid as well as a Lewis base to control the direction of a carbonyl function and enhances the reactivity of an enone by forming an activated malonate [10, 11].

La(O-*i*-Pr)$_3$ + 3 (binol) + 3 NaO-*t*-Bu → La-Na-BINOL complex (LSB) **1** (12.1)

cyclohexenone + BnO(CO)CH(CH$_3$)(CO)OBn —LSB (10 mol %), 91%, 92% ee→ 3-substituted cyclohexanone (CO$_2$Bn, CO$_2$Bn)

The first catalytic asymmetric tandem Michael-aldol reactions were also achieved by the Al-Li-binol complex (ALB), which was prepared from LiAlH$_4$ and binol. The ALB catalysts gave the Michael adducts in up to 99% ee (Eq. (12.2)) [12]. Mechanistic and calculation studies on ALB revealed that ALB is a heterobimetallic complex which acts as a multifunctional catalyst.

cyclopentenone + EtO(CO)CH(CH$_3$)(CO)OEt + Ph(CH$_2$)$_2$CHO —ALB (10 mol %), rt, 36 h, 64%, 91% ee→ aldol-Michael adduct (OH, Ph, CO$_2$Et, CO$_2$Et) (12.2)

In the presence of a catalytic amount of [(*R*)-1,1′-bi-2-naphthalenediolato(2)-O,O′]oxotitanium **2**, silyl enol ethers derived from thioesters reacted with cyclopentenone to afford the corresponding Michael adducts in high yields and up to 90% ee (Eq. (12.3)) [6].

cyclopentenone + CH$_2$=C(OSi^{t}BuMe$_2$)SCHPh$_2$ —**2** (Ti=O) (20 mol %), toluene, –78 °C→ (F$^-$), H$_2$O, 75%, 90% ee → 3-(CH$_2$COSCHPh$_2$)cyclopentanone (12.3)

The rubidium salt of L-proline **3** catalyzed the asymmetric Michael reaction of malonate and nitroalkane with enones. High enantiomeric excesses were realized when di(t-butyl) malonate was added to the acyclic (E)-enones in the presence of CsF (Eq. (12.4)) [14, 15].

O + $CH_2(CO_2t\text{-}Bu)_2$ —[N H, CO_2Rb **3** (20 mol %), CsF (20 mol %); $CHCl_3$, rt; 65%, 88% ee]→ O, $CH(CO_2t\text{-}Bu)_2$ (12.4)

The asymmetric Michael reaction of 2-cyanopropionates with vinyl ketones or acrolein in the presence of 0.1 mol% of a rhodium catalyst prepared in situ from $Rh(acac)(CO)_2$ and trans-chelating chiral bisphosphine ligand (S,S)-(R,R)-PhTRAP **4** gave the optically active Michael adducts in 89–97% ee and high yields (Eq. (12.5)) [16–18]. The catalyst was also applicable to the reaction of N-methoxy-N-methyl-2-cyanopropionamide and (1-cyanoethyl)phosphonates with PhTRAP **4** to give the adducts in high enantioselectivities.

R, O (R = H, Me, Ph) + NC, O, N, Me, Me, OMe —[**4**: Me, H, Fe, PPh_2, PPh_2, Fe, Me, H; PhTRAP/Rh (0.1 mol %); benzene, 3 °C; 89–94% ee]→ R, O, O, N, Me, Me, CN, OMe (12.5)

The simple chiral lithium alkoxide **5** was applied to the Michael reaction of methyl phenylacetate with methyl acrylate to give the corresponding adduct in 84% ee (Eq. (12.6)) [19]. The catalytic enantioselective Michael reaction was also effected by the chiral alkoxide **5**.

Ph CO_2Me + CO_2Me —[Me, Ph, H_2N, O–Li **5**; THF, –78 °C, 60 h; 68%, 84% ee]→ Ph, CO_2Me, CO_2Me (12.6)

12.3 Reaction of Organometallic Reagents

12.3.1 Reaction of Organolithium Reagents Using External Chiral Ligands

Due to the high nucleophilicity of the organolithium reagent, a bulky reagent such as dithiane used to be the Michael donor [20]. Recent progress has been made based on the use of a bulky carbonyl masking group for protection. Asymmetric Michael-type addition of organolithium reagents to α,β-unsaturated carbonyl compounds has recently been raised to a useful level by using an external chiral ligand, especially chiral diether **6** and chiral natural diamine, (–)-sparteine **7**. The diether **6** was readily prepared by dimethylation of the chiral stilbene diol, which was prepared by an AD-mix reaction of stilbene in high ee and high yield.

The ligand **6** for the organolithium reagent has been used to perform a high-level catalytic asymmetric Michael-type addition reaction with hindered α,β-unsaturated esters [21, 22] as well as naphthyl esters [23]. The chiral diether **6** shows high efficiency with aryllithium reagents and (–)-sparteine **7** with alkyllithium reagents (Eq. (12.7)). Therefore, the use of these ligands is complementary to realizing high enantioselectivity. The catalytic turnover of **7** is superior to that of **6** [24]. The use of a poorly coordinating solvent such as toluene or ether is essential for high enantioselectivity, because of formation of the tight lithium-ligand-chelated complex as a chiral nucleophile. The ligands **6** and **7** are recoverable for reuse in high yield.

Ph Ph MeO OMe **6** **7** t-Bu Me O O t-Bu OMe + RLi toluene, –78 °C Me CO_2BHA R
R = Ph: **6**, 99%, 84% ee
R = Bu: **7**, 80%, 97% ee (12.7)

(–)-Sparteine **7** is also an excellent chiral ligand for the asymmetric Michael addition of chirally fixed organolithiums (Eq. (12.8)) [25]. The choice of ligand for the lithium cation provides control of 1,2- vs 1,4-addition of organolithium species to cycloalkenones. Furthermore, in these addition reactions, two contiguous stereocenters were constructed with high diastereo- and enantioselectivities.

PMP N Boc Ph Li + **7** toluene-TMSCl –78 °C 82%, 92% ee Ph Boc N PMP (12.8)

Chiral diether **6** has shown the greatest efficiency for the asymmetric Michael-type addition of organolithium reagents to α,β-unsaturated *N*-cyclohexylimines (Eqs. (12.9) and (12.10)) [26]. After hydrolysis, β-substituted aldehydes are obtained with excellent enantioselectivity. The sense of enantiofacial selection has been interpreted through lithium-coordinated complex formation between organolithium, imine and chiral diether **6**. From the favored complex, the R group of organolithium reagent is then transferred to the less hindered face of the double bond of the unsaturated imine.

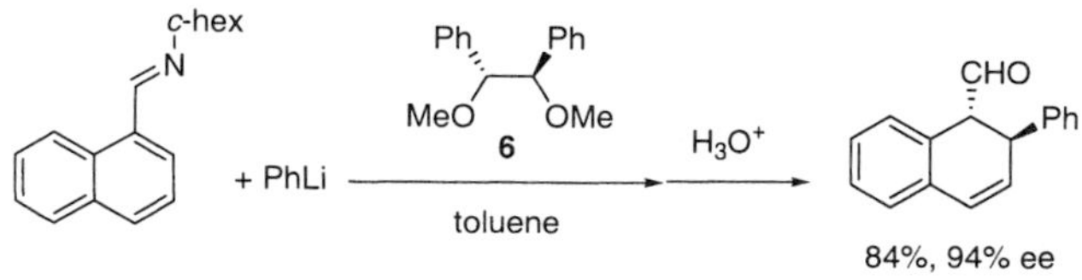

(12.9)

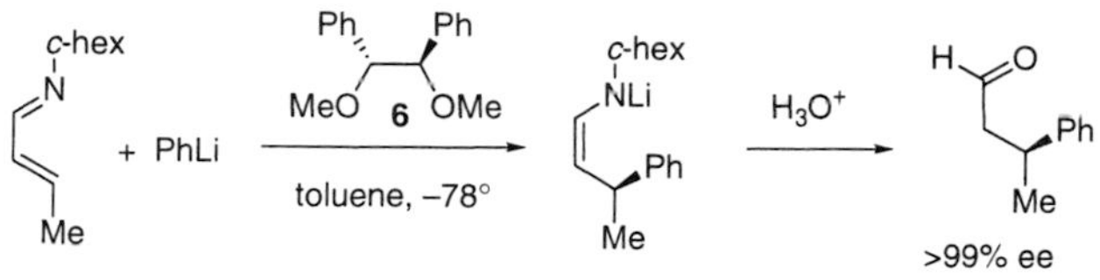

(12.10)

The regioselectivity, that is 1,4- vs 1,2-addition, is directed mainly by a larger LUMO coefficient of the corresponding reaction site [27]. Change of the cyclohexyl group of the imine moiety to an aromatic group leads to a larger coefficient at the imine carbon. The reaction with an imine bearing an aryl group on the imine nitrogen results in selective 1,2-addition, not the Michael type-conjugate addition. Catalytic asymmetric 1,2-addition reaction of the imine with an organolithium reagent was catalyzed by **6** or **7** to provide the corresponding chiral amine in high ee (Eq. (12.11)) [28–30].

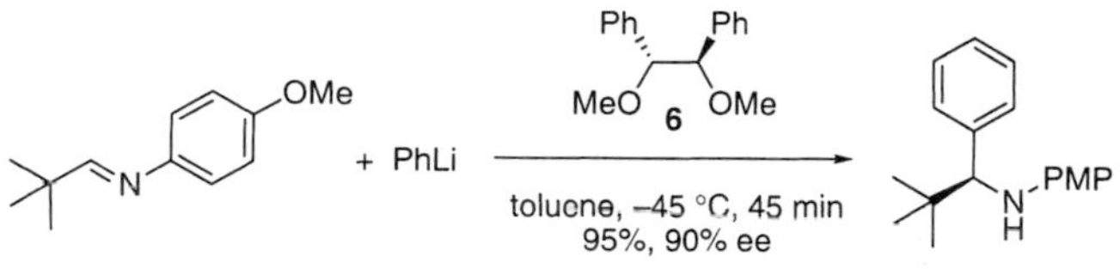

(12.11)

The first prominent catalytic asymmetric Michael-type addition reaction of an organolithium reagent was shown by the reaction of 1-naphthyllithium with 1-fluoro-2-naphthylaldehyde imine in the presence of **6** to afford the binaphthyls in high ee. Only catalytic amounts of **6** (0.05 mol%) effects the reaction to give 82% ee, in which an enantioselective Michael-type addition-elimination mechanism is operative (Eq. (12.12)) [31].

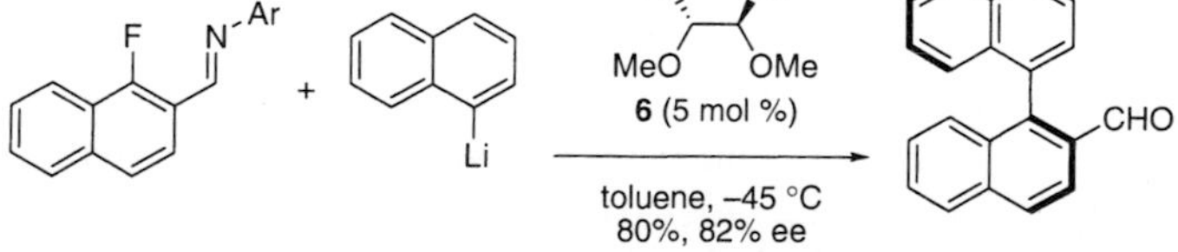

(12.12)

A chromium complex of benzaldehyde imine is also a good substrate for the Michael-type addition of organolithium reagents mediated by a stoichiometric amount of the chiral diether **6** in toluene to give the corresponding product in up to 93% ee (Eq. (12.13)) [32].

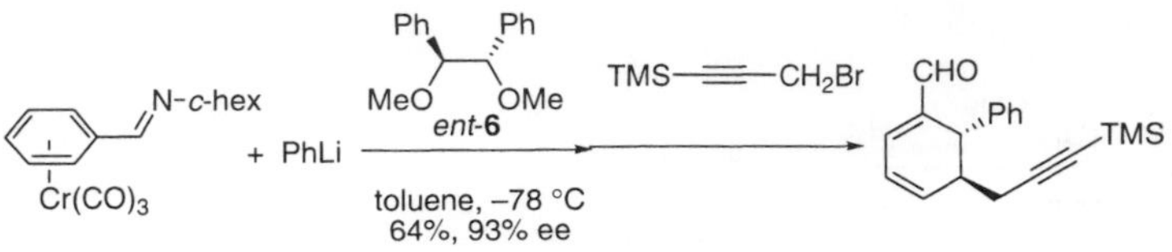

(12.13)

12.3.2 Reaction of Organocopper Reagent

12.3.2.1 Reaction of chiral heteroorganocuprates

The most reliable Michael donor is an organocopper reagent. The organocopper reagents for the asymmetric conjugate addition are classified into two categories: one is the chiral heterocuprate, and the other is organocopper coordinated by the chiral external ligands such as phosphines, sulfides, and oxazolines.

12.3.2.2 Chiral alkoxycuprates

Treatment of copper(I) salt with organolithium or Grignard reagents in the presence of an alcoholate of a chiral alcohol generates the chirally modified alkoxycuprate. Although the enantioselectivity was not high in the early attempts [33, 34], the use of *N*-methylprolinol as a chiral alkoxide source opened up a new route. The asymmetric conjugate addition reaction of methylmagnesium bromide with chalcone provided a relatively good enantioselectivity of 68% [35]. The reaction was optimized to afford the addition product in 88% ee [36].

Breakthrough was brought by using an ephedrine-derived chiral aminoalcohol **8** to effect conjugate addition of organolithiums with over 90% enantioselectivity (Eq. (12.14)) [37]. The relationships between cluster structure and enantiofacial selection are a matter of discussion [38]. The sense of the observed enantiofacial selection was rationalized by the model **9**.

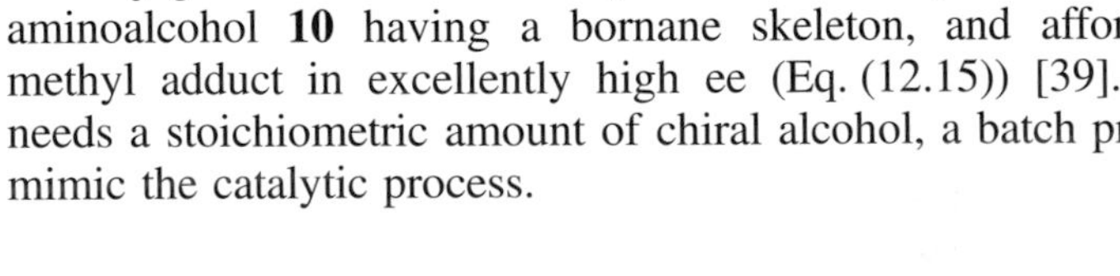

(12.14)

Conjugate addition of methyllithium with cyclic alkenone was mediated by aminoalcohol **10** having a bornane skeleton, and afforded the corresponding methyl adduct in excellently high ee (Eq. (12.15)) [39]. Although the reaction needs a stoichiometric amount of chiral alcohol, a batch process was applicable to mimic the catalytic process.

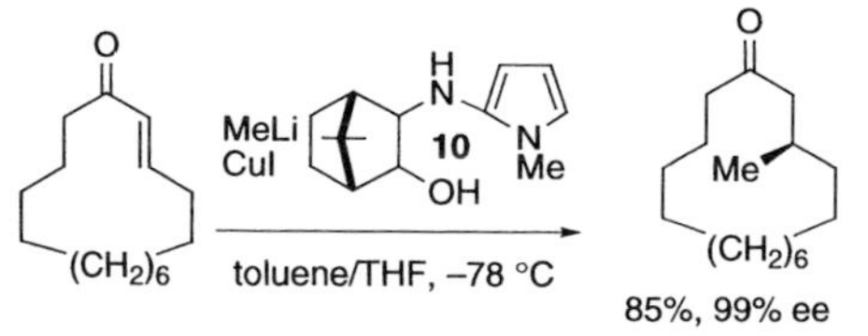

(12.15)

12.3.2.3 Chiral amidocuprates

Treatment of organolithium or Grignard reagents with copper(I) salt in the presence of a lithium or magnesium amide of a chiral amine generates the chirally modified amidocuprates. The relatively high enantioselectivity was first reported by using a chiral amine **11** and copper iodide. The reaction of phenyllithium with cyclohexenone afforded the adduct in 50% ee (Eq. (12.16)) [40]. A more simple prolinol-derived amine (**12**) is interesting, affording either enantiomer by choosing bromide or thiocyanate as a copper source in over 82% ee (Eq. (12.17)) [41]. A linear amine **13** was also designed to effect the addition of organolithium reagent with cycloalkenone to provide the adduct in up to 97% ee (Eq. (12.18)) [42, 43]. Observed enantiofacial selection was interpreted by the model **14** assuming a dimer structure in which the presence of the phenyl group of **13** blocks the bottom face to lead a top face reaction. The dimeric structure was supported by the observation of an amplification effect. Conjugate addition of isopropenyllithium to 2-methylcyclopentenone was mediated by a prolinol-derived amine (**15**) to afford the adduct which was a key intermediate for the asymmetric synthesis of (+)-confertin (Eq. 12.19) [44].

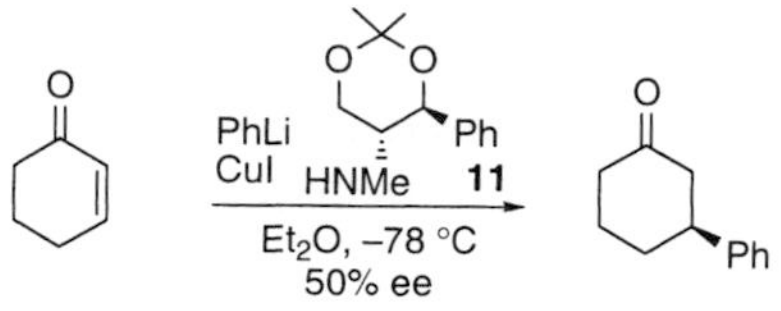

(12.16)

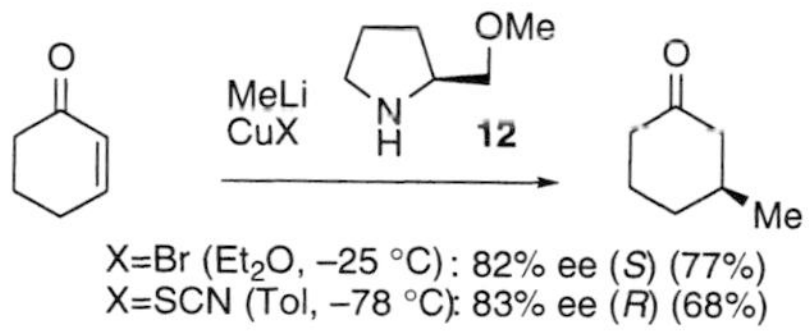

(12.17)

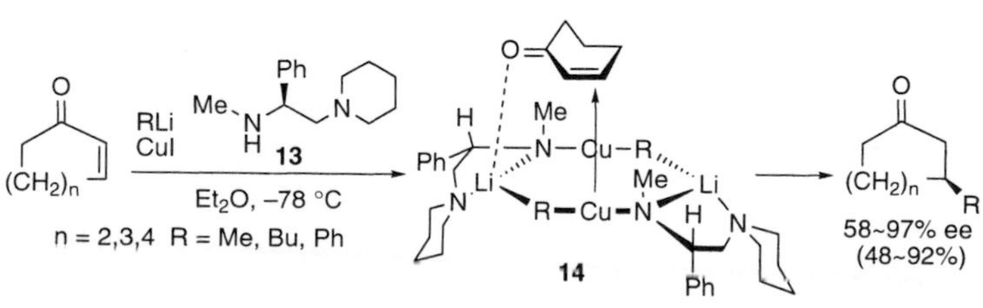

(12.18)

(12.19)

CuSCN, 15; Et_2O, −100 °C; 88%, 88% ee

The first epoch-making catalytic process was developed by using a chiral copper amide derived from an amine **16** in 1990. With 3 mol% of aminotroponeimine (**16**) and copper, butylmagnesium chloride reacted with cyclohexenone to afford the corresponding adduct in 74% ee (Eq. (12.20)) [45].

(12.20)

16 (3 mol %), BuMgCl, $Ph_2{}^tBuSiCl$; HMPA/THF, −78 °C; 57%, 74% ee

12.3.2.4 Chiral thiocuprates

Chirally modified thiocuprates are used mostly in the catalytic process. The success is probably due to the high affinity of the sulfur atom to copper and the great stability of the sulfur-copper bond.

Thiocuprates can be generated from organolithium or Grignard reagents and copper(I) salt in the presence of a lithium or magnesium thiolate of a chiral thiol. It is remarkable that the reaction of Grignard reagents is catalyzed by a catalytic amount of chiral copper thiolates **17–19** to afford the corresponding adduct in high ee (Eqs. (12.21) to (12.23)) [46–48].

(12.21)

MeMgI, **17** (9 mol%); Et_2O, 0 °C; 97%, 76% ee

RMgCl, **18** (5~10 mol%), THF / HMPA, –78 °C

n = 1, 2, 3 R = Bu, iPr — 16~87% ee (30~71%)

(12.22)

19 (4 mol%), BuMgBr CuS, Et_2O, –78 °C, 92%, 60% ee

(12.23)

12.3.2.5 Reaction of homoorganocoppers using external chiral ligands

Generally, a homoorganocopper reagent has two different metals in the cluster. Their chiral modification needs a chiral ligand whose heteroatoms coordinate selectively to copper and other metals. The first approach along this line was reported by Kretchmer, who used chiral natural diamine **7** as a ligand for methylcopper in the reaction with cyclohexenone to afford the adduct in only 6% ee [49].

The breakthrough in the stoichiometric reaction was brought about by Leyendecker in 1983 by using hydroxyprolinol-derived sulfide **20** bearing three coordinating sites, as shown in **21** [50]. The reaction of dimethylcopper lithium with chalcone gave the product in 94% ee (Eq. (12.24)). In 1991, Alexakis introduced chiral phosphines, e.g. **22**, as the ligands in the reaction of the medium order cuprate with cycloalkenones in the presence of lithium bromide to afford the product in 76–95% ee (Eq. (12.25)) [51].

Me_2CuLi, **20**, Et_2O, –50 °C, >90%, 94% ee

(12.24)

$R_5Cu_3Li_2$, LiBr, **22**, THF, –78 - –30 °C

n=1,2,3 R=Et, Bu, $tBuO(CH_2)_4$ — 76~>95% ee (72~83%)

(12.25)

Based on the metal-differentiating coordination concept, proline-derived bidentate amidophosphines **23–25** were developed. The carbonyl oxygen and phosphorus atoms of the ligand selectively coordinate to lithium and copper atoms of an organocopper species, which discriminate the reaction face of the complex as shown in **26**. The reaction of dimethylcopper lithium with chalcone gave the adduct in 84% ee [52]. Enantioselectivity was later improved to 90% with the more bulky amidophosphine **25** based on the model **26** (Eq. (12.26)) [53].

Me_2CuLi; **23**: R = *t*-Bu; **24**: R = Me_2N; −20 °C; Et_2O : 79%, 84% ee (*S*); THF : 72%, 50% ee (*R*)

23 R^1 = H: 84% ee
25 R^1 = Me: 90% ee

26

(12.26)

The metal-selective coordination was supported by NMR analysis [54]. The reaction with cycloalkenone was also highly efficient to give the Michael adduct in up to 95% ee by the reaction of lithium cyanocuprate in the presence of lithium bromide (Eq. (12.27)) [55]. However, the catalytic version of the lithium cyanocuprate was unsuccessful. On the other hand, magnesium cyanocuprate prepared from a Grignard reagent was highly effective, affording the products in up to 98% ee. It is noteworthy that the same chiral ligand gave the products with the reversed absolute configuration by exchanging the lithium with the magnesium [56].

cuprate; Et_2O, −78 °C (HMPA / TMSCl)

$(CH_2)_n$, n = 1, 2, 3

RLi/CuCN/LiBr/**23** (R = Me, Et, Bu): 74~95% ee (*R*) (46~99%)

RMgCl/CuCN/**24** (R = Et, Pr, Bu, Hex, $PhCH_2$): 53~98% ee (*S*) (61~98%)

(12.27)

Catalytic asymmetric reaction was realized by using 8 mol% of copper iodide and 32 mol% of the chiral amidophosphine **24** to afford the *β*-substituted cycloalkanones in 72–94% ee (Eq. (12.28)) [57]. The amidophosphine is recoverable for reuse in high yield.

RMgCl
24 (32 mol %)
CuI (8 mol %)
Et_2O, –78 °C

n = 2, 3 X = CH_2, O
R = Pr, Bu, Hex, $Ph(CH_2)_2$

72~94% ee (*S*)
(66~92%)

(12.28)

Chiral ferrocenylphosphine oxazoline **27** was also introduced as a ligand for the copper catalyst. The reaction of Grignard reagents with cyclohexenone afforded the product in 83% ee (Eq. (12.29)) [58].

+ BuMgCl

CuI
27 (12 mol %) 10 mo l%
Et_2O, –78 °C
97%, 83% ee

(12.29)

12.3.2.6 Reaction of organozinc using external chiral ligands

Asymmetric conjugate additions of organozincs to enones in the presence of chiral ligands are a rapidly developing and exciting area in recent conjugate addition chemistry. Alexakis discovered the copper-catalyzed asymmetric conjugate addition of diethylzinc to cyclohexenone using **22**, giving 3-ethylcyclohexanone in 32% ee.

Since the bisphosphine as well as a monophosphine greatly accelerate the copper-catalyzed reaction [59], a survey of the known diphosphine was carried out to find that 0.5% of copper(II) triflate and 0.5% of phosphine are sufficient, though enantioselectivity was at most 44% [60]. Chiral phosphite ligand bearing tartrate moiety **28** accelerated the reaction [61], but the ee was not so satisfactory at 40% (Eq. (12.30)) [62]. Chiral thiazolidinone **29** was developed as a chiral ligand to afford the product in 63% ee [63].

Binaphthol-based phosphorus amidite **30** was developed by Feringa to afford 3-ethylcyclohexanone in over 98% ee [64, 65]. However, high enantioselectivity is limited to cyclohexenone, and rather poor selectivity was observed in the reaction of cyclopentenone (10% ee) and cycloheptenone (53% ee). Symmetric aminophosphine ligand **31** was synthesized and the reaction with cyclohexenone was examined in the presence of 5 mol% of copper triflate to afford the product in 55% ee [66]. Amide-phosphine **32** was examined in the reaction to afford the product in 35% ee. Higher selectivity (64%) was observed in the reaction of 4,4-dimethylcyclohexenone [67, 68].

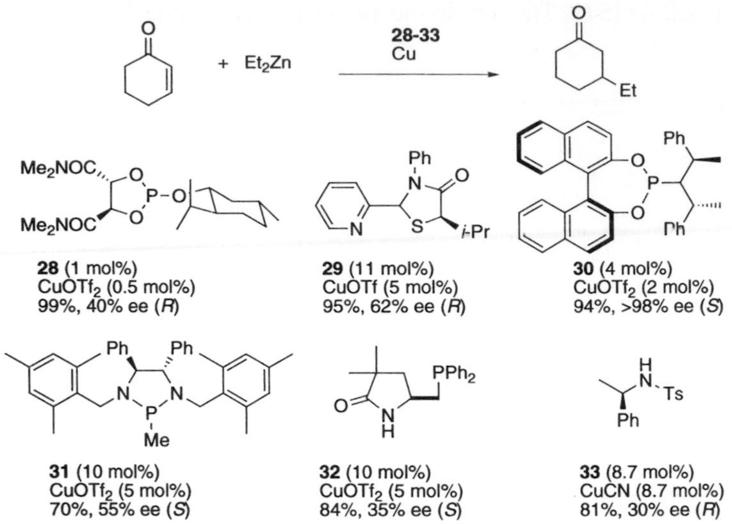

(12.30)

Based on the reaction of diorganozinc with cycloalkenone catalyzed by N-monosubstituted sulfonamide and copper(I) [69], the effect of chiral sulfonamide **33** was examined. It was found that catalytic amounts of both sulfonamide and copper(I) are necessary to catalyze the reaction, but ee was at most 32% [70].

The asymmetric conjugate addition of diethylzinc with chalcone was also catalyzed by nickel and cobalt complex (Eq. (12.31)) [71]. A catalytic process was achieved by using a combination of 17 mol% of an aminoalcohol **34** and nickel acetylacetonate in the reaction of diethylzinc and chalcone to provide the product in 90% ee [72, 73]. Proline-derived chiral diamine **35** was also effective, giving 82% ee [74]. Camphor-derived tridentate aminoalcohol **36** also catalyzes the conjugate addition reaction of diethylzinc in the presence of nickel acetylacetonate to afford the product in 83% ee [75]. Similarly, the ligand **37**-cobalt acetylacetonate complex catalyzes the reaction to afford the product in 83% ee [76].

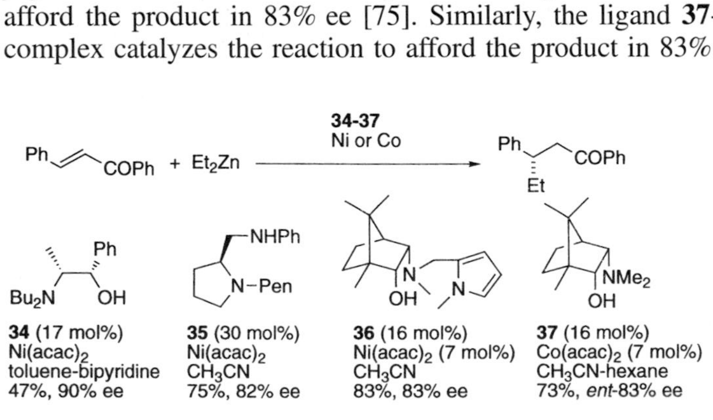

(12.31)

12.4 Reaction of Other Organometals Using External Chiral Ligands

The prominent asymmetric Michael-type addition reaction of arylborane was realized using binap-Rh catalyst **38** (Eq. (12.32)) [77, 78].

Reaction of trimethylaluminum [79, 80] with cyclohexadienone was catalyzed by the oxazoline ligand **39** and copper(I) triflate complex to afford high selectivity (Eq. (12.33)) [81, 82].

PPh_2 Rh **38** PPh_2

CO_2R + $PhB(OH)_2$ → (BINAP-transition metal) CO_2R Ph >95% ee (12.32)

OMe, O, OMe N, *i*-Pr **39** (20 mol%) CuOTf 5 mol%

O + Me_3Al → (THF, 0 °C; 93%, 62% de) O (12.33)

12.5 Recent Michael-Type Reactions Using Chirally Modified α,β-Substituted Carbonyl Compounds

The Michael-type addition reaction of nucleophilic reagents with chirally modified α,β-substituted carbonyl compounds constitutes the established methodology for the preparation of β-substituted carbonyl compounds. The disadvantage of this type of asymmetric Michael reaction is the loading and disloading process of the chiral auxiliary on the Michael acceptor. However, this type of the reaction has been well documented to give the adduct with a high level of diastereoselectivity [83, 84].

References

1. Perlmutter, P. *Conjugate Addition Reactions in Organic Synthesis*, Tetrahedron Organic Chemistry Series Vol. 9, 1992, Pergamon Press, Oxford.
2. Tomioka, K.; Koga, K. In *Asymmetric Synthesis* Vol. 2, Academic Press, New York, 1983, Chapter 7; Posner, G.H. *ibid.* Chapter 8; Lutomoski, K.A.; Meyers, A.I. in *Asymmetric Synthesis* Vol. 3, Academic Press, New York, 1984, Chapter 3.
3. Tomioka, K. *Synthesis* **1990**, 541–549.
4. Rossiter, B.E.; Swingle, N.M. *Chem. Rev.* **1992**, *92*, 771–806.
5. Noyori, R. *Asymmetric Catalysis in Organic Synthesis,* John Wiley and Sons, Inc., New York, 1994.
6. Seyden-Penne, J. *Chiral Auxiliaries and Ligands in Asymmetric Synthesis*, John Wiley and Sons, Inc., New York, 1995.
7. Kanai, M.; Nakagawa, Y.; Tomioka, K. *J. Syn. Org. Chem. Jpn.*, **1996**, *54*, 474–480.
8. Krause, N. *Angew. Chem. Int. Ed. Engl.* **1998**, *37*, 283–285.
9. Hayashi, T.; Tomioka, K.; Yonemitsu, O. (Eds.) *Asymmetric Synthesis, Graphical Abstracts and Experimental Methods*, Kodansha and Gordon and Breach Science Publishers, 1998, Tokyo.
10. Sasai, H.; Arai, T.; Shibasaki, M. *J. Am. Chem. Soc.* **1994**, *116*, 1571–1572.
11. Sasai, H.; Arai, T.; Satow, Y.; Houk, K.N.; Shibasaki, M. *J. Am. Chem. Soc.* **1995**, *117*, 6194–6196.
12. Arai, T.; Sasai, H.; Aoe, K.; Date, T.; Shibasaki, M. *Angew. Chem. Int. Ed. Engl.* **1996**, *35*, 104–105.
13. Kobayashi, S.; Suda, S.; Yamada, M.; Mukaiyama, T. *Chem. Lett.* **1994**, 97–100.
14. Yamaguchi, M.; Shiraishi, T.; Hirama, M. *Angew. Chem., Int. Ed. Engl.* **1993**, *32*, 1176–1177.
15. Yamaguchi, M.; Shiraishi, T.; Hirama, M. *J. Org. Chem.* **1996**, *61*, 3520–3527.
16. Sawamura, M.; Hamashima, H.; Ito, Y. *J. Am. Chem. Soc.* **1992**, *114*, 8295–8297.
17. Sawamura, M.; Hamashima, H.; Ito, Y. *Tetrahedron* **1994**, *50*, 4439–4445.
18. Sawamura, M.; Hamashima, H.; Shinoto, H.; Ito, Y. *Tetrahedron Lett.* **1995**, *36*, 6479–6482.
19. Kumamoto, T.; Aoki, S.; Nakajima, M.; Koga, K. *Tetrahedron Lett.* **1994**, *35*, 1431–1432.
20. Tomioka, K.; Sudani, M.; Shinmi, Y.; Koga, K. *Chemistry Lett.* **1985**, 329–332.
21. Asano, Y.; Iida, A.; Tomioka, K. *Tetrahedron Lett.*, **1997**, *38*, 8973–8976.
22. Xu, F.; Tillyer, R.D.; Tschaen, D.M.; Grabowski, E.J.J.; Reider, P.J. *Tetrahedron Asymmetry* **1998**, *9*, 1651–1654.
23. Tomioka, K.; Shindo, M.; Koga, K. *Tetrahedron Lett.* **1993**, *34*, 681–682.
24. Asano, Y.; Iida, A.; Tomioka, K. *Chem. Pharm. Bull.* **1998**, *46*, 184–186.
25. Park, Y.S.; Weisenburger, G.A.; Beak, P. *J. Am. Chem. Soc.* **1997**, *119*, 10537–10538.
26. Tomioka, K.; Shindo, M.; Koga, K. *J. Org. Chem.* **1998**, *63*, 9351–9357.
27. Tomioka, K.; Okamoto, T.; Kanai, M.; Yamataka, H. *Tetrahedron Lett.* **1984**, *35*, 1891–1892.
28. Denmark, S.E.; Nicaise, O.J.-C. *J. Chem. Soc., Chem. Commun.* **1996**, 999–1004.
29. Taniyama, D.; Hasegawa, M.; Tomioka, K. *Tetrahedron: Asymmetry,* **1999**, *10*, 221–224.
30. Inoue, I.; Shindo, M.; Koga, K.; Kanai, M.; Tomioka, K. *Tetrahedron: Asymmetry,* **1995**, *6*, 2527–2533.
31. Shindo, M.; Koga, K.; Tomioka, K. *J. Am. Chem. Soc.* **1992**, *114*, 8732–8733.
32. Amurrio, D.; Khan, K.; Kundig, E.P.J. *Org. Chem.* **1996**, *61*, 2258–2259.
33. Zweig, J.S.; Luche, J.L.; Barreiro, E.; Crabbé, P. *Tetrahedron Lett.* **1975**, 2355–2358.
34. Huché, M.; Berlan, J.; Pourcelot, G.; Cresson, P. *Tetrahedron Lett.* **1981**, *22*, 1329–1332.
35. Mukaiyama, T.; Imamoto, T. *Chem. Lett.* **1980**, 45–46.
36. Leyendecker, F.; Laucher, D. *Nouv. J. Chim.* **1985**, *9*, 13–19.
37. Corey, E.J.; Naef, R.; Hannon, F.J. *J. Am. Chem. Soc.* **1986**, *108*, 7114–7116.
38. Dieter, R.K.; Lagu, B.; Deo, N.; Dieter, J.W. *Tetrahedron Lett.* **1990**, *31*, 4105–4108.
39. Tanaka, K.; Ushio, H.; Kawabata, Y.; Suzuki, H. *J. Chem. Soc., Perkin I* **1991**, 1445–1450.
40. Bertz, S.H.; Dabbagh, G.; Sundararajan, G. *J. Org. Chem.* **1986**, *51*, 4953–4959.
41. Dieter, R.K.; Tokles, M. *J. Am. Chem. Soc.* **1987**, *109*, 2040.
42. Swingle, N.M.; Reddy, K.V.; Rossiter, B.E. *Tetrahedron* **1994**, *50*, 4455–4466.

43. Miano, G.; Rossiter, B.E. *J. Org. Chem.* **1995**, *60*, 8424–8427.
44. Quinkert, G.; Müller, T.; Königer, A.; Schortheis, O.; Sickenberger, B.; Düner, G. *Tetrahedron Lett.* **1992**, *33*, 3469–3472.
45. Ahn, K.-H.; Klassen, R.B.; Lippard, S.J. *Organometallics* **1990**, *9*, 3178–3181.
46. van Klaveran, M.; Lambert, F.; Eijkelkamp, D.J.F.M.; Grove, D.M.; van Koten, G. *Tetrahedron Lett.* **1994**, *35*, 6135–6138.
47. Zhou, Q.-L.; Pfaltz, A. *Tetrahedron* **1994**, *50*, 4467–4478.
48. Spescha, M.; Rihs, G. *Helv. Chim. Acta* **1993**, *76*, 1219–1230.
49. Kretchmer, R.A. *J. Org. Chem.* **1972**, *37*, 2744–2747.
50. Leyendecker, F.; Laucher, D. *Idem. Nouv. J. Chim.* **1985**, *9*, 13–19.
51. Alexakis, A.; Mutti, S.; Normant, J.F. *J. Am. Chem. Soc.* **1991**, *113*, 6332–6334.
52. Kanai, M.; Koga, K.; Tomioka, K. *Tetrahedron Lett.* **1992**, *33*, 7193–7196.
53. Nakagawa, Y.; Kanai, M.; Nagaoka, Y.; Tomioka, K. *Tetrahedron* **1998**, *54*, 10295–10307.
54. Kanai, M.; Koga, K.; Tomioka, K. *J. Chem. Soc., Chem. Commun.* **1993**, 1248–1249.
55. Kanai, M.; Nakagawa, Y.; Tomioka, K. *Tetrahedron* **1999**, *55*, 3831–3842.
56. Kanai, M.; Tomioka, K. *Tetrahedron Lett.* **1995**, *36*, 4273–4274.
57. Kanai, M.; Nakagawa, Y.; Tomioka, K. *Tetrahedron* **1999**, *55*, 3843–3854.
58. Stangeland, E.L.; Sammakia, T. *Tetrahedron* **1997**, *53*, 16503–16510.
59. Alexakis, A.; Vastra, J.; Mangeney, P. *Tetrahedron Asymmetry* **1997**, *8*, 7745–7748.
60. Alexakis, A.; Burton, J.; Vastra, J.; Mangeney, P. *Tetrahedron Asymmetry* **1997**, *8*, 3987–3990.
61. Berrisford, D.J.; Bolm, C.; Sharpless, K.B. *Angew. Chem. Int. Ed. Engl.* **1995**, *34*, 1050–1064.
62. Alexakis, A.; Vastra, J.; Burton, J.; Mangeney, P. *Tetrahedron Asymmetry* **1997**, *8*, 3193–3196.
63. de Vries, A.H.M.; Hof, R.P.; Staal, D.; Kellogg, R.M.; Feringa, B.L. *Tetrahedron Asymmetry* **1997**, *8*, 1539–1543.
64. de Vries, A.H.M.; Meetsma, A.; Feringa, B.L. *Angew. Chem. Int. Ed. Engl.* **1996**, *35*, 2374–2376.
65. Feringa, B.L.; Pineschi, M.; Arnold, L.A.; Imbos, R.; de Vries, A.H.M. *Angew. Chem. Int. Ed. Engl.* **1997**, *36*, 2620–2623.
66. Mori, T.; Kosaka, K.; Nakagawa, Y.; Nagaoka, Y.; Tomioka, K. *Tetrahedron Asymmetry* **1998**, *9*, 3175–3178.
67. Tomioka, K.; Nakagawa, Y. *Heterocycles* **2000**, *53*, 95–98.
68. Nakagawa, Y.; Matsumoto, K.; Tomicka, K. *Tetrahedron*, in press.
69. Kitamura, M.; Miki, T.; Nakano, K.; Noyori, R. *Tetrahedron Lett.* **1996**, *37*, 5141–5144.
70. Wendish, V.; Sewald, N. *Tetrahedron Asymmetry* **1997**, *8*, 1253–1257.
71. Jansen, J.F.G.A.; Feringa, B.L. *J. Chem. Soc., Chem. Commun.*, **1989**, 741–742.
72. Soai, K.; Hayasaka, T.; Ugajin, S.; Yokoyama, S. *J. Chem. Soc., Chem. Commun.* **1989**, 516–517.
73. Soai, K.; Okudo, M.; Okamoto, M. *Tetrahedron Lett.* **1991**, *32*, 95–98.
74. Asami, M.; Usui, K.; Higuchi, S.; Inoue, S. *Chem. Lett.* **1994**, 297–300.
75. de Vries, A.H.M.; Imbos, R.; Feringa, B.L. *Tetrahedron Asymmetry* **1997**, *8*, 1467–1473.
76. de Vries, A.H.M.; Feringa, B.L. *Tetrahedron Asymmetry* **1997**, *8*, 1377–1378.
77. Takaya, Y.; Ogasawara, M.; Hayashi, T.; Sakai, M.; Miyaura, N. *J. Am. Chem. Soc.* **1998**, *120*, 5579.
78. Takaya, Y.; Ogasawara, M.; Hayashi, T. *Tetrahedron Lett.* **1998**, *39*, 8479–8482.
79. Westermann, J.; Nickisch, K. *Angew. Chem. Int. Ed. Engl.* **1993**, *32*, 1368–1370.
80. Kabbara, J.; Fleming, S.; Nickisch, K.; Neh, H.; Westermann, J. *Tetrahedron* **1995**, *51*, 743–754.
81. Takemoto, Y.; Kuraoka, S.; Hamaue, N.; Aoe, K.; Hiramatsu, H.; Iwata, C. *Tetrahedron* **1996**, *52*, 14177–14188.
82. Takemoto, Y.; Kuraoka, S.; Ohra, T.; Yonetoku, Y.; Iwata, C. *J. Chem. Soc., Chem. Commun.* **1996**, 1655–1656.
83. Tsuge, H.; Takumi, K.; Nagai, T.; Okano, T.; Eguchi, S.; Kimoto, H. *Tetrahedron* **1997**, *53*, 823–838.
84. Didiuk, M.T.; Johannes, C.W.; Morken, J.P.; Hoveyda, A.H. *J. Am. Chem. Soc.* **1995**, *117*, 7097–7104.

13 Stereoselective Radical Reactions

Mukund P. Sibi and Tara R. Ternes

13.1 Introduction

Synthetic chemists faced with the challenge of developing a stereoselective route to a target molecule often resort to bond construction strategies involving ionic or concerted reactions. Only in the last two decades have free radical reactions come into prominence in this context. In the 1980s, many selective free radical-mediated cyclizations were developed. There was an explosion of new stereoselective methodologies using free radical intermediates in the last decade. Chiral auxiliary-mediated methods for stereocontrol in both acyclic as well as cyclic systems were demonstrated during the period 1980–1995 [1]. This culminated in the publication of several key reviews [2] and an authoritative monograph by three leading experts [3]. In the last five years, another milestone was reached in that enantioselective free radical methods under catalytic conditions were reported [4]. Thus, stereoselective bond construction by radical methods is now commonplace. There are several advantages to free radical methods over their ionic counterparts. These are (1) reactions under neutral conditions, (2) compatibility with radical acceptors containing functional groups such as carbonyls, enol ethers, and enamines, (3) compatibility with Lewis acids, (4) no necessity for protection of alcohol and amine functional groups, (5) compatibility with protic solvents, (6) potential for reaction in aqueous systems, and (7) ease of quaternary center formation.

This review focuses on free radical-mediated stereoselective bond construction in which the carbonyl group plays a key role. Reaction at the carbonyl group as well as on carbons alpha and beta are described. The general reaction characteristics of these reactive intermediates are as follows. The acyl radicals are nucleophilic in character and thus they react easily with electrophilic acceptors. On the other hand, radicals on carbon alpha to the carbonyl are electrophilic in nature and their reactivity matches with nucleophilic partners. The majority of reactions at carbon beta to the carbonyl are in α,β-unsaturated systems and in these the beta carbon is electrophilic.

The review highlights the chemistry described in the past five years with a small emphasis on work from our laboratory. The chemistry of ketyl radicals has been dealt with only in a cursory fashion and readers are advised to consult recent seminal reviews in this area [5]. Due to space limitations, the chemistry of malonyl radicals are also not covered. Readers should consult important contributions from Prof. Snider and a recent review [6].

13.2 Reactions at the Carbonyl Carbon: Stereoselective Acyl and Ketyl Radical Reactions

13.2.1 Acyclic Diastereoselection

Few examples exist in the literature concerning the stereoselective addition of acyl radicals to a radical acceptor in an acyclic manner. Equation (13.1) shows the efficient 1,2-asymmetric induction in the addition of aliphatic or aromatic acyl radicals to chiral acyclic alkenes **1** [7]. The corresponding α-hydroxy ketones **3** were produced with high *syn* selectivity (Table 13-1). This acyl radical addition is very exothermic, and it is hypothesized that Hammond's postulate can be invoked to predict a transition state that is very close in energy to the starting alkene **1**. The X-ray structure of **1** was then used to rationalize the stereochemical outcome of this radical addition by determination of the least sterically hindered path for the approaching radical.

OY SMe, R, SO_2Tol, **1** + R_1CHO **2** —hυ, PhCOPh→ OY SMe, R, SO_2Tol, R_1, O **3** (13.1)

Table 13-1. Photocycloadditions of acyl radicals.

Entry	R	Y	R_1	Time/h	Yield/%	*syn:anti*
1	Me	H	Et	2	92	83:17
2	Me	Ac	Et	2	77	91:9
3	*i*-Pr	Ac	Et	4	89	96:4
4	Me	H	Ph	2	64	83:17
5	Me	Ac	Ph	2	86	93:7
6	*i*-Pr	Ac	Ph	5.5	80	95:5

13.2.2 Diastereoselective Cyclizations of Acyl Radicals

Two strategies are currently available for obtaining high diastereoselectivity in acyl radical cyclizations. The first strategy involves taking advantage of chirality remote from the acyl radical to favor one cyclic transition state over another through steric or electronic considerations. This strategy is commonly employed in natural product synthesis where multiple stereocenters exist. The second, more recent, strategy uses chiral diols to first "protect" the acyl radical precursor such that when the radical is formed it is essentially a chiral acyl radical equivalent. The chirality of the ketal radical then determines the stereochemical outcome of the cyclization.

13.2.2.1 Diastereoselective cyclizations towards natural products

Natural products which contain cyclic ethers, such as mucocin and gamberic acids A–D, have been accessed using intramolecular acyl radical cyclizations onto α,β-unsaturated esters or sulfones [8, 9]. Equation (13.2) illustrates the general reaction scheme for the formation of 5, 6 and 7 membered cyclic ethers [10]. Formation of the acyl radical can be carried out from the acylselenide precursor by using AIBN or triethylborane as the radical initiator and tin or silicon hydride as the chain carrier. The *syn* product **5** is obtained as the major product in high yield (85–94%) and good diastereoselectivity (up to >19:1). Higher selectivities are obtained by increasing the size of the R group and on using $(TMS)_3SiH$ as the hydrogen atom source. Preference for a cyclic transition state **7** where the α,β-unsaturated ester occupies a pseudoequatorial position of the ring accounts for the predominant formation of the *syn* product. Performing this transformation at lower temperatures (room temperature or lower) favors the cyclization over decarbonylation, which is a common problem in reactions involving acyl radicals. A similar strategy has been employed in the cyclization of acyl radicals to enamines and enol thioethers as well as enol ethers in the formation of the corresponding heterocycles [11].

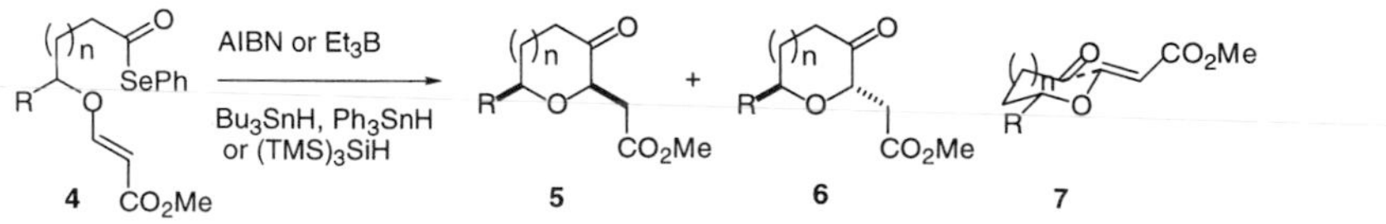

(13.2)

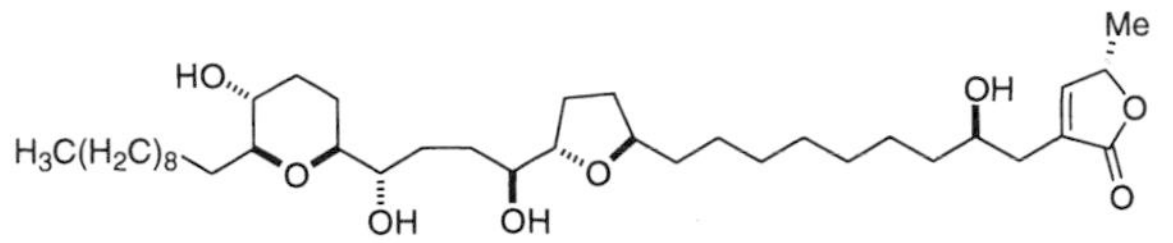

Mucocin

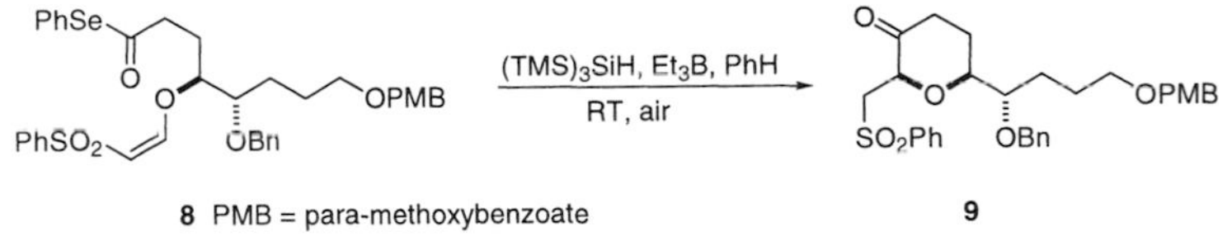

8 PMB = para-methoxybenzoate

(13.3)

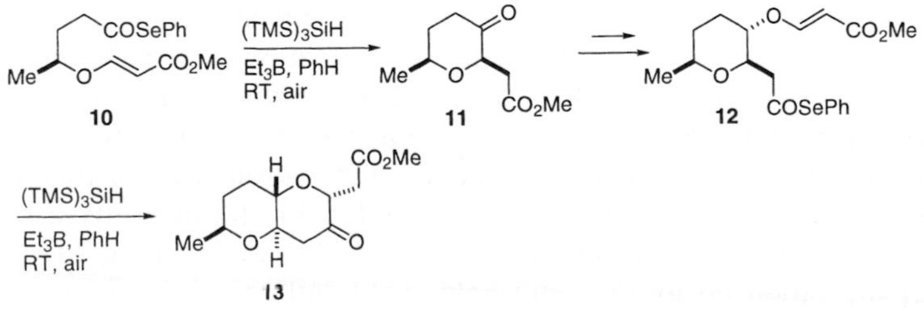

This methodology has been extended to the synthesis of the C-16 to C-26 fragment of the natural product mucocin in 81% yield and 15:1 diastereoselectivity as shown in Eq. (13.3). This cyclization can also be carried out sequentially. The methodology was extended towards the synthesis of polycyclic natural products [9]. Both cyclization steps proceeded in greater than 90% yield with diastereomer ratios of 6:1 for the formation of **11** and >19:1 for the formation of **13**.

The acyl radical cyclizations have been cleverly applied to the synthesis of complex natural and unnatural products (Eq. (13.4)). In one example, tandem radical cyclization of **14** provides access to the steroidal skeleton. The reaction proceeds through a 6-*endo-trig* cyclization to form the A/B ring and a macrocyclization/transannulation to establish the C/D ring [12]. More recently, a cascade cyclization initiated by an acyl radical has allowed for the establishment of fourteen chiral centers in a single step [13]. Once again the formation of the polycyclic compound **17** involves sequential 6-*endo-trig* reactions.

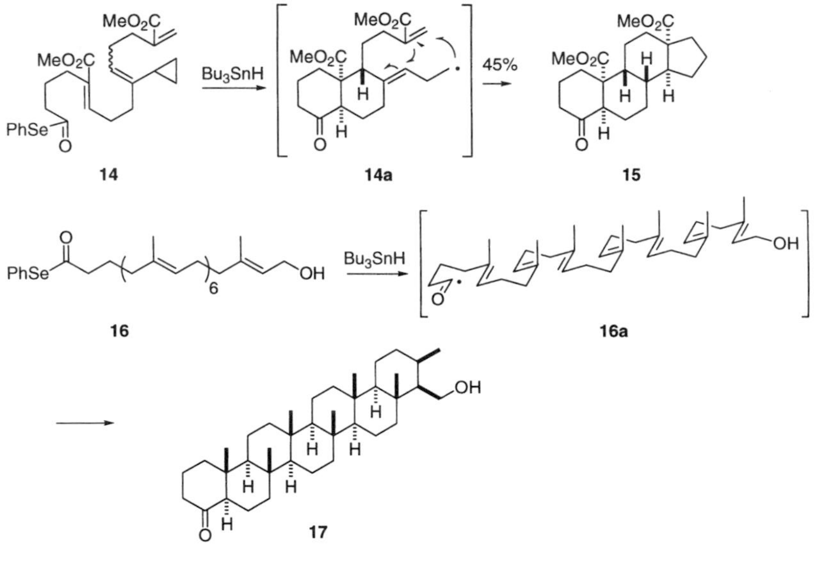

(13.4)

13.2.2.2 Chiral acyl radical equivalents

Chiral diols **18a–c** have been employed as easily removable chiral auxiliaries in the formation of chiral acyl radical equivalents [14, 15]. The reaction of carbonyl compounds with these diols produces acetals containing an additional 6-membered ring which is rigidly held in a *trans*-decalin conformation. The rigid *trans*-fused bicyclic system is desirable such that boat and twist-boat conformers observed with acyclic chiral diols are disfavored. The reaction starts with the formation of a vinyl radical followed by an intramolecular 1,5-hydrogen atom transfer resulting in a dioxanyl radical. This then undergoes cyclization followed by a hydrogen atom transfer to furnish the products **20** and **21** as shown in Eq. (13.5). Products **20** and **21** are produced in 72% and 65% yield and enantiomeric excess of the hydrolyzed ketone products of 52% and 42% respectively. Enantiomeric excesses of up to 95% were obtained using **18c** as the chiral diol.

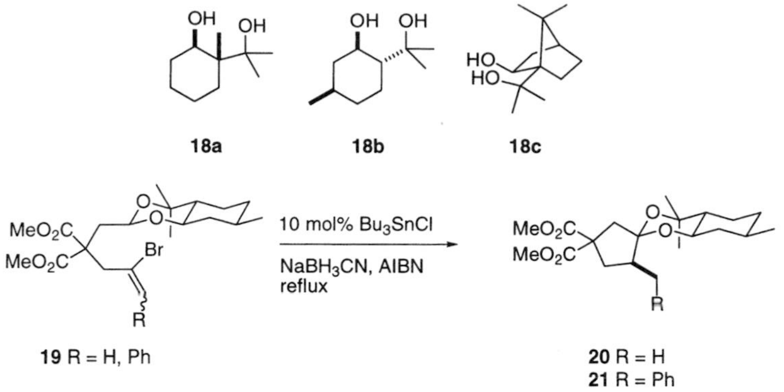

(13.5)

13.2.3 Diastereoselective Cyclizations of Ketyl Radicals

Samarium diiodide is commonly used to reductively generate ketyl radicals through the donation of a single electron. The resulting ketyl radical can then be trapped by a number of radical-trapping agents such as acrylates to produce an assortment of coupled products. This premise has been utilized in conjunction with the observation that chromium tricarbonyl complexes of arenes effectively shield one face of an aromatic ring [16]. Reactions can then occur selectively onto the opposite face of the arene from the chromium complex. Equation (13.6) illustrates how these two concepts were applied in the diastereomeric synthesis of spirocyclic compounds [17]. The overall reaction is a two-electron reduction sequence, which yields **26** as a single diastereomer.

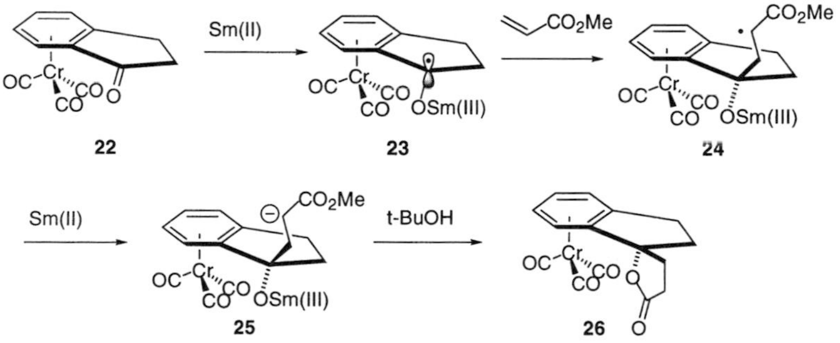

(13.6)

13.2.4 Diastereoselective Photocycloadditions to Carbonyl Compounds

The Paterno-Buchi reaction is the photocycloaddition of an alkene with an aldehyde or ketone to form oxetanes. This transformation has been shown to proceed through a biradical intermediate, and up to three new stereocenters can be formed as a result of this reaction. A general mechanism for the reaction between an aldehyde and a chiral enol silyl ether is shown in Eq. (13.7) [18]. Allylic 1,3-strain is cited as the control element in reactions of this type, and diastereomeric ratios of >95:5 are reported for products **30** containing four contiguous stereocenters. Examples of photocyclizations of amino acid derivatives proceeding through biradical intermediates have been reported [19].

27 + **28** —hυ→ [**29**] → **30** (13.7)

13.3 Stereocontrol α- to the Carbonyl

13.3.1 Acyclic Diastereoselection

13.3.1.1 Reductions

Stereoselective reductions of diastereomeric mixtures of α-halo-β-alkoxyesters have been utilized as a synthetic strategy towards *anti*-aldol products. Selective hydrogen atom transfer is achieved via 1,2-stereoinduction [20]. As shown in Eq. (13.8), it has been demonstrated that high selectivities can be obtained at low temperatures in systems where R is large, but diastereoselectivity falls dramatically when the size of R is reduced [21].

31 —Bu_3SnH→ *syn*-**32** + *anti*-**33** (13.8)

R	Y	Temp.	Ratio syn/anti	Yield (%)
Ph	Me	-78 °C	1:30	90
i-Pr	Me	-78 °C	1:8	66

It was later found that by linking R and Y as in the cyclic ether **34**, diastereoselectivities of the subsequent reductions could be substantially increased (Eq. (13.9)) [22]. It is suggested that the increased *anti* selectivity of the exocyclic radical is due to destabilization of the transition state **37** leading to *syn* products rather than stabilization of the transition state leading to *anti* products.

Bu_3SnH, Et_3B, - 30 °C, 93%

34 → **35** + **36** (1:52)

37

(13.9)

It was also found that the addition of chelating Lewis acids reversed the sense of stereoinduction in similar types of reductions yielding products with high levels of *syn* selectivity [23]. In the absence of a chelating Lewis acid, a transition state can be envisaged as in **39** where allylic 1,3-strain, dipole-dipole repulsion and stabilization through hyperconjugation favors an *anti* conformation with respect to the carbonyl and the alkoxy substituent. Tin hydride can then approach from the face opposite the phenyl ring resulting in products of *anti* configuration. A 70% yield and 20:1 *anti*:*syn* ratio are obtained using the conditions illustrated in Eq. (13.10).

38 → (Bu_3SnH, Et_3B, -78 °C) → [**39**] → *anti*-**40**

(13.10)

On the other hand, a bidentate Lewis acid holds the two Lewis basic oxygens in a *syn* conformation forming a 6-membered chelated transition state **42**. This chelate reverses the favored approach of the incoming tin hydride, leading to *syn* products with high diastereoselectivity. Equation (13.11) illustrates the formation of **43** in a 70% yield and 28:1 *syn*:*anti* ratio.

41 → (Bu_3SnH, MgI_2 or $MgBr_2 \cdot Et_2O$, Et_3B, -78 °C) → [**42**] → *syn*-**43**

(13.11)

Higher selectivities are generally obtained in the presence of a Lewis acid as the reactive conformation is essentially locked into place. The reaction in the absence of a Lewis acid, in contrast, is prone to rotamer control problems as the *s-trans* to *s-cis* interconversion can result in lower selectivities depending on the size of substituents and reaction temperatures. Larger substituents and lower reaction temperatures counteract the *s-trans* to *s-cis* rotation.

13.3.1.2 Allylations

Diastereoselective radical allylations have been studied in many different contexts, and a plethora of information exists regarding stereocontrol in these reactions. Allylations have been performed using the traditional trapping and β-elimination sequence occurring typically with allylstannanes as well as a stepwise atom transfer/elimination sequence found to occur with allylsilanes. Stereochemistry is commonly controlled through the use of chiral auxiliaries or by 1,2-induction, and functionalized *anti*-aldol and amino acid products are available using this established methodology.

Excellent yields and diastereoselectivities have been obtained in allylations using a new oxazolidinone chiral auxiliary derived from diphenylalaninol [24]. The use of oxazolidinone chiral auxiliaries was sparked by the application of Lewis acids to radical reactions. Bidentate Lewis acids are used to favor one rotamer (**44**) out of a possible four by forming a chelated intermediate with the two carbonyl groups and through steric interactions imparted by the 4-substituent of the oxazolidinone (Eq. (13.12)). Trapping with the allylstannane can then occur on the face opposite the bulky oxazolidinone-4-substituent.

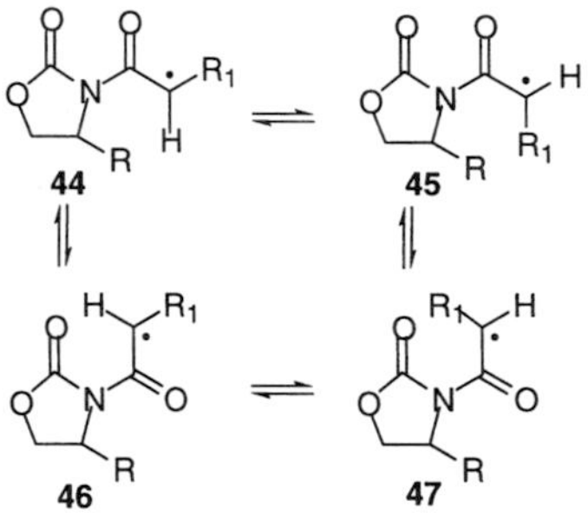

44 ⇌ 45
46 ⇌ 47

(13.12)

Several mono- and multidentate Lewis acids were tested, and these results are tabulated in Table 13-2 (Eq. (13.13)). As to be expected, single-point-binding Lewis acids such as $BF_3{\cdot}Et_2O$ gave poor selectivity (entry 2). It was found that $MgBr_2{\cdot}Et_2O$ gave the highest selectivities even at room temperature (entries 4 and 5). Lanthanide triflates, however, showed poor selectivity at low temperature (entry 6).

The effect of the 4-substituent on the oxazolidinone ring was also examined. Results shown in Table 13-3 indicate that aromatic substituents provide the highest selectivities, and the diphenylmethyl (entry 1) and tritylmethyl auxiliaries (en-

(13.13)

Table 13-2. Diastereoselective allylations using oxazolidinone auxiliaries.

Entry	Lewis acid [a])	Reaction conditions	% Yield [b])	Ratio (RS:RR) [c])
1	None	–78 °C, CH_2Cl_2, 3 h	93	1:1.8
2	$BF_3 \cdot OEt_2$	–78 °C, CH_2Cl_2, 2.5 h	85	1:1.4
3	$Zn(OTf)_2$	–78 °C, CH_2Cl_2/THF (1:1), 2 h	86	7.6:1
4	$MgBr_2$	–78 °C, CH_2Cl_2, 2 h	94	≥100:1
5	$MgBr_2$	25 °C, CH_2Cl_2, 2 h	94	30:1
6	$Yb(OTf)_3$ [d])	–78 °C, CH_2Cl_2/THF (1:1), 2 h	64 (14)	5:1

[a]) 2 eq of Lewis acid was used in all reactions unless otherwise noted.
[b]) Isolated yield and yield in parentheses is for the reduction product.
[c]) Diastereomer ratios were determined by ^{1}H NMR (400 MHz).
[d]) 1 eq of the Lewis acid was used.

try 3) give the best results. The traditional Evans auxiliary derived from phenyl alaninol and valinol gave moderate and low selectivities respectively (entries 2 and 4). A model that accounts for the observed selectivity is shown in **52**. In this model, the metal coordinates to both carbonyl oxygens and allylic 1,3-strain favors the *s-cis* rotamer of the intermediate radical. Allylstannane addition to the chelated intermediate takes place from the face opposite the bulky oxazolidinone 4-substituent.

Table 13-3. Effect of oxazolidinone-4-substituent on selectivity.

Entry	Substrate [a])	Reaction conditions	Yield, %	Ratio [b, c, d])
1	**48** (R)	$MgBr_2$ (2 eq), –78 °C, CH_2Cl_2, 2 h	94	≥100:1 (1:1.8) (S)
2	**50** (R)	$MgBr_2$ (2 eq), –78 °C, CH_2Cl_2, 2 h	84	17:1 (1:1) (S)
3	**50a** (R)	$MgBr_2$ (2 eq), –78 °C, CH_2Cl_2, 2 h	94	>100:1 (1:1) (S)
4	**50b** (S)	$MgBr_2$ (2 eq), –78 °C, CH_2Cl_2, 2 h	85	2.8:1 (1:1)[e])
5	**50c** (S)	$MgBr_2$ (2 eq), –78 °C, CH_2Cl_2, 2 h	92	3:1 (1.5:1)[e])

[a]) The absolute stereochemistry of the auxiliary is indicated in parentheses.
[b]) The ratio in parenthesis is for reactions without Lewis acid additive.
[c]) The absolute stereochemistry of the newly formed stereocenter.
[d]) Diastereomer ratios were determined by ^{1}H NMR (400 MHz).
[e]) The absolute stereochemistry of the products was not determined.

Highly diastereoselective allylations were also achieved in a slightly different manner through radical addition to chiral oxazolidinone acrylate and trapping with allylstannane [25]. In reactions with α,β-unsaturated substrates, the Lewis acid functions not only to control the rotamer populations but also to increase reactivity at the β-carbon. After initial addition of the radical, an intermediate is generated at the α-position similar to the sequence shown in Eq. (13.14). This intermediate could also be trapped readily with allylstannane. It was found that lanthanide Lewis acids such as ytterbium triflate provide the highest selectivities (>100:1) in the tandem addition of isopropyl radical and trapping with allylstannane. A variety of radical precursors (Eq. (13.15), Table 13-4) were employed to evaluate the scope of this methodology. Excellent diastereoselectivities resulted from alkyl radicals and acyl radicals, but the methoxymethyl radical appeared to interfere with the Lewis acid, in particular $MgBr_2$. Higher selectivities were attainable with $Yb(OTf)_3$ as a Lewis acid, presumably as a reflection of its higher coordinating ability.

$SnBu_3$, -78 °C, Et_3B/O_2

50 R = CH_2Ph
50a R = CH_2CPh_3
50b R = $CHMe_2$
50c R = Ph

51 **52**

(13.14)

Lewis acid, RX, $SnBu_3$, CH_2Cl_2, Et_3B/O_2, -78 °C

53 **54**

(13.15)

Table 13-4. Conjugate addition/trapping experiments.

Entry	RX	Lewis acid (eq)	Yield	Ratio [a, b]
1	EtI	$MgBr_2$ (2)	93	>100:1
2	EtI	$Yb(OTf)_3$ (1)	90	>100:1
3	*i*-PrI	$MgBr_2$ (2)	85	>100:1
4	$MeOCH_2Br$	$MgBr_2$ (2)	50	1.8:1
5	$MeOCH_2Br$	$Yb(OTf)_3$ (1)	70	58:1
6	MeCOBr	$MgBr_2$ (2)	55	50:1

[a]) Diastereomer ratios were determined by ^{1}H NMR (400 MHz).
[b]) The absolute stereochemistry of the products was not determined.

Strongly electron-deficient β-ketoamidyl radicals on chiral oxazolidine auxiliaries have also been shown to trap allylstannanes with high levels of selectivity (Eq. (13.16), Table 13-5) [26]. High diastereoselectivity in these reactions is obtained with bulky R groups on the chiral auxiliary and by lower reaction temperatures.

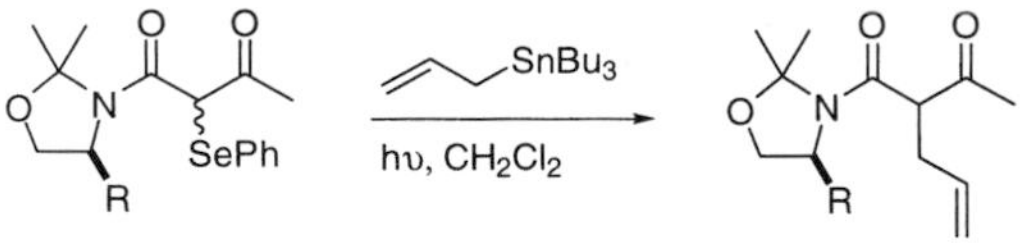

(13.16)

Table 13-5. Diastereoselective allylation of oxazolidines.

Entry	R	Temperature (°C)	dr
1	*i*-Pr	0	2.6:1
2	CH_2Ph	0	2.5:1
3	CH_2Ph	–78	3.1:1
4	*t*-Bu	80	13:1
5	*t*-Bu	25	24:1
6	*t*-Bu	–78	>100:1

Functionalized *anti*-aldol products can be obtained through the stereoselective allylation of *β*-hydroxy or *β*-alkoxy esters. Once again, addition of Lewis acid additives enhanced the selectivities in these reactions by favoring transition states similar to **59** which can be trapped leading to *anti* products selectively (Eq. (13.17)). It was found that unprotected *β*-hydroxy esters can be alkylated via 1,2-stereoinduction by forming an aluminum alcoholate (**59**) analogous to chelates obtained using bidentate Lewis acids [27]. The cyclic radical intermediate is then trapped by the allylstannane from the face opposite the R group. Selectivities up to 32 : 1 *anti* : *syn* and 98% isolated yield have been obtained using this methodology.

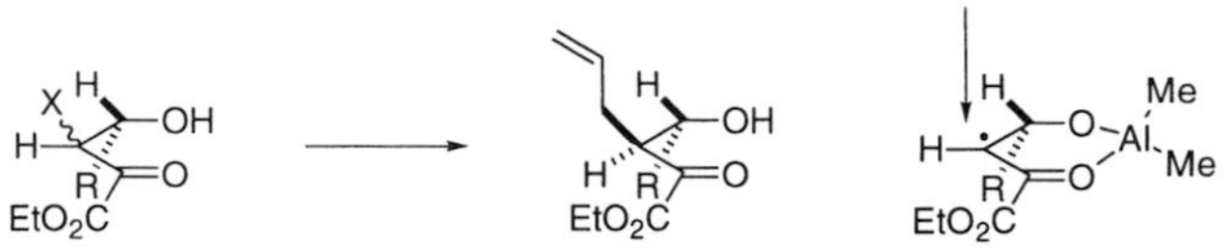

(13.17)

Allylations of *β*-alkoxy esters were demonstrated under atom and group transfer conditions using allylsilanes and chelating Lewis acid conditions [28]. Equation (13.18) illustrates the basic mechanism for this reaction. Once again, selectivities greater than 100 : 1 favoring the *anti* product and yields of 87% are reported. Results from this study suggest that inclusion of Lewis acids may help facilitate the atom transfer step.

Chiral auxiliary-mediated diastereoselective allylations of *α*-bromoglycine derivatives **65** have also been established. 8-Phenylmenthol has been successfully employed as a chiral auxiliary in glycine allylations (Eq. (13.19)) [29]. The captodative radical intermediate generated in this reaction benefits from the observation that *α*-amino acid radicals prefer an *s-cis* geometry about the single bond, presum-

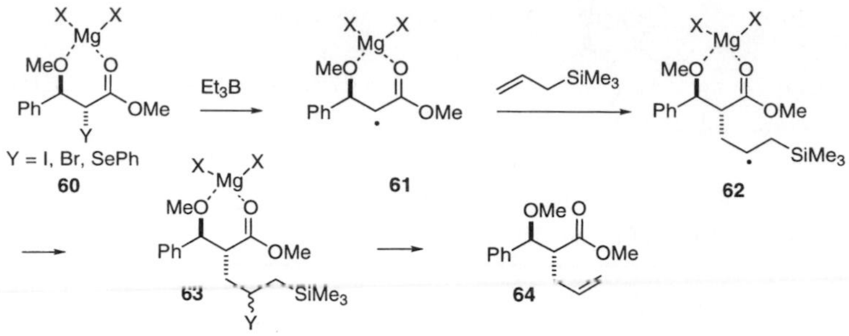

(13.18)

ably through hydrogen bonding or electronic effects. Having controlled the single bond rotation, facial selectivity can then simply be imparted by the chiral auxiliary. Even at 80 °C the diastereomeric excesses of 86% (2*S*) for **66** can be attained in this reaction. The creation of two new chiral centers was also attempted by trapping with butenyl stannane. The *α* carbon retained its preference for (*S*) configuration in a 93:7 ratio (**67**). Less stereocontrol was exerted over the chiral center formed at the *β* carbon, however, providing only a 3:2 mixture of (*R*:*S*).

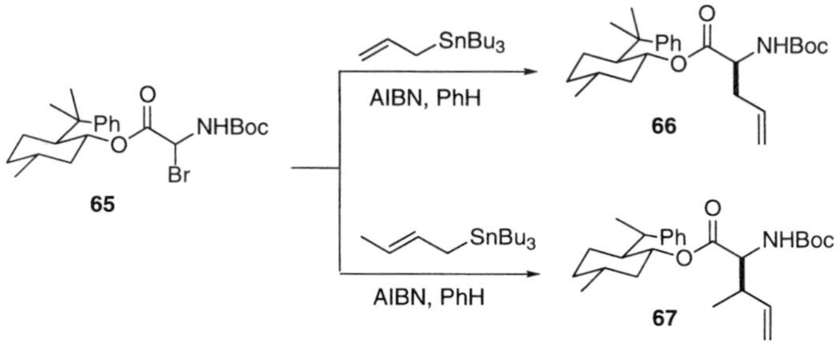

(13.19)

A similar allylation has been reported with the chiral auxiliary on the nitrogen of the bromoglycine attached as an imide rather than as an ester (Eq. (13.20)) [30]. A valinol-derived oxazolidinone chiral auxiliary **68** was employed under bidentate Lewis acid conditions ($ZnCl_2$). It was found that $ZnCl_2$ functions as a radical initiator as well as a Lewis acid in these reactions. The best example provides the allylglycine **69** derivative in 85% yield and the newly formed stereocenter in a ratio of 87:13 (*R*:*S*) with a reaction time of 1 h at –78 °C.

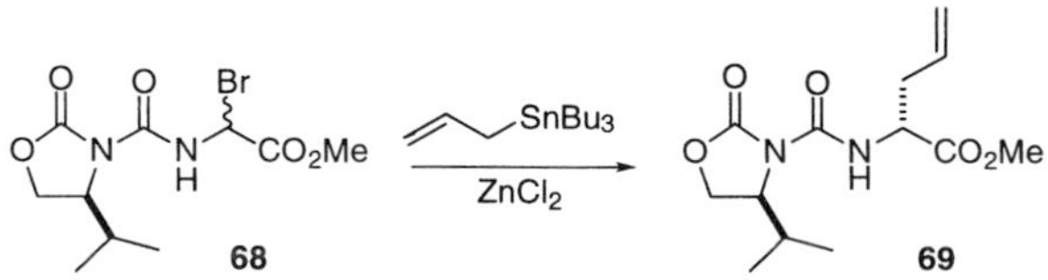

(13.20)

Intramolecular hydrogen bonding has been exercised as a control element in the allylation of α-acyl radicals derived from a series of amino acid derivatives [31]. Equation (13.21) shows that good to excellent levels of diastereoselectivity are achieved not only via 1,2-induction, but also 1,3-, 1,4- and 1,5-asymmetric inductions. Hydrogen bonding occurs between the amine hydrogen and the carbonyl oxygen, favoring a more rigid reactive conformation.

SnBu3; AIBN, hυ

70 → **71** n = 1; major isomer + **72a**, n = 2; *anti*-major; **72b**, n = 3; *syn*-major (13.21)

13.3.1.3 Conjugate addition and reduction to install the α center

A conjugate addition and trapping strategy has emerged as a viable method for forming a chiral center α to a carbonyl by diastereoselective hydrogen atom transfer after initial radical addition to an acrylate. An example shown in Eq. (13.22) uses a C-2 symmetric pyrrolidine chiral auxiliary to induce facial selectivity in the hydrogen atom transfer step [32]. This particular example afforded 89% yield and 25 : 1 preference for **74**.

PhSH; AIBN, Δ

73 → **74** (13.22)

Other applications of this strategy include addition and trapping sequences with *exo* double bonds to install the α centers. This methodology has been applied to the formation of *anti*-aldol adducts as well as substituted α-amino acids. Stereocenters located in adjacent positions on the ring provide facial bias for selective hydrogen atom transfer. Equation (13.23) illustrates how this was applied towards the formation of α-amino acids [33].

It was found that a variety of radical precursors could be added, and that the specific electronic and steric effects exerted on the resulting radical effected the diastereoselectivity of the hydrogen atom transfer. Increasing the size of R group appeared to increase the selectivity of the trap. For instance, reaction with *t*-butyl radical and tributyltin hydride gave the highest selectivity, >98 : 2 (70% yield), for the *trans* product (**77**). Reactions with electron-deficient radicals suffered from low yields and decreased selectivity. Results also indicate that reactions with tributyltin hydride produced higher selectivities but lower yields than those performed with mercury hydrides. A similar strategy was employed using photoinduced prochiral alcohol and ester radicals (**78**) to establish the *α* center as well as the *γ* center (Eq. (13.23)) [34]. Once again, good control of *α* selectivity was achieved, but little stereocontrol was exerted at the *γ* center.

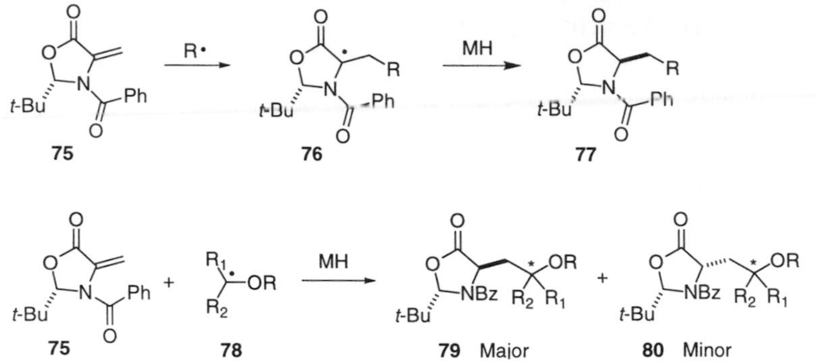

(13.23)

13.3.1.4 Radical addition to imines

It has been shown that oxime ethers (**81**) can function as radical acceptors and that the intermediate aminyl radical formed is stabilized by the lone pair on the adjacent oxygen atom [35]. This supposition has also been applied towards the stereoselective synthesis of *α*-amino acids using Oppolzer's camphor sultam chiral auxiliary as shown in Eq. (13.24) [36]. It was also established that under these conditions alkyl radical addition to glyoxylic oxime ethers was facile. With bulky radicals, such as *t*-butyl, selectivities up to >98:2 could be obtained. Lewis acid additives such as $BF_3 \cdot Et_2O$ led to increased product yields, but no effect was found on levels of diastereoselectivity in these reactions. Product from ethyl radical addition was a by-product in most cases (up to 22%), a consequence of using triethylborane as an initiator. The nitrogen-oxygen single bond in the product can be reductively cleaved by treatment with $Mo(CO)_6$ and the chiral auxiliary can be hydrolyzed to yield the free amino acids with no racemization. Rationalization of the resultant stereochemistry in these radical additions is shown in **83**. Dipole-dipole repulsion forces the *s-cis* conformation and the radical approaches from the less hindered face allowing for formation of the (2*R*) product.

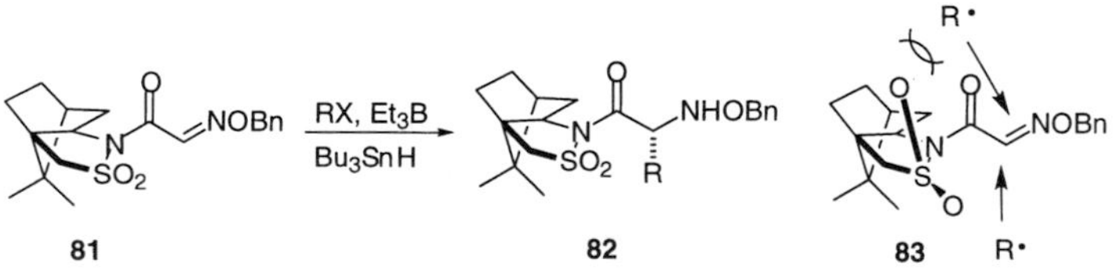

(13.24)

13.3.2 Acyclic Enantioselection

13.3.2.1 Reductions

Achieving high levels of enantioselectivity *α* to a carbonyl in radical chemistry remains an elusive goal. Chiral ligands and chiral tin hydride reagents have recently been utilized towards the formation of carbon-hydrogen bonds in an enantioselective manner. Enantiomeric excesses of up to 62% and yields of 88% have been obtained in the reduction of **84** using chiral diamine ligand **85**, MgI_2 as a Lewis acid and tributyltin hydride (Eq. (13.25)) [37]. It is suggested that the Lewis acid is coordinated to the carbonyl oxygen, the ether oxygen, as well as the chiral ligand. The chiral environment provided by this association determines from which face the tributyltin hydride will deliver the hydrogen atom.

84 → (MgI_2, Bu_3SnH, (S)-**85**) → (-)-**86** (13.25)

Another approach to enantioselective reductions involves reactions with chiral tin hydrides. The helical chirality of the binaphthyl group has been taken advantage of in the design of chiral tin reagents. An example of an enantioselective reduction using chiral tin hydride **88** is shown in Eq. (13.26) [38]. The reduced products are formed in low enantiomeric excesses (41% ee) and low chemical yields (often under 50%). These factors and the difficulty in synthesizing the chiral tin hydride reagents serve to diminish the utility of these types of enantioselective reductions thus far.

87 + **88** → (Et_3B, -78 °C) → **89** (13.26)

13.3.2.2 Allylations

Greater success has been achieved in the enantioselective formation of carbon-carbon bonds *α* to a carbonyl via enantioselective allylation using chiral Lewis acids. Once again, radical addition to acrylates and trapping with allyltributyltin has been applied as a strategy in enantioselective allylations. Oxazolidinone acrylates (**90**) are good templates for the formation of bidentate chelates with chiral bisoxazoline ligand (**92**)/Lewis acid complexes. Equation (13.27) shows that good selectivities (up to 90% ee) result using $Zn(OTf)_2$ as a Lewis acid and when R is large (*t*-butyl) [39, 40]. The configuration of the product **91** can be rationalized using a model similar to that used for Diels-Alder reactions, where the zinc metal adopts a tetrahedral geometry with two bidentate donors: the oxazolidinone substrate and the ligand [41]. One face of the prochiral radical in an *s-cis* conformation is shielded by the phenyl group on the ligand. This allows the allylstannane to selectively approach the unhindered face of the radical.

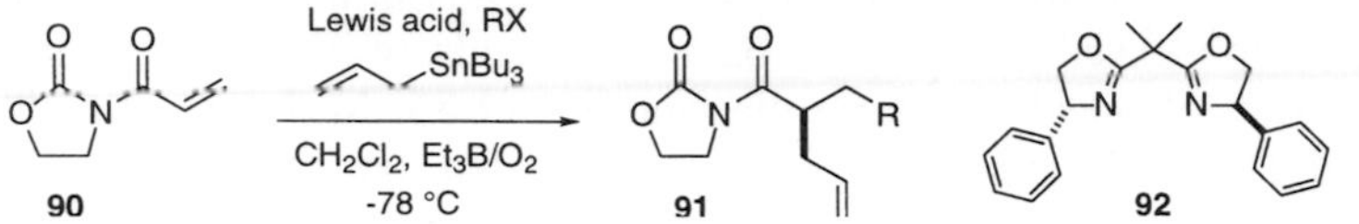

(13.27)

A similar type of allylation has been reported by generating the α radical directly from the corresponding halide, also using an oxazolidinone substrate and a chiral zinc Lewis acid (Eq. (13.28)) [34]. Enantioselectivities in this latter case are slightly lower (76% ee) than those obtained in the addition and trapping experiments, but the trend showing that larger R groups provide increased levels of selectivity still holds true.

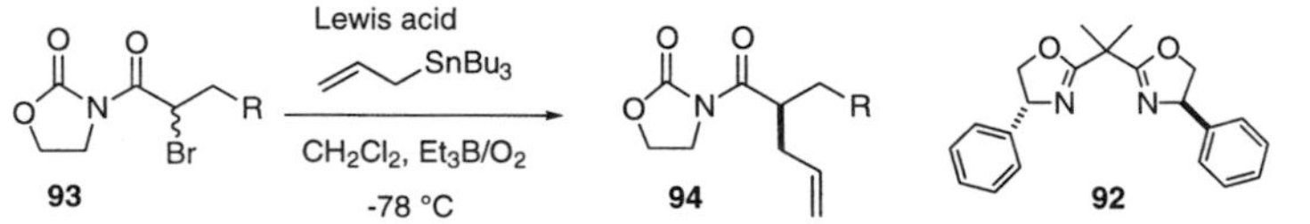

(13.28)

The intermediate α-amide radical generally prefers an *s-cis* orientation (**95**), minimizing allylic strain. However, when R is small, allylic strain is decreased and the *s-trans* conformation **96** is more accessible. Additionally, trapping from the wrong face of the chiral Lewis acid/radical complex is more likely with small R (Eq. (13.29)). The lower selectivity generally observed with the α-bromide substrates is presumably due to the fact that the α-halo amide carbonyl is not as good a donor as the α,β-unsaturated amide carbonyl and that this may be adversely affecting interactions with the chiral Lewis acid. No selectivity was observed using these conditions when R=H.

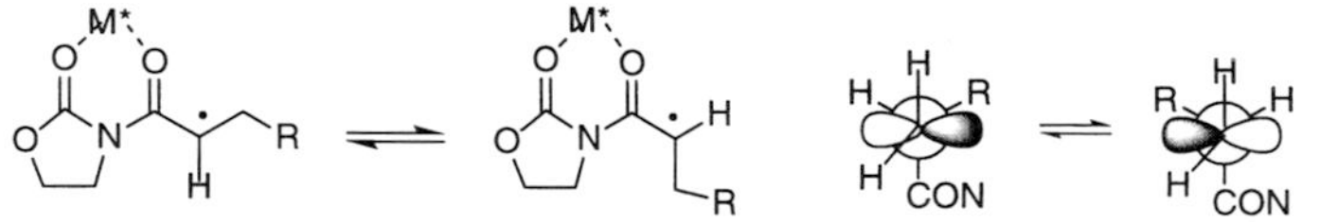

(13.29)

The combination of aluminum Lewis acids and chiral diols has allowed for modest enantioselectivities in similar reactions using oxazolidinone templates and α-iodopropionate substrate as in Eq. (13.30) [42]. The 4,4-dimethyl substituent on the oxazolidinone (**99**) favors the *s-cis* conformation for the intermediate radical. Enantioselectivities of 32 and 34% and yields above 90% were obtained for allylations with R=H and R=Me respectively.

(13.30)

Radical reactions are also valuable strategies for the formation of quaternary carbon centers. An enantioselective variant of this has recently come to light utilizing aluminum as a Lewis acid complexed to a chiral binol ligand (**103**) in the allylation of α-iodolactones **101** (Eq. (13.31), Table 13-6) [43]. It was established that diethyl ether as an additive in these reactions dramatically increases product enantioselectivities (compare entries 1 and 2, Table 13-6). Catalytic reactions were also demonstrated (entry 3) with no appreciable loss of selectivity. A proposed model for how diethyl ether functions to enhance selectivity in the enantioselective formation of these quaternary chiral centers is shown in **104**.

(13.31)

Table 13-6. Enantioselective allylations: effect of additives.

Entry	R	Chiral LA (eq)	Additive	Yield (%)[a]	ee (%)[b]	Config.[c]
1	Me	1.0	none	72	27	R
2	Me	1.0	Et_2O	84	81	R
3	Me	0.2	Et_2O	81	80	R
4	CH_2OMe	1.0	Et_2O	85	82	R
5	CH_2OBn	1.0	Et_2O	76	91	R

[a]) Isolated yields. [b]) Enantiomeric excesses of the allylated product were determined by chiral HPLC analysis. [c]) Absolute stereochemistry determined to be (*R*) by CD spectrum of 2-methyl-2-propyl-4-methoxyindanone.

13.3.3 Diastereoselective Cyclizations

13.3.3.1 General considerations

Intramolecular free radical cyclizations are commonly employed for the formation of 5- or 6-membered ring heterocycles that can be extended to the total synthesis of natural products. Often, 5-*exo-trig* cyclizations are facile, and conditions typically utilized in these reactions are Bu_3SnH/AIBN in refluxing benzene. Previously, it was shown that radical cyclizations of α-halo amides proceeded to the desired product only when tertiary amides were used with bulky protecting groups (Eq. (13.32)) [44]. It was presumed that the bulky protecting group facilitated these reactions by favoring the *anti* conformer **106**, which is properly positioned to undergo cyclization.

105 ⇌ 106 (13.32)

It was recently discovered, however, that the bulky protecting group was unnecessary and that efficient 5-*exo-trig* cyclizations were also possible for secondary amides [45]. It was found that variation of substitution at the α carbon, ipso to the radical formed, and at the acceptor alkene also influenced the efficiency of these cyclizations (Eq. (13.33)). Substitution as in **107** allowed for the formation of **108** as the *trans-trans* adduct in 40–56% yield. Higher yields were obtained in refluxing toluene. Minor products included the simple reduction product of the halogen and the other diastereomers, which account for about 25% of the overall yield.

107 → (Bu_3SnH, AIBN; benzene or toluene) major-**108** (13.33)

It has been demonstrated that Lewis acids function to control rotamer populations of *N*-enoyloxazolidinones leading to selective intramolecular cyclizations [46]. Results indicate that organotin halides, the ubiquitous by-products of many radical reactions, can function as Lewis acids and alter the course of these reactions. Whereas strong Lewis acids such as $MgBr_2$ or Et_2AlCl gave only reduction product, weakly coordinating organotin halides still provided enough coordination to favor the *syn* conformer and allow for the 5-*exo* cyclization to take place. Equation (13.34) illustrates how this is applied in the formation of bicyclic oxazolidinones. Diastereoselectivities of >97% are observed with yields in the range of 80–90%. Reactions also worked well with $(TMS)_3SiH$ as the hydrogen atom donor and organotin halides as an additive.

109 → (Bu$_3$SnH, AIBN; Benzene, reflux) → major-110 (13.34)

Another slightly different approach to diastereoselective radical cyclizations uses group transfer methodology in order to access chiral tertiary alcohol moieties commonly found in natural products (Eq. (13.35)) [47]. The reaction occurs *anti* to the bulky *t*-butyl group, resulting in the formation of the major product **112**.

111 → (hν) → 112 (77%) + 113 (15%) + 114 (8%) (13.35)

An example of 1,3-asymmetric induction has been illustrated in the copper-mediated addition of electron-deficient radicals to alkenes [48]. The reaction is shown as in Eq. (13.36). The mechanism involves a single-electron transfer from copper, which forms the copper(I) halide as a by-product. This reaction also uses atom-transfer methodology to obtain halogen transfer at the γ position (**116**), which then readily lactonizes with the ester to form the product **117**.

115 → ($(CH_2)_8CO_2CH_3$ alkene; Cu powder; NaI, 120 °C, 5-9 h) → 116 → *syn*-117 (13.36)

It is noted that this reaction proceeds in the complete absence of solvent. The *syn* product is favored through relative 1,3-induction, reaching the highest levels of selectivity (2 : 1) when substituents are large. A model that accounts for the preferential formation of *syn* products via destabilization of the intermediate leading to *anti* products is shown in **118** (Eq. (13.37)).

118 → 119 → *anti*-120 (13.37)

Radical polymerization of acrylates can be used to make low-molecular-weight oligomers under high dilution conditions with a relatively large concentration of chain transfer reagent. These conditions were extended to the intramolecular cyclization of two acrylate entities tethered by a chiral diol in the formation of a remote stereocenter and a medium-sized ring (Eq. (13.38)) [49].

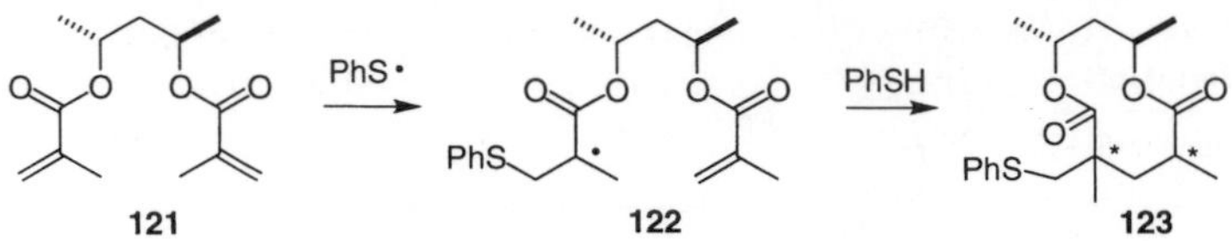

(13.38)

13.3.3.2 Protection as chiral acetal

5-*Exo* radical cyclizations of bromoacetals that contain allylic chiral centers are often referred to as Ueno-Stork reactions [50]. It was recently discovered that stereochemistry in these cyclizations could also be controlled by the acetal center, as shown in Eq. (13.39) [51]. A chair-like transition state (**125**) is envisaged where the alkene occupies a pseudoequatorial position leading to predominantly *cis* product **126**. Yields of approximately 70% and *cis*:*trans* ratios of up to 98:2 are obtained.

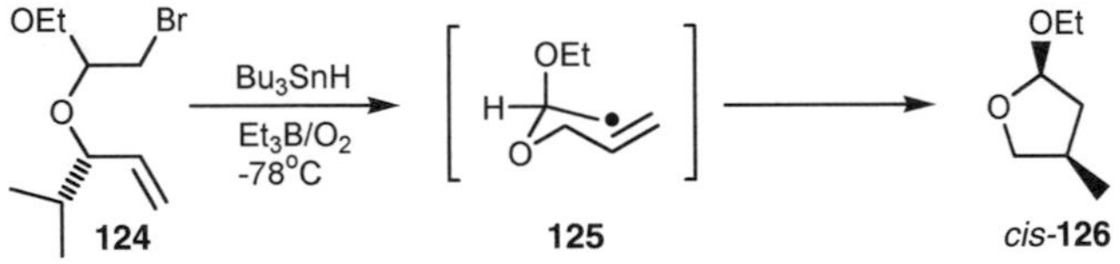

(13.39)

Sugars have also been used as chiral auxiliaries in acetal formation for diastereoselective radical cyclizations [52]. In Eq. (13.40) a chiral acetal is utilized to control the stereochemistry of a 5-*exo-dig* cyclization resulting in the formation of quaternary carbon-based stereocenters. Product **129** is formed as a single diastereomer in 35% yield. An allylic strain model is proposed to account for the stereochemical outcome of this reaction.

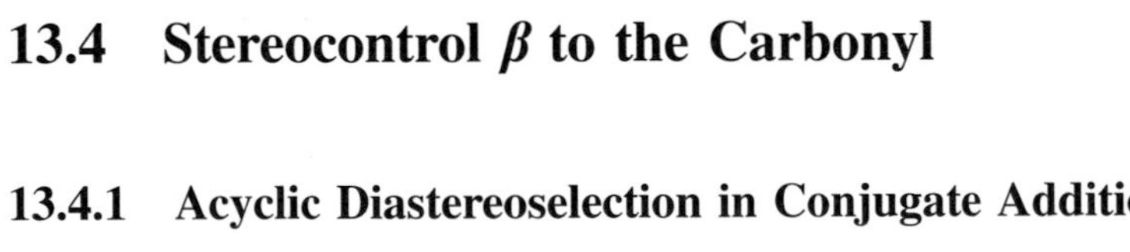

(13.40)

13.4 Stereocontrol β to the Carbonyl

13.4.1 Acyclic Diastereoselection in Conjugate Additions

Diastereoselective conjugate radical additions to control stereochemistry at the β center has been approached in a number of ways. Chiral auxiliaries have been exploited on the radical trap as well as the radical itself, producing products with high diastereoselectivities in both cases. 1,2-Selectivity has been induced by the application of chiral sulfoxides as α-sulfinyl enones, and relative stereochemistry can be controlled by allylic strain. Many types of β-functionalized products are available using this diastereoselective radical methodology including functionalized *anti*-aldol and α-amino acid products.

The realization of acyclic diastereoselectivity at the β center via conjugate radical additions has been approached in an analogous fashion to that utilized for selective allylations at the α center. Once again, starting with 4-diphenylmethyl-*N*-enoyloxazolidinone as the substrate, rotamer control is accessed through use of the appropriate bidentate Lewis acid to chelate both the carbonyl oxygens, and the *s-cis* rotamer is favored through steric interactions with the bulky diphenylmethyl portion of the chiral auxiliary. With the reactive rotamer **132** locked into place, β radical addition can occur selectively to the least sterically hindered face of the olefin. Examples of β-selective conjugate radical additions are shown in Eq. (13.41) Table 13-17 [53].

i-PrI, Bu_3SnH
Lewis acid
Et_3B/O_2, -78 °C

130 **131** (13.41)

Top face addition

132

Table 13-7. Diastereoselective conjugate additions using oxazolidinone auxiliaries.

Entry	R	Lewis acid (eq)	Solvent	Yield (%)	Ratio [a])
1	Me	none	CH_2Cl_2/Hex	60	1.3:1
2	Me	$BF_3{\cdot}OEt_2$ (2)	CH_2Cl_2/Hex	80	1.3:1
3	Me	$MgBr_2$ (2)	CH_2Cl_2/Hex/ether	90	6:1
4	Me	$Yb(OTf)_3$ (2)	CH_2Cl_2/Hex/THF	93	25:1
5	Me	$Yb(OTf)_3$ (0.3)	CH_2Cl_2/Hex/THF	90	20:1
6	Me	$Yb(OTf)_3$ (2)	CH_2Cl_2/Hex/THF/H_2O [b])	90	20:1
7	Ph	$MgBr_2$ (2)	CH_2Cl_2/Hex/ether	90	20:1
8	Ph	$Yb(OTf)_3$ (2)	CH_2Cl_2/Hex/THF	89	45:1

[a]) Diastereomer ratios were determined by ^{1}H NMR (400 MHz). [b]) Ten equivalents of water relative to substrate were added.

After the evaluation of several Lewis acids some highlights emerged. Lanthanide Lewis acids provided the highest levels of selectivity (entries 4 and 8) as did appropriate selection of the R group (compare entries 4 and 8). The use of substoichiometric Lewis acids had negligible effects on stereoselectivity or reaction efficiency (entry 5), and limited amounts of water did not hinder either the progress or the selectivity in this reaction (entry 6). It must also be noted that the level of selectivity achieved through free radical conjugate addition to oxazolidinone substrates rivals if not surpasses that obtained through ionic methodology [54]. The high yields and diastereoselectivities can be explained by the chelation model **132**. The bulky diphenylmethyl portion of the chiral auxiliary functions to block the back face of the enoyl fragment, leaving the top face exposed for facile radical addition. In addition, the Lewis acid in β additions not only functions to control the rotamer population, it also activates the β carbon towards nucleophilic alkyl radical addition, making this type of Lewis acid-mediated reaction amenable for catalytic reactions.

In a similar manner, conjugate radical additions to differentially protected fumarates allowed access to functionalized succinates [55]. Regioselectivity in these radical additions is provided by preferential Lewis acid activation of the β carbon by coordination of the imide carbonyl. The general scheme is illustrated in Eq. (13.42) (Table 13-8).

COOEt, 133; i-PrI, Bu_3SnH, Lewis acid, Et_3B/O_2, -78 °C → major-**134** + minor-**135** (13.42)

Table 13-8. Diastereoselective conjugate additions: desymmetrization of fumarates.

Entry	Lewis acid (eq)	Solvent	Yield (%)	dr[a])	**134**:**135**[b])
1	none	CH_2Cl_2	92	1.6:1	11:1
2	$BF_3{\cdot}OEt_2$ (1)	CH_2Cl_2	86	1.2:1	9:1
3	$Sm(OTf)_3$ (1)	CH_2Cl_2/THF, 4:1	95	29:1	>100:1
4	$Er(OTf)_3$ (1)	CH_2Cl_2/THF, 4:1	90	33:1	>100:1
5	$Er(OTf)_3$ (0.2)	CH_2Cl_2/THF, 4:1	88	3:1	11:1
6	$Yb(OTf)_3$ (1)	CH_2Cl_2/THF, 4:1	91	10:1	80:1

[a]) Diastereomeric ratios were determined by ^1H 400 MHz NMR of the crude reaction mixture. [b]) Regioselectivity was determined by ^1H 400 MHz NMR analysis of the crude reaction mixture.

This conjugate addition proceeds in excellent yields. High regio- and diastereoselectivity is observed with lanthanide and other Lewis acids (entries 3, 4 and 6), but little selectivity is seen in the absence of a chelating Lewis acid (entries 1 and 2) and with substoichiometric amounts of Lewis acid (entry 5). It had previously been established that radical addition to crotonates and cinnamates could be accomplished with catalytic amounts of Lewis acid. However, the fumarate substrate is orders of magnitude more reactive, and hence radical addition to the uncomplexed substrate presumably competes, leading to lower observed selectivity.

This methodology was combined with established *syn*-selective aldol methodology in the formation of trisubstituted butyrolactone natural products, (–)-roccellaric acid **137** and (–)-nephrosteranic acid **138** (Eq. (13.43)). These natural products were synthesized in four steps in 53% and 42% overall yields respectively.

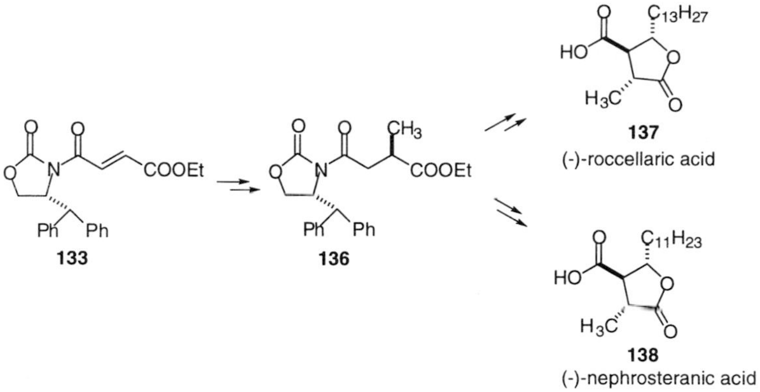

(13.43)

Diastereoselective β additions of stannyl radicals have also been achieved through the use of a chiral auxiliary and Lewis acid activation (Eq. (13.44)) [56]. It is observed that the *s-trans* conformation **140** is favored, and stannyl radical addition occurs to the face opposite the phenyl ring to give products **141**. One-point-binding Lewis acids such as $BF_3{\cdot}Et_2O$ were found to activate the substrate for stannyl radical addition. Without Lewis acid activation no reaction resulted. It was also apparent that larger R groups on the substrate gave higher selectivities; up to 19:1 with approximately 80% yield. Applications of chiral β-stannyl esters include cyclopropane synthesis and as reactive species in other organometallic reactions.

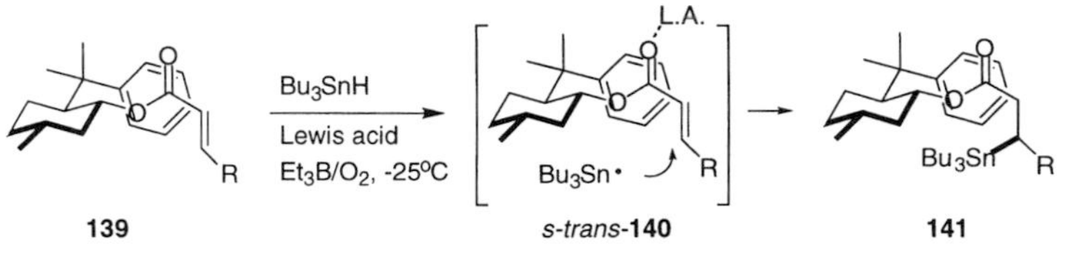

(13.44)

Stereoselective β additions have also been controlled by the introduction of chiral sulfinyl auxiliaries at the α position of alkenes. Several advantages of sulfinyl auxiliaries are pointed out. The use of bulky substituents on the chiral sulfoxide functions to shield one face of the β position of the alkene, allowing for control of radical addition. The dipole of the sulfoxide can be used to control rotamer populations with and without Lewis acids. In the absence of a chelating Lewis acid, dipole-dipole repulsion orients the sulfoxide and carbonyl groups in an *anti* conformation. On the other hand, bidentate Lewis acids can chelate these two Lewis basic functionalities and keep them oriented in a *syn* conformation. Finally, the sulfinyl group is easily removed or chemically transformed into other functionality.

Conjugate radical additions to α-sulfinyl cyclopentenones has met with much success (Eq. (13.45)) [57]. Yields >90% and selectivities >98:2 were obtained for the product **143**. Since the addition of nucleophilic alkyl radicals to electron-poor olefins is an exothermic process, the ground state conformation of the starting cyclopentenone substrate determines the likely path for the approaching radical in the transition state. Single crystal X-ray analysis of **143** confirmed the *anti* conformation of the sulfoxide oxygen and the carbonyl oxygen.

R_3B

142 major-**143** Ar = 2,4,6-(*i*-Pr)$_3$C$_6$H$_2$ (13.45)

This methodology was extended to the addition of functionalized carbon-based radicals generated by photo-irradiation of alcohols in the presence of benzophenone [58]. The general reaction mechanism is shown in Eq. (13.46). High diastereoselectivities of >98 : 2 and yields of 97–99% are obtained. This photo-induced addition of hydroxyalkyl radicals also proceeded with high selectivity and yield with acyclic substrates, contrary to what was observed with simple alkyl radicals.

hυ [PhCOPh]‡ PhCOPh PhĊPh H–O

144 **145** **146** (13.46)

Chiral auxiliaries have also been applied to the radicals themselves in the formation of chiral hydroxyalkyl radical equivalents [59]. Once again, stereocontrol is accessed through the use of chiral acetals, which are readily available in the form of sugars. Typical reactions of this type are shown in Eq. (13.47). First, the thiohydroxamate ester **148** is prepared so that radical intermediate **149** can be formed photolytically via Barton's radical decarboxylation protocol [60]. The chiral radical **149** can then be trapped by methyl acrylate in a 61% yield with an 11 : 1 diastereomeric preference for γ-substituted **150**.

DCC, CH_2Cl_2; hυ, -78 °C; 1. methyl acrylate (CO_2Me), 2. Ra-Ni

147 **148** **149** **150** (13.47)

Asymmetric aldol-type products are also accessible using chiral acetal methodology (Eq. (13.48)) [61]. Conjugate addition of a chiral radical (**151**) to 2-nitropropene **152** results in aldol products **154** following conversion of the resultant nitro thioether to a ketone. Maximum diastereoselectivities obtained using this methodology are 35:1 at –78 °C, with maximum yields just under 80%. The use of chiral dihydropyrans for formation of chiral acetals provides higher selectivity than acetals derived from sugars. The free aldol product can be obtained without dehydration following treatment with thiophenol and $BF_3{\cdot}Et_2O$.

R*O radical **151** + NO_2 alkene **152** → OR* NO_2 SPyr Me **153** —15% aq. $TiCl_3$, buffer 48°C→ OR* O Me **154** (13.48)

R* = chiral auxiliary

Relative stereochemistry can be established through allylic strain present in transition states. In this context, conjugate radical additions to β-substituted dehydroamino acid derivatives have been examined [62]. Allylic strain provides selectivity of the ensuing hydrogen atom trap, and relative stereochemistry between the α and β centers is established as shown in Eq. (13.49). The *anti* product **156** is modestly favored by a ratio of 2.3:1, and the combined yield of **156** and **157** is 41%.

155 (COOBn) —Bu_3SnH, AIBN, *t*-Bu-I, hυ→ *anti*-**156** (*t*-Bu, COOBn) + *syn*-**157** (*t*-Bu, COOBn) (13.49)

The facial selectivity of additions of *gem*-dihalocyclopropanes to electron-deficient olefins was recently studied (Eq. (13.50)) [63]. Results indicate that 2,3-*cis*-disubstituted dihalocyclopropanes **158** react with olefins **159**, with a high preference for the formation of the *exo*-adduct. Yields are typically in the range of 50% and *exo*:*endo* ratios of >99:1 are observed.

158 (Br, Br) + **159** (CO_2Me) —Bu_3SnH, AIBN, benzene, reflux→ *exo*-**160** (Br, CO_2Me) + *endo*-**161** (CO_2Me, Br) (13.50)

13.4.2 Acyclic Enantioselection

The success in diastereoselective Lewis acid-mediated conjugate radical additions using chiral oxazolidinones led to the development of enantioselective variants. The achiral template selection was based on literature precedents for rotamer control (*s-cis* vs *s-trans*) and the requirement for a bidentate Lewis acid chiral ligand combination for obtaining selectivity (Eq. (13.51)).

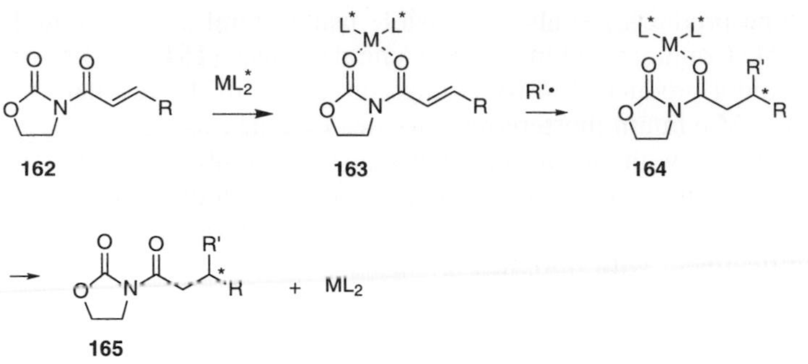

(13.51)

Chiral bisoxazoline ligands were initially chosen for these experiments (Eq. (13.52)). It was found that simple bisoxazoline ligands **167** derived from amino acids led to good yields but moderate selectivities of products **168** when catalytic amounts of chiral Lewis acid were used (67% ee with 20 mol% of the catalyst and 86% chemical yield).

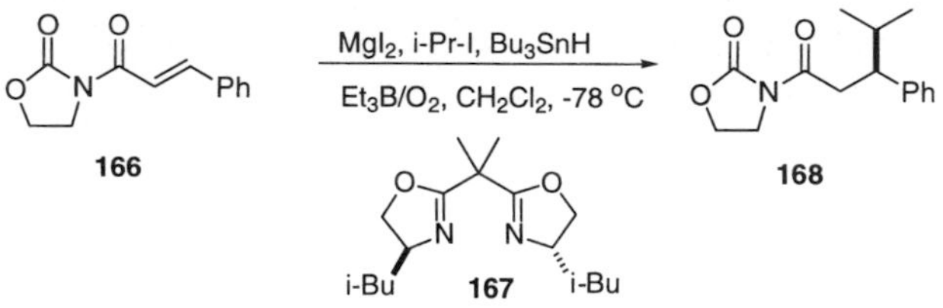

(13.52)

After some modifications to the bisoxazoline ligand, a practical and efficient route was now available for catalytic enantioselective conjugate radical additions [64]. These included changing the bite angle of the ligand by adding a cyclopropane functionality at the one-carbon bridge and making the ligand more rigid by using aminoindanols instead of phenyl glycinol in the formation of the ligand [65]. Equation (13.53) (Table 13-9) illustrates how the new cyclopropyl bisoxazoline ligand **169** improved the practical utility of conjugate radical additions.

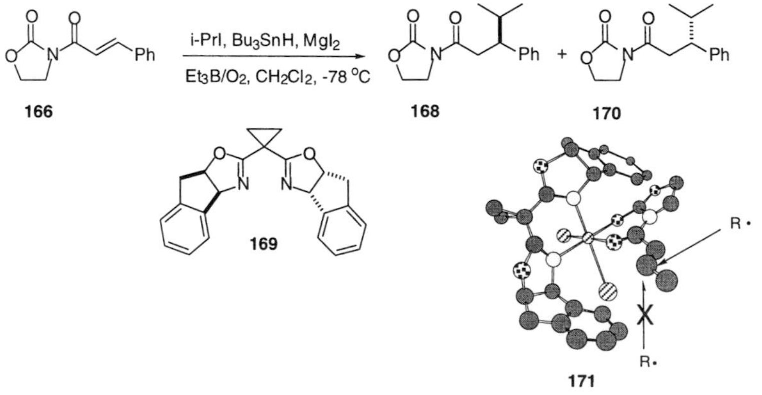

(13.53)

Table 13-9. Enantioselective conjugate additions.

Entry	Lewis acid (mol%)	T (°C)	yield (%) [a])	ee (%) [b])	er [c])
1	100	−78	88	93	28:1
2	30	−78	91	97	66:1
3	30	−20	93	95	39:1
4	30	25	87	93	28:1

[a]) Yields are for isolated and purified materials. [b]) ee's were determined by chiral HPLC analysis using a Chiralcel OD column. [c]) Enantiomeric ratios above 10:1 are rounded off to the nearest integer. The absolute stereochemistry of the product was established by hydrolysis to the known carboxylic acid and comparison of the sign of its rotation.

An octahedral model **171** is proposed and is consistent with the observed absolute stereochemistry of the product. In the case of the amino indanol-derived ligands, the ligand-MgI_2-substrate complex adopts an octahedral geometry where the two iodides have a *cis* orientation and the more Lewis-basic carbonyl oxygen is *trans* to the iodide [66]. The ring constraint and the larger bite angle in **171** provides for optimal face shielding in the radical addition and thus accounts for the high levels of enantioselectivity. Attack of the radical on the least hindered *re*-face of the substrate accounts for the observed absolute stereochemistry of the product (4S,5R-**171** gave *R* product).

The effect of varying the achiral template on enantioselective conjugate additions was also studied (Eq. (13.54)) [67]. *N*-Acylpyrazoles, templates capable of forming 5-membered chelates with a two-point-binding Lewis acid have been evaluated in conjugate radical additions. Addition of isopropyl radical to **172** using a chiral Lewis acid prepared from $Zn(OTf)_2$ and **169** gave **173** with moderate enantioselectivity (51% ee). In comparison, isopropyl radical addition to **166**, an oxazolidinone-derived substrate, using $Zn(OTf)_2$ and **169** gave product of opposite configuration (84% yield, 51% ee). Model **174** or a square planar model accounts for the product configuration. These experiments illustrate that achiral templates are thus convenient handles for the formation of products of opposite configuration.

H3C CH3 Ph **172** —i-PrI, Bu_3SnH, Et_3B/O_2 / $Zn(OTf)_2$/**169**→ H3C CH3 Ph **173**

OTf L* Zn OTf N O CH3 Ph **174** L* = Ligand **169**

(13.54)

13.4.3 Diastereoselective Cyclizations: Intramolecular Conjugate Addition

A wealth of information exists regarding intramolecular conjugate additions in the formation of cyclic compounds. Due to space limitations, only a small sample of these reactions will be discussed here. For a more thorough treatment the reader should consult a recent review on this subject [68].

Chiral auxiliaries such as 8-phenylmenthol have been employed to control facial selectivity of intramolecular conjugate cyclizations (Eq. (13.55)) [69]. With no other additives there is little preference, however, for either the *s-cis* or the *s-trans* rotamer. A noteworthy approach to this problem has been through the use of bulky aluminum-based Lewis acids, which coordinates the carbonyl oxygen and favors radical addition to the *s-trans* rotamer [70]. Not only was alkyl aluminum reagent useful as a chelating agent, it was also found to function as a radical initiator through homolytic cleavage of the carbon-aluminum bond [71]. Cyclization with MAD and no other radical initiator provided **176** in 90% yield and a 93 : 7 diastereomeric preference.

COOR*
MAD, Bu_3SnH
Toluene, -78 °C, dry air
CO_2R^*
175 R* = 8-phenylmenthyl
major-**176**
MAD =
Me, Al, *t*-Bu, *t*-Bu, *t*-Bu, *t*-Bu, Me, Me
L.A.
s-trans-**177**

(13.55)

Intramolecular conjugate radical additions are common strategies towards the synthesis of cyclic natural products. For instance, in the synthesis of dactomelynes prochiral halogenated radicals add to a β-alkoxyacrylate to form disubstituted tetrahydropyranyl rings (Eq. (13.56)) [72]. A transition state is adopted as in **179** where the bromide prefers an orientation away from the steric congestion of the ring. Compound **180** was obtained as a single diastereomer and the radical cyclization step proceeded in 75% yield.

Free-radical cyclizations onto β carbons have also been utilized in the formation of the tricyclic core ring structure of alkaloids gelsemine and oxogelsemine (Eq. (13.57)) [73]. Tricyclic compound **183** was formed as a single compound in 87% yield.

Examples of stereoselective cyclizations involving heteroatom radicals are rare. Tandem oxy radical cyclization and diastereoselective hydrogen atom transfer reactions, however, have also been studied (Eq. (13.58)) [74]. These reactions take advantage of chirality at the γ carbon to induce anti-β cycloaddition. Hydrogen

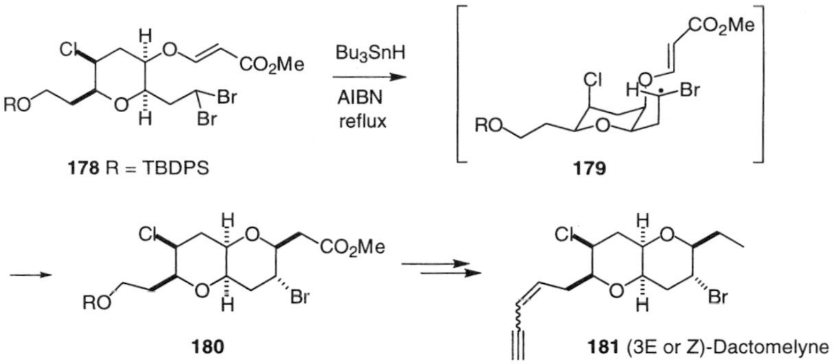

(13.56)

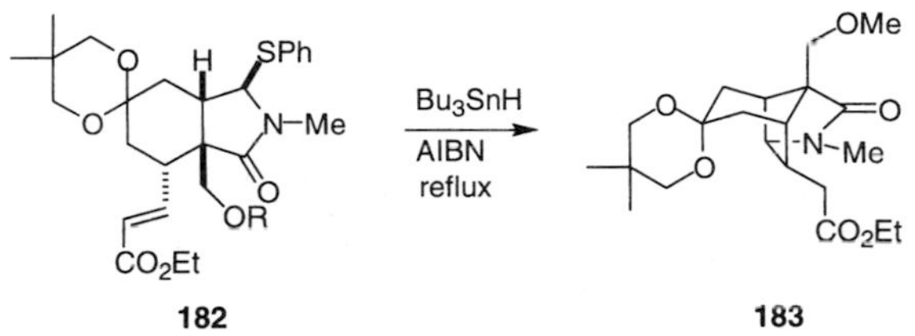

(13.57)

atom transfer is predicted to occur through transition state **187** to produce 2,5,8-*anti* product **186**. Modest yields of 50–60% were typically observed and *anti*/*syn* ratios reached a maximum of 22 : 1.

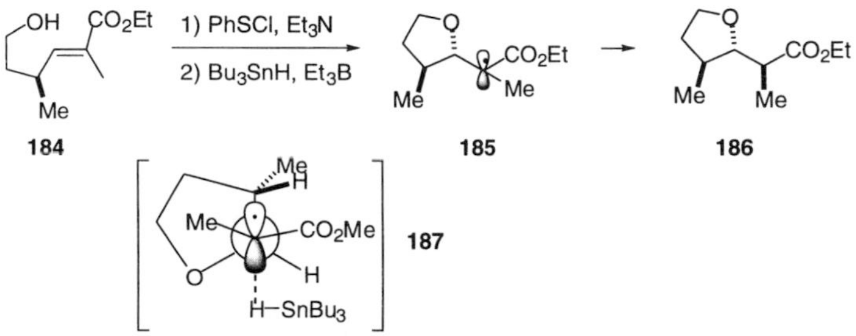

(13.58)

13.4.4 Enantioselective Cyclizations

Few examples have been reported demonstrating enantioselective cyclization methodology. One known example, however, is similar to the diastereoselective cyclization of **175**, which uses a menthol-derived chiral auxiliary and a bulky aluminum Lewis acid (see Eq. (13.55)). The enantioselective variant simply utilizes an achiral template **188** in conjunction with a bulky chiral binol-derived aluminum Lewis acid **189** (Eq. (13.59)) [75]. Once again the steric bulk of the chiral aluminum Lewis acid complex favors the *s-trans* rotamer of the acceptor olefin. Facial selectivity of the radical addition can then be controlled by the chiral Lewis acid. The highest selectivity (48% ee) was achieved with 4 equivalents of chiral Lewis acid, providing a yield of 63%.

188 R = cyclohexyl

Bu_3SnH, Et_3B
CH_2Cl_2, -78 °C, dry air

189

major-**190**

(13.59)

13.5 Conclusions

This review has detailed stereoselective bond construction strategies using free radicals with an emphasis on literature from the past five years. The carbonyl group plays a key role in the reactions described. Tremendous achievements have been made in this area, and the future is equally rosy for new and exciting accomplishments.

References

1. (a) Giese, B. *Radical in Organic Synthesis. Formation of Carbon-Carbon Bond*, Pergamon, Oxford, **1986**. (b) Giese, B. *Angew. Chem., Int. Ed. Engl.* **1989**, *28*, 969.
2. (a) Porter, N.A.; Giese, B.; Curran, D.P. *Acc. Chem. Res.* **1991**, *24*, 296–301. (b) Smadja, W. *Synlett.* **1994**, 1–26. (c) Beckwith, A.L.J. *Chem. Soc. Rev.* **1993**, 143–151.
3. Curran, D.P.; Porter, N.A.; Giese, B. *Stereochemistry of Radical Reactions*, VCH, Weinheim, **1995**. For an excellent recent review on Lewis acid mediated radical reactions see: Renaud, P.; Gerster, M. *Angew. Chem., Int. Ed. Engl.* **1998**, *37*, 2562–2579.
4. Sibi, M.P.; Porter, N.A. *Acc. Chem. Res.* **1999**, *32*, 163–171.
5. Molander, G.A.; Harris, C.I. *Chem. Rev.* **1996**, *96*, 307–363.
6. Snider, B.B. *Chem. Rev.* **1996**, *96*, 339–363.
7. Ogura, K.; Arai, T.; Kayano, A.; Akazome, M. *Tetrahedron Lett.* **1999**, *40*, 2537–2540.
8. Evans, P.A.; Roseman, J.D. *Tetrahedron Lett.* **1997**, *38*, 5249–5252.
9. Evans, P.A.; Roseman, J.D.; Garber, L.T. *J. Org. Chem.* **1996**, *61*, 4880–4881.
10. Evans, P.A.; Roseman, J.D. *J. Org. Chem.* **1996**, *61*, 2252–2253.
11. Evans, P.A.; Roseman, J.D. *Tetrahedron Lett.* **1995**, *36*, 31–34.
12. Handa, S.; Pattenden, G.; Li, W.-S. *Chem. Commun.* **1998**, 311–312.
13. Handa, S.; Pattenden, G. *J. Chem. Soc., Perkin Trans. 1* **1999**, 843–845.
14. Bertrand, M.P.; Crich, D.; Nouguier, R.; Samy, R.; Stien, D. *J. Org. Chem.* **1996**, *61*, 3588–3589. For other examples of dioxolanyl radicals in synthesis see: Rychnovsky, S.D.; Skalitzky, D.J. *Synlett.* **1995**, 555–556; Batsanov, A.S.; Begley, M.J.; Fletcher, R.J.; Murphy, J.A.; Sherburn, M.S. *J. Chem. Soc. Perkin Trans. 1* **1995**, 1281–1294.
15. Stien, D.; Samy, R.; Nouguier, R.; Crich, D.; Bertrand, M.P. *J. Org. Chem.* **1997**, *62*, 275–286.
16. Taniguchi, N.; Uemura, M. *Tetrahedron Lett.* **1997**, *38*, 7199–7202.
17. Merlic, C.A.; Walsh, J.C. *Tetrahedron Lett.* **1998**, *39*, 2083–2086.
18. Bach, T.; Jödicke, K.; Kather, K.; Fröhlich, R. *J. Am. Chem. Soc.* **1997**, *119*, 2437–2445.
19. Wyss, C.; Batra, R.; Lehmann, C.; Sauer, S.; Giese, B. *Angew. Chem. Int. Ed. Engl.* **1996**, *35*, 2529–2531.

20. For a discussion on 1,2-stereoinduction involving *β*-oxy radicals see: Hassler, C.; Batra, R.; Giese, B. *Tetrahedron Lett.* **1995**, *36*, 7639–7642 and Giese, B.; Bulliard, M.; Dickhaut, J.; Halbach, R.; Hassler, C.; Hoffmann, U.; Hinzen, B.; Senn, M. *Synlett.* **1995**, 116–118.
21. Guindon, Y.; Yoakim, C.; Lemieux, R.; Boisvert, L.; Delorme, D.; Lavallée, J.-F. *Tetrahedron Lett.* **1990**, *31*, 2845–2848.
22. Guindon, Y.; Faucher, A.-M.; Bourque, É.; Caron, V.; Jung, G.; Landry, S.R. *J. Org. Chem.* **1997**, *62*, 9276–9283.
23. Guindon, Y.; Rancourt, J. *J. Org. Chem.* **1998**, *63*, 6554–6565.
24. Sibi, M.P.; Ji, J. *Angew. Chem., Int. Ed. Engl.* **1996**, *35,* 190–192.
25. Sibi, M.P.; Ji, J. *J. Org. Chem.* **1996**, *61*, 6090–6091.
26. Rosenstein, I.J.; Tynan, T.A. *Tetrahedron Lett.* **1998**, *39*, 8429–8432.
27. Gerster, M.; Audergon, L.; Moufid, N.; Renaud, P. *Tetrahedron Lett.* **1996**, *37*, 6335–6338.
28. Guindon, Y.; Géurin, B.; Chabot, C.; Ogilvie, W. *J. Am. Chem. Soc.* **1996**, *118*, 12528–12535.
29. Hamon, D.P.; Massey-Westropp, R.A.; Razzino, P. *Tetrahedron* **1995**, *51*, 4183–4194.
30. Yamamoto, Y.; Onuki, S.; Yumoto, M.; Asao, N. *J. Am. Chem. Soc.* **1994**, *116*, 421–422.
31. Hanessian, S.; Yang, H.; Schaum, R. *J. Am. Chem. Soc.* **1996**, *118*, 2507–2508. For another example of 1,2-asymmetric induction controlled through hydrogen bonding see: Kündig, E.P.; Xu, L.-H.; Romanmens, P. *Tetrahedron Lett.* **1995**, *36*, 4047–4050.
32. Taber, D.F.; Gorski, G.J.; Liable-Sands, L.M.; Rheingold, A.L. *Tetrahedron Lett.* **1997**, *38*, 6317–6318.
33. Axon, J.R.; Beckwith, A.L.J. *J. Chem. Soc., Chem. Commun.* **1995**, 549–550. For reactions at a center remote to the carbonyl see: Beaulieu, F., Arora, J., Veith, U., Taylor, N.J., Chapell, B.J., Snieckus, V. *J. Am. Chem. Soc.* **1996**, *118*, 8727–8728.
34. Pyne, S.G.; Schafer, K. *Tetrahedron* **1998**, *54*, 5709–5720. For reactions with lactones see: Piber, M.; Leahy, J.W. *Tetrahedron Lett.* **1998**, *39*, 2043–2046.
35. Booth, S.E.; Jenkins, P.R.; Swain, C.J.; Sweeny, J.B. *J. Chem. Soc., Perkin Trans. 1* **1994**, 3499–3508.
36. Miyabe, H.; Ushiro, C.; Naito, T. *J. Chem. Soc., Chem. Commun.* **1997**, 1789–1790.
37. Murakata, M.; Tsutsui, H.; Hoshino, O. *J. Chem. Soc., Chem. Commun.* **1995**, 481–482.
38. Nanni, D.; Curran, D.P. *Tetrahedron: Asymmetry* **1996**, *7*, 2417–2422.
39. Wu, J.H.; Radinov, R.; Porter, N.A. *J. Am. Chem. Soc.* **1995**, *117*, 11029–11030.
40. Wu, J.H.; Zhang, G.; Porter, N.A. *Tetrahedron Lett.* **1997**, *38*, 2067–2070.
41. (a) Corey, E.J.; Ishihara, K. *Tetrahedron Lett.* **1992**, *33*, 6807–6810. (b) Evans, D.A.; Miller, S.J.; Lectka, T. *J. Am. Chem. Soc.* **1993**, *115*, 6460–6461.
42. Fhal, A.-R.; Renaud, P. *Tetrahedron Lett.* **1997**, *38*, 2661–2664.
43. (a) Murakata, M.; Jono, T.; Mizuno, Y.; Hoshino, O. *J. Am. Chem. Soc.* **1997**, *119*, 11713–11714. (b) Murakata, M.; Jono, T.; Hoshino, O. *Tetrahedron: Asymmetry* **1998**, *9*, 2087–2092.
44. (a) Sato, T.; Wada, Y.; Nishimoto, M.; Ishibashi, H.; Ikeda, M. *J. Chem. Soc., Perkin Trans.1* **1989**, 879–886. (b) Stork, G.; Mah, R. *Heterocycles* **1989**, *28*, 723–727.
45. Bryans, J.S.; Large, J.M.; Parsons, A.F. *Tetrahedron Lett.* **1999**, *40*, 3487–3490.
46. Sibi, M.P.; Ji, J. *J. Am. Chem. Soc.* **1996**, *118*, 3063–3064.
47. Abazi, S.; Rapado, L.P.; Schenk, K.; Renaud, P. *Eur. J. Org. Chem.* **1999**, 477–483.
48. Metzger, J.O.; Mahler, R.; Francke, G. *Liebigs Ann./Recueil* **1997**, 2303–2313.
49. Sugimura, T.; Nagano, S.; Tai, A. *Tetrahedron Lett.* **1997**, *38*, 3547–3548.
50. (a) Ueno, Y.; Chino, K.; Watanabe, M.; Moriya, O.; Okawara, M.J. *J. Am. Chem. Soc.* **1982**, *104*, 5564–5566. (b) Ueno, Y.; Moriya, O.; Chino, K.; Watanabe, M.; Okawara, M.J. *J. Chem. Soc., Perkin Trans. 1* **1986**, 1351–1356. (c) Stork, G.; Mook, R.; Biller, S.A.; Rychnovsky, S.D. *J. Am. Chem. Soc.* **1983**, *105*, 3741–3742.
51. Villar, F.; Renaud, P. *Tetrahedron Lett.* **1998**, *39*, 8655–8658.
52. McCague, R.; Pritchard, R.G.; Stoodley, R.J.; Williamson, D.S. *Chem. Commun.* **1998**, 2691–2692.
53. Sibi, M.P.; Jasperse, C.P.; Ji, J. *J. Am. Chem. Soc.* **1995**, *117*, 10779–10780.
54. For conjugate addition using oxazolidinone auxiliaries, see: Nicolas, E.; Russel, K.C.; Hruby, V.J. *J. Org. Chem.* **1993**, *58*, 766–770.
55. Sibi, M.P.; Ji, J. *Angew. Chem., Int. Ed. Engl.* **1997**, *36*, 274–276.
56. Nishida, M.; Nishida, A.; Kawahara, N. *J. Org. Chem.* **1996**, *61*, 3574–3575.

57. Toru, T.; Watanabe, Y.; Mase, N.; Tsusake, M.; Hayakawa, T.; Ueno, Y. *Pure and Appl. Chem.* **1996**, *68*, 711–714.
58. Mase, N.; Watanabe, Y.; Toru, T. *Bull. Chem. Soc. Jpn.* **1998**, *71*, 2957–2965.
59. Garner, P.P.; Cox, P.B.; Klippenstein, S.J. *J. Am. Chem. Soc.* **1995**, *117*, 4183–4184.
60. Barton, D.H.R.; Crich, D.; Kretzschmar, G.J. *J. Chem. Soc., Perkin Trans. 1* **1986**, 39–53.
61. Garner, P.P.; Leslie, R.; Anderson, J.T. *J. Org. Chem.* **1996**, *61*, 6754–6755.
62. Renaud, P.; Stojanovic, A. *Tetrahedron Lett.* **1996**, *37*, 2569–2572. (b) Stojanovic, A.; Renaud, P. *Helv. Chim. Acta* **1998**, *81*, 268–284.
63. Tanabe, Y.; Wakimura, K.-I.; Nishii, Y. *Tetrahedron Lett.* **1996**, *37*, 1837–1840.
64. (a) Sibi, M.P.; Ji, J.; Wu, J.H.; Gurtler, S.; Porter, N.A. *J. Am. Chem. Soc.* **1996**, *118*, 9200–9201. (b) Sibi, M.P.; Ji, J. *J. Org. Chem.* **1997**, *62*, 3800–3801.
65. Davies, I.W.; Gerena, L.; Castonguay, L.; Senanayake, C.H.; Larsen, R.D.; Verhoeven, T.R.; Reider, P.J. *J. Chem. Soc., Chem. Commun.* **1996**, 1753–1754.
66. For work on octahedral *cis*-models using iron Lewis acids see: Corey, E.J.; Imai, N.; Zhang, H.-Y. *J. Am. Chem. Soc.* **1991**, *113*, 728–729. For an octahedral model using Mg Lewis acid see: Desimoni, G.; Faita, G.; Righetti, P. P. *Tetrahedron Lett.* **1996**, *37*, 3027–3030.
67. Sibi, M.P.; Shay, J.J.; Ji, J. *Tetrahedron Lett.* **1997**, *38*, 5955–5958.
68. For a recent review see: Naito, T. *Heterocycles* **1999**, *50*, 505–541.
69. Nozaki, K.; Oshima, K.; Utimoto, K. *J. Am. Chem. Soc.* **1987**, *109*, 2547–2549.
70. Nishida, M.; Hayashi, H.; Yonemitsu, O. *Synlett.* **1995**, 1045–1046.
71. For literature precedence regarding the formation of alkyl radicals from alkylaluminum species see: Ruck, K.; Kunz, H. *Synthesis* **1993**, 1018–1028.
72. Lee, E.; Park, C.M.; Yun, J.S. *J. Am. Chem. Soc.* **1995**, *117*, 8017–8018. Also see: (a) Yuasa, Y.; Sato, W.; Shibuya, S. *Synth. Commun.* **1997**, *27*, 573–585. (b) Cossy, J.; Salle, L. *Tetrahedron Lett.* **1995**, *36*, 7235–7238.
73. Atarashi, S.; Choi, J.-K.; Ha, D.-C.; Hart, D.J.; Kuzmich, D.; Lee, C.-S.; Ramesh, S.; Wu, S. C. *J. Am. Chem. Soc.* **1997**, *119*, 6226–6241.
74. Guindon, Y.; Denis, R.C. *Tetrahedron Lett.* **1998**, *39*, 339–342.
75. Nishida, M.; Hayashi, H.; Nishida, A.; Kawahara, N. *J. Chem. Soc., Chem. Commun.* **1996**, 579–580.

14 Activation of Carbonyl and Related Compounds in Aqueous Media

Shū Kobayashi, Kei Manabe, Satoshi Nagayama

14.1 Introduction

Lewis acid activation of C=O and C=N groups has been of great interest in organic synthesis [1]. While various kinds of Lewis acid-promoted reactions have been developed and many have been applied in industry, these reactions must be carried out under strictly anhydrous conditions. The presence of even a small amount of water stops the reaction, because most Lewis acids immediately react with water rather than the substrates and decompose or deactivate, and this has restricted the use of Lewis acids in organic synthesis.

On the other hand, in the course of our investigations to develop new synthetic methods, we have found that rare earth metal triflates [$Sc(OTf)_3$, $Yb(OTf)_3$, etc.] [2] and some other metal salts can be used as water-stable Lewis acids for activation of C=O and C=N groups in water-containing solvents.

In this chapter, our research on use of the Lewis acid catalysts in carbon-carbon bond-forming reactions in aqueous solvents is overviewed.

14.2 Activation of C=O in Aqueous Media

14.2.1 Aldol Reaction

14.2.1.1 Lanthanide triflate-catalyzed aldol reactions in water-containing solvents

The titanium tetrachloride ($TiCl_4$)-mediated aldol reaction of silyl enol ethers with aldehydes was first reported in 1973 [3]. The reaction (Mukaiyama aldol reaction) is notably distinguished from the conventional aldol reactions carried out under basic conditions; it proceeds in a highly regioselective manner to afford cross aldols in high yields [4]. Since this pioneering effort, several efficient activators (trityl salts [5], clay montmorillonite [6], fluoride anions [7], etc. [8]) have been developed to realize high yields and selectivities, and now the reaction is considered to be one of the most important carbon-carbon bond-forming reactions in organic synthesis. These reactions are usually carried out under strictly anhydrous conditions. The presence of even a small amount of water causes lower yields, probably due to the rapid decomposition or deactivation of the promoters and the hydrolysis of the silyl enol ethers. Furthermore, the promoters cannot be recovered and reused because they decompose under usual quenching conditions.

On the other hand, the water-promoted aldol reactions of silyl enol ethers with aldehydes were reported in 1986 [9]. While the fact that the aldol reactions proceeded without catalyst in water was very novel, the yields and the substrate scope were not satisfactory. In 1991, we reported the first example of Lewis acid catalysis in aqueous media [10], namely the hydroxymethylation reaction of silyl enol ethers with commercial formaldehyde solution using lanthanide triflates [Ln(OTf)$_3$]. Formaldehyde is a versatile reagent, being one of the most highly reactive C1 electrophiles in organic synthesis [11]. Dry gaseous formaldehyde required for many reactions has some disadvantage because it must be generated before use from solid polymeric paraformaldehyde by way of thermal depolymerization, and it self-polymerizes easily [12]. On the other hand, commercial formaldehyde solution, which is an aqueous solution containing 37% formaldehyde and 8–10% methanol, is cheap, easy to handle, and stable even at room temperature. However, the use of this reagent is strongly restricted due to the presence of a large amount of water. For example, the $TiCl_4$-promoted hydroxymethylation reaction of silyl enol ethers was carried out by using trioxane as an HCHO source [3b, 13]. Formaldehyde/water solution could not be used because $TiCl_4$ and the silyl enol ether reacted with the water rather than the HCHO in the solution.

Lanthanide compounds were expected to act as strong Lewis acids because of their hard character and to have a strong affinity toward carbonyl oxygens [14]. Among these compounds, Ln(OTf)$_3$ compounds were expected to be some of the strongest Lewis acids because of the electron-withdrawing trifluoromethanesulfonyl group. Their hydrolysis was postulated to be slow based on their hydration energies and hydrolysis constants [15]. In fact, while most metal triflates are prepared under strictly anhydrous conditions, lanthanide triflates are reported to be prepared in aqueous solution [16, 17]. The large radius and the specific coordination number of lanthanide(III) are also unique, and we initiated research on the use of Ln(OTf)$_3$ as a water-tolerant Lewis acid.

The effects of Ln(OTf)$_3$ in the reaction of 1-phenyl-1-trimethylsiloxypropene (**1**) with commercial formaldehyde solution were examined [10]. In most cases, the reactions proceeded smoothly to give the corresponding adducts in high yields. The reactions were most effectively carried out in commercial formaldehyde solution-THF media under the influence of Yb(OTf)$_3$ (Eq. (14.1)). It is noted that only a catalytic amount of Yb(OTf)$_3$ was required to complete the reaction. The amount of the catalyst was examined, and the reaction was found to be

OSiMe$_3$ Ph **1** + HCHO aq. —[Yb(OTf)$_3$ (10 mol %), H_2O–THF]→ Ph(C=O)CH(CH$_3$)CH$_2$OH (14.1)

catalyzed by even 1 mol% of $Yb(OTf)_3$: 1 mol% (90% yield); 5 mol% (90% yield); 10 mol% (94% yield); 20 mol% (94% yield); 100 mol% (94% yield).

The use of $Ln(OTf)_3$ in the activation of aldehydes other than formaldehyde was also investigated [18]. Several examples of the present aldol reaction of silyl enol ethers with aldehydes are listed in Table 14-1. In every case, the aldol adducts were obtained in high yields in the presence of a catalytic amount of $Yb(OTf)_3$, $Gd(OTf)_3$, or $Lu(OTf)_3$ in aqueous media. Diastereoselectivities were generally good to moderate. One feature in the present reaction is that water-soluble aldehydes, for instance, acetaldehyde, acrolein, and chloroacetaldehyde, can be reacted with silyl enol ethers to afford the corresponding cross aldol adducts in high yields (entries 5–7). Some of these aldehydes are commercially supplied as water solutions and are appropriate for direct use. Phenylglyoxal monohydrate also worked well (entry 8). It is known that water often interferes with the aldol reactions of aldehydes with metal enolates and that, in the cases where such water

R^1CHO + $R^2C(OSiMe_3)=CHR^3$ —[$Yb(OTf)_3$ (10 mol %), H_2O–THF, rt]→ $R^1CH(OH)CH(R^3)C(=O)R^2$

Table 14-1. Lanthanide triflate-catalyzed aldol reactions in aqueous media.

Entry	Aldehyde	Silyl enol ether	Yield/%
1	PhCHO	1-(trimethylsilyloxy)cyclohexene	91 [a])
2	PhCHO	OSiMe3 enol ether (3-(trimethylsilyloxy)pent-2-ene)	89 [b])
3	PhCHO	OSiMe3 enol ether (isomer)	93 [c])
4	PhCHO	Ph(OSiMe3)C=CHCH3 **1**	81 [d])
5	CH_3CHO	**1**	93 [e, f])
6	CH2=CH–CHO	**1**	82 [e, g])
7	Cl–CH2–CHO	**1**	95 [h])
8	PhCOCHO•H_2O	**1**	67 [i])
9	2-HO-C6H4-CHO	**1**	81 [j, k])
10	2-pyridinecarboxaldehyde	**1**	87 [j, l])

[a]) *syn/anti* = 73/27. [b]) *syn/anti* = 63/37. [c]) *syn/anti* = 71/29. [d]) *syn/anti* = 53/47. [e]) $Gd(OTf)_3$ was used instead of $Yb(OTf)_3$. [f]) *syn/anti* = 46/54. [g]) *syn/anti* = 60/40. [h]) *syn/anti* = 45/55. [i]) *syn/anti* = 27/73. [j]) $Lu(OTf)_3$ was used instead of $Yb(OTf_3)$. [k]) *syn/anti* = 55/45. [l]) *syn/anti* = 42/58.

soluble aldehydes are employed, some troublesome purifications including dehydration are necessary. Moreover, salicylaldehyde (entry 9) and 2-pyridinecarboxaldehyde (entry 10) could be successfully employed. The former has a free hydroxy group which is incompatible with metal enolates or Lewis acids, and the latter is generally difficult to use under the influence of Lewis acids because the nitrogen atom coordinates to the Lewis acids resulting in the deactivation of the acids.

The aldol reactions of aldehydes with silyl enol ethers were also found to proceed smoothly in water-ethanol-toluene (1:7:4) [19]. Some reactions proceeded much faster in this solvent system than in THF-water. Furthermore, the new solvent system enabled continuous use of the catalyst by a very simple procedure. Although the water-ethanol-toluene system was one phase, it easily became two phases by adding toluene after the reaction was completed. The product was isolated from the organic layer by a usual work-up. On the other hand, the catalyst remained in the aqueous layer, which was used directly in the next reaction without removing water. It is noteworthy that the yields of the 2nd, 3rd, and 4th runs were comparable to that of the 1st run (Eq. (14.2)).

CHO + OSiMe3 Ph **1** — $Yb(OTf)_3$ (10 mol %), H_2O/EtOH/toluene (1 : 7 : 4), rt, 8 h → OH O Ph (N)

1st run: 86% (*syn/anti* = 38/62)
2nd run: 82% (*syn/anti* = 38/62)
3rd run: 90% (*syn/anti* = 38/62)
4th run: 82% (*syn/anti* = 39/61)

(14.2)

While continuous use of $Ln(OTf)_3$ is possible, it is also easy to recover $Ln(OTf)_3$ compounds themselves. Lanthanide triflates are more soluble in water than in organic solvents such as dichloromethane. Almost 100% of $Ln(OTf)_3$ was quite easily recovered from the aqueous layer after the reaction was completed and could be reused. The reactions are usually quenched with water and the products are extracted with an organic solvent (for example, dichloromethane). The lanthanide triflate is in the aqueous layer and removal of the water is all that is required to give the catalyst which can be used in the next reaction (Scheme 14-1). It is noteworthy that lanthanide triflates are expected to solve some severe environmental problems induced by Lewis acid-promoted reactions in industry chemistry [20].

14.2.1.2 Aldol reaction catalyzed by various metal salts

Although the element scandium is in group 3 and lies above La and Y, its use in organic synthesis is rather limited in spite of its promising properties. In the course of our investigations to search for novel Lewis acid catalysts, especially metal triflates, we focused on the element scandium.

$Sc(OTf)_3$ was found to behave as a Lewis acid catalyst in aqueous media [21]. $Sc(OTf)_3$ was stable in water and was effective in the aldol reactions of aldehydes

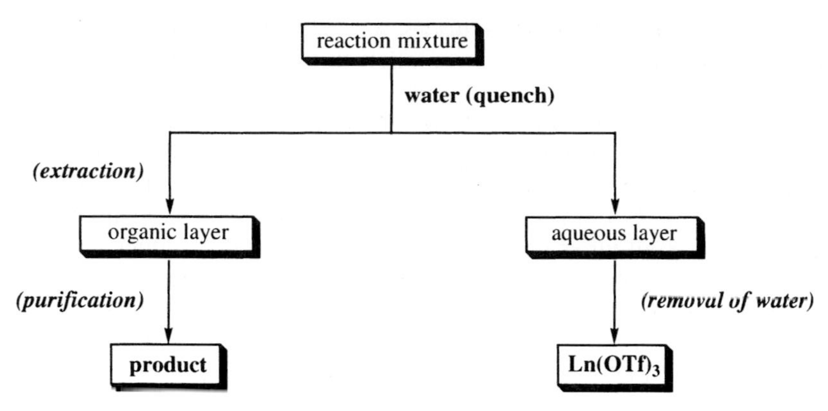

Scheme 14-1. Recover of the catalyst.

with silyl enolates in aqueous media (Table 14-2). The reactions of the usual aromatic and aliphatic aldehydes such as benzaldehyde and 3-phenylpropanal with silyl enolates were carried out in aqueous solvents, and water-soluble formaldehyde and chloroacetaldehyde were directly treated as water solutions with silyl enolates to afford the aldol adducts in good yields. Moreover, the catalyst could be recovered almost quantitatively from the aqueous layer after the reaction was completed. The recovered catalyst was also effective in the 2nd reaction, and the yield of the 2nd run was comparable to that of the 1st run.

R^1CHO + $R^2R^3C{=}C(OSiMe_3)R^4$ $\xrightarrow[H_2O\text{–}THF\ (1:9),\ rt]{Sc(OTf)_3\ (5\ mol\ \%)}$ $R^1CH(OH)CR^2R^3C(O)R^4$

Table 14-2. $Sc(OTf)_3$-Catalyzed aldol reaction in aqueous media.

Entry	Aldehyde	Silyl enol ether	Yield%
1	HCHO aq.	$CH_3CH{=}C(OSiMe_3)Ph$	84
2	$ClCH_2CHO$ aq.	$CH_3CH{=}C(OSiMe_3)Ph$	66 [a])
3	PhCHO	1-(trimethylsiloxy)cyclohexene ($OSiMe_3$)	83 [b])
4	PhCHO	$(CH_3)_2C{=}C(OSiMe_3)SMe$	92
5	$Ph(CH_2)_2CHO$	$(CH_3)_2C{=}C(OSiMe_3)SMe$	73

[a]) *syn/anti*=47/53. [b]) *syn/anti*=62/38. The diastereomer ratios were determined by ^{1}H NMR

We also screened group 1–15 metal chlorides searching for Lewis acids stable in aqueous solvents (Table 14-3) [22]. As a model, the reaction of benzaldehyde with (*Z*)-1-phenyl-1-(trimethylsiloxy)propene was selected. In the first screening, the chloride salts of Fe(II), Cu(II), Zn(II), Cd(II), In(III), and Pb(II) as well as the rare earth metals [Sc(III), Y(III), Ln(III)] gave promising yields. When the chloride salts of B(III), Si(IV), P(III), P(V), Ti(IV), V(III), Ge(IV), Zr(IV), Nb(V), Mo(V), Sn(IV), Sb(V), Hf(IV), Ta(V), W(VI), Re(V), and Tl(III) were used, decomposition of the silyl enol ether occurred rapidly and no aldol adduct was obtained. This is because hydrolysis of such metal chlorides is very fast and the silyl enol ether was protonated and then hydrolyzed to afford the corresponding ketone. On the other hand, no product or only a trace amount of the product was detected using the metal chloride salts of Li(I), Na(I), Mg(II), Al(III), K(I), Ca(II),

PhCHO + $OSiMe_3$ Ph —[MX_n (0.2 eq.); H_2O:THF = 1:9, rt, 12 h]→ Ph OH O Ph

Table 14-3. Effect of metal salts in the aldol reaction [a]).

MX_n	Yield%	MX_n	Yield%
$AlCl_3$	Trace	$InCl_3$	**68**
$ScCl_3$	**70 (78)** [b]	$In(ClO_4)_3$	14
Sc$(ClO_4)_3$	**82**	$SnCl_2$	4
$CrCl_3$	Trace	**La**$(OTf)_3$	**80**
$MnCl_2$	Trace	**Ce**$(OTf)_3$	**81**
$Mn(ClO_4)_2$	18 (40) [b]	**Pr**$(OTf)_3$	**83**
$FeCl_2$	39	**Nd**$(OTf)_3$	**78**
Fe$(ClO_4)_2$	26 (**55**) [b])	**Sm**$(OTf)_3$	**85**
$FeCl_3$	21	**Eu**$(OTf)_3$	**88**
$Fe(ClO_4)_3$	7	**Gd**$(OTf)_3$	**90**
$CoCl_2$	Trace	**Tb**$(OTf)_3$	**81**
$Co(ClO_4)_2$	17 (7) [b]	**Dy**$(OTf)_3$	**85**
$NiCl_2$	Trace	**Ho**$(OTf)_3$	**89**
$Ni(ClO_4)_2$	17 (7) [b])	**Er**$(OTf)_3$	**86**
$CuCl_2$	25	**Tm**$(OTf)_3$	**85**
Cu$(ClO_4)_2$	47 (**81**) [b])	**$YbCl_3$**	11 (**92**) [b])
$ZnCl_2$	10	**Yb**$(ClO_4)_3$	**84**
Zn$(ClO_4)_2$	46 (**57**)	**Yb**$(OTf)_3$	**92**
$GaCl_3$	Trace	**Lu**$(OTf)_3$	**84**
YCl_3	5 (**86**)	$IrCl_3$	Trace
Y$(ClO_4)_3$	**90**	$PtCl_2$	Trace
$RhCl_3$	Trace	AuCl	Trace
$PdCl_2$	Trace	$HgCl_2$	Trace
AgCl	Trace	HgCl	Trace
$AgClO_4$	42 (36) [b]	$PbCl_2$	15
$CdCl_2$	18	**Pb**$(ClO_4)_2$	**59 (65)** [b])
Cd$(ClO_4)_2$	49 (**72**) [b])	$BiCl_3$	Trace

[a]) No adduct was obtained and the starting materials were recovered when LiCl, NaCl, $MgCl_2$, PCl_3, KCl, $CaCl_2$, $RuCl_3$, $SbCl_3$, $BaCl_2$, and $OsCl_3$ were used. No adduct was obtained and the silyl enol ether was decomposed when BCl_3, $SiCl_4$, PCl_5, $TiCl_4$, VCl_3, $GeCl_4$, $ZrCl_4$, $NbCl_5$, $MoCl_5$, $SnCl_4$, $SbCl_5$, $HfCl_4$, $TaCl_5$, WCl_6, $ReCl_6$ and $TlCl_3$ were used. [b]) H_2O : EtOH : toluene = 1 : 7 : 4.

Cr(III), Mn(II), Co(II), Ni(II), Ga(III), Ru(III), Rh(III), Pd(II), Ag(I), Ba(II), Os(III), Ir(III), Pt(II), Au(I), Hg(II), and Bi(III). Some of these salts are stable in water, but have low catalytic ability. We also carried out the same aldol reaction using the corresponding perchlorates or triflates for the more promising metals (Table 14-3). It was found that Lewis acids based on Fe(II), Cu(II), Zn(II), Cd(II), and Pb(II) as well as the rare earth metals Sc(III), Y(III), and Ln(III) were both stable and active in water. Mn(II) and Ag(I) perchlorates gave moderate yields of the aldol adduct.

We noticed a correlation between the catalytic activity in water and the hydrolysis constants and exchange rate constants for substitution of inner-sphere water ligands [water exchange rate constant (WERC)]. The pK_h values (K_h=hydrolysis constant) [15, 23] and WERC [24] of the cations are shown in Table 14-4. The metal compounds which gave more than 50% yields in the aldol reaction have pK_h values from 4.3 to 10.08 *and* WERC greater than 3.2×10^6 $M^{-1}s^{-1}$. There was no exception in all the metal compounds we tested. These findings should provide a reliable indication of the catalytic ability of Lewis acids in aqueous solvents.

Table 14-4. Hydrolysis constants [a]) and exchange rate constants for substitution of inner-sphere water ligands [b]).

Element	Charge	pK_h	WERC
Li	+1	13.64	4.7×10^7
Be		—	—
B	+3	—	—
C		—	—
N		—	—
Na	+1	14.18	1.9×10^8
Mg	+2	11.44	5.3×10^5
Al	+3	1.14	1.6×10^0
Si	+4	—	—
P	+5	—	—
K	+1	14.46	1.5×10^8
Ca	+2	12.85	5×10^7
Sc	+3	4.3	4.8×10^7
Ti	+4	≤2.3	—
V	+3	2.26	1×10^3
Cr	+3	4.0	5.8×10^{-7}
Mn	+2	10.59	3.1×10^7
Fe	+2	9.5	3.2×10^6
Co	+2	9.65	2×10^5
Ni	+2	9.86	2.7×10^4
Cu	+2	7.53	2×10^8
Zn	+2	8.96	5×10^8
Ga	+3	2.6	7.6×10^2
Ge	+4	—	—
As		—	—
Rb		—	—
Sr		—	—
Y	+3	7.7	1.3×10^7
Zr	+4	0.22	—
Nb	+5	(0.6)	—
Mo	+5	—	—
Tc		—	—
Ru	+3	—	—
Rh	+3	3.4	3×10^{-8}
Pd	+2	2.3	—
Ag	+1	12	$>5\times10^6$
Cd	+2	10.08	$>1\times10^8$
In	+3	4.00	4.0×10^4
Sn	+4	—	—
Sb	+5	—	—
Cs		—	—
Ba	+2	13.47	$>6\times10^7$
Ln	+3	7.6-8.5	10^6-10^8
Hf	+4	0.25	—
Ta	+5	(-1)	—
W	+6	—	—
Re	+5	—	—
Os	+3	—	—
Ir	+3	—	—
Pt	+2	4.8	—
Au	+1	—	—
Hg	+2	3.40	2×10^9
Tl	+3	0.62	7×10^5
Pb	+2	7.71	7.5×10^9
Bi	+3	1.09	—
La	+3	8.5	2.1×10^8
Ce	+3	8.3	2.7×10^8
Pr	+3	8.1	3.1×10^8
Nd	+3	8.0	3.9×10^8
Pm		—	—
Sm	+3	7.9	5.9×10^8
Eu	+3	7.8	6.5×10^8
Gd	+3	8.0	6.3×10^7
Tb	+3	7.9	7.8×10^7
Dy	+3	8.0	6.3×10^7
Ho	+3	8.0	6.1×10^7
Er	+3	7.9	1.4×10^8
Tm	+3	7.7	6.4×10^6
Yb	+3	7.7	8×10^7
Lu	+3	7.6	6×10^7

[a] pK_h = -log K_{xy}. $x M^{z+} + y H_2O \rightleftharpoons M_x(OH)_y^{(xz-y)+} + y H^+$

$$K_{xy} = \frac{[M_x(OH)_y^{(xz-y)+}][H^+]^y}{[M^{z+}]^x} \cdot \frac{g_{xy}\, g_{H^+}{}^y}{g_{M^{z+}}{}^x\, a_{H_2O}{}^y}$$

[b] Measured by NMR, sound absorption, or multidentate ligand method.

14.2.1.3 Catalytic asymmetric aldol reaction in aqueous ethanol

Catalytic asymmetric aldol reactions provide one of the most powerful carbon-carbon bond-forming processes affording synthetically useful chiral β-hydroxy ketones and esters [25]. Chiral Lewis acid-catalyzed reactions of silyl enol ethers with aldehydes (the Mukaiyama reaction) [3] are among the most convenient and promising, and several successful examples have been reported since the first chiral tin(II)-catalyzed reactions appeared in 1990 [26]. Some common characteristics of these catalytic asymmetric reactions are the use of aprotic anhydrous solvents such as dichloromethane, toluene, and propionitrile [27], and low reaction temperatures (–78 °C) [28], which are also observed in many other catalytic asymmetric reactions.

In the course of our investigations to develop new chiral catalysts and catalytic asymmetric reactions in water, we focused on several elements whose salts are stable and work as Lewis acids in water [29]. In addition to the finding that the stability and activity of Lewis acids in water was related to hydration constants and exchange rate constants for the substitution of inner-sphere water ligands of elements (cations) (see Section 14.2.1.2), it was expected that undesired achiral side reactions would be suppressed in aqueous media and that desired enantioselective reactions would be accelerated in the presence of water (see below). Moreover, besides metal chelations, other factors such as hydrogen bonds, specific solvation, and hydrophobic interactions are anticipated to increase enantioselectivities in such media.

After the screening of chiral Lewis acids which could be used in aqueous solvents, the combination of Cu(II) triflate and bis(oxazoline) ligands [30] was found to give good enantioselectivities. Several examples of the catalytic asymmetric aldol reactions of aldehydes with silyl enol ethers are summarized in Table 14-5 [31].

R^1CHO + $OSiMe_3$ / R^2 → [$Cu(OTf)_2$ + ligand (x mol%); H_2O–EtOH (1/9), 20 h] → OH O / R^1 R^2

Table 14-5. Catalytic asymmetric aldol reactions in aqueous media.

R^1	R^2	*E/Z*	Ligand [a]) (x/mol%)	Temp/°C	Yield/%	*syn/anti*	33/% (*syn*)
Ph	Ph	*Z* [b])	**2** (20)	–10	74	3.2/1	67 [e])
Ph	Ph	*Z*	**3** (20)	0	98	2.6/1	61 [e])
Ph	Et	*Z* [b])	**2** (20)	–15	81	3.5/1	81
Ph	Et	*Z*	**2** (10)	–15	64	3.7/1	80
Ph	Et	*Z*	**2** (5)	–15	34	3.7/1	76
Ph	Et	*E* [c])	**2** (20)	–10	32	1.6/1	32
Ph	*i*-Pr	*Z* [d])	**2** (20)	0	17	4.0/1	85
Ph	*i*-Pr	*Z*	**2** (20)	5	95	4.0/1	77
p-ClPh	Ph	*Z*	**3** (20)	0	56	1.6/1	67
p-ClPh	Et	*Z*	**2** (10)	–10	88	2.6/1	76
p-ClPh	Et	*Z*	**2** (5)	–10	78	2.4/1	75
o-MeOPh	Et	*Z*	**2** (10)	–10	87	2.9/1	75
2-naphthyl	Et	*Z*	**2** (20)	–10	91	4.0/1	79
2-naphthyl	Et	*Z*	**2** (10)	–10	87	4.0/1	76
2-naphthyl	*i*-Pr	*Z*	**2** (20)	–10	97	4.0/1	81
2-furyl	Et	*Z*	**2** (20)	–10	86	4.0/1	76
2-thiophene	Et	*Z*	**2** (10)	–10	78	5.7/1	75
PhCH=CH	Et	*Z*	**2** (20)	–10	94	2.3/1	57
$Ph(CH_2)_2$	Et	*Z*	**2** (20)	–5	37	4.6/1	59
c-C_6H_{11}	Ph	*Z*	**3** (20)	0	77	4.6/1	42

[a]) **2:** R = *i*-Pr; **3:** R = CH_2Ph

[b]) *E/Z* = <1/<99.
[c]) *E/Z* = 77/23.
[d]) *E/Z* = 2/98.
[e]) (2S, 3S).

While the aldol reaction of benzaldehyde with (Z)-3-trimethylsiloxy-2-pentene proceeded smoothly in water-ethanol (1:9) at –15 °C to afford the desired adduct in a high yield with good selectivities (quant., *syn/anti*=3.3/1, *syn*=75% ee), much lower yield and selectivities were observed in ethanol (without water) under the same conditions (10% yield, *syn/anti*=2.3/1, *syn*=41% ee). Furthermore, when the reaction was performed in dichloromethane at –15 °C, the aldol adduct was obtained in 11% yield with *syn/anti*=2.1/1 [*syn*=20% ee (the opposite absolute configuration)]. From these results, we assume that desired chiral reactions are accelerated by water and that undesired achiral side reactions which proceed rapidly in aprotic solvents [32] are suppressed in aqueous media.

14.2.1.4 Aldol reaction in water without organic solvents

While aldol reactions stated above were smoothly catalyzed by the water-stable Lewis acids in aqueous media, a certain amount of organic solvent such as THF or EtOH had to be still combined with water to promote the reactions efficiently. To avoid the use of the organic solvents, we have developed a new reaction system in which $Sc(OTf)_3$ catalyzes Mukaiyama aldol reactions in pure water without any organic solvents in the presence of a small amount of a surfactant such as sodium dodecyl sulfate (SDS).

Lewis acid catalysis in micellar systems [33] was first found in the model reaction in Table 14-6 [34]. While the reaction proceeded sluggishly in the presence of 0.2 eq. $Yb(OTf)_3$ in water, remarkable enhancement of the reactivity was observed when the reaction was carried out in the presence of 0.2 eq. $Yb(OTf)_3$ in an aqueous solution of sodium dodecyl sulfate (SDS, 0.2 eq., 35 mM), and the corresponding aldol adduct was obtained in 50% yield. In the absence of the Lewis acid and the surfactant (water-promoted conditions), only 20% yield of the aldol adduct was isolated after 48 h, while 33% yield of the aldol adduct was obtained after 48 h in the absence of the Lewis acid in an aqueous solution of SDS. The amount of the surfactant also influenced the reactivity, and the yield was improved when $Sc(OTf)_3$ was used as a Lewis acid catalyst. Judging from the critical micelle concentration, micelles would be formed in these reactions, and it is noteworthy that the Lewis acid-catalyzed reactions proceeded smoothly in micellar systems [35]. It was also found that the surfactants influenced the yield, and that Triton X-100 was effective in the aldol reaction (but required long reaction time), while only a trace amount of the adduct was detected when using cetyltrimethylammonium bromide (CTAB) as a surfactant.

Several examples of the $Sc(OTf)_3$-catalyzed aldol reactions in micellar systems are shown in Table 14-7. Not only aromatic but also aliphatic and α,β-unsaturated aldehydes reacted with silyl enol ethers to afford the corresponding aldol adducts in high yields. Formaldehyde/water solution also worked well. Ketene silyl acetal **4**, which is known to be hydrolyzed very easily even in the presence of a small amount of water, reacted with an aldehyde in the present micellar system to afford the corresponding aldol adduct in a high yield.

Furthermore, we have introduced new types of Lewis acids, scandium tris(dodecyl sulfate) (**5a**) and scandium trisdodecanesulfonate (**6a**) (Chart 14-1) [36].

PhCHO + 1 → (cat. $Ln(OTf)_3$, surfactant, H_2O)

Table 14-6. Effect of $Ln(OTf)_3$ and surfactants.

$Ln(OTf)_3$/eq.	Surfactant/eq.	Time/h	Yield%
$Yb(OTf)_3$/0.2	–	48	17
$Yb(OTf)_3$/0.2	SDS/0.04	48	12
$Yb(OTf)_3$/0.2	SDS/0.1	48	19
$Yb(OTf)_3$/0.2	SDS/0.2	48	50
$Yb(OTf)_3$/0.2	SDS/0.1	48	22
$Sc(OTf)_3$/0.2	SDS/0.2	17	73
$Sc(OTf)_3$/0.1	SDS/0.2	4	88
$Sc(OTf)_3$/0.1	TritonX-100[a])/0.2	60	89
$Sc(OTf)_3$/0.1	CTAB/0.2	4	Trace

[a]) $H_3C-C(CH_3)_2-CH_2-C(CH_3)_2-C_6H_4-(OCH_2CH_2)_xOH$

R^1CHO + silyl enol ether → ($Sc(OTf)_3$ (0.1 eq.), SDS (0.2 eq.), H_2O, rt)

Table 14-7. $Sc(OTf)_3$-Catalyzed aldol reactions in micellar systems.

Aldehyde	Silyl enol ether	Yield%
PhCHO	1	88[a])
$PhCH_2CH_2CHO$	1	86[b])
PhCH=CHCHO	1	88[c])
HCHO	1	82[d])
PhCHO	$EtC(OSiMe_3)=CHCH_3$	88[e])
PhCHO	$PhC(OSiMe_3)=CH_2$	75[f, g])
PhCHO	$EtSC(OSiMe_3)=C(CH_3)_2$	94
PhCHO	$MeOC(OSiMe_3)=C(CH_3)_2$ **4**	84[g])

[a]) *syn/anti* = 50/50. [b]) *syn/anti* = 45/55. [c]) *syn/anti* = 41/59. [d]) Comercially available HCHO aq. (3 ml), **1** (0.5 mmol), $Sc(OTf)_3$ (0.1 mmol), and SDS (0.1 mmol) were combined. [e]) *syn/anti* = 57/43. [f]) $Sc(OTf)_3$ (0.2 eq.) was used. [g]) Additional silyl enolate (1.5 eq.) was charged after 6 h.

These "Lewis acid-surfactant-combined catalysts (LASCs)" were found to form stable colloidal dispersion systems with organic substrates in water and efficiently catalyze aldol reactions of aldehydes with very water-labile silyl enol ethers.

$M(O_3SO\ C_{12}H_{25})_n$

5a: M = Sc, n = 3
5b: M = Cu, n = 2

$M(O_3S\ C_{12}H_{25})_n$

6a: M = Sc, n = 3
6b: M =Yb, n = 3
6c: M = Mn, n = 2
6d: M = Co, n = 2
6e: M = Cu, n = 2
6f: M = Zn, n = 2
6g: M =Na, n = 1
6h: M = Ag, n = 1

Chart 14-1.

We also prepared several similar scandium sulfates, sulfonates, and alkylbenzenesulfonates, and these scandium salts were evaluated in a model aldol reaction (Table 14-8). As for the alkyl groups, the dodecyl groups gave the best results in all cases (sulfates, sulfonates, and alkylbenzenesulfonates). In the scandium sulfonate series, $Sc(O_3SC_{12}H_{25})_3$ gave the aldol adduct in 83% yield, while $Sc(O_3SC_{10}H_{21})_3$ and $Sc(O_3SC_{14}H_{29})_3$ afforded the products in 60% and 19% yield, respectively. The mixture of $Sc(O_3SC_{12}H_{25})_3$ and the organic substrates or $Sc(O_3SC_{10}H_{21})_3$ and the organic substrates formed stable dispersion systems, and their particle sizes were proved to be 1.1 mm and 0.7 mm, respectively. On the other hand, the mixture of $Sc(O_3SC_{14}H_{29})_3$ and the organic substrates did not form a stable dispersion system. It was indicated from these results that an excellent large hydrophobic reaction field was formed when $Sc(O_3SC_{12}H_{25})_3$ and the organic substrates were combined in water, and that the desired aldol reaction proceeded smoothly in the reaction field.

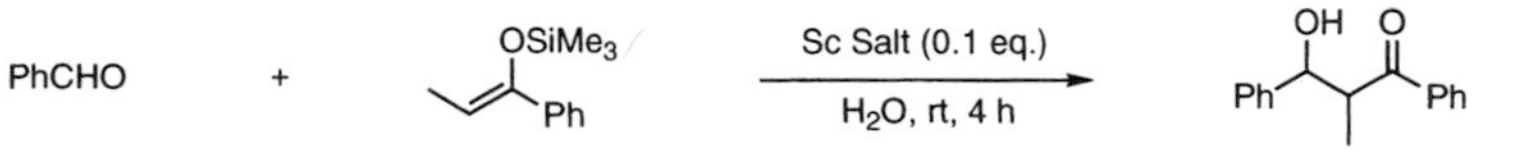

PhCHO + $OSiMe_3$ / Ph —— Sc Salt (0.1 eq.), H_2O, rt, 4 h ——> Ph, OH, O, Ph

Table 14-8. Effects of alkyl chains of the Sc salts[a]).

R	$Sc(O_3SOR)_3$	$Sc(O_3Sp\text{-}R\text{-}C_6H_4)_3$	$Sc(O_3SR)_3$
$C_{10}H_{21}$	–	55	60 (0.7 μm)[b])
$C_{11}H_{23}$	–	–	68
$C_{12}H_{25}$	92	91	83 (1.1 μm)[b])
$C_{13}H_{27}$	–	–	76
$C_{14}H_{29}$	73	33	19 (0.4 μm)[b])
$C_{16}H_{33}$	–	14	12

[a]) Numbers shown in the columns are isolated yields (%). [b]) Particle size of the dispersions.

It was also found that **5a** worked well in water rather than in organic solvents. The effect of solvents on the aldol reaction is shown in Table 14-9. While the reaction proceeded smoothly in water, it is very slow in organic solvents.

PhCHO + (OSiMe₃, Ph silyl enol ether) —5a (0.1 eq.), H_2O, rt, 4 h→ aldol product (OH, O, Ph, Ph)

Table 14-9. Effects of solvents [a]).

Solvent	Yield%	Solvent	Yield%
H_2O	92	CH_2Cl_2	3
CH_3OH	4	THF	Trace
DMF	14	Et_2O	Trace
DMSO	9	Toluene	Trace
CH_3CN	3	Hexane	4

[a]) While **5a** is dissolved in CH_3OH, DMF, DMSO, and THF, it is not dissolved or slightly dissolved in other solvents.

To investigate this LASC system in more detail, we have synthesized dodecyl sulfate and dodecanesulfonate salts with various metal cations (Chart 14-1) and studied the effects of the metal cations on catalytic ability for aldol reactions in water [37]. The catalysts (10 mol%) were used in the aldol reactions of benzaldehyde (1 eq) with thioketene silyl acetal **7** (1.5 eq) in water (Eq. (14.3)). Figure 14-1 shows the plot of GC yield (average value for two runs) versus time for the reaction catalyzed by the dodecanesulfonates (**6a–h**) at 30 °C. Remarkable effects of the metal cations on catalytic activity can be seen in Fig. 14-1. The order of catalytic activity at the initial stage of the reaction is as follows: Cu (**6e**) > Zn (**6 f**), Ag (**6h**) > Sc (**6a**), Yb (**6b**) > Na (**6g**) > Mn (**6c**), Co (**6d**). The Cu salt (**6e**) has the highest ability to catalyze the aldol reaction among the catalysts used. However, the yield of **8** did not exceed 70%, because **6e** accelerated not only the aldol reaction but also hydrolysis of thioketene silyl acetal **7**. The same trend was observed for the Zn and Ag salts (**6f**, **6h**). On the other hand, the Sc and Yb salts (**6a**, **6b**) afforded the aldol product (**8**) in >90% final yields, although the catalytic activities of **6a** and **6b** at the initial stage of the reaction were slightly lower than those of **6e**, **6f**, and **6h**. It should be noted that, in the dispersion system derived from **6a** and **6b**, the hydrolysis of thioketene silyl acetal **7** was attenuated. Especially in the case of **6a**, a small amount of **7** still remained when the aldol reaction completed. When the Na, Mn, and Co salts (**6g**, **6c**, **6d**) were used as catalysts, the aldol reactions proceeded very slowly, and the yields of **8** did not exceed 70% because of the hydrolysis of **7**.

PhCHO + **7** (OSiMe₃, SEt) —LASC (10 mol %), H_2O→ **8** (OH, O, Ph, SEt) (14.3)

During our investigations on the reactions mediated by LASCs, we have found that addition of a small amount of a Brønsted acid dramatically increases the reactivity of the aldol reaction [38]. As shown in Table 14-10, the combination of the LASC and Brønsted acids such as TsOH and HCl gave the product in better

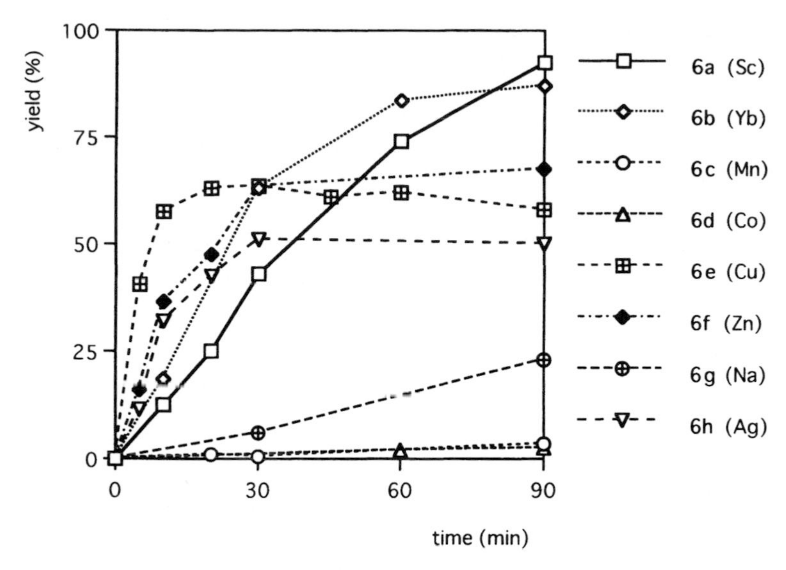

Figure 14-1. Plot of yield versus time for the aldol reactions in the presence of the dodecanesulfonate salts .

yields than the LASC or the Brønsted acids alone did. Although, from a mechanistic point of view, little is known about the real catalytic function of scandium and proton, this cooperative effect of a Lewis acid and a Brønsted acid will provide a new methodology for efficient catalytic systems in synthetic chemistry [39].

CHO + OSiMe$_3$ SEt → (LASC (10 mol%) + Bronsted Acid (10 mol%), H_2O, 23 °C, 5 min) OH O SEt

Table 14-10. LASC-Mediated aldol reactions in the presence of Brønsted acids.

LASC	Brønsted acid	Yield%
$Sc(O_3SC_{12}H_{25})_3$	–	10
–	TsOH	0
–	HCl	0
$Sc(O_3SC_{12}H_{25})_3$	**TsOH**	**55**
$Sc(O_3SC_{12}H_{25})_3$	**HCl**	**67**
$Sc(O_3SC_{12}H_{25})_3$	**HCl**	**89**[a])

[a]) 15 min.

14.2.2 Allylation Reaction

Synthesis of homoallylic alcohols by the reaction of allyl organometallics with carbonyl compounds is one of the most important processes in organic synthesis

Table 14-11. $Sc(OTf)_3$-Catalyzed allylation reactions of carbonyl compounds with tetraallyltin.

Entry	Carbonyl compound	Product	Solvent	Yield%
1	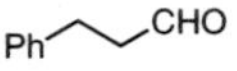 $PhCH_2CH_2CHO$	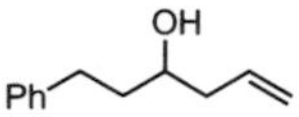	$H_2O:THF$ (1:9)	92
			$H_2O:EtOH$ (1:9)	96
			$H_2O:CH_3CN$ (1:9)	96
			EtOH	86
			CH_3CN	94
2	PhCHO	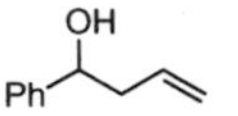	$H_2O:THF$ (1:9)	94
			CH_3CN	82
3	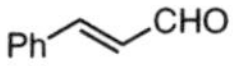 PhCH=CHCHO	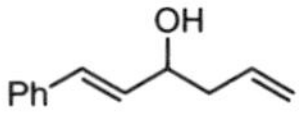	$H_2O:THF$ (1:9)	98
			CH_3CN	94
4	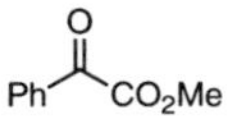 $PhCOCO_2Me$	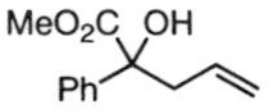	$H_2O:THF$ (1:9)	87
5	D-arabinose	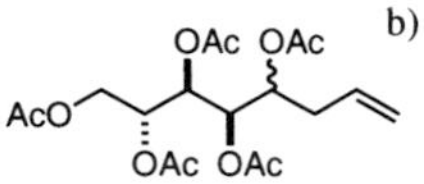 b)	$H_2O:THF$ (1:4)	81 c)
			$H_2O:EtOH$ (1:4)	89 d)
			$H_2O:CH_3CN$ (1:9)	93 e)
6	2-deoxy-D-ribose	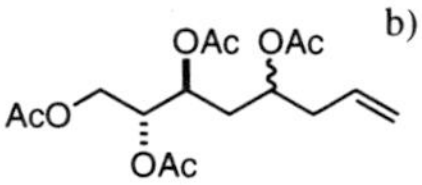 b)	$H_2O:THF$ (1:9)	89 f)
7	2-deoxy-D-glucose	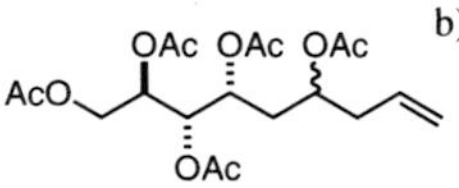 b)	$H_2O:THF$ (1:9)	88 f)
8	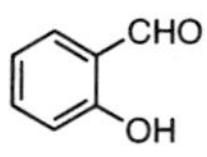	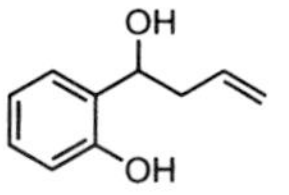	$H_2O:THF$ (1:9)	quant.
			CH_3CN	90
9	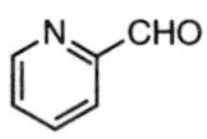	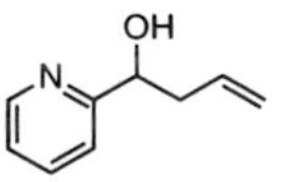	$H_2O:THF$ (1:9)	quant.
			CH_3CN	84

a) Carried out at 25 °C except for entries 6 and 7 (60 °C). b) The products were isolated after acetylation. c) *syn/anti* = 28/72. d) *syn/anti* = 27/73. e) *syn/anti* = 26/74. f) *syn/anti* = 50/50.

[40]. The allylation reactions of carbonyl compounds were found to proceed smoothly under the influence of 5 mol% of $Sc(OTf)_3$ [41] by using tetraallyltin [42] as an allylating reagent [43]. Several examples are shown in Table 14-11. The reactions proceeded smoothly in the presence of only a catalytic amount of $Sc(OTf)_3$ under extremely mild conditions, and the adducts, homoallylic alcohols, were obtained in high yields. In most cases, the reactions were successfully carried out in aqueous media. It is noteworthy that unprotected sugars reacted directly to give the adducts in high yields (entries 5–7). The allylated adducts are intermediates for the synthesis of higher sugars [44]. Under the present reaction condi-

RCHO (1.0 eq.) + (allyl)$_4$Sn (0.5 eq.) → [$Sc(OTf)_3$ (0.1 eq.), surfactant (0.2 eq.), H_2O, rt] → R–CH(OH)–CH$_2$CH=CH$_2$

Table 14-12. $Sc(OTf)_3$-Catalyzed allylation reactions in micellar systems.

RCHO	Surfactant	Yield%
 PhCH$_2$CH$_2$CHO	SDS	84 (88)[a]
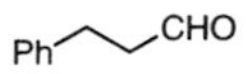 PhCH$_2$CH$_2$CHO	TritonX-100	90
PhCHO	TritonX-100	83 (79)[a]
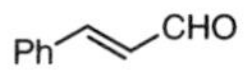 PhCH=CHCHO	TritonX-100	92 (83)[a]
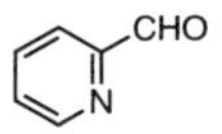 2-pyridinecarboxaldehyde	TritonX-100	77
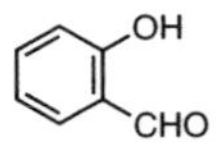 2-hydroxybenzaldehyde	SDS	85
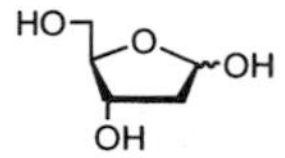 2-deoxyribose (HO, OH, OH)	SDS	99[b,c]
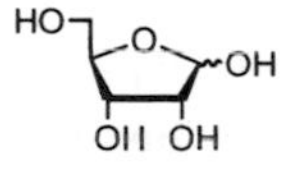 ribose (HO, OH OH, OH)	TritonX-100	95[b,d,e]
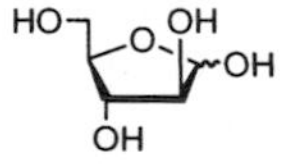 (HO, OH, OH, OH) furanose	TritonX-100	82[b,d,f]
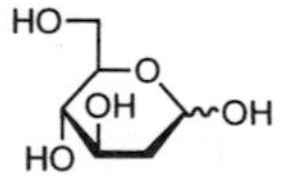 (HO, OH, HO, OH) pyranose	TritonX-100	72[b,d,g]

[a]) Tetraallyltin (0.25 eq.) was used. [b]) The products were isolated after acetylation. [c]) *syn/anti*=50/50. [d]) Carried out at 50 °C. [e]) *syn/anti*=29/71. [f]) *syn/anti*=30/70. [g]) *syn/anti*=48/52.

TritonX-100: $H_3C-C(CH_3)_2-CH_2-C(CH_3)_2-C_6H_4-(OCH_2CH_2)_xOH$

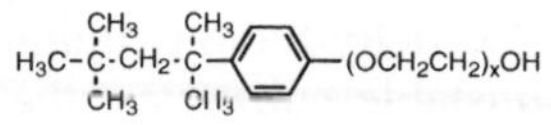

tions, salicylaldehyde and 2-pyridinecarboxaldehyde reacted with tetraallyltin to afford the homoallylic alcohols in good yields (entries 8 and 9). Under general Lewis acid conditions, these compounds react with the Lewis acids rather than the nucleophile.

Furthermore, the $Sc(OTf)_3$-catalyzed allylation reactions of aldehydes with tetraallyltin proceeded smoothly in micellar systems to afford the corresponding homoallylic alcohols in high yields (Table 14-12) [45]. The reactions were successfully carried out in the presence of a small amount of a surfactant in water without using any organic solvents.

14.3 Activation of C=N in Aqueous Media

In Lewis acid-mediated reactions of C=N groups, many Lewis acids are deactivated or sometimes decomposed by the nitrogen atoms of the starting materials or products. Furthermore, even when the desired reactions proceed, more than stoichiometric amounts of the Lewis acids are needed in most cases, because the acids are trapped by the nitrogen atoms. It is desirable from a synthetic point of view that nitrogen-containing compounds are activated by Lewis acids catalytically. Here we summarize our research efforts on catalytic activation of C=N groups in water.

14.3.1 Mannich-type Reaction

The Mannich and related reactions provide one of the most fundamental and useful methods for the synthesis of *β*-amino ketones and esters [46]. Although the classical protocols include some severe side reactions, new modifications using preformed iminium salts and imines have improved the process. Some of these materials are, however, unstable and difficult to isolate, and deaminations of the products that occur under the reaction conditions still remain as problems. The direct synthesis of *β*-amino ketones from aldehydes, amines, and silyl enolates under mild conditions is desirable from a synthetic point of view [47, 48]. Our working hypothesis is that aldehydes could react with amines in the hydrophobic micellar system in the presence of a catalytic amount of lanthanide triflate and a surfactant to produce imines, which could react with hydrophobic silyl enolates [49].

First, a model reaction of benzaldehyde, *o*-methoxyaniline, and 1-phenyl-1-trimethylsiloxyethene was performed in the presence of 5 mol% of $Sc(OTf)_3$ in an aqueous solution of SDS (SDS, 0.2 eq., 35 mM). The reaction proceeded smoothly at room temperature to afford the corresponding *β*-amino ketone derivative in 87% yield. It is noted that the dehydration (imine formation) and the coupling reaction between two water-unstable substrates, the imine and the silyl enolate, occurred successfully in water, and that only a trace amount of the product was obtained without SDS under the same reaction conditions. Other catalysts such as $Yb(OTf)_3$ and $Cu(OTf)_2$ were also found to be effective in this reaction [22]. Side reaction adducts such as deamination and aldol products were not obtained at all in aqueous media. On the other hand, when the same reaction was carried out in dichloromethane, the yield of the desired product decreased and the deamination product was obtained (Eq. (14.4)). We then tested other examples and the results are summarized in Table 14-13. Aromatic aldehydes as well as heterocyclic, α,β-unsaturated, aliphatic aldehydes, and a glyoxal worked well to afford the desired adducts in high yields. Formaldehyde also reacted smoothly. For silyl enolates, not only ketone-derived silyl enol ethers but also thioester- and ester-derived ketene silyl acetals worked well.

PhCHO + $PhNH_2$ + $CH_2{=}C(OSiMe_3)Ph$ —[$Sc(OTf)_3$ (5 mol%), CH_2Cl_2, rt]→ PhCH(NHPh)CH$_2$COPh (66%) + PhCH=CHCOPh (22%) (14.4)

The products were readily converted to free β-amino ketones and esters. Thus, treatment of the products with cerium ammonium nitrate (CAN) in acetonitrile-water (9:1) at room temperature induced smooth deprotection of the 2-methoxyphenylamino group to give the corresponding free β-amino carbonyl compounds (Eq. (14.5)) [50, 51].

PhCH(NH-2-MeOC$_6$H$_4$)C(CH$_3$)$_2$COPh —[CAN]→ PhCH(NH$_2$)C(CH$_3$)$_2$COPh, 83 %Yield (14.5)

14.3.2 Allylation Reaction

The reaction of imines with allyltributylstannane provides a useful route for the synthesis of homoallylic amines [40]. In the course of our investigations to develop new synthetic reactions in water, we have found that three-component reac-

R^1CHO + (o-OMe)$PhNH_2$ + $R^2R^3C{=}C(OSiMe_3)R^4$ $\xrightarrow[\text{SDS (0.2 eq.), } H_2O\text{, Temp.}]{Sc(OTf)_3 \text{ (5 mol \%)}}$ $R^1CH(NH(o\text{-OMe})Ph)CR^2R^3C(O)R^4$

Table 14-13. Three-component coupling reactions of aldehydes, amine, and silyl enol ethers.

Entry	R^1	Silyl enol ether	Temp. (°C)	Yield%
1	Ph	$CH_2{=}C(OSiMe_3)Ph$ **8**	rt	87
2	Ph	$CH_3CH{=}C(OSiMe_3)SEt$ **9**	rt	85
3	Ph	$CH_2{=}C(OSiMe_3)SEt$	rt	68
4	Ph	$(CH_3)_2C{=}C(OSiMe_3)OMe$	rt	73
5	2-furyl	**8**	rt	85 [a, b]
6	PhCO	**8**	rt	86
7	PhCH=CH	**8**	rt	74 [a, b]
8	$(CH_3)_2CHCH_2$	**8**	0	71 [a, b]
9	$PhCH_2CH_2$	**8**	0	75 [a, b]
10	*c*-Hex	**8**	0	79 [a, b]
11	2-furyl	**9**	rt	80
12	$(CH_3)_2CHCH_2$	**9**	0	88

[a] $Sc(OTf)_3$ (10 mol%) was used. [b]) Aldehyde (1.3 eq.) was used.

tions of aldehydes, amines and allyltributylstannane proceed smoothly in micellar systems using $Sc(OTf)_3$ as a Lewis acid catalyst.

The reaction of benzaldehyde, aniline, and allyltributylstannane in water was chosen as a model, and several reaction conditions were examined. While the reaction proceeded sluggishly in the presence of either $Sc(OTf)_3$ or SDS, 77% yield of the desired homoallylic amine was obtained in the presence of both $Sc(OTf)_3$ and SDS. It was suggested that an imine formed from the aldehyde and the amine rapidly reacted with allyltributylstannane to afford the desired adduct.

Several examples of the present three-component reactions of aldehydes, amines, and allyltributylstannane are shown in Table 14-14. The reactions proceeded smoothly in water without using any organic solvents in the presence of a small amount of $Sc(OTf)_3$ and SDS, to afford the corresponding homoallylic amines in high yields. Not only aromatic aldehydes but also aliphatic, unsaturated, and heterocyclic aldehydes worked well. It is reported that severe side reactions occur to decrease yields in reactions of imines having α-protons with allyltributylstannane in organic solvents [40]. On the other hand, aliphatic aldehydes including non-branched aliphatic aldehydes gave the homoallylic amines in high yields in water. In addition, no aldehyde adducts (homoallylic alcohols) were obtained in all cases. It was suggested that the imine formation from aldehydes and amines was very fast in the presence of both $Sc(OTf)_3$ and SDS [49], and that the selective activation of imines rather than aldehydes was achieved using $Sc(OTf)_3$ as a catalyst [52]. It is also noteworthy that using a small amount of a surfactant created efficient hydrophobic reaction fields and enabled smooth dehydration and addition reactions in water [53].

R^1CHO + R^2NH_2 + $SnBu_3$ —[$Sc(OTf)_3$ (20 mol %); SDS (0.2 eq.), H_2O, rt, 20 h]→ R^2NH, R^1

Table 14-14. Three-component reactions of aldehydes, amine, and allyltributylstannane.

Entry	R^1	R^2	Yield%
1	Ph	Ph	83
2	Ph	(*p*-Cl)Ph	90
3	Ph	(*p*-OMe)Ph	81
4	(*p*-Cl)Ph	(*p*-Cl)Ph	70
5	2-furyl	(*p*-Cl)Ph	67 (71[a]), 82[b]))
6	2-thiophene	(*p*-Cl)Ph	67[a] (78)[b])
7	$PhCH_2CH_2$	(*p*-Cl)Ph	78
8	$CH_3(CH_2)_7$	(*p*-Cl)Ph	66
9	*c*-Hex	Ph	80 (83)[b])
10	*c*-Hex	(*p*-OMe)Ph	74
11	PhCO	(*p*-Cl)Ph	70 (83)[b])
12	PhCH=CH	(*p*-Cl)Ph	80 (82)[b])

[a]) SDS 0.4 eq., [b]) SDS 0.6 eq., [c]) SDS 1.0 eq.

14.3.3 Strecker-type Reaction

The Strecker-type reaction provides one of the most efficient routes to α-amino nitriles, which are useful intermediates for the synthesis of amino acids [54] and nitrogen-containing heterocycles such as thiadiazoles and imidazoles [55]. Although classical Strecker reactions have some limitations, use of trimethylsilyl cyanide (TMSCN) as a cyanide anion source provides promising and safer routes to these compounds [54b, 56]. However, TMSCN is easily hydrolyzed in the presence of water, and it is therefore necessary to perform the reactions under strictly anhydrous conditions. In the course of our program to develop new synthetic reactions in aqueous media [34, 45], we have focused on tributyltin cyanide (Bu_3SnCN) [57]. Bu_3SnCN is stable in water and a potential cyanide source, albeit there are no reports on Strecker-type reactions using Bu_3SnCN to the best of our knowledge.

We chose valelaldehyde, diphenylmethylamine, and Bu_3SnCN as model substrates, and the Strecker-type reaction was first performed in the presence of

10 mol% of $Sc(OTf)_3$ in organic solvents. The reactions proceeded smoothly at room temperature in acetonitrile, benzene, dichloromethane, and toluene to afford the corresponding *α*-amino nitrile in high yields. Among these solvents tested, acetonitrile toluene (1:1) gave the best yield (84%). It was found that no dehydration reagent such as molecular sieves, $MgSO_4$, or drierite was needed in these reactions. We then performed the reaction in water. It was found that the model Strecker-type reaction also proceeded smoothly in water in the presence of a catalytic amount of $Sc(OTf)_3$ to give the corresponding *α*-amino nitrile in 94% yield. No surfactant was needed in this reaction. The reaction is assumed to proceed via imine formation and successive cyanation. It is noted that the dehydration process (imine formation) proceeded smoothly in water, and that the reaction rate in water was almost the same as those in organic solvents.

While the desired reaction proceeded smoothly, it was thought that use of toxic tin reagent might restrict the application of the reaction [58]. We then tried to recover the tin materials after the reaction (Scheme 14-2). The Strecker-type reaction was performed using an equimolar amount of an aldehyde and an amine, and a slight excess of Bu_3SnCN. After the reaction was completed, excess Bu_3SnCN was treated with a weak acid to form bis(tributyltin) oxide [59]. On the other hand, the adduct, an *α*-(tributylstannylamino) nitrile, was hydrolyzed by adding water to produce an *α*-amino nitrile and tributyltin hydroxide, which was readily converted to bis(tributyltin) oxide [59]. Thus, all tin sources were converted to bis(tributyltin) oxide. It was already reported that bis(tributyltin) oxide can be converted to tributyltin chloride [60] and then to Bu_3SnCN [57]. Since the catalyst, $Sc(OTf)_3$, is also recoverable and reusable [61], the present Strecker-type reactions represent a completely recyclable system.

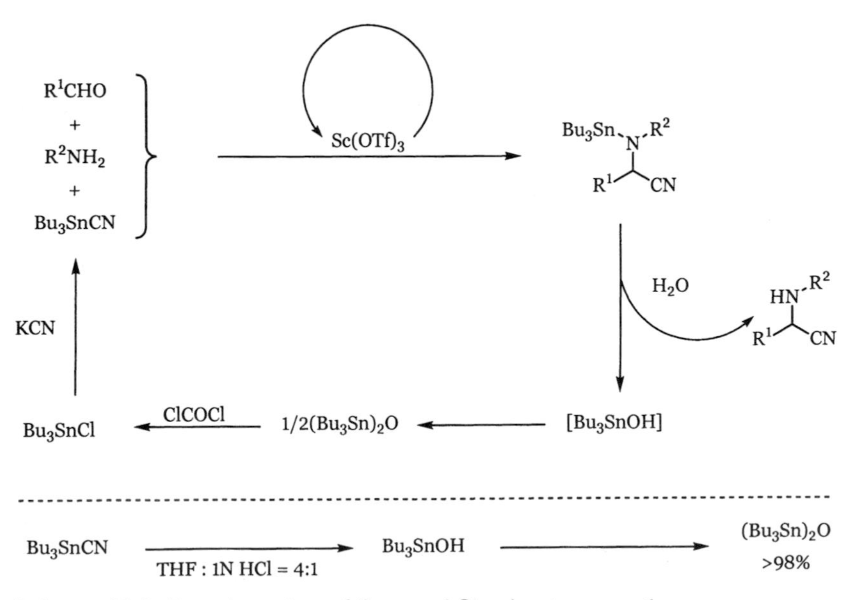

Scheme 14-2. Recycle system of the novel Strecker-type reactions.

Several examples of the Strecker-type reaction are shown in Table 14-15. In all cases, including the reactions using aromatic, aliphatic, heterocyclic, as well as α,β-unsaturated aldehydes, the reactions proceeded smoothly to afford the corresponding α-amino nitriles in high yields. Furthermore, the adducts, α-(*N*-benzhydryl)amino nitriles were readily converted to the corresponding α-amino acids [62]. The present Strecker-type reactions using other amines such as benzylamine also proceeded smoothly to afford the corresponding adducts in high yields.

$$\underset{1.0\ \text{eq}}{RCHO} + \underset{1.0\ \text{eq}}{Ph_2CHNH_2} + \underset{1.5\ \text{eq}}{Bu_3SnCN} \xrightarrow[\substack{H_2O \\ \text{rt, 20 h}}]{Sc(OTf)_3\ (10\ \text{mol}\ \%)} R\text{–}CH(CN)\text{–}NH\text{–}CHPh_2$$

Table 14-15. Streker-type reactions of aldehydes, amine, and Bu_3SnCN in water.

Entry	Solvent	Yield%
1	Ph	88
2	PhCH=CH	84
3	2-furyl	89
4	$PhCH_2CH_2$	79
5	C_4H_9	94 (94)[a]
6	*c*-Hex	94

[a]) $Sc(OTf)_3$ was reused.

14.4 Conclusions

Various metal salts such as rare earth metal triflates and copper triflate can function as Lewis acids in aqueous media. They can effectively activate aldehydes and imines in the presence of water molecules, and the first successful examples of Lewis acid-catalyzed reactions in aqueous solution have been demonstrated. Water-soluble aldehydes such as formaldehyde could be employed directly in these reactions. Moreover, the catalysts could be easily recovered after the reactions were completed and could be reused. There are many kinds of Lewis acid-promoted reactions in industrial chemistry, and treatment of large amounts of the acids left over after the reactions have induced some severe environmental problems. From the standpoints of their catalytic use and reusability, the Lewis acids described in this chapter are expected to be new types of catalysts providing some solutions for these problems.

Organic reactions in water without harmful organic solvents are of great current interest, not only because water is an environmentally benign solvent, but also because organic reactions in water display unique reactivity and selectivity. Furthermore, it is noted that water plays essential roles for reactions in organisms. The investigations on reactions in aqueous media will lead to the full understanding of the roles of water, and contribute to development of many fields of science including synthetic chemistry [63].

References

1. Schinzer, D., Ed. Selectivities in Lewis Acid Promoted Reactions, Kluwer Academic Publishers: Dordrecht, 1989.
2. Kobayashi, S. *Synlett* **1994**, 689.
3. (a) Mukaiyama, T.; Narasaka, K.; Banno, T. *Chem. Lett.* **1973**, 1011. (b) Mukaiyama, T.; Banno, K.; Narasaka, K. *J. Am. Chem. Soc.* **1974**, *96*, 7503.
4. Mukaiyama, T. *Org. React.* **1982**, *28*, 203.
5. Kobayashi, S.; Murakami, M.; Mukaiyama, T. *Chem. Lett.* **1985**, 1535.
6. (a) Kawai, M.; Onaka, M.; Izumi, Y. *Chem. Lett.* **1986**, 1581. (b) Kawai, M.; Onaka, M.; Izumi, Y. *Bull. Chem. Soc. Jpn.* **1988**, *61*, 1237.
7. (a) Noyori, R.; Yokoyama, K.; Sakata, J.; Kuwajima, I.; Nakamura, E.; Shimizu, M. *J. Am. Chem. Soc.* **1977**, *99*, 1265. (b) Nakamura, E.; Shimizu, M.; Kuwajima, I.; Sakata, J.; Yokoyama, K.; Noyori, R. *J. Org. Chem.* **1983**, *48*, 932.
8. Lanthanide(III) chlorides or some organolanthanide compounds catalyzed aldol reactions of ketene silyl acetals with aldehydes were reported. (a) Takai, K.; Heathcock, C.H. *J. Org. Chem.* **1985**, *50*, 3247. (b) Vougioukas, A.E.; Kagan, H.B. *Tetrahedron Lett.* **1987**, *28*, 5513. (c) Gong, L.; Streitwieser, A. *J. Org. Chem.* **1990**, *55*, 6235. (d) Mikami, K.; Terada, M.; Nakai, T. *J. Chem. Soc., Chem. Commun.* **1993**, 343, and references cited therein.
9. (a) Lubineau, A. *J. Org. Chem.* **1986**, *51*, 2142. (b) Lubineau, A.; Meyer, E. *Tetrahedron* **1988**, *44*, 6065.
10. Kobayashi, S. *Chem. Lett.* **1991**, 2187.
11. For example, (a) Hajos, Z.G.; Parrish, D.R. *J. Org. Chem.* **1973**, *38*, 3244. (b) Stork, G.; Isobe, M. *J. Am. Chem. Soc.* **1975**, *97*, 4745. (c) Lucast, D.H.; Wemple, J. *Synthesis* **1976**, 724. (d) Ono, N.; Miyake, H.; Fujii, M.; Kaji, A. *Tetrahedron Lett.* **1983**, *24*, 3477. (e) Tsuji, J.; Nisar, M.; Minami, I. *Tetrahedron Lett.* **1986**, *27*, 2483. (f) Larsen, S.D.; Grieco, P.A.; Fobare, W.F. *J. Am. Chem. Soc.* **1986**, *108*, 3512.
12. B.B. Snider and H. Yamamoto respectively developed formaldehyde-organoaluminum complex as formaldehyde source in several reactions. (a) Snider, B.B.; Rodini, D.J.; Kirk, T.C.; Cordova, R. *J. Am. Chem. Soc.* **1982**, *104*, 555. (b) Snider, B.B. In *Selectivities in Lewis Acid Promoted Reactions*; Schinzer, D., Ed.; Kluwer Academic Publishers, London, 1989; pp 147–167. (c) Maruoka, K., Conception, A.B., Hirayama, N., Yamamoto, H. *J. Am. Chem. Soc.* **1990**, *112*, 7422. (d) Maruoka, K.; Conception, A.B.; Murase, N.; Oishi, M.; Yamamoto, H. *J. Am. Chem. Soc.* **1993**, *115*, 3943.
13. Cf. TMSOTf-mediated aldol-type reaction of silyl enol ethers with dialkoxymethanes was also reported. Murata, S.; Suzuki, M.; Noyori, R. *Tetrahedron Lett.* **1980**, *21*, 2527.
14. Review: Molander, G.A. *Chem. Rev.* **1992**, *92*, 29.
15. Baes, Jr. C.F.; Mesmer, R.E. *The Hydrolysis of Cations*, John Wiley & Sons, New York, 1976, p. 129.
16. Thom, K.F., US Patent 3615169 (1971); CA **1972**, *76*, 5436a.
17. (a) Forsberg, J.H.; Spaziano, V.T.; Balasubramanian, T.M.; Liu, G.K.; Kinsley, S.A.; Duckworth, C.A.; Poteruca, J.J.; Brown, P.S.; Miller, J.L. *J. Org. Chem.* **1987**, *52*, 1017. See also (b) Collins, S.; Hong, Y. *Tetrahedron Lett.* **1987**, *28*, 4391. (c) Almasio, M.-C.; Arnaud-Neu, F.; Schwing-Weill, M.-J. *Helv. Chim. Acta* **1983**, *66*, 1296. Cf. (d) Harrowfield, J.M.; Kepert, D.L.; Patrick, J.M.; White, A.H. *Aust. J. Chem.* **1983**, *36*, 483.
18. Kobayashi, S.; Hachiya, I. *Tetrahedron Lett.* **1992**, 1625.
19. Kobayashi, S.; Hachiya, I.; Yamanoi, Y. *Bull. Chem. Soc. Jpn.* **1994,** *67*, 2342.
20. Haggin, J. *Chem. Eng. News* **1994**, Apr 18, 22.
21. Kobayashi, S.; Hachiya, I.; Ishitani, H.; Araki, M. *Synlett.* **1993**, 472.
22. Kobayashi, S.; Nagayama, S.; Busujima, T. *J. Am. Chem. Soc.* **1998**, *120*, 8287.
23. Yatsimirksii, K.B.; Vasil'ev, V. P. *Instability Constants of Complex Compounds*; Pergamon, New York, 1960.
24. Martell, A.E., Ed.; *Coordination Chemistry*; ACS Monograph 168; American Chemical Society, Washington, DC, 1978; Vol. 2.

25. Review: (a) Gröger, H.; Vogl, E.M.; Shibasaki, M. *Chem. Eur. J.* **1998,** *4*, 1137. (b) Nelson, S.G. *Tetrahedron: Asymmetry.*, **1998**, *9*, 357. (c) Bach, T. *Angew. Chem., Int. Ed. Engl.* **1994**, *33*, 417.
26. (a) Kobayashi, S.; Fujishita, Y.; Mukaiyama, T. *Chem. Lett.* **1990**, 1455. (b) Mukaiyama, T.; Kobayashi, S.; Uchiro, H.; Shiina, I. *Chem. Lett.* **1990**, 129.
27. Catalytic asymmetric aldol reactions in wet dimethylformamide were reported. (a) Sodeoka,M.; Ohrai, K.; Shibasaki, M. *J. Org. Chem.*, **1995,** *60*, 2648. Cf. (b) Mikami, K.; Kotera, O.; Motoyama, Y.; Sakaguchi, H. *Synlett*, **1995**, 975.
28. Some catalytic asymmetric aldol reactions were performed at higher temperatures (–20 °C–23 °C). (a) Mikami K.; Matsukawa, S. *J. Am. Chem. Soc.* **1993**, *115*, 7039. (b) Mikami, K.; Matsukawa, S. *J. Am. Chem. Soc.* **1994**, *116*, 4077. (c) Carreira, E.M.; Singer, R.A.; Lee, W. *J. Am. Chem. Soc.* **1994**, *116*, 8837. (d) Keck, G.E.; Krishnamurthy, D. *J. Am. Chem. Soc.* **1995**, *117*, 2363. See also Ref. 27.
29. Copper(II) was revealed to be one of the most promising metals. Kobayashi, S.; Nagayama, S.; Busujima, T. *Chem. Lett.* **1997**, 959.
30. (a) Fritschi, H.; Leutenegger, U.; Pfaltz, A. *Angew. Chem., Int. Ed. Engl.* **1986**, *25*, 1005. (b) Brunner, H.; Obermann, U. *Chem. Ber.* **1989**, *122*, 499. (c) Nishiyama, H.; Sakaguchi, T.; Nakamura, T.; Horihata, M.; Kondo, M.; Ito, K. *Organometallics* **1989**, *8*, 821. (d) Balavoine, G.; Clinet, J.C.; Lellouche, I. *Tetrahedron Lett.* **1989**, *30*, 5141. (e) Lowenthal, R.E.; Abiko, A.; Masamune, S. *Tetrahedron Lett.* **1990**, *31*, 6005. (f) Evans, D.A.; Woerpel, K.A.; Hinman, M.M.; Faul, M.M. *J. Am. Chem. Soc.* **1991**, *113*, 726. (g) Corey, E.J.; Imai, N.; Zhang, H.-Y. *J. Am. Chem. Soc.* **1991**, *113*, 728. (h) Müller, D.; Umbricht, G.; Weber, B.; Pfaltz, A. *Helv. Chim. Acta* **1991**, *74*, 232. (i) Evans, D.A.; Peterson, G.S.; Johnson, J.S.; Barnes, D.M.; Campos, K.R.; Woerpel, K.A. *J. Org. Chem.* **1998**, *63*, 4541. Catalytic asymmetric aldol reactions of silyl enolates with (benzyloxy)acetaldehyde or α-ketoesters using similar Cu(II) complexes in anhydrous organic solvents were reported. Low selectivities in the reaction of benzaldehyde or dihyderocinnamaldehyde with the silyl enolate derived from *t*-butyl thioacetate were indicated. See: (j) Evans, D.A.; Murry, J.A., Kozlowski, M.C. *J. Am. Chem. Soc.* **1996**, *118*, 5814. (k) Evans, D.A.; Kozlowski, M.C.; Burgey, C.S.; MacMillan, D.W.C. *J. Am. Chem. Soc.* **1997**, *119*, 7893. (l) Evans, D.A.; Kozlowski, M.C.; Murry, J.A.; Burgey, C.S.; Campos, K.R.; Connell, B.T.; Staples, R.J. *J. Am. Chem. Soc.* **1999**, *121*, 669. (m) Evans, D.A.; Burgey, C.S.; Kozlowski, M.C.; Tregay, S.W. *J. Am. Chem. Soc.* **1999**, *121*, 686.
31. Kobayashi, S.; Nagayama, S.; Busujima, T. *Chem. Lett.* **1999**, 71.
32. (a) Carreira, E.M.; Singer, R.A. *Tetrahedron Lett.* **1994**, *25*, 4323. (b) Denmark, S.E.; Chen, C.-T. *Tetrahedron Lett.* **1994**, *25*, 4327. (c) Hollis, T.K.; Bosnich, B. *J. Am. Chem. Soc.* **1995**, *117*, 4570. (d) Oishi, M.; Aratake, S.; Yamamoto, H. *J. Am. Chem. Soc.* **1998**, *120*, 8271. See also, (e) Kobayashi, S.; Nagayama, S. *J. Am. Chem. Soc.* **1997**, *119*, 10049.
33. For a review of organic reactions in micellar systems, Tascioglu, S. *Tetrahedron* **1996**, *34*, 11113.
34. Kobayashi, S.; Wakabayashi, T.; Nagayama, S.; Oyamada, H. *Tetrahedron Lett.* **1997**, *38*, 4559.
35. Diels-Alder reactions catalyzed by a Cu(II) salt in a micellar system were reported: Otto, S.; Engberts, J.B.F.N.; Kwak, J.C.T. *J. Am. Chem. Soc.* **1998**, *120*, 9517.
36. Kobayashi, S.; Wakabayashi, T. *Tetrahedron Lett.* **1998**, *39*, 5389.
37. Manabe, K.; Kobayashi, S. *Synlett* **1999**, 547.
38. Manabe, K.; Kobayashi, S. *Tetrahedron Lett.* **1999**, *40*, 3773.
39. For benzoic acid acceleration in $Yb(OTf)_3$-catalyzed allylation of aldehydes in acetonitrile, (a) Aspinall, H.C.; Greeves, N.; McIver, E.G. *Tetrahedron Lett.* **1998**, *39*, 9283. For acetic acid acceleration in $Yb(fod)_3$-catalyzed ene reaction of aldehydes with alkyl vinyl ethers, ene reaction of aldehydes with alkyl vinyl ethers, (b) Deaton, M.V.; Ciufolini, M.A. *Tetrahedron Lett.* **1993**, *34*, 2409. Yamamoto et al. reported Brønsted acid-assisted chiral Lewis acids and Lewis acid-assisted Brønsted acids which were used for catalytic asymmetric Diels-Alder reactions and protonations and stoichiometric asymmetric aza Diels-Alder reactions, aldol-type reactions of imines, and an aldol reaction. (c) Ishihara, K.; Yamamoto, H. *J. Am. Chem. Soc.* **1994**, *116*, 1561. (d) Ishihara, K.; Kurihara, H.; Yamamoto, H. *J. Am. Chem. Soc.* **1996**, *118*, 3049. (e) Ishihara, K.; Nakamura, S.; Kaneeda, M.; Yamamoto, H. *J. Am. Chem. Soc.* **1996**, *118*, 12854. (f) Ishihara, K.; Miyata, M.; Hattori, K.; Tada, T.; Yamamoto, H. *J. Am. Chem. Soc.* **1994**, *116*, 10520. (g) Yamamoto, H. *J. Am. Chem. Soc.* **1994**, *116*, 10520. (h) Ishihara, K.; Kurihara, H.; Matsumoto, M.; Yamamoto Ishihara, K.; Kurihara, H.; Matsumoto, M.; Yamamoto, H. *J. Am. Chem. Soc.* **1998**, *120*, 6920.

40. Review: Yamamoto, Y.; Asao, N. *Chem. Rev.* **1993**, *93*, 2207.
41. Hachiya, I.; Kobayashi, S. *J. Org. Chem.* **1993**, *58*, 6958.
42. For the reactions of carbonyl compounds with tetraallyltin, (a) Peet, W.G.; Tam, W. *J. Chem. Soc., Chem. Commun.* **1983**, 853. (b) Daude, G.; Pereyre, M. *J. Organometal. Chem.* **1980**, *190*, 43. (c) Harpp, D.N.; Gingras, M. *J. Am. Chem. Soc.* **1988**, *110*, 7737. See also, (d) Fukuzawa, S.; Sato, K.; Fujinami, T.; Sakai, S. *J. Chem. Soc., Chem. Commun.* **1983**, 853. Quite recently, H. Yamamoto et al. reported allylation reactions of aldehydes with tetraallyltin in the presence of hydrochloric acid. (e) Yanagisawa, A.; Inoue, H.; Morodome, M.; Yamamoto, H. *J. Am. Chem. Soc.* **1993**, *115*, 10356.
43. Lewis acid-promoted allylation reactions of carbonyl compounds with allyltrialkyltin were reported. (a) Yamamoto, Y.; Yatagai, H.; Naruta, Y.; Maruyama, K. *J. Am. Chem. Soc.* **1980**, *102*, 7107. (b) Pereyre, M.; Quintard, J.-P.; Rahm, A. *Tin in Organic Synthesis*, Butterworths, London, **1987**, p 216.
44. (a) Schmid, W.; Whitesides, G.M. *J. Am. Chem. Soc.* **1991**, *113*, 6674. (b) Kim, E.; Gordon, D.M.; Schmid, W.; Whitesides, G.M. *J. Org. Chem.* **1993**, *58*, 5500.
45. Kobayashi, S.; Wakabayashi, T.; Oyamada, H. *Chem. Lett.* **1997**, 831.
46. Kleinmann, E.F. In *Comprehensive Organic Synthesis,* Trost, B.M. Ed.; Pergamon Press: New York, 1991; Vol. 2, Chapter 4.1.
47. Kobayashi, S.; Araki, M.; Yasuda, M. *Tetrahedron Lett.* **1995**, *36*, 5773.
48. Annunziata, R.; Cinquini, M.; Cozzi, F.; Molteni, V.; Schupp, O. *J. Org. Chem.* **1996**, *61*, 8293.
49. Kobayashi, S.; Ishitani, H. *J. Chem. Soc. Chem. Commun.* **1995**, 1379.
50. Kronenthal, D.R.; Han, C.Y.; Taylor, M.K. *J. Org. Chem.* **1982**, *47*, 2765.
51. Ishitani, H.; Ueno, M.; Kobayashi, S. *J. Am. Chem. Soc.* **1997**, *119*, 7153.
52. (a) Kobayashi, S.; Nagayama, S. *J. Org. Chem.* **1997**, *62*, 232. (b) Kobayashi, S.; Nagayama, S. *J. Am. Chem. Soc.* **1997**, *119*, 10049.
53. Cf. Fendler, J.H.; Fendler, E.J. *Catalysis in Micellar and Macromolecular Systems*, Academic Press, London, 1975; *Mixed Surfactant Systems*, ed. by P.M. Holland and D.N. Rubingh, ACS, Washington, DC, 1994; *Surfactant-Enhanced Subsurface Remediatuon*, ed. by D.A. Sabayini, R.C. Knox and J.H. Harwell, ACS, Washington, DC, 1995.
54. (a) Strecker, A. *Ann. Chem. Pharm.* **1850**, *75*, 27. (b) Shafran, Y.M.; Bakulev, V.A.; Mokrushin, L.S. *Russ. Chem. Rev.* **1989**, *58*, 148.
55. (a) Weinstock, L.M.; Davis, P.; Handelsman, B.; Tull, R. *J. Org. Chem.* **1967**, *32*, 2823. (b) Matier, W.L.; Owens, D.A.; Comer, W.T.; Deitchman, D.; Ferguson, H.C.; Seïdehamel, R.J.; Young, J.R. *J. Med. Chem.* **1973**, *16*, 901.
56. (a) Ojima, I.; Inaba, S.; Nakatsugawa, K. *Chem. Lett.*, **1975**, 331. (b) Mai, K.; Patil, G. *Tetrahedron Lett.* **1984**, *25*, 4583. (c) Kobayashi, S.; Ishitani, H.; Ueno, M. *Synlett* **1997**, 115.
57. (a) Luijten, J.G.A.; van der kerk, G.J. *Investigations in the Field of Organotin Chemistry,* Tin Reserch Institute, Greenford, 1995, p. 106. (b) Tanaka, M. *Tetrahedron Lett.* **1980**, *21*, 2959. (c) Harusawa, R.; Yoneda, R.; Omori, Y.;Kurihara, T. *Tetrahedron Lett.* **1987**, *28*, 4189.
58. Davies, A.G. *Organotin Chemistry,* VCH, Weinheim, 1997.
59. Brown, J.M.; Chapman, A.C.; Harper, R.; Mowthorpe, D.J.; Davies, A.G.; Smith, P. J. *J. Chem. Soc. Dalton Trans.* **1972**, 338.
60. Davies, A.G.; Kleinschmidt, D.C.; Palan, P.R.; Vasishtha, S.C. *J. Chem. Soc.* (C) **1971**, 3972.
61. Kobayashi, S.; Hachiya, I.; Araki, M.; Ishitani, H. *Tetrahedron Lett.* **1993**, *34*, 3755.
62. Iyer, M.S.; Gigstad, K.M.; Namdev, N.D.; Lipton, M. *J. Am. Chem. Soc.* **1996**, *118*, 4910.
63. For recent reviews on organic reactions in aqueous media, see (a) Li, C.-J. *Chem. Rev.* **1993**, *93*, 2023. (b) *Organic Synthesis in Water*; Grieco, P.A. Ed.; Blacky Academic and Professional, London, 1998. (c) *Aqueous-Pase Organometallic Catalysis*, Cornils, B., Herrmann, W.A. Eds.; Wiley-VCH, Weinheim, 1998. (d) *Lanthanides: Chemistry and Use in Organic Synthesis*, S. Kobayashi, Ed.; Springer, Berlin, 1999. (e) Kobayashi, S. *Eur. J. Org. Chem.* **1999**, 15.

15 Thermo- and Photochemical Reactions of Carbonyl Compounds in the Solid State

Fumio Toda

15.1 Introduction

Most thermo- and photochemical organic reactions have been studied in solution. However, these reactions occur also in the absence of solvent and can be carried out by keeping a mixture of finely powdered reaction substrate and reagent at room temperature. Solid state photoreactions are usually carried out by irradiation of crystalline reactants. In some cases, solid state thermal reactions are accelerated by heating, shaking, irradiation with ultrasound or grinding of the reaction mixture using a mortar and pestle. Generation of local heat by grinding crystals of substrate and reagent is also helpful. In special cases, a mixture of reactant and reagent turns into a glassy material. In some cases, solid-gas, solid-liquid and even liquid-liquid reactions can also be accomplished in the absence of solvent. These solvent-free reactions are all included in this Chapter. Furthermore, in some cases, the reaction product can be separated from the reaction mixture directly by distillation, showing that no solvent is necessary throughout the reaction and separation processes. This is a genuine green, clean and economical process.

When the solid state reaction is carried out in an inclusion complex with an optically active host compound, an optically active reaction product is obtained by enantioselective reaction. This methodology is very effective for photoreactions. In some cases, a photoreaction which is inefficient in solution proceeds very efficiently as the inclusion complex.

Since this text describes the methodology of solid state reactions of carbonyl compounds using our own experimental results, not all other reported data from other workers are described.

15.2 Thermochemical Reactions

15.2.1 Baeyer-Villiger Oxidation

Some Baeyer-Villiger oxidations of ketones with *m*-chloroperbenzoic acid proceed much faster in the solid state than in solution. For example, when a mixture of powdered ketone and two equivalents of *m*-chloroperbenzoic acid was kept at room temperature, the oxidation product was obtained in the yield indicated in Table 15-1 [1]. Each yield is much higher than that obtained by the reaction in $CHCl_3$ (Table 15-1).

Table 15-1. Yields of Baeyer-Villiger oxidation products [a]).

Ketone	Reaction time	Product	Yield (%)	
			Solid state	$CHCl_3$ [b])
4-*t*Bu-cyclohexanone	30 min	5-*t*Bu-ε-caprolactone	95	94
Br–C6H4–COMe	5 days	Br–C6H4–OCOMe	64	50
$PhCOCH_2Ph$	24 h	$PhOCOCH_2Ph$	97	46
PhCOPh	24 h	PhCOOPh	85	13
Me–C6H4–COPh	24 h	Me–C6H4–OCOPh	50	12
2-Me–C6H4–COPh	4 days	2-Me–C6H4–OCOPh	39	6

a) Molar ratio of ketone and *m*-chloroperbenzoic acid is 1:2. b) The reaction was carried out with 1 g of ketone in 50 ml of $CHCl_3$.

We have shown that the movement of molecules during the host-guest complex formation process in the solid state is easy. Some host-guest complexes can be formed simply by mixing host and guest compounds in the solid state or keeping a mixture of powdered host and guest at room temperature [2]. Therefore, the ease of oxidation in the solid state is not unexpected.

15.2.2 Enantioselective Reduction of Ketone by $NaBH_4$ and $2BH_3$-$NH_2CH_2CH_2NH_2$

Reduction of ketone with $NaBH_4$ also proceeds in the solid state. A mixture of the ketone and a tenfold molar amount of $NaBH_4$ was finely powdered using an agate mortar and pestle and kept in a dry box at room temperature for five days, being stirred once a day. The reaction mixture was extracted with ether, and the dried ether solution was evaporated to give the corresponding alcohol in the yields shown in Table 15-2 [3].

Table 15-2. Reduction of ketones in the solid state by $NaBH_4$ [a]).

Ketone	Product	Yield (%)
Ph_2CO	Ph_2CH–OH	100
trans-PhCH=CHCOPh	*trans*-PhCH=CHCH(OH)Ph; $PhCH_2CH_2CH(OH)Ph$	100 (1 : 1)
4-Br-C_6H_4-COPh	4-Br-C_6H_4-CH(OH)Ph	100
2-naphthyl-COMe	2-naphthyl-CH(OH)Me	53
PhCH(OH)COPh	*meso*-PhCH(OH)–CH(OH)Ph	62
$PhCH_2COPh$	$PhCH_2CH(OH)Ph$	63
2-Cl-C_6H_4-CO-C_6H_4-4-Me	2-Cl-C_6H_4-CH(OH)-C_6H_4-4-Me	21
4-*t*Bu-cyclohexanone	4-*t*Bu-cyclohexanol	92
PhCOCON(*i*Pr)$_2$	PhCH(OH)CON(*i*Pr)$_2$	24

[a]) Reduction was carried out by keeping a 1 : 10 mixture of powdered ketone and $NaBH_4$ at room temperature for 5 days.

When the reduction of ketone is carried out as its inclusion complex with an optically active host compound, regio- and enantioselective reductions occur yielding an optically active alcohol [4]. For example, treatment of a 1 : 1 inclusion complex of (*R*)-(–)-**1** and optically active host **2a** with $NaBH_4$ in the solid state for 3 days gave (*R,R*)-(–)-**4** of 100% ee in 54% yield [3]. The corresponding reaction of a 1 : 1 complex of (*S*)-(+)-**1** and **3b** gave (*S,S*)-(+)-**4** of 100% ee. Treatment of the racemic diketone **1** with **2a** results in selective formation of an inclusion complex with (*R*)-(–)-**1**, and its decomposition gave (*R*)-(–)-**1** of 100% ee. Since the hydride attacks the carbonyl carbon at the 7-position from the side opposite to the methyl group, (*R*)-(–)-**1** should give (*R,R*)-(–)-**4** of 100% ee, as was found.

The enone moiety of (*R*)-(–)-**1** is presumably masked by forming a hydrogen bond with the hydroxyl group of **2a**, so that the other carbonyl group is reduced selectively. When the reduction of (*R*)-(–)-**1** is carried out in the presence of **2a** in a suspension in water, the diol **5** is obtained as a mixture of diastereomers.

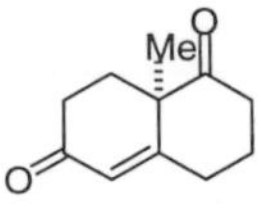

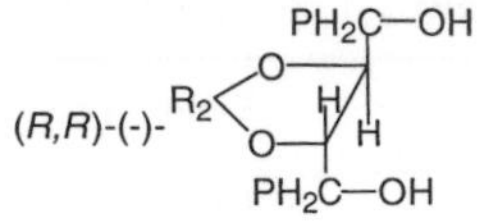

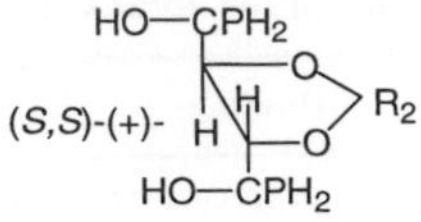

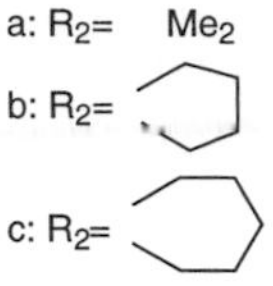

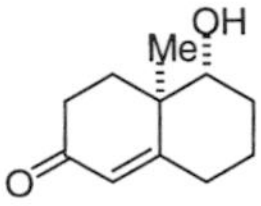

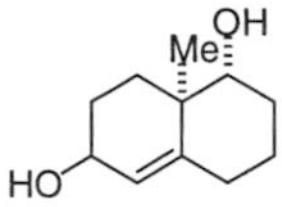

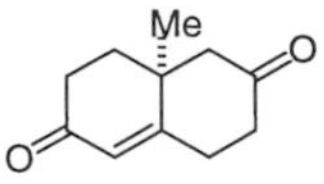

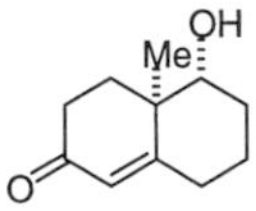

It is not easy to prepare even racemic ketoalcohol *rac*-**4**. *Rac*-**4** has been prepared by selective reduction of *rac*-**1** with $NaBH_4$ in $MeOH/ClCH_2CH_2Cl$ at –78 °C and by selective oxidation of *rac*-**5** with MnO_2 [5].

Similar reduction of a 1 : 1 complex [6] of (*R*)-(–)-**6** and **2a** with $NaBH_4$ in the solid state gave (*R,R*)-(–)-**7** of 100% ee in 55% yield.

Solid state reduction of alkyl aryl ketone **8** in an inclusion complex with the chiral host **10** with borane-ethylenediamine complex $2BH_3\text{-}NH_2CH_2CH_2NH_2$ gave optically active alcohol **9** in the optical and chemical yields summarized in Table 15-3 [7].

$NaBH_4$ reduction of **8** included in *β*-cyclodextrin also proceeded quite easily in the solid state. However, the product alcohol **9** in all cases gave only a modest level of optical purity [8].

15.2.3 Grignard Reaction

A Grignard reaction also occurred in the solid state, but some reactions gave different results from those observed in solution. In particular, reactions of ketones in the solid state gave more reduction products than addition products [9].

Dried Grignard reagents were obtained as white powders by evaporation of the solvent in vacuo from the material prepared by the usual method in ether. ^1H NMR spectra of the dried Grignard reagents in $CDCl_3$ showed the presence of

Table 15-3. Yield, optical purity, and absolute configuration of the alcohol **9** obtained by solid-solid reaction of a 1 : 1 inclusion crystal of **8** and **10** with $2\,BH_3{\cdot}NH_2CH_2CH_2NH_2$.

Ketone	Product		
	Alcohol	Yield (%)	Optical purity (% ee)
8a	(*R*)-(+)-**9a**	96	44
8b	(*R*)-(+)-**9b**	57	59
8c	(*R*)-(+)-**9c**	20	22
8d	(*R*)-(+)-**9d**	55	42
8e	(*R*)-(+)-**9e**	64	42

ArCOR **8**

ARCHR(OH) **9**

(*S,S*)-(-)- Ph–C(OH)(o-ClC₆H₄)–C≡C–C≡C–C(OH)(o-ClC₆H₄)–Ph **10**

a: Ar = Ph; R = Me
b: AR = *o*-MeC_6H_4; R = Me
c: Ar = 1-naphthyl; R = Me
d: Ar = Ph; R = Et
e: Ar = *o*-MeC_6H_4; R = Et

$$RMgX\ (\mathbf{11}) + Ph_2CO\ (\mathbf{12}) \longrightarrow Ph_2RCOH\ (\mathbf{13}) + Ph_2CHOH\ (\mathbf{14})$$

11: a: R = Me; X = I; b: R = Et; X = Br; c: R = *i*Pr; X = Br; d: R = Ph; X = Br

13: a: R = Me; b: R = Et; c: R = *i*Pr; d: R = Ph

Table 15-4. Products and yields of Grignard reactions of **11** and **12** at room temperature for 0.5 h in the solid state [a]) and in solution [b]).

Grignard reagent	Product and yield (%)			
	Solid state		Solution	
11a	–	–	**13a** (99)	–
11b	**13b** (30)	**14** (31)	**13b** (80)	**14** (20)
11c	**13c** (2)	**14** (20)	**13c** (59)	**14** (22)
11d	**13d** (59)	–	**13d** (94)	–

[a]) Only the reaction of **11a** and **12** was carried out at 50 °C since the reaction did not occur at room temperature. [b]) All the reactions in solution were carried out in ether by the usual method.

two moles of ether to one of RMgX. An ether solution of the dried Grignard reagent behaves identically to the usual Grignard reagent.

One mole of ketone and three moles of the dried Grignard reagent were finely powdered and well mixed using an agate mortar and pestle, and the mixture was

Table 15-5. Products and yields of Grignard reactions in the solid state at room temperature for 0.5 h.

Grignard reagent	Product and yield (%)		
	Ketone **15**	**17**	**18**
11b	**16a** (39)	**18a** (43)	**20a** (31)
11d	**16b** (64)	**18b** (67)	**20b** (79)

Ph(H)C=C(H)COPh **15** —11→ PhRCHCH$_2$COPh **16**

PhCOCH(OEt)Ph **17** —11→ PhRC(OH)CH(OEt)Ph **18**

Cl–C$_6$H$_4$–CHO **19** —11→ Cl–C$_6$H$_4$–CH(R)OH **20**

a : R=Et
b : R=Ph

then decomposed with aqueous NH_4Cl, extracted with ether, and the organic solution dried over Na_2SO_4. Evaporation of the solvent gave products in the yields shown in Tables 15-4 and 15-5.

Although **11a** did not react with benzophenone **12** in the solid state, other reagents (**11b–d**), did react and gave **13** and **14** in the yields shown in Table 15-4. In the case of **11b** and **11c**, more of the reduction product **14** was obtained in the solid state than in solution. A plausible interpretation for this difference is that the hydrogen radical can move more easily in the solid state than the alkyl radical.

In contrast, 1,4-addition of **11** to **15** and 1,2-addition of **11** to **17** and **19** proceeded in a similar manner to those in solution (Table 15-5).

15.2.4 Reformatsky and Luche Reactions

Reformatsky and Luche reactions with Zn provide more economical C–C bond formation methods than Grignard reactions, which use more expensive Mg metal. Both the reactions proceed efficiently in the absence of solvent [10]. These nonsolvent reactions can be carried out by a very simple procedure and give products in higher yield than with solvent.

In general, the solvent-free reactions were carried out by mixing aldehyde or ketone, organic bromo compound, and Zn-NH_4Cl using an agate mortar and pestle and by keeping the mixture at room temperature for several hours.

$$ArCHO + BrCH_2COOEt \xrightarrow[NH_4Cl]{Zn} ArCH(OH)CH_2COOEt$$

21 **22** **23**

Table 15-6. Reaction time and yields of product **23** in the nonsolvent Reformatsky reactions of **21** and **22**.

21	Ar	Reaction time (h)	Yield (%) of **23**
a	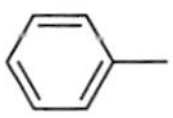 phenyl	2	91
b	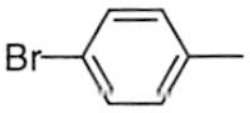 4-Br-phenyl	3	94
c	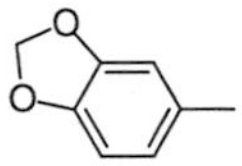 3,4-methylenedioxyphenyl	3	94
d	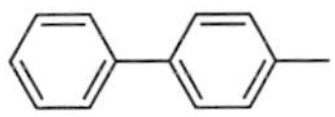 4-biphenylyl	3	83
e	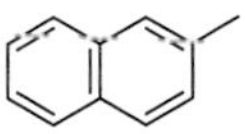 2-naphthyl	3	80

$BrCH_2CH{=}CH_2$ **24**

$CH_3(CH_2)_4CHO$ **25**

trans-$CH_3CH{=}CHCHO$ **26**

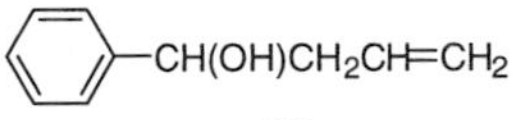

$CH_3(CH_2)_3COCH_3$ **27**

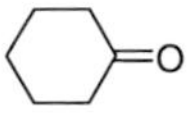

Cyclohexanone **28**

C_6H_5–$CH(OH)CH_2CH{=}CH_2$ **29**

2-Naphthyl–$CH(OH)CH_2CH{=}CH_2$ **30**

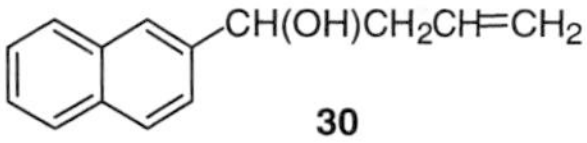

$CH_3(CH_2)_4CH(OH)CH_2CH{=}CH_2$ **31**

trans-$CH_3CH{=}CHCH(OH)CH_2CH{=}CH_2$ **32**

$CH_3(CH_2)_3C(CH_3)(OH)CH_2CH{=}CH_2$ **33**

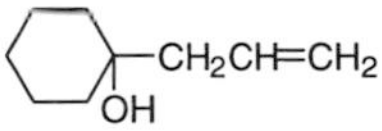

1-(OH)-1-($CH_2CH{=}CH_2$)cyclohexane **34**

Treatment of the aromatic aldehydes (**21a–e**) with ethyl bromoacetate (**22**) and Zn-NH_4Cl gave the corresponding Reformatsky reaction products **23a–e** in the yields shown in Table 15-6. The yield, for example, of **23a** obtained in the solvent-free reaction (91%) is much higher than that obtained by the reaction in dry benzene-ether solution (61–64%) [11, 12]. The nonsolvent Reformatsky reaction, which does not require the use of an anhydrous solvent, is thus advantageous.

Luche reaction also proceeds efficiently in the absence of solvent [10]. Treatment of aldehydes, **21a**, **21e**, **25**, **26**, or ketones, **27**, **28**, with 3-bromopropene

Table 15-7. Reaction time and yields of the nonsolvent Luche reaction of aldehydes and ketones with **24**.

Aldehyde or ketone	Reaction time (h)	Product	
			Yield (%)
21a	4	**29**	99
21e	4	**30**	87
25	1	**31**	83
26	1	**32**	98
27	2	**33**	89
28	2	**34**	90

(**24**) and Zn-NH_4Cl in the absence of solvent gave the corresponding Luche reaction products, **29–34**, in the yields shown in Table 15-7. It has been reported that the Luche reaction of **28** with **24** in water and DMF [13] at room temperature gives **34** in 82% and 99% yields, respectively. However, the solvent-free reaction procedure is much simpler and more economical.

15.2.5 Benzilic Acid Rearrangement

Benzilic acid rearrangement is usually carried out by heating benzil derivatives and an alkali metal hydroxide in aqueous organic solvent. This reaction also proceeds efficiently in the solid state. For example, a mixture of finely powdered benzil (**35a**) (0.5 g, 2.38 mmol) and KOH (0.26 g, 4.76 mmol) was heated at 80 °C for 0.2 h, and the reaction mixture was mixed with 3N HCl (20 ml) to give benzilic acid (**39a**) as colorless needles (0.49 g, 90% yield) [13]. Since the product is collected simply by filtration, this method is very simple and economical. Similar treatment of benzil derivatives (**35b–g**) in the solid state also gave the corresponding benzilic acid (**39b–g**) in good yields (Table 15-8) [13].

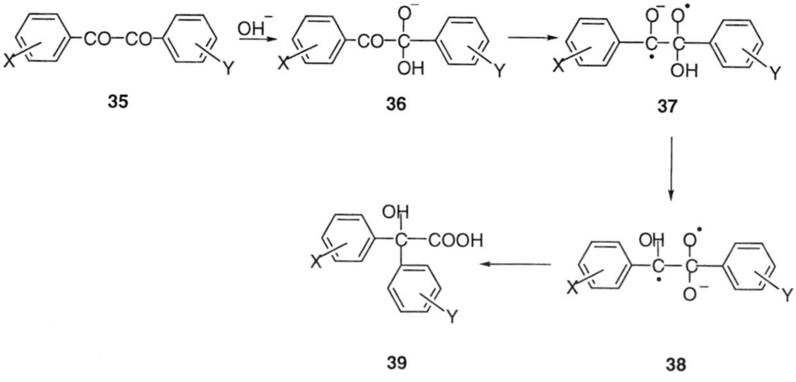

Table 15-8. Yields of benzilic acid **39** produced by treatment of benzil bolo 35 with KOH at 80 °C in the solid state.

35	X	Y	Reaction time (h)	Yield of **39** (%)
a	H	H	0.2	90
b	H	*p*-Cl	0.5	92
c	*p*-Cl	*p*-Cl	6	68
d	H	*p*-NO_2	0.1 [a])	98
e	*m*-NO_2	*m*-NO_2	0.1 [a])	72
f	H	*m*-McOH	6	91
g	*p*-MeO	*p*-MeO	6	32

[a]) Reaction was carried out at room temperature.

By ESR studies, the benzilic acid rearrangement in solution has been proven to proceed via a radical intermediate [14]. For the benzilic acid rearrangement in the solid state, a radical intermediate was also detected. For example, a freshly prepared mixture of finely powdered **35e** and KOH showed a strong ESR signal (g=2.0049), and this signal declined as the reaction proceeded.

The effect of the alkali metal hydroxide on the rate of the benzilic acid rearrangement in the solid state is different from that in solution. Its effect on the rate of rearrangement of **35a** in the solid state increases in the order: KOH > $Ba(OH)_2$ > RbOH > NaOH > CsOH. On the other hand, the rate of rearrangement of **35a** in boiling 50% aqueous EtOH and in 67% aqueous dioxane increases in the order: KOH > NaOH > $Sr(OH)_2$ > LiOH > $Ba(OH)_2$ > RbOH > CsOH, and LiOH > NaOH > CsOH > KOH, respectively [13]. The rearrangement using RbOH and $Ba(OH)_2$ proceeded faster in the solid state than in solution. However, LiOH and $Sr(OH)_2$ were inert towards the solid state rearrangement, although these reagents are effective in solution [13].

15.2.6 Azomethine Synthesis

When aniline (**40**) and benzaldehyde derivatives (**41**) are ground together at room temperature, their condensation reaction starts immediately usually with gentle heat production but without melting because azomethines have higher melting points (Table 15-9). By this method, azomethines (**42**) are obtained in quantitative yield [15].

X–C6H4–NH2 (**40**) + OHC–C6H4–Y (**41**) ⟶ X–C6H4–N=CH–C6H4–Y (**42**)

Table 15-9. Melting points of azomethines (**42**) formed quantitatively by solid-solid reaction of **40** and **41** at room temperature.

40 (Mp/°C)	**41** (Mp/°C)	**42** (Mp/°C)
X	Y	
Me (44–46)	Me (44–46)	125
Me (44–46)	Br (55–58)	136
Me (44–46)	NO_2 (105–108)	125
Me (44–46)	OH (117–119)	221
Me (44–46)	OH, *m*-Me (81–83)	116
MeO (57–60)	Cl (47–50)	124–125
MeO (57–60)	Br (55–58)	148
MeO (57–60)	NO_2 (105–108)	134
MeO (57–60)	OH (117–119)	210
MeO (57–60)	OH, *m*-Me (81–83)	132
NO_2 (148–151)	Cl (47–50)	168
Cl (68–71)	OH (117–119)	187
Br (62–64)	OH (117–119)	192
OH (188–190)	Cl (47–50)	181

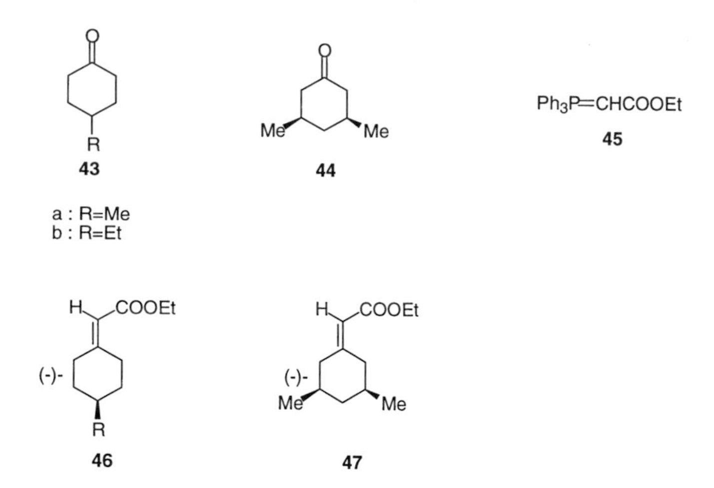

15.2.7 Enantioselective Wittig-Horner Reaction

Solid state Wittig-Horner reaction of 4-methyl- (**43a**), 4-ethyl- (**43b**) and 3,5-dimethylcyclohexanone (**44**) as their inclusion complex with optically active host compound and (carbethoxymethylene)triphenylphosphorane (**45**) gave optically active 4-methyl- (**46a**), 4-ethyl- (**46b**) and 3,5-dimethyl-1-(carbethoxymethylene)cyclohexane (**47**), respectively [16]. For example, when a mixture of the finely powdered 1 : 1 inclusion complex of **43a** with **2b** (1.5 g) and **45** (2.59 g) was kept at 70 °C, the Wittig-Horner reaction was completed within 4 h. To the reaction mixture was added ether-petroleum (1 : 1) and the precipitated solid (Ph_3PO

Table 15-10. Enantioselective Wittig-Horner reactions.

Host	Ketone	Reaction conditions		Product		
		Temperature (°C)	Time (h)		Yield (%)	Opt. purity (% ee)
2a	**43a**	70	4	**46a**	50.8	42.8
2b	**43a**	70	4	**46a**	73.0	42.3
2c	**43a**	80	4	**46a**	47.5	39.0
2b	**43b**	70	4	**46b**	72.5	45.2
2c	**43b**	80	4	**46b**	58.0	44.4
2c	**44**	80	2	**47**	58.0	56.9

and excess **45**) was removed by filtration. The crude product left after evaporation of the solvent from the filtrate was distilled in vacuo to give (–)-**46a** of 42.3% ee in 73% yield. The chiral hosts **2b** and **2c** are also effective for the enantioselective Wittig-Horner reaction of **43a** (Table 15-10). Following this procedure, **43b** and 44 gave optically active **46b** and **47**, respectively (Table 15-10) [16].

Only three enantioselective Wittig-Horner reactions using optically active reagents have been reported [17]. In comparison with these, the solid state reaction is much more simple and successful.

15.2.8 Aldol Condensations

Some aldol condensation reactions proceed more efficiently and stereoselectively in the absence of solvent than in solution [18]. When the solvent-free aldol reaction is carried out in an inclusion complex with a chiral host complex, diastereo- and enantioselective reactions occur, although enantioselectivity is not high [18].

When a slurry of *p*-methylbenzaldehyde (1.5 g, 12.5 mmol), acetophenone (1.5 g, 12.5 mmol) and NaOH (0.5 g, 12.5 mmol) was ground using a mortar and pestle at room temperature for 5 min, the mixture turned into a pale yellow solid. This solid was combined with water and filtered to give *p*-methylchalcone (2.7 g) in 97% yield (Table 15-11) [18]. When the condensation was carried out in 50% aqueous EtOH according to the reported procedure [19] for the same reaction time as above (5 min), the product was obtained only in 11% yield (Table 15-10). Other aldol reactions of benzaldehyde (**41**) and acetophenone derivatives (**48**) also proceed efficiently in the solid state (Table 15-11).

X-C₆H₄-CHO (**41**) + MeOC-C₆H₄-Y (**48**) ⟶ X-C₆H₄-CHCH₂CO-C₆H₄-Y with OH (**49**) + X-C₆H₄-CH=CHCO-C₆H₄-Y (**50**)

Table 15-11. Aldol reactions of **41** and **48** in the absence of solvent[a]) and in 50% aqueous EtOH[b]).

41	**48**	Reaction time (min)	Solvent	Yield (%)	
X	Y			**49**	**50**
H	H	30	–	10	0
			50% EtOH	0	36
p-Me	H	5	–	0	97
			50% EtOH	0	11
p-Me	Me	5	–	0	99
			50% EtOH	0	3
p-Cl	H	5	–	0	98
			50% EtOH	18	59
p-Cl	MeO	10	–	2	79
			50% EtOH	25	52
p-Cl	Br	10	–	0	81
			50% EtOH	0	92

[a]) Reaction was carried out by grinding a mixture of **41**, **48**, and NaOH using a pestle and mortar at room temperature. [b]) Reaction was carried out by keeping a solution of **41**, **48** and NaOH in 50% aqueous EtOH at room temperature.

15.2.9 Pinacol Coupling of Aromatic Aldehydes and Ketones

Pinacol coupling of aromatic aldehyde and ketones to afford α-glycols has been carried out by heating with Zn-AcOH, Mg-MgX_2, Al(Hg), Al, $TiCl_4$, $TiCl_4$-Zn, or Zn-$ZnCl_2$ [19]. These reactions are usually carried out at low temperature under an inert gas atmosphere, since the active reagents are sensitive to oxygen, and reaction at high temperature gives an olefin.

However, the coupling reaction with Zn-$ZnCl_2$ in the solid state and in 50% aqueous THF proceeds efficiently in air at room temperature to give the α-glycol. For example, a mixture of benzaldehyde (1 g), Zn powder (5 g) and $ZnCl_2$ (1 g) was kept at room temperature for 3 h. The reaction mixture was combined with 3 N HCl (5 ml) and toluene (10 ml), then filtered to remove Zn powder. From the filtrate, benzpinacol (0.48 g, 46% yield) was obtained (Table 15-12). Similar treatment of the benzaldehyde derivatives **51** gave the corresponding pinacol **52** in addition to the reduction products **53** [19].

When the reaction was carried out in 50% aqueous THF, more benzhydrol (**53**) was produced than in the solid state (Table 15-12). Since the reaction in the solid state is a high-concentration process, the intermolecular reaction of **51** would occur more easily to produce mainly the coupling product **52**. As the water content in aqueous THF decreases, the ratio of **52**:**53** increases. For example, when the coupling reaction of **51** (X=Br) with Zn-$ZnCl_2$ was carried out at room tempera-

X–C6H4–CHO (51) —Zn-ZnCl2→ X–C6H4–CH(OH)–CH(OH)–C6H4–X (52) + X–C6H4–CH2OH (53)

Table 15-12. Yield of **52** and **53** produced by treatment of **51** with Zn-$ZnCl_2$ at room temperature for 3 h in the solid state and in 50% aqueous THF.

51	Solvent	Yield (%)		*meso* : *dl* Ratio [b]) in **52**
X		**52**	**53**	
H	–	46	Trace	60:40
	A[a])	11	39	50:50
Me	–	87	Trace	70:30
	A	7	81	50:50
Cl	–	65	25	80:20
	A	16	82	50:50
Br	–	55	19	70:30
	A	27	72	50:50
CN	–	93	0	[c])
	A	90	0	[c])
Ph	–	64	2	70:30
	A	38	49	80:20

[a]) A=50% aqueous THF. [b]) The *meso* : *dl* ratio was determined by ^{1}H NMR spectroscopy. [c]) The *meso* : *dl* ratio was not determined.

ture for 3 h in THF containing 50, 20, 5 and 0% of water, the ratios of **52** (X=Br) and **53** (X=Br) were 27:72, 32:67, 49:50 and 55:19, respectively [19]. However, pinacol coupling of benzaldehydes with Mg in aqueous NH_4Cl has been reported to proceed efficiently [20].

The coupling reaction of aromatic ketones (**54**) with Zn-$ZnCl_2$ is more selective, and only the *α*-glycols (**55**) were produced (Table 15-13). This time, the reaction in aqueous THF is more effective than in the solid state. In many cases of reaction in the solid state, heating for a long time is necessary (Table 15-13). This coupling method is applicable to the acetophenone derivatives (**56**), and their coupling products (**57**) were obtained in good yields (Table 15-14) [19].

ArCOAr' (54) —Zn-ZnCl2→ Ar–C(Ar')(OH)–C(Ar')(OH)–Ar (55)

X–C6H4–COMe (56) —Zn-ZnCl2→ X–C6H4–C(Me)(OH)–C(Me)(OH)–C6H4–X (57)

Table 15-13. Yield of **55** produced by treatment of **54** with Zn-$ZnCl_2$ in the solid state and in 50% aqueous THF.

54		Solvent	Reaction time (h)	Temp. (°C)	Yield (%) of 55
Ar	Ar′				
C_6H_5	C_6H_5	–	6	rt [b])	86
		A [a])	1	rt	84
p-MeC_6H_4	*p*-MeC_6H_4	–	84	70	30
		A	3	rt	84
p-ClC_6H_4	*p*-ClC_6H_4	–	6	70	39
		A	1	rt	92
C_6H_5	*p*-ClC_6H_4	–	6	70	26 [c])
		A	2	rt	90 [c]
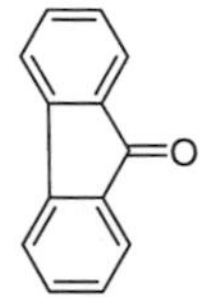		–	3	rt	92
		A	0.5	rt	94
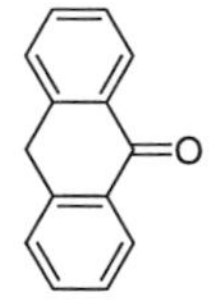		–	84	70	43
		A	1	rt	95
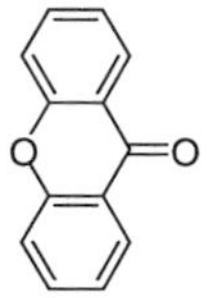		–	84	70	20
		A	2	rt	94

[a]) A=50% aqueous THF. [b]) rt=room temperature. [c]) The *meso*:*dl* ratio was not determined.

Table 15-14. Yield of **57** produced by treatment of **56** with Zn-$ZnCl_2$ in the solid state and in 50% aqueous THF.

56	Solvent	Reaction time (h)	Temp. (°C)	Yield (%) of **57** [c])
X=H	–	24	70	89
	A [a])	3	rt [b])	0
X=Br	–	3	70	77
	A	3	rt	62
X=CN	–	3	70	65
	A	3	rt	100

[a]) A=50% aqueous THF. [b]) rt=room temperature. [c]) The *meso*:*dl* ratio was not determined.

15.2.10 Dieckmann Condensation

Dieckmann condensation reactions of diesters normally should be carried out in dried solvent under reflux under an inert atmosphere [21, 22]. Furthermore, Dieckmann reactions are often carried out under high dilution conditions in order to avoid intermolecular reaction [21, 22]. However, it was found that the reaction of diethyl adipate (**58a**) or pimelate (**58b**) proceeds efficiently in the absence of solvent under air [23]. It is also possible to isolate the reaction products from the reaction mixture directly by distillation [23]. These results establish a completely solvent-free procedure throughout the reaction and work-up of the reaction mixture. This is a very clean, green, simple and economical procedure.

$(CH_2)_n$-CH_2COOEt / CH_2COOEt → $(CH_2)_n$ ring with COOEt and =O

58 **59**

a : n=2
b : n=3

After mixing **58a** and powdered Bu*t*OK for 10 min at room temperature, using a mortar and pestle, the solidified reaction mixture was kept in a desiccator for 1 h in order to complete the reaction and evaporate the Bu*t*OH formed. The dried reaction mixture was neutralized by addition of *p*-TsOH-H_2O, then distilled under reduced pressure to give **59a** in 82% yield. Similar treatment of **58b** gave **59b** in 69% yield. Instead of Bu*t*OK, other alkoxides can be used (Table 15-15).

For comparison with the solvent-free reaction, the Dieckmann condensation of **58a** and **58b** was studied using the same base in toluene under reflux. As shown in Table 15-15, there is no marked difference between the yields of the solvent-free and solution reactions. This clearly shows that the solvent-free reaction is much better than the other in terms of simplicity, cleanliness and economy.

Table 15-15. Yield of Dieckmann condensation products **59a** and **59b** in the solid state (and in toluene [b]).

Base	Yield (%) of **59a**	Yield (%) of **59b**
Bu^{t}OK	82 (98)	69 (63)
Bu^{t}ONa	74 (69)	68 (56)
EtOK	63 (41)	56 (60)
EtONa	61 (58)	60 (60)

[a]) All reactions were carried out at room temperature for 10 min and the reaction products were isolated by distillation. [b]) All reactions were carried out in toluene under reflux for 3 h.

15.2.11 Methylene Transfer from Me_2S^+-CH_2^- to Ketones

The methylene transfer reaction from ylides to ketones has been developed as a convenient synthetic method for obtaining oxiranes [24]. However, the experimental procedure is complex. For example, a THF solution of dimethylsulfonium-methylide (**61**) is obtained by treatment of trimethylsulfonium iodide (**60**) with BuLi in THF at 0 °C, and after addition of the ketone the mixture is heated at 50–55 °C under nitrogen to yield the oxirane. Throughout the reaction and separation of the product, the organic solvent is essential [25].

This methylene transfer reaction can also be accomplished without using solvent, and the reaction product isolated from the reaction mixture by distillation. For example, a mixture of propiophenone (**62a**) (0.5 g), **60** (0.9 g) and powdered Bu*t*OK (0.5 g) was heated at 60 °C for 1 h in a flask, and then the reaction mixture was distilled using a Kugelrohr apparatus at 150 °C under 18 mmHg, to give **63a** (0.4 g, 75% yield). By a similar procedure, **63b–j** were prepared from the corresponding ketones (**62b–f**), and the products were isolated by distillation to give the yields shown in Table 15-16 [25].

$Me_2\overset{+}{S}$–Me • I^- **60** $Me_2\overset{+}{S}$–$\overset{-}{CH_2}$ **61** ArC(=O)R **62** (Ar)(R)-oxirane **63**

Table 15-16. Preparation of **63** by the combination of methylene transfer reaction to **62** in the absence of solvent and by Kugelrohr distillation.

Ketone **62**			Reaction conditions		Yield (%) of **63**
	Ar	R	Temp. (°C)	Time (h)	
a	C_6H_5	Et	60	1	75
b	C_6H_5	*i*Pr	70	1	89
c	C_6H_5	cyclohexyl	70	10	91
d	*p*-MeC_6H_4	Et	70	2	86
e	*p*-BrC_6H_4	Me	70	5	64
f	2-Naphthyl	Me	70	3	86

15.2.12 Enantioselective Michael Addition Reaction

Enantioselective Michael addition of thiols to enones is a useful reaction for the synthesis of sex pheromones [26] and terpenes [27]. For example, enantioselective Michael additions of thiols to 2-cyclohexenone (**64**) and maleic acid esters in the presence of chiral bases such as cinchona alkaloids [28, 29] and optically active amino alcohols [30, 31] have been reported. It has also been found that the enantioselective Michael addition reaction proceeds efficiently in an inclusion crystal

Table 15-17. Michael addition of **65** to **64** as its inclusion crystal with **2c** in the presence of a catalytic amount of **66**.

	65 R	Reaction time (h)[a]	Product		
			67	Yield (%)	Optical purity (% ee)
a		24	(+)-**67a**	51	80
b		36	(+)-**67b**	58	78
c		12	(+)-**67c**	62	[b]
d		36	(+)-**67d**	77	74
e		12	*rac*-**67e**	100	0
f		12	*rac*-**67f**	93	0

[a]) A mixture of a 1:1 inclusion complex of **64** with **2c**, **65** and **66** was irradiated by ultrasound (28 KHz) for 1 h, and was kept at room temperature.

of enone with an optically active host compound. For example, when a mixture of the powdered 1:1 inclusion complex of **64** and **2c** (0.5 g), 2-mercaptopyridine **65a** (0.11 g) and a 40% aqueous solution of benzyltrimethylammonium hydroxide (66) was irradiated with ultrasound for 1 h at room temperature, (+)-**67a** of 80% ee (0.088 g, 51% yield) was obtained [32]. The same reaction of the inclusion complex with 2-mercaptopyridine (**65b**), 4,6-dimethyl-2-mercaptopyridine (**65c**) and 2-mercaptothiazoline (**65d**) gave the corresponding optically active Michael addition product (Table 15-17) [32].

Michael addition of **65** to 3-methyl-3-buten-2-one (**68**) in its inclusion crystal with **2c** also occurred enantioselectively. When the complex of **68** with **2c** was treated with **65a–f** in the presence of a catalytic amount of **66** under the same conditions as for the reaction shown in Table 15-17, products **69a–f** were obtained in the optical purities shown in Table 15-18. The optical purities of **69a** and **69d** were relatively higher, but those of **69b** and **69c** were lower, whilst, in the case of **69e** and **69f**, enantiocontrol was not manifested [32]. It has been reported that some of the solid state organic reactions are accelerated by an irradiation of ultrasound [33].

RSH $PhCH_2\overset{+}{N}Me_3$ $\overset{-}{O}H$ SR

64 **65** **66** **67**

$CH_2{=}C(Me)COMe$ $RSCH_2CH(Me)COMe$

68 **69**

Table 15-18. Michael addition of **65** to **68** as its inclusion crystal with **2c** in the presence of a catalytic amount of **66** [a]).

65	Product		
	69	Yield (%)	Optical purity (% ee)
65a	(+)-**69a**	76	49
65b	(+)-**69b**	93	9
65c	(+)-**69c**	78	53
65d	(+)-**69d**	89	4
65e	*rac*-**69e**	63	0
65f	*rac*-**69f**	55	0

[a]) Reactions were carried out for 23 h by keeping at room temperature after irradiation by ultrasound (28 KHz) for 1 h at room temperature.

15.2.13 Michael Addition to Chalcone in a Water Suspension Medium

Very efficient Michael addition reactions of amines, thiophenol and acetylacetate to chalcone in a water suspension medium have been developed as completely organic solvent-free reactions.

As a typical example of Michael addition of an amine to chalcone in a water suspension medium, a suspension of powdered chalcone (**70a**) in a small amount of water containing nBuNH$_2$ (**71e**) and the surfactant hexadecyltrimethylammonium bromide (**72**) was stirred at room temperature for 4 h. The reaction product was filtered and air dried to give the Michael addition product **73e** as a colorless powder in 98% yield. The filtrate containing **72** can be used again [34]. By the same procedure, Michael addition reactions of the various amines **71a–q** to **70a** were carried out and pure amineadducts were obtained in good yields (Table 15-19) [34]. The solubility of the amines in water is not related to the efficiency of the reaction. Amines (**71h–k**) which are poorly soluble in water reacted with **70a** in the water suspension as effectively as the water-soluble amines (Table 15-19).

These Michael adducts cannot be obtained in a pure state through reaction in organic solvents. For example, when a solution of **70a** and **71e** in toluene was stirred at room temperature and solvent evaporated below 40 °C in vacuo, a mixture of **73e** and **70a** was obtained as an oily material which contained **73e** in only

Table 15-19. Michael addition reactions of amires (**71**) to chalcone (**70a**) in a water suspension medium containing **72** as a surfactant.

Amine **71**			Reaction time (h)	Product **73**	
	R^1	R^2		Yield (%)[a]	
a	Me	H	48	–	–[b]
b	Et	H	48	–	–[b]
c	*n*Pr	H	48	–	–[b]
d	*i*Pr	H	48	–	–[b]
e	*n*Bu	H	4	**73**	98
f	*s*Bu	H	48	–	–[b]
g	*t*Bu	H	48	–	–[b]
h	*n*Pen	H	0.7	**73**	98
i	*n*Hex	H	0.3	**73**	96
j	*n*Oct	H	0.3	**73**	93
k	C_6H_{11}	H	48	**73**	93
l	Ph	H	48	–	–[b]
m	Et	Et	48	–	–[b]
n			5	**73**	93
o	Me		48	–	–[b]
p	Me		0.5	**73**	95
q	Me		2	**73**	96

[a]) Isolatcd yields. [b]) No reaction occurred.

R^1, R^2 — **70**

a : R^1=R^2=H
b : R^1=*m*-Me ; R^2=H
c : R^1=*p*-Me ; R^2=H
d : R^1=*p*-Cl ; R^2=H
e : R^1=*p*-Br ; R^2=H
f : R^1=*p*-MeO ; R^2=H
g : R^1=H ; R^2=*p*-Me
h : R^1=H ; R^2=*p*-Cl
i : R^1=H ; R^2=*p*-Br
j : R^1=H ; R^2=*p*-MeO
k : R^1=R^2=*p*-Me
l : R^1=R^2=*p*-Cl

$C_{16}H_{33}\overset{+}{N}Me_3 \cdot \overset{-}{Br}$ **72**

70a + R^1R^2NH (**71**) —**72**, Water→ **73**

70f + PhSH (**74**) —**72**, K_2CO_3, water→ **75**

38% yield (^{1}H NMR analysis in $CDCl_3$). Furthermore, because of the equilibrium in solution, it is difficult to isolate **73** in a pure state from the mixture. For example, ^{1}H NMR spectral analysis showed that a solution of pure **73e** (0.1 g) in toluene (1 ml) consists of 53% of **73e** and 47% of **70a**. As far as we are aware, only **73n** has been prepared previously in a pure crystalline state by the reaction of **70a** and **71n** in a sealed tube at 100 °C [34].

Michael addition of thiophenol (**74**) to *p*-methoxychalcone (**70f**) also proceeded efficiently in a water suspension medium. When a mixture of **70f**, **74**, K_2CO_3, **72** and water was stirred for 24 h at room temperature and the reaction product was filtered and air dried, the addition product **75** was obtained in 92% yield. Although similar Michael addition reactions can be carried out in solution [35], the procedure of the solid state reaction is rather simple, economical and free from pollution problems involving organic solvents [34].

The Michael addition reaction in water suspension medium is also applicable to carbon-carbon bond formation. A mixture of **70a**, methyl acetylacetate (**76**), **72** and water was stirred for 5 h at room temperature, then the reaction product was filtered and dried to give the addition product **77** in 98% yield [34]. Although a similar solvent-free Michael addition reaction of **70a** with diethyl malonate at 60 °C has been reported, organic solvent was still necessary to isolate the product from the reaction mixture [36].

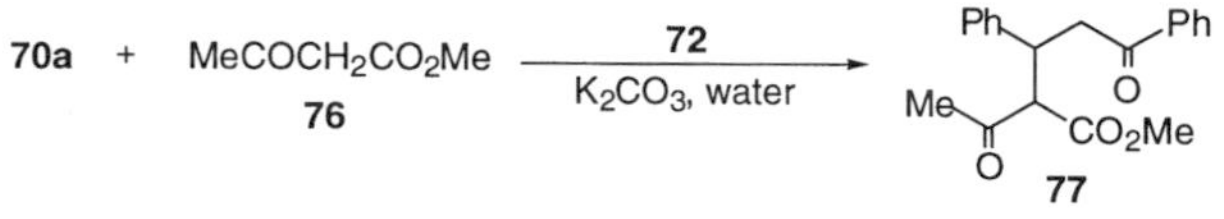

15.2.14 Epoxidation of Chalcones with NaOCl or $Ca(ClO)_2$ in a Water Suspension

The water suspension method described in Section 15.2.13 can also be applied to epoxidation reactions of chalcones **70** with NaOCl or $Ca(ClO)_2$. A mixture of **70a**, **72** and commercially available 11% aqueous NaOCl was stirred at room temperature for 24 h. The reaction product was filtered and dried to give **76a** in quantitative yield [34]. This procedure was applied to various kinds of chalcone derivatives, and **70b–j** were oxidized efficiently giving the corresponding epoxides **76b–j** respectively, in good yields (Table 15-20) [34]. In the case of **70h** and **70i**, the oxidation reaction proceeds very fast. This organic solvent-free reaction procedure is much more simple and convenient in comparison with the usual solvent procedure [37]. $Ca(OCl)_2$ can also be used for this epoxidation reaction in water suspension (Table 15-20) [34]. In the case of **70g** and **70h**, the reaction proceeds extremely quickly. However, the reaction product must be isolated from water-insoluble $Ca(OCl)_2$ by extraction with organic solvent. Furthermore, in the case of **70d–f** and **70j**, the reaction products were not extracted with ether from the reaction mixture (Table 15-20).

70 —(72; NaOCl or Ca(OCl)$_2$; water)→ 76

Table 15-20. Epoxidation reactions of chalcones in a water suspension medium containing **72** as a surfactant.

Chalcone	Product	Reagent			
70	**76**	NaOCl[a]		$Ca(ClO)_2$[b]	
		Reaction time (day)	Yield (%)	Reaction time (day)	Yield (%)
70a	**76a**	1	100	1	77
70b	**76b**	2	80	2	70
70c	**76c**	2	85	2	89
70d	**76d**	4	85	7	–[c]
70e	**76e**	2	90	7	–[c]
70f	**76f**	1	78	7	–[c]
70g	**76g**	2	36	0.04	77
70h	**76h**	0.4	90	0.1	95
70i	**76i**	0.3	99	3	78
70j	**76j**	1	93	7	–[c]
70k	**76k**	5	43	5	50
70l	**76l**	5	30	4	82

[a]) Commercially available 11% aqueous solution was used and products were collected by filtration.
[b]) Commercially available powder was used and products were collected by extraction with ether.
[c]) Reaction products were not extracted with ether from reaction mixture.

15.3 Photochemical Reactions

15.3.1 Photocyclization of Achiral Oxo Amides in their Chiral Crystals to Chiral *β*-Lactams: Generation of Chirality

In some cases, achiral molecules are arranged in a chiral form in their crystals. When the chirality can be fixed by photoreaction, this becomes a convenient asymmetric synthesis without using any further chiral source, so can be described as an absolute asymmetric synthesis. This phenomenon is also important in relation to the mechanism of generation of chirality on Earth. Several such examples of absolute asymmetric synthesis in crystals have been reported so far. In this Chapter, some of these found in the author's laboratory are described.

Recrystallization of the achiral oxo amide *N,N*-diisopropylphenylglyoxylamide (**78a**) from benzene gave colorless chiral prisms. Each crystal is chiral and shows a CD spectrum in Nujol mulls. One type of chiral crystal shows a (+)-Cotton effect and the other type shows a (–)-Cotton effect (Fig. 15-1). Crystals of (+)- and (–)-**78a** can easily be prepared in large quantities by seeding with finely powdered crystals of (+)- or (–)-**78a** during recrystallization of **78a** from benzene [35]. Measurement of the CD spectrum of chiral crystals as Nujol mulls is now well established [36, 37].

Irradiation of crystals of (+)-**78a** (200 mg) with a high-pressure mercury lamp, providing occasional grinding using an agate mortar and pestle at room temperature for 40 h, gave (+)-β-lactam 3-hydroxy-1-isopropyl-4,4-dimethyl-3-phenylazetidin-2-one (**79a**) of 93% ee in 74% yield. Irradiation of (–)-**78a** under the same conditions gave (–)-**79a** of 93% ee in 75% yield [35]. The reason why achiral **78a** molecules are easily arranged in a chiral form was clarified by X-ray analysis of a crystal. X-ray analysis showed that the two carbonyl groups are twisted around the single bond connecting these two groups as schematically depicted in Scheme 15-1 [38]. Photoreaction between the benzoylcarbonyl and the isopropyl group of (+)- and (–)-**78a** gives (+)- and (–)-**79a**, respectively (Scheme 15-1).

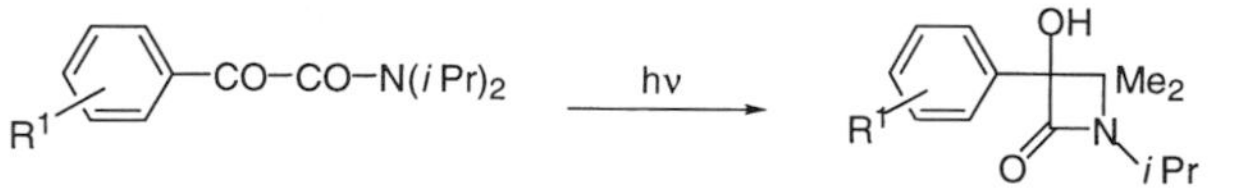

78 **79**

a : R=H
b : R=*m*-Me
c : R=*m*-Cl
d : R=*m*-Br
e : R=*o*-Me
f : R=*o*-Cl
g : R=*o*-Br
h : R=*p*-Me
i : R=*p*-Cl
j : R=*p*-Br

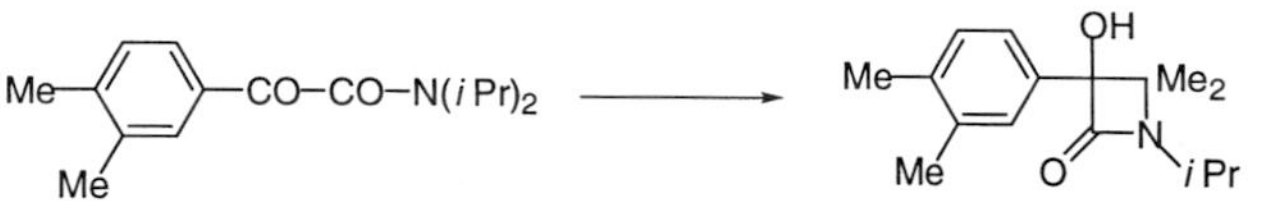

80 **81**

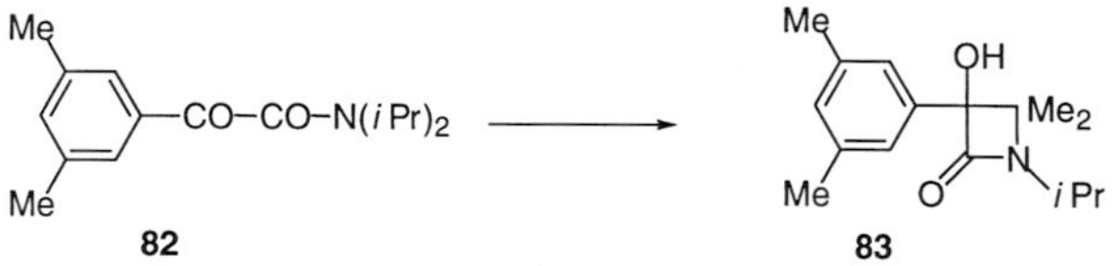

82 **83**

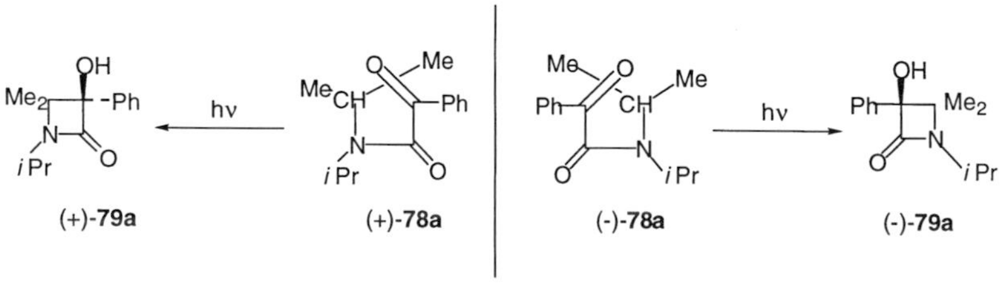

(+)-**79a** (+)-**78a** (-)-**78a** (-)-**79a**

mirror

Scheme 15-1. A mode of formation of optically active **79a** from achiral **78a** in its chiral crystal.

In this case, chiral crystals are available in bulk, and mass production of the chiral products is possible. Moreover, the present data may throw some light on the generation of optically active amino acids on Earth [39, 40]. Photocyclization of **78a** proceeds efficiently in sunlight, and hydrolysis of the optically active **79a** gives an optically active β-amino acid.

Of eleven derivatives of **78a**, **78b–j**, **80** and **82**, the compounds **78b–e** and **80** also formed chiral crystals, and their photoirradiation in the solid state gave the corresponding chiral β-lactams, **79b** (91% ee, 63% yield), **79c** (100% ee, 75% yield), **79d** (96% ee, 97% yield), **79e** (92% ee, 54% yield) and **81** (54% ee, 62% yield) in the optical and chemical yields indicated [41].

Although all meta-substituted derivatives **79b–d** formed chiral crystals, all the para-substituted ones **78h–j** did not. Photoirradiation of **78h–j** in the solid state gave racemic β-lactams **79h** (63% yield), **79i** (42% yield) and **79j** (65% yield), respectively, in the yields indicated. In the case of *o*-substituted derivatives, **78e–g**, only the *o*-methyl-substituted one **78e** formed chiral crystals, and its photoirradiation gave optically active **79e** of 92% ee in 54% yield. These data suggested that the *m*-substitution pattern is better for forming chiral crystals than the other substitution isomer. However, the *m,m*-dimethyl-substituted derivative **82** did not form chiral crystals and its photoirradiation gave racemic **83**, although the *m,p*-dimethyl derivative **80** formed chiral crystals and its photoirradiation gave optically active **81** of 54% ee in 62% yield [41].

By X-ray analysis, the chiral arrangement of **78c** molecules and achiral arrangement of **78f** and **78i** molecules in their crystals have been proven [42].

The photochemical conversion of **78d** in its chiral crystal to the optically active β-lactam **79d** was monitored by continuous measurements of CD spectra of Nujol mulls (Fig. 15-2). As the reaction proceeds, the CD spectra of **78d** decrease and new spectra due to **79d** appear.

It has been found that a combination of the two alkyl groups on the nitrogen atom of **78** is important to form chiral crystal. Of the four derivatives of **78a**, **84a–c** and **86**, only the derivative **84b** which has iPr and Et groups formed chiral crystals. The chirality of **84b** was confirmed by CD spectra measured as Nujol mulls (Fig. 15-3) [43]. The chiral arrangement of the **84b** molecule within its crystal was proven by X-ray analysis [43].

Photoirradiation of (–)- and (+)-**84b** crystals in the solid state gave (–)- and (+)-**85b** of 80% ee in 55% yield, respectively. Of course, irradiation of **84a**, **84c** and **86** gave racemic **85a**, **85c** and **87**, respectively.

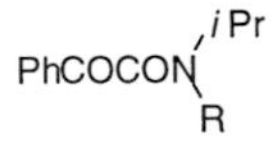

PhCOCON(*i*Pr)R

84

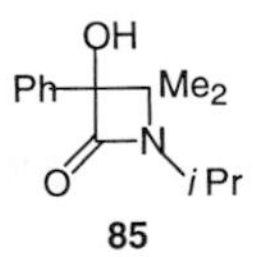

85

a : R=Me
b : R=Et
c : R=*n*Pr

PhCOCONMe$_2$

86

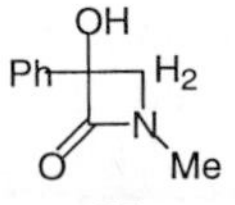

87

Even for the oxoamides which do not form chiral crystals, enantioselective photoconversion to optically active β-lactams can easily be accomplished by photoirradiation in their inclusion crystals with an optically active host compound. For example, irradiation of a 1:1 inclusion complex crystal of **86** with **10** gave (–)-**87** of 100% ee in a quantitative yield [44]. The host compound **10** is recovered and can be used again. The chiral arrangement of **86** molecules in the inclusion complex was studied by X-ray analysis [44].

Optically active hosts **2** and **3** are also useful for an enantioselective photocyclization of **86** [45]. The mechanism of the enantioselective reaction of **86** in an inclusion complex with **2** has also been studied by X-ray analysis [46].

Very interestingly, inclusion complex crystals of oxoamides with **2** can be prepared just by mixing both components in the solid state [47]. Irradiation of the complex formed by such mixing gives the optically active β-lactam. For example, mixing of **2c** and **86** in a 2:1 ratio gave their 2:1 inclusion complex and photoirradiation of the complex gave (+)-**87** of 41% ee in 48% yield. However, photoirradiation of a 2:1 inclusion complex prepared by recrystallization of **2** and **86** from solvent gave (–)-**87** of 85% ee in 39% yield. The chiral arrangement of **86** in these two complexes is reversed, although the reason for this drastic change is not clear [47].

When **2b** is used instead of **2c**, irradiation of the two 1:1 inclusion complexes of **2b** with **86** which are produced by mixing or recrystallization gave (–)-**87** (82% ee, 29% yield) and (–)-**87** (79% ee, 47% yield), respectively, in the optical and chemical yields indicated [47]. In contrast, the inclusion complex of **2a** with **86**, formed by mixing both components, on its photoirradiation gave (+)-**87** of 61% ee in 70% yield. However, an inclusion complex could not be obtained by recrystallization, despite more than 15 organic solvents being tested. In all cases, **2a** crystallized out separately.

15.3.2 Enantioselective Photocyclization of *N*-(Aryloylmethyl)-δ-valerolactams

Although *cis*- (**89**) and *trans*-7-hydroxy-1-azabicyclo[4.2.0]octan-2-one (**90**) can easily be obtained by photocyclization of *N*-(aryloylmethyl)-δ-valerolactam (**88**), this reaction proceeds nonstereoselectively to give a mixture of *rac*-**89** and *rac*-**90** [48]. Stereocontrol of this reaction is, however, easily accomplished by carrying out the reaction in an inclusion crystal with an optically active host compound such as **2** or **3**.

A suspension of powdered 1:1 inclusion complex of **88a** with **2c** (3.21 g) in water (120 ml) containing sodium alkylsulfate as a surfactant was irradiated under stirring at room temperature for 12 h. The reaction mixture was filtered to give optically active **89a** in the optical and chemical yields indicated in Table 15-21 [49]. From aqueous solution, additional (+)-**89a** was isolated (Table 15-21). By the same procedure, **88b** and **88c** gave optically active **89b** and **89c**, respectively (Table 15-21) [49]. However, **88d** was inert to irradiation in its inclusion crystal with **2c**.

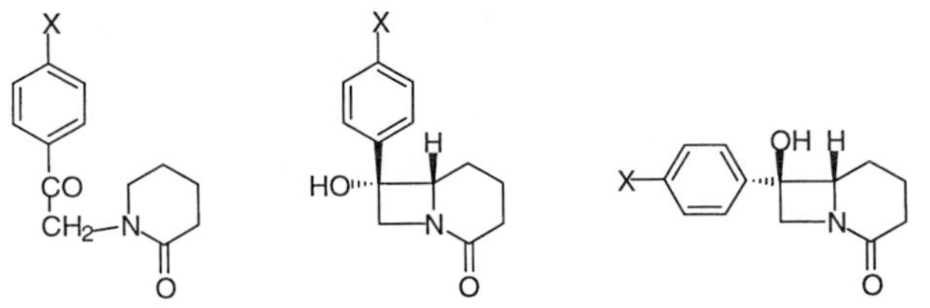

88 **89** **90**

a : X=H c : X=Br
b : X=Cl d : X=Me

Table 15-21. Photocyclization of **88** in a 1 : 1 inclusion crystal with **2c** or **3c** [a]).

88	Host	Product			
		By filtration		From aqueous layer	
		Yield (%) (% cc)		Yield (%) (% ee)	
88a	**2c**	(+)-**89a**	59 (98)	(+)-**89a**	28 (85)
88a	**3c**	(–)-**89a**	54 (99)	(–)-**89a**	28 (81)
88b	**2c**	(–)-**89b**	45 (84)	(–)-**89b**	25 (95)
88c	**2c**	(–)-**89c**	42 (98)	–	–

[a]) Irradiation was carried out in a water suspension through Pyrex filter at room temperature for 12 h by using 100-W high pressure Hg lamp under stirring.

The mechanism of the very effective enantioselective photocyclization of **88** was clarified by X-ray analysis of its inclusion crystal with **2c**. In this inclusion crystal, the **88** molecules are arranged in a chiral form so as to give only chiral **89** but not **90** upon irradiation [50].

15.3.3 Photodimerization of Enones

Although photodimerization of chalcone and its derivatives in solution have long been studied, most gave unsatisfactory results [51]. Photodimerization of chalcones in the solid state has also long been studied in order to correlate crystal-packing geometry with photochemical behavior [52]. In spite of many efforts at photodimerization by several research groups, all such attempts have failed [51]. X-ray analysis of all photochemically inert chalcones shows that the shortest contact of C=C groups of two neighboring chalcone molecules is longer than 4.2 Å. Finally, the rule “the photodimerization of chalcone does not occur when the distance between C=C groups is longer than 4.2 Å”, the so-called Schmidt’s rule, has appeared [53].

However, in an inclusion crystal of chalcone **70a** with 1,1,6,6-tetraphenylhexa-2,4-diyne-1,6-diol host compound **91**, the **70a** molecules aggregate in close positions and give as product the *syn*-head-to-tail dimer **92** in 82% yield by photoirradiation in the solid state [54]. X-ray analysis of this inclusion complex showed that two chalcone molecules are arranged in close positions so as to give the stereoisomer **92** selectively, as depicted schematically in Scheme 15-2 [55]. In the inclusion crystal, the distance between C=C groups of **70a** is 3.862 Å [55].

Very interestingly, however, photodimerizations of **70a** and its derivatives in the molten state proceeded efficiently and stereoselectively to give *rac-anti*-head-to-head dimers in all cases tested. All photodimerizations of **70a**, **70f**, **70g** and **70j** in the molten state for 24 h gave the corresponding *anti*-head-to-head dimer, **93a** (31%), **93f** (20%), **93g** (25%) and **93j** (32%) respectively [56]. This result suggests that two molecules of **70** aggregate in the liquid state so as to give the dimer **93** by photoreaction as shown in structure **94**. Since irradiation of **70a** and **70g** in solution also gives dimers **93a** and **93g**, respectively, in low yields [51], molecules of **70** would also be relatively easy to aggregate as species **94** even in solution. Since photodimerization of **70** is more efficient in the molten state than in solution, molecules of **70** must aggregate more easily as structure **94** in the molten state [56].

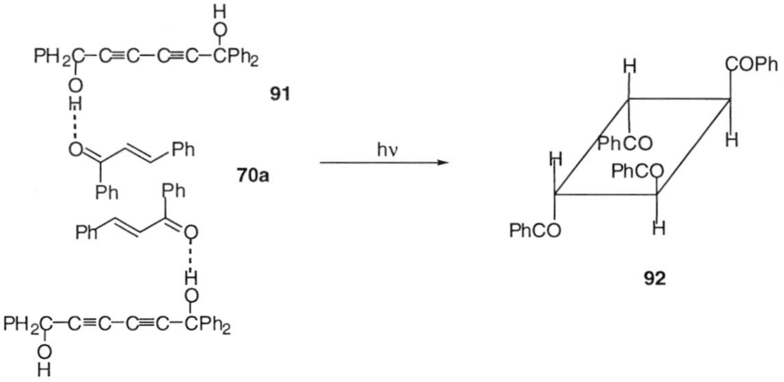

Scheme 15-2.

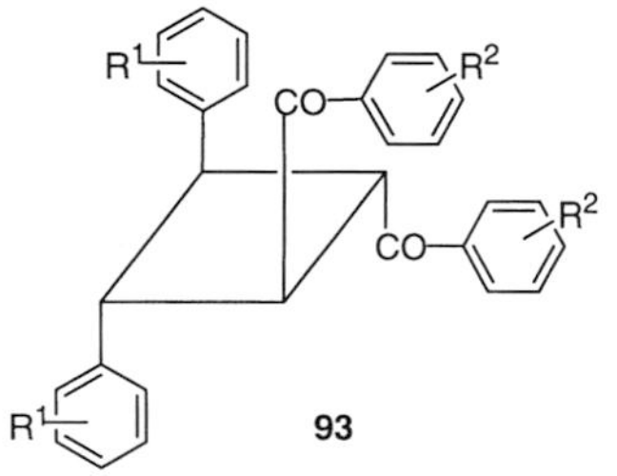

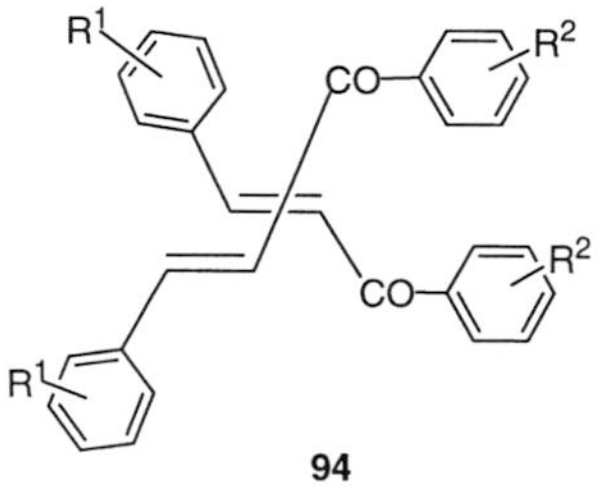

Photoirradiations of both neat and benzene solution of 2-cyclohexenone (**95**) give a complex mixture of the *syn-trans*-**96** and *anti-trans*-dimer **97**, and two other dimers of unknown structure [57]. When a 1:2 inclusion complex of the chiral host compound (–)-1,4-bis[3-(*o*-chlorophenyl)-3-hydroxy-3-phenylprop-1-

ynyl]benzene (**98**) with **95** was irradiated for 24 h, (–)-**96** of 48% ee was obtained in 75% yield [58]. Inclusion complexation of the (–)-**96** of 48% ee with **10** gave optically pure (–)-**96** [58].

Photoreactions of coumarin (**99**) in EtOH both in the absence and presence of benzophenone as a sensitizer give, respectively, a mixture of *syn*-head-to-head dimer **100** and *syn*-head-to-tail dimer **101** [59], and *anti*-head-to-head dimer **102** together with a small amount of *anti*-head-to-tail dimer **103** [60]. On the other hand, photoirradiation of a 1 : 2 inclusion complex of **10** with **99** and of a 1 : 2 inclusion complex of (*S,S*)-(–)-1,6-(2,4-dimethylphenyl)-1,6-diphenylhexa-2,4-diyne-1,6-diol (**104**) with **99** in the solid state gave **100** (74.5%) and *rac*-**103** (94%), respectively, in the yields indicated [61].

Interestingly, however, irradiation of a 1 : 1 inclusion complex of **2a** with **99** prepared by recrystallization of the two components from ethyl acetate-hexane gave (–)-**102** of 96% ee, although the same irradiation of a 1 : 1 inclusion complex of **2a** with **99** prepared by recrystallization from toluene-hexane gave **100** [61]. Photoconversion of **99** into **100** by the irradiation of the powdered 1 : 2 inclusion complex of **10** with **99** in the solid state has also been carried out recently by the Venkatesan research group [62].

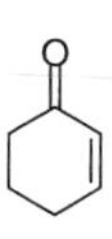

95 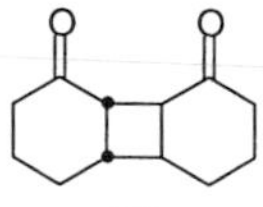 96 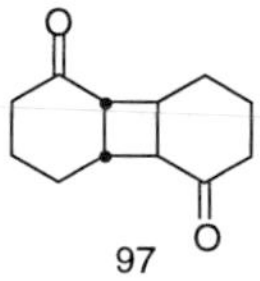 97

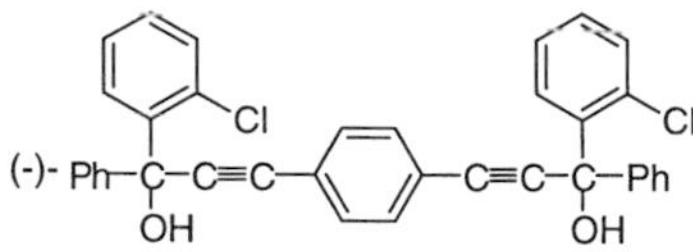

(-)- Ph–C(OH)(2-ClC6H4)–C≡C–C6H4–C≡C–C(OH)(2-ClC6H4)–Ph

98

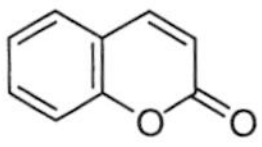

99 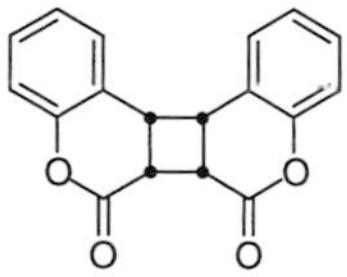 100 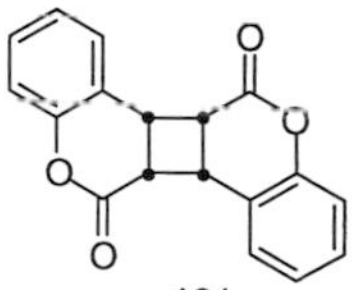 101

Ph–C(OH)(2,4-Me2C6H3)–C≡C–C≡C–C(OH)(2,4-Me2C6H3)–Ph

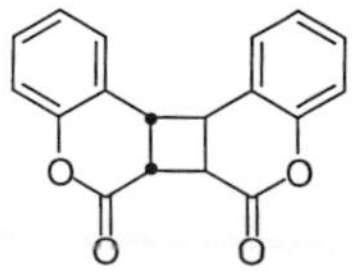

102 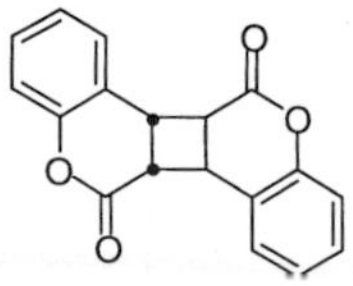 103 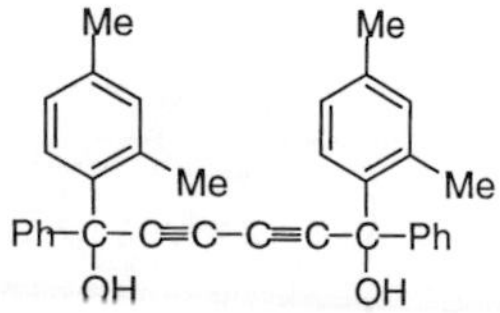 104

15.3.4 Enantioselective Photocyclization of Enones

15.3.4.1 Photoreaction of Tropolone Alkyl Ether, Cycloocta-2,4-dien-1-one and Pyridone

Enantioselective intramolecular photocyclization reactions of enones as inclusion crystals with chiral host compounds proceeds very efficiently. Several examples of such reactions are described in this Chapter.

When the photoreaction of tropolone alkyl ethers **105a** and **105b**, which produces the corresponding cyclization product **106** and its ring-opened derivative **107** [63], was carried out in a 1 : 1 inclusion complex with the chiral host **10**, then (1*S*,5*S*)-**106a** (100% ee, 11% yield) and (*S*)-**107a** (91% ee, 26% yield) and (1*S*, 5*R*)-**106b** (100% ee, 12% yield) and (*S*)-**107b** (72% ee, 14% yield) were obtained, respectively, in the optical and chemical yields indicated [64]. In the inclusion crystal with **10**, disrotatory [2+2] photoreaction of **105** would occur in the A direction, but not in the B direction, due to steric hindrance of the *o*-chlorophenyl group (Scheme 15-3), hence giving (1*S*,5*R*)-**106** but not (1*R*,5*S*)-**106** [64].

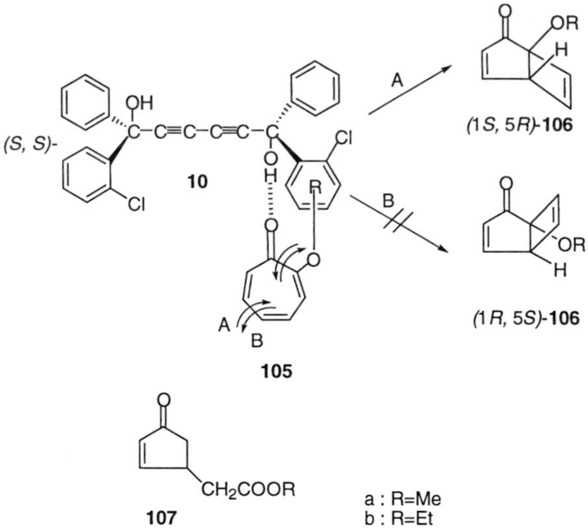

(S, S)- OH C–C≡C–C≡C–C O H Cl Cl 10 O H R O A B 105 A B OR H (1S, 5R)-106 OR H (1R, 5S)-106 O CH2COOR 107 a : R=Me b : R=Et

Scheme 15-3.

Irradiation of the 1 : 2 inclusion complex of cycloocta-2,4-dien-1-one (**108**) with **10** for 48 h gave (–)-**109** of 78% ee in 55% yield [65]. The mechanism of this enantioselective reaction has been studied by X-ray analysis of the inclusion complex [66].

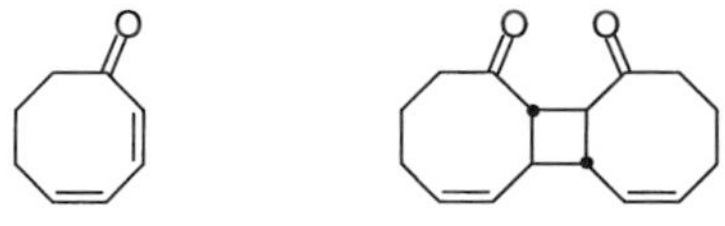

108 109

Irradiation of the inclusion complex of *N*-methylpyridine (**110a**) with **2a** gave (+)-**111a** of 100% ee in 93% yield after 15% conversion [67]. Irradiations of inclusion complexes of **110b** with the chiral host **10** or **2a** gave (–)-**111b** (100% ee, 97% yield) or (+)-**111b** (72% ee, 99% yield), after 50 and 15% conversions, respectively, in the optical and chemical yields indicated [67]. Photoreaction of fifteen other derivatives of **110** as inclusion complexes has also been studied [68].

110 **111**

a : R=H
b : R=OMe

15.3.4.2 Photoreaction of Cyclohexenone Derivatives

Irradiation of **112a–c** in their inclusion complexes with **2c** gave (+)-**113a** (99% ee, 81% yield), (+)-**113b** (99% ee, 71% yield) and (+)-**113c** (99% ee, 65% yield), respectively, in the optical and chemical yields indicated [69]. Similar irradiation of **114a–c** in their inclusion complexes with **2b** gave (–)-**115a** (99.9% ee, 18% yield), (–)-**115b** (99.9% ee, 41% yield) and (–)-**115c** (99.9% ee, 51% yield), respectively [69].

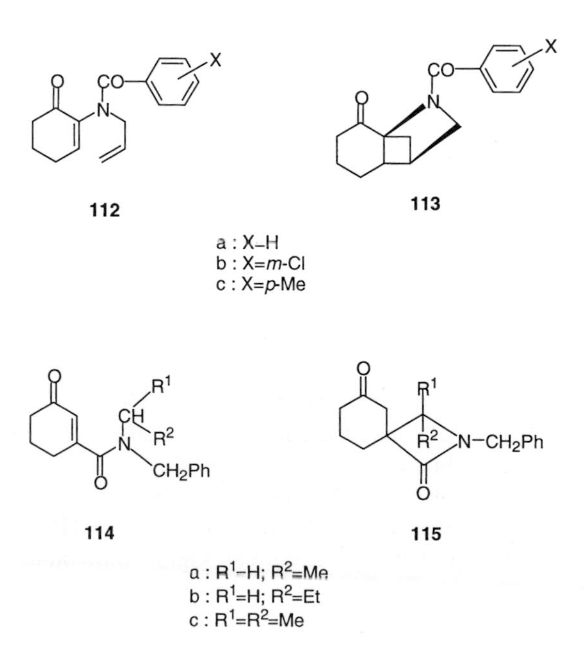

When the benzoyl group of **112** is substituted with alkyl or benzyl groups, the type of the photoreaction changes. Irradiation of **116a–h** in the inclusion complexes with the chiral hosts shown in Table 15-22 gave (–)-**117a–h**, respectively, in the optical and chemical yields indicated [70].

Table 15-22. Photocyclization reactions of **116** in 1:2 inclusion complex with **2** in a water suspension.

Host **2**	Guest **116**	Irradiation time (h)	Product		
				Yield (%)	Optical purity (% ee)
2b	**116a**	10	(–)-**117a**	32	65
2c	**116a**	10	(–)-**117a**	17	68
2b	**116b**	100	(–)-**117b**	40	14
2c	**116b**	100	(–)-**117b**	30	67
2b	**116c**	10	(–)-**117c**	38	64
2c	**116c**	10	(–)-**117c**	69	97
2b	**116d**	10	(–)-**117d**	13	28
2c	**116d**	10	(–)-**117d**	25	53
2b	**116e**	10	(–)-**117e**	90	100
2c	**116e**	10	(–)-**117e**	87	100
2c	**116f**	10	(–)-**117f**	56	100
2c	**116g**	10	(–)-**117g**	42	100
2b	**116h**	10	(–)-**117h**	53	100

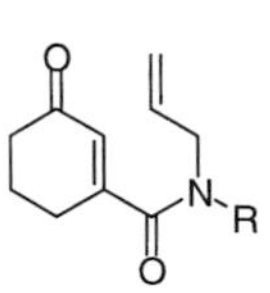

116 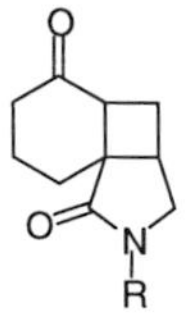 **117**

a: R=Me
b: R=Et
c: R=*n*Pr
d: R=*n*Bu
e: R= C_6H_5–CH_2–
f: R= Me–C_6H_4–CH_2–
g: R= Cl–C_6H_4–CH_2–
h: R= (Cl)C_6H_4–CH_2–

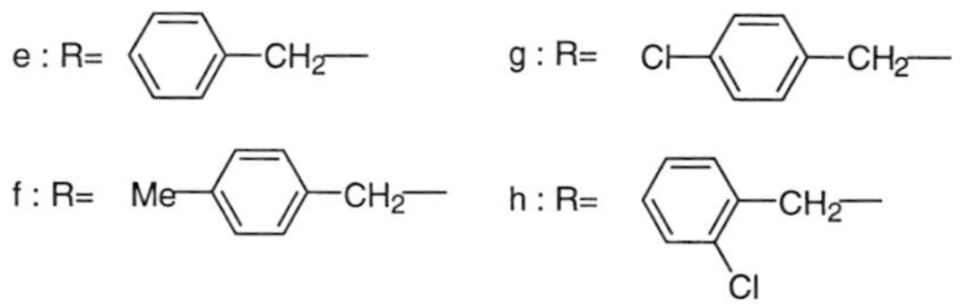

Photocyclization of 2-phenylthio-3,5,5-trimethylcyclohexen-1-one (**118**) to the dihydrobenzothiophene derivative (**119**) [71] can also be carried out enantioselectively. Photoirradiation of **118a–g** in their inclusion complexes with **2c** for 30 h in a water suspension gave (+)-**119c** (72% ee, 86% yield), (+)-**119d** (75% ee, 80% yield), (+)-**119e** (81% ee, 92% yield), (+)-**119f** (70% ee, 89% yield) and (+)-**119g** (83% ee, 91% yield), respectively, in the optical and chemical yields indicated [72].

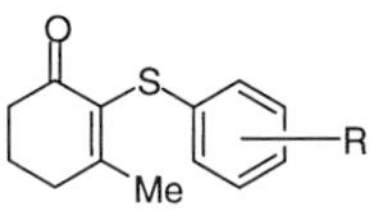

118 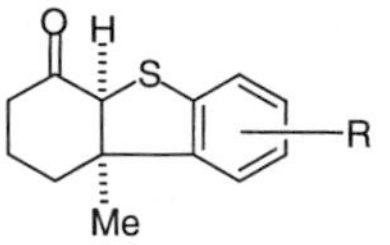 119

a : R=H e : R=*o*-Cl
b : R=*p*-Me f : R=*p*-Br
c : R=*o*-Me g : R=*o*-Br
d : R=*p*-Cl

Enantioselective photocyclization of *N*-phenyl enaminones also proceeds efficiently in inclusion complex with a chiral host. For example, irradiation of the 1 : 1 inclusion complex of **120a** with **2a** for 23 h gave (–)-**121a** of 94% ee in 43% yield. Interestingly, however, inclusion complexation of **120b** with **2c** gave two kinds of inclusion crystal, prisms and needles, and while irradiation of the former for 11.5 h gave (+)-**121b** of 87% ee in 54% yield, the latter was photochemically inert [73].

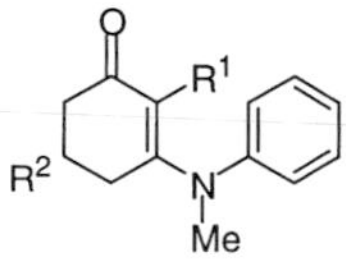

120 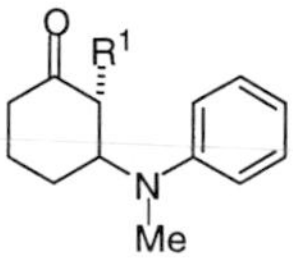 121

a : R^1=Me; R^2=H
b : R^1=H; R^2=Me

15.3.4.3 Photoreaction of Acrylanilides

The photocyclization of acrylanilide (**122**) to 3,4-dihydroquinolinone (**123**), which was first reported in 1971 [74], can also be carried out enantioselectively by using a chiral host compound. Irradiation of the finely powdered 1 : 1 inclusion complex of **122** with **26** in a water suspension gave (–)-**123** of 98% ee in 46% yield [75]. By the same procedure, optically active **125**, **127** and **129** were prepared from **124**, **126** and **128**, respectively (Table 15-23) [75].

15.3.4.4 Photoreaction of Furan-2-carboxanilides

Although the very useful synthetic route to *trans*-dihydrofuran derivatives (**131**) by photoreaction of the furan-2-carboxanilides (**130**) in MeOH has been established, the reaction gives *rac*-**131** together with some other by-products [76]. To control the reaction and to produce optically active **131**, the photoreaction of **130** was carried out in an inclusion crystal with **2**. Although **130a** did not form an inclusion crystal with **2a–c**, **130b–f** formed inclusion crystals in the ratios indicated in Table 15-24. Irradiation of these powdered inclusion crystals in the water sus-

122 **123**

124 **125**

126 **127**

128 **129**

Table 15-23. Photocyclization of anilides in 1 : 1 inclusion complexes with **2b** and **2c**.

Anilide	Host **2**	Reaction time (h)	Product		
				Yield (%)	Optical purity (% ee)
122	**2b**	150	(–)-**123**	46	98
122	**2c**	150	(+)-**123**	29	95
124	**2b**	150	(–)-**125**	65	98
124	**2c**	150	(+)-**125**	44	98
126	**2b**	50	(+)-**127**	62	70
126	**2c**	50	(–)-**127**	29	99
128	**2b**	15	(–)-**129**	64	98
128	**2c**	15	(+)-**129**	41	8

a : R=H   d : R=*i*Pr
b : R=Me   e : R=$CH_2CH=CH_2$
c : R=Et   f : R=CH_2Ph

pension gave optically active **131b–f** in the chemical and optical yields summarized in Table 15-24 [77].

Optically active **131b**, **131e** and **131f** were obtained in an almost pure optical state. In the case of **130e**, two kinds of inclusion compounds were formed, with the host **2b** having the different host: guest ratios indicated (Table 15-24). Recrystallization of **2b** and **130e** from ether gave a mixture of the 1:1 (colorless prisms) and 2:1 inclusion complexes (colorless needless), which were easily separated mechanically. Interestingly, photolysis of the 1:1 and 2:1 inclusion complexes gave (–)-**131e** and (+)-**131e**, respectively (Table 15-24). This is the first example of the formation of different enantiomers from a prochiral guest included by the same host in different ratios. To clarify the reasons, the crystal structure of the 1:1 complex was studied by X-ray analysis. It was found that **130e** was arranged in a chiral form so as to give (–)-**131e** on photoreaction [77]. Unfortunately, however, the molecular and crystal structure of the 2:1 complex could not be studied, because no suitable crystal for analysis was obtained [77]. In the 2:1 inclusion complex, **130e** molecules are probably arranged in the opposite chiral form to that of **130e** in the 1:1 complex. Nevertheless, this procedure is very convenient for the preparation of both (–)-**130e** and (+)-**130e**, since both these enantiomers can be prepared using just one enantiomeric host **2b**.

Table 15-24. Inclusion compounds of **130** with **2** and their photoreaction in a water suspension.

Host **2**	Guest	Inclusion compound Host: Guest	Irradiation time (h)	Product		
					Yield (%)	Optical purity (% ee)
2b	**130b**	1:1	40	(–)-**131b**	25	8
2c	**130c**	2:1	120	(+)-**131b**	41	99
2c	**130d**	1:1	143	(+)-**131d**	16	52
2a	**130e**	1:1	96	(+)-**131e**	20	93
2b	**130e**	1:1	77	(+)-**131e**	50	96
2b	**130e**	2:1	48	(+)-**131e**	86	98
2c	**130e**	2:1	50	(+)-**131e**	77	98
2c	**130f**	1:1	120	(–)-**131f**	72	98

References

1. Toda, F.; Yagi, M.; Kiyoshige, K. *J. Chem. Soc., Chem. Commun.* **1998**, 958.
2. Toda, F.; Tanaka, K.; Sekikawa, A. *J. Chem. Soc., Chem. Commun.* **1998**, 279.
3. Toda, F.; Kiyoshige, K.; Yagi, M. *Angew. Chem. Int. Ed. Engl.* **1989**, *28*, 320.
4. Ward, D.E.; Rhee, C.K.; Zoghaib, W.M. *Tetrahedron Lett.* **1988**, *29*, 517.
5. Sonfhrimrt, F.; Elad, D. *J. Am. Chem. Soc.* **1957**, *79*, 5542.
6. Toda, F.; Tanaka, K. *Tetrahedron Lett.* **1988**, *29*, 551.
7. Toda, F.; Mori, K. *J. Chem. Soc., Chem. Commun.* **1989**, 1245.
8. Toda, F.; Shigemasa, T. *Carbohyd. Res.* **1989**, *192*, 363.
9. Toda, F.; Takumi, H. *Chem. Express* **1989**, *4*, 507.

10. Tanaka, K.; Kishigami, S.; Toda, F. *J. Org. Chem.* **1991**, *56*, 4333.
11. Hauser, C.R.; Breslow, D.S. *Org. Synth.* **1941**, *21*, 51; Wilson, S.R.; Guazzaroni, M.E. *Org. Synth.* **1989**, *54*, 3087.
12. Shone, T.; Ishifumi, M.; Kashima, S. *Chem. Lett.* **1990**, 449.
13. Toda, F.; Tanaka, K.; Kagawa, Y.; Sakaino, Y. *Chem. Lett.* **1990**, 373.
14. Rajaguru, I.; Rzepa, H.S. *J. Chem. Soc., Perkin Trans. 2* **1987**, 1819.
15. Schmeyers, J.; Toda, F.; Boy, J.; Kaupp, G. *J. Chem. Soc., Perkin Trans. 2* **1998**, 989.
16. Toda, F.; Akehi, H. *J. Org. Chem.* **1990**, *55*, 3446.
17. Tomoskoz, I.; Iauzso, G. *Chem. Ind.* (London) **1962**, 2085; Bestmann, H.J.; Lienert, J. *J. Chem. Zeit.* **1970**, *94*, 487; Hanessian, S.; Delorme, G.D.; Beaudin, S.; Leblan, Y. *J. Am. Chem. Soc.* **1984**, *106*, 5754.
18. Toda, F.; Tanaka, K.; Hamai, K. *J. Chem. Soc., Perkin Trans. 1* **1990**, 3207.
19. Tanaka, K.; Kishigami, S.; Toda, F. *J. Org. Chem.* **1990**, *55*, 2981 and references cited therein.
20. Zhang, W.-C.; Li, C.-J. *J. Org. Chem.* **1999**, *64*, 3230.
21. Shaffer, J.P.; Bloomfield, J. *J. Org. React.* **1967**, *15*, 1.
22. Pinkney, P.S. *Org. Synth.* **1943**, Col. Vol. 2, 116.
23. Toda, F.; Suzuki, T.; Higa, S. *J. Chem. Soc., Perkin Trans. 1* **1998**, 3521.
24. Corey, E.J.; Chaykovsky, M. *J. Am. Chem. Soc.* **1965**, *87*, 1353.
25. Toda, F.; Kanemoto, K.; *Heterocycles*, **1997**, *46*, 185.
26. Trost, B.M.; Keeley, D.E. *J. Org. Chem.* **1975**, *40*, 2013.
27. Suzuki, K.; Ikegawa, A.; Mukaiyama, T. *Chem. Lett.* **1982**, 899.
28. Mukaiyama, T.; Ikegami, A.; Suzuki, K. *Chem. Lett.* **1981**, 165.
29. Yamashita, H.; Mukaiyama, T. *Chem. Lett.* **1985**, 363.
30. Helder, H.; Krends, R.; Bolt, W.; Hiemstra, H.; Wynberg, H. *Tetrahedron Lett.* **1977**, 2181.
31. Hiemstra, H.; Wynberg, H. *J. Am. Chem. Soc.* **1981**, *103*, 417.
32. Toda, F.; Tanaka, K.; Sato, J. *Tetrahedron Asymm.* **1993**, *4*, 1771.
33. Toda, F.; Tanaka, K.; Iwata, S. *J. Org. Chem.* **1989**, *54*, 3007.
34. Toda, F.; Takumi, H.; Nagami, M.; Tanaka, K. *Heterocycles* **1998**, *47*, 469.
35. Toda, F.; Yagi, M.; Soda, S. *J. Chem. Soc., Chem. Commun.* **1997**, 413.
36. Toda, F.; Miyamoto, H.; Kanemoto, K. *J. Org. Chem.* **1996**, *61*, 6490.
37. Toda, F.; Miyamoto, H.; Kikuchi, S.; Kuroda, R.; Nagami, F. *J. Am. Chem. Soc.* **1996**, *118*, 11315.
38. Sekine, A.; Hori, K.; Ohashi, Y.; Yagi, M.; Toda, F. *J. Am. Chem. Soc.* **1989**, *111*, 697.
39. Green, B.S.; Lahav, M.; Rabinovich, D. *Acc. Chem. Res.* **1979**, *12*, 191.
40. Addadi, L.; Lahav, M. *Origins of Optical Activity in Nature*; Elsevier, New York, 1979.
41. Toda, F.; Miyamoto, H. *J. Chem. Soc., Perkin Trans. 1,* **1993**, 1129.
42. Hashizume, D.; Koga, H.; Sekine, A.; Ohashi, Y.; Miyamoto, H.; Toda, F. *J. Chem. Soc., Perkin Trans. 2* **1996**, 61.
43. Toda, F.; Miyamoto, H.; Koshima, H. *J. Org. Chem.* **1997**, *62*, 1997.
44. Kaftory, M.; Yagi, M.; Tanaka, K.; Toda, F. *J. Org. Chem.* **1988**, *53*, 4391.
45. Toda, F.; Miyamoto, H.; Matsukawa, R. *J. Chem. Soc., Perkin Trans. 1* **1992**, 1461.
46. Hashizume, D.; Uekusa, H.; Ohashi, Y.; Matsugawa, R.; Miyamoto, H.; Toda, F. *Bull. Chem. Soc. Jpn.* **1994**, *67*, 985.
47. Toda, F.; Miyamoto, H.; Kanemoto, K. *J. Chem. Soc., Chem. Commun.* **1995**, 1719.
48. Quazzani-Chadi, L.; Quirion, J.-C.; Troin, Y.; Gramain, J.-C. *Tetrahedron* **1990**, *46*, 7751.
49. Toda, F.; Tanaka, K.; Kakinoki, O.; Kawakami, T. *J. Org. Chem.* **1993**, *58*, 3784.
50. Hashizume, D.; Ohashi, Y.; Tanaka, K.; Toda, F. *Bull. Chem. Soc. Jpn.* **1994**, *67,* 2383.
51. Stobbe, H.; Brener, K. *J. Prakt. Chem.* **1929**, *123*, 1; Rabinovich, D.; Schmidt, G.M. *J. Chem. Soc.* **1970**, *B*, 6.
52. Schmidt, G.M.J. *Photochemistry of the excited state, in reactivity of the photoexcited organic molecules*; J. Wiley, New York, 1967.
53. Schmidt, G.M.J. *Pure Appl. Chem.* **1926**, *27*, 647.
54. Tanaka, K. Toda, F. *J. Chem. Soc., Chem. Commun.* **1983**, 593.
55. Kaftory, M.; Tanaka, K.; Toda, F. *J. Org. Chem.* **1985**, *50*, 2154.
56. Toda, F.; Tanaka, K.; Kato, M. *J. Chem. Soc., Perkin Trans. 1* **1998**, 1315.
57. Lam, E.Y.Y.; Valentine, D.; Hammond, G.S. *J. Am. Chem. Soc.* **1967**, *89*, 3482.

58. Tanaka, K.; Kakinoki, O.; Toda, F. *J. Chem. Soc., Perkin Trans. 1* **1992**, 307.
59. Kraus, C.H.; Farid, S.; Schenck, G.O. *Chem. Ber.* **1966**, *99*, 625.
60. Shenck, G.O.; von Wilucki, I.; Krans, C.H. *Chem. Ber.* **1962**, *95*, 1409.
61. Tanaka, K.; Toda, F. *J. Chem. Soc., Perkin Trans. 1* **1992**, 943.
62. Moorthy, J.N.; Venkatesan, K. *J. Org. Chem.* **1991**, *56*, 6957.
63. Daupen, W.G.; Koch, K.; Smith, S.L.; Chapman, O.L. *J. Am. Chem. Soc.* **1963**, *85*, 2616.
64. Toda, F.; Tanaka, K. *J. Chem. Soc., Chem. Commun.* **1986**, 1429.
65. Toda, F.; Tanaka, K.; Oda, M. *Tetrahedron Lett.* **1988**, *29*, 653.
66. Fujiwara, T.; Nanba, N.; Hamada, K.; Toda, F.; Tanaka, K. *J. Org. Chem.* **1990**, *55*, 4532.
67. Toda, F.; Tanaka, K. *Tetrahedron Lett.* **1988**, *29*, 4299.
68. Kuzuya, M.; Noguchi, A.; Yokota, N.; Okuda, T.; Toda, F.; Tanaka, K. *Nippon Kagaku Kaishi* **1986**, 1753.
69. Toda, F.; Miyamoto, H.; Takeda, K.; Matsugawa, R.; Maruyama, N. *J. Org. Chem.* **1993**, *58*, 6208.
70. Toda, F.; Miyamoto, H.; Kikuchi, S. *J. Chem. Soc., Chem. Commun.* **1995**, 621.
71. Schultz, A.G. *J. Org. Chem.* **1974**, *39*, 3185.
72. Toda, F.; Miyamoto, H.; Kikuchi, S.; Kuroda, R.; Nagami, F. *J. Am. Chem. Soc.* **1996**, *118*, 11315.
73. Toda, F.; Miyamoto, H.; Tamashima, T. *J. Org. Chem.* **1999**, *64*, 2690.
74. Ogata, Y.; Takaki, K.; Ishino, I. *J. Org. Chem.* **1971**, *36*, 3975; Ninomiya, I.; Naito, T.; Tada, Y. *Heterocycles* **1984**, *22*, 237.
75. Tanaka, K.; Kakinoki, O.; Toda, F. *J. Chem. Soc., Chem. Commun.* **1992**, 1053.
76. Bates, R.B.; Kane, V.V.; Martin, A.R.; Mujumdar, R.B.; Ortega, R.; Hatanaka, Y.; Sannohe, K.; Kanaoka, Y. *J. Org. Chem.* **1987**, *52*, 3178.
77. Toda, F.; Miyamoto, H.; Kanemoto, K. *J. Org. Chem.* **1996**, *61*, 6490; Toda, F.; Miyamoto, H.; Kanemoto, K.; Tanaka, K.; Takahashi, Y.; Takenaka, Y. *J. Org. Chem.* **1999**, *64*, 2096.

Index